The Molecular and Cellular Biology of Wound Repair

Second Edition

The Molecular and Cellular Biology of Wound Repair

Second Edition

Edited by

Richard A. F. Clark

State University of New York at Stony Brook
Stony Brook, New York

PLENUM PRESS • NEW YORK AND LONDON

Library of Congress Cataloging-in-Publication Data

The molecular and cellular biology of wound repair / edited by Richard
   A.F. Clark. -- 2nd ed.
        p.   cm.
     Includes bibliographical references and index.
     ISBN 0-306-45159-X
     1. Wound healing.   I. Clark, R. A. F. (Richard A.F.)
   RD94.M65   1996
   617.1--dc20   95-47850
                                                                      CIP

Front cover: Photomicrographs of central granulation tissue in porcine cutaneous wounds at day 5 (left panel), day 7 (middle panel), and day 10 (right panel) that had been formalin-fixed and stained with Masson's trichrome. On day 5 blood vessels are vertically aligned while fibroblasts are randomly oriented in a collagen-poor provisional matrix. By day 7 much collagen (seen as blue fibrils on the original photomicrograph) has been deposited in the granulation tissue. The blood vessels continue to course through the wound vertically while the fibroblasts have assumed a horizontal alignment. At day 10 the granulation tissue has become collagen-rich and blood vessels have diminished in number. Fibroblasts have become stretched across the wound in a strict horizontal orientation as they contract the wound. This book is about the cell-matrix-cytokine interactions that drive this well-regulated progression of events called wound repair.

ISBN 0-306-45159-X

© 1996, 1988 Plenum Press, New York
A Division of Plenum Publishing Corporation
233 Spring Street, New York, N. Y. 10013

10 9 8 7 6 5 4 3 2

Contributors

Judith A. Abraham   Scios Nova Inc., Mountain View, California 94043

Monique Aumailley   Institute of Protein Biology and Chemistry, CNRS, 69367 Lyon cedex 7, France

Merton Bernfield   Joint Program in Neonatology, Harvard Medical School, Boston Children's Hospital, Boston, Massachusetts 02115

Richard A. F. Clark   Department of Dermatology, Health Sciences Center, State University of New York at Stony Brook, Stony Brook, New York 11794-8165

Alexis Desmoulière   Department of Pathology, University of Geneva, 1211 Geneva 4, Switzerland; and CNRS-URA 1459, Institut Pasteur de Lyon, 69365 Lyon cedex 7, France

Beate Eckes   Department of Dermatology, University of Cologne, D-50924 Cologne, Germany

M. W. J. Ferguson   School of Biological Sciences, University of Manchester, Manchester M13 9PT, England

Leo T. Furcht   Department of Laboratory Medicine and Pathology, Biomedical Engineering Center, University of Minnesota, Minneapolis, Minnesota 55455

Giulio Gabbiani   Department of Pathology, University of Geneva, 1211 Geneva 4, Switzerland

James Gailit   Department of Dermatology, Health Sciences Center, School of Medicine, State University of New York at Stony Brook, Stony Brook, New York 11794-8165

Richard L. Gallo   Department of Dermatology, Harvard Medical School, Boston Children's Hospital, Boston, Massachusetts 02115

Christopher Haslett   Respiratory Medicine Unit, Department of Medicine, University of Edinburgh, Royal Infirmary, Edinburgh EH3 9YW, Scotland

Carl-Henrik Heldin   Ludwig Institute for Cancer Research, Biomedical Center, S-751 24 Uppsala, Sweden

Peter Henson   Departments of Medicine and Pathology, National Jewish Center for Immunology and Respiratory Medicine, Denver, Colorado 80206

Joji Iida   Department of Laboratory Medicine and Pathology, Biomedical Engineering Center, University of Minnesota, Minneapolis, Minnesota 55455

Lloyd E. King, Jr.   Departments of Plastic Surgery, Cell Biology, and Dermatology, Vanderbilt University Medical Center, Department of Veteran's Affairs, Nashville, Tennessee 37232-2631

Michael Klagsbrun   Departments of Surgery and Pathology, Children's Hospital, and Harvard Medical School, Boston, Massachusetts 02115

Thomas Krieg   Department of Dermatology, University of Cologne, D-50924 Cologne, Germany

Joseph A. Madri   Department of Pathology, Yale University School of Medicine, New Haven, Connecticut 06510

Alain Mauviel   Departments of Dermatology and Cutaneous Biology, and Biochemistry and Molecular Biology, Jefferson Medical College, and Section of Molecular Dermatology, Jefferson Institute of Molecular Medicine, Thomas Jefferson University, Philadelphia, Pennsylvania 19107

R. L. McCallion   School of Biological Sciences, University of Manchester, Manchester M13 9PT, England

James B. McCarthy   Department of Laboratory Medicine and Pathology, Biomedical Engineering Center, University of Minnesota, Minneapolis, Minnesota 55455

John McGrath   Departments of Dermatology and Cutaneous Biology, and Biochemistry and Molecular Biology, Jefferson Medical College, and Section of Molecular Dermatology, Jefferson Institute of Molecular Medicine, Thomas Jefferson University, Philadelphia, Pennsylvania 19107

Paolo Mignatti   Department of Genetics and Microbiology, University of Pavia, 27100 Pavia, Italy

Lillian B. Nanney   Departments of Plastic Surgery, Cell Biology, and Dermatology, Vanderbilt University Medical Center, Department of Veteran's Affairs, Nashville, Tennessee 37232-2631

William C. Parks   Division of Dermatology, Department of Medicine, Washington University School of Medicine at the Jewish Hospital, St. Louis, Missouri 63110

David W. H. Riches   Division of Basic Sciences, Department of Pediatrics, National Jewish Center for Immunology and Respiratory Medicine, Denver, Colorado 80206

Daniel B. Rifkin   Department of Cell Biology and Kaplan Cancer Center, New York University Medical Center, New York, New York 10016-6402

Anita B. Roberts   Laboratory of Chemoprevention, National Cancer Institute, Bethesda, Maryland 20892-5055

Anne M. Romanic   Department of Pathology, Yale University School of Medicine, New Haven, Connecticut 06510

Sabita Sankar   Department of Pathology, Yale University School of Medicine, New Haven, Connecticut 06510

Michael B. Sporn   Laboratory of Chemoprevention, National Cancer Institute, Bethesda, Maryland 20892-5055

Jouni Uitto   Departments of Dermatology and Cutaneous Biology, and Biochemistry and Molecular Biology, Jefferson Medical College, and Section of Molecular Dermatology, Jefferson Institute of Molecular Medicine, Thomas Jefferson University, Philadelphia, Pennsylvania 19107

Howard G. Welgus   Division of Dermatology, Department of Medicine, Washington University School of Medicine at the Jewish Hospital, St. Louis, Missouri 63110

Bengt Westermark   Department of Pathology, University Hospital, S-751 85 Uppsala, Sweden

David T. Woodley   Department of Dermatology, Northwestern University, Chicago, Illinois 60611-3008

Kenneth M. Yamada   Laboratory of Developmental Biology, National Institute of Dental Research, National Institutes of Health, Bethesda, Maryland 20892-4370

Preface

It has been a great pleasure to write chapters and work with the other authors on the second edition of *The Molecular and Cellular Biology of Wound Repair*. The book has been totally revised with all chapters rewritten. Several chapters in the first edition explored inflammation. These chapters have been deleted, since other books comprehensively cover this topic. The deletions provided room to add Chapter 2 on the provisional matrix proteins, to expand Chapters 5 through 8 on growth factors, to add Chapter 9 on integrins, and to add Chapter 18 on scarring and nonscarring wound repair.

As with the first edition, I have gained an immense knowledge about the molecular and cellular biology of wound repair from my participation in the production of this book. The book has become a bridge between the basic sciences and the bedside, rather than a detailed treatise on molecular and cellular biology or a comprehensive review of animal and clinical wound studies. Thus, basic scientists who wish to determine whether their scientific endeavors might have applications in wound healing will find this volume valuable. On the other hand, clinical scientists will find that this volume gives a good scientific foundation for their clinical work.

Chapter 1 gives an overview of processes involved in cutaneous wound repair. The initial version of this overview was published 10 years ago in the *Journal of the American Academy of Dermatology* and has undergone many revisions for reviews in journals and book chapters, including the first edition of *The Molecular and Cellular Biology of Wound Repair*. This overview has once again been revised to include the highlights of data available up to and including the beginning of 1995. In Chapter 2, Kenneth Yamada and I took on the formidable task of relating the molecular structure and function of provisional matrix proteins, such as fibrinogen, fibronectin, vitronectin, thrombospondin, and tenascin, to their potential activities in wound repair. In Chapter 3, David Riches comprehensively reviews the macrophage with emphasis on lineage and phenotype modulation and how the different macrophage phenotypes might affect wound healing. In Chapter 4, Chris Haslett and Peter Henson examine how inflammation is resolved, not only in soft tissue repair, but in inflammatory processes in general. Programmed cell death is a key phenomenon in this resolution.

The growth factor section of the book gives the reader a comprehensive survey of the epidermal growth factor (EGF), fibroblast growth factor (FGF), platelet-derived growth factor (PDGF), and transforming growth factor-β (TGF-β) families and how these many diverse factors may interact in wound repair. Specifically, in Chapter 5, Lillian Nanney and Lloyd King review the seminal work of Stanley Cohn in discovering and characterizing epidermal growth factor, which led to his Nobel prize. Furthermore, they outline the ever-expanding second messenger pathways that transmit the extracellular ligand–receptor signals into the nucleus and they allude to the many studies that have been done with EGF on animals and in the clinic. In Chapter 6, Judy Abraham and Michael Klagsbrun produced a comprehensive treatise on the FGF family, including molecular biology and biochemistry of the nine family members and an in-depth review of wound healing responses to these factors in animal and clinical studies. In Chapter 7, Carl-Henrik Heldin and Bengt Westermark have written an excellent treatise on platelet-derived growth factors, explaining how the different PDGF isoforms can elicit diverse activities in cells by stimulating different receptor pairs. Last but not least, in Chapter 8, Anita Roberts and Michael Sporn have written an in-depth analysis of the mammalian TGF-β isoforms from their molecular biology and structure to their use in animal and clinical studies. Transgenic mouse studies are beginning to elucidate the precise activities and necessities of many of these growth factor molecules.

The third section of the book, entitled "New Tissue Formation: The Cutaneous Paradigm," begins with a review, in Chapter 9, of molecular and structural biology of integrin extracellular matrix receptors and how these receptors are expressed and potentially act during the wound repair process, with particular emphasis on reepithelialization. In Chapter 10, David Woodley continues with a discussion of the biology of the keratinocyte in wound repair. In Chapter 11, Joe Madri and colleagues survey the current knowledge on angiogenesis and describe how angiogenesis is closely controlled by cytokines and extracellular matrix signals. In Chapter 12, Jim McCarthy and his colleagues describe how parenchymal cell motility during wound repair and other morphogenetic events might depend on extracellular receptors other than integrins. This chapter also alludes to important new information on the controls and assembly of the actin motor apparatus in motile cells. In Chapter 13, Alexis Desmoulière and Giulio Gabbiani regale us with descriptions of myofibroblast phenotypes and functions in wound repair and fibrocontractive diseases.

The fourth and final section of the book, entitled "Essentials of Tissue Remodeling," includes chapters on proteases, extracellular matrix molecules, and scar formation. In Chapter 14, Paolo Mignatti, Daniel Rifkin, Howard Welgus, and William Parks present a comprehensive review of the important proteases involved in wound repair, including plasminogen activators, plasmin, and the metalloproteinases. In Chapter 15, Richard Gallo and Mert Bernfield provide a review on proteoglycans and their expression in cutaneous wound repair. In Chapter 16, Beate Eckes, Monique Aumailley, and Thomas Krieg summarize current knowledge about the extracellular matrix collagens and how these proteins assemble in wound repair to patch the injured tissue with a scar. In Chapter 17, Jouni Uitto, Alain Mauviel, and John McGrath survey the molecular genetics and structure of basement membrane proteins and the deficiencies in these proteins that lead to recurrent or nonhealing cutaneous wounds. These lessons of nature provide great insight into the homeostatic value of these proteins and their importance in tissue reconstruction after architecturally disruptive injury. In Chapter 18, R. L. McCallion and M. W. J. Ferguson describe repair without scarring in fetal tissue and propose ways in which adult wound healing might be modified to prevent or at least reduce scarring.

In composite, these chapters give the reader a fairly comprehensive view of the molecular and cellular biology of wound repair, in both normal and abnormal healing processes. In addition, several chapters suggest methods by which inadequate or substandard healing might be improved.

Richard A. F. Clark
Stony Brook, New York

Contents

Chapter 3

Macrophage Involvement in Wound Repair, Remodeling, and Fibrosis

David W. H. Riches

Chapter 4

Resolution of Inflammation

Christopher Haslett and Peter Henson

PART II. GROWTH FACTORS IN SOFT TISSUE HEALING

Chapter 5

Epidermal Growth Factor and Transforming Growth Factor-α

Lillian B. Nanney and Lloyd E. King, Jr.

Chapter 6

Modulation of Wound Repair by Members of the Fibroblast Growth Factor Family

Judith A. Abraham and Michael Klagsbrun

Chapter 7

Role of Platelet-Derived Growth Factor *in Vivo*

Carl-Henrik Heldin and Bengt Westermark

Chapter 8

Transforming Growth Factor-β

Anita B. Roberts and Michael B. Sporn

PART III. NEW TISSUE FORMATION: THE CUTANEOUS PARADIGM

Chapter 9

Integrins in Wound Repair

Kenneth M. Yamada, James Gailit, and Richard A. F. Clark

Chapter 10

Reepithelialization

David T. Woodley

Chapter 11

Angiogenesis

Joseph A. Madri, Sabita Sankar, and Anne M. Romanic

PART IV. ESSENTIALS OF TISSUE REMODELING

Chapter 14

Proteinases and Tissue Remodeling

Paolo Mignatti, Daniel B. Rifkin, Howard G. Welgus,
and William C. Parks

Chapter 15

Proteoglycans and Their Role in Wound Repair

Richard L. Gallo and Merton Bernfield

Chapter 16
Collagens and the Reestablishment of Dermal Integrity

Beate Eckes, Monique Aumailley, and Thomas Krieg

Chapter 17
The Dermal–Epidermal Basement Membrane Zone in Cutaneous Wound Healing

Jouni Uitto, Alain Mauviel, and John McGrath

Chapter 18

Fetal Wound Healing and the Development of Antiscarring Therapies for Adult Wound Healing

R. L. McCallion and M. W. J. Ferguson

Part I
Preliminaries

Chapter 1

Wound Repair

Overview and General Considerations

RICHARD A. F. CLARK

1. Introduction

When tissue loss disrupts normal architecture in higher vertebrate adult animals, the organ fails to regenerate. Instead, repair proceeds as a fibroproliferative response that develops into a fibrotic scar. Thus, the organ is patched rather then restored. Alterations in the normal healing processes produce even less desirable outcomes. For example, when injurious events persist or recur, inflammation is perpetuated, extending tissue damage and repair. In addition, a plethora of pathobiological states, such as diabetes, Cushing's syndrome, poor arterial perfusion, venous hypertension, poor nutrition, and sepsis, disrupt normal repair processes. Such situations often lead to nonhealing wounds or excessive fibrosis.

Over the past two decades, extraordinary advances in cellular and molecular biology have led to an enriched comprehension of the basic biological processes involved in wound repair. Clinical investigators had hoped that the great strides in basic knowledge would quickly lead to advancements in wound care that would culminate in accelerated rates of ulcer and normal wound repair, scars of greater strength, prevention of keloids and fibrosis, and ultimately substitute tissue regeneration for scar formation. While these lofty goals still have not been achieved, new scientific information continues to accumulate at an accelerating pace. It has never been more clear that today's scientific breakthroughs will lead to tomorrow's therapeutic successes. Thus, Chapters 1–17 focus on current knowledge in the molecular and cellular biology of soft tissue repair of adult animals. In contrast, Chapter 18 addresses fetal cutaneous wound repair, a process that occurs without scarring. This chapter raises the possibility that adult wound repair processes might be manipulated to prevent scarring.

Wound repair is not a simple linear process in which growth factors released by phylogistic events activate parenchymal cell proliferation and migration, but rather it is

RICHARD A. F. CLARK • Department of Dermatology, Health Sciences Center, State University of New York at Stony Brook, Stony Brook, New York 11794-8165.

The Molecular and Cellular Biology of Wound Repair, (Second Edition), edited by Richard A. F. Clark. Plenum Press, New York, 1995.

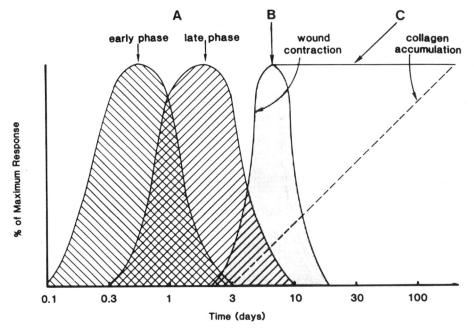

Figure 1. Phases of cutaneous wound repair. Healing of a wound has been arbitrarily divided into three phases: (A) inflammation (early and late), (B) reepithelialization and granulation tissue formation, and (C) matrix formation and remodeling. These wound repair processes are plotted along the abscissa as a logarithmic function of time. The phases of wound repair overlap considerably with one another as discussed in the text. Inflammation is divided into early and late phases denoting neutrophil-rich and mononuclear cell-rich infiltrates, respectively. In tight-skilled animals such as humans, wound contraction begins after granulation tissue is well established, as indicated by the arrow. Collagen accumulation begins shortly after the onset of granulation tissue formation, as indicated.

an integration of dynamic interactive processes involving soluble mediators, formed blood elements, extracellular matrix, and parenchymal cells. Unencumbered, these wound repair processes follow a specific time sequence and can be temporally categorized into three major groups: inflammation, tissue formation, and tissue remodeling (Fig. 1). The three phases of wound repair, however, are not mutually exclusive but rather overlapping in time. In this overview, current ideas about wound repair will be presented in a sequence that roughly follows the chronology of wound repair: inflammation, tissue formation, and tissue remodeling.

2. Inflammation: Important Preliminaries

Severe tissue injury causes blood vessel disruption with concomitant extravasation of blood constituents. Blood coagulation and platelet aggregation generate a fibrin-rich clot that plugs severed vessels and fills any discontinuity in the wounded tissue.

While the blood clot within the vessel lumen reestablishes hemostasis, the clot within the wound space provides a provisional matrix for cell migration. Platelets also have a dual function since they not only facilitate the formation of a hemostatic plug but also secrete multiple mediators, including growth factors. In fact, the coagulation pathways, as well as the activated complement pathways, and injured or activated parenchymal cells all generate numerous vasoactive mediators and chemotactic factors, which together recruit inflammatory leukocytes to the wounded site.

Infiltrating neutrophils cleanse the wounded area of foreign particles, including bacteria. If excessive microorganisms or indigestible particles have lodged in the wound site, neutrophils will probably cause further tissue damage as they attempt to clear these contaminants through the release of enzymes and toxic oxygen products. When particle clearance has been completed, the generation of granulocyte chemoattractants usually ceases. Effete neutrophils in the wound are either extruded with the eschar or phagocytosed by macrophages or fibroblasts. Peripheral blood monocytes continue to infiltrate the wound site, however, in response to specific monocyte chemoattractants. Once in tissue, monocytes progressively become activated and become macrophages. Macrophages, like platelets, release growth factors that initiate granulation tissue formation. However, unlike platelets, which release stored proteins and peptides but produce little if any of these molecules, macrophages have the ability to continually synthesize and secrete growth factors and cytokines.

2.1. Blood Coagulation

When blood extravasation accompanies tissue injury, blood clotting results from (1) surface activation of Hageman factor, (2) tissue procoagulant factor released from damaged cells, and (3) surface membrane coagulation factors and phospholipids expressed on activated platelets and endothelial cells (Furie and Furie, 1988). The critical event in all cases is the availability of a surface that promotes adsorption and activation of specific coagulation proenzymes. Surface adsorption is a prerequisite for proenzyme activation since these proteins otherwise are literally afloat in a sea of enzyme inhibitors. When the proenzymes have been adsorbed on a surface in a microenvironment relatively free of protease inhibitors, however, minute amounts of spontaneous activation are quickly amplified into the physiological response of blood clotting.

Blood coagulation terminates when stimuli for clot initiation dissipate. In addition, several intrinsic blood vessel activities limit the extent of platelet aggregation and clotting to the area proximate to the injury. These include production of prostacyclin, which inhibits platelet aggregation (Moncada et al., 1976); antithrombin III binding to thrombin, which inhibits its activity (Stern et al., 1985); generation of protein C, a potent enzyme that degrades coagulation factors V and VIII (Loedam et al., 1988); and release of plasminogen activator, which initiates clot lysis through conversion of plasminogen to plasmin (Loskutoff and Edgington, 1977).

Clot lysis is controlled as rigorously as is coagulation. The major proteolytic enzymes—plasminogen activators and plasmin—escape inactivation by fluid-phase protease inhibitors, like plasminogen activator inhibitor and α_2-antiplasmin, through

Table I. Integrin Superfamily

Integrins	Ligands
β1 family	
α1β1	Fibrillar collagen, laminin
α2β1	Fibrillar collagen, laminin
α3β1	Fibronectin (RGD), laminin-5, entactin, denatured collagens
α4β1	Fibronectin (LEDV), VCAM-1
α5β1	Fibronectin (RGD)
α6β1	Laminin
α7β1	Laminin
α8β1	Fibronectin, vitronectin
α9β1	Tenascin
αv family	
αvβ1	Fibronectin (RGD), vitronectin
αvβ3	Vitronectin (RGD), fibronectin, fibrinogen, von Willebrand factor, thrombospondin, denatured collagen
αvβ5	Fibronectin (RGD), vitronectin
αvβ6	Fibronectin, tenascin
Other ECM integrins	
αIIbβ3	Same as αvβ3
α6β4	Laminin
β2 family	
$\alpha_M\beta2$	ICAM-1, iC3b, fibrinogen, factor X
$\alpha_L\beta2$	ICAM-1, 2, and 3
$\alpha_X\beta2$	iC3b, fibrinogen

binding to the fibrin clot (Thorsen *et al.,* 1972; Castellino *et al.,* 1983) and cell surfaces (Hajjar *et al.,* 1994). Although plasminogen activator and plasmin have the ability to degrade a wide variety of extracellular matrix proteins, a specific inhibitor of plasminogen activator binds to the extracellular matrix (Salonen *et al.,* 1989) and limits matrix degradation to the microenvironment around cell surfaces.

Clearly, hemostasis is a major function of blood coagulation; however, the clot also provides a matrix scaffold for the recruitment of cells to an injured site (see Chapter 2 for details). Specifically, fibrin in conjunction with fibronectin act as a provisional matrix for the influx of monocytes (Ciano *et al.,* 1986; Lanir *et al.,* 1988) and fibroblasts (Grinnell *et al.,* 1980; Knox *et al.,* 1986; Brown *et al.,* 1993b). Presumably, migrating cells use integrin receptors that recognize fibrin, fibronectin, and vitronectin to interact with the clot matrix (Table I) (see Chapter 9 for details) (Ruoslahti, 1991; Hynes, 1992).

Blood clotting is also a part of the inflammatory response. For example, Hageman factor activation leads to generation of its fragments, bradykinin, and potent vasoactive agents (Yamamoto and Cochrane, 1981; Muller-Esterl, 1989) and to the initiation of classical and alternative complement cascades (Ghebrehiwet *et al.,* 1981; DiScipio,

Table II. Coagulation in Wound Repair

Activity	Effect
Hageman factor fragments	Vasopermeability
Bradykinin	Vasodilation
	Vasopermeability
	Pain
Complement activation	Leukocyte recruitment
	Vasopermeability
Fibrin clot	Hemostatic plug
	Reservoir of growth factors
	Provisional matrix for cell migration

1982), with the resultant generation of the anaphylatoxins C3a and C5a. The anaphylatoxins directly increase blood vessel permeability and attract neutrophils and monocytes to sites of tissue injury (Fernandez *et al.,* 1978). In addition, these substances stimulate the release of other vasoactive mediators such as histamine and leukotriene C_4 and D_4 from mast cells (Stimler *et al.,* 1982) and the release of granule constituents and biologically active oxygen products from neutrophils and macrophages (McCarthy and Henson, 1979). The activities of coagulation in wound repair are compiled in Table II.

2.2. Platelets

Successful hemostasis is dependent on platelet adhesion and aggregation. Platelets first adhere to interstitial connective tissue, then aggregate. In the process of aggregation, platelets release many mediators, including adenosine diphosphatase (ADP), and express several clotting factors on their membrane surface. Together, these platelet products facilitate coagulation and further platelet activation. The fibrin clot and locally generated thrombin act as a nidus for additional platelet adhesion and aggregation.

When activated platelets discharge their alpha-granules, several adhesive proteins including fibrinogen, fibronectin, thrombospondin, and von Willebrand factor VIII are released. The first three act as ligands for platelet aggregation, while von Willebrand factor VIII mediates platelet adhesion to fibrillar collagens and their subsequent activation (Ginsberg *et al.,* 1988; Ruggeri, 1993). Platelet adhesion to all four adhesive proteins is mediated through the platelet GPIIb/IIIa (integrin αIIbβ3) surface receptor (Ginsberg *et al.,* 1992) and other integrin extracellular matrix receptors (Ruoslahti, 1991; Hynes, 1992) (Table I) (see Chapter 9 for details). Platelet fibrinogen, once converted to fibrin by thrombin, adds to the fibrin clot. In addition, platelets release chemotactic factors for blood leukocytes (Weksler, 1992) and growth factors such as platelet-derived growth factor (PDGF) (Ross and Raines, 1990) and transforming growth factor-alpha (TGF-α) (Derynck, 1988) and -beta (TGF-β) (Sporn and Roberts,

Table III. Growth Factors in Wound Repair

Growth factor	Effect
Fibroblast growth factor-1 and -2 (FGF)	Fibroblast and epidermal cell proliferation; angiogenesis
Insulinlike growth factor (IGF)	Progression factor for cell proliferation
Keratinocyte growth factor (KGF), also known as FGF-7	Keratinocyte proliferation
Platelet-derived growth factor (PDGF) including isoforms AA, AB, and BB	Fibroblast chemotaxis, proliferation, and contraction
Transforming growth factor-α (TGF-α) and epidermal growth factor (EGF)	Reepithelialization
Transforming growth factor-β (TGF-β) including isoforms β1, β2, and β3	Fibroblast chemotaxis; extracellular matrix deposition; protease inhibitor secretion
Vascular endothelial growth factor (VGEF)	Vascular permeability; angiogenesis

1992), which promote new tissue generation (Table III) (see Chapters 5–8 for details). The effect of platelets in wound repair are listed in Table IV.

2.3. Neutrophils

Neutrophils and monocytes begin to emigrate into injured tissue concurrently, but neutrophils arrive first in great numbers partly due to their abundance in the circulation. A variety of chemotactic factors attract both cell types to the site of injury (Williams,1988) (also see Chapter 3). General leukocyte chemo attractants include fibrinopeptides cleaved from fibrinogen by thrombin; fibrin degradation products produced by plasmin degradation of fibrin; C5a arising from activated classical or alternative complement cascades; leukotriene B$_4$ released by activated neutrophils; platelet-activating factor (PAF) released from endothelial cells or activated neutrophils; formyl methionyl peptides cleaved from bacterial proteins; and PDGF and platelet factor 4 released from platelets. Besides providing the stimulus for directed migration, chemotactic factors also increase CD11/CD18 expression on the neutrophil surface (Tonnesen *et al.*, 1989). These heterodimeric complexes, in conjunction with Lewis

Table IV. Platelets in Wound Repair

Activity	Effect
Adhesion	Plug small leaks in blood vessels
Aggregation	Plug large leaks in blood vessels
	Induce coagulation
Mediator release	Vasoconstriction
	Stimulate additional platelet aggregation
	Growth factor release

factor X, mediate adherence of neutrophils to blood vessel endothelium and thereby facilitate transmigration of leukocytes through the endothelium (Albelda and Buck, 1990). Neutrophil activation by chemoattractants also stimulates release of elastase and collagenase molecules. These enzymes facilitate cell penetration through blood vessel basement membranes. Neutrophils at the wound site destroy contaminating bacteria via phagocytosis and subsequent enzymatic and oxygen radical mechanisms (Tonnesen *et al.,* 1988; Elsbach and Weiss, 1992; Klebanoff, 1992).

If substantial wound contamination has not occurred, neutrophil infiltration usually ceases within a few days. Most invading neutrophils become entrapped within the wound clot and desiccated tissue. This eschar sloughs during tissue regeneration. Neutrophils within viable tissue become senescent within a few days and are phagocytosed by tissue macrophages (Newman *et al.,* 1982). These processes mark the end of neutrophil-rich inflammation. However, substantial wound contamination will provoke a persistent neutrophil-rich inflammatory response. Bacteria, or other foreign objects, provide a surface on which alternative pathway proenzymes can adsorb, thus escaping their plasma inhibitors (Muller-Eberhard, 1992). Such surfaces continually activate the alternative pathway, which results in opsonization of foreign surfaces with C3b and generation of C3a and C5a anaphylatoxins. Additional neutrophils would be attracted to such contaminated wounds.

2.4. Monocytes

Whether neutrophil infiltrates resolve or persist, monocyte accumulation continues, stimulated by selective monocyte chemoattractants. These factors include fragments of collagen (Postlethwaite and Kang, 1976), elastin (Senior *et al.,* 1980), and fibronectin (Clark *et al.,* 1988), enzymatically active thrombin (Bar-Shavit *et al.,* 1983), and TGF-β (Wahl *et al.,* 1987). Similar to neutrophil recruitment, chemoattractants stimulate circulating monocytes to attach to the endothelium of blood vessels at the site of injury and to migrate through the blood vessel wall into the tissue stroma (Doherty *et al.,* 1987). Binding of monocytes or macrophages to specific extracellular matrix proteins through integrin receptors stimulates extracellular matrix phagocytosis and Fc- and C3b-mediated phagocytosis (Brown and Goodwin, 1988). Thereby, macrophages are armed to debride tissue through phagocytosis and digestion of pathogenic organisms, tissue debris, and effete neutrophils (Newman *et al.,* 1982). Cultured macrophages and presumably wound macrophages release enzymes such as collagenase (Campbell *et al.,* 1987) that facilitate tissue debridement. In addition, when macrophages are activated by bacterial endotoxin, substances like neutrophil-activating protein are released (Wolpe and Cerami, 1989), which recruit additional inflammatory cells.

Besides promoting phagocytosis and debridement, adherence to extracellular matrix also stimulates monocytes to undergo metamorphosis into inflammatory or reparative macrophages (see Chapter 3 for details). Adherence induces selective mRNA expression of colony-stimulating factor-1, a cytokine necessary for monocyte–macrophage survival; tumor necrosis factor-α (TNF-α), a potent inflammatory cytokine; PDGF, a

Table V. Macrophages in Wound Repair

Activity	Effect
Recruitment and maturation	Transition from circulating mono-cyte to tissue macrophage
Phagocytosis and killing of microorganisms	Wound decontamination
Phagocytosis of tissue debris	Wound debridement
Growth factor release	Autocrine and paracrine stimulation

potent chemoattractant and mitogen for fibroblasts; as well as c-fos and c-jun, transactivating factors necessary for many activation signals (Shaw *et al.,* 1990; Juliano and Haskill, 1992). mRNAs for other important macrophage cytokines are adherence-independent, e.g., TGF-β is constitutively expressed; interleukin-1 (IL-1) mRNA is stimulated by bacterial endotoxin; and human leukocyte antigen-D-related (HLA-DR) is stimulated by gamma-interferon (γ-IFN) (Shaw *et al.,* 1990, 1991). Wound macrophages, in fact, express TGF-β and PDGF mRNA as well as TGF-α and insulinlike growth factor-1 (IGF-1) mRNA (Rappolee *et al.,* 1988). Since cultured macrophages produce and secrete the peptide growth factors, IL-1 (Dinarello, 1984), PDGF (Shimokado *et al.,* 1985), TGF-β (Assoian *et al.,* 1987), TGF-α (Madtes *et al.,* 1988), and fibroblast growth factor (FGF) (Baird *et al.,* 1985), presumably wound macrophages also synthesize these protein products. Such macrophage-derived growth factors are almost certainly necessary for initiation and propagation of new tissue formation in wounds, since macrophage-depleted animals have defective wound repair (Leibovich and Ross, 1975). Thus, macrophages appear to play a pivotal role in the transition between inflammation and repair (Table V) (see Chapters 3 and 4 for details). A synopsis of growth factors and their activities is outlined in Table III and comprehensively addressed in Chapters 5–8.

3. Epithelialization: Reestablishing a Cutaneous Cover

Reepithelialization of a wound begins within hours after injury. Epithelial cells from residual epithelial structures move quickly across the wound defect (see Chapter 10 for details). It is clear that rapid reestablishment of any epithelial barrier decreases victim morbidity and mortality. In the skin, keratinocytes of the stratified epidermal sheet or hair follicle appear to move one over the other in a leapfrog fashion (Winter, 1962), whereas in the cornea, cells of the monolayer sheet appear to move in single file with the lead cells remaining in front (Fujikawa *et al.,* 1984). Stem cells for hair growth have been found in the infundibular bulge (Cotsarelis *et al.,* 1990). Since these cells reside near the epidermis, they may give rise to migrating epidermal cells that repave denuded skin.

Concomitant with migration, epithelial cells undergo marked phenotypic alteration. This metamorphosis includes retraction of intracellular tonofilaments; dissolution of most intercellular desmosomes (structures that interlink epithelial cells and thereby provide tensile strength for epithelium); and formation of peripheral cytoplasmic actin filaments (Odland and Ross, 1968; Gabbiani *et al.*, 1978). An additional manifestation of altered epidermal phenotype is the loss of tenacious binding between the epidermis and dermis. Loss of junctional adherence results from the dissolution of hemidesmosome links between the epidermis and the basement membrane (Krawczyk and Wilgram, 1973). As a consequence of these phenotypic changes, wound epidermal cells have lateral mobility and the motor apparatus for motility. The epithelial cells at the wound edge lose their apical–basal polarity and extend pseudopodia from their free basolateral sides into the wound.

Migrating wound epidermal cells do not terminally differentiate as do keratinocytes of normal epidermis. For example, migrating wound epidermal cells do not contain keratin proteins normally found in mature stratified epidermis nor do they contain filaggrin, a matrix protein in which these keratins are embedded. In contrast, cells in all layers of the migrating epidermis contain keratins normally found only in the basal cells of stratified epidermis (Mansbridge and Knapp, 1987). Nevertheless, the phenotype of migrating epidermal cells is not identical to basal cells, since the migrating cells also contain involucrin, a component of differentiated keratinocyte cell walls, and transglutaminase, an enzyme, that cross-links cell wall proteins. Involucrin and transglutaminase usually appear only in the stratum granulosum of normal epidermis. The unique phenotype of migrating wound epidermal cells is similar to the phenotype present in lesional psoriatic skin and in cultured epidermal cells (Hennings *et al.*, 1980; Mansbridge and Knapp, 1987). Induction signals for the migrating epidermal cell phenotype are not known, although low calcium concentrations impart cultured keratinocytes with a similar phenotype, while normal calcium concentrations drive terminal differentiation (Hennings *et al.*, 1980).

One to two days after injury, epithelial cells at the wound margin begin to proliferate (Krawczyk, 1971). The stimuli for epithelial proliferation during reepithelialization have not been delineated, but several possibilities exist. Perhaps the absence of neighbor cells at the wound margin signals both epithelial migration and proliferation. This "free-edge effect" has been thought to stimulate reendothelialization of large blood vessels after intimal damage (Heimark and Schwartz, 1988). Another possibility, not exclusive of the former, is local release of growth factors that induce epidermal migration and proliferation. In addition, increased expression of growth factor receptors may stimulate these processes. Leading contenders include the epidermal growth factor (EGF) family, especially TGF-α (Barrandon and Green, 1987); heparin-binding epidermal growth factor (HB-EGF) (Higashiyama *et al.*, 1991); and the FGF family (O'Keefe *et al.*, 1988; Werner *et al.*, 1992). Although growth factors may derive from macrophages or dermal parenchymal cells and act on epidermal cells through a paracrine pathway (Baird *et al.*, 1985; Rappolee *et al.*, 1988; Werner *et al.*, 1992), TGF-α, and perhaps other growth factors, originate from keratinocytes themselves and act directly on the producer cell or adjacent epidermal cells in an autocrine or juxtacrine fashion (Coffey *et al.*, 1987; Brachmann *et al.*, 1989). Many of these growth factors have been shown to stimulate reepithelialization in animal models (Brown *et al.*, 1989; Hebda *et al.*, 1990) or to be absent in models of deficient reepithelialization (Werner *et al.*, 1994), supporting the hypothesis that they are active during normal wound repair.

If the basement membrane is destroyed by injury, epidermal cells migrate over a provisional matrix consisting of type V collagen (Stenn *et al.*, 1979), fibrin, fibronectin (Clark *et al.*, 1982b), tenasin (Mackie *et al.*, 1988), and vitronectin (Cavani *et al.*, 1993), as well as type I collagen (Odland and Ross, 1968). If the basement membrane is not destroyed, fibronectin infiltrates the intact basement membrane (Fujikawa *et al.*, 1984). Fibrin and fibronectin in the provisional matrix initially originate from the circulation (Clark *et al.*, 1983). However, a few days after injury, fibronectin is deposited by wound fibroblasts, macrophages, or the migrating epidermal cells themselves (Clark *et al.*, 1983; Grimwood *et al.*, 1988; Brown *et al.*, 1993a). Furthermore, wound keratinocytes express functionally active integrin receptors for fibronectin in contrast to normal epidermal cells (Toda *et al.*, 1987). Thus, wound keratinocytes can pave the wound surface with a provisional matrix and express cell surface receptors that facilitate their migration across this matrix (Clark, 1990; Cavani *et al.*, 1993; Larjava *et al.*, 1993; Gailit *et al.*, 1994). Cultured keratinocyte migration, in fact, is accentuated on fibronectin- or type I collagen-coated surfaces compared to laminin, a major component of the basement membrane (O'Keefe *et al.*, 1985).

Often the epidermis does not simply transit over a wound coated with provisional matrix but rather dissects through the wound, separating desiccated or otherwise nonviable tissue from viable tissue (Clark *et al.*, 1982b). Epidermal movement through tissue depends on epidermal cell production of collagenase (Woodley *et al.*, 1986) and plasminogen activator (Grondahl-Hansen *et al.*, 1988). The latter enzyme activates collagenase as well as plasminogen. Interestingly, keratinocytes in direct contact with collagen greatly increase the amount of collagenase they produce compared to that produced when they reside on laminin-rich basement membrane or purified laminin (Petersen *et al.*, 1990).

Whether the driving force for epithelial cell movement is chemotactic factors, active contact guidance, loss of nearest neighbor cells, or a combination of these processes is unknown; however, migration does not depend on cell proliferation (Winter, 1972). Interestingly, TGF-β can promote the outgrowth of epidermal cells from organ cultures (Hebda, 1988) despite the fact that it is a potent inhibitor of keratinocyte proliferation *in vitro* (Shipley *et al.*, 1986). Perhaps TGF-β stimulates epidermal cell migration as it does monocyte migration (Wahl *et al.*, 1987) or perhaps TGF-β promotes epidermal cell movement through the induction of fibronectin matrix deposition (Nickoloff *et al.*, 1988; Wikner *et al.*, 1988) and fibronectin receptor expression (Gailit *et al.*, 1994).

As reepithelialization ensues, basement membrane proteins reappear in a very ordered sequence from the margin of the wound inward in a zipperlike fashion (Clark *et al.*, 1982b). Epidermal cells revert to their normal phenotype, once again firmly attaching to the reestablished basement membrane through hemidesmosomes and to the underlying neodermis through type VII collagen fibrils (Gipson *et al.*, 1988). Table VI summarizes the activities of epidermal cells during reepithelialization.

Table VI. Epidermal Cells in Wound Repair

Activity	Effect
Growth factor production	Autocrine and paracrine stimulation
Migration and proliferation	Reepithelialization
Protease release	Dissection under clot and nonviable tissue
ECM production	Provisional matrix and basement membrane formation
Terminal differentiation	Barrier function reestablished

4. Granulation Tissue: Reestablishing Dermal Integrity

New stroma, often called granulation tissue, begins to form approximately 4 days after injury. The name derives from the granular appearance of newly forming tissue when it is incised and visually examined. In fact, numerous new capillaries endow the neostroma with its granular appearance. Besides new blood vessels, granulation tissue consists of macrophages, fibroblasts, and loose connective tissue.

Cytokines, with chemoattractant, mitogenic, and other regulatory activities, presumably are probably necessary for granulation tissue induction. Net chemotaxis, cell proliferation, and phenotype modulation depends on the type and quantity of cytokines present, the activity level of the target cells, and the extracellular matrix environment (Sporn and Roberts, 1986; Damsky and Werb, 1992; Juliano and Haskill, 1992). Cytokines with potent mitogenic activities are usually referred to as growth factors (see Chapters 5–8). Low levels of some growth factors circulate in the plasma; however, activated platelets release substantial amounts of preformed growth factors into wounded areas. Arrival of peripheral blood monocytes and their activation to macrophages establish conditions for continual synthesis and release of growth factors. In addition, injured and activated parenchymal cells can synthesize and secrete growth factors.

The provisional extracellular matrix, e.g., fibrin clot, also promotes granulation tissue formation by providing scaffolding for contact guidance (fibronectin and collagen), low impedance for cell mobility (hyaluronic acid), a reservoir for cytokines (Nathan and Sporn, 1991), and direct signals to the cells through integrin receptors (Damsky and Werb, 1992).

Macrophages, fibroblasts, and blood vessels move into the wound space as a unit (Hunt, 1980), which correlates well with the proposed biological interdependence of these cells during tissue repair. That is, macrophages provide a continuing source of cytokines necessary to stimulate fibroplasia and angiogenesis, fibroblasts construct new extracellular matrix necessary to support cell ingrowth, and blood vessels carry oxygen and nutrients necessary to sustain cell metabolism.

Granulation tissue formation will be discussed further under the categories: fibroplasia, a dynamic reciprocity of fibroblasts, cytokines, and extracellular matrix; and neovascularization, a protease- and integrin-dependent event.

4.1. Fibroplasia: A Dynamic Reciprocity of Fibroblasts, Cytokines, and Extracellular Matrix

Fibroplasia consists of granulation tissue components that arise from fibroblasts. Thus, fibroplasia is an admixture of fibroblasts and extracellular matrix. A procession of cytokines, with chemotactic, mitogenic, and modulatory activities, stimulate the fibroblast response necessary to elicit fibroplasia formation (Chapters 5–8). Many of these cytokines are released from platelets and macrophages (Fig. 2); however, fibroblasts themselves can produce cytokines to which they respond in an autocrine fashion (Sporn and Roberts, 1986). Regardless of their exact origin, cytokines generated at a wound site most likely act in concert to induce fibroblast proliferation and migration into the wound space, and ECM production (also see Chapter 9). Multiple complex interactive biological phenomena occur within fibroblasts as they respond to wound cytokines, including the induction of additional cytokines (Loef *et al.*, 1986; Raines *et al.*, 1989) and modulation of cytokine receptor number or affinity (Oppenheimer *et al.*, 1983; Assoian *et al.*, 1984). *In vivo* studies support the hypothesis that growth factors are active in wound repair fibroplasia. Several studies have demonstrated that PDGF, PDGF-like peptides, TGF-α, and EGF-like peptides are present at sites of tissue repair (Grotendorst *et al.*, 1989; Matsuoka and Grotendorst, 1989; Katz *et al.*, 1991). Furthermore, purified and recombinant-derived growth factors have been shown to stimulate wound granulation tissue in normal and compromised animals (Sporn *et al.*, 1983; Lawrence *et al.*, 1986; Lynch *et al.*, 1989; Pierce *et al.*, 1989; Greenhalgh *et al.*, 1990; Mustoe *et al.*, 1991), and a single growth factor may work both directly and indirectly by inducing the production of other growth factors *in situ* (Mustoe *et al.*, 1991).

Structural molecules of the early extracellular matrix also contribute to tissue formation by providing a scaffold for contact guidance (fibronectin and collagen), low impedance for cell mobility (hyaluronic acid) (Toole, 1991), and a reservoir for cytokines (Nathan and Sporn, 1991). In addition, a dynamic reciprocity between fibroblasts and their surrounding extracellular matrix creates further complexity. That is, fibroblasts affect the extracellular matrix through new synthesis, deposition, and remodeling of the extracellular matrix (Kurkinen *et al.*, 1980; Welch *et al.*, 1990), while the extracellular matrix affects fibroblasts by regulating their function, including their ability to synthesize, deposit, and remodel the extracellular matrix (Mauch *et al.*, 1988; Grinnell, 1994; Clark *et al.*, 1995a). Thus, the interactions between extracellular matrix and fibroblasts dynamically evolve during granulation tissue development.

As fibroblasts migrate into the wound space, they initially penetrate the blood clot composed of fibrin and lesser amounts of fibronectin and vitronectin. Since cells can adhere to and detach from fibronectin substratum (Lark *et al.*, 1985) and fibronectin can bind to fibrin either noncovalently (Garcia-Pardo *et al.*, 1985) or covalently (Mosher and Johnson, 1983), fibroblasts can presumably use fibronectin matrix for move-

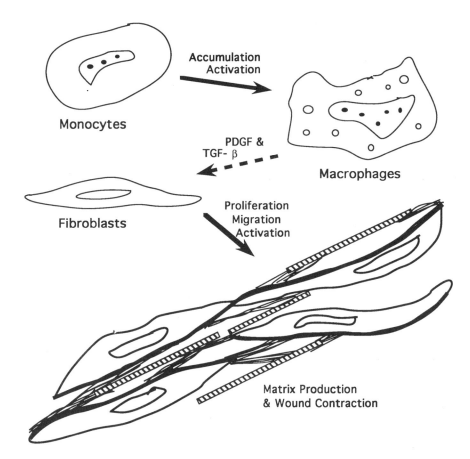

Monocytes

Accumulation
Activation

Macrophages

PDGF &
TGF- β

Fibroblasts

Proliferation
Migration
Activation

Matrix Production
& Wound Contraction

Figure 2. Initiation of fibroplasia by activated macrophages. Peripheral blood monocytes are recruited to a site of injury by a variety of specific and nonspecific chemotactic factors. Adhesion to extracellular matrix (ECM) substrata and certain cytokines activate the monocyte to become a fibrogenic macrophage that produces factors, such as PDGF and TGF-β, that stimulate fibroblast proliferation, migration, and ECM production, respectively. Fibroblasts that are thereby recruited to the wound transform into myofibroblasts, which contract the wound under PDGF and TGF-β stimulation once they have formed connection with each other and the ECM. (From Clark, 1993).

ment through the wound as they do for movement in tissue culture (Hsieh and Chen, 1983). Fibroblasts also may use fibrin and vitronectin directly as substratum for translocation since they can directly attach to both of these matrix proteins (Dejana *et al.,* 1984; Singer *et al.,* 1988). Fibroblasts bind to fibronectin, vitronectin, and fibrin through specialized cell membrane receptors of the integrin superfamily (Chapter 9) (Table I) (Ruoslahti, 1991; Hynes, 1992). The Arg-Gly-Asp-Ser (RGDS) tetrapeptide within the cell-binding domain of these proteins is critical for binding to the integrin receptors. In addition, the CSIII domain of fibronectin provides a second binding site for human dermal fibroblasts via the α4β1 integrin receptor (Gailit *et al.,* 1993). Interestingly, in studies on cultured fibroblastlike cell lines, the α4β1 integrin receptor facilitates movement, while the classic fibronectin receptor α5β1 retards movement, at least when overexpressed (Giancotti and Ruoslahti, 1990; Chan *et al.,* 1992). Importantly, wound fibroblasts produce a fibronectin that contains abundant alternative splice sites within the CSIII domain (ffrench-Constant *et al.,* 1989; Brown *et al.,* 1993a). Thus, *in vivo* expression of the α4β1 receptor may facilitate migration into a wound space, while expression of the α5β1 receptor may interrupt migration and induce the cell to reside at its existent location. In fact, α5β1 is maximally expressed in wound fibroblasts after they cease migration and establish links to the extracellular matrix (Welch *et al.,* 1990).

Both PDGF and TGF-β can stimulate fibroblasts to migrate (Seppa *et al.,* 1982; Senior *et al.,* 1985; Postlethwaite *et al.,* 1987) and can up-regulate integrin receptors (Heino *et al.,* 1989; Ahlen and Rubin, 1994; Gailit *et al.,* 1995). Therefore, these growth factors may be partially responsible for inducing a migrating fibroblast phenotype. Furthermore, the extracellular matrix alters PDGF regulation of fibroblast integrins (Xu and Clark, 1995). When fibroblasts are in a fibrin or fibronectin matrix, PDGF maximally stimulates receptors for these ligands, while maximal stimulation of collagen receptors occur when fibroblasts are in collagen gels. Hence, fibroblast responds to PDGF differently, depending on the extracellular matrix environment. *In vitro* fibroblast migration also has been observed in response to a variety of chemoattractants, including fragments of the 5th component of complement (Postlethwaite *et al.,* 1979; Senior *et al.,* 1988); types I, II, and III collagen-derived peptides (Postlethwaite and Kang, 1976); a fibronectin fragment (Postlethwaite *et al.,* 1981); elastin-derived peptides (Senior *et al.,* 1980); and interleukin 4 (IL-4) (Postlethwaite and Seyer, 1991).

Mechanistically, fibroblasts move toward a chemotactic gradient by extending lamellipodia toward the stimulus while their opposite poles remain firmly bound until released by some unknown process (see Chapter 12). This kind of cellular reorganization is similar to fibroblast movement up a surface-bound adhesion gradient, a process called *haptotaxis* (Carter, 1970; Harris *et al.,* 1980; Trinkaus, 1984). In this system, fibroblasts translocate by extending lamellipodia randomly. Each cell protrusion competes for the cell's finite surface membrane. When one lamellipodium becomes dominant, the cell spreads in that direction, pulling the cell forward and at the same time inhibiting further random protrusion activity. As the cell oozes forward and breaks old adhesions, the excess membrane at the trailing edge becomes available to resume random protrusions. Again, the most adherent extension becomes dominant and motility continues. Mechanically lifting the trailing edge of a migrating fibroblast off the substratum accelerates the advance of the leading edge of the cell (Chen, 1981). This observation supports the idea that the leading edge competes for finite cell surface membrane. In addition, new membrane from the Golgi apparatus may be inserted selectively at the leading edge of the cell (Bergmann *et al.,* 1983).

Even in the absence of an adhesion gradient, extracellular matrix fibrils strongly influence the direction of fibroblast migration, since the cells tend to align and migrate along discontinuities in substrata to which they are attached, a process called *contact guidance* (Trinkaus, 1984). For example, cultured fibroblasts plated on preformed fibronectin fibrils migrate along, not across, the fibrils (Hsieh and Chen, 1983). In fact, neutrophils also migrate faster when the chemotactic gradient is parallel to surface-bound fibrin fibrils rather than perpendicular to this axis (Wilkinson and Lackie, 1983). Thus, extracellular matrix fibers can provide additional directional information to cells stimulated to migrate by chemotactic gradients. Thus, chemotactic, haptotactic, and contact guidance signals may all influence fibroblast migration into the provisional matrix-filled wound space.

Movement into a cross-linked fibrin blood clot or any tightly woven extracellular matrix may also necessitate an active proteolytic system that can cleave a path for migration. A variety of fibroblast-derived enzymes in conjunction with serum-derived plasmin are potential candidates for this task, including plasminogen activator, interstitial collagenase [matrix metalloproteinase 1, (MMP-1)], gelatinase (MMP-2), and stromelysin (MMP-3) (Grant *et al.*, 1987; Wilhelm *et al.*, 1987; Saus *et al.*, 1988; Stetler-Stevenson *et al.*, 1989). Some chemotactic factors, such as PDGF and TGF-β, can also stimulate the production and secretion of these proteinases (Laiho *et al.*, 1986; Overall *et al.*, 1989) (see Chapter 14 for details).

Once the fibroblasts have migrated into the wound, they gradually switch their major function to protein synthesis (Welch *et al.*, 1990). Initially, the endoplasmic reticulum and Golgi apparatus become more dispersed throughout the cytoplasm of each cell as the fibroblasts begin to deposit loose extracellular matrix composed of great quantities of fibronectin (Kurkinen *et al.*, 1980; Grinnell *et al.*, 1981). Ultimately, the migratory phenotype is completely supplanted by a profibrotic phenotype characterized by abundant rough endoplasmic reticulum and Golgi apparatus filled with new collagen protein (Welch *et al.*, 1990). Since TGF-β is highly expressed in these cells (Clark *et al.*, 1995a) and can induce fibroblasts to produce great quantities of collagen (Ignotz and Massague, 1986; Roberts *et al.*, 1986), one presumes a causal relationship.

Interleukin-4 also induces a modest increase in production of types I and III collagen as well as fibronectin (Postlethwaite *et al.*, 1992). Mast cells, which are present in wounds, as well as fibrotic tissue, produce IL-4 and may contribute to collagenous matrix accumulation by releasing IL-4 in these sites. Mast cells also produce an abundance of tryptase, a serine esterase, which has recently been demonstrated to stimulate cultured fibroblast proliferation (Ruoss *et al.*, 1991).

Once an abundant collagen matrix is deposited in the wound, fibroblasts cease collagen production despite the continuing expression of TGF-β (Clark *et al.*, 1995a). Although the stimuli responsible for fibroblast proliferation and matrix synthesis during wound repair were originally extrapolated from many *in vitro* investigations over the past two decades (Derynck, 1988; Ross and Raines, 1990; Sporn and Roberts, 1992) and then confirmed by *in vivo* manipulation of wounds within the last 10 years (Sprugel *et al.*, 1987; Pierce *et al.*, 1991; Schultz *et al.*, 1991), less attention had been directed toward elucidating the signals responsible for down-regulating fibroblast proliferation and matrix synthesis until more recently. Both *in vitro* and *in vivo* studies suggest that gamma-interferon may be one such factor (Duncan and Berman, 1985; Granstein *et al.*, 1987). In addition, collagen matrix can suppress both fibroblast proliferation and fibroblast collagen synthesis (Grinnell, 1994; Clark *et al.*, 1995a). In contrast, a fibrin or fibronectin matrix has little or no suppressive effect on the mitogenic or synthetic potential of fibroblasts (Clark *et al.*, 1995a) even though baseline type I collagen mRNA is somewhat suppressed in fibrin gels (Pardes *et al.*, 1995).

Although the attenuated fibroblast activity in collagen gels is not associated with

Table VII. Fibroblasts in Wound Repair

Activity	Effect
Growth factor production	Autocrine and paracrine stimulation
Proliferation and migration	Granulation tissue formation
Protease release	Provisional matrix lysis and ECM remodeling
ECM production	Connective tissue formation
Dynamic linkage between actin bundles and ECM	Tissue contraction
Program cell death	Transition from cell-rich granulation tissue to cell-poor scar

cell death, many fibroblasts in day-10 healing wounds develop pyknotic nuclei, a cytological marker for apoptosis or programmed cell death (Williams, 1991). The signal(s) for wound fibroblast apoptosis has not been elucidated; nevertheless, this phenomenon marks the transition from a fibroblast-rich granulation tissue to a relatively acellular scar. Thus, fibroplasia in wound repair is tightly regulated, whereas functional dysregulation of these processes occurs in fibrotic diseases such as keloid formation, morphea, and scleroderma. Table VII summarizes the activities of fibroblasts during wound healing.

4.2. Neovascularization: A Protease- and Integrin-Dependent Event

Fibroplasia would halt if neovascularization failed to accompany the newly forming complex of fibroblasts and extracellular matrix. The process of new blood vessel formation is called *angiogenesis* (see Chapter 11 for details) and has been extensively studied in the chick chorioallantoic membrane and the cornea (Folkman and Shing, 1992). One or two days after implantation of angiogenic material in the cornea, the basement membrane of venules in the adjacent limbus begins to fragment, probably secondary to local endothelial cell enzyme release. *In vitro* studies support this possibility, since endothelial cells migrating through a filter impregnated with radiolabeled basement membrane collagens degrade these collagens during their transit (Kalebic *et al.*, 1983) and capillary endothelial cells grown on human amnion basement membrane release plasminogen activator and collagenase in response to angiogenic stimuli (Magnatti *et al.*, 1989). Endothelial cells from the side of the venule closest to the angiogenic stimulus begin to migrate on the second day by projecting pseudopodia through fragmented basement membranes. Subsequently, the entire endothelial cell migrates into the perivascular space and other endothelial cells follow. Endothelial cells remaining in the parent vessel begin to proliferate by the second or third day, providing a continuing source of endothelial cells for angiogenesis. Ultimately, many endothelial cells within the neovasculature proliferate. However, endothelial cells at the capillary tip do not divide (Ausprunk and Folkman, 1977). Capillary bud formation and extension can continue temporarily even after sufficient x-irradiation has been given to inhibit all DNA synthesis (Sholley *et al.*, 1978). These findings imply that angiogenic stimuli may operate through chemotaxis and that endothelial replication may be a secondary event (Folkman, 1982). Capillary sprouts eventually branch at their tips and join to form capillary loops through which blood flow begins. New sprouts then extend from these loops to form a capillary plexus.

Angiogenesis is a complex process that relies on an appropriate extracellular matrix in the wound bed as well as phenotype alteration, stimulated migration, and mitogenic stimulation of endothelial cells (see Chapter 11 for details). Endothelial cell phenotype is modified during angiogenesis (Ausprunk and Folkman, 1977), but the angiogenic phenotype is not as well delineated as the fibroblast and epidermal cell alterations described previously.

The soluble factors that stimulate angiogenesis in wound repair are not known. Many candidates exist, however, including numerous endothelial cell growth and chemotactic factors recognized by *in vitro* assays as well as angiogenic substances demonstrated by corneal or chorioallantoic membrane systems (Folkman and Klagsbrun, 1987). Angiogenic activity has been recovered from activated macrophages as well as various tissues including the epidermis and cutaneous wounds. For awhile, acidic or basic fibroblast growth factor (aFGF and bFGF) appeared to be responsible for most of these activities (Folkman and Klagsbrun, 1987). More recently, however, other molecules have been also shown to have angiogenic activity. These include TGF-α, TGF-β, TNF-α, platelet-derived endothelial cell growth factor (PD-ECGF), angiogenin, angiotropin, vascular endothelial growth factor (VEGF), interleukin-8 (IL-8), PDGF, and low-molecular-weight substances including the peptide KGHK, low oxygen tension, biogenic amines, and lactic acid (Folkman and Shing, 1992; Koch *et al.,* 1992; Battegay *et al.,* 1994; Lane *et al.,* 1994).

The VEGF stimulates endothelial cell proliferation *in vitro* and causes marked vasopermeability *in vivo* (Keck *et al.,* 1989); thus, the factor has been called vascular permeability factor (VPF) as well as VEGF. It is related to the PDGF family of growth factors (see Chapter 7) and is produced in large quantities by the epidermis during wound healing (Brown *et al.,* 1992).

Despite promoting angiogenesis *in vivo* (Roberts *et al.,* 1986; Yang and Moses, 1990), TGF-β is inhibitory to the growth and proliferation of monolayer endothelial cell growth (Baird and Durkin, 1986; Frater-Schroder *et al.,* 1986; Heimark *et al.,* 1986). However, a more recent study has demonstrated that TGF-β is a mitogen for cultured endothelial cells that have formed capillarylike tubes (Iruela-Arispe and Sage, 1993). Likewise, cultured monolayer endothelial cells make PDGF-BB, but have no receptor for this ligand. In contrast, once the cultured cells form tubes, they express PDGF-β receptor and respond to the ligand that they no longer produce (Battegay *et al.,* 1994).

Angiogenin and TNF-α also stimulate angiogenesis *in vivo,* but inhibit cultured endothelial cell monolayer growth (Folkman and Shing, 1992). Whether these molecules stimulate mitogenesis of cultured endothelial cell tubes remains to be determined. Clearly, endothelial cell response depends on the microenvironment and the cell's phenotype. Molecules may also induce angiogenesis *in vivo* by stimulating chemotaxis of endothelial cells or by recruiting monocytes or other cells to produce angiogenic factors (Weisman *et al.,* 1988).

Turning to the protein side of the angiogenesis equation, peptides within proteins may have very different activities than the intact protein. As examples of this general phenomenon, biologically active fibrinopeptides A and B and anaphylatoxins C3a and C5a are released by proteolysis from fibrinogen and the complement components C3 and C5, respectively. In angiogenesis regulation, it has recently been discovered that KGHK peptides, which can be proteolytically released from osteopoutin (SPARC), are potent angiogenic factors, while the parent protein has moderate angiostatic activity (Funk and Sage, 1993; Lane *et al.*, 1994). Other low-molecular-weight molecules that appear to potentiate angiogenesis include low oxygen tension, biogenic amines, and lactic acid, all of which are generated in the relatively hypoxic wound environment. Some agents may promote angiogenesis indirectly by injuring parenchymal cells and thereby causing the release of FGF, which is otherwise not secreted (Jackson *et al.*, 1992). Alternatively, low oxygen tension may stimulate macrophages to produce and secrete angiogenic factors (Knighton *et al.*, 1983).

Folkman and Shing (1992) have postulated that endothelial cell migration can induce proliferation. If this is true, endothelial cell chemotactic factors may be critical for angiogenesis. Adult tissue extracts and fibronectin stimulate endothelial cell migration across filters in chemotactic chambers (Glaser *et al.*, 1980; Bowersox and Sorgente, 1982); platelet-derived factors stimulate radial outgrowth from irradiated endothelial cell colonies (Wall *et al.*, 1978); and heparin as well as platelet factors stimulate phagokinetic migration of endothelial cells cultured on surfaces coated with colloidal gold particles (Azizkhan *et al.*, 1980; Bernstein *et al.*, 1982). Some factors, of course, may have both mitogenic and chemotactic activities. For example, PDGF can be either a chemotactic factor or a mitogenic factor for dermal fibroblasts (Senior *et al.*, 1985).

Besides growth factors and chemotactic factors, an appropriate extracellular matrix is also necessary for angiogenesis (also see Chapter 11). For example, aortic endothelial cell migration depends on continued collagen secretion (Madri and Stenn, 1982), is accompanied by chondroitin and dermatan sulfate proteoglycan synthesis (Kinsella and Wight, 1986), is enhanced by type IV collagen but not laminin (Herbst *et al.*, 1988), and is inhibited by fibronectin (Madri *et al.*, 1989). Three-dimensional gels of extracellular matrix proteins have an even more pronounced effect on both large and small vessel endothelial cells. Rat epididymal microvascular cells cultured in type I collagen gels with TGF-β produce capillarylike structures within 1 week (Madri *et al.*, 1988). Omission of TGF-β markedly reduces the effect. In contrast, laminin-containing gels in the absence of growth factors induce human umbilical vein and dermal microvascular cells to produce capillarylike structures within 24 hr of plating (Kubota *et al.*, 1988). Together, these studies support the hypothesis that the extracellular matrix plays an important role in angiogenesis.

Findings of several *in vivo* studies on angiogenesis are in concert with the observation that extracellular matrix profoundly affects cultured endothelial cell morphology and function. Proliferating microvascular blood vessels adjacent to and within wounds transiently deposit increased amounts of fibronectin within the vascular wall (Clark *et al.*, 1982a,c). Furthermore, capillary buds in a neovascularizing cornea are surrounded by an amorphous provisional matrix instead of a basement membrane (Ausprunk *et al.*, 1981). This amorphous matrix is reminiscent of the fibronectin-rich provisional matrix under migrating wound epithelia (Clark *et al.*, 1982b). Since cultured microvascular endothelial cells adhere to fibronectin (Clark *et al.*, 1986), the fibronectin matrix associated with activated blood vessels may act as a contact guidance system for endothelial cell movement. In support of this hypothesis, angiogenesis in the chick chorioallantoic membrane is dependent on the expression of αvβ3, an integrin that recognizes fibrin and fibronectin, as well as vitronectin (Brooks *et al.,* 1994a). Furthermore, in porcine cutaneous wounds, αvβ3 is only expressed on capillary sprouts as they invade the fibrin clot (Clark *et al.,* 1995a). *In vitro* studies, in fact, demonstrate that αvβ3 can promote endothelial cell migration on provisional matrix proteins (Leavesley *et al.,* 1993).

With the multiple bits of data outlined above, a series of events leading to angiogenesis can be hypothesized (Fig. 3). Tissue injury causes tissue cell destruction and disruption. Proteolytic enzymes released into the connective tissue degrade extracellular matrix proteins, including fibronectin. Fibronectin fragments and other degradation products from other extracellular matrix proteins attract peripheral blood monocytes to the injured site. Activated macrophages and injured tissue cells release FGF, which stimulates endothelial cells to release plasminogen activator and procollagenase. Plasminogen activator converts plasminogen to plasmin and procollagenase to active collagenase, and in concert these two proteases digest basement membrane constituents (see Chapter 14 for details). The fragmentation of the basement membrane allows endothelial cells to migrate into the injured site in response to FGF, fibronectin fragments, heparin released from disrupted mast cells, and other endothelial cell chemoattractants. As endothelial cells migrate into the fibrin–fibronectin-rich wound, they

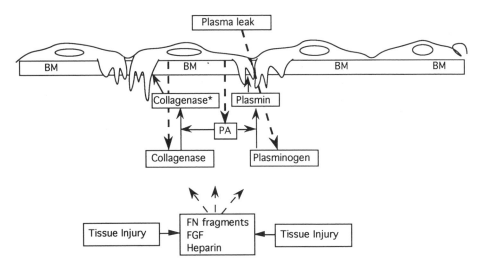

Figure 3. Initiation of angiogenesis. Fibroblast growth factor (FGF) released from injured cells stimulates the endothelial cells to release plasminogen activator (PA) and procollagenase. PA liberates plasmin from plasma-derived plasminogen and activates collagenase. Together these enzymes degrade the basement membrane (BM) beneath the stimulated endothelial cells. Endothelial cell chemoattractants such as heparin and fibronectin fragments (FNF), which have been released at the site of injury, stimulate the endothelial cells to project pseudopodia through the BM defect and subsequently to migrate into the connective tissue space. (From Clark, 1993).

Table VIII. Endothelial Cells in Wound Repair

Activity	Effect
Thrombomodulin and syndecan surface expression	Limits coagulation
Prostacyclin release	Limits platelet thrombi
Plasminogen activator release	Clot lysis
Surface expression of cell adhesion molecules	Leukocyte diapedesis
Metalloproteinase release	Basement membrane degradation
Growth factor production	Autocrine and paracrine stimulation
Migration and proliferation	Angiogenesis
ECM production	Provisional matrix and basement membrane formation
Tube formation	Blood flow

form tubes that express $\alpha v \beta 3$ integrin to facilitate adhesion and migration. The neovasculature first deposits its own provisional matrix containing fibronectin and proteoglycans, and ultimately forms a true basement membrane. The TGF-β may stimulate the fibronectin and proteoglycan synthesis as well as induce the correct endothelial cell phenotype for capillary tube formation. The FGF and other mitogens such as VEGF stimulate endothelial cell proliferation, resulting in a continual supply of endothelial cells for capillary extension. In summary, evidence has accumulated that angiogenesis is a complex process depending on at least four interrelated phenomena: cell phenotype alteration, chemoattractant-driven migration, mitogenic stimulation, and an appropriate extracellular matrix (see Chapter 11 for more details).

Within a day or two after removal of angiogenic stimuli, capillaries undergo regression as characterized by: mitochondrial swelling in the endothelial cells at the distal tips of the capillaries; platelet adherence to degenerating endothelial cells; vascular stasis; endothelial cell necrosis; and ingestion of the effete capillaries by macrophages (Ausprunk *et al.,* 1978). Although $\alpha v \beta 3$ has recently been shown to regulate apoptosis of endothelial cells in culture and in tumors (Brooks *et al.,* 1994b), $\alpha v \beta 3$ is not present on wound endothelial cells as they undergo programmed cell death, indicating another pathway of apoptosis in wound blood vessels (M. G. Tonnesen and R. A. F. Clark, unpublished observations). Endothelial cell activities that contribute to cutaneous wound repair in general are outlined in Table VIII.

5. Tissue Remodeling: Transition from Provisional Matrix to Collagenous Scar

Extracellular matrix remodeling, cell maturation, and cell apoptosis create the third phase of wound repair, which overlaps with tissue formation. In fact, remodeling of the extracellular matrix and maturation of the neoepidermis, fibroplasia, and neovasculature begin at the wound margin while granulation tissue is still invading the wound space in all but the smallest wounds (Kurkinen *et al.*, 1980). Once the wound is filled with granulation tissue and covered with a neoepidermis, fibroblasts transform into myofibroblasts, which contract the wound, and epidermal cells differentiate to reestablish the permeability barrier. Endothelial cells appear to be the first cell type to undergo apoptosis, followed by the myofibroblasts, leading gradually to a rather acellular scar. In the months following granulation tissue formation, the extracellular matrix continuously, albeit slowly, changes (Compton *et al.*, 1989).

Spatially, extracellular matrix is deposited first at the wound margin concurrent with granulation tissue development, and then more centrally as the granulation tissue grows into the wound space. However, at any given time, the extracellular matrix at the wound margin differs qualitatively and quantitatively from the extracellular matrix situated centrally. Thus, composition and structure of granulation tissue extracellular matrix depends both on the time elapsed since tissue injury and on the distance from the wound margin (Kurkinen *et al.*, 1980). The TGF-β, which is known to stimulate the production of fibronectin (Ignotz and Massagne, 1986) and collagen (Roberts *et al.*, 1986), appears in wound fibroblasts that are producing type I procollagen as judged by immunofluorescence (Clark *et al.*, 1995a). However, once the collagenous extracellular matrix is established, collagen synthesis ceases despite the continued presence of TGF-β (Clark *et al.*, 1995a). Thus, the progression of tissue formation, i.e., cell matrix maturation, at any given time and place in a wound probably depends not only on the cells, cytokines, and enzymes present in that locale but also on the extracellular matrix microenvironment. This complex interaction and feedback control of cells–cytokines–enzymes–matrix has been termed *dynamic reciprocity*.

Extracellular matrix components serve several critical functions for effective wound repair. First, during granulation tissue formation, fibronectin provides a provisional substratum for the migration and ingrowth of cells, a linkage for myofibroblasts to affect wound contraction, and a nidus for collagen fibrillogenesis. The presence of large quantities of highly hydrated hyaluronic acid in granulation tissue provides a matrix that is easily penetrated by ingrowing parenchymal cells. The early formation of types I, III, and V collagen fibrils provides nascent tensile strength for the wound. As the matrix matures over the ensuing weeks, the fibronectin and hyaluronic acid disappear; collagen bundles grow in size, increasing wound tensile strength; and proteoglycans are deposited, increasing wound resilience to deformation. It is also clear that many if not all extracellular matrix molecules can regulate cell function through interaction with specific cell surface receptors that transduce signals to the nucleus (see Chapters 2, 9, 15, and 16).

5.1. Provisional Matrix Alteration: Clot Lysis and Additional Deposition

Provisional matrix can consist of any extracellular matrix that is transiently present and has a provisional function. In wounds the fibrin clot with all its associated proteins is the classic example of a provisional matrix. Denatured collagen may also serve such a function, and several integrins have recently been found to bind this form of collagen, either directly or indirectly (Davis, 1992; Pfaff *et al.*, 1993; Tuckwell *et al.*, 1994). Once the initial fibrin matrix is cleared by proteolysis, the invading cells deposit a second provisional matrix rich in fibronectin and hyaluronan. Most if not all of these provisional matrix composites seem to promote cell proliferation and migration.

5.1.1. Fibrin Deposition and Lysis

The first extracellular matrix to be deposited in the wound space and surrounding tissue is the fibrin clot, which contains a plethora of other proteins including fibronectin, vitronectin, von Willebrand factor, and thrombospondin, as well as growth factors. This extracellular matrix has been termed the *provisional matrix* (Clark *et al.*, 1982b) (see Chapter 2). Fibrin is the major structural protein in the clot. It is derived from fibrinogen, a plasma protein circulating at 3 mg/ml. Fibrinogen has a molecular mass of 340 kDa and consists of three pairs of polypeptide chains, Aα, Bβ, and γ chains, held together by disulfide bonds. Fibrin is formed when thrombin cleaves first fibrinopeptide A and then fibrinopeptide B from fibrinogen. This process generates new amino-terminal ends in the α and β chains that interact with sites in the γ chain carboxyl-termini of adjacent fibrinogen molecules. Such interaction leads to noncovalent lateral assembly of fibrin monomers into protofibrils. Protofibrils aggregate to form fibrin fibrils that intertwine, producing a clot. For effective hemostasis, the fibrin clot must interact with platelets. This interaction is mediated by the platelet membrane receptor GPIIb-IIIa (αIIbβ3 integrin). On nonactivated platelets, this receptor selectively binds fibrinogen or fibrin. After platelet activation, however, the receptor binds von Willebrand factor, fibronectin, and vitronectin as well as fibrin(ogen) (Ginsberg *et al.*, 1992). Of note is the fact that the interaction of αIIbβ3 with a fibrinogen substrata can activate the receptor so that it recognizes other ligands (Du *et al.*, 1991). Through protein–protein interactions, platelet–protein interactions, and cell-protein interactions, the clot becomes firmly integrated in the wounded tissue where it provides a scaffold for cell migration and proliferation; a reservoir for growth factors, proteases, and protease inhibitors; and an inducer and modulator of cell function, as well as a hemostatic plug.

The fibrin clot may contain fibronectin (Mosesson and Umfleet, 1970; Clark *et al.*, 1982b), vitronectin (Preissner and Jenne, 1991), plasminogen (Castellino *et al.*, 1983), plasminogen activator (Thorsen *et al.*, 1972), plasminogen activator inhibitor (Wagner *et al.*, 1989), and thrombin (Liu *et al.*, 1979; Wilner *et al.*, 1981; Kaminski and Mc-Donagh, 1983; Siebenlist *et al.*, 1990). Furthermore, preliminary data indicate that two cytokines, TGF-β and PDGF, bind fibrinogen and fibrin. In addition, α_2-plasmin inhibitor can be reversibly cross-linked to fibrinogen or fibrin by factor XIIIa (Sakata and Aoki, 1980). As the fibrin clot undergoes degradation, these associated proteins may either be destroyed or released as active moieties. Of course, the associated plasminogen activator, plasminogen activator inhibitor, and α_2-plasmin inhibitor can exert major control over the degradation process itself.

Although fibrin can promote fibroblast migration into the matrix (Brown *et al.*, 1993b), fibrin-associated fibronectin (Mosesson and Umfleet, 1970) also appears to play a critical role for cell migration into clot (Knox *et al.*, 1986). Fibrin matrix also appears to play a complex but potentially critical role in fibroblast gene expression. For example, although fibrin depresses baseline fibroblast collagen mRNA (Pardes *et al.,* 1995), fibrin matrix is permissive to TGF-β stimulation of fibroblast collagen synthesis (Clark *et al.,* 1995a). In contrast, collagen matrix attenuates the fibroblast response to TGF-β. Furthermore, fibrin matrix also enhances the ability of PDGF to induce provisional matrix integrins α3β1 and α5β1 by stabilizing α3 and α5 mRNA (Xu and Clark, 1995). Thus, it is becoming clear that the fibrin provisional matrix promotes an early granulation tissue fibroblast phenotype. In contrast, collagen matrix promotes a relative quiescent dermal-resident fibroblast phenotype (Grinnell, 1994).

It appears that as the granulation tissue invades the fibrin clot, it is lysed (Clark *et al.,* 1995b). This is in concert with the observation that the leading edge of granulation tissue expresses proteases (Saarialho-Kere *et al.,* 1992). Concomitantly, fibroblasts begin to deposit a web of fibronectin (Welch *et al.,* 1990) immersed in a sea of hyaluronan (Bently, 1967). Thus, the wound transcends from the initial plasma-derived provisional matrix to *in situ* cell-derived provisional matrix (Clark *et al.,* 1983; Welch *et al.,* 1990), which can probably be more intimately controlled by the wound cells than the original fibrin clot.

5.1.2. Fibronectin

The clot provisional matrix contains much fibronectin (see Chapter 2). Fibronectin is a multifunctional cell adhesion protein found in blood and in a variety of tissues. Since it circulates in the blood at an appreciable concentration (0.3 mg/ml) and has an affinity for fibrin, it is thus present in clots at a substantial concentration. Specific functional domains and cell-binding sites permit fibronectin to interact with a remarkably wide range of cell types, extracellular matrix, and cytokines. Although encoded by a single gene, fibronectin exists in a number of variant forms that differ in sequence at three general regions of alternative splicing. Thus, its molecular weight varies from approximately 250 to 350 kDa. Nevertheless, it is always composed of two nearly identical chains that are linked by a single disulfhydryl bond.

Plasma fibronectin is first deposited in conjunction with fibrin, but after clot lysis, cellular fibronectin is deposited by a variety of wound cells (Clark *et al.,* 1982c, 1983; Grimwood *et al.,* 1988; Brown *et al.,* 1993a). Various spliced forms of fibronectin are found in the wound, but their differential roles have not been delineated (ffrench-Constant *et al.,* 1989; Brown *et al.,* 1993a). In general, these fibronectins probably have a multitude of functions in the wound (Clark, 1988). Clearly, fibronectin can support fibroblast, keratinocyte, and endothelial cell adhesion and movement (Grinnell and Feld, 1979; Hsieh and Chen, 1983; Clark *et al.,* 1985, 1986; O'Keefe *et al.,* 1985); and a 120-kDa fragment containing the Arg-Gly-Asp (RGD) cell binding is chemotactic for fibroblasts, endothelial cells, and monocytes (Postlethwaite *et al.,* 1981; Bowersox and Sorgente, 1982; Clark *et al.,* 1988; Doherty *et al.,* 1990). Fibronectin can also opsonize extracellular matrix debris and activate macrophages so they can more effectively phagocytize such particles and thus debride the wound (Pommier *et al.,* 1983; Brown and Goodwin, 1988). In addition to its role in cell recruitment and opsonization, fibronectin may serve as a template for collagen deposition (McDonald *et al.,* 1982).

Recently, the complex structural biology and regulatory control of fibronectin matrix assembly have become more clear (McDonald *et al.,* 1987; Morla and Ruoslahti, 1992; Mosher *et al.,* 1992; Checovich and Mosher, 1993; Somers and Mosher, 1993; Wu *et al.,* 1993; Zhang *et al.,* 1993; Hocking *et al.,* 1994; Sottile and Wiley, 1994) (also see Chapter 2).

Fibronectin acting through a variety of integrin receptors may also modulate gene expression similar to fibrin (Werb *et al.,* 1989, 1990; Xu and Clark, 1995). For example, fibroblasts that have adhered to the 120-kDa fibronectin fragment, containing the RGD cell-binding domain but not other integrin recognition sites, express metalloproteinase-1, the classic mammalian collagenase (Werb *et al.,* 1989). However, when fibroblasts reside on intact fibronectin, which contains the $\alpha4\beta1$ integrin recognition site CS-1, collagenase is not induced (Huhtala *et al.,* 1995). Hence, fibronectin transmits different signals to cells, depending on whether it is intact or fragmented. Fibronectin fragmentation, as would occur during granulation tissue invasion of the fibrin clot, might trigger a positive feedback loop by eliciting more enzyme secretion that would cause more fibronectin digestion.

After clot lysis, fibronectin is deposited by fibroblasts as a second-order provisional matrix (Welch *et al.,* 1990). The *in vivo* interrelationship of fibronectin and types I and III collagen deposition has been studied utilizing cellulose sponge implants in rats (Kurkinen *et al.,* 1980). Seven days after subcutaneous implantation, fibroblasts invaded 1–2 mm into the sponge and argyrophilic reticulin fibers, which stained uniformly and strongly for fibronectin, appeared. Type III procollagen and some type I collagen were present but trailed behind the leading edge of fibroblasts. As mentioned previously, unique spliced variants of fibronectin are deposited at this time (ffrench-Constant *et al.,* 1989). By 3–5 weeks, the fibroblasts had reached the center of the sponge where their numbers were now greatest, and the fibronectin and collagen distribution was as described in the sponge periphery at earlier times. The now more mature periphery contained birefringent collagen bundles that stained for type I collagen. Fibronectin was diminished after 5 weeks as more mature birefringent collagen fibers were formed. The appearance of fibronectin first and collagen second in healing wounds is consistent with fibronectin serving as a template for collagen fibril organization (McDonald *et al.,* 1982).

The sequence of fibronectin followed by interstitial collagen has also been observed in wounds of other mammals (Holund *et al.,* 1982) including humans (Viljanto *et al.,* 1981), in embryogenesis, and in certain pathological processes. In each, fibronectin appears to be an early component, followed by type III collagen and then type I collagen. The initial fibronectin matrix can be easily degraded by either cell or plasma proteases (Furie and Rifkin, 1980; McDonald and Kelley, 1980; Vartio *et al.,* 1981). The fibrillar collagens ultimately form fibrous bundles that greatly enhance the tissue tensile strength.

5.1.3. Hyaluronan

Hyaluronan, or hyaluronic acid (HA), is a linear polymer of repeating *N*-acetyl glucosamine-glucuronic acid disaccharides. This molecule is in the general class of polysaccharides termed glycosaminoglycans. All glycosaminoglycans are repeating disaccharides of hexosamine and an acidic sugar (see Chapter 15 for details). Most cells produce hyaluronan, but it is especially abundant around mesenchymal cells. Unlike other glycosaminoglycans it is synthesized at the plasma membrane through which the growing chain is extruded into the extracellular space (Prehm, 1983). This mechanism of synthesis allows for the production of very large molecules that can occupy an extraordinary volume since hyaluronan is extremely hydrophilic.

Hyaluronan is a major component of early granulation tissue. In open cutaneous wounds hyaluronan content increases early, falls from day 5 to 10, and then remains fairly constant, while the sulfated glycosaminoglycans, chondroitin-4-sulfate and dermatan sulfate, increase from day 5 to 7 (Bently, 1967). Fibroblasts isolated from early granulation tissue produce substantially more hyaluronic acid than fibroblasts from normal skin (Bronson et al., 1987). In addition, during regeneration and morphogenesis, HA appears at times of cell movement and mitosis and disappears at the onset of differentiation (Toole and Gross, 1971; Toole, 1972). That hyaluronan appears to promote cell movement is supported by the concomitant occurrence of hyaluronan and cell migration during both tissue repair and organ generation. Interestingly, no decrease in hyaluronan is observed during cutaneous healing of the fetal sheep (Longaker et al., 1991), and this could contribute to the lack of scarring observed in fetal skin repair (Adzick and Longaker, 1992) (see Chapter 18).

Hyaluronan probably promotes cell movement in early granulation tissue as it does in embryogenesis and morphogenesis (Toole, 1991). At least three possibilities exist, not necessarily mutually exclusively, for the role of hyaluronan in cell motility. First, HA may facilitate adhesion–disadhesion between the cell membrane and the matrix substratum during cell movement. It has been proposed that interaction of cell surface heparan sulfate and fibronectin mediates cell attachment to substratum and that accumulation of hyaluronan weakens this adhesion (Lark et al., 1985). Precise regulation of cell surface heparan sulfate and hyaluronan could result in waves of adhesion–disadhesion and cell movement. Second, since hyaluronan becomes extremely hydrated, the expanded interstitial space at sites of deposition might allow more cell recruitment and proliferation in these areas (Toole, 1981). Finally, specific cell surface receptors exist for hyaluronan, and cell movement into hyaluronan-rich areas is likely to be mediated, at least in part, by such transmembrane proteins (Toole, 1991; Stamenkovic and Aruffo, 1994). Although several distinct receptors for hyaluronan probably exist, only two have been well characterized at the molecular level. The CD44 receptor is widely distributed and has multiple isoforms that result from alternative splicing of 10 exons (Goldstein et al., 1989; Stamenkovic et al., 1989; Screaton et al., 1992; Underhill, 1992). Although the fibroblast receptor for hyaluronan-mediated motility (RHAMM) is not related to the link family of proteins as is CD44 (Hardwick et al., 1992), it does share an HA binding motif with the link family of proteins (Yang et al., 1994). These receptors seem to have different, but perhaps overlapping, cellular functions. CD44 mediates cell attachment to HA, cellular uptake, and degradation of hyaluronan and locomotion on hyaluronan substrates (Thomas et al., 1992; Underhill, 1992). RHAMM mediates cell locomotion in response to soluble HA (Turley, 1992). Compared to fibroblasts from normal skin, fibroblasts from hypertrophic scar display an increased expression of CD44 and a decreased internalization of CD44 in the presence of hyaluronan (Messadi and Bertolami, 1993). RHAMM is induced on macrophages and fibroblasts in rat skin wounds (E. A. Turley, personal communication), as well as on infiltrating macrophages in other types of tissue injury where RHAMM expression coincides with TGF-β1 secretion (Savani et al., 1994a,b). The TGF-β1 in vitro stimulates fibrosarcoma cell motility by simultaneously inducing the expression of RHAMM and the secretion of hyaluronan (Samuel et al., 1993). It is possible that TGF-β in wounds stimulates migration by a similar mechanism.

In addition to evidence that hyaluronan may facilitate cell movement, several lines of investigation suggest that hyaluronan may promote cell division. Hyaluronan production is greater during fibroblast proliferation in vitro (Hopwood and Dorfman, 1977). Furthermore, stimulated proliferation of cultured fibroblasts by serum (Tomida et al., 1974), insulin (Moscatelli and Rubin, 1975), or EGF (Lembach, 1976) results in a marked increase in hyaluronan production. Thus, the presence of hyaluronan in the extracellular matrix may be important for cell division. The mechanistic reason may be similar to that proposed for cell motility. Hyaluronan receptor has been found to be preferentially expressed on proliferating epithelial cells (Alho and Underhill, 1989), suggesting that the proliferating cells may be more interactive with hyaluronan.

As granulation tissue matures, hyaluronan is decreased through the action of tissue hyaluronidase, an enzyme that has specific endoglycosidic activity (Bertolami and Donoff, 1982). The sulfated glycosaminoglycans that replace hyaluronan are associated with a protein core and are called proteoglycans. These substances provide the tissue with more resilience than hyaluronan, but accommodate cell movement and proliferation less well. Thus, early in granulation tissue formation, fibroblasts deposit a fibronectin and hyaluronan matrix that is conducive to cell migration and proliferation and later a collagen and proteoglycan matrix that increases tissue tensile strength and resilience.

5.2. Deposition of Proteoglycans and Collagen

5.2.1. Proteoglycans

Proteoglycans contain a core protein to which at least one glycosaminoglycan chain is covalently bound (see Chapter 15 for details). However, many proteoglycans contain numerous glycosaminoglycan chains that may be of one or several types. Individual glycosaminoglycan chains can vary from as few as 10 to as many as 20,000 disaccharides. Classically, proteoglycans were named for the most prevalent glycosaminoglycan chain in their structure; however, with the cloning of several proteoglycan protein cores, specific proteoglycans have taken on specific names. The core proteins are diverse and heterogeneous, but share the capacity to bear glycosaminoglycan chains. Often, proteoglycans form a noncovalent aggregate with hyaluronan. Thus, enormous molecular versatility permits proteoglycans to have many diverse structural and organizational functions in tissues. Proteoglycans are found in cell secretory granules, at the cell surface as either intrinsic or extrinsic membrane proteins, and in the extracellular matrix. In fact, it is most logical to classify these molecules according to their resident location (see Table II in Chapter 15). Thus, proteoglycans will be noted as vesicular, extracellular, or transmembrane. Functions of these molecules include extracellular organization, growth factor storage in the extracellular matrix, promotion of growth factor receptor binding, enzyme and autocoid storage in cell granules, and regulation of blood coagulation.

Proteoglycans first act in wounds as regulators of blood coagulation and time-release capsules for mast cell and platelet products. Heparin and heparan sulfate side chains, from the mast cell vesicular proteoglycan serglycin and from the endothelial cell transmembrane proteoglycan syndecan, respectively, can interact with antithrombin III to form complexes that inhibit factor Xa and thrombin (Danielson *et al.,* 1986; Kojima *et al.,* 1992). Besides acting as an anticoagulant, serglycin heparin of mast cell granules binds histamine and a variety of cationic proteases. After mast cell granule discharge associated with injury, these biologically active factors are slowly released into the surrounding tissue through cation exchange (Wight *et al.,* 1991). Histamine dilates blood vessels and increases vasopermeability, presumably to bring more cells and nutrients to the site of injury. The mast cell proteases tryptase and chymase probably facilitate wound debridement. Importantly, tryptase is also know to activate the very potent matrix metalloproteinase stromelysin, which in turn activates procollagenase (Gruber *et al.,* 1989). Platelet factor-4 (PF-4), the prototype molecule of a newly recognized family of cytokines called chemokines (Wolpe and Cerami, 1989), binds to chondroitin sulfate and exists in platelet α-granules complexed with the chondroitin sulfate proteoglycan serglycin (Perin *et al.,* 1988). Other members of the PF-4 family may bind to serglycin within platelet and leukocyte granules. If so, the interaction may provide a reservoir for PF-4-like molecules at sites of inflammation and injury.

Extracellular matrix proteoglycans containing chondroitin-4-sulfate and dermatan sulfate increase during the second week of wound repair as hyaluronan is on the wane (Bently, 1967) and are produced by mature scar fibroblasts (Bronson *et al.,* 1988). Presumably these glycosaminoglycans are structural components of the proteoglycans decorin, biglycan, and versican (see Table II in Chapter 15). Decorin, in fact, has been shown to vary in wounds (Yeo *et al.,* 1991). Extracellular matrix proteoglycan function in wounds is not entirely clear; however, great strides have been made over the last decade in elucidating possible functions for these molecules. Clearly, versican, which is related to the cartilage proteoglycan aggrecan, contributes substantially to tissue resilience (Wight *et al.,* 1991). Several lines of *in vitro* and *in vivo* evidence also suggest that proteoglycans have the capacity to regulate collagen fibrillogenesis (McPherson *et al.,* 1988; Scott, 1993). Chondroitin-4-sulfate has been shown to accelerate polymerization of monomer collagen *in vitro* (Wood, 1960). Since this glycosaminoglycan occurs at high levels in granulation tissue (Kischer and Shetlar, 1974) but not in mature scar (Shetlar *et al.,* 1972), it may facilitate collagen deposition during the matrix formation and remodeling phase of wound healing. That both collagen synthesis (Cohen *et al.,* 1971) and chondroitin-4-sulfate levels (Shetlar *et al.,* 1972) are elevated in hypertrophic scars supports this concept. Proteoglycans, like other matrix molecules, are probably continually remodeled during wound healing, especially at epithelial– and endothelial–stroma interfaces.

In addition to their structural role, extracellular matrix proteoglycans have the capacity to regulate cell function. Versican may promote cell migration by decreasing adhesion (Yamagata *et al.*, 1993). Hyaluronan and other extracellular matrix proteoglycans also modulate cell adhesion either directly or indirectly (Lark *et al.*, 1985; Toole, 1991). Extracellular matrix proteoglycans also can regulate cell proliferation. For example, a cloned and expressed chondroitin–dermatan sulfate proteoglycan, called decorin, inhibits the growth of Chinese hamster ovary cells (Yamaguchi and Ruoslahti, 1988). Since chondroitin–dermatan sulfates appear in wounds at approximately the time cell proliferation decreases, a causal relationship may exist between the two phenomena. Previously, heparin and heparan sulfate were found to inhibit smooth muscle cell growth (Castellot *et al.*, 1981); and the synthesis and expression of sulfated glycosaminoglycans had been noted to be inversely related to cell growth (Hopwood and Dorfman, 1977). Furthermore, specific shedding of plasma membrane-bound heparan sulfate proteoglycan has been noted to occur immediately prior to cell division (Kraemer and Tobey, 1972).

The previous observations should not be taken as an indication that sulfated glycosaminoglycans and proteoglycans always down-regulate cell growth. On the contrary, FGF is bound by heparan sulfate within basement membranes and can be released in an active form when the heparan sulfate is degraded by heparanase expressed by normal or malignant cells (Vlodavsky *et al.*, 1991). Furthermore, binding of basic FGF (FGF-2) to its receptor requires prior binding either to a membrane-bound heparan sulfate proteoglycan or to free heparin glycosaminoglycans chains (Yayon *et al.*, 1991). Other cytokines that also bind heparin or heparan sulfate include FGF-1 and FGF-4 (Guimond *et al.*, 1993), FGF-7 (Reich-Slotky *et al.*, 1994), PDGF (Ross and Raines, 1990), granulocyte–macrophage colony-stimulating factor, interleukin-3 (Roberts *et al.*, 1988), pleiotrophin, a neurite-promoting factor (Merenmies and Rauvala, 1990), PF-4 (Wolpe and Cerami, 1989), TGF-β1 (McCaffrey *et al.*, 1992), and VEGF (Gitay-Goren *et al.*, 1992). The biological function of these heparin (heparan) cytokines is under intense investigation. Some interactions at the cell membrane clearly lead to cytokine receptor activation (Yayon *et al.*, 1991; Gitay-Goren *et al.*, 1992; Guimond *et al.*, 1993; Reich-Slotky *et al.*, 1994). Cytokines may also interact with the protein core of the proteoglycan as exemplified by TGF-β's interaction with decorin (Yamaguchi *et al.*, 1990). Bound TGF-β is inactive, but since the reaction is reversible, active TGF-β may be released at a later time. In summary, extracellular matrix proteoglycans may have direct effects on cell growth or may act as a repository for cytokines. The net effect on cell growth depends on the context in which the cell encounters the proteoglycan.

Cell surface proteoglycans are also modulated during wound repair and may be critical for proper cell migration, proliferation, and gene expression. The transmembrane syndecan family has received the most study in wounds. Increased expression of syndecan-1 has been observed in mouse wounds during granulation tissue formation and in hyperproliferative keratinocytes near the wound edge (Elenius *et al.*, 1991). In contrast, syndecan-1 expression is decreased during rabbit corneal reepithelialization (Gruskin-Lerner and Trinkaus-Randall, 1991). Unpublished data regarding syndecan-1 and -4 expression in human wounds is reviewed in Chapter 15. Binding of the heparan sulfate side chains of syndecan to other extracellular matrix molecules such as collagens, fibronectin, and laminin would serve to mediate cell movement or retention in the newly forming tissue. In fact, both syndecan-1 and -4 facilitate fibroblast focal contact formation (Yamagata *et al.,* 1993; Woods and Couchman, 1994). In addition, pericellular and transmembrane heparan sulfate proteoglycans facilitate FGF binding to its receptor, and thereby can have a great effect on cell response to this important growth factor (Yayon *et al.,* 1991; Guimond *et al.,* 1993; Spivak-Kroizman *et al.,* 1994).

5.2.2. Collagen

Collagens are a family of glycoproteins containing triple helices, which are found in the extracellular matrix (see Chapter 16 for details). Non-matrix proteins that contain triple helical domains, such as the complement component Clq, are not included in the collagen family (see Table II in Chapter 16). At present, there are 18 collagen types designated type I–XVIII according to their chronological order of discovery (Fukai *et al.,* 1994). These collagens can be divided into four major classes (Linsenmayer, 1991; Fukai *et al.,* 1994); fibrillar collagens with uninterrupted triple helices (types I, II, III, V, XI); collagens with interrupted triple helices that form a meshwork in the lamina densa of basement membrane (type IV); fibrillar collagens with interrupted helices (types VI and VII); and nonfibrillar collagens (types VIII–XVIII) (see Fig. 1 in Chapter 16 for a conformational grouping). Nonfibrillar collagens have been further subdivided into short-chain collagens with uninterrupted triple helices (types VIII and X), fibril-associated collagens with interrupted triple helices, often referred to as FACIT proteins (types IX, XII, XIV, and XVI), multiplexins (types XV and XVIII), and the transmembrane collagen BP180 of keratinocyte hemidesmosomes (type XVII) (see Chapter 17 for details). Fibrillar collagens types II and XI and nonfibrillar collagens types IX, X, and XII are found only in cartilage.

The principal characteristic of the fibrillar collagens is the ability of the monomeric collagen molecules to polymerize both side-by-side and end-on-end into long fibrillar aggregates (Birk *et al.,* 1990). Such fibrillar bundles constitute the major structural collagens in all connective tissues. In contrast, type IV procollagen fails to undergo proteolytic processing in the extracellular matrix, and thus retains its large globular terminal domains, which prevents fibrillar aggregation. These nonhelical domains interact, resulting in the multimolecular collagen meshwork of basement membranes (Yurchenco and Schittny, 1990) (see Chapters 16 and 17). Type VI collagen forms distinctive 100-nm periodic microfibrils intercalated between the 67-nm periodic fibrils of types I and III collagen of noncartilagenous stroma like the dermis (Bruns *et al.,* 1986). Type VII collagen forms the anchoring fibrils of epidermal basement membranes (Sakai *et al.,* 1986) (see Chapter 17 for details). Type VIII collagen is found in the connective tissue around hair follicles as well as around arterioles and venules (Sawada *et al.,* 1990).

Most studies on the collagen content of healing wounds and artificially induced granulation tissue (sponge implants) have examined types I and III collagens, since these two collagens have been characterized for some time and their supramolecular structures are well defined (see Chapter 16 for details). Bazin and Delaunay (1964) first showed, by biochemical techniques, that granulation tissue contained a collagen distinct from the normal adult dermis and rather similar to that in embryonic skin. This collagen was later recognized as type III (Epstein, 1974), which does occur in small amounts in normal dermis but is greatly increased in granulation tissue (Gabbiani *et al.*, 1976). As previously discussed, granulation tissue matrix deposition occurs in an ordered sequence of fibronectin, type III collagen, and type I collagen (Kurkinen *et al.*, 1980).

Rigid helical collagen macromolecules aggregated into fibrillar bundles gradually provide the healing tissue with increasing stiffness and tensile strength (Levenson *et al.*, 1965). After a 5-day lag, a high rate of type I collagen synthesis begins, which coincides with increased wound-breaking strength (Diegelmann *et al.*, 1975; Gabbiani *et al.*, 1976). Types I and III fibrillar collagen deposition peaks between the 7 and 14 days (Clore *et al.*, 1979). A similar time course for types I and III collagen mRNA expression has been documented (Scharffetter *et al.*, 1989; Oono *et al.*, 1993). Type V collagen also increases during granulation tissue development in parallel with tissue vascularity, thus suggesting an association between capillary endothelial cells and type V collagen (Hering *et al.*, 1983). The finding of abundant type V collagen in hypertrophic scars (Ehrlich and White, 1981), which have numerous capillaries, provides additional support for this possibility. Type VI collagen gene expression in the neo-stroma of healing cutaneous wounds has also been delineated (Oono *et al.*, 1993). As with types I and III collagen mRNAs, type VI collagen mRNA peaks between 1 and 2 weeks after injury. Collagen VI gene expression was localized to fibroblasts and endothelial cells of newly forming blood vessels. It may provide an important anchor for the neovasculature (Keene *et al.*, 1988).

Besides providing structural support and strength to the new tissue, collagen can have a profound effect on the cells within and on its matrix. For example, collagen-derived peptides act as chemoattractants for fibroblasts *in vitro* (Postlethwaite *et al.*, 1978) and may have a similar activity *in vivo*. In addition, intact collagen can alter the phenotype and function of many different cell types (Hay, 1991; Juliano and Haskill, 1992; Lin and Bissell, 1993). These effects may be mediated, in part, through activation of the integrin collagen receptors $\alpha1\beta1$ and $\alpha2\beta1$ (Staatz *et al.*, 1989; Wayner and Carter, 1989; Ignatius *et al.*, 1990). With regard to fibroblasts, collagen matrices reduce cell proliferation and collagen synthesis (Grinnell, 1994) but induce procollagenase (Unemore and Werb, 1986) and $\alpha2\beta1$ integrin expression (Klein *et al.*, 1991). Perhaps the collagen-rich extracellular matrix, which accumulates in mature granulation tissue, reduces the ability of wound fibroblasts to produce further collagenous matrix, but promotes the ability of these cells to remodel collagen-rich matrix already present. Such dynamic reciprocity between cells and extracellular matrix most likely continues until the proper balance of cells and extracellular matrix is obtained.

5.3. Wound Contraction and Extracellular Matrix Reorganization

After the initial synthesis and deposition of types I, III, and VI fibrillar collagens, myofibroblasts remodel the matrix by wound contraction. Prior to the mid-1950s, wound contraction was thought to be driven by the physicochemical forces of collagen polymerization, which intrinsically lead to collagen bundle shortening. This concept originated from investigations on clot contraction that demonstrated that contraction resulted from the realignment of fibrin fibrils in a more compact array and with a lower free energy state. However, in 1956, Abercrombie and co-workers (Abercrombie *et al.*, 1956) demonstrated that wounds in ascorbic-acid-deficient animals can close independently of collagen formation, casting doubt on the notion that physicochemical forces alone were responsible for wound contraction. Fifteen years later, Gabbiani, Majno, and their co-workers (Majno *et al.*, 1971; Gabbiani *et al.*, 1972) demonstrated that wound fibroblasts assumed some characteristics of smooth muscle cells and that these so-called myofibroblasts were responsible for the ability of granulation tissue to contract *ex vivo* in response to a variety of autocoids (see Chapter 13 for details). Thus, investigators began to focus on the myofibroblast as the engine behind wound contraction.

Fibroblasts undergo a series of phenotypic changes during granulation tissue formation that continually modify their interactions with the extracellular matrix. First, fibroblasts assume a migratory phenotype and then switch to a profibrotic phenotype during which they produce abundant types I and III collagen (Table VII) (Gabbiani *et al.*, 1978; Welch *et al.*, 1990). Subsequently, during the second and third week of healing, fibroblasts begin to assume a myofibroblast phenotype characterized by large bundles of actin-containing microfilaments disposed along the cytoplasmic face of the plasma membrane and the establishment of cell–cell and cell–matrix linkages (Welch *et al.*, 1990). The appearance of the myofibroblasts corresponds to the commencement of connective tissue compaction and the contraction of the wound. Fibroblasts link to the extracellular fibronectin matrix through $\alpha5\beta1$ and presumably other fibronectin receptors (Singer *et al.*, 1984; Welch *et al.*, 1990); to collagen matrix through $\alpha1\beta1$ and $\alpha2\beta1$ collagen receptors (Staatz *et al.*, 1989; Wayner and Carter, 1989; Ignatius *et al.*, 1990); and to each other through direct adherens junctions (Welch *et al.*, 1990). New collagen bundles in turn have the capacity to join end-to-end with collagen bundles at the wound edge and to ultimately form covalent cross-links among themselves and with the collagen bundles of the adjacent dermis (Yamauchi *et al.*, 1987; Birk *et al.*, 1989, 1990). These cell–cell, cell–matrix, and matrix–matrix links provide a network across the wound whereby the traction of fibroblasts on their pericellular matrix can be transmitted across the wound (Singer *et al.*, 1984).

Wound contraction is now ascribed to the actin-rich myofibroblasts that, in fact, are the most numerous cells in mature granulation tissue and are aligned within the wound along the lines of contraction (see Chapter 13). By contrast, neither capillaries nor macrophages are aligned along wound contraction lines. Cultured fibroblasts dispersed within a hydrated collagen gel provides a functional *in vitro* model of tissue contraction (Bell *et al.*, 1979). When serum is added to the admixture, contraction of the collagen matrix occurs over the course of a few days. When observed with time-lapse microphotography, collagen condensation appears to result from a "collection of collagen bundles" executed by fibroblasts as they extend and retract pseudopodia attached to collagen fibers (Bell *et al.*, 1983). The transmission of these traction forces across the *in vitro* collagen matrix depends on two linkage events: fibroblast attachment to the collagen matrix through the $\alpha2\beta1$ integrin receptors (Schiro *et al.*, 1991) and cross-links between the individual collagen bundles (Woodley *et al.,* 1991). This linkage system probably plays a significant role in the *in vivo* situation of wound contraction as well. In addition, cell–cell adhesions and cell–fibronectin linkages appear to provide an additional means by which the traction forces of the myofibroblast may be transmitted across the wound matrix (Gabbiani *et al.,* 1978; Singer *et al.,* 1984; Welch *et al.,* 1990).

F-actin bundle arrays, cell–cell and cell–matrix linkages, and collagen cross-links are all facets of the biomechanics of extracellular matrix contraction. The contraction process, however, probably needs a cytokine signal. In fact, cultured fibroblasts mixed in a collagen gel contract the collagen matrix only if serum is added to the medium (Bell *et al.,* 1979). Since PDGF, the major fibroblast mitogen found in serum (Ross and Raines, 1990), also had been found to stimulate fibroblast migration (Seppa *et al.,* 1982; Senior *et al.,* 1985) and arterial smooth muscle contraction (Berk *et al.,* 1986), its ability to initiate fibroblast contraction of collagen matrix was investigated. The major platelet and macrophage isoforms (AB and BB) of PDGF were found to stimulate fibroblasts to contract collagen matrix, while the major fibroblast isoform (AA) had no activity (Clark *et al.,* 1989). Since PDGF is present in wounds (Rappolee *et al.,* 1988), this factor may also provide the signal for wound contraction. Interestingly, when monocytes are cultured on a matrix *in vitro,* they develop the phenotype of a tissue macrophage and express PDGF B-chain mRNA in a biphasic time course (Shaw *et al.,*1990). The second peak of expression is at 6 days, which corresponds to the time that extracellular matrix contraction occurs in a healing wound. Perhaps tissue macrophages release a second wave of PDGF-BB or -AB approximately 1 week after cutaneous injury—a time when myofibroblasts have filled the wound and are linked to each other and to the extracellular matrix. This then may be the signal for wound contraction to commence. Thus, wound contraction represents a complex and masterfully orchestrated interaction of cells, extracellular matrix, and cytokines.

Collagen remodeling during the transition of granulation tissue to mature scar is dependent on both continued collagen synthesis and collagen catabolism. The degradation of wound collagen is controlled by a variety of collagenase enzymes from granulocytes, macrophages, epidermal cells, and fibroblasts (see Chapter 14). These collagenases are specific for particular types of collagens, but most cells probably contain two or more different types of these enzymes (Hasty *et al.,* 1986). Currently, three major metalloproteinases with collagenase activity have been well characterized: MMP-1 or interstitial collagenase, which cleaves types I, II, III, XIII, and X collagens (Grant *et al.,* 1987), MMP-2 or gelatinase, which degrades denatured collagens of all types and native types V and XI collagens (Hibbs *et al.,* 1987; Stetler-Stevenson *et al.,* 1989), and MMP-3 or stromelysin, which degrades types III, IV, V, VII, and IX collagens as well as proteoglycans and glycoproteins (Saus *et al.,* 1988; Okada *et al.,* 1989). Two other metalloproteinases that may have matrix-regulating functions are integral membrane proteins: enkapphalinase (Werb and Clark, 1989) and meprin (Butler and Bond, 1988). These activities are controlled by various inhibitor counterparts called tissue inhibitor of metalloproteinases (TIMP), which are tightly regulated during development (Brenner *et al.,* 1989) and are likely to be so during wound repair. In fact, modulation of collagenase activity and TIMP can greatly affect branching morphogenesis of the submandibular gland (Fukuda *et al.,* 1988). Cytokines such as TGF-β, PDGF, and IL-1 and the extracellular matrix itself may play an important role in the modulation of collagenase and TIMP expression *in vivo* (Werb *et al.,* 1990; Circolo *et al.,* 1991; Sporn and Roberts, 1992).

Wounds gain only about 20% of their final strength by the third week, during which time fibrillar collagen has accumulated relatively rapidly and has been remodeled coordinately with myofibroblast-driven wound contraction. Thereafter, the rate at which wounds gain tensile strength is slow, reflecting a much slower rate of collagen accumulation. In fact, the gradual gain in tensile strength has less to do with new collagen deposition than with further collagen remodeling with formation of larger collagen bundles and an alteration of intermolecular cross-links (Bailey *et al.,* 1975). Even so, wounded tissue fails to attain the same breaking strength as uninjured skin. At maximum strength a scar is only 70% as strong as intact skin (Levenson *et al.,* 1965).

References

Abercrombie, M., Flint, M. H., and James, D. W., 1956, Wound contraction in relation to collagen formation in scorbutic guinea pigs, *J. Embryol. Exp. Morph.* **4:**167–175.

Adzick, N. S., and Longaker, M. T., 1992, *Fetal Wound Healing,* Elsevier, New York.

Ahlen, K., and Rubin, K., 1994, Platelet-derived growth factor-BB stimulates synthesis of the integrin α2-subunit in human diploid fibroblasts, *Exp. Cell Res.* **215:**347–353.

Albelda, S. M., and Buck, C. A., 1990, Integrins and other cell adhesion molecules, *FASEB J.* **4:**2868–2880.

Alho, A. M., and Underhill, C. M., 1989, The hyaluronate receptor is preferentially expressed on poliferating epithelial cells, *J. Cell Biol.* **108:**1557–1566.

Assoian, R. K., Frolik, C. A., Roberts, A. B., Miller, D. M., and Sporn, M. B., 1984, Transforming growth factor-β controls receptor levels for epidermal growth factor in NRK fibroblasts, *Cell* **36:**35–41.

Assoian, R. K., Fleurdelys, B. E., Stevenson, H. C., Miller, P. J., Madtes, D. K., Raines, E. W., Ross, R., and Sporn, M. B., 1987, Expression and secretion of type β transforming growth factor by activated human macrophages, *Proc. Natl. Acad. Sci. USA* **84:**6020–6024.

Ausprunk, D. H., and Folkman, J., 1977, Migration and proliferation of endothelial cells in preformed and newly formed blood vessels during tumor angiogenesis, *Microvasc. Res.* **14:**53–65.

Ausprunk, D. H., Falterman, K., and Folkman, J., 1978, The sequence of events in the regression of corneal capillaries, *Lab. Invest.* **38:**284–294.

Ausprunk, D. H., Boudreau, C. L., and Nelson, D. A., 1981, Proteoglycans in the microvasculature. II. Histochemical localization in proliferating capillaries of the rabbit cornea, *Am. J. Pathol.* **103:**367–375.

Azizkhan, R. G., Azizkhan, J. C., Zetter, B. R., and Folkman, J., 1980, Mast cell heparin stimulates migration of capillary endothelial cells *in vitro, J. Exp. Med.* **152:**931–944.

Bailey, A. J., Bazin, S., Sims, T. J., LeLeus, M., Nicholetis, C., and Delaunay, A., 1975, Characterization of the collagen of human hypertrophic and normal scars, *Biochim. Biophys. Acta* **405:**412–421.

Baird, A., and Durkin, T., 1986, Inhibition of endothelial cell proliferation by type-beta transforming growth factor: Interactions with acidic and basic fibroblast growth factors, *Biochem. Biophys. Res. Commun.* **138:**476–482.

Baird, A., Mormede, P., and Bohlen, P., 1985, Immunoreactive fibroblast growth factor in cells of peritoneal exudate suggests its identity with macrophage growth factor, *Biochem. Biophys. Res. Commun.* **126:**358–364.

Barrandon, Y., and Green, H., 1987, Cell migration is essential for sustained growth of keratinocytes colonies: The roles of transforming growth factor-α and epidermal growth factor, *Cell* **50:**1131–1137.

Bar-Shavit, R., Kahn, A., Fenton, J. W., and Wilner, G. D., 1983, Chemotactic response of monocytes to thrombin, *J. Cell Biol.* **96:**282–285.

Battegay, E. F., Rupp, J., Iruela-Arispe, L., Sage, E. H., and Pech, M., 1994, PDGF-BB modulates endothelial proliferation and angiogenesis *in vitro* via PDGF β-receptors, *J. Cell Biol.* **125**:917–928.

Bazin, S., and Delaunay, A., 1964, Biochimie de l'inflammation. VI. Fluctuations du taux de collagene et des
proteines non fibrillaires dans differents types de foyers inflammatoires, *Am. Inst. Pasteur* **107**:163–
172.

Bell, E., Ivarsson, B., and Merrill, C., 1979, Production of a tissue-like structure by contraction of collagen
lattices by human fibroblasts of different proliferative potential *in vitro, Proc. Natl. Acad. Sci. USA*
76:1274–1278.

Bell, E., Sher, S., Hull, B., Merrill, C., Rosen, S., Chamson, A., Asselineau, D., Dubertret, L., Coulomb, B.,
Lepiere, C., Nusgens, B., and Neveux, Y., 1983, The reconstitution of living skin, *J. Invest. Dermatol.*
81(suppl.):2S–10S.

Bently, J. P., 1967, Rate of chondroitin sulfate formation in wound healing, *Ann. Surg.* **165**:186–191.

Bergmann, J. E., Kupfer, A., and Singer, S. J., 1983, Membrane insertion at the leading edge of motile
fibroblasts, *Proc. Natl. Acad. Sci. USA* **80**:1367–1371.

Berk, B. C., Alexander, R. W., Brock, T. A., and Gimbrone, J., M.A., 1986, Vasoconstriction: A new activity
for platelet-derived growth factor, *Science* **232**:87–90.

Bernstein, L. R., Antoniades, H., and Zetter, B. R., 1982, Migration of cultured vascular cells in response to
plasma and platelet-derived factors, *J. Cell Sci.* **56**:71–82.

Bertolami, C. N., and Donoff, R. B., 1982, Identification, characterization, and partial purification of
mammalian skin wound hyaluronidase, *J. Invest. Dermatol.* **79**:417–421.

Birk, D. E., Zycband, E. I., Winkelmann, D. A., and Trelstad, R. L., 1989, Collagen fibrillogenesis *in situ:*
Fibril segments are intermediates in assembly, *Proc. Natl. Acad. Sci. USA* **86**:4549–4553.

Birk, D. E., Zycband, E. I., Winkleman, D. A., and Trelstad, R. L., 1990, Collagen fibrilogenesis *in situ, NY
Acad. Sci.* **580**:176–194.

Bowersox, J. C., and Sorgente, N., 1982, Chemotaxis of aortic endothelial cells in response to fibronectin,
Cancer Res. **42**:2547–2551.

Brachmann, R., Lindquist, P. B., Nagashima, M., Kohr, W., Lipari, T., Napier, M., and Derynck, R., 1989,
Transmembrane TGF-α precursors activate EGF/TGF-α receptors, *Cell* **56**:691–700.

Brenner, C. A., Adler, R. R., Rappolee, D. A., Pederson, R. A., and Werb, Z., 1989, Genes for extracellular
matrix-degrading metalloproteases and their inhibitor, TIMP, are expressed during early mammalian
development, *Genes Dev.* **3**:848–859.

Bronson, R. E., Bertolami, C. N., and Siebert, E. P., 1987, Modulation of fibroblast growth and glycosaminoglycan synthesis by interleukin-1, *Coll. Rel. Res.* **7**:323–332.

Bronson, R. E., Argenta, J. G., and Bertolami, N., 1988, Interleukin-1 induced changes in extracellular glycosaminoglycan composition of cutaneous scar-derived fibroblasts in culture, *Coll. Rel. Res.* **8**:199–208.

Brooks, P. C., Clark, R. A. F., and Cheresh, D. A., 1994a, Requirement of vascular integrin αvβ3 for
angiogenesis, *Science* **264**:569–571.

Brooks, P. C., Montgomery, A. M. P., Rosenfeld, M., Reisfeld, R. A., Hu, T., Klier, G., and Cheresh, D. A.,
1994b, Integrin αvβ3 antagonists promote tumor regression by inducing apoptosis of angiogenic blood
vessels, *Cell* **79**:1157–1164.

Brown, E. J., and Goodwin, J. L., 1988, Fibronectin receptors of phagocytes. Characterization of the arg-gly-
asp binding proteins of human monocytes and polymorphonuclear leukocytes, *J. Exp. Med.* **167**:777–
793.

Brown, G. L., Nanney, L. B., Griffen, J., Cramer, A. B., Yancey, J. M., Curtsinger, L. J., Holtzin, L., Schultz,
G. S., Jurkiewicz, M. H., and Lynch, J. B., 1989, Enhancement of wound healing by topical treatment
with epidermal growth factor, *N. Engl. J. Med.* **321**:76–79.

Brown, L. F., Yeo, K.-T., Berse, B., Yeo, T.-K., Senger, D. R., Dvorak, H. F., and Van De Water, L., 1992,
Expression of vascular permeability factor (vascular endothelial growth factor) by epidermal keratinocytes during wound healing, *J. Exp. Med.* **176**:1375–1379.

Brown, L. F., Dubin, D., Lavigne, L., Logan, B., Dvorak, H. F., and Van De Water, L., 1993a, Macrophages
and fibroblasts express "embryonic" fibronectins during cutaneous wound healing, *Am. J. Pathol.*
142:793–801.

Brown, L. F., Lanir, N., McDonagh, J., Tognazzi, K., Dvorak, A. M., and Dvorak, H. F., 1993b, Fibroblast
migration in fibrin gel matrices, *Am. J. Pathol.* **142**(1):273–283.

Bruns, R. R., Press, W., Engvall, E., Timpl, R., and Gross, J., 1986, Type VI collagen in extracellular, 100 nm periodic filaments and fibrils: Identification by immunoelectron microscopy, *J. Cell Biol.* **103**:393–404.

Butler, P. E., and Bond, J. S., 1988, A latent proteinase in mouse kidney membranes. Characterization and relationship to meprin, *J. Biol. Chem.* **263**:13419–13426.

Campbell, E. J., Cury, J. D., Lazarus, C. J., and Welgus, H. G., 1987, Monocyte procollagenase and tissue inhibitor of metallopoteinases. Identification, characterization and regulation of secretion, *J. Biol. Chem.* **262**:15862–15868.

Carter, S. B., 1970, Cell movement and cell spreading: A passive or an active process? *Nature* **255**:858–859.

Castellino, F. J., Strickland, D. K., Morris, J. P., Smith, J., and Chibber, B., 1983, Enhancement of the streptokinase-induced activation of human plasminogen by human fibrinogen and human fibrinogen fragment D1, *Ann. NY Acad. Sci.* **408**:595–601.

Castellot, J. J., Addonizio, M. L., Rosenberg, R., and Karnovosky, M. J., 1981, Vascular endothelial cells produce a heparin-like inhibitor of smooth muscle growth, *J. Cell Biol.* **90**:372–379.

Cavani, A., Zambruno, G., Marconi, A., Manca, V., Marchetti, M., and Giannetti, A., 1993, Distinctive integrin expression in the newly forming epidermis during wound healing in humans, *J. Invest. Dermatol.* **101**:600–604.

Chan, B. M., Kassner, P. D., Schiro, J. A., Byers, R., Kupper, T. S., and Hemler, M. E., 1992, Distinct cellular functions mediated by different VLA integrin α subunit cytoplasmic domains, *Cell* **68**:1051–1060.

Checovich, W. J., and Mosher, D. F., 1993, Lysophosphatidic acid enhances fibronectin binding to adherent cells, *Arterioscler. Thromb.* **13**:1662–1667.

Chen, W.-T., 1981, Mechanism of retraction of the trailing edge during fibroblast movement, *J. Cell Biol.* **90**:187–200.

Ciano, P. S., Colvin, R. B., Dvorak, A. M., McDonagh, J., and Dvorak, H. F., 1986, Macrophage migration in fibrin gel matrices, *Lab. Invest.* **54**:62–70.

Circolo, A., Welgus, H. G., Pierce, G. F., Kramer, J., and Strunk, R. C., 1991, Differential regulation of the expression of proteinases/antiproteinases in fibroblasts. Effects of interleukein-1 and platelet-derived growth factor, *J. Biol. Chem.* **266**:12283–12288.

Clark, R. A. F., 1988, Potential roles of fibronectin in cutaneous wound repair, *Arch. Dermatol.* **124**:201–206.

Clark, R. A. F., 1990, Fibronectin matrix deposition and fibronectin receptor expression in healing and normal skin, *J. Invest. Dermatol.* **94**(Suppl):128S–134S.

Clark, R. A. F., 1993, Mechanisms of cutaneous wound repair, in: *Dermatology in General Medicine* (T. B. Fitzpatrick, A. Z. Eisen, K. Wolff, I. M. Freedberg, and K. F. Austen, eds.), pp. 473–486, McGraw Hill, New York.

Clark, R. A. F., DellaPelle, P., Manseau, E., Lanigan, J. M., Dvorak, H. F., and Colvin, R. B., 1982a, Blood vessel fibronectin increases in conjunction with endothelial cell proliferation and capillary ingrowth during wound healing, *J. Invest. Dermatol.* **79**:269–276.

Clark, R. A. F., Lanigan, J. M., DellaPelle, P., Manseau, E., Dvorak, H. F., and Colvin, R. B., 1982b, Fibronectin and fibrin provide a provisional matrix for epidermal cell migration during wound reepithelialization, *J. Invest. Dermatol.* **70**:264–269.

Clark, R. A. F., Quinn, J. H., Winn, H. J., Lanigan, J. M., DellaPelle, P., and Colvin, R. B., 1982c, Fibronectin is produced by blood vessels in response to injury, *J. Exp. Med.* **156**:646–651.

Clark, R. A. F., Quinn, H. J., Winn, H. J., and Colvin, R. B., 1983, Fibronectin beneath reepithelializing epidermis *in vivo:* Sources and significance, *J. Invest. Dermatol.* **80**(Suppl):26S–30S.

Clark, R. A. F., Folkvord, J. M., and Wertz, R. L., 1985, Fibronectin, as well as other extracellular matrix proteins, mediates human keratinocyte adherence, *J. Invest. Dermatol.* **84**:378–383.

Clark, R. A. F., Folkvord, J. M., and Nielsen, L. D., 1986, Either exogenous or endogenous fibronectin can promote adherence of human endothelial cells, *J. Cell Sci.* **82**:263–280.

Clark, R. A. F., Wikner, N. E., Doherty, D. E., and Norris, D. A., 1988, Cryptic chemotactic activity of fibronectin for human monocytes resides in the 120 kDa fibroblastic cell-binding fragment, *J. Biol. Chem.* **263**:12115–12123.

Clark, R. A. F., Folkvord, J. M., Hart, C. E., Murray, M. J., and McPherson, J. M., 1989, Platelet isoforms of platelet-derived growth factor stimulate fibroblasts to contract collagen matrices, *J. Clin. Invest.* **84**:1036–140.

Clark, R. A. F., Nielsen, L. D., Welch, M. P., and McPherson, J. M., 1995a, Collagen matrices attenuate the collagen synthetic response of cultured fibroblasts to TGF-β, *J. Cell Sci.* **108:**1251–1261.

Clark, R. A. F., Tonnesen, M. G., Gailit, J., and Cheresh, D. A., 1995b, Transient functional expression of αvβ3 on vascular cells during wound repair, *Am. J. Path.*, in press.

Clore, J. N., Cohen, I. K., and Biegelmann, R. F., 1979, Quantitation of collagen type I and III during wound healing in rat skin, *Proc. Soc. Exp. Biol. Med.* **161:**337–340.

Coffey, R. J., Derynck, R., Wilcox, J. N., Bringman, T. S., Goustin, A. S., Moses, H. L., and Pittelkow, M. R., 1987, Induction and autoinduction of TGF-α in human keratinocytes, *Nature* **328:**817–820.

Cohen, I. K., Keiser, H. R., and Sjoerdsma, A., 1971, Collagen synthesis in human keloid and hypertrophic scar, *Surg. Forum* **22:**488–489.

Compton, C. C., Gill, J. M., Bradford, D. A., Regauer, S., Galico, C. G., and O'Conner, N. E., 1989, Skin regenerated from cultured epithelial autografts on full-thickness burn wounds from 6 days to 5 years after grafting. A light, electron microscope and immunohistochemical study, *Lab. Invest.* **60:**600–612.

Cotsarelis, G., Sun, T.-T., and Lavker, R. M., 1990, Label-retaining cells reside in the bulge area of pilosebaceous unit: Implications for follicular stem cells, hair cycle, and skin carcinogenesis, *Cell* **61:**1329–1337.

Damsky, C. H., and Werb, Z., 1992, Signal transduction by integrin receptors for extracellular matrix: Cooperative processing of extracellular information, *Curr. Opin. Cell Biol.* **4:**772–781.

Danielson, A., Raub, E., Lindahl, U., and Bjork, I., 1986, Role of ternary complexes, in which heparin binds both antithrombin and proteinase, in the acceceleration of the reactions between antithrombin and thrombin or factor Xa, *J. Biol. Chem.* **261:**15467–15473.

Davis, E. D., 1992, Affinity of integrins for damaged extracellular matrix: avb3 binds to denatured collagen type I through RGD sites, *Biochem. Biophys. Res. Commun.* **182:**1025–1031.

Dejana, E., Vergara-Dauden, M., Balconi, G., Pietra, A., Cherel, G., Bonati, M. B., Larrieu, M. J., and Marguerie, G., 1984, Specific binding of human fibrinogen to cultured human fibroblasts. Evidence for the involvement of the E domain, *Eur. J. Biochem.* **139:**657–662.

Derynck, R., 1988, Transforming growth factor-α, *Cell* **54:**593–595.

Diegelmann, R. F., Rothkopf, L. C., and Cohen, I. K., 1975, Measurement of collagen biosynthesis during wound healing, *J. Surg. Res.* **19:**239–243.

Dinarello, C. A., 1984, Interleukin-1 and the pathogenesis of the acute-phase response, *N. Engl. J. Med.* **311:**1413–1418.

DiScipio, R. G., 1982, The activation of the alternative pathway C3 convertase by human plasma kallikrein, *Immunology* **45:**587–595.

Doherty, D. E., Haslet, C., Tonnesen, M. G., and Henson, P. M., 1987, Human monocyte adherence: A primary effect of chemotactic factors on the monocyte to stimulate adherence to human endothelium, *J. Immunol.* **138:**1762–1771.

Doherty, D. E., Henson, P. M., and Clark, R. A. F., 1990, Fibronectin fragments containing the RGDS cell-binding domain mediate monocyte migration into the rabbit lung, *J. Clin. Invest.* **86:**1065–1075.

Du, X. P., Plow, E. F., Frelinger, A. L., O'Toole, T. E., Loftus, J. C., and Ginsberg, M. H., 1991, Ligands activate integrin αIIbβ3 (platelet GPIIb-IIIa), *Cell* **65:**409–416.

Duncan, M. R., and Berman, B., 1985, Gamma interferon is the lymphokine and beta interferon the monokine responsible for inhibition of fibroblast collagen production and late but not early fibroblast proliferation, *J. Exp. Med.* **162:**516–527.

Ehrlich, H. P., and White, B. S., 1981, The identification of A and B collagen chains in hypertrophic scars, *Exp. Mol. Pathol.* **34:**1–8.

Elenius, K., Vainio, S., Laato, M., Salmivirta, M., Theslef, I., and Jalkanen, M., 1991, Induced expression of syndecan in healing wounds, *J. Cell. Biol.* **114:**585–595.

Elsbach, P., and Weiss, J., 1992, Oxygen-independent antimicrobial systems of phagocytosis, in: *Inflammation: Basic Principles and Clinical Correlates* (J. I. Gallin, I. M. Goldstein, and R. Snyderman, eds.), pp. 603–636, Raven Press, New York.

Epstein, E. H. J., 1974, α1(III)$_3$ human skin collagen. Release by pepsin digestion and preponderance in fetal life, *J. Biol. Chem.* **249:**3225–3231.

Fernandez, H. Å., Henson, P. M., Otani, A., and Hugli, T. E., 1978, Chemotactic response to human C3a and

C5a anaphylatoxins. I. Evaluation of C3a and C5a leukotaxis *in vitro* and under simulated *in vivo* conditions, *J. Immunol.* **120**:109–115.

ffrench-Constant, K., Van De Water, L., Dvorak, H. F., and Hynes, R. O., 1989, Reappearance of an embryonic pattern of fibronectin splicing during wound healing in the adult rat, *J. Cell. Biol.* **109**:903–914.

Folkman, J., 1982, Angiogenesis: Initiation and control, *Ann. NY Acad. Sci.* **401**:212–227.

Folkman, J., and Klagsbrun, M., 1987, Angiogenic factors, *Science* **235**:442–448.

Folkman, J., and Shing, T., 1992, Angiogenesis, *J. Biol. Chem.* **267**:10931–10934.

Frater-Schroder, M., Muller, G., Birchmeirer, W., and Bohlem, P., 1986, Transforming growth factor-beta inhibits endothelial cell proliferation, *Biochem. Biophys. Res. Commun.* **137**:295–302.

Fujikawa, L. S., Footer, C. S., Gipson, I. K., and Colvin, R. B., 1984, Basement membrane components in healing rabbit corneal epithelial wounds: Immunofluorescence and ultrastructural studies, *J. Cell Biol.* **98**:128–138.

Fukai, N., Apte, S. S., and Olsen, B. R., 1994, Nonfibrillar collagens, in: *Extracellular Matrix Components* (E. Ruoslahti and E. Engvall, eds.), pp. 3–28, Academic Press, San Diego, CA.

Fukuda, Y., Masuda, Y., Kishi, J. I., Hashimoto, Y., Hayakawa, T., Nogawa, H., and Nakanishi, Y., 1988, The role of interstitial collagens in cleft formation of mouse embryonic submandibular gland during initial branching, *Development* **103**:259–268.

Funk, S. E., and Sage, E. H., 1993, Differential effects of SPARC and cationic SPARC peptides on DNA synthesis by endothelial cells and fibroblasts, *J. Cell. Physiol.* **154**:53–63.

Furie, B., and Furie, B. C., 1988, The molecular basis of blood coagulation, *Cell* **53**:505–518.

Furie, M. B., and Rifkin, D. B., 1980, Proteolytically derived fragments of human plasma fibronectin and their localization within intact molecule, *J. Biol. Chem.* **365**:3134–3140.

Gabbiani, G., Hirschel, B. J., Ryan, G. B., Statkov, P. R., and Majno, G., 1972, Granulation tissue as a contractile organ. A study of structure and function, *J. Exp. Med.* **135**:719–734.

Gabbiani, G., Lelous, M., Bailey, A. J., and Delauney, A., 1976, Collagen and myofibroblasts of granulation tissue. A chemical, ultrastructural and immunologic study, *Virchows Arch. B Cell Pathol.* **21**:133–145.

Gabbiani, G., Chapponnier, C., and Huttner, I., 1978, Cytoplasmic filaments and gap junctions in epithelial cells and myofibroblasts during wound healing, *J. Cell Biol.* **76**:561–568.

Gailit, J., Pierschbacher, M., and Clark, R. A. F., 1993, Expression of functional α4 integrin by human dermal fibroblasts, *J. Invest. Dermatol.,* in press.

Gailit, J., Welch, M. P., and Clark, R. A. F., 1994, TGF-β1 stimulates expression of keratinocyte integrins during re-epithelialization of cutaneous wounds, *J. Invest. Dermatol.* **103**:221–227.

Gailit, J., Bueller, H., and Clark, R., 1995, Platelet-derived growth factor and inflammatory cytokines have differential effects on the expression of integrins α1β1 and α5β1 by human dermal fibroblasts, *J. Invest. Dermatol.,* in press.

Garcia-Pardo, A., Pearlstein, E., and Frangione, B., 1985, Primary structure of human plasma fibronectin. Characterization of a 31,000 dalton fragment from the COOH-terminal region containing a free sulfhydryl group and a fibrin binding site, *J. Biol. Chem.* **260**:10320–10325.

Ghebrehiwet, B., Silverberg, M., and Kaplan, A. P., 1981, Activation of classic pathway of complement by Hageman factor fragment, *J. Exp. Med.* **153**:665–676.

Giancotti, F. G., and Ruoslahti, E., 1990, Elevated levels of the α5β1 fibronectin receptor suppress the transformed phenotype of chinese hamster ovary cells, *Cell* **60**:849–859.

Ginsberg, M. H., Loftus, J. C., and Plow, E. F., 1988, Cytoadhesins, integrins, and platelets, *Thromb. Haemost.* **59**:1–6.

Ginsberg, M. H., Du, X., and Plow, E. F., 1992, Inside-out integrin signalling, *Curr. Opin. Cell Biol.* **4**:766–771.

Gipson, I. K., Spurr-Michaud, S. J., and Tisdale, A. S., 1988, Hemidesmosomes and anchoring fibril collagen appear synchronously during development and wound healing, *Dev. Biol.* **126**:253–262.

Gitay-Goren, H., Soker, S., Vlodavsky, I., and Neufeld, G., 1992, The binding of vascular endothelial growth factor to its receptors is dependent on cell surface-associated heparin-like molecules, *J. Biol. Chem.* **267**:6093–6098.

Glaser, B. M., D'Amore, P. A., Seppa, H., Seppa, S., and Schiffmann, E., 1980, Adult tissues contain chemoattractants for vascular endothelial cells, *Nature* **288**:483–484.

Goldstein, L. A., Zhou, D. F. H., Picker, L. J., Minty, C. N., Bargatze, R. F., Ding, J. F., and Butcher, E. C., 1989, A human lymphocyte homing receptor, the Hermes antigen, is related to cartilage proteoglycan core and link proteins, *Cell* **56:**1063–1072.

Granstein, R. D., Murphy, G. F., Margolis, R. J., Byrne, M. H., and Amento, E. P., 1987, Gamma interferon inhibits collagen synthesis *in vivo* in the mouse, *J. Clin. Invest.* **79:**1254–1258.

Grant, G. A., Eisen, A. Z., Marmer, B. L., Roswit, W. T., and Goldberg, G. I., 1987, The activation of human skin fibroblast procollagenase. Sequence identification of the major conversion products, *J. Biol. Chem.* **262:**5886–5889.

Greenhalgh, D. G., Sprugel, K. H., Murray, M. J., and Ross, R., 1990, PDGF and FGF stimulate wound healing in the genetically diabetic mouse, *Am. J. Pathol.* **136:**1235–1246.

Grimwood, R. E., Baskin, J. B., Nielsen, L. D., Ferris, C. F., and Clark, R. A. F., 1988, Fibronectin extracellular matrix assembly by human epidermal cells implanted into athymic mice, *J. Invest. Dermatol.* **90:**434–440.

Grinnell, F., 1994, Fibroblasts, myofibroblasts, and wound contraction, *J. Cell. Biol.* **124:**401–404.

Grinnell, F., and Feld, M. K., 1979, Initial adhesion of human fibroblasts in serum-free medium: Possible role of secreted fibronectin, *Cell* **17:**117–129.

Grinnell, F., Feld, M., and Minter, D., 1980, Fibroblast adhesion to fibrinogen and fibrin substrata: Requirement for cold-insoluble globulin (plasma fibronectin), *Cell* **19:**517–525.

Grinnell, F., Billingham, R. E., and Burgess, L., 1981, Distribution of fibronectin during wound healing *in vivo, J. Invest. Dermatol.* **76:**181–189.

Grondahl-Hansen, J., Lund, L. R., Ralfkiaer, E., Ottevanger, V., and Dano, K., 1988, Urokinase- and tissue-type plasminogen activators in keratinocytes during wound reepithelilaization *in vivo, J. Invest. Dermatol.* **90:**790–795.

Grotendorst, G. R., Soma, Y., Takehara, K., *et al.,* 1989, EGF and TGF-alpha are potent chemoattractants for endothelial cells and EGF-like peptides are present at sites of tissue regeneration, *J. Cell. Physiol.* **139:**617–623.

Gruber, B. L., Marchese, M. J., Suzuki, K., Schwartz, L. B., Okada, Y., Nagase, H., and Ramamurthy, N. S., 1989, Synovial procollagenase activation by human mast cell tryptase dependence upon matrix metalloproteinase 3 activation, *J. Clin. Invest.* **84:**1657–1662.

Gruskin-Lerner, L. S., and Trinkaus-Randall, V., 1991, Localization of integrin and syndecan *in vivo* in a corneal epithelial abrasion and keratectomy, *Curr. Eye Res.* **10:**75–85.

Guimond, S., Maccarana, M., Olwin, B. B., Lindahl, U., and Rapraeger, A. C., 1993, Activating and inhibitory heparin sequences for FGF-2 (basic FGF): Distinct requirements for FGF-1, FGF-2 and FGF-4, *J. Biol. Chem.* **268:**23906–23914.

Hajjar, K., Jacovina, A., and Chacko, J., 1994, An endothelial cell receptor for plasminogen and tissue plasminogen activator: Identity with annexin II, *J. Biol. Chem.* **269:**21191–21197.

Hardwick, C., Hoare, K., Owens, R., Holn, H. P., Hook, M., Moore, D., Cripps, V., Austen, L., Nance, D. M., and Turley, E. A., 1992, Molecular cloning of a novel hyaluron receptor that mediates tumor cell motility, *J. Cell Biol.* **117:**1343–.

Harris, A. K., Wild, P., and Stopak, S., 1980, Silicone rubber substrata: A new wrinkle in the study of cell locomotion, *Science* **208:**177–179.

Hasty, K. A., Hibbs, M. S., Seyer, J. M., Mainardi, C. L., and Kang, A. H., 1986, Secreted forms of human neutrophil collagenase, *J. Biol. Chem.* **261:**5645–5650.

Hay, E. D., 1991, Collagen and other matrix glycoproteins in embryogenesis, in: *Cell Biology of the Extracellular Matrix* (E. D. Hay, ed.), pp. 419–462, Plenum Press, New York.

Hebda, P. A., 1988, Stimulatory effects of transforming growth factor-beta and epidermal growth factor on epidermal cell outgrowth from porcine skin explant cultures, *J. Invest. Dermatol.* **91:**440–445.

Hebda, P. A., Klingbeil, C. K., Abraham, J. A., and Fiddes, J. C., 1990, Basic fibroblast growth factor stimulation of epidermal wound healing in pigs, *J. Invest. Dermatol.* **95:**626–631.

Heimark, R. L., and Schwartz, S. M., 1988, The role of cell–cell interaction in the regulation of endothelial cell growth, in: *Molecular and Cellular Biology of Wound Repair* (R. A. F. Clark and P. M. Henson, eds.), pp. 359–371, Plenum Press, New York.

Heimark, R. L., Twardzik, D. R., and S. S., 1986, Inhibition of endothelial cell regeneration by type-β transforming growth factor from platelets, *Science* **233:**1078–1080.

Heino, J., Ignotz, R. A., Hemler, M. E., Crouse, C., and Massague, J., 1989, Regulation of cell adhesion receptors by transforming growth factor-β. Concomitant regulation of integrins that share a common β1 subunit, *J. Biol. Chem.* **264**:380–388.

Hennings, H., Michael, D., Cheng, D., Steinert, P., Holbrook, K., and Yuspa, S. H., 1980, Calcium regulation of growth and differentiation of mouse epidermal cells in culture, *Cell* **19**:245–254.

Herbst, T. J., McCarthy, J. B., Tsilibary, E. C., and Furcht, L. T., 1988, Differential effects of laminin, intact type IV collagen, and specific domains of type IV collagen on endothelial cell adhesion and migration, *J. Cell Biol.* **106**:1365–1373.

Hering, T. M., Marchant, R. E., and Anderson, J. M., 1983, Type V collagen during granulation tissue development, *Exp. Mol. Pathol.* **39**:219–229.

Hibbs, M. S., Hoidal, J. R., and Kang, A. H., 1987, Expression of a metallo-proteinase that degrades native type V collagen and denatured collagens by cultured human alveolar macorophages, *J. Clin. Invest.* **80**:1644–1650.

Higashiyama, S., Abraham, J. A., Miller, J., Fiddes, F. C., and Klagsbrun, M., 1991, A heparin-binding growth factor secreted by macrophage-like cells that is related to EGF, *Science* **251**:936–939.

Hocking, D. C., Sottile, J., and McKeown-Longo, P. J., 1994, Fibronectin's III-1 module contains a conformation-dependent binding site for the amino-terminal region of fibronectin, *J. Biol. Chem.* **269**:19183–19187.

Holund, B., Clemmensen, I., Junke, R. P., and Lyon, H., 1982, Fibronectin in experimental granulation tissue, *Acta Pathol. Microbiol. Immunol. Scand.* **90**:159–165.

Hopwood, J. J., and Dorfman, A., 1977, Glycosaminoglycan synthesis by cultured human skin fibroblasts after transformation with simian virus 40, *J. Biol. Chem.* **252**:4777–4785.

Hsieh, P., and Chen, L. B., 1983, Behavior of cells seeded on isolated fibronectin matrices, *J. Cell Biol.* **96**:1208–1217.

Huhtala, P., Humphries, M. J., McCarthy, J. B., Tremble, P. M., Werb, Z., and Damsky, C. H., 1995, α4β1 and α5β1 play differential roles in metalloproteinase induction, *J. Cell Biol.* **129**:867–879.

Hunt, T. K., 1980, *Wound Healing and Wound Infection: Theory and Surgical Practice,* Appleton-Century-Crofts, New York.

Hynes, R. O., 1992, Integrins: Versatility, modulation, and signaling in cell adhesion, *Cell* **69**:11–25.

Ignatius, M. J., Large, T. H., Houde, M., Tawil, J. W., Barton, A., Esch, F., Carbonetto, S., and Reichardt, L. F., 1990, Molecular cloning of the rate integrin α1-subunit: A receptor for laminin and collagen, *J. Cell Biol.* **111**:709–720.

Ignotz, R. A., and Massague, J., 1986, Transforming growth factor-β stimulates the expression of fibronectin and collagen and their incorporation into extracellular matrix, *J. Biol. Chem.* **261**:4337–4340.

Iruela-Arispe, M., and Sage, H., 1993, Endothelial cells exhibiting angiogenesis *in vitro* proliferate in response to TGF-β1, *J. Cell Biochem.* **52**:414–430.

Jackson, A., Friedman, S., Zhan, X., Engleka, K., Forough, R., and Maciag, T., 1992, Heat shock induces the release of FGF1 from NIH 3T3 cells, *Proc. Natl. Acad. Sci. USA* **89**:10691–10695.

Juliano, R. L., and Haskill, S., 1992, Signal transduction from the extracellular matrix, *J. Cell Biol.* **120**:577–585.

Kalebic, T., Garbisa, S., Glaser, B., and Liotta, L. A., 1983, Basement membrane collagen: Degradation by migrating endothelial cells, *Science* **221**:281–283.

Kaminski, M., and McDonagh, J., 1983, Studies on the mechanism of thrombin interaction with fibrin, *J. Biol. Chem.* **258**:10530–10535.

Katz, M. H., F, A. A., Kirsner, R. S., Eaglstein, W. H., and Falanga, V., 1991, Human wound fluid from acute wounds stimulates fibroblast and endothelial cell growth, *J. Am. Acad. Dermatol.* **25**:1054–1058.

Keck, P. J., Hauser, S. D., Krivi, G., Sanzo, K., Warren, T., Feder, J., and Connolly, D. T., 1989, Vascular permeability factor, an endothelial cell mitogen related to PDGF, *Science* **246**:1309–1313.

Keene, D. R., Engvall, E., and Glanvill, R. W., 1988, Ultrastructure of type VI collagen in human skin and cartilage suggests an anchoring function for this filamentous network, *J. Cell Biol.* **107**:1995–2006.

Kinsella, M. G., and Wight, T. N., 1986, Modulation of sulfated proteoglycan synthesis by bovine aortic endothelial cells during migration, *J. Cell Biol.* **102**:679–687.

Kischer, C. W., and Shetlar, M. R., 1974, Collagen and mucopolysaccharides in the hypertrophic scar, *Connect. Tissue Res.* **2**:205–213.

Klebanoff, S. J., 1992, Oxygen metabolites from phagocytes, in: *Inflammation: Basic Principles and Clinical Correlates* (J. I. Gallin, I. M. Goldstein, and R. Snyderman, eds.), pp. 541–601, Raven Press, New York.

Klein, C. E., Dressel, D., Steinmayer, T., Mauch, C., Eckes, B., Krieg, T., Bankert, R. B., and Werber, L., 1991, Integrin α2β1 is up-regulated in fibroblasts and highly aggressive melanoma cells in three dimensional collagen lattices and mediates the reorganization of collagen I fibrils, *J. Cell Biol.* **115:**1427–1436.

Knighton, D. R., Hunt, T. K., Scheuenstuhl, H., Halliday, B. J., Werb, Z., and Banda, M. J., 1983, Oxygen tension regulates the expression of angiogenesis factor by macrophages, *Science* **221:**1283–1285.

Knox, P., Crooks, S., and Rimmer, C. S., 1986, Role of fibronectin in the migration of fibroblasts into plasma clots, *J. Cell Biol.* **102:**2318–2323.

Koch, A. E., Polverini, P. J., Kunkel, S. L., Harlow, L. A., DiPietro, L. A., Elner, V. M., Elner, S. G., and Strieter, R. M., 1992, Interleukin-8 as a macrophage-derived mediator of angiogenesis, *Science* **258:**1798–1801.

Kojima, T., Leone, C., Marchildon, G. A., Marcum, J. A., and Rosenberg, R. D., 1992, Isolation and characterization of heparan sulfate proteoglycans produced by cloned rat microvascular endothelial cells, *J. Biol. Chem.* **267:**4859–4869.

Kraemer, P. M., and Tobey, R. A., 1972, Cell-cycle-dependent desquamation of heparan sulfate from the cell surface, *J. Cell Biol.* **55:**713–717.

Krawczyk, W. S., 1971, A pattern of epidermal cell migration during wound healing, *J. Cell Biol.* **49:**247–263.

Krawczyk, W. S., and Wilgram, G. F., 1973, Hemidesmosome and desmosome morphogenesis during epidermal wound healing, *J. Ultrastruct. Res.* **45:**93–101.

Kubota, Y., Kleinman, H. K., Martin, G. R., and Lawley, T. J., 1988, Role of laminin and basement membrane in the morphological differentiation of human endothelial cells into capillary-like structures, *J. Cell Biol.* **107:**1589–1598.

Kurkinen, M., Vaheri, A., Roberts, P. J., and Stenman, S., 1980, Sequential appearance of fibronectin and collagen in experimental granulation tissue, *Lab. Invest.* **43:**47–51.

Laiho, M., Saksela, O., and Keski-Oja, J., 1986, Transforming growth factor β alters plasminogen activator activity in human skin fibroblasts, *Exp. Cell Res.* **164:**399–407.

Lane, T. F., Iruela-Arispe, M. L., Johnson, R. S., and Sage, E. H., 1994, SPARC is a source of copper-binding peptides that stimulate angiogenesis, *J. Cell Biol.* **125:**929–943.

Lanir, N., Ciano, P. S., Van de Water, L., McDonagh, J., Dvorak, A. M., and Dvorak, H. F., 1988, Macrophage migration in fibrin gel matrices II. Effects of clotting factor XIII, fibronectin, and glycosaminoglycan content on cell migration, *J. Immunol.* **140:**2340–2349.

Larjava, H., Salo, T., Haapasalmi, K., Kramer, R. H., and Heino, J., 1993, Expression of integrins and basement membrane components by wound keratinocytes, *J. Clin. Invest.* **92:**1425–1435.

Lark, M. W., Laterra, J., and Culp, L. A., 1985, Close and focal contact adhesions of fibroblasts to a fibronectin-containing matrix, *Fed. Proc.* **44:**394–403.

Lawrence, W. T., Sporn, M. B., Gorschbath, C., North, J. A., and Grotendorst, G., 1986, The reversal of an adriamycin induced healing impairment with chemoattractants and growth factors, *Ann. Surg.* **203:**142–147.

Leavesley, D. I., Schwartz, M. A., Rosenfeld, M., and Cheresh, D. A., 1993, Integrin β1- and β3-mediated endothelial cell migration is triggered through distinct signaling mechanisms, *J. Cell Biol.* **121:**163–170.

Leibovich, S. J., and Ross, R., 1975, The role of the macrophage in wound repair: A study with hydrocortisone and antimacrophage serum, *Am. J. Pathol.* **78:**71–100.

Lembach, K. J., 1976, Enhanced synthesis and extracellular accumulation of hyaluronic acid during stimulation of quiescent human fibroblasts by mouse epidermal growth factor, *J. Cell. Physiol.* **89:**277–288.

Levenson, S. M., Geever, E. F., Crowley, L. V., Oates, J. F. 3., Berard, C. W., and Rosen, H., 1965, The healing of rat skin wounds, *Ann. Surg.* **161:**293–308.

Lin, C. Q., and Bissell, M. J., 1993, Multi-faceted regulation of cell differentiation by extracellular matrix, *FASEB J.* **7:**737–743.

Linsenmayer, T. F., 1991, Collagen, in: *Cell Biology of Extracellular Matrix,* 2nd ed. (E. D. Hay, ed.), pp. 7–44, Plenum Press, New York.

Liu, C. Y., Nossel, H. L., and Kaplan, K. L., 1979, The binding of thrombin by fibrin, *J. Biol. Chem.* **254:**10421–10426.

Loedam, J. A., Meijers, J. C. M., Sixma, J. J., and Bouma, B. N., 1988, Inactivation of human factor VIII by activated protein C: Cofactor activity of protein S and protective effect of von Willebrand factor, *J. Clin. Invest.* **82:**1236–1243.

Loef, E. B., Proper, J. A., Goustin, A. S., Shipley, G. D., DiCorleto, P. E., and Moses, H. L., 1986, Induction of c-sis RNA and activity similar to platelet-derived growth factor by transforming growth factor-β: A proposed model for indirect mitogenesis involving autocrine activity, *Proc. Natl. Acad. Sci. USA* **83:**2453–2457.

Longaker, M. T., Chiu, E., Adzick, N. S., Stern, M., Harrison, M., and Stern, R., 1991, Studies in fetal wound healing V. Prolonged presence of hyaluraonic acid in fetal wound fluid, *Ann. Surg.* **213:**290–296.

Loskutoff, D. J., and Edgington, T. S., 1977, Synthesis of a fibrinolytic activator and inhibitor by endothelial cells, *Proc. Natl. Acad. Sci. USA* **74:**3903–3907.

Lynch, S. E., Colvin, R. B., and Antoniades, H. N., 1989, Growth factors in wound healing. Single and synergistic effects on partial thickness porcine skin wounds, *J. Clin. Invest.* **84:**640–646.

Mackie, E. J., Halfter, W., and Liverani, D., 1988, Induction of tenascin in healing wounds, *J. Cell Biol.* **107:**2757–2767.

Madri, J. A., and Stenn, K. S., 1982, Aortic endothelial cell migration. I. Matrix requirements and composition, *Am. J. Pathol.* **106:**180–186.

Madri, J. A., Pratt, B. M., and Tucker, A. M., 1988, Phenotypic modulation of endothelial cells by transforming growth factor-β depends upon the composition and organization of the extracellular matrix, *J. Cell Biol.* **156:**1375–1385.

Madri, J. A., Reidy, M. A., Kocher, O., and Bell, L., 1989, Endothelial cell behavior after denudation injury is modulated by transforming growth factor-β1 and fibronectin, *Lab. Invest.* **60:**755–765.

Madtes, D. K., Raines, E. W., Sakariassen, K. S., Assoian, R. K., Sporn, M. B., Bell, G. I., and Ross, R., 1988, Induction of transforming growth factor-α in activated human alveolar macrophages, *Cell* **53:**285–293.

Magnatti, P., Tsuboi, R., Robbins, E., and Rifkin, D. B., 1989, *In vitro* angiogenesis on the human amniotic membrane: Requirement for basic fibroblast growth factor-induced proteinases, *J. Cell Biol.* **108:**671–682.

Majno, G., Gabbiani, G., Hirschel, B. J., Ryan, G. B., and Statkov, P. R., 1971, Contraction of granulation tissue *in vitro:* Similarity to smooth muscle, *Science* **173:**548–550.

Mansbridge, J. N., and Knapp, A. M., 1987, Changes in keratinocyte maturation during wound healing, *J. Invest. Dermatol.* **89:**253–263.

Matsuoka, J., and Grotendorst, G. R., 1989, Two peptides related to platelet-derived growth factor are present in human wound fluid, *Proc. Natl. Acad. Sci. USA* **86:**4416–4420.

Mauch, C., Hatamochi, A., Scharffetter, K., and Krieg, T., 1988, Regulation of collagen synthesis in fibroblasts within a three-dimensional collagen gel, *Exp. Cell Res.* **178:**493–530.

McCaffrey, T. A., Falconed, D. J., and Dud, B., 1992, Transforming growth factor-β1 is a heparin-binding protein: Identification of putative heparin-binding regions and isolation of heparins with varying affinity for TGF-β1, *J. Cell. Physiol.* **152:**430–440.

McCarthy, K., and Henson, P. M., 1979, Induction of lysosomal enzyme secretion by macrophages in response to the purified complement fragments C5a and C5a des Arg, *J. Immunol.* **123:**2511–2517.

McDonald, J. A., and Kelley, D. G., 1980, Degradation of fibronectin by human leukocyte elastase, *J. Biol. Chem.* **255:**8848–8858.

McDonald, J. A., Kelley, D. G., and Broekelmann, T. J., 1982, Role of fibronectin in collagen deposition: Fab1 antibodies to the gelatin-binding domain of fibronectin inhibits both fibronectin and collagen organization in fibroblast extracellular matrix, *J. Cell Biol.* **92:**485–492.

McDonald, J. A., Quade, B. J., Broekelmann, T. J., LaChane, R., Forsman, K., Hasegawa, E., and Akiyama, S., 1987, Fibronectin's cell-adhesive domain and an amino-terminal matrix assembly domain participate in the assembly into fibroblast pericellular matrix, *J. Biol. Chem.* **262:**2957–2967.

McPherson, J. M., Sawamura, S., Condell, R. A., Rhee, W., and Wallace, D. G., 1988, The effects of heparin on the physicochemical properties of reconstituted collagen, *Collagen Rel. Res.* **8:**65–82.

Merenmies, J., and Rauvala, G., 1990, Molecular cloning of the 18-kDa growth-associated protein of developing brain, *J. Biol. Chem.* **265:**16721–16724.

Messadi, D. V., and Bertolami, C. N., 1993, CD44 and hyaluronan expression in human cutaneous scar fibroblasts, *Am. J. Pathol.* **142:**1041–1049.

Moncada, S., Gryglewski, R., Bunting, S., and Vane, J. R., 1976, An enzyme isolated from arteries transforms prostaglandin endoperoxides to an unstable substance that inhibits platelet aggregation, *Nature* **263:**663–665.

Morla, A., and Ruoslahti, E., 1992, A fibronectin self-assembly site involved in fibronectin matrix assembly: Reconstruction in a synthetic peptide, *J. Cell Biol.* **118:**421–429.

Moscatelli, D., and Rubin, H., 1975, Increased hyaluronic acid production on stimulation of DNA synthesis in chick embryo fibroblasts, *Nature* **254:**65–66.

Mosesson, M. W., and Umfleet, R., 1970, The cold insoluble globulin of human plasma. I. Purification, primary characterization, and relationship to fibrinogen and other cold insoluble fraction components, *J. Biol. Chem.* **245:**5726–5736.

Mosher, D. F., and Johnson, R. B., 1983, Specificity of fibronectin–fibrin cross-linking, *Ann. NY Acad. Sci.* **408:**583–594.

Mosher, D. F., Sottile, J., Wu, C., and McDonald, J. A., 1992, Assembly of extracellular matrix, *Curr. Opin. Cell Biol.* **4:**810–818.

Muller-Eberhard, H. J., 1992, Complement: Chemistry and pathways, in: *Inflammation: Basic Principles and Clinical Correlates* (J. I. Gallin, I. M. Goldstein, and R. Snyderman, eds.), pp. 33–61, Raven Press, New York.

Muller-Esterl, N., 1989, Kininogens, kinins and kinships, *Thromb. Haemost.* **62:**2–6.

Mustoe, T. A., Pierce, G. F., Morishima, C., and Deuel, T. F., 1991, Growth factor-induced acceleration of tissue repair through direct and inductive activities in a rabbit dermal ulcer model, *J. Clin. Invest.* **87:**694–703.

Nathan, C., and Sporn, M., 1991, Cytokines in context, *J. Cell Biol.* **113:**981–986.

Newman, S. L., Henson, J. E., and Henson, P. M., 1982, Phagocytosis of senescent neutrophils by human monocyte derived macrophages and rabbit inflammatory macrophages, *J. Exp. Med.* **156:**430–442.

Nickoloff, B. J., Mitra, R. S., Riser, B. L., Dixit, V. M., and Varani, J., 1988, A modulation of keratinocyte motility. Correlation with production of extracelluar matrix molecules in response to growth promoting and anti-proliferative factors, *Am. J. Pathol.* **132:**543–551.

O'Keefe, E. J., R. E., Jr., Payne, Russell, N., and Woodley, D. T., 1985, Spreading and enhanced motility of human keratinocytes on fibronectin, *J. Invest. Dermatol.* **85:**125–130.

O'Keefe, E. J., Chiu, M. L., and Payne, R. E., 1988, Stimulation of growth of keratinocytes by basic fibroblast growth factor, *J. Invest. Dermatol.* **90:**767–769.

Odland, G., and Ross, R., 1968, Human wound repair. I. Epidermal regeneration, *J. Cell Biol.* **39:**135–157.

Okada, Y., Konomi, H., Yada, T., Kimata, K., and Nagase, H., 1989, Degradation of type IX collagen by matrix metalloproteinase 3 (stromelysin) from human rheumatoid synovial cells, *FEBS Lett.* **244:**473–476.

Oono, T., Specks, U., Eckes, B., Majewski, S., Hunzelmann, N., Timpl, R., and Krieg, T., 1993, Expression of type VI collagen mRNA during wound healing, *J. Invest. Dermatol.* **100:**329–334.

Oppenheimer, C. L., Pessin, J. E., Massague, J., Gitomer, W., and Czech, M. P., 1983, Insulin action rapidly modulates the apparent affinity of the insulin-like growth factor II receptor, *J. Biol. Chem.* **258:**4824–4830.

Overall, C. M., Wrana, J. I., and Sodek, J., 1989, Independent regulation of collagenase, 72 kD progelatinase, and metalloendoproteinase inhibitor expression in human fibroblasts by transforming growth factor-β, *J. Biol. Chem.* **264:**1860–1869.

Pardes, J. B., Takagi, H., Martin, T. A., Ochoa, M. S., and Falanga, V., 1995, Decreased levels of alpha 1(I) procollagen mRNA in dermal fibroblasts grown of fibrin gels and in response to fibrinopeptide B, *J. Cell. Physiol.* **162:**9–14.

Perin, J.-P., Bonnet, F., Mailet, P., and Jolies, P., 1988, Characterization and N-terminal sequence of human platelet proteoglycan, *Biochem. J.* **255:**1007–1013.

Petersen, M. J., Woodley, D. T., Stricklin, G. P., and O'Keefe, E. J., 1990, Enhanced synthesis of collagenase by human keratinocytes cultured on type I or type IV collagen, *J. Invest. Dermatol.* **94:**341–346.

Pfaff, M., Aumailley, M., Specks, U., Knolle, J., Zerwes, H. G., and Timpl, R., 1993, Integrin and Arg-Gly-Asp dependence of cell adhesion to the native and unfolded triple helix of collagen type VI, *Exp. Cell Res.* **206**(1):167–176.

Pierce, G. F., Mustoe, T. A., Lingelbach, J., Masakowski, V. R., Griffin, G. L., Senior, R. M., and Deuel, T. F., 1989, Platelet-derived growth factor and transforming growth factor-β enhance tissue repair activities by unique mechanisms, *J. Cell Biol.* **109**:429–440.

Pierce, G. F., Mustoe, T. A., Altrock, B., Deuel, T. F., and Thomas, A., 1991, Role of platelet-derived growth factor in wound healing, *J. Cell. Biochem.* **45**:319–326.

Pommier, C. G., Inada, S., Fried, L. F., *et al.*, 1983, Plama fibronectin enhances phagocytosis of opsonized particles by human peripheral blood monocytes, *J. Exp. Med.* **157**:1844–1854.

Postlethwaite, A. E., and Kang, A. H., 1976, Collagen and collagen peptide-induced chemotaxis of human blood monocytes, *J. Exp. Med.* **143**:1299–1307.

Postlethwaite, A. E., and Seyer, J. M., 1991, Fibroblast chemotaxis induction by human recombinant interleukin-4: Identification by synthetic peptide analysis of two chemotactic domains residing in amino acid sequences 70–88 and 89–122, *J. Clin. Invest.* **87**:2147–2152.

Postlethwaite, A. E., Seyer, J. M., and Kang, A. H., 1978, Chemotactic attraction of human fibroblast to type I, II, and III collagens and collagen-derived peptides, *Proc. Natl. Acad. Sci. USA* **75**:871–875.

Postlethwaite, A. E., Snyderman, R., and Kang, A. H., 1979, Generation of a fibroblast chemotactic factor in serum by activation of complement, *J. Clin. Invest.* **64**:1379–1385.

Postlethwaite, A. E., Keski-Oja, J., Balian, G., and Kang, A., 1981, Induction of fibroblast chemotaxis by fibronectin. Location of the chemotactic region to a 140,000 molecular weight nongelatin binding fragment, *J. Exp. Med.* **153**:494–499.

Postlethwaite, A. E., Keski-Oja, J., Moses, H. L., and Kang, A. H., 1987, Stimulation of the chemotactic migration of human fibroblasts by transforming growth factor-β, *J. Exp. Med.* **165**:251–256.

Postlethwaite, A. E., Holness, M. A., Katai, H., and Raghow, R., 1992, Human fibroblasts synthesize elevated levels of extracellular matrix proteins in response to interleukin 4, *J. Clin. Invest.* **90**:1479–1485.

Prehm, P., 1983, Synthesis of hyaluronate in differentiated teratocarcinoma cells: Mechanism of chain growth, *Biochem. J.* **211**:191–198.

Preissner, K. T., and Jenne, D., 1991, Vitronectins. A new molecular connection in haemostasis, *Thromb. Haemost.* **66**:189–194.

Raines, E. W., Dower, S. K., and Ross, R., 1989, Interleukin-1 mitogenic activity for fibroblasts and smooth muscle cells is due to PDFG-AA, *Science* **243**:393–396.

Rappolee, D. A., Mark, D., Banda, M. J., and Werb, Z., 1988, Wound macrophages express TGF-α and other growth factors *in vivo:* Analysis by mRNA phenotyping, *Science* **241**:708–712.

Reich-Slotky, R., Bonneh-Barkay, D., Shaoul, E., Bluma, B., Svahn, C. M., and Ron, D., 1994, Differential effect of cell-associated heparan sulfates on the binding of keratinocyte growth factor (KGF) and acidic fibroblast growth factor to the KGH receptor, *J. Biol. Chem.* **269**:32279–32285.

Roberts, A. B., Sporn, M. B., Assoian, R. K., Smith, J. M., Roche, M. S., Heine, U. F., Liotta, L., Falanga, V., Kehrl, J. H., and Fauci, A. S., 1986, Transforming growth factor beta: Rapid induction of fibrosis and angiogenesis *in vivo* and stimulation of collagen formation, *Proc. Natl. Acad. Sci. USA* **83**:4167–4171.

Roberts, R., Gallagher, J., Spooncer, E., Allen, R. D., Bloomfield, F., and Dexter, R. M., 1988, Heparan sulphate bound growth factors: A mechanism for stromal cell mediated haemopoiesis, *Nature* **332**:376–378.

Ross, R. R., and Raines, E. W., 1990, Platelet-derived growth factor and cell proliferation, in: *Growth Factors: From Genes to Clinical Application* (V. R. Sara *et al.*, eds.), pp. 193–199, Raven Press, New York.

Ruggeri, Z. M., 1993, von Willebrand factor and fibrinogen, *Curr. Opin. Cell Biol.* **5**:898–906.

Ruoslahti, E., 1991, Integrins, *J. Clin. Invest.* **87**:1–5.

Ruoss, S. J., Hartmann, T., and Caughey, G. H., 1991, Mast cell tryptase is a mitogen for cultured fibroblasts, *J. Clin. Invest.* **88**:493–499.

Saarialho-Kere, U. K., Chang, E. S., Welgus, H. G., and Parks, W. C., 1992, Distinct localization of collagenase and tissue inhibitor of metalloproteinases expression in wound healing associated with ulcerative pyogenic granuloma, *J. Clin. Invest.* **90**:1952–1957.

Sakai, L., Keene, D. R., Morris, N. P., and Burgeson, R. E., 1986, Type VII collagen is a major structural component of anchoring fibrils, *J. Cell Biol.* **103:**1577–1586.

Sakata, Y., and Aoki, N., 1980, Cross-linking of α2-plasma inhibitor to fibrin by fibrin-stabilizing factor, *J. Clin. Invest.* **65:**290–297.

Salonen, E.-M., Vaheri, A., Pollanen, J., Stephens, R., Andreasen, P., Mayer, M., Dano, K., Gailit, J., and Ruoslahti, E., 1989, Interaction of plasminogen activator inhibitor (PAI-1) with vitronectin, *J. Biol. Chem.* **264:**6339–6343.

Samuel, S. K., Hurta, R. A. R., Spearman, M. A., Wright, J. A., Turley, E. A., and Greenberg, A. H., 1993, TGF-β1 stimulation of cell locomotion utilizes the hyaluronan receptor RHAMM and hyaluronan, *J. Cell Biol.* **123:**749–758.

Saus, J., Quinones, S., Otani, Y., Nagase, H., Harris, Jr., E. D., and Kurkinen, M., 1988, The complete primary structure of human matrix metallo-proteinase-3. Identity with stromelysin, *J. Biol. Chem.* **263:**6742–6745.

Savani, R. C., Wang, C., Yang, B., Zang, S., Kinsella, M. G., Wight, T. N., Stern, R., Nance, D. M., and Turley, E. A., 1994a, Migration of bovine aortic and muscle cells after wound injury: A role for hyaluronan and the hyaluronan receptor RHAMM, *J. Clin. Invest.* **95:**1158–1168.

Savani, R. C., Wang, C., Stern, R., Khalil, N., Greenberg, A. H., and Turley, E. A., 1994b, The expression and role of hyaluronan (HA) and the HA receptor RHAMM in bleomycin-induced pulmonary inflammation, *J. Exp. Med.* In Press:

Sawada, H., Konomi, H., and Hirosawa, K., 1990, Characterization of the collagen in hexagonal lattice of Descemet's membrane: Its relation to type VIII collagen, *J. Cell. Biol.,* **110:**219–227.

Scharffetter, K., Kulozik, M., Stolz, W., Lankat-Buttgereit, B., Hatamochi, A., Sohnchen, R., and Krieg, T., 1989, Localization of collagen α1(I) gene expression during wound healing by *in situ* hybridization, *J. Invest. Dermatol.* **93:**405–412.

Schiro, J. A., Chan, B. M. C., Roswit, W. R., Kassner, P. D., Pentland, A. P., Hemler, M. E., Eisen, A. Z., and Kupper, T. S., 1991, Integrin α2β1 (VLA-2) mediates reorganization and contraction of collagen matrices by human cells, *Cell* **67:**403–410.

Schultz, G., Rotatori, D. S., and Clark, W., 1991, EGF and TGF-α in wound healing and repair, *J. Cell. Biochem.* **45:**346–352.

Scott, J. E., 1993, Proteoglycan–fibrillar collagen interactions in tissues: Dermatan sulfate proteoglycan as a tissue organizer, in: *Dermatan Sulphate Proteoglycans: Chemistry, Biology, Chemical Pathology* (J. E. Scott, ed.), pp. 165–181, Portland Press, London, England.

Screaton, G. R., Bell, M. V., Jackson, D. G., Cornelis, F. B., Gerth, U., and Bell, J. I., 1992, Genomic structure of DNA encoding the lymphocyte homing receptor CD44 reveals at least 12 alternatively spliced exons, *Proc. Natl. Acad. Sci. USA* **89:**12160–12164.

Senior, R. M., Griffin, G. L., and Mecham, R. P., 1980, Chemotactic activity of elastin-derived peptides, *J. Clin. Invest.* **66:**859–862.

Senior, R. M., Huang, J. S., Griffin, G. L., and Deuel, T. F., 1985, Dissociation of the chemotactic and mitogenic activities of platelet-derived growth factor by human neutrophil elastase, *J. Cell Biol.* **100:**351–356.

Senior, R. M., Griffin, G. L., Perez, H. D., and Webster, R. O., 1988, Human C5a and C5a des arg exhibit chemotactic activity for fibroblasts, *J. Immunol.* **141:**3570–3574.

Seppa, H. E. J., Grotendorst, G. R., Seppa, S. I., Schiffmann, E., and Martin, G. R., 1982, Platelet-derived growth factor is chemotactic for fibroblasts, *J. Cell Biol.* **92:**584–588.

Shaw, R. J., Doherty, D. E., Ritter, A. G., Benedict, S. H., and Clark, R. A. F., 1990, Adherence-dependent increase in human monocyte PDGF(B) mRNA is associated with increases in c-fos, c-jun, and EGF2 mRNA, *J. Cell Biol.* **111:**2139–2148.

Shaw, R. J., Benedict, S. H., Clark, R. A. F., and King, Jr., T. E., 1991, Pathogenesis of pulmonary fibrosis in interstitial lung disease: Alveolar macrophage PDGF(B) gene activation and up-regulation by interferon gamma, *Am. Rev. Resp. Dis.* **143:**167–173.

Shetlar, M. R., Shetlar, C. L., Chien, S.-F., Linares, H. A., Dobrokovsky, M., and Larson, D. L., 1972, The hypertrophic scar. Hexosamine containing components of burn scars, *Proc. Soc. Exp. Biol. Med.* **139:**544–547.

Shimokado, K., Raines, E. W., Madtes, D. K., Barrett, T. B., Benditt, E. P., and Ross, R., 1985, A significant part of macrophage-derived growth factor consists of two forms of PDGF, *Cell* **43**:277–286.

Shipley, G. D., Pittelkow, M. R., Wille, J. J., Scott, R. E., and Moses, H. L., 1986, Reversible inhibition of normal human prokeratinocyte proliferation by type β transforming growth factor-growth inhibitor in serum-free medium, *Cancer Res.* **46**:2068–2071.

Sholley, M. M., Gimbrone, M. A. J., and Cotran, R. S., 1978, The effects of leukocyte depletion on corneal neovascularization, *Lab. Invest.* **38**:32–40.

Siebenlist, K. R., DiOrio, J. P., Budzynski, A. Z., and Mosseson, M. W., 1990, The polymerization and thrombin-binding properties of des-(Bβ1-42)-fibrin, *J. Biol. Chem.* **265**:18650–18655.

Singer, I. I., Kawka, D. W., Kazazis, D. M., and Clark, R. A. F., 1984, *In vivo* co-distribution of fibronectin and actin fibers in granulation tissue: Immunofluorescence and electron microscope studies of the fibronexus at the myofibroblast surface, *J. Cell Biol.* **98**:2091–2106.

Singer, I. I., Scott, S., Kawka, D. W., Kazazis, D. M., Gailit, J., and Ruoslahti, E., 1988, Cell surface distribution of fibronectin and vitronectin receptors depends on substrate composition and extracellular matrix accumulation, *J. Cell Biol.* **106**:2171–2182.

Somers, C. E., and Mosher, D. F., 1993, Protein kinase C modulation of fibronectin matrix assembly, *J. Biol. Chem.* **268**:22277–22280.

Sottile, J., and Wiley, S., 1994, Assembly of amino-terminal fibronectin dimers into the extracellular matrix, *J. Biol. Chem.* **269**:17192–17198.

Spivak-Kroizman, T., Lemmon, M. A., Dikic, I., Ladbury, J. E., Pinchasi, D., Huang, F., Jaye, M., Crumley, G., Schlessinger, J., and Lax, I., 1994, Heparin-induced oligomerization of FGF molecules is responsible for FGF receptor dimerization, activation and cell proliferation, *Cell* **79**:1015–1024.

Sporn, M. B., and Roberts, A. B., 1986, Peptide growth factors and inflammation, tissue repair, and cancer, *J. Clin. Invest.* **78**:329–332.

Sporn, M. B., and Roberts, A. M., 1992, Transforming growth factor-β: Recent progress and new challenges, *J. Cell Biol.* **119**:1017–1021.

Sporn, M. B., Roberts, A. B., Shull, J. H., Smith, J. M., Ward, J. M., and Sodek, J., 1983, Polypeptide transforming growth factor isolated from bovine sources and used for wound healing *in vitro, Science* **219**:1329–1331.

Sprugel, K. H., McPherson, J. M., Clowes, A. W., and Ross., R., 1987, Effects of growth factors *in vivo, Am. J. Pathol.* **129**:601–613.

Staatz, W. D., Rajpara, S. M., Wayner, E. A., Carter, W. G., and Santoro, S. A., 1989, The membrane glycoprotein Ia-IIa (VLA-2) complex mediates the Mg^{++}-dependent adhesion of platelets to collagen, *J. Cell Biol.* **108**:1917–1924.

Stamenkovic, I., and Aruffo, A., 1994, Hyaluronic acid receptors, in: *Methods in Enzymology* (E. Ruoslahti and E. Engvall, ed.), pp. 195–218, Academic Press, San Diego, CA.

Stamenkovic, I., Amiot, M., Pesando, J. M., and Seed, B., 1989, A lymphocyte molecule implicated in lymph node homing is a member of the cartilage link protein family, *Cell* **56**:1057–1062.

Stenn, K. S., Madri, J. A., and Roll, R. J., 1979, Migrating epidermis produces AB_2 collagen and requires continual collagen synthesis for movement, *Nature* **277**:229–232.

Stern, D. M., Nawroth, P. P., Marcum, J., Handley, D., Kisiel, D., Rosenberg, R., and Stern, K., 1985, Interaction of antithrombin III with bovine aortic segments, *J. Clin. Invest.* **75**:272–279.

Stetler-Stevenson, W. G., Krutzsch, H. C., Wacher, M. P., Margulies, I. M. K., and Liotta, L. A., 1989, The activation of human type IV collagenase proenzyme. Sequence identification of the major conversion product following organomercurial activation, *J. Biol. Chem.* **264**:1353–1356.

Stimler, N. P., Bach, M. K., Bloor, C. M., and Hugli, T. E., 1982, Release of leukotrienes from guinea pig lung stimulated by C5a des arg anaphylatoxin, *J. Immunol.* **128**:2247–2257.

Thomas, L., Byers, H. R., Vink, J., and Stamenkovic, I., 1992, CD44H regulates tumor cell migration on hyaluronate-coated substrate, *J. Cell Biol.* **118**:971–977.

Thorsen, S., Glas-Greenwalt, P., and Astrup, T., 1972, Difference in the binding to fibrin of urokinase and tissue plasminogen activator, *Thromb. Pathol. Haemost.* **28**:65–74.

Toda, K.-I., Tuan, T.-L., Brown, P. J., and Grinnell, F., 1987, Fibronectin receptors of human keratinocytes and their expression during cell culture, *J. Cell Biol.* **105**:3097–3104.

Tomida, M., Koyama, H., and Ono, T., 1974, Hyaluronic acid synthetase in cultured mammalian cells producing hyaluronic acid. Oscillatory change during the growth phase and suppression by 5-bromodeoxyuridine, *Biochim. Biophys. Acta* **338:**352–363.

Tonnesen, M. G., Worthen, G. S., and Johnston, R. B. J., 1988, Neutrophil emigration, activation, and tissue damage, in: *Molecular and Cellular Biology of Wound Repair* (R. A. F. Clark and P. M. Henson, eds.), pp. 149–183, Plenum Press, New York.

Tonnesen, M. G., Anderson, D. C., Springer, T. A., Knedler, A., Avdi, N., and Henson, P. M., 1989, Adherence of neutrophils to cultured human microvascular endothelial cells. Stimulation by chemotactic peptides and lipid mediators and dependence upon the Mac-1, LFA-1, p150,95 glycoprotein family, *J. Clin. Invest.* **83:**637–646.

Toole, B. P., 1972, Hyaluronate turnover during chondrogenesis in the developing chick limb and axial skeleton, *Dev. Biol.* **29:**321–329.

Toole, B. P., 1981, Glycosaminoglycans in morphogenesis, in: *Cell Biology of Extracellular Matrix* (E. D. Hay, ed.), pp. 259–294, Plenum Press, New York.

Toole, B. P., 1991, Proteoglycans and hyaluronan in morphogenesis and differentiation, in: *Cell Biology of the Extracellular Matrix* (E. D. Hay, ed.), pp. 305–341, Plenum Press, New York.

Toole, B. P., and Gross, J., 1971, The extracellular matrix of the regenerating newt limb: Synthesis and removal of hyaluronate prior to differentiation, *Dev. Biol.* **25:**57–77.

Trinkaus, J. P., 1984, *Cells into Organs. The Forces That Shape the Embryo,* Prentice-Hall, Englewood Cliffs, NJ.

Tuckwell, D. S., Ayad, S., Grant, M. E., Takigawa, M., and Humphries, M. J., 1994, Conformation dependence of integrin-type II collagen binding. Inability of collagen peptides to support $\alpha 1\beta 1$ binding, and mediation of adhesion to denatured collagen by a novel $\alpha 5\beta 1$-fibronectin bridge, *J. Cell Sci.* **107**(Pt4):993–1005.

Turley, E. A., 1992, Hyaluronan and cell locomotion, *Cancer Metastasis Rev.* **11:**21–30.

Underhill, C., 1992, CD44: The hyaluronan receptor, *J. Cell Sci.* **103:**293–298.

Unemore, E. N., and Werb, Z., 1986, Reorganization of polymerized actin: A possible trigger for induction of procollagenase in fibroblasts cultured in and on collagen gels, *J. Cell Biol.* **103:**1021–1031.

Vartio, T., Seppa, H., and Vaheri, A., 1981, Susceptibility of soluble and matrix fibronectins to degraduation by tissue proteinases, mast cell chymase and cathepsin G, *J. Biol. Chem.* **256:**471–477.

Viljanto, J., Penttinen, R., and Raekallio, J., 1981, Fibronectin in early phases of wound healing in children, *Acta Chir. Scand.* **147:**7–13.

Vlodavsky, I., Fuks, Z., Ishai-Michaeli, R., Bashkin, P., Levi, E., Korner, G., Bar-Shavit, R., and Klagsbrun, M., 1991, Extrcellular matrix-resident basic fibroblast growth factor: Implication for the control of angiogenesis, *J. Cell. Biochem.* **45:**167–176.

Wagner, O. F., Nicolosa, G., and Bachmann, F., 1989, Plasminogen activator inhibitor 1: Development of a radioimmunoassay and observations on its plasma concentration during venous occlusion and after platelet aggregation, *Blood* **70:**1645–1653.

Wahl, S. M., Hunt, D. A., Wakefield, L. M., McCartney-Francis, N., Wahl, L. M., Roberts, A. B., and Sporn, M. B., 1987, Transforming growth factor type β induces monocyte chemotaxis and growth factor production, *Proc. Natl. Acad. Sci. USA* **84:**5788–5792.

Wall, R. T., Harker, L. A., and Striker, G. E., 1978, Human endothelial cell migration. Stimulated by a released platelet factor, *Lab. Invest.* **39:**523–529.

Wayner, E. A., and Carter, W. G., 1989, Identification of multiple cell adhesion receptors for collagen and fibronectin in human fibrosarcoma cells possessing unique α and common β subunits, *J. Cell Biol.* **105:**1873–1884.

Weisman, D. M., Polverini, P. J., Kamp, D. W., and Leibovich, S. J., 1988, Transforming growth factor-beta (TGF-β) is chemotactic for human monocytes and induces their expression of angiogenic activity, *Biochem. Biophys. Res. Commun.* **157:**793–800.

Weksler, B. B., 1992, Platelets, in: *Inflammation: Basic Principle and Clinical Correlates* (J. I. Gallin, I. M. Goldstein, and R. Snyderman, eds.), pp. 727–746, Raven Press, New York.

Welch, M. P., Odland, G. F., and Clark, R. A. F., 1990, Temporal relationships of F-actin bundle formation, collagen and fibronectin matrix assembly, and fibronectin receptor expression to wound contraction, *J. Cell Biol.* **110:**133–145.

Werb, Z., and Clark, E. J., 1989, Phorbol diesters regulate expression of the membrane neutral metalloendopeptidase (EC 3.4.24.11) in rabbit synovial fibroblasts and mammary epithelial cells, *J. Biol. Chem.* **264:**9111–9113.

Werb, Z., Tremble, P. M., Behrendtsen, O., Crowley, E., and Damsky, C. H., 1989, Signal transduction through the fibronectin receptor induces collagenase and stromelysin gene expression, *J. Cell Biol.* **109:**877–889.

Werb, Z., Tremble, P., and Damsky, C. H., 1990, Regulation of extracellular matrix degradation by cell–extracellular matrix interactions, *Cell. Differ. Dev.* **32:**299–306.

Werner, S., Peters, K. G., Longaker, M. T., Fuller-Pace, F., Banda, M. J., and Williams, L. T., 1992, Large induction of keratinocyte growth factor in the dermis during wound healing, *Proc. Natl. Acad. Sci. USA* **89:**6896–6900.

Werner, S., Breeden, M., Hubner, G., Greenhalgh, D. G., and Longaker, M. T., 1994, Induction of keratinocyte growth factor expression is reduced and delayed during wound healing in the genetically diabetic mouse, *J. Invest. Dermatol.* **103:**469–475.

Wight, T. N., Heinegard, D. K., and Hascall, V. C., 1991, Proteoglycans: Structure and function, in: *Cell Biology of Extracellular Matrix* (E. D. Hay, ed.), pp. 45–78, Plenum Press, New York.

Wikner, N. E., Persichitte, K. A., Baskin, J. B., Nielsen, L. D., and Clark, R. A. F., 1988, Transforming growth factor-β stimulates the expression of fibronectin by human keratinocytes, *J. Invest. Dermatol.* **91:**207–212.

Wilhelm, S. M., Collier, I. E., Kronberger, A., Eisen, A. Z., Marmer, B. L., Grant, G. A., Bauer, E. A., and Goldberg, G. I., 1987, Human skin fibroblast stromelysin: Structure, glycosylation, substrate specificity, and differential expression in normal and tumorigenic cells, *Proc. Natl. Acad. Sci. USA* **84:**6725–6729.

Wilkinson, P. C., and Lackie, J. M., 1983, The influence of contact guidance on chemotaxis of human neutrophil leukocytes, *Exp. Cell Res.* **145:**255–264.

Williams, G. T., 1991, Programmed cell death: Apoptosis and oncogenesis, *Cell* **65:**1097–1098.

Williams, T. J., 1988, Factors that affect vessel reactivity and leukocyte emigration, in: *Molecular and Cellular Biology of Wound Repair* (R. A. F. Clark and P. M. Henson, eds.), pp. 115–183, Plenum Press, New York.

Wilner, G. D., Danitz, M. P., Mudd, M. S., Hsieh, K.-H., and Fenton II, J. W., 1981, Selective immobilization of alpha-thrombin by surface-bound fibrin, *J. Lab. Clin. Med.* **97:**403–411.

Winter, G. D., 1962, Formation of the scab and the rate of epithelialization of superficial wounds in the skin of the young domestic pig, *Nature* **193:**293–294.

Winter, G. D., 1972, Epidermal regeneration studied in the domestic pig, in: *Epidermal Wound Healing* (H. I. Maibach and D. T. Rovee, eds.), pp. 71–112, Yearbook Medical Publishing, Chicago.

Wolpe, S. D., and Cerami, A., 1989, Macrophage inflammatory proteins 1 and 2: Members of a novel superfamily of cytokines, *FASEB J.* **3:**2565–2573.

Wood, G. C., 1960, The formation of fibrils from collagen solutions. Effect of chondroitin sulfate and other naturally occurring polyanions on the rate of formation, *Biochem. J.* **75:**605–612.

Woodley, D. T., Kalebec, T., Banes, A. J., Link, W., Prunieras, M., and Liotta, L., 1986, Adult human keratinocytes migrating over nonviable dermal collagen produce collagenolytic enzymes that degrade type I and type IV collagen, *J. Invest. Dermatol.* **86:**418–423.

Woodley, D. T., Yamauchi, M., Wynn, K. C., Mechanic, G., and Briggaman, R. A., 1991, Collagen telopeptides (cross-linking sites) play a role in collagen gel lattice contraction, *J. Invest. Dermatol.* **97:**580–585.

Woods, A., and Couchman, J. R., 1994, Syndecan-4 heparan sulfate proteoglycan is a selectively enriched and widespread focal adhesion components, *Mol. Biol. Cell* **5:**183–192.

Wu, C., Bauer, J. S., Juliano, R. L., and McDonald, J. A., 1993, The α5β1 integrin fibronectin receptor, but not the α5 cytoplasmic domain, functions in an early and essential step in fibronectin matrix assembly, *J. Biol. Chem.* **268:**21883–21888.

Xu, J., and Clark, R. A. F., 1995, Extracellular matrix alters PDGF regulation of fibroblast integrins, *J. Cell Biol.*, in press.

Yamagata, M., Saga, S., Kato, M., Bernfield, M., and Kimata, K., 1993, Selective distributions of proteoglycans and their ligands in pericellular matrix of cultured fibroblasts. Implications for their roles in cell–substratum adhesion, *J. Cell Sci.* **106:**55–65.

Yamaguchi, T., and Ruoslahti, E., 1988, Expression of human proteoglycan in Chinese hamster ovary cells inhibits cell proliferation, *Nature* **336:**244–246.

Yamaguchi, T., Mann, D. M., and Ruoslahti, E., 1990, Negative regulation of transforming growth factor-β by the proteoglycan decorin, *Nature* **346:**281–284.

Yamamoto, T., and Cochrane, C. G., 1981, Guinea pig Hageman factor as a vascular permeability enhancement factor, *Am. J. Pathol.* **105:**164–175.

Yamauchi, M., London, R. E., Guenat, C., Hashimoto, F., and Mechanic, G. L., 1987, Structure and formation of a stable histidine-based trifunctional cross-link in skin collagen, *J. Biol. Chem.* **262:**11428–11434.

Yang, B., Yang, B., Savani, R. C., and Turley, E. A., 1994, Identification of a common hyaluronan binding motif in the hyaluronan binding proteins RHAMM, CD44 an link protein, *EMBO J.* **13:**286–296.

Yang, E. Y., and Moses, H. L., 1990, Transforming growth factor-β1-induced changes in cell migration, proliferation, and angiogenesis in the chicken chorioallantoic membrane, *J. Cell Biol.* **111:**731–741.

Yayon, A., Klagsbrun, M., Esko, J. D., Leder, P., and Ornitz, D. M., 1991, Cell surface, heparin-like molecules are required for binding of basic fibroblast growth factor to its high affinity receptor, *Cell* **64:**841–848.

Yeo, T.-K., Brown, L., and Dvorak, H. F., 1991, Alterations in proteoglycan synthesis common to healing wounds and tumors, *Am. J. Pathol.* **138:**1437–1450.

Yurchenco, P. D., and Schittny, J. C., 1990, Molecular architecture of basement membranes, *FASEB J* **4:**1577–1590.

Zhang, Z., Morla, A. O., Vuori, K., Bauer, J. S., Juliano, R. L., and Ruoslahti, E., 1993, The αvβ1 integrin functions as a fibronectin receptor but does not support fibronectin matrix assembly and cell migration on fibronectin, *J. Cell Biol.* **122:**235–242.

Chapter 2

Provisional Matrix

KENNETH M. YAMADA and RICHARD A. F. CLARK

1. Introduction

Fibronectin, fibrinogen, and other extracellular proteins play important roles in cell surface interactions. For example, fibronectin helps to mediate cell adhesion, embryonic cell migration, and wound repair. Each of these glycoproteins can participate in a variety of functions by use of its different specialized domains or peptide recognition sequences for binding to specific cell surface receptors or to other extracellular molecules. There has been remarkable recent progress in our understanding of the structures, domain organization, and biological roles of these multifunctional cell interaction proteins.

For each of these cell adhesion and interaction proteins, several major repeating themes have been observed as their mechanisms of action have been defined (Fig. 1). These molecules contain discrete functional polypeptide domains specialized for binding to specific cell surface receptors, as well as to other extracellular molecules. Examples include cell-adhesive domains and fibrin-binding domains. Most of these proteins have short, specific peptide sequences that are bound by cell surface receptors. Strikingly, short amino acid sequences of only three to five residues within these proteins can serve as adhesive recognition signals for cell surface receptors. Some sequences have intrinsic cell-type specificity, whereas others need a "synergy" site to provide high affinity and specificity to an otherwise less-specific adhesion site. All of these molecules can be organized into oligomers or polymers, either by formation of interchain disulfide bonds at specific residues or by noncovalent self–self association. Repetitive structural units such as fibronectin- or epidermal growth factor (EGF)-like repeats can comprise a large part of some of these proteins. Local domains of these repeats can be specialized for very different functions. Families of closely related

KENNETH M. YAMADA • Laboratory of Developmental Biology, National Institute of Dental Research, National Institutes of Health, Bethesda, Maryland 20892-4370.   RICHARD A. F. CLARK • Department of Dermatology, Health Sciences Center, State University of New York at Stony Brook, Stony Brook, New York 11794-8165.

The Molecular and Cellular Biology of Wound Repair (Second Edition), edited by Richard A. F. Clark. Plenum Press, New York, 1996

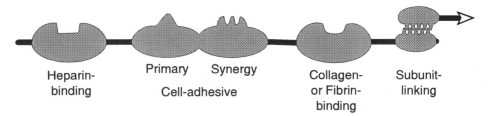

Figure 1. General conceptual model for the structure of cell-adhesive extracellular molecules, showing functional domains. Proteins of this type contain one or more domains specialized for binding to a cell surface receptor (frequently termed the cell-binding domain or the cell-adhesive domain). Most of these domains contain a short, specific amino acid sequence recognized by a cell surface receptor (primary site). As for the two major adhesive sites of fibronectin, there may be additional sequences that synergize to enhance activity by 20- to 100-fold (synergy site). Most of these proteins also contain a domain that binds to heparin and to related glycosaminoglycans, and some also contain a collagen-binding domain or other functional domains, e.g., fibrin binding. A number of the larger glycoproteins of this class also have regions involved in the formation of interchain polymers, either by intersubunit disulfide bond formation at specific cysteine residues or by means of self-assembly domains that mediate noncovalent polymerization (subunit-linking domain) to one or more similar subunits (arrow).

proteins can be generated by alternative splicing or gene duplication, e.g., to form variant fibronectins and tenascins. These extracellular matrix proteins and their receptors are regulated in location, type, and quantity during embryonic development and tissue remodeling.

These general properties of extracellular cell-adhesive proteins provide considerable sophistication to the repertoire of potential interactions between cells and the extracellular matrix, as well as between various matrix molecules. Most of the studies we review have tended to focus on only one of these proteins at a time; but it is obviously important to recall that a healing wound contains complex mixtures of extracellular matrix proteins, each of which may contribute independently and synergistically to the final effects on cell behavior. In this chapter, recent references are often cited in preference to older studies, in order to permit access to the broadest range of past references. For each protein, we also list reviews that can provide other details or perspectives on this extremely complex and rapidly expanding field.

2. Fibrinogen

2.1. Molecular Structure and Synthesis

The last step in the blood coagulation pathway involves conversion of fibrinogen to form cross-linked fibrin. Fibrinogen, a 340,000-Da hexamer composed of two Aα, two Bβ, and two γ chains, is synthesized by hepatocytes and circulates at approximately 3 gm/liter. The three fibrinogen chains are encoded by distinct genes located in close proximity within a 50 kilobase span on the long arm of chromosome 4, region q23-q32. Disulfide-linked Aα-γ and Bβ-γ dimers and Aa–Bβ-γ half molecules are intermediate species of molecular assembly (Huang *et al., 1993a*). Disulfide bonds between Aα and Bβ chains occur when Aα−Bβ-γ half molecules form. The latter then form dimers to generate mature fibrinogen. Dimerization of the two half molecules involves cysteine residues in the amino-terminal region of the three chains (Huang *et al., 1993b*).

2.2. Molecular Morphology of Fibrinogen

Electron microscopy studies of fibrinogen demonstrate a typical trinodular structure (Veklich *et al., 1993*). A central nodule represents the amino-terminal disulfide knot of the six chains (E domain); the two end nodules represent the carboxyl-terminal regions of the Bβ and γ chains (D domains) and are connected to the central nodule by a rodlike α-helical, coiled coil region. The carboxyl-terminus of the Aα chain is more flexible than the rest of the fibrinogen molecule. It is often not visible on electron microscopy since it folds back onto the E domain. The Aα carboxyl-terminal region may be involved in regulating fibrin polymerization, since isolated proteolytic fragments of the Aα chain carboxyl-terminus interfere with fibrin polymerization (Veklich *et al., 1993*).

2.3. Conversion of Fibrinogen to Fibrin

Thrombin cleavage of fibrinopeptide A and fibrinopeptide B from the amino-terminus of fibrinogen exposes four binding sites of two types that function cooperatively in fibrin polymerization (Mosesson, 1992). After the release of the 16-residue fibrinopeptide A, new α chain amino-terminal ends are generated that interact with sites in the γ chain carboxyl-termini of adjacent molecules to promote noncovalent lateral assembly. Subsequent removal of the 14-residue fibrinopeptide B creates new β chain amino-terminal ends that also interact with sites in the γ chain carboxyl-termini of adjacent molecules, reinforcing lateral aggregation and leading to thicker fiber formation. Fibrin thus polymerizes in a staggered overlapping manner into protofibrils. Subsequently, lateral protofibril associations occur, resulting in thick fibrin fibers that intertwine to produce a clot.

2.4. Interaction with Platelets

For effective hemostasis, the fibrin clot must interact with platelets. This interaction is mediated by the platelet membrane receptor GPIIb-IIIa (αIIbβ3 integrin). On nonactivated platelets this receptor selectively binds fibrinogen substrata. After platelet activation, however, the receptor binds von Willebrand factor, fibronectin, and vitronectin as well as fibrin(ogen) (Ginsberg *et al., 1992*). Of note is the fact that the interaction of αIIbβ3 with a fibrinogen substrata can activate the receptor so that it recognizes other ligands (Du *et al., 1991*). Through protein–protein interactions, platelet–protein interactions, and cell–protein interactions, the clot becomes firmly integrated in the wounded tissue where it provides a scaffold for cell migration and proliferation; a reservoir for growth factors, proteases, and protease inhibitors; modulation of cell function; and a hemostatic plug.

2.5. Intermolecular Cross-Links

The resultant clot, however, is friable, and stability is conferred on this structure by covalent cross-linking of the chains of adjacent fibrin stands. A nonserine protease known as plasma transglutaminase (factor XIIIa) produces these intermolecular cross-links by formation of ε-(γ-glutamyl)lysine [ε-(γ-glu)lys] isopeptide bonds (Pisano *et al.,* 1968). Cross-linking of γ chains within fibrils forms dimers (McKee *et al.,* 1970) that occur as reciprocal bridges between lysine at position 406 of one γ chain and glutamine at position 398 or 399 of another (Mosesson, 1992). Intermolecular cross-linking among α chains creates oligomers and larger α chain polymers; cross-links also occur between α chains and γ chains (Mosesson *et al.,* 1989; Shainoff *et al.,* 1991). Gamma chain dimerization occurs much more rapidly than α chain polymers, which can take several hours (McKee *et al.,* 1970; Mosesson *et al.,* 1989). Recent *in vitro* studies, however, indicate that in addition to γ dimers, higher-order forms of cross-linked γ chain multimers (i.e., γ trimers, γ tetramers) form slowly and progressively over a period of hours to days (Siebenlist and Mosesson, 1992). Thus, the biochemical structure of a fibrin clot undergoes dynamic evolution.

2.6. Molecules Associated with the Fibrin Clot

Fibrin clots may contain a variety of factors and may be variably degraded. Fibrin matrix may contain fibronectin (Clark *et al.,* 1982; Siebenlist and Mosesson, 1992), vitronectin (Preissner and Jenne, 1991), plasminogen (Castellino *et al.,* 1983), plasminogen activator (Thorsen *et al.,* 1972), plasminogen activator inhibitor (Wagner *et al.,* 1989), and thrombin (Liu *et al.,* 1979; Wilner *et al.,* 1981; Kaminski and McDonagh, 1983; Siebenlist *et al.,* 1990). Furthermore, we have preliminary data that two cytokines, transforming growth factor-β and platelet-derived growth factor (PDGF), bind fibrinogen and fibrin. In addition, α_2-plasmin inhibitor can be reversibly cross-linked to fibrinogen or fibrin by factor XIIIa (Sakata and Aoki, 1980).

2.7. Fibrinolysis

The fibrin clot must ultimately be removed from the site of injury to permit healing to ensue. Normal fibrinolytic activity is essential to maintain the hemostatic balance and allow beneficial clotting while controlling the formation of dangerous thrombi that might lead to vascular occlusion. In fact, either intravascular fibrin clots or clots in tissue repair undergo physiological degradation (fibrinolysis) by plasmin. First, however, plasmin must be generated from plasminogen. To this end, various tissue cells contain plasminogen activators that can be released when the cells are perturbed (see Chapter 13).

Plasminogen activators cleave plasminogen at an Arg-Val bond. Similar to other coagulation factors (e.g., Hageman factor, prekallikrein, factor X), this cleavage site is located within a peptide span that is bridged by a disulfide bound. Hence, the 92,000-Da single-chain plasminogen molecule is converted to a disulfide-linked two-chain plasmin molecule consisting of a 67,000-Da heavy chain and a 25,000-Da light chain (Robbins *et al.*, 1967). The enzymatically active site of plasmin resides in the light chain (Summaria *et al.*, 1967).

Two distinct types of plasminogen activator have been identified: urokinase and tissue plasminogen activator. Tissue plasminogen activator is released from stressed endothelium and controls intravascular clotting (Loskutoff and Edgington, 1977), while urokinase-type plasminogen activator facilitates cell invasiveness (Del Rosso *et al.*, 1990; Quax *et al.*, 1991). Urokinase plasminogen activator, like other proteases, is secreted in a latent proform. Such zymogens require activation for proteolytic activity. Once activated, enzymatic activity is strictly regulated by specific protease inhibitors. The synthesis and secretion of plasminogen activators are governed in part by cytokines and growth factors (Laiho *et al.*, 1986; Circolo *et al.*, 1991). Once secreted, the plasminogen activator zymogen binds to a cell surface receptor. Here, it is activated and brought into proximity of cell-bound plasminogen (Hajjar *et al.*, 1994). These membrane-bound complexes of plasminogen activator and plasminogen greatly facilitate plasmin generation.

Plasmin activity also can be modulated by extracellular matrix proteins (Flaumenhaft and Rifkin, 1991). For example, fibrin can bind plasminogen activator (Thorsen *et al.*, 1972) and plasminogen (Castellino *et al.*, 1983), and thereby enhance conversion of the latter to plasmin. Interestingly, the fibrin-bound plasminogen activator converts fibrin-bound plasminogen to plasmin more slowly in the context of thin fibers compared to thick fibers, suggesting intermolecular conversion (Gabriel *et al.*, 1992). Nevertheless, the most efficient plasmin generation probably occurs at the interface between cells and the fibrin clot.

Fibrinolysis is also regulated by enzyme inhibitors in the circulation and associated with fibrin and other provisional matrix proteins. In general, serine proteinases escape circulating inhibitors by binding to a surface as described above for plasmin activator and plasmin. However, plasminogen activator inhibitor and α_2-antiplasmin inhibitor also associate with the fibrin clot, and thereby can control fibrinolysis of surface-bound plasmin (Sakata and Aoki, 1980; Wagner *et al.*, 1989). Furthermore, plasminogen activator inhibitor associates with vitronectin and further limits plasmin proteolysis of the provisional matrix (Salonen *et al.*, 1989).

2.8. Fibrin Activity in Wounds

Fibrin is prominent in normal wounds and venous ulcers both clinically and histologically (Clark *et al.*, 1982). Clinically, a thick yellow adherent fibrinoid material often overlies the entire wound surface. Histologically, fibrin is the major protein of the fibrinoid eschar in both normal wounds and ulcers, is intercalated among the collagen bundles of the ulcer bed and adjacent dermis of normal wounds and ulcers, and is particularly prominent around the microvasculature (fibrin cuffs) of venous ulcers. Through understanding the pathobiology of fibrin in chronic wounds such as venous leg ulcer, we may be better able to understand the role of fibrin in the normal wound repair process.

Data over the past decade have implied that fibrin cuffs around the cutaneous microvasculature are important in the pathogenesis of venous leg ulcers, but substantiation is lacking (Falanga and Eaglestein, 1993). Sustained venous backpressure leads to distention of skin capillaries and fibrinogen leaks into the perivascular space (Burnand *et al.*, 1982). The relationship between sustained venous pressures, fibrin accumulation, and leg ulceration is not understood, although several hypotheses exist.

It has been proposed that perivascular fibrin cuffs inhibit the diffusion of oxygen from the blood vessels to the tissue, and thus impede wound repair (Browse and Burnand, 1982). Assuming that oxygen diffusion is impaired in venous leg ulcers (Falanga *et al.*, 1991), it is not clear that the fibrin cuff is responsible for the diffusion block. Histology studies demonstrate that the fibrin cuffs are discontinuous, rendering the hypothesis that the fibrin cuffs limit oxygen diffusion unlikely (Pardes *et al.*, 1990). Fibrin cuffs, nevertheless, may become highly cross-linked with $\gamma-\gamma$ multimers as well as $\alpha-\alpha$ and $\alpha-\gamma$ cross-links (Mosesson *et al.*, 1989; Shainoff *et al.*, 1991; Siebenlist and Mosesson, 1992), and as a consequence, the matrix may exhibit reduced porosity, reduced susceptibility to fibrinolysis, and altered interactions with cells and cytokines. It is well known, in fact, that old venous thrombi become highly cross-linked and resist fibrinolysis (Brommer and van Bocke, 1992).

Recently it was proposed that fibrin cuffs may entrap growth factors and thereby hinder their distribution into the wound (Falanga and Eaglestein, 1993). Growth factors, in fact, do localize to the perivascular cuff (R. A. F. Clark, unpublished observations). These entrapped growth factors may remain active and induce mesenchymal cell proliferation around the blood vessels. In fact, continuous cuffs of mesenchymal cells, which stain positive for smooth muscle actin and are embedded in a thick layer of extracellular matrix including fibrin, fibronectin, laminin, and type IV collagen, exist around the microvasculature of venous ulcers (Pardes *et al.*, 1990; Herrick *et al.*, 1992). Thus, the cuffs enveloping the microvasculature at the margin and base of venous ulcers are actually a complex interwoven meshwork of cells and matrix proteins. These cell–matrix composite cuffs, rather than fibrin cuffs, may impede diffusion of growth factors, oxygen, and other nutrients into the wound.

In addition to the effects of ectopic perivascular fibrin in venous ulcers, the fibrin in the ulcer bed interstitium may fail to support healing as it becomes excessively cross-linked (Brommer and van Bocke, 1992) and thereby only partially degraded (Bini *et al.*, 1989) while being stripped of other molecules important for wound healing. Resistance to complete fibrinolysis could be due to progressive and enhanced deposition of collagen in the clot (Mirshahi *et al.*, 1991), extensive α chain cross-linking (Francis and Marder, 1988), $\gamma-\gamma$ multimer formation (Siebenlist and Mosesson, 1992), alteration in proteases or proteases inhibitors, or to a combination of these possibilities.

Although normally cross-linked fibrin may promote fibroblast migration into the matrix (Brown *et al.,* 1993b), highly cross-linked fibrin, as found in aged clots, may inhibit cell penetration. Furthermore, as the fibrin backbone of the provisional matrix becomes more resistant to proteolysis, it may become stripped of other proteins essential for wound repair, thus becoming a barren superstructure without function. For example, fibronectin is associated with fibrin in plasma clots and is potentially critical for cell migration into clots (Knox *et al.,* 1986). Uncontrolled degradation of fibronectin within the fibrin matrix might hinder migration, even though localized, controlled proteolysis is probably essential for fibroblast migration through fibrin matrix (Knox *et al.,* 1987) as it is for tumor invasiveness (Blasi, 1993). Indeed, chronic wound fluid contains markedly elevated levels of metalloproteinases (Wysocki *et al.,* 1993) and degraded fibronectin (Grinnell *et al.,* 1992). In fact, fibronectin, which codistributes with fibrin in normal wounds (Clark *et al.,* 1982), is relatively sparse in venous ulcers (Herrick *et al.,* 1992). To compound matters, Arg-Gly-Asp (RGD)-containing fibronectin fragments released by proteolysis can increase fibroblast secretion of metalloproteinases (Werb *et al.,* 1989) and plaminogen activator, thus establishing a positive feedback loop. Widespread, chronic proteolysis may also destroy fibrin-associated cytokines critical for wound repair. Thus, with age, fibrin matrix may become more resistant to proteolysis and less capable of supporting fibroplasia.

In summary, fibrin-rich matrix may contribute to venous ulcer pathobiology because of its abnormal location (fibrin cuffs), its altered structural state, or both. Conversely, in normal wounds the provisional fibrin matrix may promote cell recruitment into the defect by providing an essential scaffold for cell migration (Ciano *et al.,* 1986; Brown *et al.,* 1993b). Furthermore, fibrinogen–fibrin, its derivatives, or other matrix constituents like fibronectin may promote cell migration (Clark *et al.,* 1988; Leavesley *et al.,* 1992) and proliferation (Michel and Harmand, 1990; Gray *et al.,* 1993) through direct coupling of extracellular matrix cell surface receptors (Damsky and Werb, 1992). Finally, the ability of fibrin matrix to bind thrombin (Liu *et al.,* 1979), transforming growth factor-beta (TGF-β), and PDGF (R. A. F. Clark, personal observations) suggests that fibrin may act as a reservoir for factors critical for cell proliferation and migration.

3. Fibronectin

3.1. Introduction

Fibronectin is a multidomain, multifunctional cell adhesion protein found in blood and in a variety of tissue extracellular matrices. Recent books have reviewed fibronectin (Carsons, 1989; Mosher, 1989; Hynes, 1990). It is a key component of the provisional matrix during wound repair (Clark and Henson, 1988), and it is often elevated in tissues during tissue remodeling and fibrosis. Major functions of fibronectin include mediating cellular adhesion, promoting cell migration and monocyte chemotaxis, and helping to regulate cell growth and gene expression. Fibronectin functions via a series of functional domains and cell-binding sites that permit it to interact with a remarkably wide range of cell types, extracellular matrix molecules, and even cytokines (see Section 3.6).

3.2. Structure and Types of Fibronectin

All fibronectin molecules appear to consist of the same basic functional domains shown in Fig. 2. Although encoded by only a single gene, fibronectins exist in a number of variant forms that differ in sequence at three general regions of alternative splicing of its precursor mRNA. Some of this alternative splicing involves cell adhesion sequences, thereby providing a posttranscriptional mechanism for potentially regulating fibronectin cell-type specificity (see Section 3.6.1.2). For example, there can be 20 different variants of the human fibronectin subunit, depending on the particular combination of spliced sites.

Fibronectin is present in blood as a soluble plasma glycoprotein at an average concentration of 0.3 g/liter; this form found in plasma is termed *plasma fibronectin*. It is thus already present in substantial concentrations in clots. *Cellular* fibronectins are produced by a wide variety of cell types, which secrete them and often organize them into extensive extracellular matrices. Cellular fibronectins vary in composition, but they characteristically contain considerably higher proportions of alternatively spliced sequences than plasma fibronectin. Healing dermal wounds contain substantially increased amounts of cellular (embryonic) fibronectins, a major source of which is wound macrophages (ffrench-Constant *et al.*, 1989; Brown *et al.*, 1993a).

3.3. Localization of Fibronectin

A large number of descriptive studies have identified fibronectin in many tissues during a host of developmental, reparative, and pathogenic events. It appears to func-

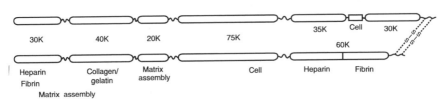

Figure 2. Modular domain structure of fibronectin. This large glycoprotein is depicted as a dimer of similar but not identical subunits linked by a pair of carboxyl-terminal disulfide bonds; the two subunits actually extend in opposite directions. The numbers indicate the approximate molecular weights of each domain (K = × 1000). The major ligand(s) bound by each domain are listed along the bottom. There are two heparin-binding domains and two major fibrin-binding domains, as well as at least two regions involved in assembly of fibronectin into a fibrillar matrix.

tion as a biological adhesive, promoter of cell migration, and stimulator of collagenous matrix formation. Summaries of such studies can be found in texts that include the first edition of this book and books focusing on fibronectin (Clark and Henson, 1988; Carsons, 1989; Mosher, 1989; Hynes, 1990).

Fibronectin (along with certain other extracellular proteins) is particularly prominent in migratory pathways for embryonic cells, such as those for gastrulation, neural crest cell migration, and precardiac mesoderm cell migration. Fibronectin is also characteristically present in loose connective tissue and in embryonic (but often not in adult) basement membranes. Large quantities of fibronectin are present in fibrin clots, where it can serve as a substrate for migrating cells during wound healing (surveyed in Chapter 1, Section 5.1.2). Fibronectin deposition is also associated with various fibrotic processes (Carsons, 1989).

3.4. Gene Structure of Fibronectin

3.4.1. The Fibronectin Gene

The fibronectin gene is relatively large, with an estimated length of about 50 kilobases. Its coding sequence is subdivided into approximately 50 exons of similar size, several of which undergo alternative splicing (Hirano *et al.,* 1983; Mosher, 1989; Hynes, 1990). Fibronectin is composed of three general types of homologous repeating units or modules, termed types I, II, and III (Fig. 3) (Kornblihtt *et al.,* 1985). There are 12 type I repeats, 2 type II repeats, and 15–17 type III repeats (reviewed in more detail in Mosher, 1989; Hynes, 1990). In the gene, each repeating module of the type I or II homology unit is encoded by a separate exon. In contrast, the larger type III repeats generally require the contribution of two exons each; however, two type III repeats exist as a single exon, and both can undergo alternative splicing (Patel *et al.,* 1987).

These modular repeating units are used as building blocks and are organized into larger structural domains with distinct functions. For example, type I modules are used in fibronectin domains that bind to fibrin alone, or to heparin as well as to fibrin, or even to collagen (Petersen *et al.,* 1989). Type II modules are found only in the collagen-binding domain, but they may not be essential for fibronectin binding to collagen (Ingham *et al.,* 1989). Type III repeats are used in domains that bind to cells or to heparin, but not to collagen (Mosher, 1989; Hynes, 1990).

The three basic protein structural modules of fibronectin are used as structural building blocks for other proteins (reviewed in Mosher, 1989). For example, repeats homologous to type I modules are also found in tissue plasminogen activator. The structure of this type of module has been determined by nuclear magnetic resonance (NMR) (Baron *et al.,* 1990). Type II modules of fibronectin are homologous to the Kringle domains of blood clotting and fibrinolytic proteins. Finally, fibronectin type III modules are found in tenascin, in cell–cell adhesion molecules such as L1, and in a surprising number of other proteins as a basic structural unit.

The promoter region of the fibronectin gene has been characterized, and it contains a cyclic AMP response element sequence responsible for fibronectin induction via

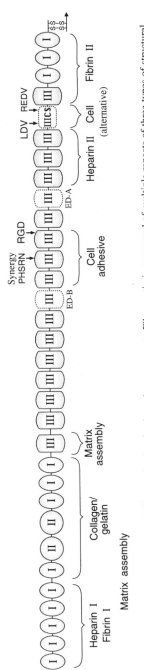

Figure 3. Detailed structural map of fibronectin showing homologous repeats. Fibronectin is composed of multiple repeats of three types of structural modules termed types I, II, or III. Dotted outlines indicate units where alternative splicing of mRNA results in the insertion or omission of type III modules termed ED-A (EIIIA) and ED-B (EIIIB), or of portions of the IIICS region. The labels along the bottom of the figure indicate functional binding domains. Arrows from above indicate the following cell interaction sites: RGD (Arg-Gly-Asp) as the primary recognition sequence in the major cell-adhesive domain; PHSRN (Pro-His-Ser-Arg-Asn) as the synergy region provided the RGD sequence with high affinity and specificity for the $\alpha_5\beta_1$ fibronectin receptor; the IIICS region, which contains the CS1 sequence containing LDV (Leu-Asp-Val) and the weaker REDV (Arg-Glu-Asp-Val) sequence. In addition, the heparin II region contains two cell-binding sequences recognized by certain cell types.

nuclear factors (Dean *et al.,* 1990; Srebrow *et al.,* 1993). A number of growth factors and cytokines can regulate fibronectin synthesis, including TGF-β (reviewed in Yamada, 1989). The mechanisms by which these factors affect gene expression, e.g., the identity of all of the transcription factors needed for regulation of fibronectin expression, remain to be determined.

3.4.2. Alternative Splicing

The patterns of alternative splicing of fibronectin are particularly interesting, since they can insert or delete cell-type-specific adhesion sites. The IIICS (also termed V) region of fibronectin shown in Fig. 3 can undergo complex patterns of alternative splicing that can produce human fibronectin molecules with five different sequence patterns in this region (reviewed in Mosher, 1989; Hynes, 1990). Two of these spliced sites, termed CS1 and CS5, contain distinct adhesive recognition sequences (see Section 3.6.1.2.). The sequence of this complex spliced region differs from the repeating structural units that comprise most of fibronectin. In addition, however, there are two other sites of alternative splicing in fibronectin that result from insertion or deletion of entire type III repeating units (Fig. 3). It is intriguing that the presence of these spliced sites, termed ED-A (EIIIA) and ED-B (EIIIB), can correlate temporally and spatially with wound repair or embryonic cell migration and morphogenesis, but their functions are not yet clear (ffrench-Constant *et al.,* 1989; Mosher, 1989; Hynes, 1990; Brown *et al.,* 1993a). They may also promote fibril formation by fibronectins (Guan *et al.,* 1990).

Differential splicing within all three of the general regions of alternative splicing in human fibronectin can produce a theoretical total of 20 different variants of each fibronectin chain of approximately 250,000 Da. Since fibronectin is a dimer, or even a multimer when associated with cell surfaces, there can be considerable complexity of fibronectin variant mixtures. Little is known about the regulation of this alternative splicing, although spliced sequences tend to be retained in cellular fibronectins, especially by certain embryonic and malignant cells (Vartio *et al.,* 1987; Oyama *et al.,* 1989) and in healing wounds (ffrench-Constant *et al.,* 1989; Brown *et al.,* 1993a). Plasma fibronectin, considerable proportions of which are produced by the liver, generally lacks ED-A and ED-B sequences, and can contain 50% or less of the IIICS type of spliced region (Hershberger and Culp, 1990). Since the fibronectin in healing wounds gradually changes from plasma fibronectin in the original clot to a much higher proportion of spliced forms derived from wound macrophages (Brown *et al.,* 1993b), it will be important to determine the biological effects of this transition.

3.4.3. Posttranslational Modifications

Fibronectin can be glycosylated, phosphorylated, and sulfated (reviewed in Mosher, 1989). Glycosylation of fibronectin is primarily of the complex N-linked oligosaccharide type. Although nonglycosylated fibronectin still displays much of the adhesive activity of the native protein, it is much more sensitive to proteases; carbohydrates therefore appear to stabilize the protein against proteolysis (Olden *et al.,* 1979). Moreover, the presence of carbohydrate moieties can modulate binding to collagen and slightly alter interactions with cells. For example, higher amounts of glycosylation cause a decreased avidity for collagen (Zhu and Laine, 1985). The wound environment is often rich in proteases, resulting in extensive degradation of fibronectin into proteolytic fragments (Grinnell and Zhu, 1994); differences in glycosylation may therefore affect this proteolysis and the activities of the fibronectin fragments it generates.

3.5. Molecular Morphology of Fibronectin

Soluble plasma fibronectin exists as a disk-shaped molecule approximately 30 nm in diameter and 2 nm thick (Sjoberg *et al.,* 1989; Benecky *et al.,* 1990). Increased ionic strength opens or unfolds the molecule to a more elongated structure. Binding of fibronectin to an artificial substrate, to certain ligands, or to cellular receptors may also cause such unfolding from a compact shape to a more linear shape, which could then theoretically open the molecule for accessibility to other ligands or to self-assembly into fibronectin fibrils (Sjoberg *et al.,* 1989; Wolff and Lai, 1989; Mosher *et al.,* 1992).

Both cellular and plasma fibronectins can become incorporated into fibronectin fibrils, but they tend to remain segregated, suggesting that alternative splicing may help regulate such supramolecular organization (Pesciotta-Peters *et al.,* 1990) (see Chapters 1 and 9 for additional review of fibronectin fibrillogenesis and matrix assembly). When incorporated into fibrils, fibronectin becomes cross-linked covalently by disulfide bonds and by transglutaminase cross-linking (reviewed in Mosher, 1989). Such cross-linking stabilizes the fibrils.

3.6. Organization of Functional Domains

Each fibronectin polypeptide chain consists of a series of structural and functional domains (Fig. 2). These domains have been defined by proteolytic fragmentation studies and by recombinant DNA analyses.

3.6.1. Cell Interaction Sequences

Fibronectin contains six or more peptide sites capable of mediating cell adhesion. They are located in three general regions: the central cell-binding domain, the alternatively spliced IIICS region, and the heparin-binding domain. In most cases, the sequences involved in recognition by cell surface receptors are relatively short, suggesting the importance of linear amino acid sequences either alone or in combination to mediate processes such as cell adhesion. Ongoing studies, however, also seem to indicate the importance of other polypeptide information in fibronectin that is necessary for function of the short, linear sequences; this requirement for a second part of the protein could be due to the need for another recognition sequence(s) that binds to the receptor, or to a requirement for a protein framework that can present a specific conformation of the recognition sequence to the receptor.

3.6.1a. Central Cell-Binding Domain. Most cells can adhere to fibronectin, at least in part by binding to the centrally located "cell-binding" domain (Figs. 2, 3). Some cells, such as fibroblasts, can bind to isolated 20- to 120-kDa proteolytic fragments containing this domain nearly as well as they can to intact fibronectin (i.e., equal molar activities) (Akiyama *et al.,* 1985, 1994). A crucial sequence in this domain is Arg-Gly-Asp-Ser (abbreviated RGDS, using the single-letter amino acid code) (Pierschbacher and Ruoslahti, 1984, 1987; Yamada and Kennedy, 1984). The first three amino acids are particularly important, and they define the now-classic RGD recognition motif. Deletion of this sequence, or even mutation of the aspartate to a glutamate residue (which retains the same charge) in an otherwise intact domain, leads to the loss of nearly all adhesive activity (Obara *et al.,* 1988). RGD-containing peptides specifically inhibit a variety of migratory processes *in vivo,* including gastrulation, neural crest cell migration, and experimental metastasis (reviewed in Mosher, 1989; Hay, 1991).

Further studies, however, indicate major roles for a second, "synergy" sequence. Deletion of the RGD sequence does not completely abolish adhesive activity, consistent with the existence of a second interaction site in this domain (Obara *et al.,* 1988; Bowditch *et al.,* 1991). Conversely, the estimated affinity of interaction of RGD-containing fibronectin peptides with cell surface fibronectin receptors is much lower than that of larger fragments (Akiyama and Yamada, 1985). Mutagenesis and monoclonal antibody inhibition experiments in fact define a second critical region of this domain located roughly 10,000 Da toward the amino-terminus of the protein, and competitive inhibition studies confirm that this region contains a second binding region that functions in synergy with the RGD sequence (Aota *et al.,* 1991; Bowditch *et al.,* 1991; Nagai *et al.,* 1991). Both $\alpha5\beta1$ and $\alpha IIb\beta3$ respond to this overall synergy region, using overlapping recognition sites. The key sequence for $\alpha5\beta1$ has been narrowed down to Pro-His-Ser-Arg-Asn (PHSRN), of which Arg_{1379} appears to be the only crucial amino acid (Aota *et al.,* 1994). A synthetic peptide containing PHSRN and amino-terminal amino acids also contained synergy recognition activity for the major platelet integrin $\alpha IIb\beta3$, but substitution analysis indicated that Asp_{1373} and Arg_{1374} were the only crucial amino acids for interactions with this integrin (Bowditch *et al.,* 1994). Nevertheless, recent studies with recombinant polypeptides rather than synthetic peptides now indicate that this platelet integrin receptor has very similar sequence requirements for only Arg_{1379} (R. Bowditch, personal communication). The two integrins may therefore not differ substantially in their amino acid sequence specificity, although they do differ in the magnitude of their requirement for this synergy site, which is much higher for the $\alpha5\beta1$ integrin.

It has been proposed that this synergistic system can account for specificity and for full affinity of fibronectin toward the $\alpha5\beta1$ integrin receptor (reviewed in Yamada, 1989). However, the potential importance of the three-dimensional conformation of the RGD sequence by itself as a determinant of specificity for fibronectin receptors remains uncertain (in contrast to the much stronger evidence for importance of RGD conformation for vitronectin receptors). These biochemical distinctions are important for both evolutionary and functional considerations, since classical biological questions such as the basis of adhesive specificity could be explained either by the use of unique combinations of such sequences or by more lock-and-key mechanisms requiring unique conformations of these sequences (reviewed in Yamada, 1989). It will therefore be important to extend current structural information on RGD-containing type III modules (Leahy *et al.*, 1992; Main *et al.*, 1992) to three-dimensional structural determination of the fully active combination of RGD and synergy sites.

In wounds, proteolytic fragments of fibronectin and other molecules are often present due to the action of wound proteases (Grinnell *et al.*, 1992; Grinnell and Zhu, 1994). Such proteolytic fragments can display biological activities that are absent or suppressed in the intact parent molecule. The central cell-adhesive domain of fibronectin displays potent chemotactic activity for monocytes, even though the intact molecule has little activity (Clark *et al.*, 1988; Wikner and Clark, 1988; Doherty *et al.*, 1990). Similarly, a fragment containing this central cell-adhesive domain can stimulate protease secretion by certain fibroblasts, even though the intact fibronectin molecule does not (Werb *et al.*, 1989). A possibly similar disparity between effects of the intact molecule and a peptide fragment is also seen in an embryonic system, where RGD peptides but not fibronectin can induce cell–cell aggregation and compaction of segmental plate cells, possibly through the induction of cell surface expression of the N-cadherin molecule (Lash *et al.*, 1987; J. Lash, personal communication). These processes provide additional mechanisms for cells to detect and respond to wound conditions that generate fragments of fibronectin, independent of growth factors.

3.6.1.2. Alternatively Spliced Adhesion Sites. The alternatively spliced sequences of the IIICS region are of particular interest because they can encode cell-type-specific adhesion sequences; fibronectin in dermal wounds contains enhanced amounts of such alternatively spliced sequences (ffrench-Constant *et al.*, 1989). Although cells such as fibroblasts cannot adhere well to this region of fibronectin, cells derived from the embryonic neural crest and activated lymphocytes can adhere readily. For example, embryonic neural crest cells, neurons from sympathetic and sensory ganglia, and melanoma cells adhere, as do activated T cells and certain B lymphocytes. All of these cells adhere to one or both of the two cell-binding sequences in this region by means of the $\alpha 4\beta 1$ integrin receptor (see Chapter 9). Alternatively, spliced IIICS sequences can be used by neural crest cells for migration in conjunction with several other sites elsewhere in the protein (Dufour *et al.*, 1988).

The stronger of the two peptide adhesion sequences in this region is unusually potent. A 25-residue synthetic peptide corresponding to this site, termed CS1, still retains 40% of the total molar activity of fibronectin itself (Humphries *et al.*, 1987), which is an unusually high retention of functional activity by a synthetic peptide. In contrast, RGD-containing peptides appear to be 100–200 times less active for recognition by the $\alpha 5\beta 1$ fibronectin receptor on a molar basis (Akiyama and Yamada, 1985). The minimal essential peptide sequence required for function of this CS1 region is the tripeptide Leu-Asp-Val (Komoriya *et al.*, 1991). Nevertheless, other information elsewhere in the CS1 peptide region appears to be required for full activity of this region, perhaps due to synergy of adjacent peptide sequences with the minimal Leu-Asp-Val sequence.

An independent sequence that is also alternatively spliced consists of the sequence Arg-Glu-Asp-Val (Humphries *et al.*, 1986). This sequence is two orders of magnitude less active than the CS1 sequence, but it appears to display identical cell-type and receptor specificity for the α4β1 integrin receptor. The existence of two alternatively spliced regions known to interact with α4β1 integrins provides a potential means for regulating the strength of these interactions by their extent of expression, which might be important for processes such as rates of cell migration or neurite extension (Humphries *et al.,* 1988).

An interesting speculation concerning the nature of adhesive recognition sequences comes from inspection of these sequences compared with the canonical Arg-Gly-Asp sequence, all of which appear to contain a critical Asp residue: It has been proposed that this residue participates in a general binding mechanism based on cation binding to a functional site contributed by both this Asp and other sequences located in the integrin receptor (Loftus *et al.,* 1990; Tuckwell *et al.,* 1992). Experiments in which this sequence is mutated in either proteins or synthetic peptides are consistent with this hypothesis to date, but whether it is the underlying mechanism for this class of adhesive interactions remains to be determined. An interesting recent alternative mechanism postulates a two-step cation displacement model in which divalent cation is first bound to the integrin, but is then displaced or replaced by the ligand during the binding process (D'Souza *et al.,* 1994).

If provided in high enough amounts, either the RGD sequence or the CS1 peptide sequence immobilized alone on a substrate can produce cell spreading and promote microfilament bundle organization (Singer *et al.,* 1987; K. Yamada, unpublished results). Even cell proliferation can be regulated by either the RGD or CS1 domains. Binding of T lymphocytes to either the central cell-binding domain of fibronectin containing RGD or to the Leu-Asp-Val (LDV)-containing CS1 peptide can help induce proliferation. In the intact fibronectin molecule, both sites can contribute to mitogenesis (Nojima *et al.,* 1990). These and other findings suggest the intracellular sequelae of binding to different fibronectin-adhesive sequences can be similar. In striking contrast, however, the induction of collagenase in rabbit synovial fibroblasts by binding to the central cell-binding domain is inhibited by concurrent binding to the CS1 sequence (Huhtala *et al.,* 1995). In this latter case, their cell surface integrin receptors, α5β1 and α4β1, respectively, mediate surprisingly different signals. It will be important to understand the differences between these proposed distinct signaling pathways.

3.6.1.3. Heparin-Binding Sites. Although binding of fibronectin to cell surface heparan sulfate proteoglycan can contribute to fibronectin-mediated adhesion, a more direct interaction between cell surface receptors and this type of domain may occur. Several peptide sequences from the high-affinity heparin-binding domain that can bind heparin directly can also mediate cell attachment (Fig. 3, site 5), and the receptor used for this interaction appears to be the α4β1 integrin (McCarthy *et al.,* 1988, 1990). The relative strengths, cell-type specificities, and functional interrelationships of these sites to the central cell-binding domain sites and to the alternatively spliced sites remain to be determined for various cell types.

3.6.2. Fibrin-Binding Domains

Fibronectin interacts with fibrin in clots, contributing to the thickening of fibrin fibers (Nair and Dhall, 1991). Biologically, fibronectin may play important roles in the capacity of cells to interact with fibrin. Depending on the cell system, fibronectin is required for either cell adhesion or cell migration into fibrin clots (Grinnell *et al.*, 1980; Knox *et al.*, 1986). In both studies, covalent cross-linking of fibronectin to fibrin via factor XIII transglutaminase was proposed to be involved in mediating the effect. The roles of fibronectin and transglutaminase cross-linking in wound healing therefore deserve investigation. Although adhesion or migration into fibrin clots required cross-linked fibronectin, fibrin gel contraction did not (Tuan and Grinnell, 1989).

Fibronectin contains at least two fibrin-binding domains; a third is detectable after proteolysis of the protein (Figs. 2, 3) (Mosher, 1989). The major fibronectin-binding site for fibrin is in the amino-terminal domain and is formed by a combination of the 4th and 5th type I repeating units ("fingers") of fibronectin (Matsuka *et al.*, 1994). Thus, the requirements for interaction of the amino-terminal domain with fibrin differ from those for binding to fibroblast cell surfaces or to *Staphylococcus*, which requires all 5 type I repeating units (Sottile *et al.*, 1991). Although the 4th and 5th type I repeats together bound to fibrin, each of these units alone did not bind, indicating that the pair interacts in some fashion to form the binding site or pocket (Matsuka *et al.*, 1994).

The three-dimensional structure of this fibrin-binding region has been determined by two-dimensional NMR, providing insight into how these individual units form a structured complex and providing candidate surface-exposed amino acid residues as potential contributors to the binding site (Williams *et al.*, 1994). These type I repeats of fibronectin are homologous to the finger domain of tissue plasminogen activator (TPA). Competition between binding of these related domains in TPA and fibronectin to fibrin might be an important regulator of fibrinolysis, and, in fact, inhibition of fibrin-dependent plasmin generation by TPA in the presence of fibronectin has been observed (Beckmann *et al.*, 1991).

The interaction of fibronectin with fibrin may also be important for mediating macrophage clearance of fibrin released into the circulation after trauma or in inflammatory states (reviewed by Blystone and Kaplan, 1993). The binding of fibronectin to the macrophage cell surface is of high affinity ($K_d = 2$–7×10^{-8} M for 29-kDa and 70-kDa amino-terminal fragments) involving about 80,000 sites per cell. A 67-kDa binding protein has been isolated that may serve as this high-affinity receptor (Blystone and Kaplan, 1992).

3.6.3. Collagen (Gelatin)-Binding Domain

Although this domain initially appeared to interact with native, nondenatured collagen, model studies indicate that the interaction may be truly significant only at subphysiological ionic strength (Ingham *et al.*, 1985). This domain binds far more effectively to denatured collagen (e.g., gelatin) than to native collagen, and thus its interactions with collagens in general may be due to its binding to unfolded regions of the collagen triple helix. Interestingly, the most preferred binding site on type I collagen is in the same region cleaved by bacterial collagenase. This region less stringently conforms to a triple helix, and thus might be preferentially bound by a fibronectin domain that recognizes only denatured collagen chains.

The evolutionary conservation of the gelatin-binding function in fibronectins from a wide variety of species suggests its biological importance. The marked preference of this domain to bind to unfolded collagen chains, however, is puzzling when considering its potential biological function *in vivo*. Although fibronectin can mediate or modulate cell adhesion to collagen *in vitro,* it is not known whether this function is important *in vivo*. In fact, cells can adhere directly to native collagen via receptors such as the $\alpha2\beta1$ integrin. It is conceivable that the physiological function of this gelatin-binding domain relates less to cell adhesion per se and more to binding and clearance of denatured collagenous materials from blood or tissue. It has even been reported that fibronectin is needed for the organization of collagen fibrils *in vitro* (McDonald *et al.,* 1982), and it may also control the organization of a variety of other extracellular matrix components.

The polypeptide regions of fibronectin capable of mediating its binding to collagen have been variously identified as including a type II repeat (Owens and Baralle, 1986) or only type I repeats (Ingham *et al.,* 1989). The results obtained to date seem to indicate that several parts of the collagen-binding region contribute to full avidity of binding (McDonald and Kelley, 1980), although the relative contributions to binding of combining independent binding sites as opposed to producing a better conformation of the binding region remains to be determined.

3.6.4. Heparin-Binding Domains

Fibronectin contains two heparin-binding domains (Figs. 2, 4), which are thought to interact most often with heparan sulfate proteoglycans. The binding of heparin by intact fibronectin is of relatively high affinity, with two classes of affinities with $K_d = 10^{-7}$ to 10^{-9} M (Yamada *et al.,* 1980). The specific heparin-binding domains are located toward opposite ends of the protein, and they differ in both affinities and sensitivity to calcium ion. The strongest heparin-binding site is in the carboxyl-

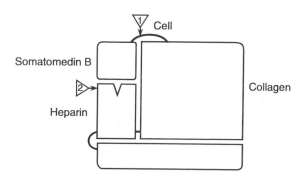

Figure 4. Schematic model of the major native form of vitronectin. This glycoprotein usually exists as a relatively globular, monomeric protein containing an amino-terminal region that can be proteolytically cleaved to release the plasma protein somatomedin B. A cell-binding sequence (numeral 1) consists of the amino acid sequence Arg-Gly-Asp-(Val). A putative collagen-binding domain can be cleaved proteolytically from an adjacent part of the protein at protease-sensitive connecting strands indicated by the thicker lines. The heparin-binding domain (numeral 2) is usually cryptic (V-shaped groove), and it generally becomes exposed or available only after denaturation or unfolding of the molecule.

terminal third of the protein (Fig. 3). A weaker binding domain is located in the amino-terminal end of the protein. This latter domain also binds to fibrin and can modulate cell spreading. Heparin or heparan binding by this amino-terminal domain can be regulated by the extracellular concentration of calcium. Inhibition of binding occurs when levels of Ca^{2+} rise above average physiological concentrations in blood; thus, tissue sites with high local divalent cation concentrations could have decreased binding by this domain (Hayashi and Yamada, 1982).

As reviewed above, heparin-binding regions of fibronectin can also mediate cell adhesion *in vitro*. They have additional activities based on their glycosaminoglycan-binding activity, as demonstrated using fibronectin fragments. Certain cells cannot efficiently organize microfilament bundles and focal adhesions unless a heparin-binding domain is present, supporting the hypothesis that cell surface heparan sulfate is important for part of fibronectin's effects on cell behavior (Izzard *et al.*, 1986; Woods *et al.*, 1986). This activity of the fibronectin heparin-binding domain on focal adhesions can be mimicked by a soluble peptides from this site containing the PRARI (Pro-Arg-Ala-Arg-Ile) sequence (Woods *et al.*, 1993). The formation of focal adhesions involves protein kinase C as part of a putative signaling pathway (Woods and Couchman, 1992). It is of interest from an evolutionary standpoint that heparin-binding activity in fibronectin arose in two structurally diverse domains. One domain consists of type III repeats lacking any disulfide bonds, whereas the other domain is based on the type I repeat and contains five internal disulfide bonds, forming a compact domain.

Although intact fibronectin binds relatively weakly to chondroitin sulfate, fragments containing the high-affinity heparin-binding domain can bind to this glycosaminoglycan as strongly as to heparin (Barkalow and Schwarzbauer, 1994). These results indicate that the overall structure of fibronectin affects its interactions with such molecules, and that proteolysis such as that which occurs in wound fluid can generate or expose cryptic binding sites for chondroitin sulfate, a widely distributed glycosaminoglycan.

3.6.5. Matrix-Assembly Regions

The assembly of fibronectin into fibrils and matrices is likely to be an important process in wound repair. Such fibrils form the natural *in vivo* substrates for cell adhesion and migration, which is not provided by soluble plasma fibronectin until it is immobilized. Normal fibronectin fibrillogenesis appears to require at least two processes: (1) association of one fibronectin molecule with another via specific sites, and (2) $\alpha5\beta1$ integrin binding with possible contributions from some other cell surface receptor. Several distinct regions are involved in the organization of secreted fibronectin molecules into extracellular fibrils. This matrix-assembly process, however, appears to be an active event involving contributions from receptors on living cells, rather than occurring by simple self-assembly (McDonald, 1988). Fibronectin matrix assembly, thus, occurs in a two-step sequence: first, fibronectin binds to a receptor system, then it is converted to a detergent-insoluble form during fibrillogenesis (Mosher *et al.*, 1992).

Not surprisingly, the central cell-binding domain of fibronectin seems to be involved in matrix assembly: the process of *in vitro* fibronectin matrix assembly is inhibited by antibodies against the cell-binding domain, by antibodies against the $\alpha5\beta1$ fibronectin receptor, or by competition with excess free proteolytic fragments containing this region (McDonald *et al.*, 1987; Fogerty *et al.*, 1990). Antibodies against either the RGD region or the synergy region of the cell-adhesive domain will inhibit matrix assembly, whereas antibodies that bind between these sites do not (Nagai *et al.*, 1991). Thus, both fibronectin elements recognized by $\alpha5\beta1$ are involved in matrix assembly. Besides preventing formation of a fibronectin-based matrix, inhibiting this matrix assembly process can block events such as embryonic cell migration during gastrulation (Darribere *et al.*, 1990).

In addition to the cell-binding domain, the amino-terminal end of the protein is surprisingly also implicated in the matrix assembly process. Fragments of 70 kDa or, somewhat less effectively, the amino-terminal domain of 25 kDa, can be bound to cells in the first step of matrix assembly (McKeown-Longo and Mosher, 1985; McDonald *et al.*, 1987). An excess of the 70-kDa domain can competitively inhibit incorporation of fibronectin into fibrils. This fragment, however, cannot complete assembly by itself, remaining detergent-soluble even while intact fibronectin becomes assembled. Linking 70-kDa monomers into a dimer via the last 51 amino acids of fibronectin to form a disulfide-bonded dimer allows incorporation of these fragments into a preexisting fibronectin matrix (Sottile and Wiley, 1994). Because they could not form a matrix in the absence of intact fibronectin, another fibronectin–fibronectin-binding site appears needed for full-fledged *de novo* matrix assembly.

An additional specific binding site involved in fibronectin self-assembly has been found in the first type III repeating unit of fibronectin (the III-1 module). This region contains a conformation-dependent binding site for the 70-kDa amino-terminal region (Chernousov *et al.*, 1991; Hocking *et al.*, 1994). A synthetic 31-amino acid peptide from this region can mediate fibronectin binding, and has been found to promote the assembly of soluble fibronectin into fibrils (Morla and Ruoslahti, 1992). Function of this region is essential for *in vivo* formation of fibronectin fibrils in embryos (Darribere *et al.*, 1992).

Interestingly, plasma and cellular (i.e., containing alternatively spliced ED-A and ED-B sites) forms of fibronectin tend not to coassemble in fibrils. They instead segregate into short alternating stripes of self–self assembly on fibrils (Pesciotta-Peters *et al.*, 1990).

The cell surface receptors involved in matrix assembly include $\alpha5\beta1$ but not $\alpha v\beta1$, with possible contributions from some other $\beta1$ integrin (Fogerty *et al.*, 1990; Wu *et al.*, 1993; Zhang *et al.*, 1993; Akiyama *et al.*, 1994). Other nonintegrins may also be involved, although more studies are needed to clarify this point (Limper *et al.*, 1991; Moon *et al.*, 1994). Moreover, gangliosides appear to considerably augment fibronectin matrix assembly (Spiegel *et al.*, 1986), perhaps because they can bind effectively ($K_d = 10^{-8}$ M) to the same 25- to 31,000-Da amino-terminal domain implicated in matrix assembly (Thompson *et al.*, 1986).

The cell-catalyzed or cell-mediated mechanism of normal fibronectin assembly contrasts markedly with collagen fibrillogenesis, which proceeds dramatically in the absence of cells *in vitro*. Nevertheless, fibronectin fibrils are major components of the extracellular matrix produced by cells in tissue culture, as well as *in vivo* pathways of embryonic cell migration, e.g., for gastrulation and neural crest cell migration. The requirement for a cellular matrix-assembly system permits more precise regulation of the sites of fibronectin fibril formation than if the fibrils were self-assembling, and probably accounts for the fact that even though high levels of fibronectin circulate in blood (0.3 g/liter), it remains freely soluble until incorporated into blood clots or tissue matrices. Recent studies suggest that fibronectin matrix assembly can be modulated by regulators such as protein kinase C and lysophosphatidic acid (Checovich and Mosher, 1993; Somers and Mosher, 1993).

3.6.6. Other Molecular and Functional Associations of Fibronectin

Like other extracellular matrix molecules, fibronectin can serve as a reservoir for the binding of growth factors and cytokines. For example, TGF-β can bind tightly to purified fibronectin and it can be recovered by acid extraction (Fava and McClure, 1987). Moreover, tumor necrosis factor-alpha (TNF-α) binds to fibronectin's amino-terminal domain, retaining high activity while immobilized (Alon et al., 1994; Hersh-koviz et al., 1994). Such extracellular matrix-bound factors may provide an immo-bilized store that can be held in close proximity to cells for cell activation or may be released after proteolysis of the matrix protein carrier.

A novel bifunctional activity of fibronectin and its amino-terminal fragment has been described for adipocyte differentiation in vitro. Intact fibronectin inhibited differ-entiation of an adipocyte cell line almost completely, in a process requiring the cell-binding domain and $\alpha5\beta1$. In contrast, fibronectin fragments and purified 24-kDa amino-terminal domain markedly stimulated differentiation (Fukai et al., 1993). Whether this activity is due to binding other molecules, such as cytokines or extracellu-lar ligands, or to direct activity of this fragment remains to be established.

Fibronectin can also bind molecules such as complement proteins, bacteria, yeast, DNA, and denatured actin. The biological roles of these interactions remain to be established conclusively, although one function may be as a scavenger or opsonic molecule that promotes the clearing of such materials from blood by the reticuloen-dothelial system (reviewed in Carsons, 1989).

4. Vitronectin

4.1. Introduction

Vitronectin (S-protein) is a multifunctional glycoprotein found in plasma, extra-cellular matrices, and fibrin clots. First identified as a cell attachment factor with high avidity for glass substrates (vitro = glass), vitronectin has been investigated under the names serum spreading factor, S-protein, and epibolin. Besides mediating cell adhesion and migration (e.g., epiboly), vitronectin can also protect cells from cytolytic destruc-tion by released activated complement complexes and can protect thrombin from inactivation by antithrombin III (reviewed by Tomasini and Mosher, 1991; Felding-Habermann and Cheresh, 1993).

4.2. Structure and Location

The vitronectin molecule is a 75,000-Da monomer that can be proteolytically clipped to a form consisting of two chains, 65,000 Da and 10,000 Da, linked by a disulfide bond. Plasma contains both forms in varying ratios in different individuals, at a total concentration of 0.2–0.4 g/liter (Kubota *et al.,* 1988). The form of vitronectin in tissues remains to be determined.

The vitronectin gene is relatively small, extending only 3 kilobases in length and containing eight exons (Jenne and Stanley, 1985; Suzuki *et al.,* 1985; Seiffert *et al.,* 1993). Exon 2 encodes the sequence for somatomedin B. Proteolysis of vitronectin apparently releases this polypeptide, whose function remains unknown. Neither highly purified somatomedin B nor vitronectin has growth factor activity (Barnes *et al.,* 1984). Levels of vitronectin can be regulated by TGF-β1 and other cytokines (Koli *et al.,* 1991).

Vitronectin is also deposited on fibers in the extracellular matrices of a variety of tissues, where it sometimes colocalizes with fibronectin. Although vitronectin is absent from the skin of children (Dahlback *et al.,* 1989), in adult skin, it is located at the periphery of dermal elastic fibers in a noncovalent association and it appears to accumulate as a function of age (Hintner *et al.,* 1991). However, vitronectin does not necessarily colocalize with fibrillin in microfibrils. Vitronectin is also associated with amyloid deposits in dermis (Dahlback *et al.,* 1993).

4.3. Cell Attachment Domain

The cell attachment activity of vitronectin is encoded by exon 3 and is based on the sequence Arg-Gly-Asp (Jenne and Stanley, 1985; Suzuki *et al.,* 1985). Cell attachment to vitronectin does not appear to require either the somatomedin B or heparin-binding domains, and mutation of this the Arg-Gly-Asp sequence leads to a loss of cell adhesive activity (Zhao and Sane, 1993). A wide variety of cell types will adhere to vitronectin. In fact, most cell-adhesive activity of serum, used at 5–10% for tissue culture, can be attributed to vitronectin [fibronectin becomes more important only at low concentrations of serum (Grinnell and Phan, 1983; Knox, 1984; Bale *et al.,* 1989)].

Attachment to vitronectin occurs via any of several vitronectin receptors in the integrin family, including αvβ3, αIIbβ3, and αvβ5. It is important to note that these receptors may also bind to fibronectin and in some cases to other proteins. For example, αIIbβ3, a major platelet aggregation receptor, binds to fibronectin, vitronectin, von Willebrand factor, and fibrinogen, all of which contain the canonical Arg-Gly-Asp (RGD) recognition sequence. Potential binding sites on αvβ3 and αIIbβ3 have been identified by cross-linking experiments using synthetic peptide ligands (D'Souza *et al.,* 1990). Current data suggest that both α and β chains contribute to formation of the binding pocket. These receptors exhibit variations of a binding specificity that is strongly focused on the RGD sequence itself. Unlike the α5β1 receptor, αvβ3 and αIIbβ3 can readily bind to immobilized synthetic peptides containing RGD. There is good evidence that the local sequence and conformation of such peptides contribute substantially to binding. Nevertheless, even the classically pure RGD-recognizing receptor, αIIbβ3 on the platelet, appears to require additional sequence information for binding to fibronectin (Bowditch *et al.*, 1991). It remains to be determined whether other sequences in vitronectin and other cooperative or competitive ligands contribute to vitronectin receptor binding.

In cultured cells, vitronectin–vitronectin receptor complexes are prominently located at focal contacts with the culture plate (focal adhesions) in association with intracellular cytoskeletal elements such as vinculin, talin, paxillin, and the ends of actin microfilament bundles (Burridge *et al.*, 1988; Dejana *et al.*, 1988; Singer *et al.*, 1988). The *in vivo* relevance of this striking transmembrane complex will be interesting to unravel. Besides forming strong adhesions to vitronectin, certain cells can migrate on vitronectin substrates. The migration of polymorphonuclear neutrophils on vitronectin depends on unusual repeated cycles of transiently increased intracellular calcium ion concentration. The relatively high intracellular calcium apparently reduces integrin-mediated adhesion to vitronectin via a process that requires calcineurin, a calmodulin-dependent phosphatase (Hendey *et al.*, 1992).

4.4. Heparin Binding and Other Binding Activities

Vitronectin contains a cryptic heparin-binding domain that is exposed (along with certain antibody epitopes) after adsorption onto a surface or after denaturation by agents such as urea or guanidine (reviewed in Tomasini and Mosher, 1991). This property has been used as the basis for a simple, efficient purification protocol (Yatohgo *et al.*, 1988). In the absence of denaturing treatments, only a small fraction of plasma or serum vitronectin will bind heparin (2% and 7%, respectively). The vitronectin form that binds heparin consists of large aggregates, which have similar cell adhesion activities as monomeric vitronectin (Izumi *et al.*, 1989). Conformationally altered forms of vitronectin can form multimers and thereby form a mutivalent adhesion complex (Zlatopolsky *et al.*, 1992). Activated vitronectin binds to heparin with high affinity ($K_d = 10^{-8}$ M), and can neutralize heparin anticoagulant activity (Preissner and Muller-Berghaus, 1987). The expression of heparin-binding activity after activation by denaturants is accompanied by the exposure of new epitopes. The current interpretation of these findings is that the vitronectin molecule is normally folded to conceal this region (Fig. 4), and unfolding exposes this cryptic site.

Urea-treated vitronectin also binds to native collagen types I–VI, although this binding is minimal at physiological salt concentrations (Izumi *et al.*, 1988). The biological relevance of this interaction, therefore, remains to be determined. Vitronectin appears to have at least two collagen-binding domains. One is located close to the heparin-binding domain in the carboxyl half, while the other is in the amino-terminal half of vitronectin (Ishikawa-Sakurai and Hayashi, 1993).

Vitronectin also binds plasminogen activator inhibitor (PAI-1) and stabilizes its activity or even activates mutated forms (Keijer *et al.*, 1991; Tomasini and Mosher, 1991). These functions may help regulate the localization and activity of this important inhibitor of plasminogen activators, proteases implicated in a host of tissue-remodeling events including implantation, cell migration, and tumor cell invasion.

5. Thrombospondin

5.1. Introduction

Thrombospondin is a large, multifunctional glycoprotein that is released when platelets are activated, but it is also secreted continuously by a variety of other cell types during cell growth (reviewed by Lawler, 1986; Frazier, 1987; Asch and Nachman, 1989; Mosher, 1990; Frazier, 1991; Bornstein, 1992; Lahav, 1993; Lawler *et al.,* 1993). It interacts with cells and binds to several extracellular matrix molecules. Major activities of this protein include mediating or inhibiting cell adhesion and regulation of growth of certain cells. A causal role for thrombospondin in the regulation of cell growth has been suggested from experiments with smooth muscle cells, in which antibodies against thrombospondin inhibit cell proliferation by arresting cells in the G1 phase of the cell cycle. Heparin-mediated inhibition of the interaction of thrombospondin with these cells also inhibits proliferation (Majack *et al.,* 1988).

5.2. Structure and Location

Thrombospondin is a trimeric glycoprotein composed of three identical 140,000-Da subunits joined together by a region of interchain disulfide bonds (Fig. 5). The molecule contains an amino-terminal heparin-binding domain that binds avidly to heparin (K_d = 80 nM). The next structural feature on each chain is the site of interchain disulfide bonding, which links each subunit into a trimer. The center of the molecule consists of a relatively linear domain, and the protein terminates in a large carboxyl-terminal domain that binds Ca^{2+}. Thrombospondin is the most abundant protein of platelet α-granules, but it is also a component of a variety of extracellular matrices, is located in embryonic basement membranes, around epithelial cells, and is associated with peripheral nerves, myoblasts, and chondroblasts. After differentiation, levels of this protein decrease (O'Shea and Dixit, 1988).

There are at least four thrombospondin genes, which share a variety of homologous regions including EGF-like and calcium-binding motifs and the carboxyl-terminal domain. Phylogenetic analysis suggests that the primordial thrombospondin gene duplicated about 600 million years ago into the four known branches of the family, which indicates their separate existence throughout the evolution of the animal kingdom (Frazier, 1991; Bornstein, 1992; Lawler *et al.,* 1993). The four thrombospondin genes are expressed differentially in embryonic development (Iruela-Arispe *et al.,* 1993). Thrombospondin 1, but not 2 or 3, is expressed in neural tube, head mesenchyme, and megakaryocytes. Thrombospondin 2 is generally confined to connective tissue as well as myoblasts. Thrombospondin 3 is restricted to brain, cartilage, and lung. Thrombospondin 4 is concentrated in heart and skeletal muscle. These major differences in localization during development strongly suggest different functions, as well as emphasizing the importance of characterizing separately the potential role of each different thrombospondin gene in wound healing.

Thrombospondin synthesis can be stimulated by TGF-β, PDGF, heparin, and heat

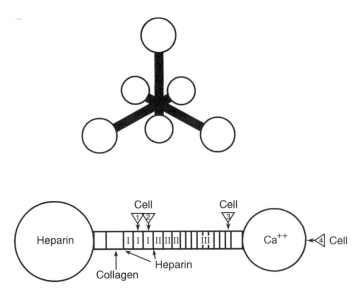

Figure 5. Schematic models of the overall morphology and internal structure of thrombospondin. The top diagram shows the general shape of the glycoprotein thrombospondin-1, which is a trimer of identical subunits bonded together by disulfide bonds located near the smaller globular domain. The diagram at the bottom shows the structure of one subunit of thrombospondin-1 or -2. Globular domains are attached to ends of a rodlike segment consisting of three types of repeating unit termed type I, type II, and type III. There are at least four cell-binding sites in thrombospondin: the first two are in type I repeats and contain the Val-Thr-Cys-Gly sequence; they may bind to CD36 or platelet gpIV. The third cell-binding cite is in a C-terminal type III repeating unit and contains the sequence Arg-Gly-Asp-. The fourth is in the carboxyl-terminal domain. In addition, the heparin-binding domain is reported to be a strong adhesion site in certain cells. Other binding sites include a procollagenlike domain thought to bind collagen and a calcium-binding domain.

shock (Ketis *et al.*, 1988; Penttinen *et al.*, 1988; Lyons-Giordano *et al.*, 1989). Analysis of the promoter region of the thrombospondin gene has been initiated to elucidate this complexity of regulation (Donoviel *et al.*, 1988; Laherty *et al.*, 1989). Conversely, however, thrombospondin also has been shown to activate latent TGF-β to its fully functional form by a novel mechanism (Schultz-Cherry and Murphy-Ullrich, 1993).

Thrombospondin-1 appears at the edge of wounds within 12 hr, and is maximal after 1–2 days (Reed *et al.*, 1993). Its mRNA is found in the thrombus, but not at the wound edge. Thrombospondin generally inhibits angiogenesis *in vitro* and *in vivo*, although enhancement was reported in a serum-free extracellular matrix assay (Iruela-Arispe *et al.*, 1991; Reed *et al.*, 1993; Nicosia and Tuszynski, 1994). Down-regulation of thrombospondin-1 in endothelial cells by antisense RNA can prevent this inhibition (DiPietro *et al.*, 1994).

5.3. Cell-Binding Sites

Thrombospondin can bind to cell surface receptors, mediating or modulating cell adhesion. Depending on the cell type, this interaction can lead to cell spreading or aggregation, or even to negative effects on strength of adhesion such as an inhibition of cell adhesion to fibronectin. In particular, thrombospondin can modulate platelet aggregation (Asch and Nachman, 1989; Mosher, 1990). Moreover, substrates coated with thrombospondin mediate attachment and spreading of keratinocytes (Varani *et al.*, 1988). In contrast, however, thrombospondin inhibits the adhesion of endothelial cells to substrates coated with serum or fibronectin (Lahav, 1988). Furthermore, thrombospondin can partially inhibit adhesion of cultured fibroblasts to fibronectin, producing a decrease in focal contacts (Murphy-Ullrich and Hook, 1989). Activity can be localized to heparin-binding peptides from thrombospondins-1 and -2 (Murphy-Ullrich *et al.*, 1993). This complexity of effects of thrombospondin may be due at least in part to the existence of at least two distinct cell interaction sites on this protein.

Thrombospondin-1 contains at least five sites that can mediate cell adhesion (Mosher, 1990; Frazier, 1991; Bornstein, 1992; Tuszynski *et al.*, 1992; Lahav, 1993; Lawler *et al.*, 1993). As described in more detail in the following section, the amino-terminal heparin-binding domain can mediate attachment of certain cells. CSVTCG and CSTSCG sequences contained in the region with fibronectinlike type I repeats can function in cell adhesion, platelet aggregation, and tumor cell metastasis. The RGDA sequence in the last fibronectinlike type III repeats is another functional site. This type III repeat also binds calcium. Cells and platelets also can bind the carboxyl-terminal domain. Cell binding for melanoma cells has been localized within the carboxyl-terminal domain to the sequences RFYVVM and IRVVM. Although three amino acids of these peptides are identical, they can synergize in combination for competitive inhibition of cell attachment to this domain.

Strong cell adhesive interactions with thrombospondin are mediated by the amino-terminal heparin-binding domain, which can bind to cell surface sulfated lipids (sulfatides) or to membrane-inserted heparan sulfate proteoglycan (Fig. 5) (Roberts and Ginsburg, 1988; Kaesberg *et al.*, 1989). This heparin-binding domain can also mediate the incorporation of thrombospondin into extracellular matrices (Prochownik *et al.*, 1989). Addition of exogenous heparin or antibodies to this domain can often block cell adhesion. Thrombospondin-mediated inhibition of angiogenesis can be attributed to two to three independent sites: the heparin-binding Trp-Ser-X-Trp peptide sequences in the fibronectinlike type I repeats and in the amino-terminal domain, as well as the procollagenlike region (Tolsma *et al.*, 1993; Vogel *et al.*, 1993).

Nevertheless, other studies using different adhesion assays or cells suggest that other regions of thrombospondin can also mediate cell adhesion. For example, keratinocytes and hematopoietic progenitor cells do not adhere to the 25,000-Da amino-terminal heparin-binding domain, but instead appear to use carboxyl-terminal sequence(s) in the large residual fragment (Varani *et al.*, 1988; Long and Dixit, 1990). The best-characterized alternative adhesion site for certain cells is the Arg-Gly-Asp sequence present in a predicted calcium-sensitive loop region of the protein near the carboxy-terminal end of the connecting rod (Fig. 5). In some assays, attachment of several cell types to thrombospondin could be inhibited by a synthetic peptide containing this adhesive recognition sequence (Lawler *et al.*, 1988). This site probably accounts for the binding of thrombospondin by the integrin receptors $\alpha v \beta 3$ and $\alpha IIb \beta 3$ (Lawler and Hynes, 1989).

5.4. Heparin- and Fibronectin-Binding Domains

Thrombospondin is secreted by cells and eventually becomes incorporated into a fibrillar matrix surrounding cells. Depending on the cell type and time in culture, it can be organized into 100- to 300-nm spherical granules together with heparan sulfate proteoglycan (Veklich *et al.*, 1993) in patterns distinct from fibronectin, or it can become organized into fibrils that colocalize with fibronectin. The latter organization appears to require a preexisting matrix as a scaffolding for assembly and involves one or more heparin-binding domains. Such domains also bind to fibronectin (Dardik and Lahav, 1989). The major heparin-binding region is a globular domain at the amino-terminus of thrombospondin. This region binds heparin with relatively high affinity (K_d = 7–8 $\times$ 10^{-8} M). In mutagenesis studies, heparin binding and incorporation of this domain into the extracellular matrix can occur as long as it retains a critical intrachain disulfide bridge region immediately adjacent to the heparin-binding domain (Prochownik *et al.*, 1989).

Like tenascin, thrombospondin can alter cell interactions with substrates by disrupting focal adhesions. This adhesion-labilizing activity can be localized to a 19-amino acid sequence corresponding to thrombospondin residues 17–35 in the heparin-binding domain. This peptide is active at 0.1 μM, and its activity is blocked by heparin and heparan sulfate. Consistent with involvement of heparin-binding activity in its function, modification of lysine residues in the peptide leads to loss of function (Murphy-Ullrich *et al.*, 1993).

Thrombospondin from endothelial cells binds to fibronectin with modest affinity (K_d = 0.7 $\times$ 10^{-7} M) using two distinct domains: a 70,000-Da core fragment similar to that in platelet thrombospondin (K_d = 3 $\times$ 10^{-7} M) and another domain of 27,000 Da that is reportedly unique to endothelial cell thrombospondin (K_d = 9 $\times$ 10^{-7} M). Heparin competitively inhibits binding of intact thrombospondin and to the 27,000-Da fibronectin fragment (Dardik and Lahav, 1989). Thrombospondin binds to fibronectin via the GGWSHW sequence within the second type I repeat. A synthetic GGWSHW peptide inhibits thrombospondin binding to fibronectin, but not direct cell adhesion to fibronectin substrates (Sipes *et al.*, 1993).

5.5. Other Binding Interactions

Some growth-modulating activity of thrombospondin appears to be due to bound TGF-β (Murphy-Ullrich *et al.*, 1992). The bound TGF-β inhibits endothelial cell growth. The TGF-β can be from exogenous sources or endogenously produced by the cells. In the latter case, thrombospondin appears to bind and function as a novel physiological regulator of TGF-β activation. Under these conditions, TGF-β has a maximal activity at concentrations of only 0.9 μM.

Thrombospondin binds to and inhibits several important proteases. It binds to plasmin in a 1 : 1 molar complex and serves as a slow tight-binding inhibitor. Thrombospondin also inhibits urokinase plasminogen activator, but it stimulates tissue plasminogen activator and had no effect in assays of α-thrombin or factor Xa (Hogg *et al.*, 1992). Thrombospondin may therefore be an important novel modulator of fibrinolysis.

In addition, thrombospondin binds to neutrophil elastase with a 1 : 3 stoichiometry (1 elastase molecule per each of the 3 thrombospondin subunits) with a site-binding affinity of $K_d = 6 \times 10^{-8}$ M. Thrombospondin functions as a tight-binding competitive inhibitor of elastase, perhaps via two reactive centers in the calcium-binding type 3 domains of thrombospondin. Because of the importance of neutrophil elastase in degradation of connective tissue components during inflammatory processes, thrombospondin may play critical roles in modulating tissue remodeling and repair (Hogg *et al.*, 1993).

Thrombospondin binds to more than a dozen other molecules (see recent tabulation by Lahav, 1993). Besides fibronectin and heparin, it can bind to the extracellular matrix molecules collagen, laminin, and osteonectin/SPARC (secreted protein acidic and rich in cysteine), as well as to fibrin and fibrinogen. It can also bind to syndecan as well as sulfated lipids.

Some bacteria can take advantage of binding sites on extracellular matrix proteins to establish sites of pyogenic infection. *Staphylococcus aureus* can bind to thrombospondin in calcium-dependent fashion with relatively high affinity ($K_d = 6 \times 10^{-9}$ M). This interaction provides a potentially important mechanism for staphylococcal adherence to blood clots or extracellular matrices to establish sites of infection. Thrombospondin is also a mediator of adhesion of malaria-infected red blood cells to endothelial cells, which appears particularly important in cerebral malaria (Aikawa *et al.*, 1990).

5.6. Receptors for Thrombospondin

The membrane glycoprotein CD36 is a major cell surface receptor for thrombospondin, although certain cells appear to use the integrin $\alpha v \beta 3$, other integrins, a 50- to 60-kDa glycoprotein, or a heparin-modulated system (Asch *et al.*, 1991, 1992; Savill *et al.*, 1992; Silverstein *et al.*, 1992; Adams and Lawler, 1993; Tuszynski *et al.*, 1993; Yabkowitz *et al.*, 1993). CD36 appears to bind to thrombospondin in a two-step process. First, a region containing residues 139–155 binds to thrombospondin, which triggers a change in the latter molecule to reveal a second site. This second, conformation-dependent site consists of the thrombospondin pentapeptide SVTCG, which binds to CD36 residues 87–99 with high affinity (Leung *et al.*, 1992; Asch *et al.*, 1993; Li *et al.*, 1993). The CD36 residue 87–99 region has a consensus protein kinase C phosphorylation site. Dephosphorylation of CD36 results in increased binding to thrombospondin, accompanied by a loss of CD36 binding to collagen. Conversely, phosphorylation mediated by protein kinase C of this extracellular site results in a loss of binding to thrombospondin and enhanced collagen binding (Asch *et al.*, 1993). Puzzlingly, the same thrombospondin pentapeptide in the hexapeptide CSVTCG bound to Sepharose also identifies another putative receptor of 50,000 and 60,000 Da from human tumor cells (Tuszynski *et al.*, 1993).

6. Tenascin

6.1. Introduction

Tenascins comprise a family of unusually large extracellular matrix molecules that presently include tenascins -C, -R, and -X. Tenascin-C (also known as cytotactin, J1–200/220, or hexabrachion) exists as a striking, six-armed, star-shaped extracellular complex of about 1.2 million Da consisting of similar subunits linked by interchain disulfide bonds (Fig. 6) (reviews include the following references: Erickson and Bourdon, 1989; Chiquet-Ehrismann, 1990; Hoffman *et al.*, 1990; Erickson, 1993). Tenascin-R was known previously as restrictin or J1–160/180 (Norenberg *et al.*, 1992). The other tenascin family member is tenascin-X (Bristow *et al.*, 1993). Tenascin is an

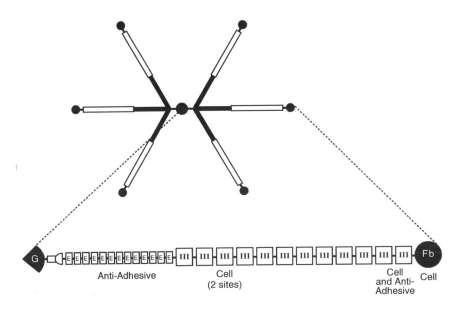

Figure 6. Schematic models of the overall morphology and internal structure of tenascin-C. This unusually large glycoprotein is a disulfide-linked complex composed of six similar subunits joined at their amino-termini to form a central globular domain (top). The structure of a single tenascin subunit is shown at the bottom. It starts with part of the central globular domain (G) that apparently receives contributions from multiple chains (indicated by pie-shaped segment), then includes heptad repeats and a prominent series of EGF repeats (E) associated with a polypeptide region that mediates antiadhesive function. Next are variable numbers of fibronectin type III repeats. Of the type III repeats that are always present and do not undergo alternative splicing, the third type III unit contains a region of relatively weak cell-adhesive activity based on an Arg-Gly-Asp- sequence and a stronger adhesive site in the same type III repeat. The seventh plus eighth invariant type III repeats contain a region of cell-adhesive function as well as an antiadhesive site. In between, variable numbers of type III units are inserted or deleted in isoforms due to alternative slicing of precursor tenascin mRNA. The carboxy-terminus consists of a fibrinogenlike knob (Fb) located at the distal end of each arm of the intact hexagonal molecule, which contains yet another cell-binding site.

unusual cell interaction protein, capable of mediating both adhesive and repulsive (antiadhesive) interactions, as well as binding to certain proteoglycans and fibronectin.

Tenascin is particularly interesting because its patterns of expression are tightly regulated. Tenascin expression is closely associated with morphogenetic events, including embryonic migration and induction, wound healing, and tumorigenesis (Chiquet-Ehrismann, 1990; Koukoulis *et al.*, 1991; Whitby and Ferguson, 1991; Whitby *et al.*, 1991; Erickson, 1993). Surprisingly, mice in which the tenascin gene was inactivated by homologous recombination lack any tenascin-C, and yet developed apparently normally, although subtle changes remain to be evaluated (Saga *et al.*, 1992). It appears likely that one of the other tenascin isoforms was able to substitute for most functions of the missing tenascin-C. Their apparent complexities and potential overlapping of biological functions may permit the tenascins to regulate tissue remodeling despite a missing isoform.

6.2. Structure and Tissue Location of Tenascin

Native tenascin molecules are predominantly six-subunit oligomers in which interchain disulfide bonds near the amino-terminus link subunits of about 190,000 to 230,000 Da into a large, six-armed complex or hexabrachion (Fig. 6) (Erickson and Bourdon, 1989; Chiquet-Ehrismann, 1990; Hoffman *et al.*, 1990; Erickson, 1993). Tenascin is characterized by structural units homologous to those of other proteins: it includes 13 epidermal growth factor-like repeats, 8–16 fibronectin type III repeats, and a globular carboxyl-terminus homologous to fibrinogen that includes a calcium-binding region (Fig. 6). Tenascin undergoes alternative splicing of its precursor mRNA to produce variant molecules, which result from the removal of sequences encoding various type III fibronectin repeats (Jones *et al.*, 1989; Spring *et al.*, 1989; Chiquet-Ehrismann, 1990; Prieto *et al.*, 1992; Aukhil *et al.*, 1993; Erickson, 1993; Tucker *et al.*, 1994). Although its complete three-dimensional structure remains to be determined, a predominance of beta structure is suggested by its far-UV circular dichroism spectrum (Taylor *et al.*, 1989). The structure of the type III fibronectin domain containing an RGD sequence has been determined, and it is very similar to the homologous domain in fibronectin (Leahy *et al.*, 1992).

Tenascin is frequently found at sites of tissue remodeling. In embryonic development, it is expressed in developing brain and in mesenchyme associated with epithelial–mesenchymal interactions, and its pattern of localization may correlate with pathways of migration (Chiquet-Ehrismann, 1990; Hay, 1991; Erickson, 1993). *In vitro* analyses of tenascin effects on neural crest cell migration reveal that crest cells migrate more rapidly on tenascin than on fibronectin or on basal lamina isolated from embryonic retina. Interestingly, crest cells on tenascin appear rounded and lack lamellipodia, whereas they are more flattened (and presumably more adhesive) on the other two substrates. When added to medium, however, tenascin becomes an inhibitor of cell migration on basal lamina. These differing effects suggest that tenascin functions to decrease cell-substrate interactions and to weaken tractional forces on migrating cells, thereby stimulating the rate of forward cell translocation (Halfter *et al.*, 1989). Tenascin can permit axonal extension while inhibiting glial cell migration, which may induce the compaction of nerve fascicles (Wehrle-Haller and Chiquet, 1993).

In adults, tenascin reappears at the edges of healing wounds, particularly beneath migrating, proliferating epidermal cells at the dermal–epidermal junction; its distribution differs from that of fibronectin, which is also thought to be involved in migration during wound healing (Mackie *et al.*, 1988; Whitby and Ferguson, 1991; Whitby *et al.*, 1991; Herrick *et al.*, 1992; Luomanen and Virtanen, 1993; Kanno and Fukuda, 1994). Tenascin first appears at approximately 2 days after wounding, forming increasingly intense patterns. However, it is absent from scar tissue (Betz *et al.*, 1993). In comparisons of fetal, neonatal, and adult murine lip wounds, tenascin emergence preceded fibronectin and varied from 1 to 24 hr. Its rate of appearance paralleled the rate of wound healing (Whitby and Ferguson, 1991; Whitby *et al.*, 1991). Since the appearance of tenascin preceded cell migration, it may function to promote rapid closure of fetal wounds. Levels of tenascin are also increased during nerve regeneration and in association with various tumors (Erickson and Bourdon, 1989; Chiquet-Ehrismann, 1990; Hoffman *et al.*, 1990; Koukoulis *et al.*, 1991).

Tenascin synthesis is stimulated by serum and by cytokines including TGF-β (Pearson *et al.*, 1988; Erickson, 1993). Tenascin levels can be induced by more than 100-fold by factors such as fibroblast growth factors, interleukins, and TNF-α, depending on the cell type (Rettig *et al.*, 1994). These findings suggest that tenascin regulation *in vivo* is due at least in part to growth factors associated with wounds, morphogenetic events, or tumors.

6.3. Cell Interactions of Tenascin

Some cells such as endothelial cells can adhere to tenascin or its fragments. The interaction with intact tenascin is weak, unlike adhesion to fibronectin, and it does not strengthen over time (Lotz *et al.*, 1989). A major cell-binding site of tenascin involved in such adhesion has been localized by studies with antibody inhibition and proteolytic fragments, and it is present near the carboxy-terminus in the 10th and 11th fibronectin type III repeats (Fig. 6) (Friedlander *et al.*, 1988; Spring *et al.*, 1989). A monoclonal antibody against this region can block cell adhesion to tenascin, supporting the importance of this site in tenascin-mediated cell attachment. A second adhesion site exists in the globular fibrinogenlike domain at the carboxyl-terminal end of the protein. Besides this site, there is an Arg-Gly-Asp (RGD) sequence in the third fibronectin type III repeat that can also function in some cells by interaction with the αvβ3 integrin receptor (Bourdon and Ruoslahti, 1989; Prieto *et al.*, 1992; Joshi *et al.*, 1993). As described in Section 6.5, an apparently distinct non-RGD adhesion site also exists in this third type III repeat.

However, tenascin also can exert a striking antiadhesive activity. Mixing tenascin with fibronectin when substrates are prepared for adhesion assays results in an inhibition of fibronectin function; tenascin can also inhibit adhesion to laminin and to the GRGDS peptide (Chiquet-Ehrismann *et al.*, 1988; Lotz *et al.*, 1989). The mechanism of this inhibition is not yet clear. A distinct antiadhesive region involved in this function can be mapped on the tenascin molecule to the region of EGF-like repeats adjacent to the amino-terminal globular domain (Chiquet-Ehrismann *et al.,* 1988; Spring *et al.,* 1989). Such antiadhesive activities may be of particular importance for neuronal outgrowth. In some cases, however, effects may be due to simple steric inhibition of cell access to adhesion proteins, since tenascin can also block access of antibodies to the substrate (Lightner and Erickson, 1990).

Inhibition of cell interactions by tenascin can also be demonstrated *in vivo.* Exogenous tenascin injected into developing amphibian gastrulas causes an arrest of gastrulation, an effect attributable to inhibition of mesodermal cell migration (Riou *et al.,* 1990). Further studies will be necessary to clarify the role of antiadhesive molecules such as tenascin during embryonic development (Keynes and Cook, 1990) and potentially in wound repair.

6.4. Interactions of Tenascin with Other Extracellular Molecules

Tenascin binds to a specific chondroitin sulfate proteoglycan, and this interaction is dependent on divalent cations. The localization of this interacting pair of molecules in embryos is similar but not always coincident; in nervous tissue, tenascin is synthesized by glia, whereas its binding proteoglycan is specifically synthesized by neurons (Hoffman *et al.,* 1988). Major chondroitin sulfate proteoglycans of nervous tissue, neurocan and phosphacan, have been shown to interact with tenascin and to have differing effects on cell adhesion (Grumet *et al.,* 1994). Remarkably, one chondroitin sulfate proteoglycan of brain is a variant form of the receptor tyrosine phosphatase β, and the extracellular domain of this transmembrane tyrosine phosphatase can bind specifically to tenascin (Barnea *et al.,* 1994). This interaction may play roles in nervous system signaling in development.

Tenascin also binds weakly to fibronectin, and in fact it can heavily contaminate cell surface fibronectin preparations that are not subjected to gel filtration. The interaction is readily reversible, but it appears to be specific in that tenascin was not found to bind to laminin or to collagens (Lightner and Erickson, 1990). This interaction may be functionally important, since it can permit inhibition of cell interactions with fibronectin (Chiquet-Ehrismann *et al.,* 1988).

6.5. Cell Surface Receptors for Tenascin

As reviewed in Section 6.3, integrins that interact with the RGD site can be involved in cellular interactions with tenascin. A second type of integrin interaction is displayed by the novel integrin $\alpha 9\beta 1$, which binds instead to a non-RGD site in the third fibronectin type III repeat of tenascin (Yokosaki *et al.,* 1994). A third type of cell surface ligand for tenascin is the neuronal cell adhesion molecule contactin/F11, which binds to tenascin-C and -R via its immunoglobulinlike domains (Zisch *et al.,* 1992). In addition, however, a novel, unexpected high-affinity "receptor" for tenascin has been identified as annexin II (Chung and Erickson, 1994). Although the latter molecule is present primarily as an intracellular molecule, it also appears to be released by cells to serve as a cell surface receptor for tenascin. Future studies will need to clarify which receptors are important for particular biological functions such as adhesion or signaling, as well as which other cell surface receptors are involved in mediating the anti-adhesive activity of tenascin.

References

Adams, J. C., and Lawler, J., 1993, Diverse mechanisms for cell attachment to platelet thrombospondin, *J. Cell Sci.* **104**:1061–1071.

Aikawa, M., Iseki, M., Barnwell, J. W., Taylor, D., Oo, M. M., and Howard, R. J., 1990, The pathology of human cerebral malaria, *Am. J. Trop. Med. Hyg.* **43**:30–37.

Akiyama, S. K., and Yamada, K. M., 1985, Synthetic peptides competitively inhibit both direct binding to fibroblasts and functional biological assays for the purified cell-binding domain of fibronectin, *J. Biol. Chem.* **260**:10402–10405.

Akiyama, S. K., Hasegawa, E., Hasegawa, T., and Yamada, K. M., 1985, The interactions of fibronectin fragments with fibroblastic cells, *J. Biol. Chem.* **260**:13256–13260.

Akiyama, S. K., Aota, S., and Yamada, K. M., 1994, Function and receptor specificity of a minimal 20 kilodalton cell adhesive fragment of fibronectin. *Cell Adhesion Commun.* **3**:13–25.

Alon, R., Cahalon, L., Hershkoviz, R., Elbaz, D., Reizis, B., Wallach, D., Akiyama, S. K., Yamada, K. M., and Lider, O., 1994, TNF-α binds to the N-terminal domain of fibronectin and augments the β1-integrin-mediated adhesion of CD4$^+$ T lymphocytes to the glycoprotein, *J. Immunol.* **152**:1304–1313.

Aota, S., Nagai, T., and Yamada, K. M., 1991, Characterization of regions of fibronectin besides the arginine-glycine-aspartic acid sequence required for adhesive function of the cell-binding domain using site-directed mutagenesis, *J. Biol. Chem.* **266**:15938–15943.

Aota, S., Nomizu, M., and Yamada, K. M., 1994, The short amino acid sequence Pro-His-Ser-Arg-Asn in human fibronectin enhances cell-adhesive function, *J. Biol. Chem.* **269**:24756–24761.

Asch, A. S., and Nachman, R. L., 1989, Thrombospondin: Phenomenology to function, *Prog. Hemost. Thromb.* **9**:157–176.

Asch, A. S., Tepler, J., Silbiger, S., and Nachman, R. L., 1991, Cellular attachment to thrombospondin. Cooperative interactions between receptor systems, *J. Biol. Chem.* **266**:1740–1745.

Asch, A. S., Silbiger, S., Heimer, E., and Nachman, R. L., 1992, Thrombospondin sequence motif (CSVTCG) is responsible for CD36 binding, *Biochem. Biophys. Res. Commun.* **182**:1208–1217.

Asch, A. S., Liu, I., Briccetti, F. M., Barnwell, J. W., Kwakye-Berko, F., Dokun, A., Goldberger, J., and Pernambuco, M., 1993, Analysis of CD36 binding domains: Ligand specificity controlled by dephosphorylation of an ectodomain, *Science* **262**:1436–1440.

Aukhil, I., Joshi, P., Yan, Y., and Erickson, H. P., 1993, Cell- and heparin-binding domains of the hexabrachion arm identified by tenascin expression proteins, *J. Biol. Chem.* **268**:2542–2553.

Bale, M. D., Wohlfahrt, L. A., Mosher, D. F., Tomasini, B., and Sutton, R. C., 1989, Identification of vitronectin as a major plasma protein adsorbed on polymer surfaces of different copolymer composition, *Blood* **74**:2698–2706.

Barkalow, F. J., and Schwarzbauer, J. E., 1994, Interactions between fibronectin and chondroitin sulfate are modulated by molecular context, *J. Biol. Chem.* **269**:3957–3962.

Barnea, G., Grumet, M., Milev, P., Silvennoinen, O., Levy, J. B., Sap, J., and Schlessinger, J., 1994, Receptor tyrosine phosphatase β is expressed in the form of proteoglycan and binds to the extracellular matrix protein tenascin, *J. Biol. Chem.* **269**:14349–14352.

Barnes, D. W., Foley, T. P., Shaffer, M. C., and Silnutzer, J. E., 1984, Human serum spreading factor: Relationship to somatomedin B, *J. Clin. Endocrinol. Metab.* **59**:1019–1021.

Baron, M., Norman, D., Willis, A., and Campbell, I. D., 1990, Structure of the fibronectin type 1 module, *Nature* **345**:642–646.

Beckmann, R., Geiger, M., de Vries, C., Pannekoek, H., and Binder, B. R., 1991, Fibronectin decreases the stimulatory effect of fibrin and fibrinogen fragment FCB-2 on plasmin formation by tissue plasminogen activator, *J. Biol. Chem.* **266**:2227–2232.

Benecky, M. J., Kolvenbach, C. G., Wine, R. W., DiOrio, J. P., and Mosesson, M. W., 1990, Human plasma fibronectin structure probed by steady-state fluorescence polarization: Evidence for a rigid oblate structure, *Biochemistry* **29**:3082–3091.

Betz, P., Nerlich, A., Tubel, J., Penning, R., and Eisenmenger, W., 1993, Localization of tenascin in human skin wounds: An immunohistochemical study, *Int. J. Legal Med.* **105**:325–328.

Bini, A., Fenoglio Jr., J. J., Mesa-Tejada, R., Kudryk, B., and Kaplan, K. L., 1989, Identification and distribution of fibrinogen, fibrin and fibrin(ogen) degradation products in atherosclerosis by monoclonal antibodies, *Arteriosclerosis* **9**:109–121.

Blasi, F., 1993, Urokinase and urokinase receptor: A paracrine/autocrine system regulating cell migration and invasiveness, *BioEssays* **15**:105–111.

Blystone, S. D., and Kaplan, J. E., 1992, Isolation of an amino-terminal fibronectin-binding protein on human U937 cells and rat peritoneal macrophages, *J. Biol. Chem.* **267**:3968–3975.

Blystone, S. D., and Kaplan, J. E., 1993, The role of fibronectin in macrophage fibrin binding: A potential mechanism for high affinity, high capacity clearance of circulating fibrin, *Blood Coagul. Fibrinolysis* **4**:769–781.

Bornstein, P., 1992, Thrombospondins: Structure and regulation of expression, *FASEB J.* **6**:3290–3299.

Bourdon, M. A., and Ruoslahti, E., 1989, Tenascin mediates cell attachment through an RGD-dependent receptor, *J. Cell Biol.* **108**:1149–1155.

Bowditch, R. D., Halloran, C. E., Aota, S., Obara, M., Plow, E. F., Yamada, K. M., and Ginsberg, M. H., 1991, Integrin αIIbβ3 (platelet GPIIb-IIIa) recognizes multiple sites in fibronectin, *J. Biol. Chem.* **266**:23323–23328.

Bowditch, R. D., Hariharan, M., Tominna, E. F., Smith, J. W., Yamada, K. M., Getzoff, E. D., and Ginsberg, M. H., 1994, Identification of a novel integrin binding site in fibronectin. Differential utilization by β3 integrins, *J. Biol. Chem.* **269**:10856–10863.

Bristow, J., Tee, M. K., Gitelman, S. E., Mellon, S. H., and Miller, W. L., 1993, Tenascin-X: A novel extracellular matrix protein encoded by the human XB gene overlapping P450c21B, *J. Cell Biol.* **122**:265–278.

Brommer, E. J. P., and van Bocke, L. J. H., 1992, Composition and susceptibility to thrombolysis of human arterial thrombi and the influence of their age, *Blood Coagul. Fibrinolysis* **3**:717–725.

Brown, L. F., Dubin, D., Lavigne, L., Logan, B., Dvorak, H. F., and Van De Water, L., 1993a, Macrophages and fibroblasts express "embryonic" fibronectins during cutaneous wound healing, *Am. J. Pathol.* **142**:793–801.

Brown, L. F., Lanir, N., McDonagh, J., Tognazzi, K., Dvorak, A. M., and Dvorak, H. F., 1993b, Fibroblast migration in fibrin gel matrices, *Am. J. Pathol.* **142**(1):273–283.

Browse, N. L., and Burnand, K. G., 1982, The cause of venous ulceration, *Lancet* **2**:243–245.

Burnand, K. G., Clemenson, G., Whimster, I., Gaunt, J., and Browse, N. L., 1982, The effect of sustained venous hypertension on the skin capillaries of the canine hind limb, *Br. J. Surg.* **69**:41–44.

Burridge, K., Fath, K., Kelly, T., Nuckolls, G., and Turner, C., 1988, Focal adhesions: Transmembrane junctions between the extracellular matrix and the cytoskeleton, *Annu. Rev. Cell Biol.* **4**:487–525.

Carsons, S. E. (ed.), 1989, *Fibronectin in Health and Disease,* CRC Press, Boca Raton, FL.

Castellino, F. J., Strickland, D. K., Morris, J. P., Smith, J., and Chibber, B., 1983, Enhancement of the streptokinase-induced activation of human plasminogen by human fibrinogen and human fibrinogen fragment D1, *Ann. NY Acad. Sci.* **408**:595–601.

Checovich, W. J., and Mosher, D. F., 1993, Lysophosphatidic acid enhances fibronectin binding to adherent cells, *Arterioscler. Thromb.* **13**:1662–1667.

Chernousov, M. A., Fogerty, F. J., Koteliansky, V. E., and Mosher, D. F., 1991, Role of the I-9 and III-1 modules of fibronectin in formation of an extracellular fibronectin matrix, *J. Biol. Chem.* **266**:10851–10858.

Chiquet-Ehrismann, R., 1990, What distinguishes tenascin from fibronectin, *FASEB J.* **4**:2598–2604.

Chiquet-Ehrismann, R., Kalla, P., Pearson, C. A., Beck, K., and Chiquet, M., 1988, Tenascin interferes with fibronectin action, *Cell* **53**:383–390.

Chung, C. Y., and Erickson, H. P., 1994, Cell surface annexin II is a high affinity receptor for the alternatively spliced segment of tenascin-C, *J. Cell Biol.* **126:**539–548.

Ciano, P. S., Colvin, R. B., Dvorak, A. M., McDonagh, J., and Dvorak, H. F., 1986, Macrophage migration in fibrin gel matrices, *Lab. Invest.* **54:**62–70.

Circolo, A., Welgus, H. G., Pierce, G. F., Kramer, J., and Strunk, R. C., 1991, Differential regulation of the expression of proteinases/antiproteinases in fibroblasts. Effects of interleukein-1 and platelet-derived growth factor, *J. Biol. Chem.* **266:**12283–12288.

Clark, R. A. F., and Henson, P. M. (eds.), 1988, *The Molecular and Cellular Biology of Wound Repair,* Plenum Press, New York.

Clark, R. A. F., Lanigan, J. M., DellaPelle, P., Manseau, E., Dvorak, H. F., and Colvin, R. B., 1982, Fibronectin and fibrin provide a provisional matrix for epidermal cell migration during wound reepithelialization, *J. Invest. Dermatol.* **70:**264–269.

Clark, R. A. F., Wikner, N. E., Doherty, D. E., and Norris, D. A., 1988, Cryptic chemotactic activity of fibronectin for human monocytes resides in the 120 kDa fibroblastic cell-binding fragment, *J. Biol. Chem.* **263:**12115–12123.

Dahlback, K., Lfberg, H., Alumets, J., and Dahlback, B., 1989, Immunohistochemical demonstration of age-related deposition of vitronectin (S-protein of complement) and terminal complement complex on dermal elastic fibers, *J. Invest. Dermatol.* **92:**727–733.

Dahlback, K., Wulf, H. C., and Dahlback, B., 1993, Vitronectin in mouse skin: Immunohistochemical demonstration of its association with cutaneous amyloid, *J. Invest. Dermatol.* **100:**166–170.

Damsky, C. H., and Werb, Z., 1992, Signal transduction by integrin receptors for extracellular matrix: Cooperative processing of extracellular information, *Curr. Opin. Cell Biol.* **4:**772–781.

Dardik, R., and Lahav, J., 1989, Multiple domains are involved in the interaction of endothelial cell thrombospondin with fibronectin, *Eur. J. Biochem.* **185:**581–588.

Darribere, T., Guida, K., Larjava, H., Johnson, K. E., Yamada, K. M., Thiery, J. P., and Boucaut, J.-C., 1990, *In vivo* analyses of integrin β1 subunit function in fibronectin matrix assembly, *J. Cell Biol.* **110:**1813–1823.

Darribere, T., Koteliansky, V. E., Chernousov, M. A., Akiyama, S. K., Yamada, K. M., Thiery, J. P., and Boucaut, J. C., 1992, Distinct regions of human fibronectin are essential for fibril assembly in an *in vivo* developing system, *Dev. Dyn.* **194:**63–70.

Dean, D. C., McQuillan, J. J., and Weintraub, S., 1990, Serum stimulation of fibronectin gene expression appears to result from rapid serum-induced binding of nuclear proteins to a cAMP response element, *J. Biol. Chem.* **265:**3522–3527.

Dejana, E., Colella, S., Conforti, G., Abbadini, M., Gaboli, M., and Marchisio, P. C., 1988, Fibronectin and vitronectin regulate the organization of their respective Arg-Gly-Asp adhesion receptors in cultured human endothelial cells, *J. Cell Biol.* **107:**1215–1223.

Del Rosso, M., Fibbi, G., Dini, G., Grappone, C., Pucci, M., Caldini, R., Fimiani, M., Lotti, T., and Pancones, E., 1990, Role of specific membrane receptors in urokinase-dependent migration of human keratinocytes, *J. Invest. Dermatol.* **94:**310–316.

DiPietro, L. A., Nebgen, D. R., and Polverini, P. J., 1994, Down-regulation of endothelial cell thrombospondin 1 enhances *in vitro* angiogenesis, *J. Vasc. Res.* **31:**178–185.

Doherty, D. E., Henson, P. M., and Clark, R. A. F., 1990, Fibronectin fragments containing the RGDS cell-binding domain mediate monocyte migration into the rabbit lung, *J. Clin. Invest.* **86:**1065–1075.

Donoviel, D. B., Framson, P., Eldridge, C. F., Cooke, M., Kobayashi, S., and Bornstein, P., 1988, Structural analysis and expression of the human thrombospondin gene promoter, *J. Biol. Chem.* **263:**18590–18593.

D'Souza, S. E., Ginsberg, M. H., Burke, T. A., and Plow, E. F., 1990, The ligand binding site of the platelet integrin receptor GPIIb- IIIa is proximal to the second calcium binding domain of its α subunit, *J. Biol. Chem.* **265:**3440–3446.

D'Souza, S. E., Haas, T. A., Piotrowicz, R. S., Byers-Ward, V., McGrath, D. E., Soule, H. R., Cierniewski, C., Plow, E. F., and Smith, J. W., 1994, Ligand and cation binding are dual functions of a discrete segment of the integrin β3 subunit: Cation displacement is involved in ligand binding, *Cell* **79:**659–667.

Du, X. P., Plow, E. F., Frelinger, A. L., O'Toole, T. E., Loftus, J. C., and Ginsberg, M. H., 1991, Ligands activate integrin αIIbβ3 (platelet GPIIb-IIIa), *Cell* **65:**409–416.

Dufour, S., Duband, J.-L., Humphries, M. J., Obara, M., Yamada, K. M., and Thiery, J. P., 1988, Attachment, spreading and locomotion of avian neural crest cells are mediated by multiple adhesion sites on fibronectin molecules, *EMBO J.* **7**:2661–2671.

Erickson, H. P., 1993, Tenascin-C, tenascin-R and tenascin-X: A family of talented proteins in search of functions, *Curr. Opin. Cell Biol.* **5**:869–876.

Erickson, H. P., and Bourdon, M. A., 1989, Tenascin: An extracellular matrix protein prominent in specialized embryonic tissues and tumors, *Annu. Rev. Cell Biol.* **5**:71–92.

Falanga, V., and Eaglestein, W. H., 1993, The "trap" hypothesis of venous ulceration, *Lancet* **341**:1006–1008.

Falanga, V., McKenzie, A., and Eaglstein, W. H., 1991, Heterogeneity in oxygen diffusion around venous ulcers, *J. Dermatol. Surg. Oncol.* **17**:336–339.

Fava, R. A., and McClure, D. B., 1987, Fibronectin-associated transforming growth factor, *J. Cell. Physiol.* **131**:184–189.

Felding-Habermann, B., and Cheresh, D. A., 1993, Vitronectin and its receptors, *Curr. Opin. Cell Biol.* **5**:864–868.

ffrench-Constant, K., Van de Water, L., Dvorak, H. F., and Hynes, R. O., 1989, Reappearance of an embryonic pattern of fibronectin splicing during wound healing in the adult rat, *J. Cell Biol.* **109**:903–914.

Flaumenhaft, R., and Rifkin, D. B., 1991, Extracellular matrix regulation of growth factor and protease activity, *Curr. Opin. Cell Biol.* **3**:817–823.

Fogerty, F. J., Akiyama, S. K., Yamada, K. M., and Mosher, D. F., 1990, Inhibition of binding of fibronectin to matrix assembly sites by anti-integrin ($\alpha 5 \beta 1$) antibodies, *J. Cell Biol.* **111**:699–708.

Francis, C. W., and Marder, V. J., 1988, Increased resistance to plasmin degradation of fibrin with highly cross-linked α-polymer chains formed at high factor XIII concentrations, *Blood* **71**:1361.

Frazier, W. A., 1987, Thrombospondin: A modular adhesive glycoprotein of platelets and nucleated cells, *J. Cell Biol.* **105**:625–632.

Frazier, W. A., 1991, Thrombospondins, *Curr. Opin. Cell Biol.* **3**:792–799.

Friedlander, D. R., Hoffman, S., and Edelman, G. M., 1988, Functional mapping of cytotactin: Proteolytic fragments active in cell–substrate adhesion, *J. Cell Biol.* **107**:2329–2340.

Fukai, F., Iso, T., Sekiguchi, K., Miyatake, N., Tsugita, A., and Katayama, T., 1993, An amino-terminal fibronectin fragment stimulates the differentiation of ST-13 preadipocytes, *Biochemistry* **32**:5746–5751.

Gabriel, D. A., Muga, K., and Boothroyd, E. M., 1992, The effect of fibrin structure on fibrinolysis, *J. Biol. Chem.* **267**:24259–24263.

Ginsberg, M. H., Du, X., and Plow, E. F., 1992, Inside-out integrin signalling, *Curr. Opin. Cell Biol.* **4**:766–771.

Gray, A. J., Bishop, J. E., Reeves, J. T., and Laurent, G. J., 1993, Aα and Bβ chains of fibrinogen stimulate proliferation of human fibroblasts, *J. Cell Sci.* **104**:409–413.

Grinnell, F., and Phan, T. V., 1983, Deposition of fibronectin on material surfaces exposed to plasma: Quantitative and biological studies, *J. Cell. Physiol.* **116**:289–296.

Grinnell, F., and Zhu, M., 1994, Identification of neutrophil elastase as the proteinase in burn wound fluid responsible for degradation of fibronectin, *J. Invest. Dermatol.* **103**:155–161.

Grinnell, F., Feld, M., and Minter, D., 1980, Fibroblast adhesion to fibrinogen and fibrin substrata: Requirement for cold-insoluble globulin (plasma fibronectin), *Cell* **19**:517–525.

Grinnell, F., Ho, C. H., and Wysocki, A. J., 1992, Degradation of fibronectin and vitronectin in chronic wound fluid: Analysis by cell blotting, immunoblotting, and cell adhesion assays, *J. Invest. Dermatol.* **98**:410–416.

Grumet, M., Milev, P., Sakurai, T., Karthikeyan, L. Bourdon, M., Margolis, R. K., and Margolis, R. U., 1994, Interactions with tenascin and differential effects on cell adhesion of neurocan and phosphacan, two major chondroitin sulfate proteoglycans of nervous tissue, *J. Biol. Chem.* **269**:12142–12146.

Guan, J. L., Trevithick, J. E., and Hynes, R. O., 1990, Retroviral expression of alternatively spliced forms of rat fibronectin, *J. Cell Biol.* **110**:833–847.

Hajjar, K., Jacovina, A., and Chacko, J., 1994, An endothelial cell receptor for plasminogen and tissue plasminogen activator: Identity with annexin II, *J. Biol. Chem.* **269**:21191–21197.

Halfter, W., Chiquet-Ehrismann, R., and Tucker, R. P., 1989, The effect of tenascin and embryonic basal lamina on the behavior and morphology of neural crest cells *in vitro*, *Dev. Biol.* **132**:14–25.

Hay, E. D. (ed.), 1991, *Cell Biology of Extracellular Matrix,* Plenum Press, New York.

Hayashi, M., and Yamada, K. M., 1982, Divalent cation modulation of fibronectin binding to heparin and to DNA, *J. Biol. Chem.* **257:**5263–5267.

Hendey, B., Klee, C. B., and Maxfield, F. R., 1992, Inhibition of neutrophil chemokinesis on vitronectin by inhibitors of calcineurin, *Science* **258:**296–299.

Herrick, S. E., Sloan, P., McGurk, M., Freak, L., McCollum, C. N., and Ferguson, M. W. J., 1992, Sequential changes in histologic pattern and extracellular matrix deposition during the healing of chronic venous ulcers, *Am. J. Pathol.* **141:**1085–1095.

Hershberger, R. P., and Culp, L. A., 1990, Cell-type-specific expression of alternatively spliced human fibronectin IIICS mRNAs, *Mol. Cell. Biol.* **10:**662–671.

Hershkoviz, R., Cahalon, L., Miron, S., Alon, R., Sapir, T., Akiyama, S. K., Yamada, K. M., and Lider, O., 1994, TNF-α associated with fibronectin enhances phorbol myristate acetate- or antigen-mediated integrin-dependent adhesion of CD4$^+$ T cells via protein tyrosine phosphorylation, *J. Immunol.* **153:**554–565.

Hintner, H., Dahlback, K., Dahlback, B., Pepys, M. B., and Breathnach, S. M., 1991, Tissue vitronectin in normal adult human dermis is non-covalently bound to elastic tissue, *J. Invest. Dermatol.* **96:**747–753.

Hirano, H., Yamada, Y., Sullivan, M., de Crombrugghe, B., Pastan, I., and Yamada, K. M., 1983, Isolation of genomic DNA clones spanning the entire fibronectin gene, *Proc. Natl. Acad. Sci. USA* **80:**46–50.

Hocking, D. C., Sottile, J., and McKeown-Longo, P. J., 1994, Fibronectin's III-1 module contains a conformation-dependent binding site for the amino-terminal region of fibronectin, *J. Biol. Chem.* **269:**19183–19187.

Hoffman, S., Crossin, K. L., and Edelman, G. M., 1988, Molecular forms, binding functions, and developmental expression patterns of cytotactin and cytotactin-binding proteoglycan, an interactive pair of extracellular matrix molecules, *J. Cell Biol.* **106:**519–532.

Hoffman, S., Crossin, K. L., Jones, F. S., Friedlander, D. R., and Edelman, G. M., 1990, Cytotactin and cytotactin-binding proteoglycan. An interactive pair of extracellular matrix proteins, *Ann. NY Acad. Sci.* **580:**288–301.

Hogg, P. J., Stenflo, J., and Mosher, D. F., 1992, Thrombospondin is a slow tight-binding inhibitor of plasmin, *Biochemistry* **31:**265–269.

Hogg, P. J., Owensby, D. A., Mosher, D. F., Misenheimer, T. M., and Chesterman, C. N., 1993, Thrombospondin is a tight-binding competitive inhibitor of neutrophil elastase, *J. Biol. Chem.* **268:**7139–7146.

Huang, S., Cao, Z., and Davie, E. W., 1993a, The role of amino-terminal disulfide bonds in the structure and assembly of human fibrinogen, *Biochem. Biophys. Res. Commun.* **190:**488–495.

Huang, S., Mulvihill, E. R., Farrell, K. H., Chung, D. W., and Davie, E. W., 1993b, Biosynthesis of human fibrinogen, *J. Biol. Chem.* **268:**8919–8926.

Huhtala, P., Humphries, M. J., McCarthy, J. B., Tremble, P. M., Werb, Z., and Damsky, C. H., 1995, Cooperative signaling by α5β1 and α4β1 integrins regulates metalloproteinase gene expression in fibroblasts adhering to fibronectin, *J. Cell Biol.* **129:**867–879.

Humphries, M. J., Akiyama, S. K., Komoriya, A., Olden, K., and Yamada, K. M., 1986, Identification of an alternatively spliced site in human plasma fibronectin that mediates cell type-specific adhesion, *J. Cell Biol.* **103:**2637–2647.

Humphries, M. J., Komoriya, A., Akiyama, S. K., Olden, K., and Yamada, K. M., 1987, Identification of two distinct regions of the type III connecting segment of human plasma fibronectin that promote cell type-specific adhesion, *J. Biol. Chem.* **262:**6886–6892.

Humphries, M. J., Akiyama, S. K., Komoriya, A., Olden, K., and Yamada, K. M., 1988, Neurite extension of chicken peripheral nervous system neurons on fibronectin: Relative importance of specific adhesion sites in the central cell-binding domain and the alternatively spliced type III connecting segment, *J. Cell Biol.* **106:**1289–1297.

Hynes, R. O., 1990, *Fibronectins,* Springer-Verlag, New York.

Ingham, K. C., Landwehr, R., and Engel, J., 1985, Interaction of fibronectin with C1q and collagen. Effects of ionic strength and denaturation of the collagenous component, *Eur. J. Biochem.* **148:**219–224.

Ingham, K. C., Brew, S. A., and Migliorini, M. M., 1989, Further localization of the gelatin-binding determinants within fibronectin. Active fragments devoid of type II homologous repeat modules, *J. Biol. Chem.* **264:**16977–16980.

Iruela-Arispe, M. L., Bornstein, P., and Sage, H., 1991, Thrombospondin exerts an antiangiogenic effect on cord formation by endothelial cells *in vitro, Proc. Natl. Acad. Sci. USA* **88:**5026–5030.

Iruela-Arispe, M. L., Liska, D. J., Sage, E. H., and Bornstein, P., 1993, Differential expression of thrombospondin 1, 2, and 3 during murine development, *Dev. Dyn.* **197:**40–56.

Ishikawa-Sakurai, M., and Hayashi, M., 1993, Two collagen-binding domains of vitronectin, *Cell Struct. Funct.* **18:**253–259.

Izumi, M., Shimo-Oka, T., Morishita, N., Ii, I., and Hayashi, M., 1988, Identification of the collagen-binding domain of vitronectin using monoclonal antibodies, *Cell Struct. Funct.* **13:**217–225.

Izumi, M., Yamada, K. M., and Hayashi, M., 1989, Vitronectin exists in two structurally and functionally distinct forms in human plasma, *Biochim. Biophys. Acta* **990:**101–108.

Izzard, C. S., Radinsky, R., and Culp, L. A., 1986, Substratum contacts and cytoskeletal reorganization of BALB/c 3T3 cells on a cell-binding fragment and heparin-binding fragments of plasma fibronectin, *Exp. Cell Res.* **165:**320–336.

Jenne, D., and Stanley, K. K., 1985, Molecular cloning of S-protein, a link between complement, coagulation and cell-substrate adhesion, *EMBO J.* **4:**3153–3157.

Jones, F. S., Hoffman, S., Cunningham, B. A., and Edelman, G. M., 1989, A detailed structural model of cytotactin: Protein homologies, alternative RNA splicing, and binding regions, *Proc. Natl. Acad. Sci. USA* **86:**1905–1909.

Joshi, P., Chung, C. Y., Aukhil, I., and Erickson, H. P., 1993, Endothelial cells adhere to the RGD domain and the fibrinogen-like terminal knob of tenascin, *J. Cell Sci.* **106:**389–400.

Kaesberg, P. R., Ershler, W. B., Esko, J. D., and Mosher, D. F., 1989, Chinese hamster ovary cell adhesion to human platelet thrombospondin is dependent on cell surface heparan sulfate proteoglycan, *J. Clin. Invest.* **83:**994–1001.

Kaminski, M., and McDonagh, J., 1983, Studies on the mechanism of thrombin interaction with fibrin, *J. Biol. Chem.* **258:**10530–10535.

Kanno, S., and Fukuda, Y., 1994, Fibronectin and tenascin in rat tracheal wound healing and their relation to cell proliferation, *Pathol. Int.* **44:**96–106.

Keijer, J., Ehrlich, H. J., Linders, M., Preissner, K. T., and Pannekoek, H., 1991, Vitronectin governs the interaction between plasminogen activator inhibitor 1 and tissue-type plasminogen activator, *J. Biol. Chem.* **266:**10700–10707.

Ketis, N. V., Lawler, J., Hoover, R. L., and Karnovsky, M. J., 1988, Effects of heat shock on the expression of thrombospondin by endothelial cells in culture, *J. Cell Biol.* **106:**893–904.

Keynes, R., and Cook, G., 1990, Cell–cell repulsion: Clues from the growth cone? *Cell* **62:**609–610.

Knox, P., 1984, Kinetics of cell spreading in the presence of different concentrations of serum or fibronectin-depleted serum, *J. Cell Sci.* **71:**51–59.

Knox, P., Crooks, S., and Rimmer, C. S., 1986, Role of fibronectin in the migration of fibroblasts into plasma clots, *J. Cell Biol.* **102:**2318–2323.

Knox, P., Crooks, S., Scaife, M. C., and Patel, S., 1987, Role of plasminogen, plasmin, and plasminogen activators in the migration of fibroblasts into plasma clots, *J. Cell. Physiol.* **132:**501–508.

Koli, K., Lohi, J., Hautanen, A., and Keski-Oja, J., 1991, Enhancement of vitronectin expression in human HepG2 hepatoma cells by transforming growth factor-beta 1, *Eur. J. Biochem.* **199:**337–345.

Komoriya, A., Green, L. J., Mervic, M., Yamada, S. S., Yamada, K. M., and Humphries, M. J., 1991, The minimal essential sequence for a major cell type-specific adhesion site (CS1) within the alternatively spliced type III connecting segment domain of fibronectin is leucine-aspartic acid-valine, *J. Biol. Chem.* **266:**15075–15079.

Kornblihtt, A. R., Umezawa, K., Vibe-Pedersen, K., and Baralle, F. E., 1985, Primary structure of human fibronectin: Differential splicing may generate at least 10 polypeptides from a single gene, *EMBO J.* **4:**1755–1759.

Koukoulis, G. K., Gould, V. E., Bhattacharyya, A., Gould, J. E., Howeedy, A. A., and Virtanen, I., 1991, Tenascin in normal, reactive, hyperplastic, and neoplastic tissues: Biologic and pathologic implications, *Hum. Pathol.* **22:**636–643.

Kubota, K., Katayama, S., Matsuda, M., and Hayashi, M., 1988, Three types of vitronectin in human blood, *Cell Struct. Funct.* **13:**123–128.

Lahav, J., 1988, Thrombospondin inhibits adhesion of endothelial cells, *Exp. Cell Res.* **177:**199–204.

Lahav, J., 1993, The functions of thrombospondin and its involvement in physiology and pathophysiology, *Biochim. Biophys. Acta* **1182:**1–14.

Laherty, C. D., Gierman, T. M., and Dixit, V. M., 1989, Characterization of the promoter region of the human thrombospondin gene. DNA sequences within the first intron increase transcription, *J. Biol. Chem.* **264:**11222–11227.

Laiho, M., Saksela, O., and Keski-Oja, J., 1986, Transforming growth factor β alters plasminogen activator activity in human skin fibroblasts, *Exp. Cell Res.* **164:**399–407.

Lash, J. W., Linask, K. K., and Yamada, K. M., 1987, Synthetic peptides that mimic the adhesive recognition signal of fibronectin: Differential effects on cell–cell and cell–substratum adhesion in embryonic chick cells, *Dev. Biol.* **123:**411–420.

Lawler, J., 1986, The structural and functional properties of thrombospondin, *Blood* **67:**1197–1209.

Lawler, J., and Hynes, R. O., 1989, An integrin receptor on normal and thrombasthenic platelets that binds thrombospondin, *Blood* **74:**2022–2027.

Lawler, J., Weinstein, R., and Hynes, R. O., 1988, Cell attachment to thrombospondin: The role of Arg-Gly-Asp, calcium, and integrin receptors, *J. Cell Biol.* **107:**2351–2361.

Lawler, J., Duquette, M., Urry, L., McHenry, K., and Smith, T. F., 1993, The evolution of the thrombospondin gene family, *J. Mol. Evol.* **36:**509–516.

Leahy, D. J., Hendrickson, W. A., Aukhil, I., and Erickson, H. P., 1992, Structure of a fibronectin type III domain from tenascin phased by MAD analysis of the selenomethionyl protein, *Science* **258:**987–991.

Leavesley, D. I., Ferguson, G. D., Wayner, E. A., and Cheresh, D. A., 1992, Requirement of the integrin β3 subunit for carcinoma cell spreading or migration on vitronectin and fibrinogen, *J. Cell Biol.* **117:**1101–1107.

Leung, L. L., Li, W. X., McGregor, J. L., Albrecht, G., and Howard, R. J., 1992, CD36 peptides enhance or inhibit CD36–thrombospondin binding. A two-step process of ligand–receptor interaction, *J. Biol. Chem.* **267:**18244–18250.

Li, W. X., Howard, R. J., and Leung, L. L., 1993, Identification of SVTCG in thrombospondin as the conformation-dependent, high affinity binding site for its receptor, CD36, *J. Biol. Chem.* **268:**16179–16184.

Lightner, V. A., and Erickson, H. P., 1990, Binding of hexabrachion (tenascin) to the extracellular matrix and substratum and its effect on cell adhesion, *J. Cell Sci.* **95:**263–277.

Limper, A. H., Quade, B. J., LaChance, R. M., Birkenmeier, T. M., Rangwala, T. S., and McDonald, J. A., 1991, Cell surface molecules that bind fibronectin's matrix assembly domain, *J. Biol. Chem.* **26:**9697–9702.

Liu, C. Y., Nossel, H. L., and Kaplan, K. L., 1979, The binding of thrombin by fibrin, *J. Biol. Chem.* **254:**10421–10425.

Loftus, J. C., O'Toole, T. E., Plow, E. F., Glass, A., Frelinger, A. L., and Ginsberg, M. H., 1990, A β3 integrin mutation abolishes ligand binding and alters divalent cation-dependent conformation, *Science* **249:**915–918.

Long, M. W., and Dixit, V. M., 1990, Thrombospondin functions as a cytoadhesion molecule for human hematopoietic progenitor cells, *Blood* **75:**2311–2318.

Loskutoff, D. J., and Edgington, T. S., 1977, Synthesis of a fibrinolytic activator and inhibitor by endothelial cells, *Proc. Natl. Acad. Sci. USA* **74:**3903–3907.

Lotz, M. M., Burdsal, C. A., Erickson, H. P., and McClay, D. R., 1989, Cell adhesion to fibronectin and tenascin: Quantitative measurements of initial binding and subsequent strengthening response, *J. Cell Biol.* **109:**1795–1805.

Luomanen, M., and Virtanen, I., 1993, Distribution of tenascin in healing incision, excision and laser wounds, *J. Oral Pathol. Med.* **22:**41–45.

Lyons-Giordano, B., Brinker, J. M., and Kefalides, N. A., 1989, Heparin increases mRNA levels of thrombospondin but not fibronectin in human vascular smooth muscle cells, *Biochem. Biophys. Res. Commun.* **162:**1100–1104.

Mackie, E. J., Halfter, W., and Liverani, D., 1988, Induction of tenascin in healing wounds, *J. Cell Biol.* **107:**2757–2767.

Main, A. L., Harvey, T. S., Baron, M., Boyd, J., and Campbell, I. D., 1992, The three-dimensional structure of the tenth type III module of fibronectin: An insight into RGD-mediated interactions, *Cell* **71**:671–678.

Majack, R. A., Goodman, L. V., and Dixit, V. M., 1988, Cell surface thrombospondin is functionally essential for vascular smooth muscle cell proliferation, *J. Cell Biol.* **106**:415–422.

Matsuka, Y. V., Medved, L. V., Brew, S. A., and Ingham, K. C., 1994, The NH2-terminal fibrin-binding site of fibronectin is formed by interacting fourth and fifth finger domains. Studies with recombinant finger fragments expressed in *Escherichia coli, J. Biol. Chem.* **269**:9539–9546.

McCarthy, J. B., Chelberg, M. K., Mickelson, D. J., and Furcht, L. T., 1988, Localization and chemical synthesis of fibronectin peptides with melanoma adhesion and heparin binding activities, *Biochemistry* **27**:1380–1388.

McCarthy, J. B., Skubitz, A. P., Qi, Z., Yi, X. Y., Mickelson, D. J., Klein, D. J., and Furcht, L. T., 1990, RGD-independent cell adhesion to the carboxy-terminal heparin-binding fragment of fibronectin involves heparin-dependent and -independent activities, *J. Cell Biol.* **110**:777–787.

McDonald, J. A., 1988, Extracellular matrix assembly, *Annu. Rev. Cell Biol.* **4**:183–207.

McDonald, J. A., and Kelley, D. G., 1980, Degradation of fibronectin by human leukocyte elastase. Release of biologically active fragments, *J. Biol. Chem.* **255**:8848–8858.

McDonald, J. A., Kelley, D. G., and Broekelmann, T. J., 1982, Role of fibronectin in collagen deposition: Fab1 antibodies to the gelatin-binding domain of fibronectin inhibits both fibronectin and collagen organization in fibroblast extracellular matrix, *J. Cell Biol.* **92**:485–492.

McDonald, J. A., Quade, B. J., Broekelmann, T. J., LaChance, R., Forsman, K., Hasegawa, E., and Akiyama, S., 1987, Fibronectin's cell-adhesive domain and an amino-terminal matrix assembly domain participate in the assembly into fibroblast pericellular matrix, *J. Biol. Chem.* **262**:2957–2967.

McKee, P. A., Mattock, P., and Hill, R. L., 1970, Subunit structure of human fibrinogen, soluble fibrin, and cross-linked insoluble fibrin, *Proc. Natl. Acad. Sci. USA* **66**:738–743.

McKeown-Longo, P. J., and Mosher, D. F., 1985, Interaction of the 70,000-mol-wt amino fragment of fibronectin with the matrix-assembly receptor of fibroblasts, *J. Cell Biol.* **100**:364–374.

Michel, D., and Harmand, M. F., 1990, Fibrin seal in wound healing: Effect of thrombin and [Ca^{2+}] on human skin fibroblast growth and collagen production, *J. Dermatol. Sci.* **1(5)**:325–33.

Mirshahi, M., Azzarone, B., Soria, J., Mirshahi, F., and Soria, C., 1991, The role of fibroblasts in organization and degradation of a fibrin clot, *J. Lab. Clin. Med.* **117(4)**:274–81.

Moon, K. Y., Shin, K. S., Song, W. K., Chung, C. H., Ha, D. B., and Kang, M. S., 1994, A candidate molecule for the matrix assembly receptor to the N-terminal 29 kDa fragment of fibronectin in chick myoblasts, *J. Biol. Chem.* **269**:7651–7657.

Morla, A., and Ruoslahti, E., 1992, A fibronectin self-assembly site involved in fibronectin matrix assembly: Reconstruction in a synthetic peptide, *J. Cell Biol.* **118**:421–429.

Mosesson, M. W., 1992, The roles of fibrinogen and fibrin in hemostasis and thrombosis, *Sem. Hematol.* **29**:177–188.

Mosesson, M. W., Siebenlist, K. R., Amrani, D. L., and DiOrio, J. P., 1989, Identification of covalently linked trimeric and tetrameric D domains in cross-linked fibrin, *Proc. Natl. Acad. Sci. USA* **86**:1113.

Mosher, D. F., 1989, *Fibronectin,* Academic Press, San Diego, CA.

Mosher, D. F., 1990, Physiology of thrombospondin, *Annu. Rev. Med.* **41**:85–97.

Mosher, D. F., Sottile, J., Wu, C., and McDonald, J. A., 1992, Assembly of extracellular matrix, *Curr. Opin. Cell Biol.* **4**:810–818.

Murphy-Ullrich, J. E., and Hook, M., 1989, Thrombospondin modulates focal adhesions in endothelial cells, *J. Cell Biol.* **109**:1309–1319.

Murphy-Ullrich, J. E., Schultz-Cherry, S., and Hook, M., 1992, Transforming growth factor-beta complexes with thrombospondin, *Mol. Biol. Cell* **3**:181–188.

Murphy-Ullrich, J. E., Gurusiddappa, S., Frazier, W. A., and Hook, M., 1993, Heparin-binding peptides from thrombospondins 1 and 2 contain focal adhesion-labilizing activity, *J. Biol. Chem.* **268**:26784–26789.

Nagai, T., Yamakawa, N., Aota, S., Yamada, S. S., Akiyama, S. K., Olden, K., and Yamada, K. M., 1991, Monoclonal antibody characterization of two distant sites required for function of the central cell-binding domain of fibronectin in cell adhesion, cell migration, and matrix assembly, *J. Cell Biol.* **114**:1295–1305.

Nair, C. H., and Dhall, D. P., 1991, Studies on fibrin network structure: The effect of some plasma proteins, *Thromb. Res.* **61**:315–325.

Nicosia, R. F., and Tuszynski, G. P., 1994, Matrix-bound thrombospondin promotes angiogenesis *in vitro, J. Cell Biol.* **124**:183–193.

Nojima, Y., Humphries, M. J., Mould, A. P., Komoriya, A., Yamada, K. M., Schlossman, S. F., and Morimoto, C., 1990, VLA-4 mediates CD3-dependent CD4+ T cell activation via the CS1 alternatively spliced domain of fibronectin, *J. Exp. Med.* **172**:1185–1192.

Norenberg, U., Wille, H., Wolff, J. M., Frank, R., and Rathjen, F. G., 1992, The chicken neural extracellular matrix molecule restrictin: Similarity with EGF-, fibronectin type III-, and fibrinogen-like motifs, *Neuron* **8**:849–863.

Obara, M., Kang, M. S., and Yamada, K. M., 1988, Site-directed mutagenesis of the cell-binding domain of human fibronectin: Separable, synergistic sites mediate adhesive function, *Cell* **53**:649–657.

Olden, K., Pratt, R. M., and Yamada, K. M., 1979, Role of carbohydrate in biological function of the adhesive glycoprotein fibronectin, *Proc. Natl. Acad. Sci. USA* **76**:3343–3347.

O'Shea, K. S., and Dixit, V. M., 1988, Unique distribution of the extracellular matrix component thrombospondin in the developing mouse embryo, *J. Cell Biol.* **107**:2737–2748.

Owens, R. J., and Baralle, F. E., 1986, Mapping the collagen-binding site of human fibronectin by expression in *Escherichia coli, EMBO J.* **5**:2825–2830.

Oyama, F., Murata, Y., Suganuma, N., Kimura, T., Titani, K., and Sekiguchi, K., 1989, Patterns of alternative splicing of fibronectin pre-mRNA in human adult and fetal tissues, *Biochemistry* **28**:1428–1434.

Pardes, J. B., Tonnesen, M. G., Falanga, V., Eaglestein, W. H., and Clark, R. A. F., 1990, Skin capillaries surrounding chronic venous ulcers demonstrate smooth muscle hyperplasia and increased laminin and type IV collagen, *Clin. Res.* **38**:628A.

Patel, R. S., Odermatt, E., Schwarzbauer, J. E., and Hynes, R. O., 1987, Organization of the fibronectin gene provides evidence for exon shuffling during evolution, *EMBO J.* **6**:2565–2572.

Pearson, C. A., Pearson, D., Shibahara, S., Hofsteenge, J., and Chiquet-Ehrismann, R., 1988, Tenascin: cDNA cloning and induction by TGF-beta, *EMBO J.* **7**:2977–2982.

Penttinen, R. P., Kobayashi, S., and Bornstein, P., 1988, Transforming growth factor beta increases mRNA for matrix proteins both in the presence and in the absence of changes in mRNA stability, *Proc. Natl. Acad. Sci. USA* **85**:1105–1108.

Pesciotta-Peters, D. M., Portz, L. M., Fullenwider, J., and Mosher, D. F., 1990, Co-assembly of plasma and cellular fibronectins into fibrils in human fibroblast cultures, *J. Cell Biol.* **111**:249–256.

Petersen, T. E., Skorstengaard, K., and Vibe-Pedersen, K., 1989, Primary structure of fibronectin, in: *Fibronectin* (D. F. Mosher, ed.), pp. 1–24, Academic Press, New York.

Pierschbacher, M. D., and Ruoslahti, E., 1984, Cell attachment activity of fibronectin can be duplicated by small synthetic fragments of the molecule, *Nature* **309**:30–33.

Pierschbacher, M. D., and Ruoslahti, E., 1987, Influence of stereochemistry of the sequence arg-gly-asp-xaa on binding specificity in cell adhesion, *J. Biol. Chem.* **262**:17294–17298.

Pisano, J. J., Finlayson, J. S., and Peyton, M. P., 1968, Cross-link in fibrin polymerized by factor XIII: ϵ-(γ-glutamyl)lysine, *Science* **160**:892–893.

Preissner, K. T., and Jenne, D., 1991, Vitronectins. A new molecular connection in haemostasis, *Thromb. Haemost.* **66**:189–194.

Preissner, K. T., and Muller-Berghaus, G., 1987, Neutralization and binding of heparin by S protein/vitronectin in the inhibition of factor Xa by antithrombin III. Involvement of an inducible heparin-binding domain of S protein/vitronectin, *J. Biol. Chem.* **262**:12247–12253.

Prieto, A. L., Andersson-Fisone, C., and Crossin, K. L., 1992, Characterization of multiple adhesive and counteradhesive domains in the extracellular matrix protein cytotactin, *J. Cell Biol.* **119**:663–678.

Prochownik, E. V., O'Rourke, K., and Dixit, V. M., 1989, Expression and analysis of COOH-terminal deletions of the human thrombospondin molecule, *J. Cell Biol.* **109**:843–852.

Quax, P. H. A., Pedersen, N., Masucci, M. T., Weening-Werhoeff, E. J. D., Dano, K., Verheijen, J., and Blasi, F., 1991, Complementation between urokinase-producing and receptor-producing cells in extracellular matrix degradations, *Cell Regul.* **2**:793–803.

Reed, M. J., Puolakkainen, P., Lane, T. F., Dickerson, D., Bornstein, P., and Sage, E. H., 1993, Differential expression of SPARC and thrombospondin 1 in wound repair: Immunolocalization and *in situ* hybridization, *J. Histochem. Cytochem.* **41:**1467–1477.

Rettig, W. J., Erickson, H. P., Albino, A. P., and Garin-Chesa, P., 1994, Induction of human tenascin (neuronectin) by growth factors and cytokines: Cell type-specific signals and signalling pathways, *J. Cell Sci.* **107:**487–497.

Riou, J. F., Shi, D. L., Chiquet, M., and Boucaut, J.-C., 1990, Exogenous tenascin inhibits mesodermal cell migration during amphibian gastrulation, *Dev. Biol.* **137:**305–317.

Robbins, K. C., Summaria, L., Hseih, B., and Shah, R., 1967, The peptide chains of human plasmin. Mechanism of activation of human plasminogen to plasmin, *J. Biol. Chem.* **242:**2333–2342.

Roberts, D. D., and Ginsburg, V., 1988, Sulfated glycolipids and cell adhesion, *Arch. Biochem. Biophys.* **267:**405–415.

Saga, Y., Yagi, T., Ikawa, Y., Sakakura, T., and Aizawa, S., 1992, Mice develop normally without tenascin, *Genes Dev.* **6:**1821–1831.

Sakata, Y., and Aoki, N., 1980, Cross-linking of α2-plasma inhibitor to fibrin by fibrin-stabilizing factor, *J. Clin. Invest.* **65:**290–297.

Salonen, E.-M., Vaheri, A., Pollanen, J., Stephens, R., Andreasen, P., Mayer, M., Dano, K., Gailit, J., and Ruoslahti, E., 1989, Interaction of plasminogen activator inhibitor (PAI-1) with vitronectin, *J. Biol. Chem.* **264:**6339–6343.

Savill, J., Hogg, N., Ren, Y., and Haslett, C., 1992, Thrombospondin cooperates with CD36 and the vitronectin receptor in macrophage recognition of neutrophils undergoing apoptosis, *J. Clin. Invest.* **90:**1513–1522.

Schultz-Cherry, S., and Murphy-Ullrich, J. E., 1993, Thrombospondin causes activation of latent transforming growth factor-beta secreted by endothelial cells by a novel mechanism, *J. Cell Biol.* **122:**923–932.

Seiffert, D., Poenninger, J., and Binder, B. R., 1993, Organization of the gene encoding mouse vitronectin, *Gene* **134:**303–304.

Shainoff, J. R., Urbanic, D. A., and DiBello, P. M., 1991, Immunoelectrophoretic characterization of the cross-linking of fibrinogen and fibrin by factor XIIIa and tissue transglutaminase, *J. Biol. Chem.* **166:**6429–6437.

Siebenlist, K. R., and Mosesson, M. W., 1992, Factors affecting γ-chain multimer formation in cross-linked fibrin, *Biochemistry* **31:**936.

Siebenlist, K. R., DiOrio, J. P., Budzynski, A. Z., and Mosesson, M. W., 1990, The polymerization and thrombin-binding properties of des-(Bβ1–42)-fibrin, *J. Biol. Chem.* **265:**18650–18655.

Silverstein, R. L., Baird, M., Lo, S. K., and Yesner, L. M., 1992, Sense and antisense cDNA transfection of CD36 (glycoprotein IV) in melanoma cells. Role of CD36 as a thrombospondin receptor, *J. Biol. Chem.* **267:**16607–16612.

Singer, I. I., Kawka, D. W., Scott, S., Mumford, R. A., and Lark, M. W., 1987, The fibronectin cell attachment sequence Arg-Gly-Asp-Ser promotes focal contact formation during early fibroblast attachment and spreading, *J. Cell Biol.* **104:**573–584.

Singer, I. I., Scott, S., Kawka, D. W., Kazazis, D. M., Gailit, J., and Ruoslahti, E., 1988, Cell surface distribution of fibronectin and vitronectin receptors depends on substrate composition and extracellular matrix accumulation, *J. Cell Biol.* **106:**2171–2182.

Sipes, J. M., Guo, N., Negre, E., Vogel, T., Krutzsch, H. C., and Roberts, D. D., 1993, Inhibition of fibronectin binding and fibronectin-mediated cell adhesion to collagen by a peptide from the second type I repeat of thrombospondin, *J. Cell Biol.* **121:**469–477.

Sjoberg, B., Eriksson, M., Osterlund, E., Pap, S., and Osterlund, K., 1989, Solution structure of human plasma fibronectin as a function of NaCl concentration determined by small-angle X-ray scattering, *Eur. Biophys. J.* **17:**5–11.

Somers, C. E., and Mosher, D. F., 1993, Protein kinase C modulation of fibronectin matrix assembly, *J. Biol. Chem.* **268:**22277–22280.

Sottile, J., and Wiley, S., 1994, Assembly of amino-terminal fibronectin dimers into the extracellular matrix, *J. Biol. Chem.* **269:**17192–17198.

Sottile, J., Schwarzbauer, J., Selegue, J., and Mosher, D. F., 1991, Five type I modules of fibronectin form a functional unit that binds to fibroblasts and *Staphylococcus aureus*, *J. Biol. Chem.* **266:**12840–12843.

Spiegel, S., Yamada, K. M., Hom, B. E., Moss, J., and Fishman, P. H., 1986, Fibrillar organization of fibronectin is expressed coordinately with cell surface gangliosides in a variant murine fibroblast, *J. Cell Biol.* **102:**1898–1906.

Spring, J., Beck, K., and Chiquet-Ehrismann, R., 1989, Two contrary functions of tenascin: Dissection of the active sites by recombinant tenascin fragments, *Cell* **59:**325–334.

Srebrow, A., Muro, A. F., Werbajh, S., Sharp, P. A., and Kornblihtt, A. R., 1993, The CRE-binding factor ATF-2 facilitates the occupation of the CCAAT box in the fibronectin gene promoter, *FEBS Lett.* **327:**25–28.

Summaria, L., Hsieh, B., Groskopf, W. R., and Robbins, K. C., 1967, The isolation and characterization of the S-carboxymethyl (light) chain derivative of human plasmin. The location of the active site on the light chain, *J. Biol. Chem.* **242:**5046–5052.

Suzuki, S., Oldberg, A., Hayman, E. G., Pierschbacher, M. D., and Ruoslahti, E., 1985, Complete amino acid sequence of human vitronectin deduced from cDNA. Similarity of cell attachment sites in vitronectin and fibronectin, *EMBO J.* **4:**2519–2524.

Taylor, H. C., Lightner, V. A., Beyer, W. F., McCaslin, D., Briscoe, G., and Erickson, H. P., 1989, Biochemical and structural studies of tenascin/hexabrachion proteins, *J. Cell. Biochem.* **41:**71–90.

Thompson, L. K., Horowitz, P. M., Bentley, K. L., Thomas, D. D., Alderete, J. F., and Klebe, R. J., 1986, Localization of the ganglioside-binding site of fibronectin, *J. Biol. Chem.* **261:**5209–5214.

Thorsen, S., Glas-Greenwalt, P., and Astrup, T., 1972, Difference in the binding to fibrin of urokinase and tissue plasminogen activator, *Thromb. Pathol. Haemost.* **28:**65–74.

Tolsma, S. S., Volpert, O. V., Good, D. J., Frazier, W. A., Polverini, P. J., and Bouck, N., 1993, Peptides derived from two separate domains of the matrix protein thrombospondin-1 have anti-angiogenic activity, *J. Cell Biol.* **122:**497–511.

Tomasini, B. R., and Mosher, D. F., 1991, Vitronectin, *Prog. Hemost. Thromb.* **10:**269–305.

Tuan, T. L., and Grinnell, F., 1989, Fibronectin and fibrinolysis are not required for fibrin gel contraction by human skin fibroblasts, *J. Cell. Physiol.* **140(3):**577–583.

Tucker, R. P., Spring, J., Baumgartner, S., Martin, D., Hagios, C., Poss, P. M., and Chiquet-Ehrismann, R., 1994, Novel tenascin variants with a distinctive pattern of expression in the avian embryo, *Development* **120:**637–647.

Tuckwell, D. S., Brass, A., and Humphries, M. J., 1992, Homology modelling of integrin EF-hands. Evidence for widespread use of a conserved cation-binding site, *Biochem. J.* **285:**325–331.

Tuszynski, G. P., Rothman, V. L., Deutch, A. H., Hamilton, B. K., and Eyal, J., 1992, Biological activities of peptides and peptide analogues derived from common sequences present in thrombospondin, properdin, and malarial proteins, *J. Cell Biol.* **116:**209–217.

Tuszynski, G. P., Rothman, V. L., Papale, M., Hamilton, B. K., and Eyal, J., 1993, Identification and characterization of a tumor cell receptor for CSVTCG, a thrombospondin adhesive domain, *J. Cell Biol.* **120:**513–521.

Varani, J., Nickoloff, B. J., Riser, B. L., Mitra, R. S., O'Rourke, K., and Dixit, V. M., 1988, Thrombospondin-induced adhesion of human keratinocytes, *J. Clin. Invest.* **81:**1537–1544.

Vartio, T., Laitinen, L., Nrvnen, O., Cutolo, M., Thornell, L. E., Zardi, L., and Virtanen, I., 1987, Differential expression of the ED sequence-containing form of cellular fibronectin in embryonic and adult human tissues, *J. Cell Sci.* **88:**419–430.

Veklich, Y. I., Gorkun, O. V., Medved, I. V., Nieuwenhuizen, W., and Weisel, J. W., 1993, Carboxyl-terminal portions of the α chains of fibrinogen and fibrin, *J. Biol. Chem.* **268:**13577–13585.

Vogel, T., Guo, N. H., Krutzsch, H. C., Blake, D. A., Hartman, J., Mendelovitz, S., Panet, A., and Roberts, D. D., 1993, Modulation of endothelial cell proliferation, adhesion, and motility by recombinant heparin-binding domain and synthetic peptides from the type I repeats of thrombospondin, *J. Cell Biochem.* **53:**74–84.

Wagner, O. F., Nicolosa, G., and Bachmann, F., 1989, Plasminogen activator inhibitor 1: Development of a radioimmunoassay and observations on its plasma concentration during venous occlusion and after platelet aggregation, *Blood* **70:**1645–1653.

Wehrle-Haller, B., and Chiquet, M., 1993, Dual function of tenascin: Simultaneous promotion of neurite growth and inhibition of glial migration, *J. Cell Sci.* **106:**597–610.

Werb, Z., Tremble, P. M., Behrendtsen, O., Crowley, E., and Damsky, C. H., 1989, Signal transduction through the fibronectin receptor induces collagenase and stromelysin gene expression, *J. Cell Biol.* **109:**877–889.

Whitby, D. J., and Ferguson, M. W., 1991, The extracellular matrix of lip wounds in fetal, neonatal and adult mice, *Development* **112:**651–668.

Whitby, D. J., Longaker, M. T., Harrison, M. R., Adzick, N. S., and Ferguson, M. W., 1991, Rapid epithelialisation of fetal wounds is associated with the early deposition of tenascin, *J. Cell Sci.* **99:**583–586.

Wikner, N. E., and Clark, R. A. F., 1988, Chemotactic fragments of fibronectin, *Methods Enzymol.* **162:**214–222.

Williams, M. J., Phan, I., Harvey, T. S., Rostagno, A., Gold, L. I., and Campbell, I. D., 1994, Solution structure of a pair of fibronectin type 1 modules with fibrin binding activity, *J. Mol. Biol.* **235:**1302–1311.

Wilner, G. D., Danitz, M. P., Mudd, M. S., Hsieh, K.-H., and Fenton II, J. W., 1981, Selective immobilization of α-thrombin by surface-bound fibrin, *J. Lab. Clin. Med.* **97:**403–411.

Wolff, C., and Lai, C. S., 1989, Fluorescence energy transfer detects changes in fibronectin structure upon surface binding, *Arch. Biochem. Biophys.* **268:**536–545.

Woods, A., and Couchman, J. R., 1992, Protein kinase C involvement in focal adhesion formation, *J. Cell Sci.* **101:**277–290.

Woods, A., Couchman, J. R., Johansson, S., and Hook, M., 1986, Adhesion and cytoskeletal organisation of fibroblasts in response to fibronectin fragments, *EMBO J.* **5:**665–670.

Woods, A., McCarthy, J. B., Furcht, L. T., and Couchman, J. R., 1993, A synthetic peptide from the COOH-terminal heparin-binding domain of fibronectin promotes focal adhesion formation, *Mol. Biol. Cell* **4:**605–613.

Wu, C., Bauer, J. S., Juliano, R. L., and McDonald, J. A., 1993, The $\alpha5\beta1$ integrin fibronectin receptor, but not the $\alpha5$ cytoplasmic domain, functions in an early and essential step in fibronectin matrix assembly, *J. Biol. Chem.* **268:**21883–21888.

Wysocki, A. B., Staiano-Coico, L., and Grinnell, F., 1993, Wound fluid from chronic leg ulcers contains elevated levels of metalloproteinases MMP-2 and MMP-9, *J. Invest. Dermatol.* **101:**64–68.

Yabkowitz, R., Dixit, V. M., Guo, N., Roberts, D. D., and Shimizu, Y., 1993, Activated T-cell adhesion to thrombospondin is mediated by the $\alpha4\beta1$ (VLA-4) and $\alpha5\beta1$ (VLA-5) integrins, *J. Immunol.* **151:**149–158.

Yamada, K. M., 1989, Fibronectin structure, functions and receptors, *Curr. Opin. Cell Biol.* **1:**956–963.

Yamada, K. M., and Kennedy, D. W., 1984, Dualistic nature of adhesive protein function: Fibronectin and its biologically active peptide fragments can autoinhibit fibronectin function, *J. Cell Biol.* **99:**29–36.

Yamada, K. M., Kennedy, D. W., Kimata, K., and Pratt, R. M., 1980, Characterization of fibronectin interactions with glycosaminoglycans and identification of active proteolytic fragments, *J. Biol. Chem.* **255:**6055–6063.

Yatohgo, T., Izumi, M., Kashiwagi, H., and Hayashi, M., 1988, Novel purification of vitronectin from human plasma by heparin affinity chromatography, *Cell Struct. Funct.* **13:**281–292.

Yokosaki, Y., Palmer, E. L., Prieto, A. L., Crossin, K. L., Bourdon, M. A., Pytela, R., and Sheppard, D., 1994, The integrin $\alpha9\beta1$ mediates cell attachment to a non-RGD site in the third fibronectin type III repeat of tenascin, *J. Biol. Chem.* **269:**26691–26696.

Zhang, Z., Morla, A. O., Vuori, K., Bauer, J. S., Juliano, R. L., and Ruoslahti, E., 1993, The $\alpha v\beta1$ integrin functions as a fibronectin receptor but does not support fibronectin matrix assembly and cell migration on fibronectin, *J. Cell Biol.* **122:**235–242.

Zhao, Y., and Sane, D. C., 1993, The cell attachment and spreading activity of vitronectin is dependent on the Arg-Gly-Asp sequence. Analysis by construction of RGD and domain deletion mutants, *Biochem. Biophys. Res. Commun.* **192:**575–582.

Zhu, B. C., and Laine, R. A., 1985, Polylactosamine glycosylation on human fetal placental fibronectin weakens the binding affinity of fibronectin to gelatin, *J. Biol. Chem.* **260:**4041–4045.

Zisch, A. H., D'Alessandri, L., Ranscht, B., Falchetto, R., Winterhalter, K. H., and Vaughan, L., 1992, Neuronal cell adhesion molecule contactin/F11 binds to tenascin via its immunoglobulin-like domains, *J. Cell Biol.* **119:**203–213.

Zlatopolsky, A. D., Chubukina, A. N., and Berman, A. E., 1992, Heparin-binding fibronectin fragments containing cell-binding domains and devoid of hep2 and gelatin-binding domains promote human embryo fibroblast proliferation, *Biochem. Biophys. Res. Commun.* **183:**383–389.

Chapter 3

Macrophage Involvement in Wound Repair, Remodeling, and Fibrosis

DAVID W. H. RICHES

1. Introduction

The process of wound repair has as its ultimate goal the restoration of normal aseptic tissue structure and function following injury. Although injury can take many forms, e.g., surgical trauma, burns, immunologically mediated injury, and so forth, the general sequence of events that are activated in response to injury and that lead to successful wound repair show striking similarity irrespective of the initial injurious insult. The sequence comprises (1) the activation of the coagulation system, leading to a cessation of blood flow and the formation of a provisional matrix; (2) the local generation of a variety of soluble chemotactic factors formed from preformed plasma proteins that attract inflammatory cells to the site of injury; (3) the sequential influx of neutrophils and monocytes, leading to wound sterilization; (4) the debridement of damaged connective tissue matrix; (5) the initiation of neovascularization; and (6) the stimulation of mesenchymal cell proliferation and connective tissue matrix remodeling. However, while in many tissues and situations, this generalized sequence of events leads to the restoration of normal tissue structure and functions, in some tissues, such as in adult skin, repair is invariably associated with scarring caused as a result of abundant collagen synthesis by fibroblasts that proliferate and differentiate within the provisional matrix. While this is generally acceptable in the case of the skin, excessive tissue fibrosis during repair of other tissues, for example, as a consequence of injury to the lung or liver parenchyma, results in a dramatic and frequently fatal loss of function as a consequence of scarring. Thus, understanding what distinguishes these two outcomes may allow treatment strategies to be developed to ameliorate tissue fibrosis in susceptible or "at-risk" individuals.

DAVID W. H. RICHES • Division of Basic Sciences, Department of Pediatrics, National Jewish Center for Immunology and Respiratory Medicine, Denver, Colorado 80206.

The Molecular and Cellular Biology of Wound Repair (Second Edition), edited by Richard A. F. Clark. Plenum Press, New York, 1996

In both situations, an inflammatory response is initiated within minutes of application of the injurious insult. The first blood leukocytes to be attracted and demobilized at the site of tissue injury are neutrophils, whose numbers increase steadily before peaking after 24–48 hr. The main function of these cells appears to be to kill bacteria that may have been introduced into the tissue during injury. Depletion of circulating neutrophils in guinea pigs by treatment with antineutrophil serum was found to have no demonstrable effect on the subsequent healing of experimentally induced aseptic wounds (Simpson and Ross, 1971, 1972). Thus, unless a wound becomes infected, the neutrophil does not appear to be an essential player in the process of wound repair.

As the number of neutrophils begin to decline, the macrophage population increases to replace the neutrophil as the predominant professional wound phagocyte. For many years, the prime function of these cells was thought principally to be the removal and degradation of injured tissue debris, in anticipation of the reparative process. However, studies reported by Leibovitch and Ross in 1975 (Leibovitch and Ross, 1975) showed that the combined depletion of circulating blood monocytes and local tissue macrophages in guinea pigs resulted not only in a severe retardation of tissue debridement, but also in a marked delay in fibroblast proliferation and subsequent wound fibrosis. These data indicated for the first time that macrophages play a vital role in the orchestration and execution of both the degradative and reparative phases of wound healing. As a result of further investigations during the ensuing two decades, it is now recognized that macrophages serve as an important source of mesenchymal cell growth factors that stimulate the proliferation of fibroblasts, smooth muscle cells, and endothelial cells. The reduced oxygen tension and other intrinsic and possibly extrinsic factors within the avascular wound are thought to stimulate wound macrophages to secrete cytokines that stimulate angiogenesis, thereby initiating neovascularization (i.e., the directed outgrowth of new capillaries) of the wound space. Finally, in response to a variety of stimuli that are present in wounds, macrophages play a critical role as a source of growth factors and cytokines that control the synthesis of connective tissue proteins, especially collagen, by other cell types during the reparative phase of the inflammatory response.

The objective of this chapter is to provide an overview of the multiple roles of macrophages in both the degradative and reparative phases of wound healing. Specifically, discussion will include the origin of wound monocytes and macrophages, the signals involved in their emigration from the vasculature, and the mechanisms underlying the differentiation of monocytes into macrophages and of subsequent macrophage phenotypic differentiation. The involvement of macrophages in tissue debridement will be reviewed with particular reference to the secreted and intracellular proteolytic enzymes that are thought to be preeminently involved in this process. In consideration of the reparative phase, the focus will be on three aspects of macrophage involvement in repair, namely (1) the characteristics and conditions under which mesenchymal growth factors are expressed, (2) the role of the macrophage in angiogenesis, and (3) the potential of the macrophage to stimulate connective tissue matrix synthesis. Unfortunately, but out of necessity, this chapter intends to be heuristic rather than comprehensive and thus will focus on underlying and emerging concepts.

2. Origin and Kinetics of Macrophages

The ontogeny of mononuclear phagocytes has been studied in great depth during the past three decades. Studies in parabiotic rats by Volkman and Gowans (1965a,b) provided the first indications of the bone marrow origin of mononuclear phagocytes. Studies in the mouse by van Furth and his colleagues (van Furth and Cohn, 1968; van Furth *et al.*, 1970, 1973, 1985a) revealed the nature and turnover kinetics of the two monocytic progenitors, namely, monoblasts and promonocytes (Goud *et al.*, 1975; Goud and van Furth, 1975; van Furth *et al.*, 1973). These investigations have been complemented by the characterization of mononuclear phagocyte lineage growth factors including colony-stimulating factor-1 (CSF-1) and granulocyte–macrophage colony-stimulating factor (GM-CSF) and their respective receptors. More recent approaches using transgenic mice that overexpress specific myeloid growth factors (Lang *et al.*, 1987) or in which specific myeloid growth factor "knockout" mice have been developed have allowed a fuller appreciation of the redundancy of these systems as well as several unexpected findings such as the occurrence of alveolar proteinosis in GM-CSF knockout mice (Stanley *et al.*, 1994). Together, these studies have revealed much about the characteristics and conditions of growth of mononuclear phagocytes and their precursors both under steady-state conditions as well as in the face of an inflammatory insult.

2.1. Monocyte Biogenesis and Circulation

A considerably simplified scheme of monocyte development is presented in Fig. 1. Monocytes are ultimately derived from a pluripotential stem cell that, to date, has never been isolated or characterized. The pluripotential stem cell is thought to give rise to

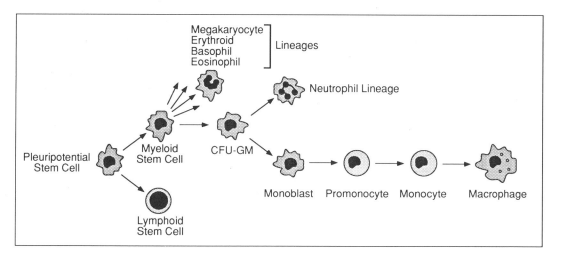

Figure 1. Schematic pathway of monocyte development from the pluripotential stem cell to the circulating blood monocyte.

either a myeloid stem cell, a lymphoid stem cell, or to undergo self-renewal. The myeloid stem cell, sometimes referred to as CFU-GEMM (colony forming unit—granulocyte, erythrocyte, macrophage, megakaryocyte) gives rise to progenitor cells of the mega-karyocytic, erythroid, and granulocytic–monocytic lineages (CFU-GM). In the presence of specific growth factors, this latter progenitor gives rise to colonies of neutrophils, monocytes, or mixed colonies of both cell types. Cell culture studies in the mouse have defined the first detectable precursor cell derived during monocytic development from the CFU-GM as the monoblast, a moderately sized (10–12 μm), weakly phagocytic, esterase-positive cell (Goud et $al.$, 1975). The monoblast gives rise to two promonocytes with a cell cycle time of approximately 11.5 hr (Goud and van Furth, 1975). While still in the bone marrow, promonocytes divide, with a $t_{1/2}$ of 16.2 hr (van Furth et $al.$, 1973), to give rise to two monocytes that are released into the blood where they circulate before randomly leaving the vasculature, under steady-state conditions, with a $t_{1/2}$ of 17.4 hr (van Furth et $al.$, 1973). For more detail the reader is referred to Metcalf (1989).

2.2. Monocyte Influx versus Local Production

Although it was initially thought that circulating blood monocytes gave rise to the resident mononuclear phagocytes distributed throughout the body, other investigators have shown that a large proportion of resident tissue macrophages replicate in $situ$, and thus may not be absolutely dependent on circulating blood monocytes for repletion. The duality in the origin of tissue macrophages appears to vary according to tissue/organ type and whether the measurements were made under steady-state or inflammatory conditions. For example, current evidence for the lung suggests that a major proportion of the pulmonary alveolar macrophage population (70%) may be sustained, under steady-state conditions, by local macrophage proliferation. This conclusion is based on several lines of evidence, including: (1) the maintenance of the size of the alveolar macrophage population during periods of monocytopenia induced either by systemic hydrocortisone administration (Lin et $al.$, 1982) or by external or internal bone marrow irradiation (Sawyer et $al.$, 1982; Tarling and Coggle, 1982); (2) the large reduction of the pulmonary alveolar macrophage population following irradiation of the thorax (Tarling and Coggle, 1982); (3) by a more accurate appreciation of the turnover time of pulmonary alveolar macrophages (Coggle and Tarling, 1984); and (4) by an analysis of the alveolar macrophage populations in parabiotic mice (Sawyer, 1986). By contrast, the Kupffer cells of the liver appear to be more dependent on the circulating blood monocyte for the repletion of senescent cells under steady-state conditions with perhaps only about 8% being derived by local division (van Furth et $al.$, 1985b).

2.3. Monocyte Influx under Inflammatory Conditions

Unlike the question of the origin of resident tissue macrophages, it is now abundantly clear that under inflammatory conditions, circulating blood monocytes are actively recruited into organs and tissues, where they further differentiate into macrophages. This has been shown to be true for the inflammatory response in the peritoneal cavity following intraperitoneal injection of mild, sterile inflammatory agents, such as latex spherules and newborn calf serum (van Furth *et al.,* 1973); in the lung following the intrabronchial instillation of BCG (Blusse van Oud Alblas *et al.,* 1983); and in the liver following the intravenous administration of inflammatory stimuli such as zymosan particles (Bouwens and Wisse, 1985) or glucan (Deimann and Fahimi, 1980). However, after the initial influx of circulating blood monocytes, there also frequently appears to be an induction of mitotic activity in local macrophages. Daems (1980) has suggested that under steady-state conditions, blood monocytes rapidly migrate through the vasculature of tissues and organs for the purposes of surveillance, and only under inflammatory or other insult conditions are retained in the tissues and differentiate into macrophages in order to assist in the activities of the resident tissue macrophages. An obvious implication of this idea is that it potentially allows the relatively immature and adaptable circulating blood monocyte to differentiate into a macrophage whose functional properties are determined by the prevailing conditions (e.g., stimuli such as bacteria, activated complement components, connective tissue breakdown products, oxygen tension, etc.) it encounters at the site of demobilization. This question of monocyte maturation and differentiation will be dealt with in more depth in Section 4.2.

As a tissue, the skin contains very few macrophages, although during an inflammatory response the numbers of cells increase dramatically. Evidence that the increase in macrophage numbers during an inflammatory response (wound healing) was derived from circulating blood monocytes was initially presented by Leibovitch and Ross (1975), who found that the systemic administration of hydrocortisone (an inducer of monocytopenia) resulted in a 66% reduction in the number of macrophages associated with experimentally induced skin wounds in guinea pigs. Locally administered antimacrophage serum had no effect on the number of circulating monocytes, nor did it affect the number of wound macrophages. Similar conclusions have been drawn concerning the bone marrow origin of macrophages in healing wounds in humans by Stewart *et al.* (1981). By observing the karyotype (sex chromosome markers) of macrophages and fibroblasts emigrating from explants of a 5-day skin wound in a 9-year-old girl who had recently received a bone marrow graft from her 11-year-old brother, it was found that the fibroblasts had the karyotype of the recipient, while the macrophages bore the karyotype of the donor. These findings suggest that the macrophages of the healing wound were derived from the bone marrow, whereas the fibroblasts are presumed to be of local tissue origin.

In an attempt to gain information about the origin and kinetics of macrophages in the skin under inflammatory conditions in the mouse, van Furth *et al.* (1985b) measured the rate of appearance of mononuclear phagocytes onto a subcutaneously inserted glass coverslip acting as an inflammatory stimulus. Removal and examination of the coverslips 3- to 6-hr postinsertion revealed a large influx of granulocytes. However, these cells were later (6–48 hr) replaced by mononuclear cells that microscopically and functionally resembled blood monocytes. The number of mononuclear phagocytes that became attached to the glass coverslips correlated directly with the time the coverslips had been left in place. On the basis first of similarities in the [³H]thymidine-labeling indices of circulating blood monocytes and the mononuclear phagocytes that became attached to the glass coverslips, and second, on the fact that the induction of mono-cytopenia (by systemic administration of hydrocortisone) prior to the insertion of the glass coverslips resulted in a large reduction in the accumulation of mononuclear phagocytes onto the surface of the coverslip, these investigators also concluded that the increase in the number of macrophages during this type of inflammatory response in the skin was almost exclusively (>99%) due to an influx of circulating blood mono-cytes, with little or no proliferation of local tissue macrophages. Thus, dependent on the site of injury and subsequent repair, different proportions of monocyte-derived macro-phages and resident tissue macrophages may contribute to the tissue degradation and repair.

3. Monocyte Migration

3.1. Multiplicity of *in Vivo* Monocyte Chemotactic Factors

From what has been discussed above, it is clear that macrophage involvement in injury and repair is strongly dependent on an influx of circulating blood monocytes. The emigration of these cells from the vasculature to the extravascular compartment appears to be affected by many different endogenous and exogenous chemotactic agents that are initially generated from preformed precursor proteins and then later expressed locally by resident and infiltrating cells present within the evolving inflam-matory site. In this context, chemotactic factors that specifically attract neutrophils or monocytes likely provides the basis for the predominance of one type of inflammatory cell over another that is characteristically seen during the inflammatory response. Disruption of capillaries will lead to the extravasation of blood plasma and formed elements, platelet aggregation and activation, activation of the coagulation cascade, and the generation of activated complement components and kinins. Many of these ele-ments behave as strong chemotactic factors for inflammatory cells and likely provide the signals for the initial waves of these cells to migrate into an inflammatory site. Once demobilized at an inflammatory site, resident and infiltrating cells, especially mononuclear phagocytes, express an extensive repertoire of chemokines, which attract and activate additional inflammatory cells, thereby amplifying the initial inflammatory response and later providing signals to specifically attract the mononuclear phagocytes that orchestrate and execute the reparative phase. The chemokine family of low-molecular-weight chemotactic factors is subdivided into two groups, α and β (for an excellent review, see Baggiolini *et al.,* 1994). Members of the α-chemokine famly, which are characterized by a C-X-C motif, are generally specific for neutrophils and are exemplified by interleukin-8 (IL-8), growth-related protein (GRO), and melanoma growth stimulating activity (MGSA). Macrophages secrete IL-8 in response to a vari-ety of bacterial products and cytokines, including lipopolysaccharide (LPS), tumor necrosis factor-alpha (TNF-α), and IL-1β (Strieter *et al.*, 1990a,b). By contrast, representatives of the β-chemokine family, which are characterized by a C-C motif, generally act on mononuclear phagocytes and include factors such as monocyte chemoattractant protein-1 (MCP-1) and macrophage inflammatory protein-1α (MIP-1α). Both α- and β-chemokines interact with specific heptahelical G-protein-linked receptors, which signal both the directed migratory response in receptive cells and the activation of a number of inflammatory functions, including reactive oxygen intermediate production and lysosomal enzyme secretion. As will be discussed in more depth (Section 5), this latter function is of some significance to the amplification of the inflammatory response since secreted lysosomal enzymes may initiate degradation of the connective tissue matrix, thereby generating an additional group of monocyte-specific chemotactic factors. The dominant monocyte chemotactic factors implicated as attractants of this cell type *in vivo* are summarized in Table I.

Table I. Major Classes of Mononuclear Phagocyte Chemoattractants

Chemoattractant	Target	Reference
Plasma-protein-derived chemoattractants		
C5a, C5a desArg	Neutrophils, monocytes	Marder *et al.* (1985)
Fibrinopeptides	Monocytes	Kay *et al.* (1973)
IgG-proteolytic fragments	Monocytes	Ishida *et al.* (1978)
α-Thrombin	Monocytes	Bar-Shavit *et al.* (1983)
Cell-derived chemoattractants		
Leukotriene B$_4$	Neutrophils monocytes	Ford *et al.* (1980)
Monocyte-chemoattractant proteins (MCP) 1, 2, and 3	Monocytes	Kunkel *et al.* (1991)
RANTES	Monocytes, lymphocytes	Schall *et al.* (1990)
Macrophage inflammatory protein (MIP) 1β	Monocytes, lymphocytes	Sherry *et al.* (1988)
Macrophage inflammatory protein (MIP) 1α	Monocytes, lymphocytes, fibroblasts	Sherry *et al.* (1988)
Platelet factor 4	Monocytes	Deuel *et al.* (1981)
PDGF	Neutrophils, monocytes, fibroblasts	Pierce *et al.* (1989)
TGF-β	Neutrophils, monocytes, fibroblasts	Pierce *et al.* (1989)
Extracellular connective tissue matrix-derived chemoattractants		
Collagen/collagen fragments	Monocytes, fibroblasts	Postlethwaite and Kang (1976)
Elastin/elastin fragments	Neutrophils, monocytes	Senior *et al.* (1980, 1984)
Fibronectin fragments	Monocytes	Norris *et al.* (1982)
Bacterial-derived chemoattractants		
Formyl methionyl peptides	Neutrophils, monocytes	Snyderman and Fudman (1980)
N-acetylmuramyl-L-alanyl-D-isoglutamine	Neutrophils, monocytes	Ogawa *et al.* (1983)

3.2. Complement Component Fragments C5a and C5a des Arg

Activation of either the classical or alternative pathway of the complement system results in cleavage of the N-terminal region of the α chain of C5, thereby generating C5a, a profoundly basic 11-kDa glycopeptide possessing potent inflammatory properties. The inflammatory properties of C5a are predominantly due to its ability to behave both as an anaphylatoxin (releasing histamine and other granule constituents from mast cells), as well as a chemotactic factor for neutrophils and monocytes. When injected into skin, nanogram quantities of human C5a induce a rapid (within seconds) wheal-and-flare reaction, which peaks in intensity 10–30 min after injection but which essentially disappears 60 min after injection (Yancey et al., 1985). The changes in vascular permeability induced by C5a injection are due to a direct effect on the vascular endothelium, as well as a vasodilatory effect caused by C5a-dependent prostaglandin production (Wedmore and Williams, 1981). After its formation, however, C5a is rapidly hydrolyzed to C5a des Arg by serum carboxypeptidase N, thereby largely inactivating its anaphylatoxic properties, while inhibiting its activity as a neutrophil chemotactic agent by a factor of 10- to 15-fold. Interestingly, and of some significance to the migration of circulating blood monocytes to sites of inflammation, the conversion of C5a to C5a des Arg apparently has no effect on the activity of this molecule as a monocyte chemotactic factor (Marder et al., 1985). In addition to their chemotactic properties, C5a and C5a des Arg also stimulate the in vitro adherence of purified human peripheral blood monocytes to monolayers of human microvascular endothelial cells (Doherty et al., 1987), a process essential to the emigration of monocytes through blood vessel walls.

3.3. β-Chemokines

Although not exclusively monocyte chemotactic factors, members of the β-chemokine family show some specificity for this cell lineage as well as for lymphoid cells (Baggiolini et al., 1994). Most thoroughly studied is MCP-1. Originally identified as the fibroblast competence factor JE, MCP-1 has been shown to be produced by many different cell types, including fibroblasts, macrophages, and endothelial cells. Expression of the MCP-1 gene is stimulated by an equally diverse group of inflammatory agents, including interferon gamma (IFN-γ), LPS, TNF-α and IL-1β (Kunkel et al., 1991). Production of MCP-1 and increased expression of its mRNA have been detected in macrophages in chronic inflammatory settings (Koch et al., 1992). The protein is recognized by a heptahelical heterotrimeric G-protein-linked surface receptor that initiates Ca^{2+}-dependent transmembrane signaling in responsive cells (Rollins et al., 1991). Monocyte-specific β-chemokines are particularly important chemotactic factors since they have the potential to promote the selective accumulation of monocytes and macrophages during the later stages of the inflammatory response when neutrophil numbers begin to wane. Unfortunately, little attention has been given to the role of these important chemotactic factors in repair processes.

3.4. Cytokines and Growth Factors

Both platelet-derived growth factor (PDGF) and transforming growth factor-beta (TGF-β) are potent chemotactic factors for monocytes, although both lack specificity, and thus, many other cells types, including neutrophils, respond similarly (Pierce *et al.*, 1989). In addition to being a monocyte chemotactic factor, TGF-β also stimulates the directed migration of macrophages. By contrast, PDGF receptor expression is absent in macrophages, and these cells fail to respond to PDGF (P. W. Noble and D. W. H. Riches, unpublished observations). Since both proteins are secreted from the granules of platelets during the early response of tissues to injury, this cell type may be an important source of inflammatory cell chemotactic factors to promote initial cell migration to the site of injury. Later (as discussed in more depth in Sections 6.4.1 and 6.4.2), other cell types including macrophages express both PDGF and TGF-β, thereby amplifying the inflammatory response.

3.5. Connective Tissue Matrix Proteins

Purified components of the connective tissue matrix itself also have been found to express chemotactic activity *in vitro* for neutrophils and monocytes, as well as for fibroblasts. These components include major structural proteins such as collagen and elastin, as well as less abundant proteins such as fibronectin and laminin. Interestingly, small proteolytic cleavage fragments of these proteins also have been found to be chemotactic for these cell types.

Type I collagen, the major structural protein of skin, has been found to be chemotactic for monocytes and fibroblasts, though not for neutrophils (Postlethwaite and Kang, 1976; Postlethwaite *et al.*, 1978). This property is expressed by native collagen, as well as by cyanogen bromide and proteolytically derived collagen fragments varying in molecular size from small oligopeptides (3–10 amino acids) to larger fragments the size of ovalbumin (43 kDa). Senior *et al.* (1980) have found that oligopeptides (14–20 kDa) derived by digestion of human aortic elastin and bovine ligament elastin with human neutrophil elastase are also preferentially chemotactic for monocytes. The chemotactic activity of elastin has been shown to be associated with the repeating peptide sequence Val-Gly-Val-Ala-Pro-Gly (Senior *et al.*, 1984). Fibronectin derived both from plasma and synthesized locally by a variety of cell types, including mononuclear phagocytes (Carre *et al.*, 1994), is deposited at sites of injury and becomes associated with the provisional matrix. As discussed elsewhere in this volume, fibronectin exhibits multiple functions, including behaving as an anchor protein for fibroblasts and mononuclear phagocytes and in directing the migration of epidermal cells during wound reepithelialization (Clark *et al.*, 1982; Donaldson and Mahan, 1983). Proteolytic fragments derived from plasma fibronectin also serve as potent chemotactic factors for circulating blood monocytes, though not for neutrophils or lymphocytes.

If connective tissue fragments are to be chemotactically active *in vivo*, this raises the question of how are they generated, since it implies that some degradation of the connective tissue matrix must take place before monocytes can be attracted to the area. One possibility is that neutrophils, whose accumulation at an inflammatory site generally precedes that of monocytes, promote some limited degradation of the connective tissue matrix, thereby generating monocyte chemotactic factors. Consistent with this notion is the finding that neutrophil granules contain both elastase (Murphy *et al.,* 1977) and collagenase (Robertson *et al.,* 1972) activities that are capable of degrading insoluble collagen and elastin to soluble (and therefore potentially chemotactic) fragments. Human granulocyte elastase also has been reported to degrade fibronectin (McDonald and Kelley, 1980). However, in apparent contradiction of this hypothesis, Simpson and Ross (1971, 1972) found that neutrophil depletion of guinea pigs with antineutrophil serum had no effect on the migration of monocytes to a site of inflammation in the skin.

3.6. Macrophage Secretion of Chemotactic Factors

Not only do macrophages respond to generic and specific chemotactic factors, they are also a major source of chemotactic factors that attract other inflammatory cells, mesenchymal cells, and cells of the immune system. As will be discussed later, many of these agents are pleiotropic growth factors and cytokines that influence the inflammatory response and subsequent repair in many ways. For example, macrophages secrete PDGF and TGF-β, which act as monocyte chemotactic factors, as well as profoundly influencing matrix remodeling and angiogenesis (see Section 5), while IL-8, an α-chemokine produced by macrophages (Carre *et al.,* 1991; Strieter *et al.,* 1990a) and other cells, also plays a role in angiogenesis (Koch *et al.,* 1986; Strieter *et al.,* 1992). Nevertheless, despite these potentially confounding issues, macrophages serve an important role in attracting additional inflammatory cells into an inflammatory site, thereby amplifying the inflammatory response.

3.7. Interactions with Multiple Chemotactic Factors

We have discussed the nature and characteristics of just a few of the multiple chemotactic factors that are generated during an inflammatory response *in vivo.* Evidence presented over a decade ago (Cianciolo and Snyderman, 1981; Falk and Leonard 1980, 1981) showed that cells migrating into a nucleopore filter in response to one chemotactic factor could then migrate out of the filter in the reverse direction when the chemotactic factor was removed from the lower chamber and a second chemotactic factor was placed in the upper chamber. Furthermore, stimulus-specific desensitization to one stimulus (*N*-formyl-methionyl-leucyl-phenylalanine) did not inhibit the migration of cells to another stimulus (activated serum) when chemotactic factors were employed at low concentrations, although high doses of single chemotactic agents will induce a state of cross-desensitization to other stimuli (Cianciolo and Snyderman, 1981). These data suggested that the ability of monocytes to migrate toward specific chemotactic factors is due to the expression of multiple receptors for specific chemotaxins. Since chemotactic factor receptor expression does not appear to be restricted to specific subpopulations of monocytes, the implications of these findings are (1) that multiple chemotactic factors may be generated during an inflammatory response for which the monocyte expresses a repertoire of surface receptors, and (2) that the responding monocyte population is relatively homogeneous in this respect.

4. Monocyte Maturation and Differentiation

4.1. Monocyte–Macrophage Maturation

Monocytes rapidly differentiate into macrophages upon arrival at a site of injury. Often referred to as "inflammatory" or "responsive" macrophages, these cells have been studied extensively *in vitro* and can be obtained in relatively large quantities by irritation of the peritoneal cavity of laboratory animals with sterile stimuli such as thioglycollate medium, proteose peptone, mineral oil, heterologous serum, and endotoxin, or by the expansion of bone marrow progenitor cells in liquid culture. Many of the functions of these cells are suggestive of their presumed function in the inflammatory response, i.e., as a predominent mediator of tissue debridement. Notable examples include increased levels of lysosomal enzymes, enhanced capacities for endocytosis, the ability to ingest particulate material via CD11b/CD18, and the induction of secretion of the neutral proteases plasminogen activator and elastase. The role of some of these altered capacities in the process of tissue debridement will be discussed in Section 5.

The mechanisms responsible for driving the differentiation of monocytes into inflammatory macrophages remain poorly understood. *In vitro* studies have indicated that several serum constituents may be implicated (Musson, 1983), including $1\alpha,25$-dihydroxyvitamin D_3 (Proveddini *et al.,* 1986; Tanaka *et al.,* 1983). In addition, adherence has been reported by Shaw *et al.* (1990) to play an important role in the differentiation of monocytes into macrophages. In the context of wound repair, an attractive theory for the extravascular differentiation of monocytes has stemmed from experiments in which the effect of fibronectin binding to monocytes was investigated. Circulating blood monocytes bind fibronectin via the fibronectin receptor, a $\beta1$ integrin. Engagement of fibronectin receptors by surface-bound ligands was found to induce several functional changes within the monocyte, including the phagocytic activation of receptors for C3b and iC3b (Wright *et al.,* 1983) and the ability to secrete plasminogen activator and elastase following nonspecific phagocytic stimulation (Bianco, 1983). Both of these qualities are associated with inflammatory or responsive macrophages such as those obtained by irritation of the peritoneal cavity of the mouse with thioglycollate medium or proteose peptone. Thus, as has been suggested by Hosein *et al.* (1985), the mere process of binding of monocytes to connective tissue fibronectin may, by itself, be sufficient to drive the differentiation of monocytes into responsive macrophages. The finding that soluble fibronectin binds minimally to monocytes whereas surface-bound or particulate fibronectin binds more effectively is suggestive of a control mechanism to limit monocyte differentiation to areas of fibronectin deposition. This mechanism, of course, does not exclude other possible mechanisms for the induction of monocyte differentiation, which probably involve the autocrine and/or paracrine effects of endogenously derived factors and/or cytokines.

4.2. Macrophage Phenotypic Differentiation

Although being traditionally perceived as a scavenger cell involved in the uptake and degradation of foreign agents, senescent cells, and tissue debris, the macrophage is now acknowledged to contribute to many diverse aspects of host defense, including the recognition and elimination of virally and neoplastically transformed cells, the destruction of pathogenic microorganisms and obligate intracellular parasites, the processing and presentation of antigens to the immune system, as well as the aforementioned vital roles in the degradative and reparative phases of the inflammatory response. One of the major issues in contemporary macrophage biology has been delineating the mechanistic basis underlying the generation of macrophage functional diversity. Two general hypotheses, illustrated in Fig. 2, have been advanced to explain the heterogeneity

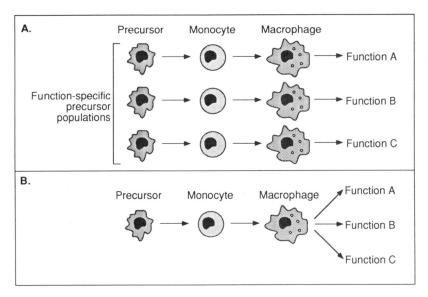

Figure 2. Prototypical pathways leading to the development of macrophage functional diversity. (A) Function-specific monocyte precursors are viewed as giving rise to function-specific subpopulations of monocytes and ultimately to function-specific subpopulations of macrophages. (B) Monocytes are viewed as a relatively homogeneous population of cells that respond to the stimuli and conditions that prevail at the site to which they have been attracted. In this hypothesis, macrophage functional diversity is viewed as being induced in an adaptive fashion. Abundant data support this latter hypothesis (panel B).

exhibited by these cells. The possibility that diversity is generated by functionally distinct and committed subpopulations of macrophages, while being conceptually attractive, has received little experimental support. The alternative hypothesis is that macrophages are pleuripotential cells that adapt themselves to the stimuli or conditions that prevail at the site to which they have been attracted, and thus functional heterogeneity in this hypothesis is viewed as an adaptive response of a relatively homogeneous population of precursor cells.

Support for the adaptation concept was first reported in the late 1970s with the observation by Russell *et al.* (1977) and by Ruco and Meltzer (1978) that monocyte-derived peritoneal exudate macrophages could respond *in vitro* to immune and bacterial stimuli by differentiating to express a new functional activity, cytocidal activity, that endows these cells with the capacity to recognize and destroy transformed target cells. During this transition, we and others have shown that macrophages are induced to express a number of gene products that collectively contribute to the new functional activity. Included in the genes whose expression is up-regulated during macrophage cytocidal differentiation are inducible nitric oxide synthase (iNOS) (Ding *et al.*, 1988; Lowenstein *et al.*, 1992; Xie *et al.*, 1992), complement component Bf (Riches *et al.*, 1988; Riches and Underwood, 1991), and IFN-β (Riches and Underwood, 1991). Importantly, however, the process of macrophage cytocidal differentiation is accompanied by the down-regulation of expression of a number of constitutively expressed genes, including the mannose receptor (Imber *et al.*, 1982), the secretory protein apolipoprotein E (Werb and Chin, 1983), and the genes encoding a number of lysosomal acid hydrolases (Riches, 1988; Riches and Henson, 1986). As might be expected, the reduced expression of these gene products results in a diminution or cessation of basic scavenging properties of macrophages as reflected by a reduction of both ligand uptake and intralysosomal degradation of ingested material (De Whalley and Riches, 1991). Thus, the development of cytocidal activity results in a cessation of other macrophage functional activities.

By analogy to the development of cytocidal activity, we have proposed that other complex macrophages functions, such as those that underlie the role of this cell in both the debridement and reparative phases of the inflammatory response, may also be explained on the basis of distinct phenotypic states defined by the expression of distinct but restricted patterns of gene expression (Laszlo *et al.*, 1993). A notable feature of macrophages that accumulate *in vivo* in response to particulate inflammatory stimuli, such as zymosan particles and β-1,3-glucan, is the presence of dramatically elevated levels of a number of lysosomal acid hydrolases such as β-hexosaminidase and β-glucuronidase (Deinmann and Fahima, 1980; Meister *et al.*, 1977; Schorlemmer *et al.*, 1977; Sugimoto *et al.*, 1978). *In vitro* exposure of macrophages to many of these stimuli has been shown to stimulate the increased synthesis of these enzymes over a similar time course to that seen *in vivo* (Lew *et al.*, 1986, 1991). Thus, in contrast to cytocidal activation of macrophages, during which the synthesis of lysosomal acid hydrolases is down-regulated (Riches, 1988; Riches and Underwood, 1991), exposure of macrophages to inflammatory stimuli results in an up-regulation of lysosomal enzyme synthesis and content. Although these findings could be considered distinct

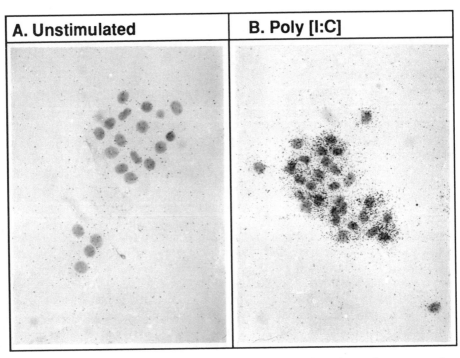

Figure 3. *In situ* hybridization of unstimulated and poly [I:C]-stimulated monolayers of mouse macrophages with a complement component Bf cRNA probe. Following stimulation with poly [I:C], the majority of the macrophage population are seen to be expressing Bf-specific transcripts, indicating relative uniformity in the responding population.

responses of separate subpopulations of macrophages, we have recently presented data that strongly argue against this possibility. As shown in Fig. 3, using *in situ* hybridization, we have shown that exposure of macrophages to poly [I:C], a stimulus that induces cytocidal differentiation, induces the uniform expression of the complement component Bf transcript (a gene product whose expression accompanies macrophage cytocidal differentiation) within the macrophage population (Laszlo *et al.*, 1993). Thus, distinct macrophage stimuli induce distinct patterns of gene expression that appear to be an adaptive response of the macrophage population rather than a response of distinct macrophage subpopulations. Figure 4 illustrates three mutually exclusive patterns of gene expression that are currently under investigation in our laboratory.

These findings raised the question of how macrophage phenotypic states may be controlled. Given the fact that macrophage responses are expressed in a tightly controlled fashion, we questioned if macrophage phenotypic states may be regulated in a mutually antagonistic fashion to preclude the possibility of the simultaneous expression of multiple states. Induction of the cytocidal state with poly [I:C], while stimulating the expression of genes associated with this state, led to a state of unresponsiveness in

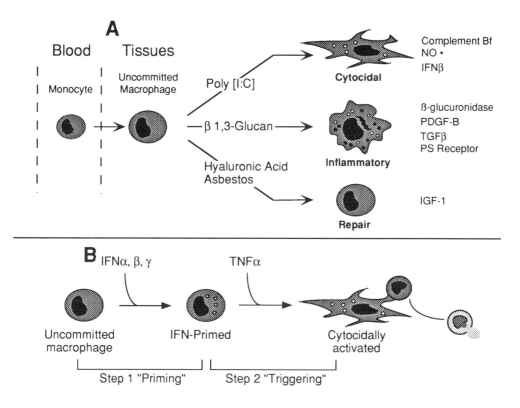

Figure 4. Prototypic pathways of macrophage phenotypic development and the groups of gene products associated with each response. (A) Expression of cytocidal activity following stimulation with poly [I:C] is accompanied by the expression of the complement component Bf, IFN-β, and the formation of nitric oxide catalyzed by the enzyme-inducible nitric oxide synthase. In response to stimulation with the inflammatory stimulus β 1,3-glucan, macrophages express increased levels of the lysosomal enzyme β-glucuronidase, the growth factors PDGF-B and TGF-β, and a receptor that recognizes phosphatidylserine and appears to be involved in the uptake of apoptotic neutrophils and lymphocytes. Stimulation with the matrix component hyaluronic acid or the fibrogenic stimulus chrysotile asbestos fibers leads to increased expression of the fibroblast growth factor, IGF-I. These pathways of development are expressed in a mutually restricted fashion. (B) Macrophage functional development is generally regulated in at least two steps. In the first step, cells are primed and exhibit increased sensitivity to a secondary ("triggering") stimulus. In the case of cytocidal activation, as illustrated here, priming is achieved by interferons α, β, or γ. Full activation is achieved in response to reception of a second signal provided by TNF-α. Neither the priming stimulus nor the triggering stimulus are individually effective in promoting activation.

which the cells failed to respond to β 1,3-glucan (Laszlo *et al.,* 1993). Conversely, exposure of macrophages to β 1,3-glucan led to a state of anergy in which the cells failed to respond to poly [I:C] (Laszlo *et al.,* 1993). These findings may have significant implications in the role of macrophages in host defense. For example, the introduction of a tumor inoculum into naive mice initiates a macrophage-rich inflammatory response at the site of injection. However, the macrophages isolated from such progressing tumors do not express cytocidal activity. By contrast, macrophages isolated from regressing tumors are spontaneously cytocidal against tumor target cells (Russell *et al.,* 1977). This finding suggests that at some pivotal point between tumor progression and tumor regression, a change in functional phenotype occurs consistent with a change from an inflammatory phenotype to a cytocidal phenotype. Interestingly, and as will be discussed in more depth later, interferons, which are required to initiate the pathway of cytocidal differentiation, express anti-inflammatory activity in a number of animal models of acute and chronic inflammation as well as in rheumatoid arthritis (Heremans and Billiau, 1989; Heremans *et al.,* 1987a,b; Mecs and Koltai, 1982; Shiozawa *et al.,* 1992). In addition, the induction of inflammatory gene expression is dependent on autocrine- and/or paracrine-acting TGF-β, which antagonizes cytocidal activation (see Section 6.4.1.3). Phenotypic anergy is only transiently expressed, and following decay of a given phenotype, macrophages regain the capacity to respond to subsequent stimulation both to the same stimulus as well as to other stimuli (Laszlo *et al.,* 1993). These findings suggest that at the single-cell level, an infiltrating macrophage may have the potential to contribute to both tissue degradation and repair in a sequential fashion.

5. The Role of Mononuclear Phagocytes in Tissue Debridement

Put very simply, the extracellular connective tissue matrix comprises a cross-linked supporting framework of collagen fibrils that confers the property of tensile strength to the tissue. To this is added a cross-linked network of elastin fibers that provide the tissue with elastic properties. Finally, the mesh of collagen and elastin fibrils is saturated with a filler substance composed of proteoglycans (which consist of a protein backbone to which are attached long-chain glycosaminoglycans) and other glycoproteins such as laminin and fibronectin that mediate the interaction between matrix and adhesive cells via integrins. Successful breakdown of the connective tissue matrix requires enzymes that are capable of degrading these major constituents. There is now abundant evidence to implicate mononuclear phagocytes as major effector cells involved in the removal and degradation of damaged connective tissue at inflammatory sites prior to, or during, the initiation of tissue reconstruction. As mentioned in Section 4.1, inflammatory macrophages synthesize and secrete a repertoire of neutral and acid pH-optimum proteases and glycosidases that play instrumental roles in this degradative process. These include activities capable of degrading collagen and elastin, as well as the ground substance proteoglycans and their constituent glycosaminoglycans. In addition, the fibroblast is an important source of several collagenolytic metalloproteinases. It seems likely that some cleavage of the extracellular connective tissue matrix is initiated extracellularly or at the cell–substratum interface, thereby generating smaller fragments of tissue debris, which can subsequently be ingested by macrophages and degraded intracellularly within the lysosomal system.

5.1. Monocyte Involvement in Matrix Degradation

In general, monocytes contain a mixed complement of tissue-degrading enzymes that overlap with those of neutrophils, macrophages, and fibroblasts and consist of metalloproteinases, serine, and cysteine proteases and acid hydrolases, all of which appear to act both intracellularly and extracellularly. Collagen degradation by monocytes appears to be mediated by a procollagenase that bears immunologic and biochemical similarities to the procollagenase of fibroblasts and alveolar macrophages, i.e., interstitial collagenase or matrix metalloproteinase-1 (MMP-1) (Campbell *et al.,* 1987). However, the levels of procollagenase secreted by monocytes are low by comparison with those of macrophages, even when the cells are stimulated with LPS or phorbol diesters (Campbell *et al.,* 1987). Monocytes and the promonocytic cell line U937 also secrete the 92-kDa type IV collagenase (MMP-9) (Welgus *et al.,* 1990), which functions to support further degradation of cleaved native collagen, and hence is important in the degradation of the damaged or partially degraded collagen found at sites of injury. The activities of monocyte MMP-1 and MMP-9 are opposed by tissue inhibitors of metalloproteinases (TIMP), of which TIMP-1 is constitutively secreted in abundant levels by monocytes (Campbell *et al.,* 1987) and which may serve to limit the activity of secreted metalloproteases. Given the fact that inflammatory cells are ubiquitously attached to elements of the surrounding connective tissue matrix and that matrix degradation is mediated at sites of cell–substratum contact that are inaccessible to antiproteases, the abundance of antiproteases within an inflammatory site would seem important to control the release of proteases from damaged or dead cells or to inhibit the activity of proteases that diffuse away from the cell of origin.

Unlike the macrophage collagen-degrading metalloproteinases, monocytes degrade elastin via serine and cysteine proteases. These include an elastase that is biochemically and antigenically indistinguishable from human neutrophil elastase (Sandhaus *et al.,* 1983), neutral pH-optimum cathepsin G (Campbell *et al.,* 1989), and cathepsin L, an acid pH-optimum cysteine protease (Etherington *et al.,* 1988; Shapiro *et al.,* 1991). In addition, plasma-derived antiproteases inhibit serine and cysteine proteases, thereby controlling their activity. Monocytes are also capable of secreting preformed lysosomal acid hydrolases, although the relative abundance of these enzymes is low compared with macrophages.

Based on the available data, the role of the monocyte in connective tissue matrix breakdown during debridement is difficult to specify precisely. The monocyte does not appear to secrete stromelysin (MMP-3) (Campbell *et al.,* 1991), a metalloproteinase that degrades proteoglycans and is required to activate the interstitial procollagenase, MMP-1. Moreover, the level of synthesis of MMP-1 is small in comparison with that of macrophages and fibroblasts. However, monocytes do secrete at least one gelatinase (92-kDa MMP-9) that can degrade damaged or cleaved collagens. Thus, it is possible that upon entering a site of injury, monocytes participate in the initial debridement of the site but do not contribute in any substantial fashion to remodeling, a process that requires the coordinated and controlled balance between matrix-degrading metalloproteinases and matrix synthesis. In addition, as monocytes differentiate into macro-

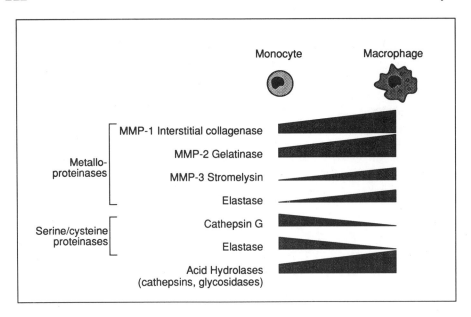

Figure 5. Changes in the types and levels of tissue-degrading serine and cysteine proteinases, lysosomal acid glycosidases, and metalloproteinases as monocytes differentiate into macrophages.

phages, the pattern and level of expression of tissue-degrading enzymes undergoes differential regulation, as illustrated in Fig. 5.

5.2. Macrophage Involvement in Extracellular Matrix Degradation

Investigations initiated in the mid-1970s identified several secreted neutral proteinase activities, including elastase, collagenase, and plasminogen activator. Plasminogen activator itself has little proteolytic activity; but in the presence of plasminogen, the potent proteolytic enzyme plasmin is generated. The role of these enzymes in the degradation of connective tissue matrices has been studied *in vitro*, using radiolabeled extracellular connective tissue matrix synthesized by vascular smooth muscle cells, endothelial cells, or fibroblasts during *in vitro* culture as substrate (Jones and Scott-Burden 1979; Werb *et al.,* 1980). Using this system, highly purified mouse macrophage elastase was found to degrade both the elastin and glycoprotein (including laminin and fibronectin) components of the connective tissue matrix, while plasmin, a product of macrophage plasminogen activation, degraded 50–70% of the glycoprotein component, but had no effect on collagen or elastin (Werb *et al.,* 1980). However, degradation of all three components of the connective tissue matrix was accomplished when peritoneal exudate macrophages were cultured on the connective tissue matrix. Ultrastructural and biochemical observations of the macrophage–connective tissue matrix interactions have revealed that matrix is initially degraded extracellularly in the immediate vicinity of the cells, although degradation can sometimes extend several millimeters from the cells (Werb *et al.,* 1980). Following their characterization as enzyme activities, recent work has led to the purification and cDNA cloning of a number of macrophage tissue-degrading enzymes.

5.2.1. Macrophage Elastinolytic Metalloproteinase

Mouse macrophage elastase is optimally active at pH 8.0, has a molecular weight of approximately 22 kDa, and is inhibited by α_2-macroglobulin but not by α_1-antitrypsin. In addition to hydrolyzing elastin, macrophage elastase is also proteolytically active against a variety of nonelastin substrates, including fibrinogen, fibrin, fibronectin, laminin, immunoglobulins, and proteoglycans (Banda *et al.,* 1983, 1985). The cDNA encoding the enzyme was cloned in 1992 by Shapiro *et al.* (1992). The proenzyme has a molecular weight of 53 kDa and undergoes proteolytic processing to yield the active enzyme by cleavage of sequences at both the C-terminal and the N-terminal. The involvement of the human counterpart of this metalloproteinase in elastin degradation was demonstrated by Senior *et al.* (1989) using alveolar macrophages. They showed that direct macrophage–substrate contact was required for elastin degradation. Furthermore, elastin degradation was blocked by dexamethasone and cycloheximide. The recent cloning of the cDNA encoding the human enzyme has shown its precursor to be similar in size to the mouse enzyme (54 kDa) and to undergo a similar mechanism of processing to yield the 22-kDa active enzyme (Shapiro *et al.,* 1993).

5.2.2. Collagenolytic Metalloproteinases

Originally recognized as an collagenolytic activity, macrophages are now known to synthesize and secrete a number of metalloproteinases that collectively play an important role in collagen degradation. Hibbs *et al.* (1987) observed that human alveolar macrophages secrete a gelatinase upon culture *in vitro.* The gelatinase had an apparent molecular weight of 90–92 kDa (MMP-9) and showed immunologic cross-reactivity to human neutrophil gelatinase. Interestingly, the protein appeared to be a major secretory product of alveolar macrophages. In a study of human macrophage metalloproteinases, Welgus *et al.* (1990) showed that human alveolar macrophages secrete interstitial collagenase (MMP-1) and stromelysin (MMP-3), although the mechanisms underlying their expression appear distinct. Stromelysin (MMP-3) was secreted only in response to stimulation of alveolar macrophages with LPS. As a general statement, these authors suggested that as macrophages became more differentiated, they secreted a broader spectrum of metalloproteinase and in greater quantities (Welgus *et al.,* 1990). However, while macrophages can undoubtedly mediate degradation of extracellular connective tissue matrix by secreting metalloproteinases and other degradative enzymes, comparison of the level of secretion of metalloproteinases induced *in vitro* with the level of secretion of many of these enzymes by fibroblasts revealed fibroblasts to be quantitatively more active in metalloproteinase secretion (Welgus *et al.,* 1990). The issue of the role of macrophages in controlling the production of metalloproteinase by fibroblasts will be addressed later in section 5.5.

5.2.3. Activities of Secreted and Cell Surface Matrix-Degrading Enzymes

Secreted lysosomal enzymes or enzymes acting at the cell–substratum interface may be involved in the degradation of the connective tissue matrix. A significant issue regarding the role of lysosomal enzymes in the degradation of the connective tissue matrix is the question of whether the pH of the local pericellular microenvironment can be reduced sufficiently to allow the acid pH-optimal lysosomal enzymes to become active. Evidence has been presented to indicate that under certain conditions, mononuclear phagocytes can acidify their immediate local environment. This appears to be true for the chondrocyte during cartilage breakdown (Dingle, 1975) and the osteoclast during bone resorption (Blair *et al.*, 1986, 1988, 1989). Indeed, Silver *et al.* (1988) and Etherington *et al.* (1981), using microelectrodes, have shown that the pH at the attachment site of a macrophage to experimental substata can rapidly drop into the range of pH 3.6–3.7. In general, lysosomal enzymes have pH optima in the 3.5–5.5 range, and thus under these conditions lysosomal acid hydrolases would be expected to be fully active.

5.3. Intracellular Degradation of the Connective Tissue Matrix

While the *in vitro* investigations discussed above have shown that connective tissue matrices can be partially degraded extracellularly (Jones and Werb, 1980; Werb *et al.*, 1980), other studies of connective tissue matrix breakdown *in vivo* (using the mouse uterus at postpartum as a model) have shown macrophages to be actively engaged in the phagocytosis of stromal collagen fibers (Parakkal, 1972), which are considerably larger than the fragments generated in *in vitro* studies. Furthermore, by localizing the lysosomal enzyme acid phosphatase to the collagen-containing organelles, it has been shown that the collagen fibers were in the process of being digested by the lysosomal system (Parakkal, 1972). Since the collagen fibers were only discernable for 24–48 hr, it was evident that the ingested collagen was rapidly digested. The reasons for the differences in the reported sizes of the connective tissue fragments produced in these two studies are unclear. However, the studies do differ fundamentally in several ways, the most notable being that (1) the composition of the connective tissue matrices produced *in vitro* may be significantly different from those laid down *in vivo*, and (2) differences almost certainly exist in the diffusional capacity and the enzyme : substrate ratios of secreted enzymes in the two models, which may result in significant differences in the rate of extracellular and intracellular degradation.

The characteristics of two lysosomal enzymes that are implicated in the intracellular degradation of collagen fragments have been studied by Etherington (1979). Cathepsins B and N are both thiol-dependent proteinases that are optimally active at pH 3.5, with negligible activity being seen above pH 4.5. Both enzymes appear to cleave the collagen molecule in the nonhelical N-telopeptide region (Etherington, 1976), and the fragments produced by the action of cathepsins B and N are then presumed to be further degraded by other lysosomal endo- and exopeptidases. Thus, the clear indication from these studies is that macrophage-dependent degradation of the connective tissue matrix probably represents a two-step process in which some degradation is initiated extracellularly, thereby producing fragments that are subsequently ingested and digested intracellularly within the lysosomal system.

5.4. Regulation of Fibroblast Metalloproteinase Secretion by Macrophages

As mentioned earlier, accumulating evidence points to the fibroblast as being an important connective tissue cell involved in collagen breakdown under physiological and pathological conditions. The secretion of collagenase is not a constitutive property of fibroblasts, but rather is an inducible response that is dependent on *de novo* protein synthesis and which appears to be tightly controlled by factors secreted by macrophages. Early evidence of macrophage regulation of fibroblast collagenase secretion was obtained by Huybrechts-Godin *et al.* (1979), who observed that coincubation of macrophages and fibroblasts on a ^{14}C-labeled collagen film led to a more rapid and extensive degradation of the collagen matrix when compared with either fibroblasts or macrophages alone. Similarly, conditioned medium from macrophages also stimulated the degradation of the collagen films by fibroblasts. These early investigations also suggested that the secretion of the macrophage factors capable of inducing collagenase secretion by fibroblasts was developmentally regulated. Maximal induction of collagenase secretion was noted during the third day of macrophage culture and declined thereafter.

Later investigations focused on the identity of the factors responsible for the induction of fibroblast collagenase synthesis (Huybrechts-Godin *et al.*, 1985). On the basis of similarities in the physical characteristics of the purified factor, several independent laboratories concluded that the active material is IL-1β (Ito *et al.*, 1988; Unemori *et al.*, 1994). This is true not only of the factor secreted by bone marrow macrophages, but also of the mononuclear cell factor that has been shown to be involved in the control of collagenase synthesis by rheumatoid synovial fibroblasts and rabbit articular chondrocytes. More recent studies have gone on to shown that a wide range of macrophage-derived cytokines, including PDGF (Hiraoka *et al.*, 1992) and TNF-α (Chua and Chua, 1990; Ito *et al.*, 1990), are capable of inducing collagen-degrading metalloproteinase secretion by fibroblasts. However, the induction and/or catalytic activity of these enzymes is controlled by other cytokines, including IFN-γ, which suppresses the IL-1β induction of stromelysin synthesis by fibroblasts (Unemori *et al.*, 1991), and IL-6 (Ito *et al.*, 1992) and TGF-β (Wright *et al.*, 1991), which stimulate the synthesis of TIMP.

6. Macrophage Involvement in Repair, Remodeling, and Fibrosis

Experiments conducted almost two decades ago, in which mononuclear phagocyte depletion in guinea pigs was shown to be associated with a dramatic impairment of wound healing (Leibovitch and Ross, 1975), provided the first concrete evidence that macrophages actively participate in the process of tissue repair and remodeling. There is little doubt that this work set the stage for the industrious pursuit in the last two decades of the mechanisms underlying the involvement of this cell type in both wound repair and pathological tissue fibrosis. Subsequent studies in multiple experimental systems have extended these seminal studies by showing that macrophages are a vital source of growth factors and cytokines that collectively induce (1) the migration and proliferation of fibroblasts, endothelial cells, and smooth muscle cells, thereby stimulating connective tissue growth and neovascularization of injured tissue, and (2) control the synthesis of connective tissue matrix components by mesenchymal cells, especially fibroblasts.

At the time of preparation of the first edition of this book in 1986, only a limited number of macrophage-derived proteins and peptides capable of stimulating fibroblast proliferation and collagen synthesis had been characterized. However, in the intervening years, the field has undergone nothing short of a renaissance with the discovery of a large number of cytokines and growth factors that have profound effects on the proliferation and differentiation of mesenchymal cells. Using reverse transcription of mRNA and the polymerase chain reaction, Rappolee *et al.* (1988) reported that macrophages isolated from subepidermally implanted wound cylinders in mice contained the transcripts for PDGF-A, TGF-β, basic fibroblast growth factor (bFGF), insulin-like growth factor-I (IGF-I), TGF-α, and IL-1β, showing that macrophages are an important source of multiple cytokines and growth factors. A indication of the breadth of cytokines and growth factors currently known to be produced by macrophages is illustrated in Table II.

As our understanding of the diversity of activities expressed by individual cytokines and growth factors has increased, it has become abundantly clear that categorizing these molecules on the basis of their ability to stimulate one activity or another is increasingly more difficult, since most cytokines and growth factors exhibit pleiotropic activities. For example, IL-2, while being traditionally viewed as a T-cell growth factor, can also stimulate macrophages to express cytocidal activity in the presence of IFN-γ (Cox *et al.*, 1990). Conversely, PDGF, while being originally thought of (and described) as a fibroblast competence-type growth factor, is now recognized as being a potent stimulus of fibroblast and monocyte chemotaxis (Pierce *et al.*, 1989). Thus,

Table II. Major Cytokines and Growth Factors Produced by Macrophages

Cytokine/growth factor	Stimulus
TNF-α	LPS
IL-1α/IL-1β	LPS
IL-6	LPS
IL-8	LPS
IL-10	LPS
IL-12	IFN-γ
MIP-1α	LPS
MIP-1β	LPS
MIP-2	LPS
IFN-β	Double-stranded RNA, viruses, LPS
IFN-α	Double-stranded RNA, viruses, LPS
PDGF-A/B	LPS
TGF-β	LPS
TGF-α	LPS
aFGF/bFGF	Scavenger receptor ligands
Heparin-binding epidermal growth factor	Constitutive
IGF-1	TNF-α, advanced-glycosylation end-products

categorizing macrophage-derived cytokines and growth factors as fibroblast growth factors, angiogenic factors, regulators of matrix synthesis, and so on would underrepresent their potential contributions to wound repair. Therefore, in what follows, the discussion has been focused (1) on the broad activities of macrophages in repair and fibrosis, and (2) on the contrasting and/or overlapping functional activities of macrophage-derived growth factors and cytokines whose expression is known to contribute to repair and fibrosis.

6.1. Mesenchymal Cell Growth Factors

From a historical perspective, once it was recognized that macrophages were a primary orchestrator of wound repair, subsequent studies soon showed that macrophages actively secrete factors that induce the proliferation of serum-deprived fibroblasts (Leibovitch and Ross, 1976). These mesenchymal cell growth factors were collectively known as macrophage-derived growth factor (MDGF) and were shown to be synthesized and secreted rather than stored intracellularly. *In vitro* studies suggested that the secretion of MDGF is a constitutive property of monocytes and macrophages, although the likelihood of trace contamination of culture materials with LPS cannot be excluded in studies from this period. However, the level of MDGF secretion was significantly increased following stimulation of macrophages with a variety of agents, including LPS, concanavalin A (Glenn and Ross, 1981), fibronectin (Martin *et al.,* 1983), phorbol diesters (Leslie *et al.,* 1984), and the phagocytic stimuli, zymosan particles and silica (Schmidt *et al.,* 1984). Macrophage-derived growth factor was also shown to stimulate the *in vitro* proliferation of other mesenchymal cells, specifically smooth muscle cells and vascular endothelial cells (Martin *et al.,* 1981; Polverini *et al.,* 1977). Subsequent studies have shown that a number of previously characterized growth factors are contained in the macrophage and monocyte conditioned media, including PDGF (Shimokado *et al.,* 1985), IL-1β (Schmidt *et al.,* 1984), bFGF (Baird *et al.,* 1985), TGF-β (Assoian *et al.,* 1987), TGF-α (Madtes *et al.,* 1988; Rappolee *et al.,* 1988), and IGF-1 (Rappolee *et al.,* 1988).

6.2. Macrophage-Derived Angiogenic Factors

Studies also initiated in the mid-1970s showed that another significant contribution of macrophages to the process of wound repair is as a source of angiogenesis factors that stimulate neovascularization of the provisional matrix. Early evidence suggesting the involvement of macrophages in wound neovascularization came from the finding that macrophages isolated from wound fluids were potently angiogenic when injected into rat corneas (Greenberg and Hunt, 1978; Thakral *et al.,* 1979). Similarly, culture supernatants of wound macrophages were also found to be angiogenic, indicating that these factors were secreted products of macrophages. Other investigations showed that freshly explanted elicited mouse peritoneal macrophages secreted angiogenesis factors (Polverini *et al.,* 1977), suggesting perhaps that this response is a general capacity of inflammatory macrophages. The following findings are consistent with this idea: (1) recently migrated peripheral blood monocytes probably acquire the same phenotype as thioglycollate-induced inflammatory macrophages as part of their differentiation and maturation; (2) macrophages isolated from solid tumors secrete angiogenesis factors (Polverini and Leibovitch, 1984), and here, also, an inflammatory response to the tumor is usually present; and (3) macrophages isolated from human rheumatoid synovia have also been found to be potent inducers of neovascularization (Koch *et al.*, 1986), while unstimulated monocytes or macrophages (as well as neutrophils and lymphocytes) failed to secrete angiogenesis factors.

The mechanism(s) controlling the secretion of angiogenesis factors were initially studied in a number of laboratories. Banda *et al.* (1982), on the basis of their findings that rabbit bone marrow-derived macrophages secrete angiogenesis factors during periods of hypoxia, suggested that the reduced oxygen tension found in the center of a healing wound might trigger newly recruited macrophages to secrete angiogenesis factors. Of significance to the validity of their hypothesis was the finding that elimination of the oxygen gradient from the hypoxic core of a wound to the oxygenated edge inhibited wound neovascularization (Knighton *et al.*, 1981, 1983). Since the secretion of angiogenesis factors ceased when macrophages were returned to 20% oxygen growth conditions, it seemed possible that the secretion of angiogenesis factor(s) was self-limiting, i.e., the response was down-regulated once neovascularization of the wound was complete. Other conditions of induction of angiogenesis factor secretion by macrophages have been determined by *in vitro* studies. For example, Polverini *et al.* (1977) found that the phagocytosis of latex spheres by resident macrophages was sufficient to induce angiogenesis factor secretion, while Koch *et al.* (1986) have shown that incubation of purified populations of human peripheral blood monocytes secreted angiogenesis factors when exposed to high concentrations of bacterial endotoxin (5 µg/ml) or concanavalin A (25 µg/ml). As with the aforementioned characterization of mesenchymal cell growth factors, the identity of several macrophage-derived angiogenesis factors has been revealed during the past 8 years, and in most situations, they also represent previously identified growth factors and cytokines to which the ability to stimulate angiogenesis represented a new functional activity. Currently characterized macrophage-derived angiogenic factors include bFGF (Baird *et al.*, 1985), TGF-α (Madtes *et al.*, 1988), TGF-β (Wiseman *et al.*, 1988), and TNF-α (Leibovitch *et al.*, 1987). In addition, recent work has shown that the neutrophil chemotactic and activating cytokine IL-8 is also a potent angiogenic factor (Koch *et al.*, 1994; Strieter *et al.*, 1992).

6.3. Macrophage Cytokines and Growth Factors Implicated in Repair and Fibrosis

Following the initial studies showing that wound macrophages secrete factors that can increase collagen synthesis in an *in vivo* corneal assay system (Hunt *et al.*, 1984), an increasing number of macrophage-derived growth factors and cytokines have been recognized by their ability to stimulate synthesis of components of the connective tissue matrix and a significant emphasis has now been placed on understanding their mechanism of action. Many of the cytokines and growth factors secreted by macrophages are pleiotropic and influence cell proliferation, angiogenesis (in which cell proliferation is an important element), and extracellular connective tissue matrix synthesis. Of these pleiotropic factors, five currently stand out as being critical to efficient wound repair, namely, TGF-β, PDGF, IGF-I, FGF, and TGF-α, and the discussion presented considers these factors in terms of their multiple individual activities.

6.3.1. TGF-β

The TGF-β represents a group of related proteins that, based on biochemical characterization and sequence analysis of cDNA clones, currently comprises five members, namely, TGF-β1, TGF-β2, TGF-β3, TGF-β4, and TGF-β5, which exhibit identities ranging from 82% (between TGF-β1 and TGF-β4) to 64% (between TGF-β2 and TGF-β4) (Roberts and Sporn, 1991). With the exception of TGF-β4, which comprises 304 amino acids, each family member is synthesized as a latent precursor protein with a molecular weight of approximately 50 kDa and consisting of 380–412 amino acids. Following homodimerization, the precursor protein is proteolytically cleaved in the COOH-terminal to yield the mature, 25-kDa active TGF-β molecule, representing a homodimer of the COOH-terminal 112 amino acid fragments. There are also data to suggest that TGF-β1 and TGF-β2 may form a heterodimer, TGF-β1.2. The mature processed dimeric TGF-β protein, however, remains noncovalently associated with the larger remnant dimeric NH_2-portion of the precursor (74 kDa), which in turn is covalently bound to a 135-kDa TGF-β1 binding protein (Wakefield *et al.*, 1988). While in this state, TGF-β remains functionally inactive. Release of active TGF-β from the latent protein complex can be achieved by extreme acidification to pH 1.5 or by proteolysis with lysosomal cathepsin D or with plasmin (Lyons *et al.*, 1988). In addition, both the TGF-β1 binding protein and the 74-kDa precursor remnant contain N-linked oligosaccharides. Hydrolysis with endo-F or sialidase or interruption of carbohydrate interactions with either mannose-6-phosphate or sialic acid have also been reported to induce release of active dimeric TGF-β from the latent complex (Miyazono and Heldin, 1989), suggesting interactions between oligosaccharide moieties and the active molecule.

6.3.1a. Macrophages Are a Major Source of TGF-β. Studies in experimental animal models have revealed macrophages to be an important source of TGF-β following their arrival in tissues 12–24 hr after the initiation of injury. In addition, macrophages likely represent the major source of the cytokine in the fibroproliferative phase of repair. During the evolution of hepatic granulomas in LEW/N rats following the intraperitoneal injection of sonicated suspensions of streptococcal cell walls, Manthey *et al.* (1990) have shown TGF-β1 protein to be present in infiltrating macrophages and in Kupffer cells as early as 18 hr after injection. Moreover, TGF-β1 was observed to colocalize intracellularly with streptococcal cell wall antigens, suggesting a localized response of Kupffer cells and infiltrating macrophages to the ingested material. Interestingly, TGF-β1 continued to be expressed as the macrophage clusters developed into more organized granulomas over time. As the granulomas became encapsulated with fibrotic tissue, extracellular staining for TGF-β1 was detected at the interface between macrophages and the fibrotic capsule (Manthey *et al.*, 1990).

The association between macrophages, TGF-β, and increased collagen deposition in human fibrotic disorders and animal models is striking. In the bleomycin model of pulmonary fibrosis, Phan and Kunkel (1992) have shown an increase in total lung TGF-β mRNA prior to the induction of increased collagen synthesis. At the protein level, Khalil *et al.* (1989) observed that total lung TGF-β protein was increased approximately 30-fold over control animals 7 days after instillation of bleomycin. Moreover, immunohistochemical staining of TGF-β in lung sections revealed intense staining of macrophages in the alveolar interstitium and in organized clusters. At later time points, TGF-β protein was localized extracellularly in areas of increased cellularity and tissue repair. Similarly in biopsy specimens obtained from patients with idiopathic pulmonary fibrosis, Khalil *et al.* (1991) have observed an increased abundance of TGF-β in macrophages and epithelial cells compared to patients without fibrosis. Similar findings have been reported in hepatic fibrosis (Nakatsukasa *et al.,* 1990) and in the fibroproliferative aspects of joint inflammation in rheumatoid arthritis (Chu *et al.,* 1992). Thus, macrophages appear to be a major source of TGF-β during repair and fibrosis.

Only recently have studies begun to identify stimuli capable of initiating the release of active and latent TGF-β by macrophages. The expression of TGF-β is stringently regulated at both the transcriptional and translational level. In unstimulated macrophages, TGF-β mRNA is constitutively expressed but translationally repressed (Assoian *et al.,* 1987; Laszlo *et al.,* 1993; Noble *et al.,* 1993a). Work first reported by Assoian *et al.* (1987) showed that incubation of human alveolar macrophages with concanavalin A or human monocytes with LPS stimulated the synthesis and release of latent TGF-β from these cells. Since that time, a number of other stimuli have been shown to induce the release of latent and/or active TGF-β from macrophages, including the inflammatory particulate stimulus, β 1,3-glucan (Noble *et al.,* 1993a), the HIV Tat protein (Zauli *et al.,* 1992), and a number of obligate and facultative intracellular parasites such as *Toxoplasma gondii* (Bermudez *et al.,* 1993) and *Mycobacterium avium intracellulare* (Bermudez, 1993), which apparently utilize macrophage-derived TGF-β to inactivate macrophage cytocidal mechanisms, thereby allowing organism survival (see Section 6.4.1c). Control of TGF-β actions is also modulated by release of the active cytokine from the latent complex, and the macrophage has many mechanisms at its disposal to mediate this process. First, as pointed out in Section 5.1, macrophages have powerful mechanisms for generating acid both intracellularly and extracellularly. Thus, in the inflammatory milieu, especially at cell–substratum contact points, macrophages may reduce pH to a level capable of inducing some release of active TGF-β dimer from the latent complex. However, Lyons *et al.* (1988) reported that at pH 4.5, only about 20–30% of active TGF-β1 was released from the latent complex. Second, macrophages are capable of secreting a wide repertoire of neutral and acid pH-optimum proteases and glycosidases (Riches and Stanworth, 1980, 1982a,b; Riches *et al.,* 1983, 1985), which may degrade the precursor remnant to release active TGF-β. Last, several studies have shown that secreted TGF-β becomes associated with protein and glycosaminoglycan components of the connective tissue matrix and that macrophages can cleave active TGF-β from matrix in a plasmin-dependent fashion at neutral pH (Falcone *et al.,* 1993a,b). Thus, in response to a variety of stimuli and conditions, both the expression of TGF-β and its release in an active form are mediated by macrophages.

6.3.1b. Inductive Actions of TGF-β on Macrophages. In addition to its potent monocyte chemotactic activity discussed earlier (Section 3.4), TGF-β exerts at least one other important inductive effect on mononuclear phagocytes. As discussed in Section 4.2, the intravenous injection of β 1,3-glucan induces a chronic inflammatory response in the liver, culminating in severe hepatic fibrosis in experimental animals. Work conducted in this laboratory has shown that *in vitro* exposure of mouse bone marrow-derived macrophages to β 1,3-glucan resulted in the increased expression of a number of inflammatory and fibrogenic gene products, including the lysosomal hydrolase β-glucuronidase (Laszlo *et al.*, 1993; Lew *et al.*, 1991) and the growth factor PDGF-B (Laszlo *et al.*, 1993; Noble *et al.*, 1993a). In work recently reported (Noble *et al.*, 1993a), we have shown that TGF-β primes macrophages for increased expression of these inflammatory gene products. Incubation of mouse bone marrow-derived macrophages with TGF-β had no direct effect on the expression of β-glucuronidase or PDGF-B. However, pretreatment with TGF-β prior to exposure to β 1,3-glucan augmented the expression of both gene products; but, more importantly, it dramatically increased the sensitivity of the macrophage to subsequent stimulation with β 1,3-glucan such that the cells were able to respond to a concentration of β 1,3-glucan, which by itself did not stimulate the expression of either β-glucuronidase or PDGF-B. In addition, the priming activity of TGF-β extended to a nonspecific phagocytic stimulus, i.e., latex particles, which by itself did not stimulate the expression of either β-glucuronidase or PDGF-B. Thus, TGF-β primes macrophages to express increased levels of inflammatory and fibrogenic gene products in response to phagocytic stimuli. We further showed that macrophage-derived TGF-β acts in a paracrine or autocrine fashion to regulate the response of these cells to phagocytic stimuli (Noble *et al.*, 1993a). Exposure of macrophages to moderate concentrations of β 1,3-glucan stimulated the increased expression of TGF-β mRNA and the secretion of TGF-β protein into culture supernatants (Noble *et al.*, 1993a). However, exposure to β 1,3-glucan in the presence of anti-TGF-β antibody substantially inhibited the increased expression of β-glucuronidase and PDGF-B, an effect not seen with anti-PDGF-B antibody. Moreover, supernatants from β 1,3-glucan-stimulated macrophages were found to prime macrophages for increased β-glucuronidase and PDGF-B mRNA expression in a fashion that was also blocked by anti-TGF-β antibody. Collectively, these results suggest that TGF-β is produced after exposure to the particulate inflammatory stimulus β 1,3-glucan and that its production enhances the expression of inflammatory and fibrogenic gene products. Clearly, given the fact that in the early phases of the inflammatory response macrophages engage in the phagocytosis and debridement of damaged connective tissue in the presence of abundant levels of platelet-derived TGF-β, these findings have important implications for the amplification of the inflammatory response and the ensuing reparative phase.

6.3.1c. "Deactivation" of Macrophage Functions by TGF-β. As illustrated in Fig. 6, and as discussed above, TGF-β can augment certain macrophage responses associated with the degradative and reparative phases of the inflammatory response. However, a growing body of work has shown that TGF-β exerts striking inhibitory

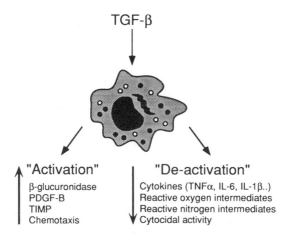

Figure 6. Opposing actions of TGF-β on monocytes and macrophages.

effects on macrophage functions, particularly on the expression of reactive oxygen intermediates and reactive nitrogen intermediates that individually and collectively are responsible for many of the cytocidal functions of macrophages with respect to the killing of bacteria, obligate intracellular organisms such as *Leishmania donovani* and *Toxoplasma gondii,* and the destruction of neoplastically transformed target cells. For example, Tsunwaski *et al.* (1988) demonstrated that pretreatment of human macrophages with TGF-β1 and TGF-β2 potently inhibited the ability of these cells to produce O_2^- in response to stimulation with phorbol myristate acetate (PMA). Using assay systems to examine the killing of amastogotes of *Leishmania donovani* and the killing of tumor target cell 1023, Nelson *et al.* (1991) showed that pretreatment of mouse macrophages with TGF-β effectively abolished killing induced by the combination of IFN-γ and LPS. Importantly, recent studies have implicated macrophage-derived TGF-β in the "deactivation" of killing mechanisms. For example, Bermudez *et al.* (1993) have shown that infection of mouse macrophages with *Toxoplasma gondii* led to the release of active TGF-β and replication of the organism within macrophages. However, infection in the presence of anti-TGF-β antibodies led to a partial inhibition of infection. Similarly, Silva *et al.* (1991) showed that macrophages secreted TGF-β in response to *Trypansoma cruzi* and that the addition of purified TGF-β to macrophage monolayers increased parasite replication. The deactivating effect of TGF-β appears to be related at least in part to down-regulation of receptors for macrophage-activating cytokines such at IFN-γ (Pinson *et al.,* 1992) and TNF-α (Bermudez *et al.,* 1993).

Taken together, these findings raise several important questions about the effects of TGF-β on macrophage functional activity in inflammation and wound repair. First, during the initial phase of the inflammatory response, when bacteria may still be present, it would seem important that the macrophage should be able to express its cytocidal activities, since infection of the inflammatory site/wound as a consequence of injury must be controlled. If so, how are the deactivating effects of TGF-β overcome since platelet-derived TGF-β is likely to be ubiquitously present? Perhaps the tissue level of TGF-β in the early phases of the inflammatory response is below the threshold to deactivate macrophage cytocidal responses and only later rises sufficiently to suppress oxygen or nitrogen free radical formation. Second, why are macrophage cytocidal functions suppressed by TGF-β? The answer to this seems more apparent since, as Tsunawaki *et al.* (1988) and Nathan (1991) have pointed out, the phagocytosis of particulate stimuli (and this may include damaged components of the connective tissue matrix) frequently elicits activation of the respiratory burst with the resulting generation of tissue and cell-damaging reactive oxygen intermediates, including O_2^-, H_2O_2, OH·, and so on. The ability of TGF-β to inhibit the generation of these tissue-damaging molecules without preventing the phagocytosis and intracellular and/or extracellular degradation of particulate stimuli seemingly would be desirable in wound repair. Third, what are the mechanisms governing these "switches" in macrophage functional activity? One mechanism involves heterologous desensitization; thus, as we have shown (Laszlo *et al.*, 1993), once the macrophage is committed to expressing a functional activity or capacity, e.g., inflammatory gene expression, it transiently loses the capacity to respond to other stimuli. Thus, during the expression of genes such as β-glucuronidase and PDGF-B, the macrophage is unable to express cytocidal activity and vice versa.

6.3.1d. Functions of TGF-β in Repair. TGF-β is a potent stimulant of connective tissue matrix accumulation, especially collagen, *in vivo* and, as might be suspected for such a pleiotropic agent, the mechanisms leading the increase in matrix abundance are many and varied and include: (1) increased collagen synthesis by fibroblasts, (2) increased synthesis of other matrix components such as glycosaminoglycans and fibronectin, (3) decreased synthesis of matrix-remodeling metalloproteinases such as collagenase, and (4) increased synthesis of TIMPs.

Roberts *et al.* (1986) have shown that subcutaneous injection of TGF-β into 1-day-old mice led to the formation of a palpable lump within 2–3 days at the site of injection, an effect that was not observed with epidermal growth factor (EGF). Histological examination of excised tissue revealed a striking induction of granulation tissue characterized by the presence of neutrophils, fibroblasts, endothelial cells, and macrophages surrounded by a newly formed collagenous network. Of some significance, the effect of injected TGF-β was reversible and self-resolving, suggesting that the availability of TGF-β is an important determinant of the duration of the fibrotic (and angiogenic) response that this agent elicits. The ability of TGF-β to stimulate synthesis of collagen by fibroblasts has also been confirmed *in vitro*. For example, exposure of fibroblasts obtained from skin and lung to TGF-β has been shown to stimulate the expression and secretion of a variety of collagen genes, including collagen types I and III (Appling *et al.*, 1989; Fine and Goldstein, 1987; Fine *et al.*, 1990; Raghow *et al.*, 1987). The increased expression of collagen genes appears to be mediated in part by increased transcription of collagen mRNAs through effects of TGF-β on nuclear factor 1 (NF1) and Sp1 binding sites in the collagen promoter (Inagaki *et al.*, 1994; Rossi *et al.*, 1988), as well as by increased stability of the procollagen mRNA (Raghow *et al.*, 1987). In addition, TGF-β has differing effects, dependent on the system studied, on the synthesis of other components of the extracellular connective tissue matrix. Using a wound-healing model in rabbits in which purified recombinant growth factors were applied to excisional ear wounds, Pierce *et al.* (1992) were unable to document a change in the level of hyaluronic acid and other glycosaminoglycans in response to TGF-β at 7 days, while a significant decrease in glycosaminoglycan content was seen at 21 days postinjury (Pierce *et al.,* 1992). However, in contrast to its effects on wound glycosaminoglycan levels, TGF-β has been shown to increase the level of fibronectin in healing wounds, and since fibronectin appears to be one avenue for fibroblast migration and proliferation in the provisional matrix, TGF-β coordinately up-regulates several mechanisms for fibroblast proliferation and matrix synthesis.

The effects of TGF-β in wound repair and fibrosis, however, are not restricted to stimulating collagen and matrix synthesis, as might be expected for such a pleiotropic cytokine. Normal turnover of the connective tissue matrix during remodeling associated with tissue growth is mediated by the concerted action of collagen-degrading metalloproteinases that mediate collagen breakdown during collagen synthesis in a controlled fashion. Work reported from several laboratories has shown that, in addition to stimulating collagen synthesis by fibroblasts, TGF-β inhibits the expression of MMP-1 and MMP-2 by fibroblasts (Overall *et al.,* 1989, 1991), a response stimulated by many agents including the growth factors PDGF (Circolo *et al.,* 1991) and IL-1β (Ito *et al.,* 1988; Unemori *et al.,* 1994). In addition, the activities of collagenase, stromelysin, and other connective tissue-degrading metalloproteinases are kept in check by a number of inhibitors, including the TIMPs. *In vitro* studies reported by Wright *et al.* (1991) have shown that TGF-β acts to stimulate the synthesis and secretion of TIMP by IL-1β-stimulated human rheumatoid synovial fibroblasts. Thus, in addition to stimulating collagen synthesis, TGF-β also prevents collagen degradation by blocking the synthesis of collagen-degrading enzymes and by increasing the synthesis of metalloproteinase inhibitors, thereby creating an imbalance between collagen synthesis and collagen degradation.

Other mechanisms whereby TGF-β may contribute to scarring and fibrosis include (1) its suppressive activities on the proliferation of epithelial cells, which, by suppressing reepithelialization of denuded basement membranes, may potentiate scarring and fibrosis that appears itself to be initiated in the absence of efficient reepithelialization; and (2) TGF-β has quite profound inhibitory activities on cells of the immune system (Kehrl *et al.,* 1986; Lotz *et al.,* 1994; Morris *et al.,* 1989; Ruegemer *et al.,* 1990). As will be discussed later, IFN-γ, a cytokine product of T helper type 1 cells (Th1 cells), has been shown to antagonize tissue collagen accumulation in several models of fibrosis. Hence, by suppressing the accumulation and/or activation of this T-cell subpopulation, TGF-β may also indirectly potentiate tissue scarring and fibrosis.

6.3.2. PDGF

The PDGF exists in either homodimeric or heterodimeric complexes as PDGF-AA, PDGF-BB, and PDGF-AB. Although platelets are able to deliver preformed PDGF during the early host response to injury, other cell types, especially macrophages, serve as a continuing source of this fibroblast competence-type growth factor during tissue repair, remodeling, and fibrosis. However, while PDGF is a potent monocyte chemotactic factor, PDGF receptor expression is not detected in macrophages (Reuterdahl *et al.*, 1993; P. W. Noble and D. W. H. Riches, unpublished observations), and thus macrophages are essentially refractory to stimulation by this growth factor. Therefore, unlike the situation with TGF-β where macrophages both express and respond to the cytokine, macrophages can only be considered a source of PDGF.

During normal wound repair, local expression of PDGF isoforms by infiltrating macrophages is greatly increased. For example, Reuterdahl *et al.* (1993), using immunofluorescence, have shown that approximately 43% of infiltrating macrophages in skin wounds express PDGF-AB/BB, whereas in normal skin, PDGF expression was restricted to peripheral nerve fibers and solitary cells in the epidermis and superficial dermis. In the exaggerated reparative response seen in idiopathic pulmonary fibrosis, Nagaoka *et al.* (1990) observed an approximately tenfold increase in PDGF-B expression by alveolar macrophages compared to normal healthy volunteers. Interestingly, there was a marked skewing of production toward PDGF-B with only about one tenth the level of expression of PDGF-A. At the tissue level, Antoniades *et al.* (1990) employed *in situ* hybridization to explore the cell types responsible for PDGF expression in idiopathic pulmonary fibrosis and showed that both alveolar macrophages and epithelial cells were significant sources of PDGF. The stimuli for PDGF expression by macrophages have not been systematically investigated, although a number of autocrine- and paracrine-acting cytokines, including IL-1β (Raines *et al.*, 1989) and IFN-γ (Shaw *et al.*, 1991) have been shown to induce PDGF expression. As mentioned earlier, work reported by this laboratory has shown that PDGF-B mRNA expression is initiated following phagocytosis of the particulate inflammatory stimulus, β 1,3-glucan (Noble *et al.*, 1993a). An extensive discussion of the role of PDGF in wound repair and its effects on mesenchymal cell proliferation is presented elsewhere in this volume.

6.3.3. IGF-I

The IGF-I is a progression-type growth factor for fibroblast proliferation that allows these and other cell types to progress through the G_1 phase of the cell cycle and to synthesize DNA. Although produced predominantly by the liver and circulating in the plasma in combination with IGF-I-binding proteins, there is now extensive evidence that IGF-I can also be expressed at increased levels by wound macrophages (Rappolee *et al.*, 1988) *in vivo* and by macrophages stimulated with a variety of fibrogenic stimuli *in vitro* (Noble *et al.*, 1993b; Rom *et al.*, 1989). For example, exposure of macrophages to chrysotile asbestos fibers stimulated the increased accumulation of IGF-I protein in culture supernatants.

Seeking to further understand the mechanism controlling IGF-I expression by macrophages, work recently conducted in this laboratory addressed the hypothesis that components of the extracellular matrix may stimulate the expression of fibrogenic growth factors, especially IGF-I. In pulmonary fibrosis and in asthma, several matrix components have been detected in cell-free brochoalveolar lavage fluid (BALF), including hyaluronic acid (Bjermer *et al.*, 1989; Nettelbladt *et al.*, 1988). Importantly, the increased abundance of hyaluronic acid in BALF is thought to precede tissue fibrosis (Nettlbladt and Hallgren, 1989). This suggested to us the possibility that hyaluronic acid may regulate IGF-I expression by macrophages. *In vitro* exposure of mouse bone marrow-derived macrophages to hyaluronic acid stimulated a two- to fourfold increase in the synthesis of IGF-I protein, with maximal stimulation being observed approximately 18–24 hr after stimulation, an effect not seen with other glycosaminoglycans, including chondriotin sulfate A or heparan sulfate (Noble *et al.,* 1993b). The interaction between hyaluronic acid and macrophages was shown to be mediated by CD44, a ubiquitous cell surface receptor for hyaluronate. In addition to stimulating the synthesis of IGF-I, hyaluronic acid stimulated the expression of TNF-α and IL-1β with a time course that preceded the expression of IGF-I. To address the possible involvement of these cytokines in the regulation of IGF-I expression, macrophages were exposed to hyaluronic acid in the presence of antibodies directed against either TNF-α or IL-1β. Antibodies directed against TNF-α but not against IL-1β inhibited the stimulation of IGF-I synthesis in response to hyaluronic acid. Furthermore, purified recombinant TNF-α was found to directly stimulate the expression of IGF-I. Taken together, these data suggest that hyaluronic acid stimulates IGF-I synthesis in a TNF-α-dependent fashion (Noble *et al.,* 1993b). In addition, given the discussion presented earlier that macrophages initially participate in the degradative phase of the inflammatory response, it is plausible that during this phase, soluble matrix components may be released that can then serve to set in motion the events that underlie the reparative phase.

6.3.4. FGF

Two functionally identical FGFs have been characterized, isolated, and cloned: namely, acidic FGF (aFGF) and basic FGF (bFGF). Both FGFs have a molecular weight of approximately 17 kDa and while lacking a conventional signal peptide are nevertheless exported to the exterior of the cell where they bind strongly to the connective tissue matrix through an interaction with heparan sulfate. Both aFGF and bFGF are potent growth factors for vascular endothelial cells, fibroblasts, and smooth muscle cells, and hence both proteins are highly angiogenic. aFGF was initially characterized in extracts of brain and pituitary and its expression for many years was thought to be restricted to these sites. By contrast, bFGF has been detected in many cells and tissues including macrophages.

Constitutive expression of bFGF by mouse peritoneal exudate macrophages was first reported by Baird *et al.* (1985), who detected the growth factor by immunocytochemistry. Similar results were obtained by Joseph-Silverstein *et al.* (1988), using a radioimmunoassay for bFGF. Analyses of FGF expression in animal models of injury and repair have revealed infiltrating macrophages to be a major cellular source of bFGF. Rappolee *et al.* (1988) detected bFGF mRNA in macrophages isolated from subepidermally implanted wound cylinders, although expression of bFGF protein was not investigated in this report. More recently, Henke *et al.* (1993) studied bFGF mRNA and protein expression in the fibroproliferative response initiated after acute lung injury in humans. Polymerase chain reaction (PCR) and Northern blot analysis of RNA isolated from alveolar macrophages revealed the presence of two species of bFGF mRNA, and biosynthetic labeling with [^{35}S]methionine followed by immunoprecipitation with anti-bFGF antibody revealed the presence of the 18-kDa protein. Of significance, immunocytochemical staining of postmortem lung tissue revealed the presence of numerous bFGF-staining macrophages in alveoli containing variable degrees of fibroplasia and vascular endothelial cell growth. Thus, alveolar macrophages appear to be an important source of bFGF in the induction of alveolar fibrosis following lung injury. Although it has generally been accepted that expression of aFGF is restricted to the brain and neuronal tissue, a recent report has also suggested that macrophages may, under certain conditions, express aFGF. In a study of FGF expression in atheromatous and nonatheromatous human arteries, Brogi *et al.* (1993) detected marked expression of aFGF in atheroma but not in control arteries. Expression of aFGF was greatest in areas of neovascularization and in macrophage-rich regions of plaque. Moreover, the expression of aFGF by macrophages was extended by showing that the human monocytic cell line THP-1 expressed aFGF mRNA following stimulation with the phorbol ester, PMA. Thus, these collective data indicate that human macrophages are capable of expressing aFGF and suggest that plaque macrophages may contribute to the neovascularization seen in human atheroma. Clearly, this area needs to be explored further to question if aFGF also contributes to macrophage-mediated neovascularization in other situations initiated by injury.

The question of how FGFs interact with responsive cells has been debated for several years. It is not yet clear how FGFs gain access to the extracellular compartment in the absence of a classical signal peptide. Possibilities include (1) a mechanism analogous to that of IL-1β, which, like FGF, lacks a signal peptide, or (2) release from dead or apoptotic cells. Once in the extracellular environment, FGF is strongly bound by elements of the connective tissue matrix and while in this state appears to be incapable of stimulating cell proliferation. For example, using immunocytochemistry, bFGF has been shown to be bound to basement membranes of blood vessels, although endothelial cell proliferation is not seen. Recently reported work suggests that macrophages may play an important role in releasing bFGF from extracellular connective tissue matrix and rendering the growth factor available for growth stimulation (Falcone *et al.*, 1993a,b). The enzyme responsible appeared to be a membrane-associated urokinaselike plasminogen activator since release of bFGF from cell-derived matrices was markedly enhanced by the addition of purified plasminogen. The released growth factor was not inactivated during its release since biologically active bFGF accumulated in culture supernatants. Interestingly, this property was not restricted to bFGF and, as alluded to earlier, matrix-bound TGF-β was released under the same conditions. Furthermore, purified soluble TGF-β, though not bFGF, TNF-α, or IL-1β, stimulated the activity of the macrophage urokinaselike plasminogen activator, indicating that TGF-β may be capable of amplifying the release of growth factors and/or cytokines from the extracellular tissue matrix (Falcone *et al.*, 1993a).

6.3.5. TGF-α

TGFα is a comparatively small macrophage-derived growth factor with a molecular weight of approximately 5–6 kDa. Sequence analysis has revealed that TGF-α shares almost 40% homology with EGF and interacts with responsive cells via the EGF receptor. TGF-α is a growth factor for endothelial cells, epithelial cells, and fibroblasts and stimulates angiogenesis *in vivo*. TGF-α has been reported to be secreted by alveolar macrophages stimulated *in vitro* with endotoxin (Madtes *et al.*, 1988) and is thought to mediate the induction of liver metallothionin following systemic injection of endotoxin in rats (Iijima *et al.*, 1989). In addition, stimulation of human tonsillar macrophages with GM-CSF resulted in a marked stimulation of TGF-α expression as detected by Northern analysis of cells and immunoassays of culture supernatants (Zhu *et al.*, 1991). Work conducted in the promyelocytic cell line HL-60 has shown that differentiation along the monocytic pathway in response to PMA results in an induction of TGF-α expression, leading to both secretion and intracellular retention of the newly synthesized protein (Davies *et al.*, 1990). Interestingly, in this study, evidence was presented to suggest that proteases mediated the release of TGF-α from the cells, since in the presence of protease inhibitors, there was a decrease in the secretion of TGF-α and a reciprocal and concurrent increase in intracellular levels.

In vitro studies have provided some important insights into the conditions under which macrophages secrete TGF-α, although few studies have investigated TGF-α production *in vivo*. Rappolee *et al.* (1988) detected TGF-α mRNA and TGF-α protein in macrophages isolated from wound cylinders, indicating that TGF-α is produced *in vivo* by macrophages during wound repair. In an analysis of TGF-α protein and mRNA expression by human adult and fetal tissues, Yasui *et al.* (1992) detected constitutive expression of TGF-α in epithelial cells of the gut, liver, pancreas, kidney, skin, adrenal, and mammary gland. Constitutive expression of TGF-α was also detected in macrophages of the lung and spleen. However, given the potential contribution of TGF-α to epithelial growth, additional studies will be required to investigate if and where TGF-α levels are detected in response to injury and during repair and fibrosis.

7. Cytokine Control of Repair Processes

With the greatly improved understanding of the macrophage growth factors and cytokines that stimulate mesenchymal cell proliferation, angiogenesis, and extracellular connective tissue matrix synthesis, attention is now being focused on identifying the mechanisms that both positively and negatively control the expression of these mediators. As might be expected for such a complex process, the mechanisms of control are anything but simple and involve multiple cell types of mesenchymal, myeloid, and lymphoid origin as well as structural motifs found in the extracellular connective tissue matrix. For the sake of simplicity and brevity, the focus of this section will be on factors known to induce repair and fibrosis and on emerging but limited data concerning the antagonism of this response.

7.1. Positive Regulation

The involvement of TNF-α in inflammation, repair, and fibrosis has been extensively studied. However, while several studies initially suggested that TNF-α serves as a growth factor for fibroblasts and as an angiogenesis factor (Leibovitch *et al.*, 1987), it is now recognized that many of these effects are indirect and require the expression of other growth factors and cytokines. For example, although TNF-α has some growth-promoting activity, it also induces the expression of PDGF by fibroblasts and IGF-I by macrophages. In addition, TNF-α induces the expression of IL-6, which has also been shown to induce the expression of PDGF. Therefore, perhaps a more accurate description of one of the roles of TNF-α is as a mediator of growth factor production.

The involvement of TNF-α in mediating tissue fibrosis was shown several years ago in the induction of pulmonary fibrosis in response to bleomycin and silica (Piguet *et al.*, 1989, 1990). In both models, TNF-α mRNA was detected in mouse lung within a few days of instillation of the stimulus. Importantly, passive immunization with anti-TNF-α antibodies completely prevented the induction of pulmonary fibrosis as measured by lung hydroxyproline levels and severely retarded the inflammatory response. Further studies are clearly required to distinguish between the potential involvement of TNF-α in (1) the genesis of the chronic inflammatory response leading to essentially irreversible epithelial damage, which may favor a fibrogenic response, and (2) the direct or indirect involvement of TNF-α in the initiation of abnormal collagen accumulation. Given the observed opposing actions of TNF-α and TGF-β, in which TNF-α amplifies the inflammatory response and TGF-β antagonizes these events (Flynn and Palladino, 1992), the former possibility would appear more likely.

An additional area for further study concerns the role of the immune system in repair and fibrosis. In normal repair processes, for example, in the skin, the involvement of the immune system seems very limited and indeed may not be required at all. However, there is increasing evidence that activation of the immune system contributes to tissue fibrosis. Piguet *et al.* (1989, 1990) showed that the combined depletion of CD4 and CD8 cells prevented the induction of pulmonary fibrosis and chronic alveolitis in mice instilled intratracheally with bleomycin. Furthermore, Hu and Stein-Streilein (1993) have developed a cell-mediated model of pulmonary fibrosis in hamsters and mice in which contact sensitivity to trinitrophenol is induced by skin painting. The pulmonary fibrosis that occurs following intratracheal instillation of the hapten is dependent on both CD4 and CD8 cells, and sensitivity can be adoptively transferred by immune spleen cells. Future studies need to be conducted to determine how the immune system mediates fibrosis in these model system. For example, do specific T-cell subset-derived cytokines favor inflammatory responses or fibrogenic responses? Interestingly, IL-4 transgenic mice experience spontaneous fibrosis of the eyelid (Dvorak *et al.*, 1994). The use of other cytokine transgenic and cytokine (and their receptor) "knockout" mice should allow a better understanding of these events, particularly with reference to how T-cell-derived cytokines control macrophage functions.

7.2. Negative Regulation

How the repair process is negatively regulated is far from clear. However, some details are beginning to emerge. In a study of the effects of IFNs and IFN inducers on the induction of pulmonary fibrosis in mice, Hyde and co-workers observed that pretreatment of animals with the IFN inducer poly [I:C] (Hyde and Giri, 1990) or with recombinant IFN-γ (Hyde *et al.*, 1988) dramatically inhibited the formation of collagen, as measured by total lung hydroxyproline. The responding cell population was not identified in these studies. In *in vitro* studies of the effects of IFNs on growth factor expression by macrophages, we have shown that IFN-β and IFN-γ abrogated the stimulation of IGF-I synthesis by mouse bone marrow-derived macrophages in response to both hyaluronic acid and TNF-α (Lake *et al.*, 1994). Interestingly, we found that coincident with the down-regulation of IGF-I synthesis, the synthesis of a gene, Bf, whose expression accompanies macrophage cytocidal activation, was up-regulated. Thus, in the absence of IFNs, TNF-α stimulated IGF-I synthesis but not Bf synthesis, while in the presence of IFNs, IGF-I synthesis ceased and the synthesis of Bf was initiated (Lake *et al.*, 1994). These data imply that IFNs mediate a "switch" from growth factor production to a pattern of gene expression consistent with cytocidal activation. Interestingly, the effects of IFNs on the suppression of growth factor expression was not restricted simply to IGF-I. In other work we also have shown that IFNs inhibit the expression of PDGF-B and TGF-β (L. Weinstein and D. W. H. Riches, unpublished observations). Clearly, additional studies are required to allow a broader and more accurate picture of the mechanisms leading to the cessation of repair and fibrosis. The immune system and mononuclear phagocyte system would appear to be logical targets for further study.

8. Concluding Comments

The experimental data that have been discussed establish that the blood-derived mononuclear phagocyte is essential to the execution of both the degradative and reparative phases of wound repair, and there is now abundant data to indicate which macrophage-derived enzymes, cytokines, and growth factors are the major players in this sequence of events. However, there remain multiple challenges in the future. We still have few clues as to why some tissues respond to injury with a complete restoration of structure and function, while other tissues respond with the formation of scar tissue. While scar formation may be desirable in some situations such as the skin, the response is clearly detrimental in other situation such as in the lung or in the liver. Is scar formation the result of an increased level of collagen synthesis by fibroblasts, a decreased level of collagenase expression and/or activity, or both? What is the involvement of the immune system? Normal wound repair occurs in the absence of significant infiltration of immune cells. What are the functions of T cells and B cells in tissue fibrosis? Mast cells are present in increased levels during tissue fibrosis. What is their role, if any? And finally, since macrophages are necessary for both repair and fibrosis, how do the different cell populations mentioned above influence macrophage activity and vice versa? These and other questions are guaranteed to keep biologists busy for many years to come.

ACKNOWLEDGMENTS. I would like to express my thanks to my friends and esteemed colleagues who collectively have made outstanding contributions to the work undertaken in my laboratory over the past 8 years. Special thanks go to Tom Bost, Philippe Carre, Ed Chan, Peter Henson, Lori Kittle, Tania Khan, Fiona Lake, Dan Laszlo, Betty Lew, Jose Martinez, Paul Noble, Linda Remigio, Carol Sable, Joe Spahn, Greg Underwood, Catherine de Whalley, and Brent Winston. This work has been funded by PHS grants HL27353 and CA50107 from the National Institutes of Health.

References

Antoniades, H. N., Bravo, M. A., Avila, R. E., Galanopoulos, T., Neville-Golden, J., Maxwell, M., and Selman, M., 1990, Platelet-derived growth factor in idiopathic pulmonary fibrosis, *J. Clin. Invest.* **86:**1055–1064.

Appling, W. D., O'Brien, W. R., Johnston, D. A., and Duvic, M., 1989, Synergistic enhancement of type I and III collagen production in cultured fibroblasts by transforming growth factor-beta and ascorbate, *FEBS Lett.* **250:**541–544.

Assoian, R. K., Fleurdelys, B. E., Stevenson, H. C., Miller, P. J., Madtes, D. K., Raines, E. W., Rooss, R., and Sporn, M. B., 1987, Expression and secretion of type beta transforming growth factor by activated human macrophages. *Proc. Natl. Acad. Sci. USA* **84:**6020–6024.

Baggiolini, M., Dewald, B., and Moser, B., 1994, Interleukin-8 and related chemotactic cytokines—CXC and CC chemokines, *Adv. Immunol.* **55:**97–179.

Baird, A., Mormede, P., and Böhlen, P., 1985, Immunoreactive fibroblast growth factor in cells of peritoneal exudate suggests its identity with macrophage-derived growth factor, *Biochem. Biophys. Res. Commun.* **126:**358–364.

Banda, M. J., Knighton, D. R., Hunt, D. R., and Werb, Z., 1982, Isolation of a nonmitogenic angiogenesis factor from wound fluid, *Proc. Natl. Acad. Sci. USA* **79:**7773–7777.

Banda, M. J., Clark, E. J., and Wegrb, Z., 1983, Selective proteolysis of immunoglobulins by mouse macrophage elastase, *J. Exp. Med.* **157:**1184–1196.

Banda, M. J., Clark, E. J., and Werb, Z., 1985, Macrophage elastase: Regulatory consequences of the proteolysis of non-elastin substrates, in: *Mononuclear Phagocytes: Characteristics, Physiology and Function,* (R. van Furth, ed.) pp. 295–300, Martinus Nijhoff, Holland, Dordrecht.

Bar-Shavit, R., Kahn, A., Fenton, J. W., and Wilner, G. D., 1983, Chemotactic response of monocytes to thrombin, *J. Cell Biol.* **96:**282–285.

Bermudez, L. E., 1993, Production of transforming growth factor-beta by *Mycobacterium avium*-infected human macrophages is associated with unresponsiveness to IFN-gamma, *J. Immunol.* **150:**1838–1845.

Bermudez, L. E., Covaro, G., and Remington, J., 1993, Infection of murine macrophages with *Toxoplasma gondii* is associated with release of transforming growth factor beta and down-regulation of expression of tumor necrosis factor receptors, *Infect. Immun.* **61:**4126–4130.

Bianco, C., 1983, Fibrin, fibronectin and macrophages, *Ann. NY Acad. Sci.* **408:**602–609.

Bjermer, L., Lundgren, R., and Hallgren, R., 1989, Hyaluron and type III procollagen peptide concentrations in bronchoalveolar lavage fluid in idiopathic pulmonary fibrosis, *Thorax* **44:**126–131.

Blair, H. C., Kahn, A. J., Crouch, E. C., Jeffrey, J. J., and Teitelbaum, S. L., 1986, Isolated osteoclasts resorb the organic and inorganic components of bone, *J. Cell Biol.* **102:**1164–1172.

Blair, H. C., Teitelbaum, S. L., Schimke, P. A., Konsek, J. D., Koziol, C. M., and Schlesinger, P. H., 1988, Receptor-mediated uptake of a mannose-6-phosphate bearing glycoprotein by isolated chicken osteoclasts, *J. Cell Physiol.* **137:**476–482.

Blair, H. C., Teitelbaum, S. L., Ghiselli, R., and Gluck, S., 1989, Osteoclastic bone resorption by a polarized vacuolar proton pump, *Science* **245:**855–857.

Blusse van Oud Alblas, A., van der Linden-Schrever, B., and van Furth, R., 1983, Origin and kinetics of pulmonary macrophage during an inflammatory reaction induced by intra-alveolar administration of aerosolized heat-killed BCG, *Am. Rev. Respir. Dis.* **128:**276–281.

Bouwens, L., and Wisse, E., 1985, Proliferation, kinetics, and fate of monocytes in rat liver during a zymosan-induced inflammation, *J. Leukocyte Biol.* **37:**531–544.

Brogi, E., Winkles, J. A., Underwood, R., Clinton, S. K., Alberts, G. F., and Libby, P., 1993, Distinct patterns of expression of fibroblast growth factors and their receptors in human atheroma and nonatherosclerotic arteries. Association of acidic FGF with plaque microvessels and macrophages, *J. Clin. Invest.* **92:**2408–2418.

Campbell, E. J., Cury, J. D., Lazarus, C. J., and Wegus, H. G., 1987, Monocyte procollagenase and tissue inhibitor of metalloproteinases. Identification, characterization, and regulation of secretion, *J. Biol. Chem.* **262:**15862–15868.

Campbell, E. J., Silverman, E. K., and Campbell, M. A., 1989, Elastase and cathepsin G of human monocytes. Quantification of cellular content, release in response to stimuli, and heterogeneity in elastase-mediated proteolytic activity, *J. Immunol.* **143:**2961–2970.

Campbell, E. J., Cury, J. D., Shapiro, S. D., Goldberg, G. I., and Welgus, H. G., 1991, Neutral proteinases of human mononuclear phagocytes. Cellular differentiation markedly alters cell phenotype for serine proteinases, metalloproteinases, and tissue inhibitor of metalloproteinases, *J. Immunol.* **146:**1286–1293.

Carre, P. C., Mortensen, R. L., King, Jr., T. E., Noble, P. W., Sable, C. L., and Riches, D. W. H., 1991, Increased expression of the interleukin-8 gene by alveolar macrophages in idiopathic pulmonary fibrosis. A potential mechanism for the recruitment and activation of neutrophils in lung fibrosis, *J. Clin. Invest.* **88:**1802–1810.

Carre, P. C., King, Jr., T. E., Mortensen, R., and Riches, D. W. H., 1994, Cryptogenic organizing pneumonia: Increased expression of interleukin-8 and fibronectin genes by alveolar macrophages, *Am. J. Resp. Cell Mol. Biol.* **10:**100–105.

Chu, C. Q., Field, M., Allard, S., Abney, E., Feldmann, M., and Maini, R. N., 1992, Detection of cytokines at the cartilage/pannus junction in patients with rheumatoid arthritis: Implications for the role of cytokines in cartilage destruction and repair, *Br. J. Rheumatol.* **31:**653–661.

Chua, C. A., and Chua, B. H., 1990, Tumor necrosis factor-alpha induces mRNA for collagenase and TIMP in human skin fibroblasts, *Connect. Tissue Res.* **25:**161–170.

Cianciolo, G. J., and Snyderman, R., 1981, Monocyte responsiveness to chemotactic stimuli is a property of a subpopulation of cells that can respond to multiple chemoattractants, *J. Clin. Invest.* **67:**60–68.

Circolo, A., Welgus, H. G., Pierce, G. F., Kramer, J., and Strunk, R. C., 1991, Differential regulation of the expression of proteinases/antiproteinases in fibroblasts. Effects of interleukin-1 and platelet-derived growth factor, *J. Biol. Chem.* **266:**12283–12288.

Clark, R. A. F., Lanigan, J. M., and Dellepella, P., 1982, Fibronectin and fibrin provide a provisional matrix for epidermal cell migration during wound reepithelialization, *J. Invest. Dermatol.* **70:**264–269.

Coggle, J. E., and Tarling, J. D., 1984, The proliferation kinetics of pulmonary alveolar macrophages, *J. Leukoc. Biol.* **35:**317–327.

Cox, G. W., Mathieson, B. J., Giardina, S. L., and Varesio, L., 1990, Characterization of IL-2 receptor expression and function on murine macrophages, *J. Immunol.* **145:**1719–1726.

Daems, W. T., 1980, Peritoneal macrophages, in: *The Reticuloendothelial System. A Comprehensive Treatise, Morphology, Vol. 1: Morphology.* (I. Carr and W. Th. Daems) pp. 57–127, Plenum Press, New York.

Davies, D., Farmer, S., and Alexander, P., 1990, Synthesis and release of TGF alpha from the myeloid leukemia cells h160 treated with phorbol ester (meeting abstract), *Br. J. Cancer* **62:**490.

De Whalley, C. V., and Riches, D. W. H., 1991, Influence of the cytocidal macrophage phenotype on the degradation of acetylated low density lipoproteins: Dual regulation of scavenger receptor activity and of intracellular degradation of endocytosed ligand, *Exp. Cell Res.* **192:**460–468.

Deimann, W., and Fahimi, H. D., 1980, Hepatic granulomas induced by glucan: An ultrastructural and peroxidase-cytochemical study, *Lab. Invest.* **43:**172–181.

Deuel, T. F., Senior, R. M., Chang, D., Griffin, G. L., Heinrikson, R. L., and Kaiser, E. T., 1981, Platelet factor 4 is chemotactic for neutrophils and monocytes, *Proc. Natl. Acad. Sci. USA* **78:**4584–4587.

Ding, A. H., Nathan, C. F., and Stuehr, D. J., 1988, Release of reactive nitrogen intermediates and reactive oxygen intermediates from mouse peritoneal macrophages. Comparison of activating cytokines and evidence for independent production, *J. Immunol.* **141:**2407–2412.

Dingle, J. T., 1975, The secretion of enzymes into the pericellular environment, *Phil. Trans. R. Soc. Lond. Biol.* **271:**315–324.

Doherty, D. E., Haslett, C., Tonneson, M. G., and Henson, P. M., 1987, Human monocyte adherence: A primary effect of chemotactic factors on the monocyte to stimulate adherence to human endothelium, *J. Immunol.* **138**:1762–1771.

Donaldson, D. J., and Mahan, J. T., 1983, Fibrinogen and fibronectin as substrates for epidermal cell migration during wound closure, *J. Cell Sci.* **39**:117–127.

Dvorak, A. M., Tepper, R. I., Weller, P. F., Morgan, E. S., Estrella, P., Monahan, E. R., and Galli, S. J., 1994, Piecemeal degranulation of mast cells in the inflammatory eyelid lesions of interleukin-4 transgenic mice. Evidence of mast cell histamine release *in vivo* by diamine oxidase-gold enzyme-affinity ultrastructural cytochemistry, *Blood* **83**:3600–3612.

Etherington, D. J., 1976, Bovine spleen cathepsin B1 and collagenolytic cathepsin: A comparative study of the properties of the two enzymes in the degradation of native collagen, *Biochem. J.* **153**:100–109.

Etherington, D. J., 1979, Proteinases in connective tissue breakdown, *Ciba Found. Symp.* **1979**:87–103.

Etherington, D. J., Pugh, D., and Silver, I. A., 1981, Collagen degradation in an experimental inflammatory lesion: Studies on the role of the macrophage, *Acta Biol. Med.* **40**:1625–1636.

Etherington, D. J., Taylor, M. A., and Henderson, B., 1988, Elevation of cathepsin L levels in the synovial lining of rabbits with antigen-induced arthritis, *Br. J. Exp. Pathol.* **69**:281–289.

Falcone, D. J., McCaffrey, T. A., Haimovitz, F. A., and Garcia, M., 1993a, Transforming growth factor-beta 1 stimulates macrophage urokinase expression and release of matrix-bound basic fibroblast growth factor, *J. Cell. Physiol.* **155**:595–605.

Falcone, D. J., McCaffrey, T. A., Haimovitz, F. A., Vergilio, J. A., and Nicholson, A. C., 1993b, Macrophage and foam cell release of matrix-bound growth factors. Role of plasminogen activation, *J. Biol. Chem.* **268**:11951–11958.

Falk, W., and Leonard, E. J., 1980, Human monocyte chemotaxis: Migrating cells are a subpopulation with multiple chemotaxin specificities on each cell, *Infec. Immun.* **29**:953–959.

Falk, W., and Leonard, E. J., 1981, Specificity and reversibility of chemotactic deactivation of human monocytes, *Infect. Immun.* **32**:464–468.

Fine, A., and Goldstein, R. H., 1987, The effect of transforming growth factor-beta on cell proliferation and collagen formation by lung fibroblasts, *J. Biol. Chem.* **262**:3897–3902.

Fine, A., Poliks, C. F., Smith, B. D., and Goldstein, R. H., 1990, The accumulation of type I collagen mRNAs in human embryonic lung fibroblasts stimulated by transforming growth factor-beta, *Connect. Tissue Res.* **24**:237–247.

Flynn, R. M., and Palladino, M. A., 1992, TNF and TGF-b: The opposite sides of the avenue? in: *Tumor Necrosis Factors. The Molecules and Their Emerging Role in Medicine,* (B. Beutler, ed.), pp. 131–144, Raven Press, New York.

Ford, H. A., Bray, M. A., Doig, M. V., Shipley, M. E., and Smith, M. J., 1980, Leukotriene B, a potent chemokinetic and aggregating substance released from polymorphonuclear leukocytes, *Nature* **286**:264–265.

Glenn, K. C., and Ross, R., 1981, Human monocyte-derived growth factor(s) for mesenchymal cells: Activation of secretion by endotoxin and concanavalin A, *Cell* **25**:603–615.

Goud, T. J. L. M., and van Furth, R., 1975, Proliferative characteristics of monoblasts grown *in vitro, J. Exp. Med.* **142**:1200–1217.

Goud, T. J. L. M., Schotte, C., and van Furth, R., 1975, Identification and characterization of the monoblast in mononuclear phagocyte colonies grown *in vitro, J. Exp. Med.* **142**:1180–1198.

Greenberg, G. B., and Hunt, T. K., 1978, The proliferative response *in vitro* of vascular endothelial and smooth muscle cells exposed to wound fluid and macrophages, *J. Cell. Physiol.* **97**:353–360.

Henke, C., Marineili, W., Jessurun, J., Fox, J., Harms, D., Peterson, M., Chiang, L., and Doran, P., 1993, Macrophage production of basic fibroblast growth factor in the fibroproliferative disorder of alveolar fibrosis after lung injury, *Am. J. Pathol.* **143**:1189–1199.

Heremans, H., and Billiau, A., 1989, The potential role of interferons and interferon antagonists in inflammatory disease, *Drugs* **38**:957–972.

Heremans, H., Billiau, A., Coutelier, J. P., and De, S. P., 1987a, The inhibition of endotoxin-induced local inflammation by LDH virus or LDH virus-infected tumors is mediated by interferon, *Proc. Soc. Exp. Biol. Med.* **185**:6–15.

Heremans, H., Dijkmans, R., Sobis, H., Vandekerckhove, F., and Billiau, A., 1987b, Regulation by interferons of the local inflammatory response to bacterial lipopolysaccharide, *J. Immunol.* **138**:4175–4179.

Hibbs, M. S., Hoidal, J. R., and Kang, A. H., 1987, Expression of a metalloproteinase that degrades native type V collagen and denatured collagens by cultured human alveolar macrophages, *J. Clin. Invest.* **80:**1644–1650.

Hiraoka, K., Sasaguri, Y., Komiya, S., Inoue, A., and Morimatsu, M., 1992, Cell proliferation-related production of matrix matalloproteinases 1 (tissue collagenase) and 3 (stromelysin) by cultured human rheumatoid synovial fibroblasts, *Biochem. Int.* **27:**1083–1091.

Hosein, B., Mosesson, M. W., and Bianco, C., 1985, Monocyte receptors for fibronectin, in: *Mononuclear Phagocytes: Characteristics, Physiology and Function* (R. van Furth, ed.), pp. 723–730, Martinus Nijhoff, Dordrecht, Holland.

Hu, H., and Stein-Streilein, J., 1993, Hapten-immune pulmonary interstitial fibrosis (HIPIF) in mice requires both CD4+ and CD8+ T lymphocytes, *J. Leukocyte Biol.* **54:**414–422.

Hunt, T. K., Knighton, D. R., Thakral, K. K., Goodson, W. H., and Andrews, W. S., 1984, Studies on inflammation and wound healing: Angiogenesis and collagen synthesis stimulated *in vivo* by resident and activated wound macrophages, *Surgery* **96:**48–54.

Huybrechts-Godin, G., Hauser, P., and Vaes, G., 1979, Macrophage–fibroblast interaction in collagenase production and cartilage degradation, *Biochem. J.* **184:**643–650.

Huybrechts-Godin, G., Peeters-Joris, C., and Vaes, G., 1985, Partial characterization of the macrophage factor that stimulates fibroblasts to produce collagenase and to degrade collagen, *Biochim. Biophys. Acta* **846:**51–54.

Hyde, D. M., and Giri, S. N., 1990, Polyinosinic-polycytidylic acid, an interferon inducer, ameliorates bleomycin-induced lung fibrosis in mice, *Exp. Lung Res.* **16:**533–546.

Hyde, D. M., Henderson, T. S., Giri, S. N., Tyler, N. K., and Stovall, M. Y., 1988, Effect of murine gamma interferon on the cellular response to bleomycin in mice, *Exp. Lung Res.* **14:**687–704.

Iijima, Y., Fukushima, T., and Kosaka, F., 1989, Involvement of transforming growth factor-alpha secreted by macrophages in metallothionein induction by endotoxin, *Biochem. Biophys. Res. Commun.* **164:**114–118.

Imber, M. J., Pizzo, S. V., Johnson, W. J., and Adams, D. O., 1982, Selective diminution of the binding of mannose by murine macrophages in the late stages of activation, *J. Biol. Chem.* **257:**5129–5135.

Inagaki, Y., Truter, S., and Ramirez, F., 1994, Transforming growth factor-beta stimulates alpha 2(I) collagen gene expression through a *cis*-acting element that contains an Sp1-binding site, *J. Biol. Chem.* **269:**14828–14834.

Ishida, M., Honda, M., and Hayashi, H., 1978, *In vitro* macrophage chemotactic generation from serum immunoglobulin G by neutrophil neutral seryl protease, *Immunology* **35:**167–176.

Ito, A., Goshowaki, H., Sato, T., Mori, Y., Yamashita, K., Hayakawa, T., and Nagase, H., 1988, Human recombinant interleukin-1 alpha-mediated stimulation of procollagenase production and suppression of biosynthesis of tissue inhibitor of metalloproteinases in rabbit uterine cervical fibroblasts, *FEBS Lett.* **234:**326–330.

Ito, A., Sato, T., Iga, T., and Mori, Y., 1990, Tumor necrosis factor bifunctionally regulates matrix metalloproteinases and tissue inhibitor of metalloproteinases (TIMP) production by human fibroblasts, *FEBS Lett.* **269:**93–95.

Ito, A., Itoh, Y., Sasaguri, Y., Morimatsu, M., and Mori, Y., 1992, Effects of interleukin-6 on the metabolism of connective tissue components in rheumatoid snovial fibroblats, *Arthritis Rheum.* **35:**1197–1201.

Jones, P. A., and Scott-Burden, T., 1979, Activated macrophages digest the extracellular matrix proteins produced by cultured cells, *Biochem. Biophys. Res. Commun.* **86:**71–77.

Jones, P. A., and Werb, Z., 1980, Degradation of connective tissue matrices by macrophages. II. Influence of matrix composition on proteolysis of glycoproteins, elastin and collagen by macrophages in culture, *J. Exp. Med.* **152:**1527–1536.

Joseph-Silverstein, J., Moscatelli, D., and Rifkin, D. B., 1988, The development of a quantitative RIA for basic fibroblast growth factor using polyclonal antibodies against the 157 amino acid form of human bFGF. The identification of bFGF in adherent elicited murine peritoneal macrophages, *J. Immunol. Methods* **110:**183–192.

Kay, A. B., Pepper, D. S., and Ewart, M. R., 1973, Generation of chemotactic activity for leukocytes by the action of thrombin of human fibrinogen, *Nature* **243:**56–57.

Kehrl, J. H., Wakefield, L. M., Roberts, A. B., Jakowlew, S., Alvarez, M. M., Derynck, R., Sporn, M. B., and

Fauci, A. S., 1986, Production of transforming growth factor beta by human T lymphocytes and its potential role in the regulation of T cell growth, *J. Exp. Med.* **163**:1037–1050.

Khalil, N., Bereznay, O., Sporn, M., and Greenberg, A. H., 1989, Macrophage production of transforming growth factor beta and fibroblast collagen synthesis in chronic pulmonary inflammation, *J. Exp. Med.* **170**:727–737.

Khalil, N., O'Connor, R. N., Unruh, H. W., Warren, P. W., Flanders, K. C., Kemp, A., Bereznay, O. H., and Greenberg, A. H., 1991, Increased production and immunohistochemical localization of transforming growth factor-beta in idiopathic pulmonary fibrosis, *Am. J. Respir. Cell Mol. Biol.* **5**:155–162.

Knighton, D. R., Silver, I. A., and Hunt, T. K., 1981, Regulation of wound-healing angiogenesis—Effect of oxygen gradients and inspired oxygen concentration, *Surgery* **90**:262–270.

Knighton, D. R., Hunt, T. K., Scheuenstuhl, H., Halliday, B. J., Werb, Z., and Banda, M. J., 1983, Oxygen tension regulates the expression of angiogenesis factor by macrophages, *Science* **221**:1283–1285.

Koch, A. E., Polverini, P. J., and Leibovitch, S. J., 1986, Induction of neovascularization by activated human monocytes, *J. Leukoc. Biol.* **39**:233–238.

Koch, A. E., Kunkel, S. L., Harlow, L. A., Johnson, B., Evanoff, H. L., Haines, G. K., Burdick, M. D., Pope, R. M., and Strieter, R. M., 1992, Enhanced production of monocyte chemoattractant protein-1 in rheumatoid arthritis, *J. Clin. Invest.* **90**:772–779.

Koch, A. E., Kunkel, S. L., Harlow, L. A., Mazarakis, D. D., Haines, G. K., Burdick, M. D., Pope, R. M., and Strieter, R. M., 1994, Macrophage inflammatory protein-1 alpha. A novel chemotactic cytokine for macrophages in rheumatoid arthritis, *J. Clin. Invest.* **93**:921–928.

Kunkel, S. L., Standiford, T., Kasahara, K., and Strieter, R. M., 1991, Stimulus specific induction of monocyte chemotactic protein-1 (MCP-1) gene expression, *Adv. Exp. Med. Biol.* **305**:65–71.

Lake, F. R., Noble, P. W., Henson, P. M., and Riches, D. W. H., 1994, Functional switching of macrophage responses to TNFα by interferons. Implications for the pleiotropic activities of TNFα, *J. Clin. Invest.* **93**:1661–1669.

Lang, R. A., Metcalf, D., Cuthbertson, R. A., *et al.,* 1987, Transgenic mice expressing a hemopoietic growth factor gene (GM-CSF) develop accumulations of macrophages, blindness, and a fatal syndrome of tissue damage, *Cell* **51**:675–686.

Laszlo, D. J., Henson, P. M., Weinstein, L., Remigio, L. K., Sable, C., Noble, P. W., and Riches, D. W. H., 1993, Development of functional diversity in mouse macrophages. Mutual exclusion of two phenotypic states, *Am. J. Pathol.* **143**:587–597.

Leibovitch, S. J., and Ross, R., 1975, The role of the macrophage in wound repair. A study with hydrocortisone and antimacrophage serum, *Am. J. Pathol.* **78**:71–91.

Leibovitch, S. J., and Ross, R., 1976, A macrophage-dependent factor that stimulates the proliferation of fibroblasts *in vitro, Am. J. Pathol.* **84**:5001–5013.

Leibovitch, S. J., Polverini, P. J., Shepard, H. M., Wiseman, D. M., Shively, V., and Nuseir, N., 1987, Macrophage-induced angiogenesis is mediated by tumour necrosis factor-a, *Nature* **329**:630–632.

Leslie, C. C., Musson, R. A., and Henson, P. M., 1984, Production of growth factor activity for fibroblasts by human nonocyte-derived macrophages, *J. Leukoc. Biol.* **36**:143–160.

Lew, D. B., Leslie, C. C., Riches, D. W. H., and Henson, P. M., 1986, Induction of macrophage lysosomal hydrolase synthesis and secretion by beta-1,3-glucan, *Cell. Immunol.* **100**:340–350.

Lew, D. B., Leslie, C. C., Henson, P. M., and Riches, D. W. H., 1991, Role of endogenously derived leukotrienes in the regulation of lysosomal enzyme expression in macrophages exposed to beta 1,3-glucan, *J. Leukoc. Biol.* **49**:266–276.

Lin, H., Kuhn, C., and Chen, D., 1982, Effects of hydrocortisone acetate on pulmonary alveolar macrophage colony-forming cells, *Am. Rev. Respir. Dis.* **125**:712–715.

Lotz, M., Ranheim, E., and Kipps, T. J., 1994, Transforming growth factor beta as endogenous growth inhibitor of chronic lymphocytic leukemia B cells, *J. Exp. Med.* **179**:999–1004.

Lowenstein, C. J., Glatt, C. S., Bredt, D. S., and Snyder, S. H., 1992, Cloned and expressed macrophage nitric oxide synthase contrasts with the brain enzyme, *Proc. Natl. Acad. Sci. USA* **89**:6711–6715.

Lyons, R. M., Keski-Oja, J., and Moses, H. L., 1988, Proteolytic activation of latent transforming growth factor-b from fibroblast conditioned medium, *J. Cell Biol.* **106**:1659–1665.

Madtes, D. K., Raines, E. W., Sakariassen, K. S., Assoian, R. K., Sporn, M. B., Bell, G. I., and Ross, R.,

1988, Induction of transforming growth factor-alpha in activated human alveolar macrophages, *Cell* **53**:285–293.

Manthey, C. L., Allen, J. B., Ellingworth, L. R., and Wahl, S. M., 1990, *In situ* expression of transforming growth factor beta in streptococcal cell wall-induced granulomatous inflammation and hepatic fibrosis, *Growth Factors* **4**:17–26.

Marder, S. R., Chenoweth, D. E., Goldstein, I. M., and Perez, H. D., 1985, Chemotactic responses of human peripheral blood monocytes to the complement-derived peptides C5a and C5a des Arg, *J. Immunol.* **134**:3325–3331.

Martin, B. M., Gimbrone, M. A., Unanue, E. R., and Cotran, R. S., 1981, Stimulation of nonlymphoid mesenchymal cell proliferation by a macrophage-derived growth factor, *J. Immunol.* **126**:1510–1515.

Martin, B. M., Gimbrone, M. A., Majeau, G. R., Unanue, E. R., and Otran, R. S., 1983, Stimulation of human monocyte/macrophage-derived growth factor (MDGF) production by plasma fibronectin, *Am. J. Pathol.* **111**:367–373.

McDonald, J. A., and Kelley, D. G., 1980, Degradation of fibronectin by human leukocyte elastase, *J. Biol. Chem.* **255**:8848–8858.

Mecs, I., and Koltai, M., 1982, *In vivo* hyporesponsiveness induced by Sendai virus in CFLP mice, *Acta Virol. (Praha)* **26**:346–352.

Meister, H., Heyman, B., Schafer, H., and Haferkamp, O., 1977, Role of *Candida albicans* in granulomatous reactions. II. *In vivo* degradation of *C. albicans* in hepatic macrophages in mice, *J. Infect. Dis.* **135**:235–242.

Metcalf, D., 1989, The molecular control of cell division, differentiation commitment and maturation in haemopoietic cells, *Nature* **339**:27–30.

Miyazono, K., and Heldin, C.-H., 1989, Interaction between TGF-b1 and carbohydrate structures in its precursor renders TGF-b1 latent, *Nature* **338**:158–160.

Morris, D. R., Kuepfer, C. A., Ellingsworth, L. R., Ogawa, Y., and Rabinovitch, P. S., 1989, Transforming growth factor-beta blocks proliferation but not early mitogenic signaling events in T-lymphocytes, *Exp. Cell Res.* **185**:529–534.

Murphy, G., Reynolds, J. J., Bretz, U., and Baggiolini, M., 1977, Collagenase is a component of the specific granules of human neutrophil leucocytes, *Biochem. J.* **162**:195–197.

Musson, R. A., 1983, Human serum induces maturation of human monocytes *in vitro, Am. J. Pathol.* **111**:331–340.

Nagaoka, I., Trapnell, B. C., and Crystal, R. G., 1990, Up-regulation of platelet-derived growth factor-A and -B gene expression in alveolar macrophages of individuals with idiopathic pulmonary fibrosis, *J. Clin. Invest.* **85**:2023–2027.

Nakatsukasa, H., Nagy, P., Evarts, R. P., Hsia, C. C., Marsden, E., and Thorgeirsson, S. S., 1990, Cellular distribution of transforming growth factor-b1 and procollagen types I, III, and IV transcripts in carbon tetrachloride-induced rat liver fibrosis, *J. Clin. Invest.* **85**:1833–1843.

Nathan, C. F., 1991, Coordinate actions of growth factors in monocytes/macrophages, in: *Peptide Growth Factors and Their Receptors II* (M. B. Sporn and A. B. Roberts, eds.), pp. 427–462, Springer-Verlag, New York.

Nelson, B. J., Ralph, P., Green, S. J., and Nacy, C. A., 1991, Differential susceptibility of activated macrophage cytotoxic effector reactions to the suppressive effects of transforming growth factor-beta 1, *J. Immunol.* **146**:1849–1857.

Nettelbladt, O., Bergh, J., Schenholm, M., Tengblad, A., and Hallgren, R., 1988, Accumulation of hyaluronic acid in the alveolar interstitial tissue in bleomycin-induced alveolitis in the rat, *Am. Rev. Respir. Dis.* **140**:1028–1032.

Nettlbladt, O., and Hallgren, R., 1989, Hyaluron (hyaluronic acid) in bronchoalveolar lavage fluid during the development of bleomycin-induced alveolitis in the rat, *Am. Rev. Respir. Dis.* **140**:1028–1032.

Noble, P. W., Henson, P. M., Carre, P. C., and Riches, D. W. H., 1993a, TGFβ primes macrophages to express inflammatory gene products in response to particulate stimuli by an autocrine/paracrine mechanism, *J. Immunol.* **151**:979–989.

Noble, P. W., Lake, F. R., Henson, P. M., and Riches, D. W. H., 1993b, Hyaluronate activation of CD44 induces insulin-like growth factor-1 expression by a tumor necrosis factor-alpha-dependent mechanism in murine macrophages, *J. Clin. Invest.* **91:**2368–2377.

Norris, D. A., Clark, R. A. F., Swigart, L. M., Huff, J. C., Weston, W. L., and Howell, S. E., 1982, Fibronectin fragment(s) are chemotactic for human peripheral blood monocytes, *J. Immunol.* **129:**1612–1618.

Ogawa, T., Kotani, S., Kusumoto, S., and Shiba, T., 1983, Possible chemotaxis of human monocytes by *N*-acetylmuramyl-L-ananyl-D-isoglutamine, *Infec. Immun.* **39:**449–451.

Overall, C. M., Wrana, J. L., and Sodek, J., 1989, Transforming growth factor-beta regulation of collagenase, 72 kDa-progelatinase, TIMP and PAI-1 expression in rat bone cell populations and human fibroblasts, *Connec. Tissue Res.* **20:**289–294.

Overall, C. M., Wrana, J. L., and Sodek, J., 1991, Transcriptional and posttranscriptional regulation of 72-kDa gelatinase/type IV collagenase by transforming growth factor-beta 1 in human fibroblasts. Comparisons with collagenase and tissue inhibitor of matrix metalloproteinase gene expression, *J. Biol. Chem.* **266:**14064–14071.

Parakkal, P. F., 1972, Macrophages: The time course and sequence of their distribution in the post-partum uterus, *J. Ultrastruct. Res.* **40:**284–291.

Phan, S. H., and Kunkel, S. L., 1992, Lung cytokine production in bleomycin-induced pulmonary fibrosis, *Exp. Lung Res.* **18:**29–43.

Pierce, G. F., Mustoe, T. A., Lingelbach, J., Masakowski, V. R., Griffin, G. L., Senior, R. M., and Deuel, T. F., 1989, Platelet-derived growth factor and transforming growth factor-b enhance tissue repair activities by unique mechanisms, *J. Cell Biol.* **109:**429–440.

Pierce, G. F., Tarpley, J. E., Yanagihara, D., Mustoe, T. A., Fox, G. M., and Thomason, A., 1992, Platelet-derived growth factor (BB homodimer), transforming growth factor-b1, and basic fibroblast growth factor in dermal wound healing. Neovessel and matrix formation and cessation of repair, *Am. J. Pathol.* **140:**1375–1388.

Piguet, P. F., Collart, M. A., Grau, G. E., Kapanci, Y., and Vassalli, P., 1989, Tumor necrosis factor/cachectin plays a key role in bleomycin-induced pneumopathy and fibrosis, *J. Exp. Med.* **170:**655–663.

Piguet, P. F., Collart, M. A., Grau, G. E., Sappino, A. P., and Vassalli, P., 1990, Requirement of tumour necrosis factor for development of silica-induced pulmonary fibrosis, *Nature* **344:**245–247.

Pinson, D. M., LeClaire, R. D., Lorsbach, R. B., Parmely, M. J., and Russell, S. W., 1992, Regulation by transforming growth factor-beta 1 of expression and function of the receptor for IFN-gamma on mouse macrophages, *J. Immunol.* **149:**2028–2034.

Polverini, P. J., and Leibovitch, S. J., 1984, Induction of neovascularization *in vivo* and endothelial proliferation *in vitro* by tumor-associated macrophages, *Lab. Invest.* **51:**635–642.

Polverini, P. J., Cotran, R. S., Gimbrone, M. A., and Unanue, E. R., 1977, Activated macrophages induce vascular proliferation, *Nature* **269:**804–806.

Postlethwaite, A. E., and Kang, A. H., 1976, Collagen- and collagen peptide-induced chemotaxis of human blood monocytes, *J. Exp. Med.* **143:**1299–1307.

Postlethwaite, A. E., Seyer, J. M., and Kang, A. H., 1978, Chemotactic attraction of human fibroblasts to type I, II, and III collagens and collagen-derived peptides, *Proc. Natl. Acad. Sci. USA* **75:**871–875.

Proveddini, D. M., Deftos, L. J., and Manolagas, S. C., 1986, 1,25-dihydroxyvitamin D_3 promotes *in vitro* morphologic and enzymatic changes in normal human monocytes consistent with their differentiation into macrophages, *Bone* **7:**23–28.

Raghow, R., Postlethwaite, A. E., Keski, O. J., Moses, H. L., and Kang, A. H., 1987, Transforming growth factor-beta increases steady state levels of type I procollagen and fibronectin messenger RNAs posttranscriptionally in cultured human dermal fibroblasts, *J. Clin. Invest.* **79:**1285–1288.

Raines, E. W., Dower, S. K., and Ross, R., 1989, Interleukin-1 mitogenic activity for fibroblasts and smooth muscle cells is due to PDGF-AA, *Science* **243:**393–396.

Rappolee, D. A., Mark, D., Banda, M. J., and Werb, Z., 1988, Wound macrophages express TGF-alpha and other growth factors *in vivo*: Analysis by mRNA phenotyping, *Science* **241:**708–712.

Reuterdahl, C., Sundberg, C., Rubin, K., Funa, K., and Gerdin, B., 1993, Tissue localization of b receptors for platelet-derived growth factor and platelet-derived growth factor B chain during wound repair in humans, *J. Clin. Invest.* **91:**2065–2075.

Riches, D. W. H., 1988, The multiple roles of macrophages in wound healing, in: *The Molecular and Cellular biology of Wound Repair,* lsted. (R. A. F. Clark and P. M. Henson, eds.), pp. 213–239, Plenum Press, New York.

Riches, D. W. H., and Henson, P. M., 1986, Bacterial lipopolysaccharide suppresses the production of catalytically active lysosomal acid hydrolases in human macrophages, *J. Cell Biol.* **102:**1606–1614.

Riches, D. W. H., and Stanworth, D. R., 1980, Primary amines induce selective release of lysosomal enzymes from mouse macrophages, *Biochem. J.* **188:**933–936.

Riches, D. W. H., and Stanworth, D. R., 1982a, Evidence for a mechanism for the initiation of acid hydrolase secretion by macrophages that is functionally independent of alternative pathway complement activation, *Biochem. J.* **202:**639–645.

Riches, D. W. H., and Stanworth, D. R., 1982b, Weak-base-induced lysosomal secretion by macrophages: An alternative trigger mechanism that is independent of complement activation, *Adv. Exp. Med. Biol.* **155:**313–323.

Riches, D. W. H., and Underwood, G. A., 1991, Expression of IFNβ during the triggering phase of macrophage cytocidal activation. Evidence for an autocrine/paracrine role in the regulation of this state, *J. Biol. Chem.* **266:**24785–24792.

Riches, D. W. H., Watkins, J. L., and Stanworth, D. R., 1983, Biochemical differences in the mechanism of macrophage lysosomal exocytosis initiated by zymosan particles and weak bases, *Biochem. J.* **212:**869–874.

Riches, D. W. H., Watkins, J. L., Henson, P. M., and Stanworth, D. R., 1985, Regulation of macrophage lysosomal secretion by adenosine, adenosine phosphate esters, and related structural analogues of adenosine, *J. Leukoc. Biol.* **37:**545–557.

Riches, D. W. H., Henson, P. M., Remigio, L. K., Catterall, J. F., and Strunk, R. C., 1988, Differential regulation of gene expression during macrophage activation with a polyribonucleotide. The role of endogenously derived IFN, *J. Immunol.* **141:**180–188.

Roberts, A. B., and Sporn, M. B., 1991, The transforming growth factor-bs, in: *Peptide Growth Factors and Their Receptors I,* (M. B. Sporn and A. B. Roberts, ed.), pp. 419–472, Springer-Verlag, New York.

Roberts, A. B., Sporn, M. B., Assoian, R. K., Smith, J. M., Roche, N. S., Wakefield, L. M., Heine, U. I., Liotta, L. A., Falanga, V., Kehrl, J. H., and Fauci, A. S., 1986, Transforming growth factor b: Rapid induction of fibrosis and angiogenesis *in vivo* and stimulation of collagen formation *in vitro, Proc. Natl. Acad. Sci. USA* **83:**4167–4171.

Robertson, P. B., Ryel, R. B., Taylor, R. E., Shyu, K. W., and Fullmer, H. M., 1972, Collagenase: Localization in polymorphonuclear leukocyte granules in the rabbit, *Science* **177:**64–65.

Rollins, B. J., Watz, A., and Baggiolini, M., 1991, Recombinant human MCP-1/JE induces chemotaxis, calcium flux and the respiratory burst in human monocytes, *Blood* **78:**1112–1122.

Rom, W. N., Basset, P., Fells, G. A., Nukiwa, T., Trapnell, B. C., and Crystal, R. G., 1989, Alveolar macrophages release an insulin-like growth factor-1-type molecule, *J. Clin. Invest.* **82:**1685–1693.

Rossi, P., Karsenty, G., Roberts, A. B., Roche, N. S., Sporn, M. B., and de Crombrugghe, B., 1988, A nuclear factor 1 binding site mediates the transcriptional activation of a type I collagen promoter by transforming growth factor-beta, *Cell* **52:**405–414.

Ruco, L. P., and Meltzer, M. S., 1978, Macrophage activation for tumor cytotoxicity: Development of macrophage cytotoxic activity requires completion of a sequence of short-lived intermediary reactions, *J. Immunol.* **121:**2035–2042.

Ruegemer, J. J., Ho, S. N., Augustine, J. A., Schlager, J. W., Bell, M. P., McKean, D. J., and Abraham, R. T., 1990, Regulatory effects of transforming growth factor-beta on IL-2- and IL-4-dependent T cell-cycle progression, *J. Immunol.* **144:**1767–1776.

Russell, S. W., Doe, W. F., and McIntosh, A. T., 1977, Functional characterization of a stable, non-cytolytic stage of macrophage activation in tumors, *J. Exp. Med.* **146:**1511–1520.

Sandhaus, R. A., McCarthy, K. M., Musson, R. A., and Henson, P. M., 1983, Elastinolytic proteinases of the human macrophage, *Chest* **83S:**60S–62S.

Sawyer, R. T., 1986, The ontogeny of pulmonary alveolar macrophages in parabiotic mice, *J. Leukoc. Biol.* **40:**347–354.

Sawyer, R. T., Strausbauch, P. H., and Volkman, A., 1982, Resident macrophage proliferation in mice depleted of blood monocytes by Strontium-89, *Lab. Invest.* **46:**165–170.

Schall, T. J., Bacon, K., Toy, K. J., and Goeddel, D. V., 1990, Selective attraction of monocytes and T lymphocytes of the memory phenotype by cytokine RANTES, *Nature* **347**:669–671.

Schmidt, J. A., Oliver, C. N., Lepe-Zuniga, J. L., Green, I., and Gery, I., 1984, Silica-stimulated macrophages release fibroblast proliferation factors identical to interleukin 1, *J. Clin. Invest.* **73**:1461–1472.

Schorlemmer, H. U., Davies, P., Hylton, W., Gugig, M., and Allison, A. C., 1977, The selective release of lysosomal acid hydrolases from mouse peritoneal macrophages by stimuli of chronic inflammation, *Br. J. Exp. Pathol.* **58**:315–326.

Senior, R. M., Griffin, G. L., and Mecham, R. P., 1980, Chemotactic activity of elastin-derived peptides, *J. Clin. Invest.* **66**:859–862.

Senior, R. M., Griffin, G. L., Mecham, R. P., Wrenn, D. S., Prasad, K. U., and Urry, D. W., 1984, Val-Gly-Val-Ala-Pro-Gly, a repeating peptide in elastin, is chemotactic for fibroblasts and monocytes, *J. Cell Biol.* **99**:870–874.

Senior, R. M., Connolly, N. L., Cury, J. D., Welgus, H. G., and Campbell, E. J., 1989, Elastin degradation by human alveolar macrophages. A prominent role of metalloproteinase activity, *Am. Rev. Respir. Dis.* **139**:1251–1256.

Shapiro, S. D., Campbell, E. J., Welgus, H. G., and Senior, R. M., 1991, Elastin degradation by mononuclear phagocytes, *Ann. NY Acad. Sci.* **624**:69–80.

Shapiro, S. D., Griffin, G. L., Gilbert, D. J., Jenkins, N. A., Copeland, N. G., Welgus, H. G., Senior, R. M., and Ley, T. J., 1992, Molecular cloning, chromosomal localization, and bacterial expression of a murine macrophage metalloelastase, *J. Biol. Chem.* **267**:4664–4671.

Shapiro, S. D., Kobayashi, D. K., and Ley, T. J., 1993, Cloning and characterization of a unique elastolytic metalloproteinase produced by human alveolar macrophages, *J. Biol. Chem.* **268**:23824–23829.

Shaw, R. J., Doherty, D. E., Ritter, A. G., Benedict, S. H., and Clark, R. A. F., 1990, Adherence-dependent increase in human monocyte PDGF(B) mRNA as associated with increases in c-fos, c-jun and EGR2 mRNA, *J. Cell Biol.* **111**:2139–2148.

Shaw, R. J., Benedict, S. H., Clark, R. A. F., and King, Jr., T. E., 1991, Pathogenesis of pulmonary fibrosis in interstitial lung disease: Alveolar macrophage PDGF(B) gene activation an up-regulation by interferon gamma, *Am. Rev. Respir. Dis.* **143**:167–173.

Sherry, B., Tekamp, O. P., Gallegos, C., Bauer, D., Davatelis, G., Wolpe, S. D., Masiarz, F., Coit, D., and Cerami, A., 1988, Resolution of the two components of macrophage inflammatory protein 1, and cloning and characterization of one of those components, macrophage inflammatory protein 1 beta, *J. Exp. Med.* **168**:2251–2259.

Shimokado, K., Raines, E. W., Madtes, D. K., Barrett, T. B., Benditt, E. P., and Ross, R., 1985, A significant part of macrophage-derived growth factor consists of at least two forms of PDGF, *Cell* **43**:277–288.

Shiozawa, S., Shiozawa, K., Kita, M., Kishida, T., Fujita, T., and Imura, S., 1992, A preliminary study on the effect of alpha-interferon treatment on the joint inflammation and serum calcium in rheumatoid arthritis, *Br. J. Rheumatol.* **31**:405–408.

Silva, J. S., Twardzik, D. R., and Reed, S. G., 1991, Regulation of *Trypanosoma cruzi* infections *in vitro* and *in vivo* by transforming growth factor beta (TGF-beta), *J. Exp. Med.* **174**:549–545.

Silver, I. A., Murrills, R. J., and Etherington, D. J., 1988, Microelectrode studies on the acid microenvironment beneath adherent macrophages and osteoclasts, *Exp. Cell Res.* **175**:266–276.

Simpson, D. M., and Ross, R., 1971, Effects of heterologous antineutrophil serum in guinea pigs. Hematologic and ultrastructural observations, *Am. J. Pathol.* **65**:79–102.

Simpson, D. M., and Ross, R., 1972, The neutrophilic leukocyte in wound repair a study with antineutrophil serum, *J. Clin. Invest.* **51**:2009–2023.

Snyderman, R., and Fudman, E. J., 1980, Demonstration of a chemotactic factor receptor on macrophages, *J. Immunol.* **124**:2754–2757.

Stanley, E., Lieschke, G. J., Grail, D., Metcalf, D., Hodgson, G., Gall, J. A., Maher, D. W., Cebon, J., Sinickas, V., and Dunn, A. R., 1994, Granulocyte/macrophage colony-stimulating factor-deficient mice show no major perturbation of hematopoiesis but develop a characteristic pulmonary pathology, *Proc. Natl. Acad. Sci. USA* **91**:5592–5596.

Stewart, R. J., Duley, J. A., Dewdney, J., Allardyce, R. A., Beard, M. E. J., and Fitzgerald, P. H., 1981, The

wound fibroblast and macrophage. II. Their origin studied in a human after bone marrow transplantation, *Br. J. Surg.* **68**:129–131.

Strieter, R. M., Chensue, S. W., Basha, M. A., Standiford, T. J., Lynch, J. P., Baggiolini, M., and Kunkel,
S. L., 1990a, Human alveolar macrophage gene expression of interleukin-8 by tumor necrosis factor-
alpha, lipopolysaccharide, and interleukin-1 beta, *Am. J. Respir. Cell. Mol. Biol.* **2**:321–326.

Strieter, R. M., Chensue, S. W., Standiford, T. J., Basha, M. A., Showell, H. J., and Kunkel, S. L., 1990b,
Disparate gene expression of chemotactic cytokines by human mononuclear phagocytes, *Biochem.
Biophys. Res. Commun.* **166**:886–891.

Strieter, R. M., Kunkel, S. L., Elner, V. M., Martonyi, C. L., Koch, A. E., Polverini, P. J., and Elner, S. G.,
1992, Interleukin-8. A corneal factor that induces neovascularization, *Am. J. Pathol.* **141**:1279–1284.

Sugimoto, M., Dannenberg, A. M., Wahl, L. M., Ettinger, W. H., Hastie, A. T., Daniels, D. C., Thomas, C. R.,
and Brahy, L. D., 1978, Extracellular hydrolytic enzymes of rabbit dermal tuberculous lesions and
tuberculin reactions collected in skin chambers, *Am. J. Pathol.* **90**:583–606.

Tanaka, H., Abe, E., Miyaura, C., Shiina, Y., and Suda, T., 1983, Iα,25-dihydroxyvitamin D$_3$ induces
differentiation of human promyelocytic leukemia cells (HL-60) into monocyte-macrophages but not
into granulocytes, *Biochem. Biophys. Res. Commun.* **117**:86–92.

Tarling, J. D., and Coggle, J. E., 1982, Evidence for the pulmonary origin of alveolar macrophages, *Cell
Tissue Kinet.* **15**:577–584.

Thakral, K. K., Goodson, W. H., and Hunt, T. K., 1979, Stimulation of wound blood vessel growth by wound
macrophages, *J. Surg. Res.* **26**:430–436.

Tsunawaki, S., Sporn, M., Ding, A., and Nathan, C., 1988, Deactivation of macrophages by transforming
growth factor-beta, *Nature* **334**:260–262.

Unemori, E. N., Bair, M. J., Bauer, E. A., and Amento, E. P., 1991, Stromelysin expression regulates
collagenase activation in human fibroblasts. Dissociable control of two metalloproteinases by
interferon-gamma, *J. Biol. Chem.* **266**:23477–23482.

Unemori, E. N., Ehsani, N., Wang, M., Lee, S., McGuire, J., and Amento, E. P., 1994, Interleukin-1 and
transforming frowth factor-alpha: Synergistic stimulation of metalloproteinases, PGE2, and proliferation in human fibroblasts, *Exp. Cell Res.* **210**:166–171.

van Furth, R., and Cohn, Z. A., 1968, The origin and kinetics of mononuclear phagocytes, *J. Exp. Med.*
128:415–435.

van Furth, R., Hirsch, J. G., and Fedorko, M. E., 1970, Morphology and peroxidase cytochemistry of mouse
promonocytes, monocytes and macrophages, *J. Exp. Med.* **132**:794–805.

van Furth, R., Diesselhoff-den Dulk, M. M. C., and Mattie, H., 1973, Quantitative study on the production
and kinetics of mononuclear phagocytes during an acute inflammatory reaction, *J. Exp. Med.* **138**:1314–
1330.

van Furth, R., Diesselhoff-den Dulk, M. M. C., Sluiter, W., and van Dissel, J. T., 1985a, New perspectives on
the kinetics of mononuclear phagocytes, in: *Mononuclear Phagocytes: Characteristics, Physiology and
Function,* (R. van Furth and Mijhoff), pp. 201–208, Dordrecht, Holland.

van Furth, R., Nibbering, P. H., van Dissel, J. T., and Diesselhoff-den Dulk, M. M. C., 1985b, The characterization, origin, and kinetics of skin macrophages during inflammation, *J. Invest. Dermatol.* **85**:398–402.

Volkman, A., and Gowans, J. L., 1965a, The origin of macrophages from bone marrow in the rat, *Br. J. Exp.
Pathol.* **46**:62–70.

Volkman, A., and Gowans, J. L., 1965b, The production of macrophages in the rat, *Br. J. Exp. Pathol.* **46**:50–
61.

Wakefield, L. M., Smith, D. M., Flanders, K. C., and Sporn, M. B., 1988, Latent transforming growth factor-
b from human platelets, *J. Biol. Chem.* **263**:7646–7654.

Wedmore, C. V., and Williams, T. J., 1981, Control of vascular permeability by polymorphonuclear leukocytes in inflammation, *Nature* **289**:646–650.

Welgus, H. G., Campbell, E. J., Cury, J. D., Eisen, A. Z., Senior, R. M., Wilhelm, S. M., and Goldberg, G. I.,
1990, Neutral metalloproteinases produced by human mononuclear phagocytes. Enzyme profile, regulation, and expression during cellular development, *J. Clin. Invest.* **86**:1496–1502.

Werb, Z., and Chin, J. R., 1983, Endotoxin suppresses expression of apolipoprotein E by mouse macrophages
in vivo and in culture. A biochemical and genetic study, *J. Biol. Chem.* **258**:10642–10648.

Werb, Z., Banda, M. J., and Jones, P. A., 1980, Degradation of connective tissue matrices by macrophages. I.

Proteolysis of elastin, glycoproteins and collagen by proteinases isolated from macrophages, *J. Exp. Med.* **152**:1340–1357.

Wiseman, D. M., Polverini, P. J., Kamp, D. W., and Leibovich, S. J., 1988, Transforming growth factor-beta (TFGβ) is chemotactic for human monocytes and induces their expression of angiogenic activity, *Biochem. Biophys. Res. Commun.* **157**:5788.

Wright, J. K., Cawston, T. E., and Hazleman, B. L., 1991, Transforming growth factor beta stimulates the production of the tissue inhibitor of metalloproteinases (TIMP) by human synovial and skin fibroblasts, *Biochim. Biophys. Acta.* **1094**:207–210.

Wright, S. D., Craigmyle, L. S., and Silverstein, S. C., 1983, Fibronectin and serum amyloid P component stimulate C3b- and C3bi-mediated phagocytosis in cultured human monocytes, *J. Exp. Med.* **158**:1339–1343.

Xie, Q.-W., Cho, H. J., Calaycay, J., Mumford, R. A., Swiderek, K. M., Lee, T. D., Ding, A., Troso, T., and Nathan, C., 1992, Cloning and characterization of inducible nitric oxide synthase from mouse macrophages, *Science* **256**:225–228.

Yancey, K. B., Hammer, C. H., Harvath, L., Renfer, L., Frank, M. M., and Lawley, T. J., 1985, Studies of human C5a as a mediator of inflammation in normal human skin, *J. Clin. Invest.* **75**:486–495.

Yasui, W., Ji, Z. Q., Kuniyasu, H., Ayhan, A., Yokozaki, H., Ito, H., and Tahara, E., 1992, Expression of transforming growth factor alpha in human tissues: Immunohistochemical study and Northern blot analysis, *Virchows Arch. A Pathol. Anat. Histopathol.* **421**:513–519.

Zauli, G., Davis, B. R., Re, M. C., Visani, G., Furlini, G., and La Placa, M., 1992, Tat protein stimulates production of transforming growth factor-beta 1 by marrow macrophages: A potential mechanism for human immunodeficiency virus-1-induced hematopoietic suppression, *Blood* **80**:3036–3043.

Zhu, J. Q., Wu, J., Zhu, D. X., Scharfman, A., Lamblin, G., and Han, K. K., 1991, Recombinant human granulocyte macrophage colony-stimulating factor (rhGM-CSF) induces human macrophage production of transforming growth factor-alpha, *Cell. Mol. Biol.* **37**:413–419.

Chapter 4

Resolution of Inflammation

CHRISTOPHER HASLETT and PETER HENSON

1. Introduction

Inflammation is a highly effective component of the innate response of the body to infection or injury. The inflammatory response is an important consequence of injury and one that normally leads to repair and restoration of function. Indeed, until the last two or three decades, inflammation was perceived as an entirely beneficial host response. From his experiences on the battle fields of Europe, John Hunter stated "Inflammation is itself not to be considered as a disease but as a salutary operation consequent either to some violence or to some disease" and Eli Metchnikoff (1968), the father of modern inflammatory cell biology, emphasized this concept in his work. The clear-cut implication of these observations is the close connection between the inflammatory process and repair. It has become apparent, however, that inflammation can contribute to the pathogenesis of a large number of diseases. For example, neutrophil infiltration, if not controlled, contributes to tissue injury and necrosis. A relationship between inflammation and scarring has been recognized in such disorders as adult respiratory distress syndrome, fibrosing alveolitis in the lung, hepatitic cirrhosis, glomerulonephritis, as well as infected cutaneous wounds. Extensive scarring or fibrosis of any organ can cause catastrophic loss of function of that organ.

These observations emphasize the importance of the processes by which inflammation normally resolves. However, little attention has been paid to this area. In the past two decades there has been a great deal of research into the early stages of evolution of the inflammatory response, but so many chronic inflammatory diseases (e.g., fibrosing alveolitis) may not present until the pathogenic events have progressed far beyond the early initiation stages. It is just as reasonable to suppose that understanding how inflammation normally resolves will not only provide important insights into the circumstances leading to the persistent inflammation that characterizes most in-

CHRISTOPHER HASLETT • Respiratory Medicine Unit, Department of Medicine, University of Edinburgh, Royal Infirmary, Edinburgh EH3 9YW, Scotland. PETER HENSON • Departments of Medicine and Pathology, National Jewish Center for Immunology and Respiratory Medicine, Denver, Colorado 80206.

The Molecular and Cellular Biology of Wound Repair (Second Edition), edited by Richard A. F. Clark. Plenum Press, New York, 1996

flammatory diseases but will also suggest novel therapies directed at promoting those mechanisms favoring resolution. In his treatise on acute inflammation, Hurley (1983) considered that the acute inflammatory response might terminate by development of chronic inflammation, suppuration, scarring, or by resolution. It is reasonable to suppose that all the alternatives to resolution are nonideal and could contribute to disease processes, particularly in organs whose function depends on the integrity of delicate exchange membranes. The remainder of this chapter therefore represents a somewhat theoretical consideration of some of the processes that are likely to occur in the resolution of inflammation and speculation concerning how a better understanding of these mechanisms will help elucidate the pathogenesis of inflammatory disease and suggest novel anti-inflammatory therapy.

In order for tissues to return to normal during the resolution of inflammation, all of the processes occurring during its evolution must be reversed. Thus, in the simplest model of a self-limited inflammatory response, such as might occur in response to the presence (e.g., by instillation of bacteria into the alveolar airspace) these must include: removal of the inciting stimulus and dissipation of the mediators so generated; cessation of granulocyte emigration from blood vessels; restoration of normal microvascular permeability; limitation of granulocyte secretion of potentially histotoxic and pro-inflammatory agents; and cessation of the emigration of monocytes from blood vessels and their maturation into inflammatory macrophages, as well as removal of extravasated fluid, proteins, bacterial and cellular debris, granulocytes, and macrophages. Finally, any damage to the tissues themselves must be repaired. Experiments *in vitro* and *in vivo* suggest that neutrophils and monocytes are capable, under some circumstances, of emigrating between endothelial cell and epithelial cell monolayers without necessarily causing injury to these "barrier cells." However, it is also clear even at sites of "beneficial" inflammation, such as streptococcal pneumonia, that there may be quite extensive endothelial and epithelial injury and even areas of complete destruction and denudation of these cell layers; but the capacity of streptococcal pneumonia to resolve implies that this injury must not be sufficient in degree or extent to inhibit mechanisms necessary for repair. Therefore, during the resolution of inflammation, there must also exist very effective mechanisms for repair of damage and reconstitution of resident tissue cells. With the completion of resolution and repair events, the stage should be set for full recovery of normal tissue architecture and function.

Each of these events will be considered in the following discussion, but factors relevant to the behavior of neutrophil and eosinophil granulocytes in the resolution of inflammation will receive most attention. The neutrophil is the archetypal acute inflammatory cell. It is essential for host defense, but it is also implicated in the pathogenesis of a wide range of inflammatory diseases (Malech and Gallin, 1988). It is usually the first cell to arrive at the scene of tissue perturbation, and a number of key inflammatory events including monocyte emigration (Doherty *et al.,* 1988) and the generation of inflammatory edema (Wedmore and Williams, 1981) appear to depend on the initial accumulation of this cell type. Neutrophils contain a variety of agents with the capacity not only to injure tissues (Weiss, 1989) but also to cleave matrix proteins into chemotactic fragments (Vartio *et al.,* 1981) with the potential to amplify inflammation by attracting more cells, and they have recently been shown to contain a granule component, the chemotactic protein CAP37 (Spitznagel, 1990), which is a specific monocyte chemotaxin. Eosinophils play an important role in host defense against worms and other parasites, yet paradoxically they are also specifically implicated in the pathogenesis of allergic diseases. The presence of large numbers of eosinophils in tissues is often associated with a local fibrogenic response. Termination of granulocyte emigration from blood vessels and their subsequent clearance from inflamed sites are obvious prerequisites for inflammation to resolve and are important events to consider in the control of inflammatory tissue injury generally. Moreover, gaining further knowledge of the mechanisms controlling the cessation of the emigration of these cells and their disposal may suggest new therapeutic opportunities for manipulating inflammation and promoting mechanisms that favor resolution rather than the persistence of inflammation.

2. Removal or Disappearance of Mediators

During the resolution of inflammation, the powerful mediators initiating the response must somehow be removed, inactivated, or otherwise rendered impotent. The "downside" of mediator biology has received much less attention than mechanisms involved in their initial generation, and it is likely that different mechanisms may be utilized for different mediators. For example, thromboxane A_2 and endothelial-derived relaxing factor (nitric oxide) are labile factors that are spontaneously unstable. Platelet-activating factor (PAF) and C5a are inhibited *in vitro* by appropriate inactivating enzymes (Berenberg and Ward, 1973), and some chemotactic cytokines such as interleukin-8 are thought to become inactivated by binding to other cells, e.g., erythrocytes. Reduction of mediator efficacy might occur by local reduction of their concentration consequent upon dilution during the generation of inflammatory edema. Mediator efficacy may also be reduced by attenuation of target cell responsiveness, for example, in the down-regulation of receptors that occurs during desensitization of neutrophils to high concentrations of a variety of inflammatory mediators (Henson *et al.,* 1981; Colditz and Movat, 1984). It is also likely *in vivo* that locally generated factors may exert opposite effects, for example, neutrophil-immobilizing factor would tend to counteract the chemotactic effects of locally generated chemotactic peptides. In cytokine biology, much attention has been paid to agents that initiate or amplify inflammation; but, by analogy with the proteins involved in the blood coagulation cascade, the whole system is kept under close control by very effective inhibitors and other negative influences. Similar agents operative in inflammation include interleukin-1 receptor antagonist (IL-1ra), yet the inhibitory "partners" of most cytokines and chemotactic peptides have yet to be described. The final requirement for the success of most of the above mechanisms is that the production of mediators at the site must cease.

It is thus likely that control of a single, complex function such as neutrophil chemotaxis in response to a chemotactic peptide, e.g., C5a or interleukin-8, is influenced at a number of points and by a number of factors, including the concentration of mediators, the concentration of their inhibitors or inactivators, possible desensitizing mechanisms, and the effects of other locally generated agents with negative influences on chemotaxis. The redundancy of the inflammatory response *in vivo* must also be taken into consideration. Not only may single mediators exert multiple effects under different circumstances, but important events may be provoked by agents from different mediator families. For example, C5a, leukotriene B_4, interleukin-8, ENA-78, and probably many more factors are likely to exert neutrophil chemotactic effects *in vivo*. In order to gain a dynamic perspective of the resolution of inflammation, it will therefore be necessary to consider how a variety of important mediators may act in concert at the inflamed site and seek to appreciate the integrated impact of negative and positive stimuli on dynamic events *in situ*. Thus, the overall propensity for inflammation to persist would be expected to cease when the balance of mediator effects tips toward the inhibitory rather than the stimulatory, presumably as a result of the combination of at least some of the possible mechanisms considered above.

3. Cessation of Granulocyte and Monocyte Emigration

Until quite recently it was considered that the differential rate of emigration of granulocytes and monocytes at the inflamed site was mainly due to a slower responsiveness of monocytes to "common" chemotactic factors, e.g., C5a. In the light of new discoveries in chemokine and adhesive molecule biology, however, it is reasonable to suggest that the emigration through microvascular endothelium of specific leukocytes in different pathological circumstances is likely to be caused by the combined effects of the local release of cell-specific chemokines and the utilization of different components of the adhesive molecule repertoire that control adhesion of inflammatory cells to endothelial cells. For example, interleukin-8 is a specific neutrophil chemotaxin, and transcapillary neutrophil migration is likely to be mediated by the adhesive interaction between leukocyte adhesion molecules and their counterparts such as intercellular adhesion molecule (ICAM-1) and CD31 on the endothelial surface, whereas specific chemotactic peptides such as macrophage chemotactic protein-1 (MCP-1) and the use of alternative adhesive molecule interactions between such as very late antigen-4 VLA-4 on the monocyte surface and VCAM-1 on the endothelium may be utilized to achieve specific monocyte emigration. Since VCAM-1 tends to be expressed later than E-selectin by stimulated endothelial cells, sequential emigration of leukocytes may also be influenced by the time course of endothelial adhesin expression. Experiments *in vitro* suggest that eosinophils may also have the capacity to use the vascular cell adhesion molecule-1 (VCAM-1)/VLA-4 adhesive axis, but PAF, "Rantes," and interleukin-5 are important in their initial attraction and stimulation.

The factors controlling cessation of inflammatory cell emigration remain obscure. Evolution and resolution of inflammation are dynamic processes, and simple histological techniques may inadequately represent these events. Because poorly understood factors such as cell removal rates may also exert major influences on the number of cells observed in "static" histological sections, the study of neutrophil emigration kinetics requires the careful monitoring of labeled populations of cells. When intravenous pulses of radiolabeled neutrophils were used to define the emigration profiles of neutrophils from blood into acutely inflamed skin (Colditz and Movat, 1984), joints (Haslett *et al.*, 1989a), or lung (Clark *et al.*, 1989), it was found that neutrophil influx ceased remarkably early, in contrast with the greatly prolonged influx that occurred in an inflammatory model that progressed to chronic tissue injury and scarring (Haslett *et al.*, 1989). Indeed, in experimental streptococcal pneumonia, where the histological appearance at 48 hr would suggest massive and continued neutrophil influx (Fig. 1), we were able to show that neutrophil migration to the site had ceased at least 24 hr previously (Fig. 2). Cessation of granulocyte emigration occurring so soon in the evolution of acute inflammation may therefore represent one of the earliest resolution events, and a number of hypothetical mechanisms could be responsible:

1. As mentioned above, locally generated chemotactic factor inhibitors could inactivate neutrophil chemotactic factors. Agents with the capacity to inactive C5a activity *in vitro* have been isolated in plasma (Berenberg and Ward, 1973), but these factors have not been characterized and quantified at inflamed sites, and a plasma-derived inactivator is unlikely to account for cessation of neutrophil emigration in situations where extravascular protein leakage is minimal or absent.

2. "Deactivation" or desensitization of neutrophils to high concentrations of inflammatory mediators may lead to extravasated neutrophils becoming unresponsive to further chemotactic factor stimulation (Ward and Becker, 1967). This might be expected to occur at the center of an inflamed site where the concentration of chemotaxins would be expected to be highest, but it seems unlikely that this mechanism could be involved in the cessation of neutrophils entering the site.

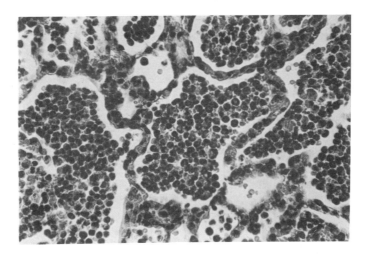

Figure 1. Histological section of experimental streptococcal pneumonia at 48 hr after the onset of disease (× ca. 400). There are large numbers of extravasated neutrophils and monocytes in the alveolar spaces.

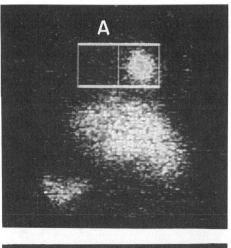

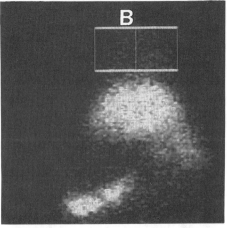

Figure 2. (A) External 24-hr g-camera scintigram of a rabbit that had received intravenous ^{111}In-labeled neutrophils 6 hr after the bronchoscopic introduction of *Streptococcus pneumoniae* into the right upper lobe. This is a major emigration of labeled cells to the lung. (B) In contrast the 24-hr scintigram of a rabbit which had received intravenous ^{111}In-labeled neutrophils 24 hr after streptococcal pneumonia instillation shows that neutrophil emigration to the lung has ceased.

3. A negative feedback loop might operate whereby neutrophils that have already accumulated exert an influence that prevents more neutrophils entering from the bloodstream. For localized inflammatory reactions in the lung, a neutrophil-dependent inhibition of local blood flow (and thus of delivery of new inflammatory cells to the site) has been demonstrated—a process that appears to involve mediators such as PAF and thromboxane (Hellewell *et al.*, 1991).

4. Cessation of neutrophil emigration may simply occur as a result of dissipation or removal of chemotactic factors from the inflamed site.

5. The layers of endothelial and epithelial cells that normally permit neutrophils to

emigrate during the initiation of inflammation could alter to form a "barrier" to further neutrophil emigration. It is well recognized that neutrophils can migrate between endothelial cells (Hurley, 1963) and epithelial cells (Milks and Cramer, 1984) without causing obvious injury. This process, known as *diapedesis* or *transmigration,* involves complex intercellular adhesive mechanisms together with the opening of intercellular endothelial and endothelial junctions by mechanisms that are as yet obscure.

Which of these hypothetical events are important *in vivo* is by no means clear. In a skin model of inflammation, it appeared that a desensitization mechanism (Colditz and Movat, 1984) was operating, and in some forms of human disease involving persistent inflammation there have been suggestions that chemotactic factor inhibitory agents may be defective. However, in experimental arthritis we found no evidence for a desensitization mechanism or for a chemotactic factor inhibitory mechanism (Haslett *et al.,* 1989a). Cessation of neutrophil emigration into the joint coincided with loss of chemoattractants from the joint space. Loss of chemoattractants was not dependent on cellular accumulation at the site, an observation providing evidence against a simple negative feedback mechanism (Haslett *et al.,* 1989a). Although the mechanism responsible for the loss of chemotaxin was not identified, these observations suggest that the local generation and removal of chemoattractants are likely to be centrally important in the persistence and cessation of neutrophil emigration.

Neutrophil surface adhesive molecules are up-regulated very rapidly upon neutrophil exposure to chemotaxins such as C5a and interleukin-8. It is now thought that L-selectin on the surface of the neutrophil is important in the initial interaction with endothelial cells under the conditions of shear stress that exists *in vivo,* whereas the leukocyte integrins, e.g., CD11b/CD18 (Mac-1), are particularly important in the second phase of "tight" adhesion necessary for capillary transmigration. Neutrophil adhesive molecules must then uncouple to permit the next stage of migration to proceed. Molecular mechanisms controlling the "turn-on" and "turn-off" signals of the integrins and other surface molecules are now the subject of detailed scrutiny.

The endothelium also plays an active role in these events. Neutrophil adhesins interact with counterreceptors on the endothelium, e.g., E-selectin, ICAM-1, ICAM-2, and P-selectin. It appears that endothelial P-selectin and E-selectin are involved in the initial neutrophil adhesion (P-selectin and E-selectin), whereas the link between ICAM-1 on the endothelium and thee leucocyte integrin CR3/Mac-1 integrin on the neutrophil surface is likely to be important in the second stage of adhesion and transmigration. Endothelial adhesive molecules are markedly up-regulated by factors such as interleukin-1 and tumor necrosis factor (TNF), which are generated by local cells, particularly macrophages, during the initiation of inflammation.

There has been no detailed *in vivo* research on changes of adhesion molecule expression during the termination of neutrophil emigration, but in experimental arthritis it is clear that the inflamed site will permit a further wave of neutrophil emigration in response to a second inflammatory stimulus (Haslett *et al.,* 1989a). Therefore, any "barriers" to cell adhesion or transmigration existing at the time of cessation of neutrophil emigration must be readily reversible, presumably by the further action of newly generated inflammatory cytokines, which induce renewed expression/activation of endothelial surface adhesive molecules together with parallel effects on neutrophil locomotion and the expression/activation of neutrophil surface adhesive molecules. Thus, it is still reasonable to suggest that the detailed identification of mechanisms controlling the local generation and dissipation of agents that promote chemotaxis and up-regulation and activation of adhesive molecules is essential for our understanding of the processes of termination or persistence of neutrophil emigration at inflamed sites.

Much less is understood of the control of eosinophil and monocyte emigration *in situ,* although similar principles would be applicable to the identification of mechanisms involved in the cessation of their emigration.

4. Restoration of Normal Microvascular Permeability

There are circumstances in which an inflammatory response can be generated experimentally in the lung without detectable leak of plasma proteins from the microvessels. From classical ultrastructural studies, it is apparent that neutrophil migration to inflamed sites is not necessarily associated with overt endothelial or epithelial injury (Majno and Palade, 1961). Nevertheless, in "real" acute inflammation, such as experimental pneumococcal pneumonia (Larsen *et al.,* 1980) or suppurative wounds, there is clear morphological evidence of endothelial injury ranging from cytoplasmic vaculation to areas of complete denudation and fluid leakage into the surrounding tissue. However, the endothelial and epithelial cells must retain the capacity for complete repair as the inflammation resolves. Since many inflammatory diseases, e.g., the adult respiratory distress syndrome (ARDS), are characterized by severe and persistent endothelial and epithelial injury and there is evidence of at least a degree of inevitable endothelial and epithelial injury in examples of "beneficial inflammation," this may represent a pivotal point at which the loss of the normal controls of tissue injury and repair might represent a major mechanism in the development of inflammatory disease. Although the underlying processes are poorly understood, repair is likely to occur by a combination of local cell proliferation to bridge gaps and the recovery of some cells from sublethal injury. Little is known of how endothelial cells recovery from sublethal injury, but epithelial cells (Parsons *et al.,* 1987) *in vitro* appear to be able to recover from hydrogen peroxide-induced injury by a mechanism that requires new protein synthesis. Such cytoprotective mechanisms have received little study. Similarly, it is known that endothelial monolayers, deliberately "wounded" *in vitro,* have a remarkable capacity to reform, yet little is known of the underlying mechanisms (Haudenschild and Schwartz, 1979).

5. Control of Inflammatory Cell Secretion

Injury to tissues is not mediated by the accumulation of inflammatory cells alone; rather, it is their secretion of potentially injurious constituents that produces the damage. Tight control and ultimately cessation of inflammatory cell synthesis and secretion are likely to be important in the limitation of inflammatory tissue injury and the resolution of inflammation. Although there has been much recent study of the initiation and up-regulation of phagocyte secretion *in vitro* (Henson *et al.,* 1988), little is understood of how secretion is down-regulated or of how these processes are controlled *in vivo.* Secretion *in situ* is likely to be modulated by the balance between stimulatory and inhibitory mediators. The simplest mechanism for termination of secretion, that is, the cell exhausting its secretory potential, is unlikely since cells removed from inflamed sites retain major capacity for further synthesis and secretion upon stimulation *ex vivo* (Zimmerli *et al.,* 1986). Other factors that may contribute to down-regulation or termination of secretion are the exhaustion of internal energy supplies, receptor desensitization and down-regulation, dissipation of stimuli, and finally death or removal of the cell itself.

In a short-lived, terminally differentiated cell like the neutrophil, which normally has a blood half-life of about 6 hr, the ultimate demise of the cell could itself represent an important mechanism in the irreversible down-regulation of its secretory function. We have recently observed that aging neutrophils and eosinophils undergo programmed cell death or apoptosis (see Section 6.3 onward). During apoptosis, the neutrophil retains its granule enzymes and an intact membrane but loses the ability to secrete granule contents in response to external stimulation with inflammatory mediators (Whyte *et al.,* 1993b). The apoptotic neutrophil undergoes surface changes by which it becomes recognized as "senescent self" and phagocytosed by inflammatory macrophages. Apoptosis therefore provides a mechanism that renders the neutrophil inert and functionally isolated from inflammatory mediators in its microenvironment, thus greatly limiting the destructive potential of the neutrophil before it is removed by local phagocytes.

6. The Clearance Phase of Inflammation

Once extravasated inflammatory cells have completed their defense tasks for the host, and inciting agents, e.g., bacteria, have been effectively destroyed or rendered impotent, the site must then be cleared of fluid, proteins, antibodies, and debris. Then the key cellular players of inflammation—granulocytes and inflammatory macrophages—must be removed before the tissues return to normality.

6.1. Clearance of Fluid, Proteins, and Debris

Most fluid is probably removed via the lymphatic vessels, although reconstitution of normal hemodynamics may contribute by restoring the balance of hydrostatic and osmotic forces in favor of net fluid absorption at the venous end of the capillary. Proteolytic enzymes in plasma exudate and inflammatory cell secretions are likely to break down any fibrin clot at the inflamed site, and products of this digestion are likely to be drained by the lymphatics which become widely distended as the removal of fluid and proteins increases.

The macrophage may also play an important role in this phase. It can remove fluids (which might contain a variety of proteins) by pinocytosis. In activated inflammatory macrophages, pinocytosis can occur at a rate such that 25% of the cell surface is reused each minute (Steinmann *et al.*, 1976). Inflammatory macrophages also have a hugely increased phagocytic potential. They can recognize opsonized and nonopsonized particles and they express cell surface receptors for a wide variety of altered and damaged cells and proteins. The critical role of macrophages in the clearance phase of inflammation was first recognized by Metchnikoff more than a century ago, and we are now just beginning to elucidate the molecular mechanisms of some of his seminal observations.

6.2. The Clearance of Extravasated Granulocytes

Although we have been aware for some time of the tissue-damaging potential of a wide variety of neutrophil contents, the fate of this cell *in situ* has not received much attention until very recently. There is no evidence that extravasated neutrophils return to the bloodstream or that lymphatic drainage provides an important disposal route, and it is generally agreed that the bulk of neutrophils meet their fate at the inflamed site. It was widely assumed that the majority of neutrophils inevitably disintegrate locally before their fragments are removed by macrophages (Hurley, 1983). However, if this were the rule, healthy tissues would inevitably be exposed to large quantities of potentially damaging neutrophil contents. Although a number of pathological descriptions have favored neutrophil necrosis as a major mechanism operating in inflammation, many of these examples have derived from tissue disease states rather than from examples of more "benign," self-limited inflammation. Furthermore, there has been evidence for over a century of an alternative fate for extravasated neutrophils, based on the classical observations of Metchnikoff (1968). He was the first to catalogue the cellular events of the evolution and resolution of acute inflammation in vital preparations. Rather than neutrophil necrosis as the major mechanism, he described an alternative process whereby intact senescent neutrophils were removed by local macrophages. Over the ensuing decades there have been a number of sporadic reports in both health and disease of macrophage phagocytosis of neutrophils, and of particular relevance to the resolution of inflammation is the clinical phenomenon of "Reiter's cells"— neutrophil-containing macrophages that have been described in the cytology of synovial fluid from the inflamed joints of patients with Reiter's disease and other forms of acute arthritis (Pekin *et al.*, 1967; Spriggs *et al.*, 1978). In experimental peritonitis, where it is possible to sample the inflammatory exudate with ease, it appears that macrophage ingestion of apparently intact neutrophils is the dominant mode of neutrophil removal from the inflamed site (Chapes and Haskill, 1983). The mechanisms underlying these *in vivo* observations have only recently been addressed *in vitro*. Newman *et al.* (1982) showed that human neutrophils harvested from peripheral blood and "aged" overnight were recognized and ingested by inflammatory macrophages (but not by monocytes), whereas freshly isolated neutrophils were not ingested. This suggested that, during aging, a time-related process must have been associated with changes in the neutrophil surface, leading to its recognition as "nonself" or "senescent self." The development of improved methods for harvesting and culturing human neutrophils with minimal activation and avoiding cell losses from aggregation allowed us to study in detail the changes occurring in cultured neutrophils. We have found that aging granulocytes constitutively undergo apoptosis or programmed cell death and that this process is responsible for the recognition of intact senescent neutrophils by macrophages (Savill *et al.*, 1989a).

6.3. Necrosis versus Apoptosis

From the work of Wyllie and his colleagues it is now recognized that the death of nucleated cells can be classified into at least two distinct types: necrosis or accidental death, and apoptosis (programmed cell death) (Kerr *et al.*, 1972; Wyllie *et al.*, 1980). Necrosis can be observed where tissues are exposed to gross insults such as high concentrations of toxins or hypoxia. It is characterized by rapid loss of membrane function and abnormal permeability of the cell membrane and can be observed by failure to exclude vital dyes such as trypan blue. There is early disruption of organelles including lysosomal disintegration and irreversible damage to mitochondria. The stimuli inducing necrosis usually affect large numbers of contiguous cells, and the widespread release of lysosomal contents can cause local tissue injury and the initiation or amplification of a local inflammatory response. By contrast, apoptosis occurs *in situ*, in situations where death is predictable or indeed physiological, such as the removal of unwanted cells during embryological remodelling, involution of the thymus, and a number of other situations in which cell turnover is physiologically rapid, e.g., crypt cells in the gut epithelium. Recognizing the widespread importance of this process in tissue kinetics, Wyllie and his colleagues sought a suitably descriptive "neologism." With the help of the Professor of Classics at Aberdeen University, they coined the term *apoptosis*, meaning the "the falling off, as of leaves from a tree" in ancient Greek. This had an appealing analogy with leaf fall during Autumn, a carefully programmed and regulated event in which the loss of individual leaves occurs in a random fashion and the overall process is not detrimental to the host.

In the many physiological and pathophysiological situations where apoptosis is now recognized, the process occurs with remarkably reproducible structural and broad biochemical changes, implying a common underlying series of molecular mechanisms (Wyllie *et al.*, 1980). During apoptosis, cells shrink and there are major changes in the cell surface, which becomes featureless with the loss of microvillae and with the development of deep invaginations in the surface. However, the membrane remains intact and continues to exclude vital dyes, and organelles such as mitochondria and lysosomes remain intact until very late in the process, although the endoplasmic reticulum appears to undergo characteristic, marked dilatation, which, on light microscopy, may give the appearance of vacuoles in the cytoplasm. The ultrastructural changes in the nucleus are most characteristic, with condensation of chromatin into dense, crescent-shaped aggregates and prominence of the nuceolus (Fig. 3). Apoptotic cells are very swiftly ingested by phagocytes *in vivo,* such that in tissue sections of the

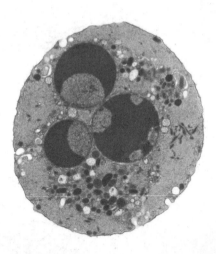

Figure 3. Electron micrograph ($\times$ ca. 11,000) of an apoptotic human neutrophil showing the characteristic chromatin aggregation, prominent nucleolus, and dilated cytoplasmic vacuoles. Note that the cell membrane is intact and the granule structure appears normal. (Electron micrograph taken by Jan Henson.)

remodeling embryo, apoptotic cells are usually seen contained within other cells. Usually macrophages are responsible for their ingestion, but other "semiprofessional" phagocytes, e.g., epithelia and fibroblastlike cells, can also participate. The speed and efficiency of clearance of cells undergoing apoptosis, together with the fact that it occurs in cells at random within the population, renders this mode of cell death much less conspicous than necrosis in histological sections. It is quite remarkable that in embryonic remodeling or during thymus involution whole tracts of tissue can be removed by this process over a few hours without causing local tissue injury or inciting an inflammatory response (thus avoiding a potentially catastrophic effect on the developing embryo).

A key biochemical feature of apoptosis is internucleosomal cleavage of chromatin in a pattern indicative of endogenous endonuclease activation (Wyllie, 1981). This creates low-molecular-sized fragments of chromatin that are integers of the 180 base pairs of DNA associated with a nucleosome. When DNA extracted from apoptotic cells is subjected to agarose gel electrophoresis, this results in a characteristic "ladder" pattern of DNA fragments. Over the past decade a number of laboratories have been pursuing the endonuclease(s) responsible for this particular feature of apoptosis. Activities have been isolated that cleave DNA in a characteristic fashion. These appear to be dependent on calcium and magnesium, but are inhibited by zinc. Final characterization has remained elusive; while a number of candidates have been put forward (Shi *et al.,* 1992; Pietsch *et al.,* 1993), there is as yet no agreement on the molecular nature of the endonuclease concerned.

It is now recognized that the process of apoptosis can be influenced by external mediators that are to some extent cell lineage restricted but can also be modulated by internal genetic influences. However, the biochemical processes integrating the external influences, the genetic influences, the nuclease responsible for chromatin cleavage, and the cell surface changes responsible for cell removal are yet to be determined.

7. Apoptosis in Aging Granulocytes

Neutrophils harvested from blood or from acutely inflamed human joints remain intact and continue to exclude trypan blue for up to 24-hr "aging" in culture (Savill *et al.*, 1989b). With time in culture there is a progressive increase in the proportion of cells exhibiting the light microscopical features of apoptosis, confirmed by electron microscopy and by the chromatin cleavage ladder pattern indicative of endogenous endonuclease activation (Savill *et al.*, 1989b). These changes occur in parallel with an increase in the proportion of human monocyte-derived macrophages ingesting cultured neutrophils in a phagocytic assay. That apoptosis is the change in the aging neutrophil population determining recognition was confirmed using counterflow centrifugation to separate aged neutrophils populations into fractions with varying proportions of apoptotic cells. Macrophage recognition was directly related to the proportion of cells showing apoptosis in each fraction taken from the same time point (Savill *et al.*, 1989b). Neutrophil apoptosis is obvious at 2–3 hr in culture, and by 10 hr up to 50% of the cells show apoptotic morphology. By 24 hr in culture, the majority of neutrophils display features of apoptosis, but there is minimal (usually less than 3%) evidence of cell necrosis or release of the granule contents (Savill *et al.*, 1989b). Nevertheless, apoptotic neutrophils are not indestructible. Beyond 24 hr in culture, there is a progressive increase in the percentage of cells that fail to exclude trypan blue and spontaneous release of granule enzymes occurs. However, when neutrophils are cultured beyond 24 hr in the presence of macrophages, the removal of apoptotic cells is so effective that no trypan blue positive cells are seen and there is no release of granule materials into the surrounding medium (Kar *et al.*, 1993).

Eosinophils undergo apoptosis at a much slower apparent constitutive rate than neutrophils, with the first apoptotic cells being obvious at about 72 hr and maximum percentage apoptosis by about 96 hr (Stern *et al.*, 1992). Up to 96 hr, there is minimal evidence of eosinophil necrosis assessed by trypan blue exclusion, but thereafter the percentage of necrotic cells steadily increases. However, the time difference between onset of apoptosis and onset of necrosis suggests that the bulk of apoptotic eosinophils can exist in culture for up to 2 or 3 days before undergoing necrosis and ultimate disintegration. As in the neutrophil, eosinophil apoptosis is responsible for macrophage recognition of the intact senescent cell (Stern *et al.*, 1992).

The speed with which macrophages recognize, ingest, and destroy apoptotic neutrophils *in vitro* is quite remarkable. Individual macrophages can ingest several neutrophils (see Fig. 4), and once ingested there is extremely rapid degradation of the neutrophils, such that in electron microscopic studies it is necessary to fix macrophages

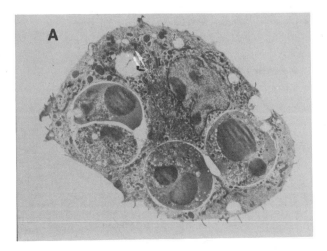

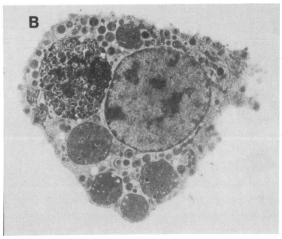

Figure 4. (A) A macrophage that *in vitro* has ingested four apoptotic neutrophils within minutes. (B) Ingested apoptotic neutrophils have degraded to the point at which they are barely recognizable.

within minutes of the initial interaction in order to demonstrate recognizable neutrophils within the phagocytes. Thereafter, ingested cells are no longer recognizable. This may help explain why the dynamic contribution of this process to cell and tissue kinetics has not been fully appreciated until recently. However, there are now several examples demonstrating clear histological evidence of a role for apoptosis in the *in vivo* removal of granulocytes in acute inflammation. These include acute arthrides (Savill *et al.*, 1989b), neonatal acute lung injury (Grigg *et al.*, 1991), and experimental pneumococcal pneumonia during its resolution phase (Fig. 5). Histological evidence of eosinophil apoptosis and ingestion by local macrophages has been described in dexamethasone-treated experimental eosinophilic jejunitis.

Given the proinflammatory potential of neutrophils and their contents, there are

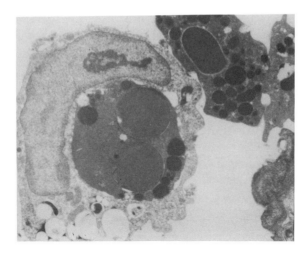

Figure 5. An electron micrograph of resolving experimental streptococcal pneumonia showing a macrophage that contains an apoptotic neutrophil.

now several lines of *in vitro* experimental evidence to support the hypothesis that apoptosis provides an injury-limiting neutrophil clearance mechanism in tissues that would tend to promote resolution rather than persistence of inflammation:

1. During the process of neutrophil apoptosis, there is marked loss of a number of neutrophil functions, including chemotaxis, superoxide production, and secretion of granule enzymes upon deliberate external neutrophil stimulation. These data suggest that apoptosis may lead to "shutting off" of neutrophil functions, resulting in it becoming functionally isolated from external stimuli that would otherwise trigger responses that could damage tissue (Whyte *et al.*, 1993b). This mechanism could be important if fully mature and competent phagocytes are not immediately available in the vicinity of the neutrophil undergoing apoptosis.

2. Neutrophils undergoing apoptosis are very rapidly taken up by macrophages. In these model systems apoptotic neutrophils retain their enzyme contents and are ingested while still intact, thus preventing the leakage of granule enzymes that would occur should the cell disintegrate before or during uptake by macrophage. This is emphasized by preliminary *in vitro* data of a simple model in which macrophages and neutrophils are cocultured. If macrophage uptake of apoptotic neutrophils is blocked (with colchicine, for example) (Kar *et al.*, 1993), rather than being ingested the apoptotic cells then disintegrate and release toxic contents such as myeloperoxidase and elastase before their cellular fragments are taken up by macrophages.

3. The usual response of macrophages to the ingestion of particles *in vitro* is to release proinflammatory mediators, enzymes, and cytokines. However, it has been found that even maximal uptake of apoptotic neutrophils fails to stimulate the release of mediators (Meagher *et al.*, 1993). This was not simply the result of a toxic inhibitory effect of the apoptotic neutrophil on the macrophage since phagocytes that have in-

gested apoptotic cells were able to generate maximum release of potential mediators when subsequently stimulated by opsonized zymozan. Moreover, when apoptotic neutrophils were deliberately opsonized prior to ingestion, macrophages did respond by the release of thromboxane (Meagher *et al.,* 1995). Furthermore, when granulocytes are cultured beyond apoptosis to a point when they fail to exclude trypan blue, their ingestion by macrophages induces the normal release of mediators. From these experiments we can conclude that (1) it is recognition of the senescent granulocyte in the apoptotic morphology rather than the necrotic morphology that determines the lack of macrophage proinflammatory response, and (2) this lack of macrophage response is not a function of the apoptotic particle itself, but relates to the mechanism by which the apoptotic cell is normally ingested. These observations provided considerable impetus for our work on the molecular mechanisms by which macrophages recognize and ingest apoptotic cells.

7.1. Mechanisms Whereby Macrophages Recognize Apoptotic Neutrophils

Early work by Duval and Wyllie (Duvall *et al.,* 1985), using various sugars to inhibit the interactions between macrophages, had suggested that phagocytes possess a lectin mechanism capable of recognizing sugar residues on the apoptotic thymocyte surface exposed by loss of sialic acid. This mechanism does not appear to be involved in the macrophage recognition of apoptotic granulocytes; but these findings stimulated our early work demonstrating that recognition of apoptotic neutrophils occurred by a "charge-sensitive" mechanism, inhibitable by cationic molecules such as amino sugars and amino acids and directly influenced by minor changes of pH, in a fashion suggesting involvement of negatively charged residues on the apoptotic neutrophil surface (Savill *et al.,* 1989b). These data implied that low pH and the presence of cationic molecules might be expected to adversely influence the clearance of apoptotic cells at inflamed sites. This was of particular interest, since a number of granulocyte-derived products such as elastase, myeloperoxidase, and eosinophil-derived major basic protein from eosinophils are known to be highly cationic and have been detected in significant amounts in the tissues. Furthermore, in situations where inflammation is chronic or where there is abscess formation, interstitial pH may be very low.

The amino sugar inhibition pattern helped define at least one set of macrophage cell surface molecules involved in the clearance of apoptotic neutrophils: (1) Since amino sugars are known to inhibit the functions of certain members of the integrin family, this led to a detailed series of investigations using a range of monoclonal antibodies directed against candidate integrins in the $\beta2$ and $\beta3$ family, leading to the implication of macrophage surface $\alpha v \beta3$ in the recognition of apoptotic neutrophils and also apoptotic lymphocytes (Savill *et al.,* 1990). (2) This amino sugar inhibition pattern was previously described in platelet–platelet interactions occurring via thrombospondin and thrombospondin receptors on their surfaces. This led to work that now implicates CD36 on the macrophage surface (Savill *et al.,* 1992a), and it is thought that thrombospondin itself may serve as an intracellular-bridging molecule between the

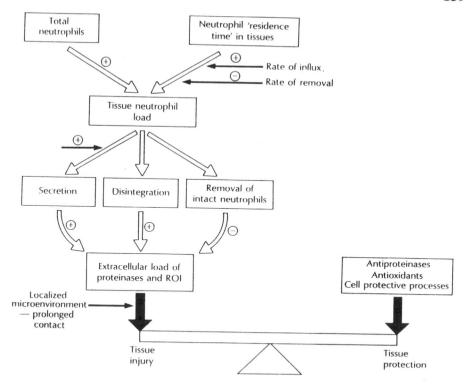

Figure 6. A model of possible surface mechanisms by which macrophage recognize apoptotic cells. (From Savill *et al.*, 1993. Reprinted with permission.)

macrophage and the apoptotic cell surface. Our present model of this recognition mechanism so far is depicted in Fig. 6. The moiety on the surface of the apoptotic cell that is responsible for apoptotic cell recognition in this system has not yet been identified.

Parallel studies addressing the putative ionic sites on the apoptotic cell surface that might be involved in macrophage recognition led to the observation that "receptors" on macrophages can recognize exposed phosphatidylserine (PS) residues on the surface of apoptotic cells (Fadok *et al.*, 1992a). Cell membranes normally exhibit asymmetry with regard to the distribution of phospholipid classes. PS is normally confined to the inner leaflet of the membrane lipid bilayer. During apoptosis, however, we have shown that the normal control elements that limit to this site are lost and the phospholipid is permanently expressed on the membrane surface (in a fashion that may be analogous to that occurring during the sickling of erythrocytes). It appears that the main difference between these two recognition systems relates to the utilization of alternative recognition mechanisms by different subpopulations of macrophages (Fadok *et al.*, 1992b). The "*in vivo*" significance of these observations is as yet uncertain, but the redundancy in recognition mechanisms serves further to emphasize the importance of this removal process.

The definition of cell surface molecules involved in macrophage uptake of apoptotic neutrophils suggests mechanisms by which this function might be regulated. Recent data show that a number of cytokines, including granulocyte–macrophage colony-stimulating factor (GM-CSF), interleukin-1 β, TNF-α, and gamma interferon promote macrophage uptake of apoptotic neutrophils. As an alternative regulatory process, phagocytosis of digestable particles by relatively undifferentiated macrophages induces expression of the PS-recognition mechanism and thus of the ability to remove apoptotic cells (Fadok *et al.*, 1993). This particular macrophage maturation pathway is controlled by the autocrine/paracrine actions of transforming growth factor-beta (TGF-β) and PAF (Noble *et al.*, 1993). Since these mediators/cytokines may amplify inflammation by recruiting leukocytes to the inflamed sites through their effects on endothelial cell adhesive molecule expression and activation, it is possible that their capacity to promote resolution by enhancing macrophage removal of apoptotic neutrophils may represent an "anticipatory" control mechanism for the removal of cells subsequent to their exudation.

7.2. Clearance of Apoptotic Granulocytes by Cells Other Than Macrophages

It is well recognized histologically in embryonic remodeling and in thymus involution that while apoptotic cells are usually taken up by local macrophages, they may also be seen within epithelial cells or fibroblastlike cells. We therefore compared the ability of monolayers of fibroblasts, endothelial cells, and epithelial cells from a variety of sources to recognize apoptotic neutrophils *in vitro*. In these experiments only the fibroblast appeared to recognize and ingest apoptotic neutrophils (Hall *et al.*, 1994). The fibroblast has long been recognized as a "semiprofessional" phagocyte capable of ingesting latex beads, dye particles, and mast cell granules. The significance of fibroblast phagocytosis of senescent neutrophils is uncertain, but the fibroblast appears to employ recognition mechanisms differing from the macrophage in that fibroblasts appear to utilize a sugar–lectin recognition mechanism in addition to the integrin mechanism described in the macrophage–neutrophil system (Hall *et al.*, 1994). More recently, Savill *et al.* (1992) have shown that renal mesangial cells, also recognized as semiprofessional phagocytes, have the capacity to take up large numbers of apoptotic neutrophils.

The significance of these observations is uncertain. It is possible that uptake of apoptotic neutrophils by resident cells, including fibroblasts, serves as a clearance mechanism before extravasated monocytes have fully matured into inflammatory macrophages capable of recognizing and ingesting apoptotic cells. Alternatively, it may serve as a "backup" mechanism should the macrophage disposal mechanism be overwhelmed by waves of neutrophil apoptosis. However, since the fibroblast is responsible for scar tissue matrix protein secretion, it is possible that this possible clearance route is an "undesirable" alternative, particularly if the uptake of apoptotic neutrophils should cause fibroblast replication and secretion of collagen.

7.3. Regulation of Granulocyte Apoptosis by External Mediators: A Control Point for Granulocyte Survival?

Recent histological observations of resolving pulmonary inflammation suggested that extravasated neutrophils undergo apoptosis at a slower rate than neutrophils derived from peripheral blood. In experimental pneumonia at 48 hr (see Fig. 1), large numbers of neutrophils without any significant evidence of apoptosis were seen. The use of radiolabeled pulses of neutrophils delivered intravenously during the evolution of this model suggested that neutrophil emigration from the blood to the inflamed lung had largely ceased by 16 hr. This implied that the bulk of neutrophils observed at 48 hr had been present for at least 24 hr, yet the $t_{1/2}$ of neutrophils in blood is about 4–5 hr. These observations suggested that factors present at the inflamed site might have retarded the inherent rate of neutrophil apoptosis. We have now shown that the rate at which neutrophils undergo apoptosis *in vitro* is inhibited by a variety of inflammatory mediators (Lee *et al.*, 1993), including endotoxic lipopolysaccharide, C5a, and GM-CSF (which is particularly potent). If, as seems likely, apoptosis controls the tissue longevity of neutrophils by leading to macrophage removal of unwanted cells, this might represent an important mechanism controlling the "tissue load" of inflammatory cells *in situ*. Experiments with eosinophils *in vitro* show that GM-CSF inhibits eosinophil apoptosis; but interleukin-5 is also extremely potent in this regard, whereas it has no effect on neutrophil longevity.

Intracellular mechanisms governing apoptosis are as yet poorly understood. However, there are indications that internal controls in granulocytes may differ from lymphoid cells. In thymocytes, elevation of intracellular calcium $[Ca^{2+}]_i$ concentration by calcium ionophores induces apoptosis, and apoptosis induced by other stimuli, such as gluocorticoids, is associated with raises in $[Ca^{2+}]_i$ (McConkey *et al.*, 1989). In neutrophils spontaneously undergoing apoptosis, however, there were no such rises in $[Ca^{2+}]_i$, and agents increasing $[Ca^{2+}]_i$ caused dramatic slowing of neutrophil apoptosis without inducing necrosis (Whyte *et al.*, 1993a). Furthermore, treatment of aging neutrophils with the chelating agents MAPTAM and BAPTA, which bind intracellular calcium, is associated with an increase in the rate of neutrophil apoptosis (Whyte *et al.*, 1993a). Rises in intracellular calcium are known to occur when neutrophils are primed or activated *in vitro*, and this is likely to represent at least part of the mechanism underlying the retardation of neutrophil apoptosis observed after external stimulation with inflammatory mediators. Further, when neutrophils are aged in the presence of inhibitors of protein synthesis, e.g., cycloheximide, there is an acceleration of the constitutive rate of apoptosis. This is in marked contrast to corticosteroid-induced thymocyte apoptosis, which is inhibited by cycloheximide. These observations suggest the existence of a protein synthesis-dependent, apoptosis-inhibitory factor in neutrophils. It is hypothesized that external inflammatory mediators inhibit the rate of neutrophil apoptosis by causing a rise in intracellular calcium, which subsequently acts on downstream processes including mRNA and protein synthesis of this putative inhibitory factor(s).

It is now clear that apoptosis in a variety of cell types can be influenced by proto-oncogene expression. The best worked out system is *bcl*-2 (Fanidi *et al.*, 1992; Vaux *et al.*, 1992), the expression of which is associated with inhibition of apoptosis in a variety of cell types; but its significance in the neutrophil is uncertain. Expression of c-*myc* in fibroblasts and lymphoid cells appears to induce apoptosis (Evan *et al.*, 1992), and more recently p53 and Rb-1 have also been implicated. It remains to be established whether the products of these genes are relevant for the control of inflammatory cell apoptosis, and it is presently unclear how the genetic controls are linked with second messenger controls and the final eeffector events including endonuclease activation, down-regulation of cell function, and cell surface changes responsible for phagocyte recognition.

Clearly, we are only just beginning to scratch the surface of the internal mechanisms of apoptosis, but ultimately it may be possible to regulate granulocyte apoptosis for therapeutic benefit. It is intriguing that two such closely related cells as the neutrophil and eosinophil should appear to possess different inherent rates of apoptosis that are further influenced by different external mediators. These observations suggest that at in the future it may be possible to specifically induce eosinophil apoptosis without influencing neutrophil apoptosis. This would presumably result in the tissue clearance of eosinophils by the nonphlogistic mechanisms "that nature intended" and yet leave other inflammatory cells available for host defense purposes.

7.4. A Role for Granulocyte Apoptosis in the Control of Inflammation?

Granulocyte apoptosis could play at least two important roles in the control of inflammation. First, by controlling the functional longevity and tissue removal of unwanted granulocytes and providing a pivotal point at which inflammatory cytokines and growth factors exert their controls on inflammatory cell longevity, it is likely that (together with neutrophil influx) neutrophil apoptosis is a critical determinant of the overall tissue load of inflammatory cells (see Fig. 7). Second, it is reasonable to suggest that neutrophil removal (whether by apoptosis or by necrosis) is an important factor in the control of inflammation. Inflammatory disease is now thought to result from a quantitative imbalance between potentially damaging and protective mechanisms, as exemplified by the proteinase–antiproteinase theory of emphysema. Whether neutrophils meet their fate by a mechanism that involves removal of the whole cell or whether they meet their fate via a mechanism that results in disintegration and disgorgement of their potentially histotoxic and proinflammatory contents is therefore critical. We have given detailed consideration to the observations that apoptosis may serve to keep potentially injurious granule contents within the cell membrane, while at the same time the cell becomes unable to respond by degranulation in response to external stimuli; and finally the intact cell is removed by a novel phagocytic recognition mechanism that determines that macrophages fail to release proinflammatory mediators during macrophage recognition and phagocytosis. This is not to suggest that apoptosis is the only mechanism for removal of neutrophils at an inflamed site; examples of neutrophil necrosis are also seen in such diseases as systemic vasculitis where light microscopic features of neutrophil necrosis and disintegration have been de-

Apoptotic cell

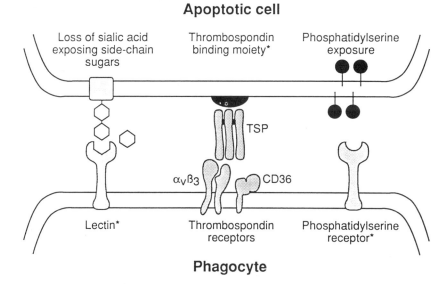

Figure 7. The "balance" between the capacity of inflammatory cells to injure tissues and tissue defense mechanisms, including pivotal points at which factors involved in the emigration and clearance of granulocytes may impinge on the equation. (From Savill *et al.*, 1993. With permission.)

scribed in tissues close to inflamed vessels and called *leukocytoclastic vasculitis*. It is possible therefore that the balance between the degree of neutrophil apoptosis and the degree of neutrophil necrosis at an inflamed site may represent a pivotal point in the control of tissue injury and in the propensity of an inflamed site to resolve or to progress.

At the present time the proper *in vivo* validation of this hypothesis presents a number of problems. It is difficult to make even semiquantitative estimates of the degree of apoptosis and necrosis at inflamed sites, since we do not have specific surface markers to enable us to develop tools for quantifying with confidence the rates of these processes *in situ*. This difficulty is compounded by the rapidity with which apoptotic cells are known to be recognized and ingested *in vitro* and degraded to the point at which the cell of origin is no longer easily recognizable. Moreover, at the present time we can only speculate what the actual consequences of neutrophil disintegration *in situ* would be. Attention has been drawn to the enormous injurious potential of a wide variety of granulocyte contents that would inevitably be disgorged during granulocyte activation and necrosis. It is also recognized that there are tissue defense mechanisms including antioxidants and antiproteinases that normally serve to shield tissues from the effects of these toxic agents. However, there is abundant evidence that this antiproteinase shield can be overcome, particularly in circumstances that generate very high localized concentrations of the damaging agent, such as the intercellular microenvironment between cells adherent to matrix or other cells. Furthermore, potent antiproteinases can be rendered ineffective in the presence of oxidants generated locally by inflammatory cells. Moreover, a variety of toxic neutrophil and eosinophil contents do not have obvious tissue inhibitors or regulators, such as the nonenzymatic, cationic antimicrobial proteins of neutrophils and the major basic protein and eosinophil cationic protein content of eosinophils. Neutrophil enzymes can cleave chemotactic fragments from complement and from matrix proteins such as fibronectin while the granule protein CAP37 also has monocyte-specific chemoattractant properties. Thus, neutrophil contents have the capacity to recruit more leukocytes to an inflamed site, amplifying and prolonging the response.

Therefore, with knowledge presently available it seems reasonable to suggest that granulocyte necrosis at inflamed sites can be regarded as a deleterious and undesirable mode of clearance that, by contrast with apoptosis, is likely to favor persistence of inflammatory tissue injury rather than resolution of inflammation.

8. The Fate of Macrophages

Although monocytes have the capacity to undergo apoptosis *in vitro,* the mechanisms involved in the clearance of macrophages that have completed their functions in host defense and clearance of inflammation remain entirely obscure.

9. Summary and Conclusions

9.1. Resolution Mechanisms and the Control of Inflammation

On preliminary evidence, a speculative scheme can be proposed whereby a stereotyped sequence of inflammatory events results in the resolution of inflammation: Injury to endothelial and epithelial cell "barriers" is minimized but is associated with leakage of fluid and protein; reconstitution of normal microvascular permeability occurs by the reforming of cell junctions and the regeneration of cell sheets; neutrophil influx ceases early in the evolution of acute inflammation and relates to cessation of local chemoattractant generation and dissipation; monocytes mature into inflammatory macrophages to remove proteins and other debris; neutrophil secretion is restricted and the aged cell undergoes apoptosis, which controls neutrophil longevity and determines the macrophage removal of the intact senescent cell without stimulating macrophage release of proinflammatory mediators; and excess tissue macrophages are removed by unknown processes. There are several steps in this scheme that could go awry and lead to circumstances favoring the development of persistent inflammation. For example, if the macrophage fails to develop the appropriate receptors for removing apoptotic cells, neutrophils would eventually become necrotic and disgorge their damaging contents, or alternatively they may be taken up by local fibroblasts, possibly with a profibrotic response. Once a chronic inflammatory state begins to develop, there is evidence that the local pH falls. This would tend to inhibit macrophage recognition of apoptotic neutrophils, as would the continued accumulation of inflammatory cell cationic products such as elastase, major basic protein, and so forth, which would contribute to the inhibition of macrophage clearance of apoptotic neutrophils.

9.2. Potential Therapeutic Application of Resolution Mechanisms of Inflammation

Hurley (1983) recognized that there were several mechanisms other than resolution whereby acute inflammation could terminate. These include chronic inflammation, scarring, and abscess formation. By comparison with resolution, all of these termination events clearly must be regarded as detrimental to organ function. As we begin to learn more about the mechanisms involved in the resolution of inflammation, it may be possible to "divert" inflammatory processes down resolution pathways rather than one of the other less desirable pathways. More specifically, with increasing knowledge of apoptosis and its internal mechanisms as well as the specific mechanisms available for the removal of apoptotic cells, it may be possible to use this "controlled" process of cell suicide or programmed cell death to remove specific inflammatory cells at particular pathogenetic stages when they are critical to the disease process.

ACKNOWLEDGMENTS. This work was supported by the Medical Research Council of Great Britain in the form of a program grant and a number of preceding project grants and fellowships. Further support was obtained from the National Asthma Campaign and the British Lung Foundation. Dr. Henson is supported by NIH Grant GM48211. Thanks are due to close colleagues who participated in the work and who were a constant source of stimulation, including John Savill in particular, Moira Whyte, Laura Meagher, and Ian Dransfield. The manuscript was typed by Mrs. J. McMahon and Mrs. B. Sebern.

References

Berenberg, J. L., and Ward, P. A., 1973, Chemotactic factor of inactivator in normal human serum, *J. Clin. Invest.* **52**:1200–1207.

Chapes, S. K., and Haskill, S., 1983, Evidence for granulocyte-mediated macrophage activation after *C. parvum* immunization, *Cell Immunol.* **75**:367–377.

Clark, R. J., Jones, H. A., Rhodes, C. G., and Haslett, C., 1989, Non-invasive assessment in self-limited pulmonary inflammation by external scintigraphy of [111]indium-labelled neutrophil influx and by measurement of the local metabolic response with positron emission tomography, *Am. Rev. Respir. Dis.* **139**:A58.

Colditz, I. G., and Movat, H. Z., 1984, Desensitisation of acute inflammatory lesions to chemotaxins and endotoxin, *J. Immunol.* **133**:2163–2168.

Doherty, D. E., Downey, G. P., Worthen, G. S., Haslett, C., and Henson, P. M., 1988. Monocyte retention and migration in pulmonary inflammation, *Lab. Invest.* **59**:200–213.

Duvall, E., Wyllie, A. H., and Morris, R. G., 1985, Macrophage recognition of cells undergoing programmed cell death, *Immunology* **56**:351–358.

Evan, G. I., Wyllie, A. H., Gilbert, G. S., Littlewood, T. D., Land, H., Brooks, M., Waters, C. M., Penn, L. Z., and Hancock, D. C., 1992, Induction of apoptosis in fibroblasts by c-myc protein, *Cell* **69**:119–128.

Fadok, V., Savill, J. S., Haslett, C., Bratton, D. L., Doherty, D. E., Campbell, P. A., and Henson, P. M., 1992a, Different populations of macrophages use either the vitronectin receptor or the phosphatidylserine receptor to recognise and remove apoptotic cells, *J. Immunol.* **149**:4029–4035.

Fadok, V. A., Voelker, D. R., Campbell, P. A., Cohen, J. J., Bratton, D. L., and Henson, P. M., 1992b, Exposure of phosphatidylserine on the surface of apoptotic lymphocytes triggers specific recognition and removal by macrophages, *J. Immunol.* **148**:2207–2216.

Fadok, V. A., Laszlo, D. J., Noble, P. W., Weinstein, L., Riches, D. W. H., and Henson, P. M., 1993, Particle digestibility is required for induction of the phosphatidylserine recognition mechanism used by murine macrophages to phagocytose apoptotic cells, *J. Immunol.* **151**:4274–4285.

Fanidi, A., Harrington, E. A., and Evan, G. I., 1992, Co-operative interaction between *c-myc* and *bcl-2* proto-oncogenes, *Nature***359**:554–556.

Grigg, J. M., Savill, J. S., Sarraf, C., Haslett, C., and Silverman, M., 1991, Neutrophil apoptosis and clearance from neonatal lungs, *Lancet* **338**:720–722.

Hall, S. E., Savill, J. S., Henson, P. M., and Haslett, C., 1994, Apoptotic neutrophils are phagocytosed by fibroblasts with participation of the fibroblast vitronectin receptor and involvement of a mannose/fucose-specific lectin, *J. Immunol* **153**:3218–3227.

Haslett, C., Jose, P. J., Giclas, P. C., Williams, T. J., and Henson, P. M., 1989a, Cessation of neutrophil influx in C5a-induced acute experimental arthritis is associated with loss of chemoattractant activity from joint spaces, *J. Immunol.* **142**:3510–3517.

Haslett, C., Shen, A. S., Feldsien, D. C., Allen, D., Henson, P. M., and Cherniack, R. M., 1989b, [111]Indium-labelled neutrophil flux into the lungs of bleomycin-treated rabbits assessed non-invasively by external scintigraphy, *Am. Rev. Respir. Dis.* **140**:756–763.

Haudenschild, C. L., and Schwartz, S. M., 1979, Endothelial regeneration. II. Restitution of endothelial continuity, *Lab. Invest.* **41**:407–418.

Hellewell, P. G., Henson, P. M., Downey, G. P., and Worthen, G. S., 1991, Control of local blood flow in pulmonary inflammation: Role for neutrophils, PAF, and thromboxane, *J. Appl. Physiol.* **70**:1184–1193.

Henson, P. M., Schwartzmann, N. A., and Zanolari, B., 1981, Intracellular control of human neutrophil secretion. II. Stimulus specificity of desensitisation induced by six different soluble and particulate stimuli, *J. Immunol.* **127**:754–759.

Henson, P. M., Henson, J. E., Fittschen, C., Kimani, G., Bratton, D. L., and Riches, D. H. W., 1988, Phagocytic cells: Degranulation and secretion, in: *Inflammation, Basic Principals and Clinical Correlates* (J. I. Gallin, ed.), pp. 363–390, Raven Press, New York.

Hurley, J., 1963, An electron microscopic study of leukocyte emigration and vascular permeability in rat skin, *Aust. J. Exp. Biol. Med. Sci.* **41**:171–179.

Hurley, J. V., 1983, Termination of acute inflammation. I. Resolution, in: *Acute Inflammation,* 2nd ed. Churchill Livingstone, pp. 109–117, London.

Kar, S., Ren, Y., Savill, J. S., and Haslett, C., 1993, Inhibition of macrophage phagocytosis *in vitro* of aged neutrophils increases release of neutrophil contents, *Clin Sci* (Abstract). **85**:27.

Kerr, J. F. R., Wyllie, A. H., and Currie, A. R., 1972, Apoptosis: A basic biological phenomenon with wide-ranging implications in tissue kinetics, *Br. J. Cancer***26**:239–257.

Larsen, G. L., McCarthy, K., Webster, R. O., Henson, J. E., and Henson, P. M., 1980, A differential effect of C5a and C5a des arg in the induction of pulmonary inflammation, *Am. J. Pathol.* **100**:179–192.

Lee, A., Whyte, M. B. K., and Haslett, C., 1993, Prolonged *in vitro* lifespan and functional longevity of neutrophils induced by inflammatory mediators acting through inhibition of apoptosis, *J. Leuk. Biol.* **54**:283–288.

Majno, E., and Palade, G. E., 1961, Studies on inflammation. I. The effect of histamine and serotonin on vascular permeability: An electron microscopic study, *J. Biol. Phys. Biochem. Cytol.* **11**:571–605.

Malech, H. D., and Gallin, J. I., 1988, Neutrophils in human diseases, *N. Engl. Med. J.* **37**:687–694.

McConkey, D. J., Nicotera, P., Hartzell, P., Bellomo, G., Wyllie, A. H., and Orrenius, S., 1989, Glucocorticoids activate a suicide process in thymocytes through an elevation of cytosolic Ca^{2+} concentration, *Arch. Biochem. Biophys.* **269**:365–370.

Meagher, L. C., Savill, J. S., Baker, A., Fuller, R. W., and Haslett, C., 1993, Phagocytosis of apoptotic neutrophils does not induce macrophage release of thromboxane B_2, *J. Leuk. Biol.* in press.

Metchnikoff, E., 1968, *Lectures on the Comparative Pathology of Inflammation. Lecture VII. Delivered at the Pasteur Institute in 1891,* Dover, New York.

Milks, L., and Cramer, E., 1984, Transepithelial electrical resistance studies during *in vitro* neutrophil migration, *Fed. Proc.* **43**:477.

Newman, S. L., Henson, J. E., and Henson, P. M., 1982, Phagocytosis of senescent neutrophils by human monocyte-derived macrophages and rabbit inflammatory macrophages, *J. Exp. Med.* **156**:430–442.

Noble, P. W., Henson, P. M., Lucas, C., Mora-Worms, M., Carre, P., and Riches, D. W. H., 1993, Transforming growth factor β TGFβ primes murine macrophages to express inflammatory gene products in response to particulate stimuli by an autocrine/paracrine mechanism, *J. Immunol.* **151**:979–989.

Parsons, P. E., Sugahara, K., Cott, G. R., Mason, R. J., and Henson, P. M., 1987, The effect of neutrophil migration and prolonged neutrophil contact on epithelial permeability, *Am. J. Pathol.* **129**:302–312.

Pekin, T., Malinin, T., and Zwaifler, R., 1967, Unusual synovial fluid findings in Reiter's syndrome, *Ann. Intern. Med.* **66**:677–684.

Pietsch, M. C., Polzar, B., Stephan, H., Crompton, T., MacDonald, H. R., Mannherz, H. G., and Tschopp, J., 1993, Characterisation of the endogenous deoxyribonuclease involved in nuclear DNA degradation during apoptosis (programmed cell death), *EMBO J.* **12**:371–377.

Savill, J. S., Wyllie, A. H., Henson, J. E., Henson, P. M., and Haslett, C., 1989a, Macrophage phagocytosis of aging neutrophils in inflammation—programmed cell death leads to its recognition by macrophages, *J. Clin. Invest.* **83**:865–875.

Savill, J. S., Henson, P. M., and Haslett, C., 1989b, Phagocytosis of aged human neutrophils by macrophages is mediated by a novel "charge sensitive" recognition mechanism, *J. Clin. Invest.* **84**:1518–1527.

Savill, J. S., Dransfield, I., Hogg, N., and Haslett, C., 1990, Macrophage recognition of "senescent self." The vitronectin receptor mediates phagocytosis of cells undergoing apoptosis, *Nature* **342**:170–173.

Savill, J. S., Hogg, N., and Haslett, C., 1992a, Thrombospondin co-operates with CD36 and the vitronectin receptor macrophage in recognition of aged neutrophils, *J. Clin. Invest.* **90**:1513–1529.

Savill, J. S., Smith, J., Sarraf, C., Ren, Y., Abbott, F., and Rees, A., 1992b, Glomerular mesangial cells and inflammatory macrophages ingest neutrophils undergoing apoptosis, *Kidney Int.* **42**:924–936.

Savill, J. S., Fadok, V., Hersar, P. M., and Haslett, C., 1993, Phagocyte recognition of cells undergoing apoptosis, *Immunol. Today* **14**:131–136.

Shi, Y., Glynn, J. M., Guilbert, L. J., Cotter, T. G., Bissonnette, R. P., and Green, D. R., 1992, Role for *c-myc* in activation-induced cell death in T cell hybridomas, *Science* **257**:212–215.

Spitznagel, J. K., 1990, Antibiotic proteins of neutrophils, *J. Clin. Invest.* **86**:1851–1854.

Spriggs, R. S., Boddington, M. M., and Mowat, A., 1978, Joint fluid cytology in Reiter's syndrome, *Ann. Rheum. Dis.* **37**:557–560.

Steinmann, R. M., Brodie, S. E., and Cohn, Z. A., 1976, Membrane flow during pinocytosis—a sterological analysis, *J. Cell Biol.* **68**:665–687.

Stern, M., Meagher, L., Savill, J., and Haslett, C., 1992, Apoptosis in human eosinophils. Programmed cell death in the eosinophil leads to phagocytosis by macrophages and is modulated by IL-5, *J. Immunol.* **148**:3543–3549.

Vartio, T., Seppa, H., and Vaheri, A., 1981, Susceptibility of soluble and matrix fibronectins to degraduation by tissue proteinases, mast cell chymase and cathepsin G, *J. Biol. Chem.* **256**:471–477.

Vaux, D. L., Cory, S., and Adams, J. M., 1992, Bcl-2 gene promotes haemopoitic cell survival and cooperates with *c-myc* to immortalise pre-B cells, *Nature* **335**:440–442.

Ward, P. A., and Becker, E. L., 1967, The deactivation of rabbit neutrophils by chemotactic factor and the nature of the activatable esterase, *J. Exp. Med.* **127**:693–709.

Wedmore, C. V., and Williams, T. J., 1981, Control of vascular permeability by polymorphonuclear leukocytes in inflammation, *Nature* **289**:646–650.

Weiss, S. J., 1989, Tissue destruction by neutrophils, *N. Engl. Med. J.* **320**:365–376.

Whyte, M. K. B., Meagher, L. C., Hardwick, S. J., Savill, J. S., and Haslett, C., 1993a, Transient elevations of cytosolic free calcium retard subsequent apoptosis in neutrophils *in vitro, J. Clin. Invest.* **92**:446–455.

Whyte, M. K. B., Meagher, L. C., MacDermott, J., and Haslett, C., 1993b, Down-regulation of neutrophil function by apoptosis: A mechanism for functional isolation of neutrophils from inflammatory mediator stimulation, *J. Immunol.* **150**:5124–5134.

Wyllie, A. H., 1981, Glucocorticoid-induced thymocyte apoptosis is associated with endogenous endo-nuclease activation, *Nature* **284:**555–558.

Wyllie, A. H., Kerr, J. F. R., and Currie, A. R., 1980, Cell death: The significance of apoptosis, *Int. Rev. Cytol.* **68:**251–306.

Zimmerli, W., Seligmann, B., and Gallin, J. I., 1986, Exudation primes human and guinea pig neutrophils for subsequent responsiveness to the chemotactive peptide N-formyl methionyl leucyl phenylamine and increases complement C3bi receptor expression, *J. Clin. Invest.* **77:**925–933.

Part II
Growth Factors in Soft Tissue Healing

Chapter 5

Epidermal Growth Factor and Transforming Growth Factor-α

LILLIAN B. NANNEY and LLOYD E. KING, JR.

1. Introduction

Since its discovery in 1963 as a polypeptide that accelerated the *in vivo* maturation of epithelial tissues (Cohen and Elliott, 1963), epidermal growth factor (EGF) has been assumed to be important in regulating epidermal growth, differentiation, and repair. Over the last 30 years, the concepts inherent to EGF and its related growth factors have merged with that of lymphokines, interferons, and cytokines to become synonymous with small peptide factors that regulate multiple functions of cells and tissues. The regulatory signals induced by binding of soluble and matrix protein-bound growth factors/cytokines to specific receptors is termed *signal transduction*. Although *in vitro* and *in vivo* biochemical and genetic studies of signal transduction are unfolding at a rapid pace, these complex signaling mechanisms have not yet been evaluated in the more complicated milieu within cutaneous wounds. To provide an appropriate background for this chapter on EGF and its common factors involved in wound healing, a brief description of this expanding family of EGF-like molecules, their cellular interactions, and emerging signal transduction mechanisms is described below.

2. Background

An ever-growing subfamily of molecules with EGF homologies has been identified by biochemical and molecular methods. Molecules with EGF-like regions include transforming growth factor-alpha (TGF-α) (Derynck, 1986), amphiregulin (Shoyab *et al.*, 1989), vaccinia virus growth factor (Brown *et al.*, 1985), neu differentiation factor (NDF) (Peles *et al.*, 1992; Wen *et al.*, 1992), heparin-binding (HB) EGF-like growth

LILLIAN B. NANNEY and LLOYD E. KING, JR. • Departments of Plastic Surgery, Cell Biology, and Dermatology, Vanderbilt University Medical Center, Department of Veteran's Affairs, Nashville, Tennessee 37232-2631.

The Molecular and Cellular Biology of Wound Repair (Second Edition), edited by Richard A. F. Clark. Plenum Press, New York, 1996

Table I. EGF Subfamily and Proteins with EGF Motifs

Amphiregulin	EGF	Heparin-binding EGF-like protein
Beta cellulin	TGF-α	Myxoma virus growth factor
NDF/heregulin	cripto	Vaccinia virus growth factor
Extracellular Matrix Molecules		
Aggrecan		
Fibrillin		
Laminin		
Tenascin		
Adhesion Molecules		
P-Selectin		
L-Selectin		
E Selectins		
Coagulation Factors		
Factor IX	Protein C	Plasminogen activator, tissue type tPA
Factor X	Protein S	Plasminogen activator, urokinase type uPA

References: Carpenter (1993); Hayaski *et al.* (1994); Nawa *et al.* (1994); Nowak *et al.* (1994); Smith *et al.* (1994); Zhong *et al.* (1994); Medved *et al.* (1994); Watanabe *et al.* (1994); Graves *et al.* (1994); Sasada *et al.* (1993); Heavner *et al.* (1993); Kansas *et al.* (1994); Lightner (1994); Moller *et al.* (1994).

molecules (Higashiyama *et al.*, 1991), and betacellulin (Shing *et al.*, 1993) (Table I). EGF sequences have also been reported in extracellular matrix molecules such as aggrecan (Moller *et al.*, 1994), tenascin (Taylor *et al.*, 1989; Nies *et al.*, 1991), and the laminin molecule (Panayotou *et al.*, 1989) (Table I). Two additional groups of proteins, the adhesion molecules and coagulation factors, are also known to have EGF modular domains (Table I).

Interactions of these potential ligands with different members of the EGF receptor (EGF-R) family are much better characterized *in vitro* than *in vivo;* therefore, much remains to be learned of how each may play a role in cutaneous wound repair. In addition to this family of EGF-like ligands, other growth factors/cytokines such as the TGF-βs, platelet-derived growth factors (PDGFs), insulinlike growth factors (IGFs), and tumor necrosis factor (TNF-α), which have their own discrete receptors, can also indirectly modify or transmodulate this common receptor, the EGF-R (Massague, 1985; Olashaw *et al.*, 1986; Krane *et al.*, 1991; Donato *et al.*, 1992). As these competing cytokine functions and interactions are defined, we are certain to gain insight into the singular processes such as cell migration, division, secretion, and differentiation that comprise wound repair.

3. The Common Receptor Pathway for EGF-like Ligands

The common receptor, EGF-R, is a part of a subfamily of receptor tyrosine kinase molecules. To date, at least four members of this EGF-R subfamily have been identified in mammalian tissues (Holmes *et al.*, 1992; Plowman *et al.*, 1993) (Table II). Only the EGF-R and *neu*(c-*erbB*-2) have been localized in normal human skin (Nanney *et*

Table II. EGF Receptor Tyrosine Kinase Subfamily

	Mol. wt. (kDa)	Binding/activation of receptor					
		EGF	TGFα	Amphiregulin	gp30	NeuDF	HER4DF
EGF-R	170	+	+	+	+	−	−
HER2/erbB2	185	−	−	−	+	+	−
HER3/erbB3	160		−	−	−	−	−
HER4/erbB4	180	−	−	−	−	−	+

References: Carpenter (1993); Callaghan *et al.* (1993); Plowman *et al.* (1993); Qian *et al.* (1994).

al., 1984; Maguire *et al.,* 1989). Although EGF-R remains as the prototypic cytokine receptor, much remains to be learned about how, when, where, why, and which EGF-like ligands interact with one or more EGF-R subfamily members. Some of these interactions may be restricted embryologically and be cell type or species specific.

A complete description of the exon–intron structure of the EGF receptor has only been described for the chicken EGF-R (Callaghan *et al.,* 1993) that is encoded by the proto-oncogene c-*erbB*-1. This form of the EGF-R is composed of 28 exons and spans over 75 kilobases (kb) that encode for the four previously identified domains: a ligand-binding (LB) domain; the transmembrane (TM) domain; the catalytic protein tyrosine kinase (PTK) domain; and the C-terminal regulatory (REG) domain. The REG domain consists of a calcium internalization (CAIN) subdomain, autophosphorylation domains that are also involved in binding with proteins containing *src* homology domains (for reviews, see Carpenter, 1992, 1993), and unspecified sites that interact with other proteins involved in signal transduction pathways such as G-proteins, *ras,* and mitogen-activated protein (MAP) kinase pathways (Fig. 1) (for reviews, see Carpenter,

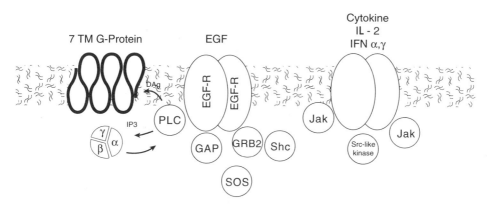

Figure 1. EGF-related signal transduction pathways. The three major signal transduction pathways mediating EGF-receptor-induced biological activation include (1) G-protein pathways, (2) MAP kinase pathway, and (3) JAK-Stat pathways.

Table III. Molecules Directly Involved in EGF Receptor Signal Transduction

Molecule	Mol. wt. (kDa)	SH2/SH3 domains	PY	Function
Membrane lipids, metabolism and cytoskeleton				
PLCγ-1	145	+	+	PI 4,5-P2 → 1,4,5-IP3 + DAG
DAG kinase	86	−	+	DAG → PA
pp85	85	+	?	Regulatory subunit, PI-3
α Adaptin	100	−	−	Interacts with coated pits
Ezerin	80	−	+	Cytoskeletal component
Annexin I	35	−	+	Phospholipid-binding protein
PTP1D	65–68	+	+	PTPase
ras GTP pathway				
pp190	190	−	+	Associates with GAP
pp62	62	−	+	Associates with GAP
GAP	20	+	+	Activates ras GTPase
Grb-2	25	+	−	Binds SOS, Shc-2
Shc-2	66	+	+	Binds SOS, Grb-2, EGF-R
Jak/Stat pathway (p91)				
Jak1	130	−	+	Associates with cytokine receptors
Jak2	130	−	+	and phosphorylates Stat proteins
Tyk2	135	−	+	
Stat1α	91	+	+	Interacts with nuclear/DNA sites
Stat1β	84	+	+	
Stat2	113	+	+	
Stat3	92	+	+	

Abbreviations: SH2,3 Src homology domain 2,3; PY, tyrosine phosphorylation; GAP, GTPase activating protein; PA, phosphotidic acid; DAG, diacylglycerol. References: Carpenter (1992, 1993); Soler *et al.* (1993); Sorokin and Carpenter (1993); Hall (1994); Darnell *et al.* (1994); Zhong *et al.* (1994).

1992, 1993; Blenis, 1993; Egan and Weinberg, 1993; Rozakis-Adcock *et al.*, 1993; Williams *et al.*, 1993; Hall, 1994). EGF-R itself has been subjected to site-directed mutagenesis to define the functions of these domains, but other subfamily members have been less well studied (for reviews, see Cadena and Gill, 1992; Soler *et al.*, 1994).

The EGF-R-related signal transduction mechanisms that activate the intrinsic tyrosine kinase and dimerization of EGF-R have been intensely studied (for reviews, see King *et al.*, 1991; Carpenter, 1993; Cadena and Gill, 1992; Sorokin *et al.*, 1994). The role of the autophosphorylation sites on the EGF-R in regulating its tyrosine kinase activity and binding of proteins containing *src* homology domains has provided important insights into signal transduction by many different types of receptors (Carpenter, 1992; Soler *et al.*, 1994; Larner *et al.*, 1993; Ruff-Jamison *et al.*, 1993; Sadowski *et al.*, 1993; Darnell *et al.*, 1994). However, the mechanisms whereby substrates for EGF-R induce tyrosine phosphorylation (Table III) and mediate the plasma membrane, cytosolic, mitochondrial, and nuclear signals regulating cell growth, differentiation, and migration are just now becoming more clearly understood (Rozakis-Adcock *et al.*, 1993; Williams *et al.*, 1993). Figures 1–4 show simplified diagrams of EGF-

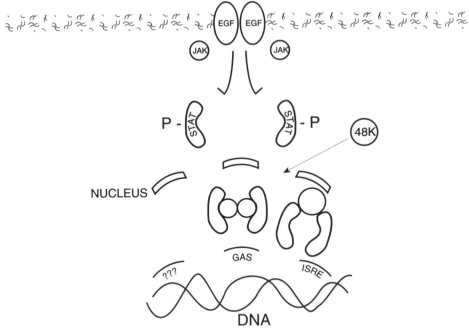

Figure 2. JAK-Stat signal transduction pathways. The Janus kinase (JAK) mediates a rapid response pathway to the nucleus via Stat protein family members and specific DNA binding sites such as GAS and ISRS.

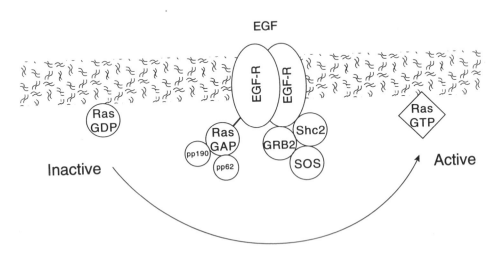

Figure 3. Ras-GTP signal transduction pathway. The *ras* protein GTP pathway is activated by hydrolysis of GTP to GDP, mediated by a complex of proteins shown in this figure.

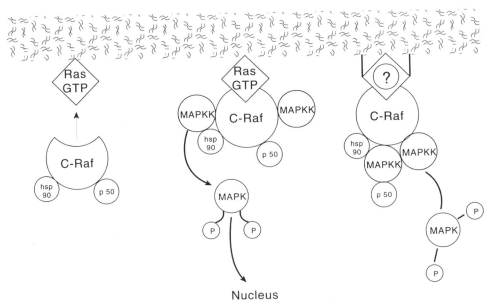

Figure 4. MAP kinase signal transduction pathway. The MAP kinase pathway also interacts with cytosolic and nuclear receptors as shown in the figure.

R-activated signal transduction mechanisms. Since this area is very dynamic, the reader is referred to recent articles on varying aspects of these pathways (Aronson, 1992; Egan and Weinberg, 1993; Pelech and Sanghera, 1992; Hall, 1994; Darnell *et al.*, 1994). Recently, a rapid, more direct mechanism to relay EGF-R-activated signal transduction to the nucleus, different from the pathway described in Figs. 3 and 4, was proposed (Fu and Zhang, 1993; Lutticken *et al.*, 1994; Darnell *et al.*, 1994) (JAK Stat pathway, Fig. 2).

In the EGF-R-mediated signal transduction pathway, more emphasis has been placed on tyrosine kinases than serine/threonine kinases and phosphotyrosyl phosphatases (PTPases). Although a PTPase activity was documented in the original studies of EGF-R-induced tyrosine kinase activity (Carpenter *et al.*, 1978), interactions of tyrosine kinases with PTPases were not intensely studied until a large family of PTPases was identified by cloning and sequencing (Fisher *et al.*, 1991). Currently, PTPases have been localized in normal human skin (Gunaratne *et al.*, 1994), but have not been examined in the wound-healing setting. What role(s) PTPases may play in wound healing should be a fruitful area for study, since it is now appreciated that stimulatory signals mediated through tyrosine kinase receptors such as the EGF-R must be attenuated with regulatory molecules such as the PTPases. *In vitro* studies have recently shown that EGF stimulates substrate-selective protein tyrosine phosphatase activity (Hernandez-Sotomayor *et al.*, 1993) and mitogen-activated protein kinase regulates the EGF-R through activation of a tyrosine phosphatase (Griswold-Prenner *et al.*, 1993).

Powerful new techniques such as the polymerase chain reaction (PCR), *in situ* hybridization, and biological studies of transgenic mice are now available to define with greater specificity how growth factors and cytokines regulate wound healing. Overexpression of growth factors such as TGF-α (Sandgren *et al.,* 1990; Dominey *et al.,* 1993) and inactivation of signal transduction (Blessing *et al.,* 1994) were recently employed to study how such mutations affect normal development and wound repair. Such studies in transgenic mice have already provided major insights and surprises. As it was supposed that TGF-α was a major regulator of keratinocyte growth (Coffey *et al.,* 1987; Pittelkow *et al.,* 1993), it came as a surprise that transgenic mice with a deletion of TGF-α grew and developed with no discernible effect on wound healing, although the untraumatized normal skin exhibited wavy hairs (Luetteke *et al.,* 1993). Similarly, transgenic "knockout" mice that do not express tenascin had normal-appearing skin, hair, and teeth and did not exhibit grossly abnormal wound healing (Saga *et al.,* 1992). Conversely, transgenic mice overexpressing TGF-α did show gross and microscopic abnormalities (papillomas) in the skin in response to cutaneous trauma (Dominey *et al.,* 1993). Targeted disruption of EGF receptor in transgenic mice produced hair follicle abnormalities (Threadgill *et al.,* 1995; Sibilin and Wagner, 1995). As expected, overexpression of inhibitory molecules of the TGF-β family (bone morphogenetic protein-4) did produce mice with severe deficiencies in keratinocyte growth (Blessing *et al.,* 1993). Thus experiments with overexpression and elimination of modulatory molecules can provide valuable information. These *in vivo* experiments may suggest the intrinsic importance of EGF-related molecules and their common receptor forms. Biological redundancy in this critical pathway would act to minimize otherwise lethal effects of inherited (or induced) mutations (Brookfield *et al.,* 1992; Erickson, 1993a). The challenge then remains to utilize *in vivo* models such as transgenic mice with selective impairment of various elements of the signal transduction to test potentially therapeutic wound-healing strategies (Threadgill *et al.,* 1995; Sibilin and Wagner, 1995). Until then, feasibility studies using various growth factors or cytokines will continue to be based on *in vitro* models prior to moving to the more costly and time-consuming *in vivo* studies on human wounds.

4. Biological Relevance of EGF-R Pathways in Human Wound Healing

Although *in vivo* studies related to wound healing and the EGF-R pathways in other species had been previously reported (for review, see King *et al.,* 1991), the first prospective, double-blind trial demonstrating clinical efficacy of topical EGF in human cutaneous wound repair was not published until 1989 (Brown *et al.,* 1989). This study documented a modest acceleration in the rate of skin resurfacing in partial thickness, donor site wounds following twice daily topical applications of EGF. Enthusiasm for this report was out of proportion to its clinical utility since donor sites are not problematic wounds. By documenting that "normal" wounds (surgically created under ideal conditions) could be induced to heal faster by topically delivered EGF, this paper put to rest the old clinical bias that normal wound repair could not be accelerated. This study also confirmed what most investigators involved in wound healing already knew to be true: The modest effect of topical EGF in wounds was a reflection of the complex interactions of cells, growth factors/cytokines, type of wound, and individual patient differences. The clinical challenge remains of how to target the EGF-R and related cytokine pathways so that acute and chronic human cutaneous wounds achieve satisfactory healing despite the presence of underlying disease states such as diabetes or other adverse wound-healing circumstances.

5. Regulation of EGF-R-Related Pathways in Normal Human Skin

The vast number of *in vitro* studies evaluating the EGF-R-related pathways continue to race ahead of the clinical studies addressing cytokine participation within the wound-healing setting. It is difficult in biological trials to mimic *in vitro* models of EGF-R signal transduction pathways that provide such seemingly clear-cut results. These difficulties in defining a precise function for EGF-R or any other cytokine *in vivo* are related to a number of variables such as age, anatomical site, proliferative state, degree of differentiation, preexisting cutaneous and systemic abnormalities, temporal intervention after injury, type of injury, and other undefined genetic and environmental factors. To date, the *in vivo* relevance and regulation of EGF-R have been examined in normal and abnormal skin using the currently available technologies. These techniques have been limited to immunoreactive localization of EGF-R and its substrates, *in situ* localization of the mRNA of interest, and therapeutic trials. Results of such studies and their interpretation are provided below.

5.1. EGF-R in Normal Fetal and Neonatal Human Skin

One way to gain insight into the participation of EGF-R and its ligands (Table I) in wound repair is to consider the known roles for this pathway during embryogenesis of human skin and its appendages. Normal embryological processes may recapitulate certain wound repair events necessary for the reestablishment of epithelial integrity and function. The original work with EGF injections into neonatal mice clearly showed that EGF could affect epithelial structures by modulating the normal developmental process (Cohen and Elliott, 1963). Since all known functions of EGF and related molecules are mediated by the EGF-R, immunolocalization of EGF-R and radioactive EGF binding sites during embryonic development of skin and its appendages were performed in rodents (Green and Couchman, 1984) and humans (Nanney *et al.*, 1984, 1990b) to gain insights by spatial localization. *In utero* and even in the neonatal human epidermis, EGF-R localization was normally distributed throughout in all nucleated layers of the epidermis and was not restricted to the germinative population of keratinocytes (Nanney *et al.*, 1990a,b). This same pattern of detection in all layers of the epidermis was also noted in rapidly proliferating epidermal diseases (Nanney *et al.*, 1986, 1992a) and even in normal hyperproliferative adult interfollicular epidermis (Fig. 5B) (Wenczak *et*

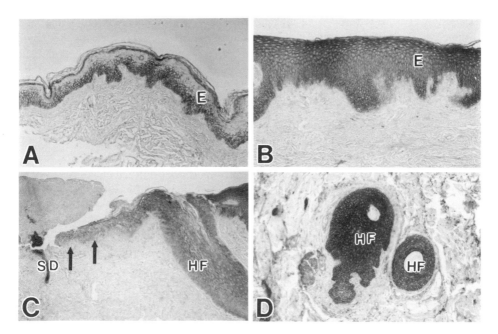

Figure 5. Immunoreactive EGF-R in paraffin embedded samples of human skin × 45. (A) Normal human skin shows EGF-R restricted to the basal/spinous population in the epidermis (E). (B) Hypertrophic epidermis near the edge of a human burn wound at postburn day 7 shows EGF-R throughout all nucleated cell layers of the epidermis (E). (C) The migrating and proliferating epithelium at the wound edge (arrows) as well as the deeper surviving epithelial cells from a hair follicle (HF) and eccrine sweat duct show immunoreactive EGF-R at postburn day 3. (D) Deeper surviving epithelial cells from hair follicles (HF) show positive staining for EGF-R at postburn day 4.

al., 1992); however, in the majority of normal adult specimens, EGF-R was confined primarily to the basal cell layer (Figs. 5A and 7A). This EGF-R distribution presumably reflects the increased mitotic rate in neonatal epidermis. Although the presence of EGF-R in all epidermal layers seems to correlate best with proliferation (Figs. 5A and 7A) (Green *et al.*, 1983; Stoscheck *et al.*, 1992), the *in vivo* evidence directly linking EGF-R to keratinocyte mitosis remains unproven although indirectly implicated (Ellis *et al.*, 1990; Nanney *et al.*, 1992a; Wenczak *et al.*, 1992). Other possibilities beyond mitotic functions need to be explored. For example, disturbances in the permeability barrier that may mimic trauma may be the trigger for this altered EGF-R distribution. Certainly psoriatic lesions, neonatal skin, and wounded epidermis, lesions in which the permeability barrier functions are disturbed, exhibit this common pattern of persistent EGF-R throughout all epidermal layers. Increased EGF-R can likewise be indirectly correlated with parakeratosis since this general pattern is also inherent to focal regions of parakeratosis in psoriatic lesions (Nanney *et al.*, 1986) and normal oral mucosa.

A third distribution pattern has also been reported for EGF-R. When EGF-R localization was examined in restrictive dermopathy, a condition with abnormal dermal proliferation, EGF-R was found to be restricted to the basal layer as compared to the expression of EGF-R throughout all epidermal layers normally expected at that age (Nanney *et al.,* 1990a). Therefore, developmental and disease-related events that regulate EGF-R cannot be predicted based on a simple model of EGF-R being expressed only in mitotically active cells. This provides additional support for the theories that the EGF pathway is not restricted to epidermal proliferation but extends into the realm of mesenchymal–epidermal interaction.

5.2. EGF-R in Normal Adult Human Skin

As has been reviewed above, in normal adult unperturbed thin and thick skin, EGF-R is spatially located primarily within the basal epidermal population (Figs. 5A and 7A). This localization of EGF-R in adult skin was confirmed by both immunoreactive EGF-R and by autoradiography of radioactive EGF (Nanney *et al.,* 1984). Ten years later, this spatial localization was again confirmed, but this time by the new technique of reverse transcriptase *in situ* polymerase chain reaction (Patel *et al.,* 1994). These original morphological studies of human tissue sections served to support the assumption that EGF-R was predominantly found in germinative cells since EGF was a well-characterized mitogenic agent of responsive cells *in vitro.* Detection of EGF-R in the renewing populations of hair follicles (Figs. 5C and 7C, D), sebaceous glands, and sweat glands was confirmatory as well. However, the highest concentration of EGF-R was detected on eccrine sweat ducts that were not known to be an actively dividing cell population (Fig. 5A). Such unexpected localizations were originally suspected to be due to experimental artifacts, but have proven to be correct. These EGF receptor sites suggest that EGF-R is spatially positioned on deeply located keratinocytes (Figs. 5C,D and 7C,D), cells that are of prime importance for the resurfacing of human partial-thickness wounds.

6. Functional Roles of EGF-R in Normal and Wounded Human Skin

The presence or absence of EGF-R on specific keratinocyte populations in the interfollicular epidermis and within epidermal appendages implies regulatory and biological significance for this cytokine pathway. Since appendages are situated deep in the dermis and subcutaneous tissues, this population of keratinocytes is more protected from the environment and therefore serves as a ready source of cells to replace epidermis lost during wounding. Induction of these cells to migrate, divide, differentiate, and reestablish an effective permeability barrier is critical to achieve satisfactory wound healing. EGF-R-mediated events appear to play vital *in vivo* roles in all of these four critical steps in wound healing.

6.1. EGF-R Roles in Reforming Epidermal Permeability Barrier in Wounded Human Skin

Resurfacing or reepithelializing a cutaneous wound is the process that is most often considered in achieving satisfactory wound closure. Inherent in this concept is that the barrier function of the surface epidermis must be reestablished to prevent fluid loss or invasion by microorganisms. To be a truly effective barrier, the epidermis must, in addition to refilling the space lost due to trauma with keratinocytes ("fill up the hole"), also reestablish the percutaneous or epidermal permeability barrier ("seal the leaky roof"). How EGF-R participates in reestablishing an effective permeability barrier in human skin is not well understood. Reports of perturbations of epidermal barrier function are only beginning to be correlated with initiation of the cytokine cascade in human skin (Nickoloff and Naidu, 1994).

In first-degree wounds including abrasions in which a defective permeability barrier is present there may be no apparent involvement of the dermis or immunocytes initially or even after replacement of the barrier. To address the role of the EGF-R pathway in this process, a model of superficial injury with minimal dermal or immunocyte involvement—tape stripping of mouse tail skin—was examined. The degree of injury and the loss of the permeability barrier in this model had been well documented to be dependent on the number of times the mouse tail was stripped. When mouse tail epidermis was wounded, it responded within 24 hr by increasing the number of EGF-R (Fig. 6). This modulation in EGF-R following trauma was predictable. Clinical trials had reported a biological response following topical EGF to human skin (Brown *et al.,* 1989) and confirmed in cutaneous wounding experiments with pigs (Brown *et al.,* 1986; Nanney, 1990). In traumatized mouse tail epidermis, the peak response of EGF-R expression occurred within 24–72 hr and returned to baseline 4–14 days after tape stripping (Stoscheck *et al.,* 1992). This increase in EGF-R occurred 24 hr prior to

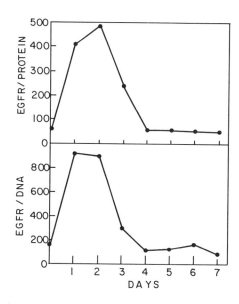

Figure 6. Tails of Balb/c mice were tape-stripped to remove the outer epidermal layers. Wounded tails were removed at various days after injury and EGF-R was quantified using an ELISA procedure. (A) Values are expressed as EGF-R per protein levels and (B) EGF-R per DNA levels. (From Stoscheck *et al.,* 1992. Reprinted with permission.)

increases in mRNA, DNA, and total protein. Thus EGF receptors show an increase in response to wounding. Presumably topical EGF directly binds to this receptor to mediate the biological acceleration of epidermal repair. These studies did not examine other biochemical events or the immediate, early gene response to tape stripping so that the molecular events related to EGF-R in response to trauma are not yet defined.

6.2. EGF-R-Mediated Keratinocyte Differentiation in Human Skin Wounds

Topical applications of EGF to human donor site wounds (Brown *et al.*, 1989) and venous leg ulcers (Falanga *et al.*, 1992) appear to accelerate wound healing or at least decrease wound size. These effects of EGF (and TGF-α) are not assumed to be limited to epidermal effects. However, an implicit assumption of these studies is that reepithelialization, including keratinocyte differentiation and normalization of EGF-R localization, proceeds normally in the damaged epidermis and over denuded dermis.

Topical EGF can induce a biological response in other epidermal settings. For example, following daily EGF treatments with very high doses of EGF (levels that are actually inhibitory in the wound-healing setting), keratinocyte differentiation was altered, a decrease in epidermal proliferation was noted, and return of the psoriatic phenotype of human psoriasis grafts on nude mice to that seen in normal epidermis was documented (Nanney *et al.*, 1992b). The abnormally differentiating psoriatic epidermis histologically normalized, with a loss of the parakeratotic stratum corneum and concomitant return of the EGF-R to its normal basal layer distribution. Other indirect evidence is also available that shows this same correlation of EGF-R expression in rapidly proliferating epidermis (Ellis *et al.*, 1991; Nanney *et al.*, 1992a) and the normally parakeratotic mucosal epithelial of humans (L. B. Nanney, unpublished observations). *In vitro* studies have recently suggested that EGF-like ligands can regulate keratin 8 expression, further implicating the involvement of the EGF pathway on keratinocyte phenotypes (Cheng *et al.*, 1993).

6.3. EGF-R-Mediated Keratinocyte Proliferation in Human Skin Wounds

Various methods have been used to determine whether EGF-R-expressing cells are the same cells that comprise the proliferating population due to trauma or the presence of hyperproliferative skin diseases. Labeling with [3H]thymidine, anti-Br-DU, and anti-PCNA (proliferating cell nuclear antigen) antibody methods provide quantitative data to determine the mitotic index and epidermal turnover rates. Colocalization of EGF-R substrates such as annexin-1 (Fava *et al.*, 1993) and PLC-γ-1 (phospholipase C) (Nanney *et al.*, 1992b) also provide some assurances that the patterns seen do or do not correlate with EGF-R expression. Although identical correlations of these substrates with morphological localization of EGF-R have not always been observed, in general there is excellent correlation in hyperproliferative epidermis and cutaneous wounds when examined in detail. For example, the tape stripping of the mouse tail epidermis model has provided strong support for a direct role of EGF-R in epidermal proliferation (see Section 6.1). In this model and in the topical EGF treatment of human psoriatic skin grafted onto nude mice the correlation of EGF-R expression throughout all epidermal layers with increased keratinocyte proliferation and delayed keratinocyte differentiation is very strong.

6.4. EGF-R-Mediated Keratinocyte Migration and Adhesion in Human Skin Wounds

Although the tape-stripping model in conjunction with enzyme-linked immunosorbent assay (ELISA) detection methods showed that the EGF-R reached peak levels during early wound repair (Stoscheck *et al.*, 1992), the injury in this model was restricted to the epidermis. It is somewhat unique among wound-healing models in that it heals both in the absence of a marked dermal inflammatory response and in the absence of epidermal migration. To examine the role of EGF-R in more severely damaged human skin, skin excised from human burn wounds during routine skin-grafting procedures was examined. These burn wounds ranged in severity from superficial partial thickness to viable margins of full-thickness burns. These studies support the conclusion that the localization of EGF-like ligands and the EGF-R was spatially and temporally regulated in these burn wounds. Although these studies were performed in retrospect to the human clinical trials with topical EGF (Brown *et al.*, 1989), the data did suggest that the EGF could be acting directly since the appropriate receptor was found within healing wounds.

In these human burn tissues, EGF-R was localized in the migrating and proliferating epithelial tips present at the wound edges (Figs. 5C, D and 7C). EGF-R was also detected in epithelial islands arising from the surviving remnant of hair follicles and sweat ducts found deep within the burned dermis (Wenczak *et al.*, 1992) (Fig. 7D). Autoradiographic studies using radioactive EGF confirm that an excess of available EGF receptors are present in healing keratinocytes populations (Fig. 7C, D). Spatially, the EGF-R was first observed to diminish in the migrating epithelial tips and only later in the adjacent proliferating epidermal keratinocytes in postburn days 4–15. These findings or localization patterns were very reminiscent of the loss of immunoreactive EGF-R in the downward migration of the epithelium that invaginates to form human hair follicle *in utero* (Nanney *et al.*, 1990b). Whether this reduction in EGF-R reflects down-regulation by specific EGF-like ligands or is due to the indirect effects of other cytokines or autocoids is unknown, but may account for the intensity and duration of keratinocyte migration, proliferation, and differentiation.

In vitro studies with keratinocytes suggest that arachidonic acid metabolites appear to function in EGF signal transduction (Peppelenbosch *et al.*, 1993). EGF-induced cortical actin polymerization was induced by lipoxygenase metabolism, whereas stress fiber breakdown was mediated by cycloxygenase metabolites. Thus arachidonic acid metabolites appear to provide a novel mechanism for directing EGF-induced cytoskeletal changes. Cytoskeletal change is certainly an integral component of the migratory

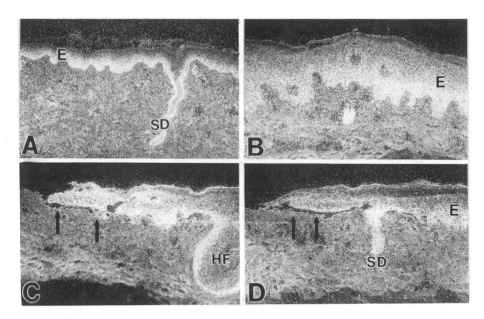

Figure 7. Autoradiographic binding studies using radioactive EGF were conducted in accordance with previously published techniques (Nanney *et al.*, 1984). Sections were slightly overexposed at 4 weeks. (×45). (A) Normal skin shows intense binding in the germinative population of the epidermis (E) and strong label in an eccrine sweat duct (SD). Some label is also present in the dermis, but to a less extent than the basal epidermal labeling. (B) Hypertrophic epidermis (E) near the edge of a burn wound shows intense labeling throughout all the nucleated layers. (C) At postburn day 4, strong binding for EGF is apparent in the migrating/proliferating tip (arrows) of epithelial cells that have grown out from an adjacent hair follicle (HF). (D) At postburn day 4, moderate label is present in the migrating/proliferating tip (arrows) that has grown out from a surviving eccrine sweat duct (SD).

events during reepithelialization following trauma. Other studies both *in vitro* and *in vivo* have also implicated the EGF-R tyrosine kinase pathway in migratory functions. An increase in a substrate known as focal adhesion kinase was localized within migrating keratinocytes following stimulation with EGF (Gates *et al.*, 1994). These studies all continue to implicate the EGF receptor pathway in migratory events, but these data only represent preliminary pieces of the puzzle.

Other ligands along the EGF-R pathway (TGF-α and vaccinia virus growth factor) likewise have been examined in wound healing and found to stimulate reepithelialization (Schultz *et al.*, 1987). Endogenous immunoreactive TGF-α is reportedly present in the healing epithelial edges throughout the first 2 weeks of wound healing (Wenczak and Nanney, 1993). These findings correlate well with the appearance of other EGF-related molecules such as tenascin (Mackie, 1988; Betz *et al.*, 1993; Lightner, 1994) and heparin-binding EGF-like molecules (Marikovsky *et al.*, 1993), which are also induced in healing wounds. This apparent redundancy in EGF-R-related ligands (Table I) may relate to the dedifferentiation of cells as they recapitulate embryological stages of skin development during wound repair. Alternatively, the presence of multiple EGF-R ligands ensures that a relative deficiency in one molecule will not preclude wound closure. The possibility that other ligands known to bind to HER-2–4 (erbB-2–4, Table II) and possible other as yet unknown variants of the EGF-R may also participate in wound healing should be considered likely, but has not been studied in any detail. If EGF-like ligands affect the migrating epithelial tip, then topical EGF or TGF-α should affect the *in vivo* outgrowth of keratinocytes in a wound.

Evidence that this effect does occur was provided by treating porcine excisional wounds with EGF. The most dramatic stimulation of epithelialization was noted during the migratory phases of wound repair (Nanney, 1990). Data from the topical EGF trial in human donor site wounds (Brown *et al.*, 1989) also showed the greatest statistically significant difference between placebo and EGF treatments during the migratory phase or earliest time points. Thus in both studies, the epidermal effects of EGF diminished as wound healing progressed. These *in vivo* findings are compatible with *in vitro* results showing that EGF and TGF-α stimulated keratinocyte migration and prolonged keratinocyte survival (Barrandon and Green, 1987; Hebda, 1988).

6.5. EGF-R Interactions in Normal and Abnormal Dermis

6.5.1. Fibroblastic Responses Regulated by the EGF-R Pathway

The *in vivo* role of EGF-R in nonepithelial cells have not been well defined due to low levels of EGF-R detected in these cells in normal dermis (Nanney *et al.*, 1984). Nevertheless, all evidence points toward involvement of this cytokine pathway in the interactions that occur between mesenchymal and epithelial cells. Newer and more sensitive techniques such as reverse transcriptase PCR will undoubtedly lead to a better understanding of endogenous levels in the connective tissue environment.

Despite these limitations, many years ago it became apparent that cultured fibroblasts express EGF receptors. A predictable down-regulation phenomenon could be measured in response to EGF (Carpenter and Cohen, 1976). Although the addition of EGF to sponge-induced granulomas in rats produced an increase in the net amount of collagen, this effect was due to an increase in fibroblast proliferation and was not due to an increase in expression of type I or II procollagen genes (Laato *et al.*, 1987).

Addition of EGF to fibroblastic culture media produces measurable increases in the synthesis of a wide variety of other matrix proteins ranging from the glycosaminoglycans (Lembach, 1976) to the degradative matrix proteins such as collagenase (Colige *et al.*, 1992) and stromelysin (Kerr *et al.*, 1992). It was recently shown that EGF, through posttranscriptional mechanisms, could regulate gene expression for collagenase and stromelysin in cultured human fibroblasts (Delany and Brinckerhoff, 1992). It should also be pointed out that other cells within the wound environment, namely keratinocytes, can also respond to EGF by increasing their expression of degradative matrix proteins such as 92-kDa gelatinases (Shing *et al.*, 1993). Thus in cultured cell populations, particularly fibroblasts, many of the necessary extracellular matrix components that would be essential for dermal remodeling following trauma are known to be regulated by the EGF–EGF-R pathway.

While other cytokines such as TGF-β and PDGF are perhaps better known for their dramatic *in vivo* fibrotic responses (see Chapters 5 and 7), a few papers have nonetheless documented some positive EGF-induced dermal effects during wound repair. While using a porcine excisional model, Nanney (1990) showed a dose-responsive increase in the thickness of granulation tissue following daily topical applications of EGF. Acceleration of tensile strength in incisional wounds treated with EGF was likewise documented (Brown *et al.*, 1988; Perry *et al.*, 1993). These positive outcomes following sustained release of EGF confirmed the earlier work by Buckley *et al.* (1985). This latter study, which evaluated granulation tissue inside a subcutaneous wound chamber, was the first to suggest that continuous drug delivery of EGF was a requirement for successful acceleration of reparative events. While all four of these studies employed pharmacological delivery methods, a host of recent papers have uncovered several possible natural mechanisms for continuous local stimulation of the EGF receptor pathway within the dermis.

The initial doses of EGF and TGF-α as well as other cytokines (TGF-β, PDGF) are undoubtedly delivered to the wound environment by platelet degranulation within the provisional fibrin clot (Ben-Ezra *et al.*, 1990; Oka and Orth, 1983; Assoian *et al.*, 1983). As mentioned above, at least three of the EGF-like molecules are temporally induced by cells within the granulation tissues. These cytokines include heparin-binding EGF molecules as well as EGF-like sequences in tenascin-c and the laminin molecule, which are present at the epithelial–mesenchymal junction. Tenascin-c expression in wounded human dermis has already been indirectly linked with epidermal proliferation (for reviews, see Schalkwijk *et al.*, 1991; Lightner, 1994). Immunohistochemical localization of tenascin in human skin wounds shows a pericellular distribution around fibroblastic cells as early as 2–3 days postinjury (Betz *et al.*, 1993). Certainly these EGF-like ligands are spatially situated for a sustained delivery to cells in the mesenchymal milieu during the remodeling phase of wound repair (Engel, 1989).

Recent studies have also expanded the role of the fibroblast in wound repair. For years, these cells were considered to be the target cells for cytokine stimuli. Recently, another member of the basic fibroblast growth factor (bFGF) family known as keratinocyte growth factor (KGF) has been shown to have significant impact on wound healing (Werner *et al.*, 1992). While the KGF molecules are produced in the granulation tissue by the fibroblast, they act in a paracrine fashion to stimulate the nearby keratinocytes in hair follicles and on the surface (Pierce *et al.*, 1994). Although KGF can bind to its specific KGF receptor located on keratinocytes, KGF at least *in vitro* can indirectly stimulate keratinocyte proliferation by induction of TGF-α (Dlugosz *et al.*, 1994). This is yet another example of the complex relationships that exist between the EGF-R pathway and a host of other cytokines.

6.5.2. Macrophage Responses Regulating the EGF-R Pathway

Overwhelming evidence suggests that the macrophage is a pivotal participant in cutaneous wound repair. This cellular population migrates into the wound during the first 24- to 48-hr period (Leibovich and Ross, 1975), recruited by a host of chemotactic factors including but not limited to TGF-β (Grotendorst *et al.*, 1989). Macrophages seem to function as local cytokine factories for heparin-binding EGF-like molecules (Higashiyama *et al.*, 1991), EGF, and TGF-α (Rappolee *et al.*, 1988), as well as many other cytokines. These products are secreted into the wound environment where their levels have been documented to date in wound fluid collected from surgical drain tubes (Dvonch *et al.*, 1992) and from porcine excisional wounds (Marikovsky *et al.*, 1993).

Macrophage secretory products are by no means limited to cytokines. These cells actively participate in the remodeling phase of repair by a host of discrete mechanisms. To date, *in vivo* studies using either PDGF or TGF-β treatments indicate that the macrophage responds by increasing expression of structural matrix proteins, including collagens I and III, elastin, and fibronectin (Quaglino *et al.*, 1990; Pierce *et al.*, 1991), and decreasing expression of degradative matrix proteins such as stromelysin (Quaglino *et al.*, 1989). Macrophages within human burn tissues also show an enhanced expression of matrix metalloproteinases such as collagenase following trauma (Stricklin *et al.*, 1993); however, in the complicated clinical setting, the probable cytokine stimuli and pathways controlling the equilibrium between matrix synthesis and degradation have not been identified.

6.5.3. Endothelial Responses to the EGF Pathway

All of the aforementioned studies suggest a physiological significance for the EGF pathway in nearly all of the cell types (keratinocytes, fibroblasts, macrophages) involved in complex wound repair. Endothelial cells and smooth muscle cells are no exception. To date, *in vivo* studies in the corneal wound-healing model have indicated that both TGF-α as well as EGF can stimulate angiogenesis (Schreiber *et al.*, 1986). The response of endothelial cells to the EGF pathway has also been confirmed in the porcine excisional model, although it should be noted that the magnitude of the angiogenic response following EGF treatment was considerably weaker than bFGF, TGF-α, or TNF-α treatments (Roesel and Nanney, 1995). Recently, keratinocyte-produced TGF-α was shown to induce tubulogenesis of human microvascular endothelial cells, suggesting a possible paracrine loop whereby the TGF-α pathway may participate in wound repair (Ono *et al.*, 1992). Heparin-binding EGF-like growth factor derived from macrophages is also known to stimulate smooth muscle cell migration (Higashiyama *et al.*, 1993) and may well serve as an endogenous cytokine source within the wound environment.

At this point it seems appropriate to expand beyond the strict confines of wound repair to compare the cytokine stimuli within wound healing to the stromal responses for cytokines that occur within cutaneous neoplasms. The concept that tumors are really wounds that did not heal was proposed in 1986 by Dvorak. Angiogenic ingrowth, fibroblastic responses, and immune modulations are common events in both of these *in vivo* circumstances. Moreover, these processes are correspondingly responsive to cytokines. The *in vitro* stimulation of capillary endothelium by EGF–TGF-α was documented years ago (Gospodarowicz *et al.*, 1978). In tumors, the overexpression of the EGF–TGF-α pathway and the ability of endothelial cells to respond in a paracrine fashion to these signals can obviously trigger capillary ingrowth, which is associated with a poor prognostic outcome (Weidner *et al.*, 1991). Conversely, the typical skin ulcer may well represent the opposite end of the growth spectrum where there is a deficiency in EGF-like ligands or a deficiency elsewhere along the common EGF-R pathway. The resulting cutaneous ulcer may exhibit poor capillary ingrowth followed by nutritional deficiencies that perpetuate the status quo. It should be emphasized that presently very little is known about the endogenous cytokine pathways in chronic wounds, although the chronic wound-healing setting is certain to be the focus of future evaluations. At present, evidence continues to amass that provides strong evidence that the EGF-R pathway is one of the integral pathways influencing many of the major events of wound repair.

ACKNOWLEDGMENTS. This work was supported by the National Institutes of Health GM40437, AR41943, DK 26518 and the Department of Veterans Affairs.

References

Assoian, R. K., Komoriya, A., Meyers, C. A., Miller, D. M., and Sporn, M. B., 1983, Transforming growth factor-beta in human platelets. Identification of a major storage site, purification, and characterization, *J. Biol. Chem.* **258**:7155–7160.

Barrandon, Y., and Green, H., 1987, Cell migration is essential for sustained growth of keratinocyte colonies: The roles of transforming growth factor-α and epidermal growth factor, *Cell* **50**:1131–1137.

Ben-Ezra, J., Sheibani, K., Hwang, D. L., and Lev-Ran, A., 1990, Megakaryocyte synthesis is the source of epidermal growth factor in human platelets, *Am. J. Pathol.* **137**:755–759.

Betz, P., Nerlich, A., Tubel, J., Penning, R., and Eisenmenger, W., 1993, Localization of tenascin in human skin wounds—an immunohistochemical study, *Int. J. Leg. Med.* **105**:325–328.

Blenis, J. P., 1993, Signal transduction via the MAP kinases, *Proc. Natl. Aca. Sci. USA* **90**:5889–5892.

Blessing, M., Nanney, L. B., King, L. E., Jones, C. M., and Hogan, B. L., 1993, Transgenic mice as a model to study the role of TGF-beta-related molecules in hair follicles, *Genes Dev.* **7**:204–215.

Brookfield, J., 1992, Can genes by truly redundant? *Curr. Biol.* **2**:553–554.

Brown, G. L., Curtsinger, L., Brightwell, J. R., Ackerman, D. M., Tobin, G. R., Polk, H. C., George-Nascimento, C., Valenzula, P., and Schultz, G. S., 1986, Enhancement of epidermal regeneration by biosythetic epidermal growth factor, *J. Exp. Med.* **163**:1319–1324.

Brown, G. L., Curtsinger, L., White, M., Mitchell, R., Pietsch, J., Nordquist, R., von Fraunhofer, A., and Schultz, G. S., 1988, Acceleration of tensile strength of incisions treated with EGF and TGF-β, *Ann. Surg.* **208**:788–794.

Brown, G. L., Nanney, L. B., Griffen, J., Cramer, A. B., Yancey, J. M., Curtsinger, L. J., Holtzin, L., Schultz, G. S., Jurkiewicz, M. J., and Lynch, J. B., 1989, Enhancement of wound healing by topical treatment with epidermal growth factor, *N. Engl. J. Med.* **321**:76–79.

Brown, J. P., Twardzik, D. R., Marguardt, H. J., and Todaro, G. J., 1985, Vaccinia virus encodes a polypeptide homologous to epidermal growth factor and transforming growth factor-α, *Nature* **313**:491–492.

Buckley, A., Davidson, J. M., Kamerath, C. D., Wolt, T. B., and Woodward, S. C., 1985, Sustained release of epidermal growth factor accelerates wound repair, *Proc. Natl. Acad. Sci. USA* **82**:7340–7344.

Cadena, D. L., and Gill, G. N., 1992, Receptor tyrosine kinases, *FASEB J.* **6**:2332–2337.

Callaghan, T., Antczak, M., Flinkinger, T., Raines, M., Myers, M. K., and Kuug, H-J., 1993, A complete description of the EGF-receptor exon structure: Implication in oncogenic activation and domain evolution, *Oncogene* **8**:2939–2948.

Carpenter, G., 1992, Receptor tyrosine kinase substrates; *src* homology domains and signal transduction, *FASEB J.* **6**:3283–3289.

Carpenter, G., 1993, Intracellular signalling from the epidermal growth factor receptor, *FORUM Trends Exp. Clin. Med.* **3**:616–634.

Carpenter, G., and Cohen, S., 1976, ^{125}I-Labelled human epidermal growth factor. Binding, internalization and degradation in human fibroblasts, *J. Cell Biol.* **71**:159–171.

Carpenter, G., King, L. E., and Cohen, S., 1978, Epidermal growth factor stimulates phosphorylation in membrane preparations *in vitro, Nature* **276**:409–410.

Cheng, C., Tennenbaum, T., Dempsey, P. J., Coffey, R. J., Yuspa, S. H., and Dlugosz, A. A., 1993, Epidermal growth factor receptor ligands regulate keratin 8 expression in keratinocytes, and transforming growth factor-α mediates the induction of keratin 8 by the *v-ras*[Ha] oncogene, *Cell Growth Differ.* **4**:317–327.

Coffey, R. J., Derynck, R., Wilcox, J. N., Bringman, T. S., Gouskin, A. S., Moses, H. L., and Pittelkow, M. R., 1987, Production and autoinduction of transforming growth factor-a in human keratinocytes, *Nature* **328**:817–819.

Cohen, S., and Elliott, G. A., 1963, The stimulation of epidermal keratinization by a protein isolated from the submaxillary gland of the mouse, *J. Invest. Dermatol.* **40**:1–5.

Colige, A. C., Lambert, C. A., Nusgens, B. V., and Lapiere, C. M., 1992, Effect of cell–cell and cell–matrix interactions on the response of fibroblasts to epidermal growth factor *in vitro.* Expression of collagen type I, collagenase, stromelysin and tissue inhibitor of metalloproteinases, *Biochem. J.* **285**:215–221.

Darnell, J. E., Kerr, I. M., and Stark, G. R., 1994, Jak-Stat pathways and transcriptional activation in response to IFNs and other extracellular signaling proteins, *Science* **264**:1415–1420.

Delany, A. M., Brinkerhoff, C. E., 1992, Post-transcriptional regulation of collagenase and stromelysin gene expression by epidermal growth factor and dexamethasone in cultured human fibroblasts, *J. Biol. Biochem.* **50**:400–410.

Derynck, R., 1986, Transforming growth factor-α: Structure and biological activities, *J. Cell Biol.* **32**: 293–304.

Dlugosz, A. A., Cheng, C., Denning, M. F., Dempsey, P. J., Coffey, R. J., and Yuspa, S. H., 1994, Keratinocyte growth factor receptor ligands induce transforming growth factor alpha expression and activate the epidermal growth factor receptor signalling pathway in cultural epidermal keratinocytes, *Cell Growth Diff.* **5**:1283–1292.

Dominey, A. M., Wang, X., King, Jr., L. E., Nanney, L. B., Gagne, T. A., Sellheyer, C., Bundman, D. S., Longley, M. A., Rothnagel, J. A., Greenhalgh, D. A., and Roop, D. R., 1993, Targeted over-expression of transforming growth factor-α in the epidermis of transgenic mice elicits hyperproliferation, hyerkeratosis, and differentiation and spontaneous squamous papillomas, *Cell Growth Differ.* **4**:1071–1082.

Donato, N. J., Rosenblum, M. G., and Steck, P. A., 1992, Tumor necrosis factor regulates tyrosine phosphorylation on epidermal growth factor receptors in A431 carcinoma cells: Evidence for a distinct mechanism, *Cell Growth Differ.* **3**:259–268.

Dvoneh, V. M., Murhpey, R. J., Matsuoka, J., and Grotendorst, C. R., 1992, Changes in growth factor levels in human wound fluid, *Surgery* **112**:18–23.

Dvorak, H. F., 1986, Tumors: wounds that do not heal: Similarities between tumor stroma generation and wound healing, *N. Eng. J. Med.* **315**:1650–1659.

Egan, S. E., and Weinberg, R. A., 1993, The pathway of signal achievement, *Nature,* **365**:781–783.

Ellis, D. L., Nanney, L. B., and King, L. E., 1990, Increased epidermal growth factor receptors in seborrheic keratoses and acrochordons of patients with the dysplastic nevus syndrome, *J. Am. Acad. Dermatol.* **23**:1070–1077.

Engel, J., 1989, EGF-like domains in extracellular matrix proteins: localized signals for growth and differentiation? *FEBS Letters* **251**:1–7.

Erickson, H. P., 1993a, Gene knockouts of c-*src*, transforming growth factor-β suggest superfluous, nonfunctional expression of proteins, *J. Cell Biol.* **120**:1070–1081.

Erikson, H. P., 1993b, Tenascin-c, tenascin-R and tenascin-x: A family of talented proteins in search of functions, *Curr. Opin. Cell. Biol.* **5**:869–876.

Falanga, V., Eaglstein, W. H., Bucalo, B., Katz, M. H., Harris, B., and Carson, P., 1992, Topical use of human recombinant epidermal growth factor (h-EGF) in venous ulcers, *J. Dermatol. Surg. Oncol.* **18**:604–606.

Fava, R. A., Nanney, L. B., Wilson, D., and King, Jr., L. E., 1993, Annexin-1 localization in human skin:

Possible association with cytoskeletal elements in keratinocytes of the stratum spinosum, *J. Invest. Dermatol.* **101**:732–737.

Fisher, E. H., Charbonneau, H., and Tonks, N. K., 1991, Protein tyrosine phosphatase: A diverse family of intracellular and transmembrane enzymes, *Science* **253**:401–406.

Fu, X.-Y., and Zhang, J. J., 1993, Transcription factor p91 interacts with the epidermal growth factor receptor and mediates activation of the c-fos gene promoter, *Cell* **74**:1135–1145.

Gates, R. E., King, L. E., Hanks, S. K., and Nanney, L. B., 1994, Potential role for focal adhesion kinase in migrating and proliferating keratinocytes near epidermal wounds and in culture, *Cell Growth Differ.* **5**:891–899.

Gospodarowicz, D., Brown, D. D., Birdwell, C. R., and Zetter, B. R., 1978, Control of proliferation of human vascular endothelial cells; characterization of the response of human umbilical vein endothelial cells to fibroblast growth factor, epidermal growth factor, and thrombin, *J. Cell Biol.* **77**:774–788.

Graves, B. J., Crowther, R. L., Chandran, C., Rumberger, J. M., Li, S., Huang, K-S., Presky, D. H., Familletti, P. C., Wolitzky, B. A., and Burns, D. K., 1994, Insight into E-selectin/ligand interaction from the crystal structure and mutagenesis of the lec/EGF domains, *Nature* **367**:532–534.

Green, M. R., and Couchman, J. R., 1984, Distribution of epidermal growth factor receptors in rat tissues during embryonic skin development, hair formation, and the adult hair growth cycle, *J. Invest. Dermatol.* **83**:118–123.

Green, M. R., Basketter, D. A., Couchman, J. R., and Rees, D. A., 1983, Distribution and number of epidermal growth factor receptors in skin is related to epithelial cell growth, *Dev. Biol.* **100**:506–512.

Griswold-Prenner, I., Carlin, C. R., and Rosner, M. R., 1993, Mitogen-activated protein kinase regulates the epidermal growth factor receptor through activation of a tryosine phosphatase, *J. Biol. Chem.* **268**:13050–13054.

Grotendorst, G. R., Smale, G., and Pencer, D., 1989, Production of transforming growth factor-β by human peripheral blood monocytes and neutrophils, *J. Cell. Physiol.* **140**:396–402.

Gunaratne, P., Stoscheck, C. M., Gates, R. E., Nanney, L. B., and King, Jr., L. E., 1994, Protein tyrosyl phosphatase-1B is expressed by normal human epideramis, keratinocytes and A-431 cells, and de-phosphorylates substrates of the epidermal growth factor receptor, *J. Invest. Dermatol.,* **103**:701–706.

Hall, A., 1994, A biochemical function for *ras*—at last, *Science* **264**:1413–1463.

Hayaski, T., Nichioka, J., Shingekiyo, T., Saito, Saito, S., and Suzuki, K., 1994, Protein S Tokushima: Abnormal molecule with a substitution of Glu for Lys-155 in the second epidermal growth factor-like domain of proteins, *Blood* **83**:683–690.

Heavner, G. A., Falcone, M., Kruszynski, M., Epps, L.; Mervic, M., Riexinger, D., and Mcever, R. P., 1993, Peptides from multiple regions of the lectin domain of P-selectin inhibiting neutrophil adhesion, *Int. J. Peptide Protein Res.* **42**:484–489.

Hebda, P. A., 1988, Stimulatory effects of transforming growth factor-beta and epidermal growth factor on epidermal cell outgrowth from porcine skin explant cultures, *J. Invest. Dermatol.* **91**:440–445.

Hernandez-Sotomayor, S. M. T., Arteaga, C. L., Soler, C., and Carpenter, G., 1993, Epidermal growth factor stimulates substrate-selective protein-tyrosine-phosphatase activity, *Proc. Natl. Acad. Sci. USA* **90**: 7691–7695.

Higashiyama, S., Abraham, J. A., Miller, J. I., Diffes, J. C., and Klagsbrun, M., 1991, A heparin-binding growth factor secreted by macrophage-like cells that is related to EGF, *Science* **251**:936–939.

Higashiyama, S., Abraham, J. A., and Klagsbrun, M., 1993, Heparin-binding EGF-like growth factor stimulation of smooth muscle cell migration: Dependence on interactions with cell surface heparan sulfate, *J. Cell Biol.* **122**:933–940.

Holmes, W. E., Sliwkowski, M. X., Akita, R. W., Henzel, W. J., Lee, J., Park, J. W., Yansura, D., Abadi, N., Raab, H., Lewis, G. D., Shepard, M., Kuang, W., Wood, W. I., Goeddel, V., and Vandlen, R. L., 1992, Identification of heregulin, a specific activator of p185erbB2, *Science* **256**:1205–1210.

Kansas, G. S., Saunders, K. B., Ley, K., Zakrzewicz, A., Gibson, R. M., Furie, B. C., Furie, B., and Tedder, T. F., 1994, A role for the epidermal growth factor-like domain of p-selectin in ligand recognition and cell adhesion, *J. Cell Biol.* **124**:609–618.

Kerr, L. D., Magun, B. E., and Matrisian, L. M., 1992, The role of c-Fos in growth factor regulation of stromelysin/transin gene expression, *Matrix Suppl.* **1**:176–183.

King, L. E., Stoscheck, C. M., Gates, R., and Nanney, L. B., 1991, Epidermal growth factor and transforming growth factor-α, in: *Physiology, Biochemistry and Molecular Biology of the Skin* (L. Goldsmith, ed.), pp. 329–350, Oxford University Press, New York.

Krane, J. F., Murphy, D. P., Carter, D. M., and Kruger, J. G., 1991, Synergistic effects of EGF and insulin-like growth factor I/stomatomedin C (IGF-1) on keratinocyte proliferation may be mediated by IGF-1 transmodulation of the EGF receptor, *J. Invest. Dermatol.* **96:**4199–4204.

Laato, M., Kahari, V. M., Niinikoski, J., and Vuorio, E., 1987, Epidermal growth factor increases collagen production in granulation tissue by stimulation of fibroblast proliferation and not by activation of procollagen genes, *Biochem. J.* **247:**385–388.

Larner, A. C., David, M., Feldman, G. M., Igarashi, K., Hackett, R. H., Webb, D. S. A., Sweitzer, S. M., Petricoin, E. F., and Finbloom, D. S., 1993, Tyrosine phosphorylation of DNA binding proteins by multiple cytokines, *Science* **261:**1730–1733.

Leibovich, S. J., and Ross, R., 1975, The role of the macrophage in wound repair. A study of hydrocortisone and antimacrophage serum, *Am. J. Pathol.* **78:**71–91.

Lembach, K. J., 1976, Enhanced synthesis and extracellular accumulation of hyaluronic acid during stimulation of quiescent human fibroblasts by mouse epidermal growth factor, *J. Cell. Physiol.* **89:**277–288.

Lightner, V. A., 1994, Tenascin: Does it play a role in epidermal morphogenesis and homeostasis? *J. Invest. Dermatol.* **102:**273–277.

Luetteke, N. C., Hu Qiu, T., Peiffer, R. L., Oliver, P., Smithies, O., and Lee, D. C., 1993, TGFα deficiency results in hair follicle and eye abnormalities in targeted and waved-1 mice, *Cell* **73:**263–278.

Lutticken, C., Wegenka, U. M., Yuan, J., Buschmann, J., Schindler, C., Ziemiecki, A., Harpur, A. G., Wilks, A. F., Yasukawa, K., Taga, T., Kishimoto, T., Barbieri, G., Pellegrini, S., Sendtner, M., Heinrich, P. C., and Horn, F., 1994, Association of transcription factor APRF and protein kinase jak1 with the interleukin-6 signal tranducer gp 130, *Science* **263:**89–92.

Mackie, E. J., Halfter, W., and Liverani, D., 1988, Induction of tenascin in healing wounds, *J. Cell Biol.* **107:**2757–2767.

Maguire, H. C., Jaworsky, C., Cohen, J. A., Hellman, M., Weiner, D. B., and Greene, M. I., 1989, Distribution of neu (c-erbB-2) protein in human skin, *J. Invest. Dermatol.* **92:**786–790.

Marikovsky, M., Breuing, K., Yu Liu, P., Eriksson, E., Higashiyama, S., Farber, P., Abraham, J., and Klagsbrun, M., 1993, Appearance of heparin-binding EGF-like growth factor in wounds fluid as a response to injury, *Proc. Natl. Acad. Sci. USA* **90:**3889–3893.

Massague, J., 1985, Transforming growth factor-β modulates the high affinity receptors of epidermal growth factor and transforming growth factor-α, *J. Cell Biol.* **100:**1500–1514.

Medved, L. V., Vysotchin, A., and Ingham, K. C., 1994, Ca^{++}-dependent interactions between Gla and EGF domains in human coagulation factor IX, *Biochemistry* **33:**478–485.

Moller, J. J., Ingemann-Hansen, T., and Poulsen, J. H., 1994, The epidermal growth factor-like domain of the human cartilage large aggregating proteoglycan, aggrecan: Increased serum concentration in rheumatoid arthritis, *Br. J. Rheumatol.* **33:**44–47.

Nanney, L. B., 1990, Epidermal and dermal effects of epidermal growth factor during wound repair, *J. Invest. Dermatol.* **94:**624–629.

Nanney, L. B., Magid, M., Stoscheck, C. M., and King, Jr., L. E., 1984, Comparison of epidermal growth factor binding and receptor distribution in normal human epidermis and epidermal appendages, *J. Invest. Dermatol.* **83:**385–393.

Nanney, L. B., Stoscheck, C. M., Magid, M., and King, Jr., L. E., 1986, Altered [125I]epidermal growth factor binding and receptor distribution in psoriasis, *J. Invest. Dermatol.* **86:**260–265.

Nanney, L. B., King, Jr., L. E., and Dale, B. A., 1990a, Epidermal growth factor receptors in genetically induced hyperproliferative skin disorders, *Pediar. Dermatol.* **7:**256–265.

Nanney, L. B., Stoscheck, C. M., King, Jr., L. E., Underwood, R. A., and Holbrook, K. A., 1990b, Immunolocalization of epidermal growth factor receptors in normal developing human skin, *J. Invest. Dermatol.* **94:**742–748.

Nanney, L. B., Ellis, D. L., Levine, J., and King, L. E., 1992a, Epidermal growth factor receptors in idiopathic and virally induced skin diseases, *Am. J. Pathol.* **140:**915–925.

Nanney, L. B., Gates, R. E., Todderud, G., King, Jr., L. E., and Carpenter, G., 1992b, Altered distribution of phospholipase Cγ 1 in benign hyperproliferative epidermal diseases, *Cell Growth Differ.* **3:**233–239.

Nanney, L. B., Yates, R. A., and King, Jr., L. E., 1992c, Modulation of epidermal growth factor receptors in psoriatic lesions during treatment with topical EGF, *J. Invest. Dermatol.* **98**:296–301.

Nawa, K., Ono, M., Fujiwara, J., Sugiyama, N., Uchiyama, T., and Marumotot, Y., 1994, Monoclonal antibodies against human thrombomodulin whose epitope is located in epidermal growth factor-like domains, *Biochem. Biophys. Acta* **1205**:162–170.

Nickoloff, B. J., and Naidu, Y., 1994, Perturbation of epidermal barrier function correlates with initiation of cytokine cascade in human skin, *J. Am. Acad. Dermatol.* **30**:535–546.

Nies, D. E., Hemesath, T. J., Kim, J. H., Gulcher, J. R., and Stefansson, K., 1991, The complete cDNA sequence of human hexabrachion (tenascin). A multidomain protein containing unique epidermal growth factor repeats, *J. Biol. Chem.* **266**:2818–2823.

Nowak, U. K., Cooper, A., Saunders, D., Smith, R. A. G., and Dobson, C. M., 1994, Unfolding studies of the protease domain of urokinase-type plasminogen activator: The existence of partly folded states and stable subdomains, *Biochemistry* **33**:2951–2960.

Oka, Y., and Orth, D. N., 1983, Human plasma epidermal growth factor urogastrone is associated with blood platelets, *J. Clin. Invest.* **72**:249–259.

Olashaw, N. E., O'Keefe, E. J., and Pledger, W. J., 1986, Platelet-derived growth factor modulates epidermal growth factor receptors by a mechanism distinct from that of phorbol esters, *Proc. Natl. Acad. Sci. USA* **83**:3834–3839.

Ono, M., Okamura, K., Nakayama, Y., Tomita, M., Sata, Y., Komatsu, Y., and Kuwano, M., 1992, Induction of human microvascular endothelial tubular morphogenesis by human keratinocytes: Involvement of transforming growth factor-alpha, *Biochem. Biophys. Res. Commun.* **189**:601–609.

Panayotou, G., Aumailley, M., Timpl, R., and Engel, J., 1989, Domains of laminin with growth factor activity, *Cell* **56**:93–101.

Patel, V. G., Shum-Siu, A., Heniford, B. W., Wieman, T. J., and Hendler, F. J., 1994, Detection of epidermal growth factor receptor mRNA in tissue sections from biopsy specimens using *in situ* polymerase chain reaction, *Am. J. Pathol.* **144**:7–14.

Pelech, S. L., and Sanghera, J. S., 1992, MAP kinases: Charting the regulatory pathways, *Science* **257**:1355–1356.

Peles, E., Bacus, S. S., Koski, R. A., Lu, H. S., Wen, D., Ogden, S. G., Levy, R. B., and Yarden, Y., 1992, Isolation of the Neu/HER-2 stimulatory ligand: A 44 kd glycoprotein that induces differentiation of mammary tumor cells, *Cell* **69**:205–216.

Peppelenbosch, M. P., Tertoolen, L. G., Hage, W. J., and DeLaat, S. W., 1993, Epidermal growth factor-induced actin remodeling is regulated by 5-lipoxygenase and cyclooxygenase products, *Cell* **74**:565–575.

Perry, L. C., Connors, A. W., Matrisian, L. M., Nanney, L. B., Charles, P. D., Reyes, D. P., Kerr, L. D., and Fisher, J., 1993, Role of TGFβ1 and EGF in the wound healing process: An *in vivo* biochemical evaluation, *J. Wound Repair Regen.* **1**:41–46.

Pierce, G. F., Berg, J. V., Rudolph, R., Tarpley, J., and Mustoe, T. A., 1991, Platelet-derived growth factor-BB and transforming growth factor-β selectively modulate glycosaminoglycans, collagen, and myofibroblasts in excisional wounds, *Am. J. Pathol.* **138**:629–646.

Pierce, G. F., Yanagishara, D., Klopchin, K., Danilenko, D. M., Hsu, E., Kenney, W. C., and Morris, C. F., 1994, Stimulation of all epithelial elements during skin regeneration by keratinocyte growth factor, *J. Exp. Med.* **179**:831–840.

Pittelkow, M. R., Cook, P. W., Shipley, G. D., Derynck, R., and Coffey, R. J., 1993, Autonomous growth of human keratinocytes requires epidermal growth factor receptor occupancy, *Cell Growth Differ.* **4**:513–521.

Plowman, G. D., Culouscou, J., Whitney, G. S., Green, J. M., Carlton, G. W., Foy, L., Neubauer, M. G., and Shoyab, M., 1993, Ligand-specific activation of HER4/p180, a fourth member of the epidermal growth factor receptor family, *Proc. Natl. Acad. Sci. USA* **90**:1746–1750.

Qian, X., Dougall, W. C., Hellman, M. E., and Greene, M. I., 1994, Kinase-deficient neu proteins suppress epidermal growth factor function and abolish cell transformation, *Oncogene* **9**:1507–1514.

Quaglino, Jr., D., Nanney, L. B., Kennedy, R., and Davidson, J. M., 1990, Transforming growth factor-beta stimulates wound healing and modulates extracellular matrix gene expression in pig skin. I. Excisional wound model, *Lab. Invest.* **63**:307–319.

Rappolee, D. A., Mark, D., Banda, M. J., and Werb, Z., 1988, Wound macrophages express TGFβ and other growth factors *in vivo:* Analysis by mRNA phenotyping, *Science* **241**:708–712.

Roesel, J. F., and Nanney, L. B., 1995, Assessment of differential cytokine effects on angiogenesis using an *in vivo* model of cutaneous wound repair, *J. Surg. Res.* **58**:449–459, 1995.

Rozakis-Adcock, M., Fernley, R., Wade, J., Pawson, T., and Bowtell, D., 1993, The SH2 and SH3 domains of mammalian Grb2 couple the EGF receptor to the Ras activator mSos1, *Nature* **363**:83–85.

Ruff-Jamison, S., Chen, K., and Cohen, S., 1993, Induction by EGF and interferon-γ of tyrosine phosphorylated DNA binding proteins in mouse liver nuclei, *Science* **261**:1733–1736.

Sadowski, H. B., Shuai, K., Darnell, J. E., and Gilman, M. Z., 1993, A common nuclear signal transduction pathway activated by growth factor and cytokine receptors, *Science* **261**:1739–1744.

Saga, Y., Yagi, T., Ikawa, Y., Sakakura, T., and Aizawa, S., 1992, Mice develop normally without tenascin, *Genes Dev.* **6**:1821–1831.

Sandgren, E. P., Luetteke, N. C., Palmiter, R. D., Brinster, R. L., and Lee, D. C., 1990, Overexpression of TGFα in transgenic mice: Induction of epithelial hyperplasia, pancreatic metaplasia, and carcinoma of the breast, *Cell* **61**:1121–1129.

Sasada, R., Ono, Y., Taniyama, Y., Shing, Y., Folkman, J., and Igarishi, K., 1993, Cloning and expression of cDNA encoding human betacellulin, a new member of the EGF family, *Biochem. Biophys. Res. Commun.* **190**:1173–1179.

Schalkwijk, J., Steijten, P. M., van Vlijmen-Willems, I. M. J. J., Oosterling, B., Mackie, E. J., and Verstraeten, A. A., 1991, Tenascin expression in human dermis related to epidermal proliferation, *Am. J. Pathol.* **139**:1143–1150.

Schreiber, A. B., Winkler, M. E., and Derynck, R., 1986, Transforming growth factor-α: A more potent angiogenic mediator than epidermal growth factor, *Science* **232**:1250–1252.

Schultz, G. S., White, M., Mitchell, R., Brown, G., Lynch, J., Twardzik, D. R., and Todaro, G. J., 1987, Epithelial wound healing enhanced by transforming growth factor-α and vaccinia growth factor, *Science* **235**:350–352.

Shing, Y., Christofori, G., Hanahan, D., Ono, Y., Sasada, R., Igarashi, K., and Folkman, J., 1993, Betacellulin: A mitogen from pancreatic B cell tumors, *Science* **259**:1604–1607.

Shima, I., Sasaguri, Y., Nakano, R., Yamana, H., Fujita, H., Kakegawa, T., Morimatsu, M., 1993, Production of matrix metalloproteinase 9192KDa gelatinase by human oesophageal squamous cell carcinoma in response to epidermal growth factor, *Brit. J. Can.* **67**:721–727.

Shoyab, M., Plowman, G. D., McDonald, V. L., Bradley, J. G., and Todaro, G. J., 1989, Structure and function of human amphiregulin: A member of the epidermal growth factor family, *Science* **243**:1074–1076.

Sibilia, M., Wagner, E. F., 1995, Strain-dependent epithelial defects in mice lacking the EGF receptor, *Science* **269**:234–238.

Smith, B. O., Lowning, K. A., Dudgeon, T. J., Cunningham, M., Driscoll, P. C., and Campbell, I. D., 1994, Secondary structure of fibronectin type 1 and epidermal growth factor modules from tissue-type plasminogen activator by nuclear magnetic resonance, *Biochemistry* **33**:2422–2429.

Soler, C., Beguinot, L., Sorkin, A., and Carpenter, G., 1993, Tyrosine phosphorylation of *ras* GTPase-activating protein does not require association with the epidermal growth factor receptor, *J. Biol. Chem.* **268**:22010–22019.

Soler, C., Beguinot, L., and Carpenter, G., 1994, Individual epidermal growth factor receptor auto-phosphorylation sites do not stringently define association motifs for several SH-2 containing proteins, *J. Biol. Chem.* **269**:12320–12324.

Sorokin, A., and Carpenter, G., 1993, Interaction of activated EGF receptors with coated pit adaptins, *Science* **261**:612–615.

Sorokin, A., Lemmon, M. A., Ullrich, A., and Schlessinger, J., 1994, Stabilization of an active dimeric form of the epidermal growth factor receptor by introduction of an inter-receptor disulfide bond, *J. Biol. Chem.* **269**:9752–9759.

Stoscheck, C. M., Nanney, L. B., and King, Jr., L. E., 1992, Quantitative determination of EGF-R during epidermal wound healing, *J. Invest. Dermatol.* **99**:645–649.

Stricklin, G. P., Li, L., Jancic, V., Wenczak, B. A., and Nanney, L. B., 1993, Localization of mRNAs representing collagenase and TIMP in sections of healing human burn wounds, *Am. J. Pathol.* **143:**1657–1666.

Taylor, H. C., Lightner, V. A., Beyer, W. F., McCaslin, D., Briscoe, G., and Erickson, H. P., 1989, Biochemical and structural studies of tenascin hexabrachion proteins, *J. Cell. Biochem.* **41:**71–90.

Threadgill, D. W., Dlugosz, A. A., Hansen, L. A., Tennenbaum, T., Lichti, U., Yee, D., LaMantia, C., Mourton, T., Herrup, K., Harris, R. C., Baarnard, J. A., Yuspa, S. H., Coffey, R. J., Magnuson, T., 1995, Targeted disruption of mouse EGF receptor: Effects of genetic background on mutant phenotype, *Science* **269:**230–234.

Watanabe, T., Shintani, A., Nakata, M., Shing, Y., Folkman, J., Igarashi, K., and Sasada, R., 1994, Recombinant human betacellulin, *J. Biol. Chem.* **269:**9966–9973.

Weidner, N., Semple, J. P., Welch, W. R., and Folkman, J., 1991, Tumor angiogenesis and metastasis-correlation in invasive breast carcinoma, *N. Engl. J. Med.* **324:**1–8.

Wen, D., Peles, E., Cupples, R., Suggs, S. V., Bacus, S. S., Luo, Y., Trail, G., Hu, S., Silbiger, S. M., Levy, R. B., Koski, R. A., Lu, H. S., and Yarden, Y., 1992, Neu differentiation factor: A transmembrane glycoprotein containing an EGF domain and an immunoglobulin homology unit, *Cell* **69:**559–572.

Wenczak, B. A., and Nanney, L. B., 1993, Correlation of TGFα and EGF-R with proliferating cell nuclear antigen in human burn wounds, *J. Wound Repair Regen.* **1:**219–230.

Wenczak, B. A., Lynch, J. B., and Nanney, L. B., 1992, Epidermal growth factor receptor distribution in burn wounds. Implications for growth factor-mediated repair, *J. Clin. Invest.* **90:**2392–2401.

Werner, S., Peters, K. G., Longaker, M. T., Fuller-Pace, F., Banda, J. J., and Williams, L. T., 1992, Large induction of keratinocyte growth factor expression in the dermis during wound healing, *Proc. Natl. Aca. Sci. USA* **89:**6896–6900.

Williams, R., Sanghera, J., Wu, F., Carbonaro-Hall, D., Campbell, D. L., Warburton, D., Pelech, S., and Hall, F., 1993, Identification of a human epidermal growth factor receptor-associated protein kinase as a new member of the mitogen-activated protein kinase/extracellular signal-regulated protein kinase family, *J. Biol. Chem.* **268:**18213–18217.

Zhong, Z., Wen, Z., and Darnell, J. E., 1994, Stat3: A STAT family member activated by tyrosine phosphorylation in response to epidermal growth factor and interleukin-6, *Science* **264:**95–98.

Chapter 6

Modulation of Wound Repair by Members of the Fibroblast Growth Factor Family

JUDITH A. ABRAHAM and MICHAEL KLAGSBRUN

1. Overview

The fibroblast growth factors (FGFs) comprise a family of at least nine structurally homologous polypeptides that are found in a variety of cells and tissues (Baird and Böhlen, 1990; Brem and Klagsbrun, 1993; Burgess and Maciag, 1989; Folkman and Klagsbrun, 1987; Klagsbrun, 1989; Klagsbrun and D'Amore, 1991; Klagsbrun and Folkman, 1990; Rifkin and Moscatelli, 1989; Tanaka *et al.*, 1992; Miyamoto *et al.*, 1993). This family includes acidic FGF (aFGF), basic FGF (bFGF), *int-2* protein, HST/K-FGF, FGF-5, FGF-6, keratinocyte growth factor (KGF), androgen-induced growth factor (AIGF), and glia-activating factor (GAF) (see Table I). These growth factors have been enumerated as FGF-1 through FGF-9, respectively, in order to simplify the nomenclature (Baird and Klagsbrun, 1991). However, since the name KGF is still widely used, the designation "KGF/FGF-7" will be used here in referring to this factor. The aFGF (FGF-1) and bFGF (FGF-2) proteins are the most extensively characterized FGF family members in terms of detailed knowledge concerning structure and biological activity.

Although the FGF family members have structural homologies of up to 35–55%, they are not homogeneous in their biological properties. For example, FGF-1 and FGF-2 are mitogens for a wide variety of cell types including vascular endothelial cells, vascular smooth muscle cells, fibroblasts, and keratinocytes (Burgess and Maciag, 1989; Klagsbrun, 1989); KGF/FGF-7, on the other hand, appears to be a highly specific mitogen for epithelial cells alone (Finch *et al.*, 1989). Another difference involves protein secretion. FGF-1 and FGF-2 lack consensus signal peptides;

JUDITH A. ABRAHAM • Scios Nova Inc., Mountain View, California 94043.   MICHAEL KLAGSBRUN • Departments of Surgery and Pathology, Children's Hospital, and Harvard Medical School, Boston, Massachusetts 02115.

The Molecular and Cellular Biology of Wound Repair (Second Edition), edited by Richard A. F. Clark. Plenum Press, New York, 1996

Table I. The FGF Family

Family member	Common name(s)	Mol. mass (kDa)	Originally found in	Target[a]
FGF-1	Acidic FGF, aFGF, HBGF-1	18	Adult tissue (neural)	Fibroblasts, EC, SMC, keratinocytes
FGF-2	Basic FGF, bFGF, HBGF-2	18	Most adult tissue	Fibroblasts, EC, SMC, keratinocytes
FGF-3	INT-2	27, 31 32.5	Site of MMTV integration; breast carcinoma	?[b]
FGF-4	HST, K-FGF	23	Human stomach tumor; Kaposi's sarcoma	Fibroblasts, EC
FGF-5	FGF5	29	Bladder carcinoma; hepatoma	Fibroblasts, EC
FGF-6	FGF6	21, 22 24	Homology to FGF-4 gene	Fibroblasts
FGF-7	Keratinocyte growth factor, KGF	22	Epithelial tissue stromal cells	Keratinocytes
FGF-8	Androgen-induced growth factor	28, 32	Shionogi carcinoma cells	SC-3 androgen-dependent carcinoma cells
FGF-9	Glia-activating factor	25, 29 30	Glioma cells	Glial cells

[a]For FGF-1 through FGF-7, mitogenic targets potentially relevant to wound healing are listed. Abbreviations used: EC, endothelial cells; SMC, smooth muscle cells.
[b]*Xenopus* FGF-3 has been shown to be mitogenic for mouse keratinocytes and mouse mammary epithelial cells, but mitogenic activity has so far not been demonstrated directly for mammalian forms of FGF-3.

consistent with this feature, these two FGFs are for the most part cell-associated and not secreted. In contrast, FGF-3, FGF-4, FGF-5, FGF-6, KGF/FGF-7, and FGF-8 all possess signal peptides and are secreted. FGF-9 is unusual in that it does not possess a classical signal sequence, but is nevertheless released into conditioned medium.

A range of studies has indicated that members of the FGF family might be able to influence—or even be natural mediators of—the process of dermal wound repair. Several members of the family have been shown to be produced by cells associated with wound healing such as inflammatory cells (macrophages, T lymphocytes), vascular endothelial cells, and dermal fibroblasts (Baird *et al.*, 1985; Blotnick *et al.*, 1994; Ross, 1993; Schweigerer *et al.*, 1987; Vlodavsky *et al.*, 1987a; Kandel *et al.*, 1991). These FGFs are capable of stimulating the proliferation and migration of cells involved in such dermal wound-healing steps as granulation tissue formation (fibroblasts), angiogenesis (endothelial cells), and reepithelialization (keratinocytes). In particular, fibroblasts are stimulated to migrate and proliferate by FGF-1 and FGF-2; angiogenesis is stimulated by FGF-1, FGF-2, FGF-4, and FGF-5; and keratinocytes have been shown to migrate in response to KGF/FGF-7 and to proliferate in response to FGF-1, FGF-2, and KGF/FGF-7 (Burgess and Maciag, 1989; Klagsbrun, 1989; Delli-Bovi *et al.*, 1987; Zhan *et al.*, 1988; O'Keefe *et al.*, 1988; Shipley *et al.*, 1989; Tsuboi *et al.*, 1993; Lobb *et al.*, 1985). Extracellular matrix synthesis and degradation can also be affected by the presence of FGFs (Gross *et al.*, 1983; Buckley-Sturrock *et al.*, 1989; Mignatti *et al.*,

1989; Flaumenhaft and Rifkin, 1991; Sato and Rifkin, 1988).

In vivo, elevated levels of FGF-1, FGF-2, FGF-5, and KGF/FGF-7 mRNA transcripts have been detected at sites of dermal injury (Werner *et al., 1992a*), as has FGF-2 protein (Whitby and Ferguson, 1991a; Kurita *et al.,* 1992; Gibran *et al.,* 1994; Grayson *et al.,* 1993; Cooper *et al.,* 1994). The presence of a neutralizing anti-FGF-2 antibody has been shown to delay granulation tissue formation (Broadley *et al.,* 1989a), while transgenic expression of a dominant negative mutant form of one of the FGF receptors causes a delay in wound reepithelialization (Werner *et al.,* 1994b). In addition, FGF-1, FGF-2, FGF-4, and KGF/FGF-7 have all been shown to accelerate tissue repair steps in at least some types of animal models when recombinant versions of these proteins were applied exogenously. Results have also been reported for several small clinical trials of FGF-2 for wound healing. Despite the sometimes striking effects of FGF-2 in animals, however, these human trials have so far shown either no or fairly small benefit from FGF-2 treatment of various types of wounds (Robson *et al.,* 1992; Mazué *et al.,* 1991; Greenhalgh and Rieman, 1994).

2. The FGF Family

2.1. FGF-1 (aFGF) and FGF-2 (bFGF)

2.1.1. Structural Properties

FGF-1 and FGF-2 share about 55% protein sequence identity and have similar biological activity profiles (Baird and Böhlen, 1990; Brem and Klagsbrun, 1993; Burgess and Maciag, 1989; Folkman and Klagsbrun, 1987; Klagsbrun, 1989; Klagsbrun and D'Amore, 1991; Klagsbrun and Folkman, 1990; Rifkin and Moscatelli, 1989; Esch *et al.,* 1985; Gimenez-Gallego *et al.,* 1985). Both are single-chain, nonglycosylated polypeptides that contain commonly 154 amino acids (molecular mass of about 18 kDa). However, 22- to 25-kDa forms of FGF-2, initiated on CUG start codons rather than the typical AUG start codon, have been identified as well (Prats *et al.,* 1989; Florkiewicz and Sommer, 1989). FGF-1 has three cysteine residues while FGF-2 has four. Two of the cysteines within each factor lie at positions that are conserved in all nine members of the FGF family. Unlike the case with some other mitogenic proteins such as epidermal growth factor, the cysteine residues in FGF-1 and FGF-2 are spaced such that they do not form intramolecular disulfide bonds (Zhu *et al.,* 1991), and the two growth factors are not inactivated by exposure to sulfhydryl-reducing agents (Klagsbrun and Shing, 1985).

2.1.2. Biological Properties Relevant to Wound Healing

FGF-1 and FGF-2 have numerous biological activities, some of which are potentially important for modulating wound-healing events. For example, these two growth factors are mitogenic and chemotactic for vascular endothelial cells *in vitro* and are angiogenic *in vivo* (Baird and Böhlen, 1990; Brem and Klagsbrun, 1993; Burgess and

Maciag, 1989; Folkman and Klagsbrun, 1987; Klagsbrun, 1989; Klagsbrun and D'Amore, 1991; Klagsbrun and Folkman, 1990; Rifkin and Moscatelli, 1989). FGF-mediated angiogenesis activity has been demonstrated in the chick chorioallantoic membrane (CAM) (Lobb et al., 1985; Shing et al., 1985), in the rabbit cornea (Lobb et al., 1985; Shing et al., 1985), in sponges implanted subcutaneously into rats (Davidson et al., 1985), and in gelfoam implants in the peritoneal cavities of rats (Thompson et al., 1989). In addition, FGF-2 stimulates collateral blood vessel formation in myocardial infarcts (Yanagisawa-Miwa et al., 1992), and FGF-1 stimulates angiogenesis in arteries (Nabel et al., 1993). The ability of FGF-2 to stimulate plasminogen activator and collagenase activity in endothelial cells has been suggested to facilitate basement membrane breakdown and the migration of endothelial cells through stroma (Mignatti et al., 1989).

FGF-1 and FGF-2 are also mitogenic for fibroblasts in vitro. In addition, fibroblasts cultured from wound granulation tissue respond chemotactically to FGF-2 (Buckley-Sturrock et al., 1989), and FGF-2 stimulates the migration of fibroblasts at the periphery of a wound boundary made by scratching a monolayer (Schreier et al., 1993). Furthermore, FGF-2 stimulates granulation-tissue-derived fibroblasts to produce collagenase, suggesting a potential role for FGF-2 in triggering collagenase-catalyzed restructuring of collagen during wound repair (Buckley-Sturrock et al., 1989). Finally, a role for FGF-1 and FGF-2 in reepithelialization has been suggested in that both are mitogens for human keratinocytes, with FGF-1 being the more potent mitogen of the two (Shipley et al., 1989; O'Keefe et al., 1988).

2.1.3. Biosynthesis and Release by Wound-Associated Cell Types

FGF-1 and FGF-2 are synthesized by a number of major cell types involved in the wound-healing process, including inflammatory cells such as monocytes/macrophages (Baird et al., 1985) and T lymphocytes (both $CD4^+$ and $CD8^+$) (Blotnick et al., 1994), vascular endothelial cells (Vlodavsky et al., 1987a; Schweigerer et al., 1987), and dermal fibroblasts (Kandel et al., 1991). However, given the fact that FGF-1 and FGF-2 lack classical secretory signal peptides (Jaye et al., 1986; Abraham et al., 1986) and in general are not secreted by cultured cells, the mechanism by which these factors might be released into the wound environment from their cell of origin remains unresolved. On the other hand, suggestive evidence exists that FGF-1 and FGF-2 can indeed be released from cells; this evidence includes the observation that FGF-2 has been found deposited in extracellular matrix (Vlodavsky et al., 1987b; Baird and Ling, 1987) and the observation that FGF receptors are abundant on many cell surfaces, implying a paracrine role for these growth factors. Possible mechanisms of signal-less FGF-1 and FGF-2 export that have been explored include exocytosis (Mignatti et al., 1992), temperature-dependent release involving homodimer formation (Jackson et al., 1992), and ATP-driven translocators (Kuchler and Thorner, 1992).

There is also evidence that the FGFs could be released as a result of damage to cells (McNeil et al., 1989; McNeil, 1993; Muthukrishnan et al., 1991). McNeil and colleagues developed an in vitro model in which endothelial cells were damaged mechanically by scraping, resulting in the disruption of plasma membranes (transient permeabilization) while preserving cell viability. Under these conditions, FGF-2 activity was released from the wounded endothelial cells but not from underlying extracellular matrix. These results thus suggest that dermal wounding itself could trigger the release of FGF protein preexisting in cells in the wound area.

2.1.4. Interactions with Heparin and Heparan Sulfate Proteoglycan

FGF-1 and FGF-2 are both characterized by their strong interaction with heparin-like molecules (Shing *et al.*, 1984; Maciag *et al.*, 1984). Despite its anionic pI, FGF-1 was found to bind tightly to columns of immobilized heparin, eluting with 1 M NaCl. FGF-2 binds more tightly to immobilized heparin than any other known growth factor and is eluted with 1.6–1.8 M NaCl, a property that has greatly facilitated its purification. In addition, heparin stabilizes FGF-1 and FGF-2 and protects them from heat, acid, and proteolytic degradation (Brem and Klagsbrun, 1993; Klagsbrun, 1989; Gospodarowicz and Cheng, 1986; Sommer and Rifkin, 1989; Rosengart *et al.*, 1988, Saksela *et al.*, 1988). The biological relevance of heparin binding was demonstrated by experiments showing that FGF-1 and FGF-2 also bind to heparan sulfate proteoglycans (HSPG) on cell surfaces (Moscatelli, 1987) and in extracellular matrix (Vlodavsky *et al.*, 1987b; Baird and Ling, 1987; Moscatelli, 1987; Bashkin *et al.*, 1989; Folkman *et al.*, 1988; Klagsbrun, 1990; Sakaguchi *et al.*, 1991). It has been suggested that FGF-1 and FGF-2 are sequestered or "stored" in the extracellular matrix as part of a highly stable FGF–HSPG complex and are released when needed by a combination of proteases and heparinases (Vlodavsky *et al.*, 1991). The role of cell surface HSPG in modulating FGF binding to high-affinity receptors and FGF mitogenicity will be discussed below.

One type of HSPG that might be relevant to wound healing is the syndecan family (Bernfield *et al.*, 1994). This family consists of four cell surface proteoglycans, each containing a protein core as well as both heparan sulfate and chondroitin sulfate side chains. Syndecan-1, the first HSPG to be described in this family, has been the most intensively studied. During the healing of cutaneous wounds, increased amounts of syndecan-1 have been found on the cell surfaces of migrating and proliferating epidermal cells and on hair follicle keratinocytes adjacent to wound margins (Elenius *et al.*, 1991). Syndecan induction was noted within a day of wounding and was most pronounced at the time of intense cell proliferation. Syndecan-1 expression is also induced in granulation tissue during wound healing. Syndecan-1 has been shown to bind FGF-2, raising the possibility that in wound healing the syndecans might serve to modulate FGF-2 as well as other heparin-binding growth factor activity at the level of the cell surface and/or extracellular matrix (Elenius *et al.*, 1992). More recently, a factor has been purified from wound fluid that stimulates syndecan-1 expression at the surface of mesenchymal cells (Gallo *et al.*, 1994).

2.2. FGF-3 (INT-2)

FGF-3 was originally identified as the product of a cellular oncogene activated by integration ("int") of the mouse mammary tumor virus into the mouse genome (Dickson and Peters, 1987). The FGF-3 gene encodes a predicted 27-kDa primary translation product of about 240 amino acids that includes a putative atypical signal peptide sequence (Dixon *et al.*, 1989; Dickson *et al.*, 1990). The structural homology of the predicted FGF-3 protein to FGF-1 and FGF-2 is 38 and 44%, respectively. The FGF-3 gene is normally silent in the adult mammary gland; FGF-3 instead appears to play a role in the developing embryo. Mouse FGF-3 is transcribed as a complex set of mRNAs as early as embryonic day 6.5 (Wilkinson *et al.*, 1988, 1989). Expression continues until parturition in a highly localized and temporally regulated manner. Unlike mammalian forms of FGF-1 and FGF-2, it has not been possible to demonstrate directly mitogenic activity for mouse FGF-3. A role in growth has been implicated, however, in that transgenic mice that express the FGF-3 gene under the control of a mouse mammary tumor virus promoter develop mammary hyperplasias (Ornitz *et al.*, 1991; Muller *et al.*, 1990). In addition, amplification of the FGF-3 gene has been found in a variety of human tumors, particularly in breast carcinomas and squamous cell carcinomas of the head and neck region (Zhou *et al.*, 1988; Meyers *et al.*, 1990; Somers *et al.*, 1990). FGF-3 is also expressed in Kaposi's sarcoma tumors (Huang *et al.*, 1993).

Recently, the homologue of the FGF-3 gene from *Xenopus laevis* has been isolated. COS-1 cells transfected with a cDNA derived from this *Xenopus* gene (XFGF-3 gene) express 31- and 27-kDa products that are secreted and bind to heparin-Sepharose (Kiefer *et al.*, 1993). The conditioned medium of XFGF-3-transfected cells induces transient transformation of NIH 3T3 cells and is mitogenic for mouse mammary epithelial cells and for mouse epidermal keratinocytes. Thus, XFGF-3 appears to be mitogenic under conditions where mouse FGF-3 is not, despite an 82% structural homology.

2.3. FGF-4 (HST/K-FGF)

The second oncogene to be identified that encodes an FGF-like protein was the FGF-4 oncogene, which was isolated from two sources simultaneously. One source was NIH-3T3 cells transfected with Kaposi's sarcoma DNA, hence the name Kaposi's FGF (K-FGF) (Delli-Bovi *et al.*, 1987, 1988). The other source was NIH-3T3 cells transfected with DNA from a *human stomach tumor*, hence the name *hst* (Yoshida *et al.*, 1987). FGF-4 shares 43, 38, and 40% sequence identity, respectively, with FGF-1, FGF-2, and FGF-3. The FGF-4 gene encodes a 206-amino-acid primary translation product that contains a hydrophobic signal peptide sequence. NIH-3T3 cells transfected with the FGF-4 gene secrete a heparin-binding, 176-amino-acid (23-kDa) protein that is glycosylated (Delli-Bovi *et al.*, 1988; Moscatelli and Quarto, 1989; Basilico *et al.*, 1989). FGF-4 is mitogenic for fibroblasts and endothelial cells; however, the addition of heparin is needed for endothelial cell growth factor activity. A truncated form of FGF-4 lacking the N-terminal 36 amino acids (as a result of mutation of the single N-linked glycosylation site) is approximately five times more mitogenic for BALB/c 3T3 cells, has higher affinity for FGF receptors, and has increased heparin-binding affinity (elution with 1.3–1.5 M NaCl rather than 1.1 M NaCl) compared to wild-type FGF-4 (Bellosta *et al.*, 1993).

FGF-4 has many of the properties of FGF-2 *in vivo*. It accelerates wound healing (see

Section 4.4.) and induces formation of mesoderm in *Xenopus* oocytes (Basilico and Moscatelli, 1992). FGF-4 is also angiogenic in model systems. For example, when single-cell suspensions of fetal rat brain (removed at embryonic days 13 and 14) were infected with retroviral vectors encoding the FGF-4 oncogene and subsequently implanted into the caudate putamen, the grafts exhibited abundant capillary proliferation and capillary angiomas (Brüstle *et al.*, 1992). Expression of FGF-4 was detected in neural cells adjacent to areas of vascular proliferation. On the other hand, despite its oncogenic properties, FGF-4 did not cause tumor formation in the brain in these experiments.

The FGF-4 gene is rarely expressed in adult cells or in adult tissues. It has been shown, however, to be expressed during midstage mouse embryogenesis (Delli-Bovi *et al.*, 1989; Niswander and Martin, 1992). NIH-3T3 cells transfected with the FGF-4 gene are transformed in culture and are tumorigenic, ostensibly by an autocrine loop of FGF-4 interacting with its FGF receptor. Suramin reverses this transformed phenotype (Moscatelli and Quarto, 1989; Yayon and Klagsbrun, 1990). As might be expected for the product of an oncogene, injection of a recombinant retrovirus encoding FGF-4 into mice causes tumor formation (aggressive fibrosarcomas) (Talarico *et al.*, 1993), and FGF-4 expression has been demonstrated in a variety of solid tumors (reviewed in Brem and Klagsbrun, 1993). Interestingly, although FGF-4 was first isolated from cells transfected with Kaposi's sarcoma DNA, it has not been detected in the secreted material from cultured Kaposi's sarcoma cells.

2.4. FGF-5

The gene for FGF-5 was isolated by transfection of human bladder tumor DNA into NIH-3T3 cells (Zhan *et al.*, 1988; Bates *et al.*, 1991). This gene encodes a signal-sequence-containing, 267-residue (29-kDa) protein with a 40–50% homology to FGF-1 and FGF-2 (Bates *et al.*, 1991). FGF-5 is a heparin-binding glycoprotein and is a potent mitogen for endothelial cells and fibroblasts (Zhan *et al.*, 1988; Bates *et al.*, 1991; Werner *et al.*, 1991). It also appears to be an embryonic muscle-derived trophic factor that supports the survival of embryonic chick motoneurons (Hughes *et al.*, 1993).

Messenger RNA transcripts for FGF-5 are found in nearly all phases of embryogenesis (Hébert *et al.*, 1990). They are also found in the neurons of adult brains (Haub *et al.*, 1990), in growing normal human fibroblasts (Werner *et al.*, 1991), in hair follicles (Hébert *et al.*, 1994), and in skin (Werner *et al.*, 1992a). Consistent with the oncogenic properties of FGF-5, this protein has been found to be secreted from bladder carcinoma, endometrial carcinoma, and human hepatoma cell lines (reviewed in Brem and Klagsbrun, 1993).

2.5. FGF-6

The FGF-6 gene was isolated from a mouse DNA plasmid library by screening with FGF-4 gene sequences (Marics *et al.*, 1989; de Lapeyiere *et al.*, 1990; Coulier *et al.*, 1991, 1994). The FGF-6 gene encodes a 208-amino-acid protein containing a hydrophobic signal sequence; this protein shares about 70% sequence identity with FGF-4 in a "core" homology region. *In vitro* translation of FGF-6 has yielded 21- to 24-kDa proteins. Transfection with the FGF-6 gene leads to transformation of NIH-3T3 cells and the appearance of foci. Recombinant FGF-6 protein has been expressed in bacteria and is heparin-binding. This protein is highly mitogenic for BALB/c 3T3 cells, but, in contrast to FGF-1, FGF-2, and FGF-4, it is far less mitogenic for aortic endothelial cells. The addition of heparin is required even to get the weak mitogenic signal for the endothelial cells. It thus appears that the effects of FGF-6 are more limited than those of FGF-1 and FGF-2. FGF-6 mRNA is present primarily in adult skeletal muscle and during embryogenesis, particularly postimplantation. FGF-6 represses myogenesis, as does FGF-1 and FGF-2, and it has been suggested that FGF-6 could play a role when muscle is damaged.

2.6. KGF/FGF-7

KGF/FGF-7 is synthesized by stromal cells derived from epithelial tissues of embryonic, neonatal, and adult sources (Finch *et al.*, 1989; Rubin *et al.*, 1989). The KGF/FGF-7 gene encodes a primary translation product of 194 amino acids that possesses a signal peptide and has a 39% structural homology to FGF-2. The 22-kDa KGF/FGF-7 protein is secreted and binds heparin. KGF/FGF-7 differs markedly from the other FGF family members in target cell specificity in that it is a highly specific mitogen for epithelial cells with no mitogenic activity for endothelial cells or fibroblasts. However, this factor is not produced by epithelial cells, but is rather the product of fibroblasts, and KGF/FGF-7 protein synthesis in fibroblasts can be potently induced by interleukin-1 (Chedid *et al.*, 1994). Thus KGF/FGF-7 appears to have the unique property of being a stromal cell-derived, paracrine mediator of epithelial cell proliferation. Besides its mitogenic activity, KGF/FGF-7 stimulates keratinocyte migration and an increase in keratinocyte urokinase-type plasminogen activator activity (Tsuboi *et al.*, 1993). It has been suggested that the keratinocyte-derived plasminogen activator activity might serve to stimulate proteolytic degradation of extracellular matrix by plasmin.

2.7. FGF-8 (AIGF) and FGF-9 (GAF)

FGF-8 (Tanaka *et al.*, 1992) and FGF-9 (Miyamoto *et al.*, 1993) are very recently identified members of the FGF family, and therefore the available information regarding these growth factors is limited. The conditioned medium of an androgen-dependent cancer line (SC-3) derived from a mouse mammary carcinoma was found to contain an autocrine heparin-binding growth factor mitogenic for SC-3 cells. This activity was inhibited by anti-FGF-2 antibodies. However, upon cloning, a novel FGF-like gene with an open reading frame encoding 215 amino acids (including a signal peptide) was discovered that has a 30–40% homology to members of the FGF family. Purification of FGF-8 resulted in the isolation of 28- and 32-kDa proteins equally active on SC-3 cells.

An interesting property of FGF-8 is that the expression of its mRNA in SC-3 cells is induced by testosterone. FGF-8 appears to be responsible for changing SC-3 cells from a normal epithelial cell phenotype to a transformed one.

More recently, FGF-9 was purified from the conditioned medium of a human glioma cell line and subsequently found to be expressed in the brain and kidney. The FGF-9 cDNA encodes a polypeptide of 208 amino acids with a structural homology to FGF family members of about 30%. FGF-9 lacks a classical signal sequence, but nevertheless appears to be secreted. Purification of recombinant FGF-9 expressed in Chinese hamster ovary (CHO) cells yielded 25-, 29-, and 30-kDa proteins. To date, FGF-9 has been shown to be mitogenic for glial cells, suggesting that FGF-9 may act as an autocrine transforming growth factor for these cells.

2.8. FGF Receptors

A family of FGF receptors has been identified (Kornbluth *et al.*, 1988; Lee *et al.*, 1989; Ruta *et al.*, 1989; Dionne *et al.*, 1990; Houssaint *et al.*, 1990; Johnson *et al.*, 1990; Mansukhani *et al.*, 1990; Pasquale, 1990; Keegan *et al.*, 1991; Partanen *et al.*, 1991; Miki *et al.*, 1991; reviewed in Partanen *et al.*, 1992). The family includes four homologous protein tyrosine kinases with K_ds of about 2×10^{-11}: FGFR-1, originally known as flg; FGFR-2, originally known as bek; FGFR-3, originally known as cek2; and FGFR-4. In addition, a unique cysteine-rich FGFR of unknown function that is not homologous to the tyrosine kinase FGF receptors has been described (Burrus *et al.*, 1992; Olwin *et al.*, 1994). The FGFR tyrosine kinases share about 60–70% sequence identity and a similar overall structure. In particular, they can each contain up to three immunoglobulin loops in the extracellular region, and the cytoplasmic tyrosine kinase domain in each is interrupted by a short insert. The second and third immunoglobulin loops, which are sites of FGF binding, and the tyrosine kinase domains are particularly conserved. The first immunoglobulin loop appears to be dispensible for ligand binding.

There are numerous splice variants among the four FGF receptors, the existence of which increases the number of possible forms of FGF receptor (Miki *et al.*, 1992; Avivi *et al.*, 1991, 1993; Bottaro *et al.*, 1990; Johnson *et al.*, 1991; Werner *et al.*, 1992b; reviewed in Partanen *et al.*, 1992). As a result of the differential splicing, forms with two rather than three immunoglobulin domains are found as well as forms with structural variations in the region of the second half of the third immunoglobulin domain loop. The four tyrosine kinase FGFR gene products and their multiple splice variants display different binding specificities for the various FGF ligands. The type IIIc form of FGFR-1 binds FGF-1, FGF-2, and (to a much lesser degree) FGF-4. On the other hand, type IIIb FGFR-1 binds FGF-1, but FGF-2 only weakly. Type IIIc FGFR-2 binds FGF-1, FGF-2, and FGF-4, but not KGF/FGF-7. In contrast, the type IIIb FGFR-2 variant, also known as the KGF receptor (KGFR), binds FGF-1, KGF/FGF-7, and (to a much lesser degree) FGF-2. Type IIIc FGFR-3 binds both FGF-1 and FGF-2, while type IIIb FGFR-3 binds exclusively FGF-1. FGFR-4 binds FGF-1 but not FGF-2. Taken together, these results indicate that all of the receptors can bind FGF-1, but the receptors differ in their ability to bind FGF-2, FGF-4, and KGF/FGF-7.

Cell surface HSPG constitutes a class of low-affinity binding sites for FGF-2, but HSPG can nonetheless be critical for the high-affinity receptor binding and optimal biological activity of FGF-2 (Klagsbrun and Baird, 1991). This conclusion was independently reached from a number of investigations. In one study, CHO cells expressing the FGFR-1 receptor but carrying a mutation in the HSPG synthesis pathway were found to be unable to bind FGF-2 (Yayon *et al.*, 1991). Addition of exogenous heparin facilitated FGF-2 binding to the FGFR-1 on these cells. Heparin has also been shown to enhance both cell-free FGF-2 binding to its receptor and FGF-2 mitogenic activity for cells mostly devoid of cell surface heparan sulfate (Ornitz *et al.*, 1992). There has been one report, however, questioning the role of heparin in mediating FGF-2 binding to its receptor in some cell types (Roghani *et al.*, 1994). In another study, it was demonstrated that heparitinase treatment of myoblasts abolished the ability of FGF-2 to stimulate myoblast proliferation, and similar treatment of Swiss 3T3 cells abolished FGF-2 mitogenic activity for those cells as well (Rapraeger *et al.*, 1991). The ability of FGF-1 to bind to high-affinity receptors on parathyroid cells is also abrogated by heparatinase treatment (Sakaguchi *et al.*, 1991). In another approach, FGF-2 has been conjugated to saporin, a toxin that kills cells if it gains access to the cytoplasm (Reiland and Rapraeger, 1993). FGF-2–saporin does not kill cells that do not express FGF receptors, even if the cells have abundant cell surface HSPG; however, the conjugate will kill these cells when they are transfected and express FGFR-1. Removal of HSPG by enzymatic digestion eliminates killing by FGF–saporin. These findings suggest that FGF-2–saporin may be endocytosed on HSPG in a complex with FGFR that is directed to the cytoplasm. HSPG also regulates neural response to FGF-1 and FGF-2 developmentally (Nurcombe *et al.*, 1993). Taken together, the above results indicate that a dual receptor system composed of cell surface HSPG and high-affinity tyrosine kinase receptors modulates the activity of FGF-1 and FGF-2. There is also evidence that FGFR-1 itself binds to HSPG as well, suggesting that HSPG may interact directly with both FGF and FGFR in the process of receptor activation (Kan *et al.*, 1993).

3. Evidence for Involvement of Endogenous FGFs in Dermal Wound Healing

To gain insight into the roles that individual FGF family members might play in the process of normal tissue repair, several groups have sought to determine whether these proteins and/or their mRNAs are present at sites of wound healing and to elucidate whether the levels of the factors are modulated during the course of healing. Various approaches to these questions have been used, including RNase protection assays, immunohistochemical staining, and the quantification of growth factor levels in wound fluids. A variety of wound-healing models have also been employed in these studies, which has sometimes complicated the comparison of results from different groups. Nonetheless, there is now considerable evidence that at least one FGF family member protein (FGF-2) is present and modulated at sites of dermal tissue injury (Whitby and Ferguson, 1991a; Kurita *et al.*, 1992; Gibran *et al.*, 1994; Grayson *et al.*, 1993; Cooper *et al.*, 1994; Chen *et al.*, 1992). Direct evidence for an actual role in wound repair for this factor has also been obtained by studying the effects when the action of FGF-2 was blocked (Broadley *et al.,* 1989a). In addition, measurements of mRNA levels after wounding (Werner *et al.,* 1992a, 1994a), along with the results of studies on transgenic mice expressing a dominant negative mutant form of the type IIIb FGFR-2 receptor (Werner *et al.,* 1994b), have strongly implicated KGF/FGF-7 as a modulator of wound reepithelialization.

3.1. Detection of FGFs at Sites of Tissue Damage

The most systematic surveys of FGF family member RNA expression at sites of dermal injury have been reported by Werner *et al.* (1992a, 1994a). In the initial study by this group (Werner *et al.,* 1992a), RNase protection assays were used to determine transcript levels in unwounded skin and in full-thickness excisional wounds made in the backs of BALB/c mice (Fig. 1). Transcripts for FGF-3, FGF-4, and FGF-6 were not detected in either the wounded or unwounded tissue. In the case of FGF-1, FGF-2, and FGF-5, on the other hand, transcripts were detected in the unwounded control skin which rose two- to tenfold in level in response to injury; these levels peaked at either 24 hr (FGF-1, FGF-5) or 5 days (FGF-2) and had returned to baseline by the seventh day postwounding. The most striking result, however, was seen with the level of the KGF/FGF-7 mRNA, which increased within 24 hr to 160-fold over the control skin level and was still 100-fold elevated in the wounds after 7 days (see Fig. 1). Transcripts for the FGF receptors FGFR-1, FGFR-2, and (at a much lower level) FGFR-3 were also detected, but their levels did not appear to differ between unwounded and wounded skin. The FGFR-2 transcript was found to be primarily in the splice variant form (type IIIb) that encodes a high-affinity receptor for FGF-1 and KGF/FGF-7. Taking these results together with the known epithelial cell specificity of KGF/FGF-7, Werner *et al.* (1992a) proposed in particular that wound-derived KGF/FGF-7 may play a key role in stimulating the migration and proliferation of epidermal keratinocytes during wound repair.

In a second study, Werner and colleagues extended their survey to analyze FGF family member transcript levels in an impaired model of wound healing: full-thickness dermal wounds in the backs of genetically diabetic (*db/db*) mice (Werner *et al.,* 1994a). Since application of exogenous FGF-2 had been shown by several groups to accelerate healing to near normal in this model (see Section 4.2.1.), the investigators reasoned that the healing delay in *db/db* mice might be associated with decreased availability of endogenous FGF(s). RNase protection assays indicated that while the KGF/FGF-7 transcript level in unwounded *db/db* skin was similar to that found in the nondiabetic BALB/c mice, the degree of induction of the transcript level after wounding was now only five- to tenfold rather than the 160-fold seen for the nondiabetic mice. The rise in the KGF/FGF-7 transcript level after wounding was also considerably delayed, and did not peak until 3–5 days after injury. Interestingly, while the basal transcript level for FGF-2 in unwounded skin (and the degree of induction upon injury) was similar in *db/db* and BALB/c mice, the basal level of the FGF-1 transcript was actually five- to tenfold higher in unwounded *db/db* skin relative to BALB/c skin, and little induction in this level was seen on injury. For both FGF-1 and FGF-2, the transcript levels peaked earlier after injury and declined more rapidly than in BALB/c mice. The FGFR-2 and FGFR-3 transcript levels also declined rapidly in the *db/db* mice soon after injury.

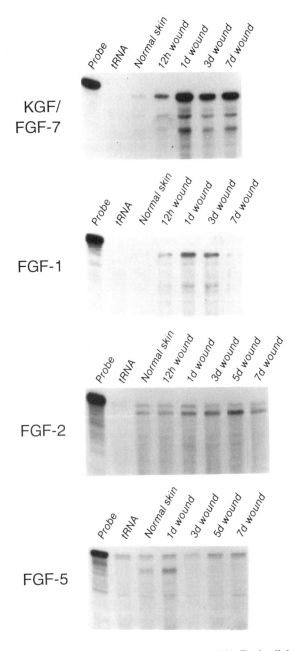

Figure 1. FGF mRNA expression in normal and wounded mouse skin. Total cellular RNA was extracted from skin at various times after full-thickness wounding. RNA was also extracted from unwounded skin (*normal skin*) for comparison. RNase protection assays were then used to quantify the mRNAs for KGF, FGF-1, FGF-2, and FGF-5 in 50-μg samples of the total cellular RNA. Autoradiographs of the protection assay gels are shown. The gels were exposed to the films for 12 hr (KGF), 3 days (FGF-5), or 5 days (FGF-1 and FGF-2). The times postwounding at which the RNA samples were extracted were 12 hr and 1, 3, 5, and 7 days. (Adapted from Werner *et al.,* 1992a, with permission.)

While it is possible that some of the transcript level differences seen by Werner *et al.* (1994a) were due to differences in strain backgrounds (BALB/c nondiabetic mice versus C57BL/KsJ diabetic mice) rather than to the consequences of the homozygous *db/db* defect, these results are nonetheless consistent with the idea that level or time-course alterations in the production of KGF/FGF-7, FGF-1, FGF-2, and/or the FGF receptors at the site of injury contribute to the wound healing impairment in *db/db* mice.

Three groups (Whitby and Ferguson, 1991a; Kurita *et al.*, 1992; Gibran *et al.*, 1994) have used immunohistochemistry to investigate FGF-2 protein localized at sites of dermal injury. Whitby and Ferguson (1991a) made vertical, full-thickness incisions in the upper lips of fetal, neonatal, and adult MF1 mice, and then examined the wounds 1 to 120 hr later. In the neonatal and adult mice, staining with an FGF-2 antibody was detected extracellularly at the surface of the wound and within the dermis adjacent to the wound at 1 and 6 hr postwounding; no staining was found at later times. The fetal wounds showed no detectable FGF-2 staining at any time point examined, which the investigators suggest may in part explain the reduced amount of capillary formation seen in the fetal versus adult mouse lip wounds (Whitby and Ferguson, 1991b).

Kurita *et al.* (1992) utilized a different mouse wound model for their immu-nohistochemical study—full-thickness, 6-mm wounds made by punch biopsy in the backs of C57BL6J mice—and obtained somewhat different results. No FGF-2 immu-noreactivity was detected in extracellular matrix spaces or in the dermal granulation tissue, but staining was found instead associated with hair bulbs at the wound edge and with basal keratinocyte layers in reepithelialized and nonwounded areas. This staining pattern was similar whether the wounds were analyzed immediately after injury or up to 21 days postinjury. The reasons for the apparent disparity between these results and the results of Whitby and Ferguson (1991a) are not clear, but possible explanations include differences in the specificity or sensitivity of the anti-FGF-2 antibodies utilized and the fact that the wounds made in the two studies were of different types and were made in different locations in the mouse. Interestingly, consistent with the RNA results of Werner *et al.* (1994a), Kurita *et al.* (1992) reported that they obtained essentially the same immunohistochemical staining pattern for FGF-2 whether they used C57BL6J nondiabetic or C57BL/KsJ *db/db* mice in their studies.

Gibran *et al.* (1994) carried out their immunohistochemical studies on human tissue samples, which were taken from either burn wound excisions or from normal skin harvested for skin grafts. In eight burn tissue samples harvested between 4 and 11 days postinjury, diffuse extracellular staining for FGF-2 was seen, similar to the stain-ing pattern reported by Whitby and Ferguson (1991a) for the mouse lip wounds at 1 or 6 hr after injury. Normal skin samples showed FGF-2 staining confined to dermal capillary basement membranes and sweat glands.

The presence of FGF-2-like protein at wound sites has also been reported by three groups (Chen *et al.*, 1992; Grayson *et al.*, 1993; Cooper *et al.*, 1994), who analyzed the growth factors present in the fluid environment at the surface of wounds. Cooper *et al.* (1994) placed hydrophilic dextranomer beads into human pressure ulcers to absorb or adsorb proteins over a 24-hr period; sandwich enzyme-linked immunosorbent assay (ELISA) analyses of the material subsequently eluted from the beads detected FGF-2 in all 20 patient samples examined, although the levels varied widely (from 47 to 697 pg/ml in the bead eluates). Variability in growth factor levels was also seen by Grayson *et al.* (1993), who analyzed fluids collected after 24 hr from a more "standardized" type of human wound: partial-thickness donor sites covered with an occlusive dressing. Using a sandwich enzyme immunoassay with a detection limit of approximately 1 ng/ml FGF-2, these investigators found FGF-2 in the wound fluid from only 5 of 13 patients. In those 5 patients, the FGF-2 levels appeared to range from about 9 to 16 ng/ml.

With regard to animal models of wound healing, Chen *et al.* (1992) examined wound fluid from full-thickness surgical excisions made in the flanks of Yorkshire White pigs. The wounds were covered with occlusive dressings, and the fluids were collected every 24 hr. When applied to human dermal fibroblasts, the wound fluids collected during the second and third 24-hr periods after wounding were found to stimulate urokinase production by the fibroblasts. Since this stimulation of protease production was blocked if the wound fluid was preincubated with an antiporcine FGF-2 antibody, Chen *et al.* (1992) concluded that FGF-2-like factors were present in the fluid.

3.2. Effects of a Neutralizing Anti-FGF-2 Antibody on Granulation Tissue Formation

The demonstrations of FGF-2 mRNA and protein at sites of dermal injury represent only indirect support for the idea that FGF-2 is involved in mediating the wound-healing process. In order to provide more concrete evidence of a role for FGF-2-like protein in tissue repair, Broadley *et al.* (1989a) made use of a neutralizing polyclonal antibody that was raised in a rabbit against human FGF-2. Immunoglobulin G (IgG) purified from the neutralizing rabbit serum was incorporated into pellets that were designed to release the protein at a controlled rate of 0.1 µg/day over 14 days. Control pellets received IgG purified from preimmune rabbit serum. Each pellet was placed in the center of a polyvinyl alcohol sponge disk, and the disks were then implanted subcutaneously under the ventral panniculus carnosus of rats. Histological analyses of the granulation tissue formed in the sponges after 7 to 9 days revealed that the material in the sponges containing the anti-FGF-2 IgG pellets exhibited strikingly less cellularity and vascularization than the granulation tissue formed in the control IgG sponges (Fig. 2). Broadley *et al.* (1989a) also showed that the DNA, protein, and collagen levels in the anti-FGF-2 sponges were reduced by about 25–35% relative to the control sponge values at day 7. Since the differences between the granulation tissues formed in the anti-FGF-2 and control sponges became less significant between 9 to 12 days after sponge implantation, the investigators interpreted their results as suggesting that endogenous FGF-2 indeed plays a role in mediating the normal rate of wound healing, but that its effects are more pronounced at earlier stages of the repair process.

With this neutralization study, as with the three immunohistochemical studies described above, it is worth noting that potential cross-reactivity of the FGF-2 antibodies with other FGF family members was only analyzed with respect to FGF-1 and (in some cases) FGF-4. The possibility thus exists that the staining seen in the immunohistochemistry studies or the growth factor activity being neutralized in the Broadley *et al.* (1989a) study could represent other FGFs besides or in addition to FGF-2.

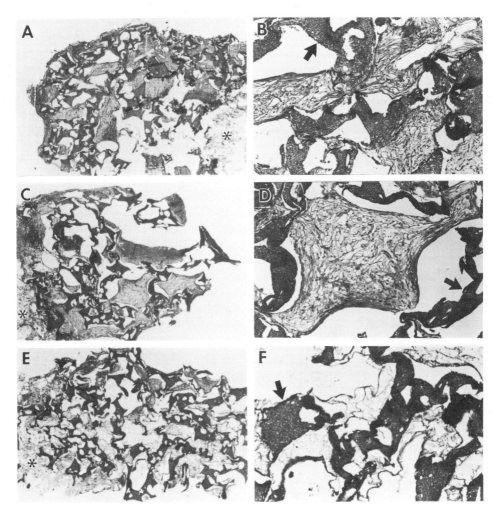

Figure 2. Histological analysis of polyvinyl alcohol sponge disks, 9 days after implantation under the panniculus carnosus on the ventral surface of rats. (A, B) Sponge containing a slow-release pellet with no additions (placebo); (C, D) sponge containing a pellet releasing preimmune IgG; (E, F) sponge containing a pellet releasing anti-FGF-2 IgG. Asterisks indicate the positions of the slow-release pellets in the sponges. The histological sections are stained with Masson's trichrome. Sponge material appears as the more uniform, darker gray areas in the panels; several such areas are indicated by arrows in panels B, D, and F. Magnification is × 25 (A, C, E) or × 100 (B, D, F). (Reprinted from Broadley *et al.*, 1989a, with permission.)

3.3. Effects of Transgenic Expression of a Dominant Negative Mutant Form of the Type IIIb FGFR-2 Receptor ("KGF Receptor")

The observation that the KGF/FGF-7 mRNA transcript level in mouse skin is markedly induced after full-thickness wounding had suggested to Werner *et al.* (1992a) that this potent keratinocyte mitogen might play a central role in the reepithelialization of wounds. To test this possibility, Werner *et al.* (1994b) took advantage of the fact that the only known high-affinity receptor for KGF/FGF-7 is the type IIIb form of FGFR-2. Transgenic mice were therefore created in which a truncated form of the type IIIb FGFR-2 protein (lacking the tyrosine kinase domains) was expressed under the control of the keratin 14 promoter. Upon binding a ligand, such truncated mutants of tyrosine kinase FGF receptors have been shown to form nonfunctional heterodimers with wild-type FGFRs, with the result that no receptor signaling occurs in response to the ligand (KGF/FGF-7 or FGF-1, in the case of type IIIb FGFR-2). In the transgenic mice made by Werner *et al.* (1994b), the keratin 14 promoter was used for the purpose of targeting expression of the truncated receptor to the undifferentiated keratinocytes in the basal layer of the epidermis.

While the transgenic mice were born with no apparent phenotypic abnormalities, Werner *et al.* (1994b) detected histological changes starting at 3 to 6 weeks after birth that included epidermal atrophy and dermal hyperthickening. When full-thickness excisional wounds were created in these mice, the reepithelialization of the wounds was found to be delayed relative to control animals. Analysis of *in vivo* 5-bromodeoxyuridine labeling of the wounded animals at day 5 after injury revealed that the number of proliferating cells in the epidermis at the wound edge in the transgenic mice was 20- to 100-fold lower than the number seen at the wound edge in control animals. These results indicate that a ligand for the type IIIb FGFR-2 receptor is indeed critical for keratinocyte proliferation in response to full-thickness wounding, supporting the proposal (Werner *et al.*, 1992a) that KGF/FGF-7 is involved in reepithelialization.

4. Effects of Exogenously Applied FGFs in Animal Models of Dermal Wound Healing

The first indication that dermal repair could be influenced by the exogenous application of FGF proteins came from the work of Buntrock *et al.* (1982a,b, 1984) (Table II). In these studies, polyvinyl rings were implanted under the dorsal skin of rats. On the day of surgery and every other day thereafter, a partially purified extract of bovine brain (now known to contain a mixture of FGF-1 and FGF-2) was applied to a point inside the ring. At 3 and 7 days after the ring implantations, Buntrock *et al.* (1982a) noted a significant increase in the wet and dry weights of the extract-treated rings, relative to vehicle-treated controls. Histological analysis (Buntrock *et al.*, 1982b) revealed in particular an increase in capillary formation in the tissue that had formed within the extract-treated rings, and the investigators therefore concluded that application of the impure preparation of FGF was primarily resulting in the stimulation of angiogenesis.

In the years since the work of Buntrock *et al.* (1982a,b, 1984), numerous wound-healing studies using highly purified preparations of FGF proteins have been published (Tables II–VI). These studies have utilized a wide range of different types of wound-healing models, including (1) granulation tissue formation in implants; (2) granulation tissue formation, epithelialization, and rate of wound closure in full-thickness excisions; (3) development of wound breaking strength in full-thickness incisions; and (4)

Table II. Studies of FGF-2 in Models of Unimpaired Healing

Reference	Wound model	Species	Dose per application and vehicle[a]	Time of application(s)[b]	Observations (vs. vehicle control)[c]
Buntrock et al. (1982a,b, 1984)	Polyvinyl ring implant	Rat	Impure extract	Days 0, 2, 4, 6	Day 3 and day 7: ↑ wet and dry weight of rings ↑ angiogenesis
Davidson et al. (1985)	Polyvinyl alcohol sponge implant	Rat	0.5 μg	Day 6	Day 9: ↑ tissue infiltration into sponges ↑ angiogenesis ↑ DNA, protein, and collagen content
Broadley et al. (1988)	Polyvinyl alcohol sponge implant	Rat	5 μg in PBS + BSA[d]	Day 3	Day 7 and day 9: ↑ DNA and protein content → % collagen; trend ↓ in total collagen
Davidson and Broadley (1991)	Polyvinyl alcohol sponge implant	Rat	1, 10, or 20 μg in PBS + BSA[d]	Day 3	Day 7 and day 9: ↑ DNA and protein content → % collagen and total collagen content
Davidson and Broadley (1991)	Polyvinyl alcohol sponge implant	Rat	1 or 2.5 μg per day in slow release pellet + heparin	Starting on day 0	Day 10: ↑ angiogenesis ↑ DNA and protein content ↑ total collagen content → % collagen
Fiddes et al. (1991)	Polyvinyl alcohol sponge implant	Rat	0.66, 3.33, or 10 μg	Day 4	Day 8–day 14: ↑ granulation tissue infiltration ↑ angiogenesis ↑ collagen content
Lazarou et al. (1989)	Polyvinyl alcohol sponge implant	Rat	0.5, 5, or 50 μg	Day 0	Day 10: → collagen content → tissue infiltration
Sprugel et al. (1987)	Porous polytetrafluoroethylene tube implant	Rat	0.1 μg in collagen + heparin	Day 0	Day 10: ↑ vascularization ↑ DNA content and fibroplasia ↑ collagen vehicle remodeling

(continued)

Table II. (*Continued*)

Reference	Wound model	Species	Dose per application and vehicle[a]	Time of application(s)[b]	Observations (vs. vehicle control)[c]
Pierce *et al.* (1992)	Full-thickness excisional dermal ulcer in ear	Rabbit	2 μg (8 μg/cm²) (mutant)[e]	Day 0	Day 7: ↑ new granulation tissue formation ↑ GAG and fibronectin content ↑ collagenase activity around wound (and ↓ collagen on days 14 and 21) ↑ angiogenesis • apparent disrupted turnoff of repair
Mustoe *et al.* (1991)	Full-thickness excisional dermal ulcer in ear	Rabbit	5 μg (20 μg/cm²) in collagen	Day 0	Day 7: ↑ reepithelialization • apparent ↑ in angiogenesis • no effect on granulation tissue formation
Mustoe *et al.* (1994)	Full-thickness excisional dermal ulcer in ear	Rabbit	5 μg (20 μg/cm²) (mutant)[e]	Day 0	Day 7: ↑ angiogenesis ↑ granulation tissue formation
Uhl *et al.* (1993)	Full-thickness excisional dermal ulcer in ear	Mouse (*hr/hr*)	0.24 or 1.35 μg (4.8 or 27 μg/cm²)	Days 1, 2, 3	Days 4–10: • no change in rate of wound closure • less time needed to achieve complete reepithelialization
Tsuboi and Rifkin (1990)	Full-thickness dorsal excisional wound	Mouse (*db/+m*)	0.5 or 5 μg (1.8 or 18 μg/cm²) in 1.5% CMC[d]	Days 0, 1, 2, 3, 4	Day 8: • no major effects on granulation tissue formation, angiogenesis, or reepithelialization • small ↑ in wound cellularity
Greenhalgh *et al.* (1990)	Full-thickness dorsal excisional wound	Mouse (*db/+m*)	0.4 or 1 μg (0.18 or 0.44 μg/cm²) in 5% PEG[d]	Days 0, 1, 2, 3, 4	Day 10: • no significant effects on granulation tissue formation, rate of wound closure, or histological appearance

Reference	Wound type	Animal	Dose	Timing	Results
Klingbeil et al. (1991)	Full-thickness excisional wound in flank	Mouse (db/+m, ob/+, or hr/hr)	2 or 20 µg (1 or 10 µg/cm²)	Day 0	Day 6–day 12: • slight (10–15%) ↑ in rate of wound closure
Stenberg et al. (1989, 1991)	Full-thickness dorsal excisional wound	Rat	100 µg (66.7 µg/cm²)	Day 0	No change in rate of wound closure ↓ breaking strength of wound scar at time of closure
McGee et al. (1988)	Full-thickness dorsal incision	Rat	0.4 µg	Day 3	Days 5, 6, 7: ↑ tensile strength • no significant change in wound collagen content
Slavin et al. (1992)	Full-thickness dorsal incision	Rat	5 or 50 µg	Day 0	Day 7: • 5 µg: no changes • 50 µg: ↓ breaking load, ↑ cellularity
Slavin et al. (1992)	Full-thickness dorsal incision	Rat	0.5 or 5 µg in collagen	Day 0	Day 7: • 0.5 µg: ↓ cellularity; no change in breaking load • 5 µg: ↑ cellularity; no change in breaking load
Slavin et al. (1992)	Full-thickness dorsal incision	Rat	5 µg in red blood cell ghosts	Day 0	Day 7: ↓ cellularity ↑ breaking load
Phillips et al. (1993)	Full-thickness dorsal incision	Rat	10 µg	Day 0	Day 7–day 21: • no change in breaking strength
Wu and Mustoe (1995)	Full-thickness incision in ear	Rabbit	20 µg (mutant)[e]	Day 0	Day 10: • no change in breaking strength
Tsuboi and Rifkin (1990)	Full-thickness dorsal incision	Mouse (db/+m)	5 µg in 1.5% CMC[d]	Day 0	Day 9: ↑ breaking strength
Hebda et al. (1990a)	Partial-thickness excisional wound	Pig	1 or 10 µg (1 or 10 µg/cm²)	Day 0	Days 5, 6, and 7: ↑ rate of reepithelialization ↑ collagen deposition ↑ neovascularization

(continued)

Table II. (*Continued*)

Reference	Wound model	Species	Dose per application and vehicle[a]	Time of application(s)[b]	Observations (vs. vehicle control)[c]
Lynch *et al.* (1989)	Partial-thickness excisional wound	Pig	0.5 µg (0.33 µg/cm²) in 3% MC[d]	Day 0	Day 7: • no significant effects on cellularity, dermal or epidermal thickness, or hydroxyproline content
LeGrand *et al.* (1993)	Partial-thickness excisional wound	Guinea pig	0.1–30 µg (0.03–10 µg/cm²) applied to dressing	Days 0, 1, 2, 3, 4	Day 7: • no significant effects on new granulation tissue thickness, epithelialization, vascularity, or collagen density
Eriksson *et al.* (1989)	Medium partial-thickness burn	Pig	0.03 µg (4.8 ng/cm²)	Daily	Day 6: ↑ reepithelialization
Cooper *et al.* (1991)	Meshed, split-thickness human skin grafted on full-thickness excision	Mouse (BALB/c *nu/nu*)	Approx. 100 ng/cm² in slow-release dressing	Days 0, 2, 4, 6	Day 8: ↓ epithelialization of interstices in grafts

[a]Unless otherwise noted, the vehicle used was an aqueous solution (most often, PBS) or was not described.
[b]Day 0 represents time of surgery (wounding).
[c]Day(s) cited represent the day(s) postwounding on which the analyses were done.
[d]Abbreviations used: PBS, phosphate-buffered saline; BSA, bovine serum albumin; CMC, carboxymethyl cellulose; PEG, polyethylene glycol; MC, methylcellulose.
[e]In "mutant" FGF-2, two cysteines were changed to serines to give the factor more stability.

Table III. Studies of FGF-2 in Models of Impaired Healing

Reference	Wound model	Species	Dose per application and vehicle[a]	Time of application(s)[b]	Observations (vs. vehicle control)[c]
Broadley et al. (1988)	Polyvinyl alcohol sponge implant in diabetic animal	Rat (strep.)[d]	5 μg in PBS + BSA[e]	Day 3	Day 9: → collagen content; ↑ DNA and protein content; ↑ cellularity
Phillips et al. (1993)	Full-thickness dorsal incision in diabetic animal	Rat (strep.)[d]	10 μg	Day 0	Day 7 and day 10: • no effect on breaking strength. Day 14: ↑ breaking strength to level of normal rats
Tsuboi and Rifkin (1990)	Full-thickness dorsal incision in diabetic animal	Mouse (db/db)	5 μg in 1.5% CMC[e]	Day 0	Day 9: ↑ breaking strength to nearly that of normal (db/+m) mice
Tsuboi and Rifkin (1990)	Full-thickness excisional wound in diabetic animal	Mouse (db/db)	0.5 or 5 μg (1.8 or 18 μg/cm²) in 1.5% CMC[e]	Day 0 or days 0, 1, 2, 3, 4	Day 8: ↑ granulation tissue, cellularity, and angiogenesis to near level seen in normal (db/+m) mice; • no change in rate of wound closure
Greenhalgh et al. (1990)	Full-thickness excisional wound in diabetic animal	Mouse (db/db)	1 μg (0.44 μg/cm²) in 5% PEG[e]	Days 0, 1, 2, 3, 4	Day 10: ↑ granulation tissue, cellularity, angiogenesis; • no significant change in rate of wound closure; Day 21: ↑ rate of wound closure; ↑ granulation tissue and cellularity
Klingbeil et al. (1991)	Full-thickness excisional wound in diabetic animal	Mouse (db/db)	2 or 20 μg (1 or 10 μg/cm²)	Day 0	Days 6–30: ↑ rate of wound closure to nearly that of normal (db/+m) mice

(continued)

Table III. (*Continued*)

Reference	Wound model	Species	Dose per application and vehicle[a]	Time of application(s)[b]	Observations (vs. vehicle control)[c]
Klingbeil et al. (1991)	Full-thickness excisional wound in obese animal	Mouse (ob/ob)	0.2–20 μg (0.1–10 μg/cm²)	Day 0	Days 3–21: ↑ rate of wound closure to nearly that of normal (ob/+) mice Day 21: ↑ neovascularization
Klingbeil et al. (1991)	Full-thickness excisional wound in steroid-treated animal	Mouse (hr/hr) (pred.)[f]	0.2–20 μg (0.1–10 μg/cm²)	Day 0	Days 5–15: ↑ rate of wound closure to nearly that of non-steroid-treated mice Day 12 ↑ granulation tissue formation ↑ neovascularization
Albertson et al. (1993)	Full-thickness excisional wound in malnourished animal	Mouse (db/+m)	1 μg (0.44 μg/cm²) in 5% PEG[d]	Days 0, 1, 2, 3, 4	Day 10: • no significant effect on wound closure or histology
Albertson et al. (1993)	Full-thickness excisional wound in malnourished, diabetic animal	Mouse (db/db)	1 μg (0.44 μg/cm²) in 5% PEG[d]	Days 0, 1, 2, 3, 4	Day 21: ↑ extent of wound closure ↑ granulation tissue formation ↑ wound cellularity
Stenberg et al. (1989, 1991)	Full-thickness excisional wound contaminated with bacteria	Rat	1, 10, or 100 μg (0.67, 6.7 or 67 μg/cm²)	Day 0	Rate of wound closure increased ↓ breaking strength of wound scar at closure (100 μg dose)
Hayward et al. (1992)	Chronic granulating wound	Rat	3000 μg (100 μg/cm²)	Days 5, 9, 12, 15, 18	Rate of wound closure increased Day 19: ↑ cellularity • no change in neovascularization or granulation tissue formation
Wu and Mustoe (1995)	Full-thickness incision in ischemic ear	Rabbit	20 μg (mutant)[g]	Day 0	Day 10: • no effect on wound breaking strength

Reference	Wound model	Animal	Dose	Day of treatment	Results
Mustoe et al. (1994)	Full-thickness excisional dermal ulcer in ischemic ear	Rabbit	5, 15, or 30 μg (20–120 μg/cm²) (mutant)[g]	Day 0	Day 7: ↑ angiogenesis • no effect on reepithelialization or new granulation tissue formation
Wu et al. (1995)	Full-thickness excisional dermal ulcer in ischemic ear	Rabbit	5, 15, or 30 μg (20–120 μg/cm²) (mutant)[g]	Day 0	Day 7: • no significant effect on reepithelialization or new granulation tissue formation
Wu et al. (1995)	Full-thickness excision in ischemic ear, + hyperbaric oxygen	Rabbit	5 μg (20 μg/cm²) (mutant)[g]	Day 0	Day 7: ↑ reepithelialization ↑ new granulation tissue formation
Uhl et al. (1993)	Full-thickness excisional dermal ulcer in ischemic ear	Mouse (hr/hr)	0.24 or 1.35 μg (4.8 or 27 μg/cm²)	Days 1, 2, 3	Day 7, day 10: ↑ rate of wound closure ↑ angiogenesis ↑ granulation tissue formation

[a] Unless otherwise noted, the vehicle used was an aqueous solution (most often, PBS) or was not described.
[b] Day 0 represents time of surgery (wounding).
[c] Day(s) cited represent the day(s) postwounding on which the analyses were done.
[d] Treated with streptozotocin to induce diabetic state.
[e] Abbreviations used: PBS, phosphate-buffered saline; BSA, bovine serum albumin; CMC, carboxymethyl cellulose; PEG, polyethylene glycol.
[f] Treated with prednisolone acetate.
[g] In "mutant" FGF-2, two cysteines were changed to serines to give the factor more stability.

Table IV. Studies of FGF-1 in Wound Healing Models

Reference	Wound model	Species	Dose per application and vehicle[a]	Time of application(s)[b]	Observations (vs. vehicle control)[c]
Mellin et al. (1992)	Full-thickness dorsal excisional wound	Mouse	0.25 μg (0.9 μg/cm²) in heparin + MSA[d]	Daily (twice per day)	Days 2, 4, and 6: ↑ rate of wound closure
Mellin et al. (1992)	Full-thickness dorsal excisional wound	Rat	0.5 μg (1.8 μg/cm²) in heparin + RSA[d]	Daily (twice per day)	Day 8: ↑ rate of wound closure ↑ granulation tissue formation ↑ angiogenesis
Mellin et al. (1992)	Full-thickness dorsal incision	Rat	2 μg in heparin + RSA[d]	Daily (once per day)	Day 10: ↑ breaking strength ↑ cellularity ↑ collagen content
Matuszewska et al. (1994)	Full-thickness dorsal excisional wound	Mouse (db/+m)	6 μg (3 μg/cm²) in heparin + 1% HEC[d]	Day 0	Days 3–16: • no change in rate of wound closure
Matuszewska et al. (1994)	Full-thickness excisional wound in diabetic animal	Mouse (db/db)	6 μg (3 μg/cm²) in heparin + 1% HEC[d]	Day 0, or days 0, 3, 7	Rate of wound closure increased ↑ granulation tissue formation ↑ angiogenesis
Mellin et al. (1995)	Full-thickness excisional wound in diabetic animal	Mouse (db/db)	1.2 or 6 μg (0.6 or 3 μg/cm²) in heparin + 1% HEC	Days 0, 3, 7	Rate of wound closure increased

[a]Unless otherwise noted, the vehicle used was an aqueous solution (most often, PBS) or was not described.
[b]Day 0 represents time of surgery (wounding).
[c]Day(s) cited represent the day(s) postwounding on which the analyses were done.
[d]Abbreviations used: MSA, mouse serum albumin; RSA, rat serum albumin; HEC, hydroxyethylcellulose.

Table V. Studies of FGF-4 in Wound Healing Models

Reference	Wound model	Species	Dose per application and vehicle[a]	Time of application(s)[b]	Observations (vs. vehicle control)[c]
Wu and Mustoe (1995)	Full-thickness incision in ear	Rabbit	20 µg	Day 0	Day 10: ↑ wound breaking strength
Wu and Mustoe (1995)	Full-thickness incision in ischemic ear	Rabbit	20 µg	Day 0	Day 10: ↑ wound breaking strength
Wu et al. (1995)	Full-thickness excisional dermal ulcer in ear	Rabbit	5 µg (20 µg/cm²)	Day 0	Day 7: ↑ reepithelialization ↑ granulation tissue formation
Wu et al. (1995)	Full-thickness excisional dermal ulcer in ischemic ear	Rabbit	5 or 15 µg (20 or 60 µg/cm²)	Day 0	Day 7: ↑ reepithelialization to level of vehicle-treated, nonischemic wounds ↑ granulation tissue formation
Hebda et al. (1990b)	Partial-thickness excisional wound	Pig	0.1–10 µg (0.1–10 µg/cm²)	Day 0	Day 4: ↑ rate of reepithelialization

[a]Unless otherwise noted, the vehicle used was an aqueous solution (most often, PBS) or was not described.
[b]Day 0 represents time of surgery (wounding).
[c]Day(s) cited represent the day(s) postwounding on which the analyses were done.

Table VI. Studies of KGF/FGF-7 in Wound Healing Models

Reference	Wound model	Species	Dose per application and vehicle[a]	Time of application(s)[b]	Observations (vs. vehicle control)[c]
Staiano-Coico et al. (1993)	Partial-thickness excisional wound	Pig	1 μg (0.08 μg/cm^2)	Day 0	Day 3: ↑ rate of reepithelialization; ↑ thickness of new epidermis
Hebda et al. (1993)	Partial-thickness excisional wound	Pig	2.8, 8.4, or 28 μg (2.8–28 μg/cm^2)	Day 0	Days 6 and 7: ↑ rate of reepithelialization; ↑ epidermal hypertrophy at wound margin; • no apparent effect on keratinocyte differentiation program
Pierce et al. (1994)	"Deep partial-thickness" excisional wound in ear	Rabbit	1 or 10 μg (4–40 μg/cm^2)	Day 0	Days 5 and/or 7: ↑ rate of reepithelialization; ↑ thickness of new epithelium; • no effect on granulation tissue; • no apparent effect on keratinocyte differentiation
Staiano-Coico et al. (1993)	Full-thickness excisional wound	Pig	1 μg (0.25 μg/cm^2)	Days 0 and 3	Days 14–28: ↑ formation of deep rete ridge pattern in epidermis; ↑ thickness of new epidermis; ↑ deposition of collagen fibers in papillary dermis; • no effect on rate of contraction
Wu et al. (1993)	Full-thickness excisional dermal ulcer in ischemic ear	Rabbit	5, 30, or 40 μg (20, 120 or 160 μg/cm^2)	Day 0	Day 7 (5 or 30 μg dose): ↑ reepithelialization; Day 10 (40 μg dose): ↑ granulation tissue formation; ↑ thickness of new epithelium

[a] In all studies listed, the growth factor was applied in an aqueous vehicle (phosphate buffered saline).
[b] Day 0 represents time of surgery (wounding).
[c] Day(s) cited represent the day(s) postwounding on which the analyses were done.

reepithelialization of partial-thickness wounds. In some cases, the wounds were made in animals with normal healing abilities; in other cases, animals were used in which wound-healing ability was impaired by the presence of a diabetic-like state, steroid treatment, malnutrition, bacterial contamination, or local tissue ischemia. As with Buntrock *et al.* (1982b), increased neovascularization after FGF application has often been noted in these studies (see Tables II–VI). These studies have also indicated that the FGFs can modulate other steps of dermal tissue repair besides angiogenesis. As summarized below, however, a review of these studies suggests that the FGF effects obtained can be influenced not only by the animal model used, but also by the dose of FGF, the vehicle, the timing of FGF treatment relative to wounding, and which FGF family member is applied.

4.1. FGF-2: Models of Unimpaired Healing

4.1.1. Implant Models

Davidson *et al.* (1985) were the first to carry out wound-healing studies with highly purified FGF-2. Like Buntrock *et al.* (1982a,b, 1984), this group utilized a "dead space" implant as a model of deep dermal healing, but in this case polyvinyl alcohol sponge disks were implanted ventrally beneath the panniculus carnosus of the rats. Six days later, after the inflammatory response triggered by the surgery had subsided, the sponges were injected with vehicle or with 0.5 μg of FGF-2 isolated from bovine cartilage. The sponges were then harvested at day 8 or day 9 postimplantation (2 or 3 days, respectively, postinjection). As in the results of Buntrock *et al.* (1982b), histological analyses of the sponges showed an increased angiogenic response in the FGF-2-treated implants. Davidson *et al.* (1985) also noted increased granulation tissue infiltration into FGF-2-treated sponges relative to controls (Fig. 3). Consistent with these histological findings, biochemical analyses of the tissue in the sponges on day 9 postimplantation showed a significantly greater accumulation of DNA, protein, and collagen in the FGF-2-treated disks versus controls. Interestingly, these effects were obtained even though the injected FGF-2 rapidly disappeared from the sponges (only about 10% of an injected dose of [^{125}I]-FGF-2 was present in the sponges 4 hr later).

In a subsequent series of reports (Davidson *et al.,* 1988; Broadley *et al.,* 1988, 1989b; Davidson and Broadley, 1991), Davidson and colleagues examined a number of variations on the sponge implant model, such as injecting the FGF-2 into the sponges on day 3 postimplantation rather than day 6, and then harvesting the sponges 4 or 6 days later (on day 7 or day 9 postimplantation). In these studies, which now used recombinant human FGF-2 (rhFGF-2), significant increases in DNA and protein content were again found in the FGF-2-treated sponges relative to vehicle-treated controls. However, in contrast to the initial study (Davidson *et al.,* 1985), the collagen content of the sponges in these subsequent studies tended to be lower in the FGF-2-treated disks relative to controls, which the investigators suggested could reflect FGF-2-induced

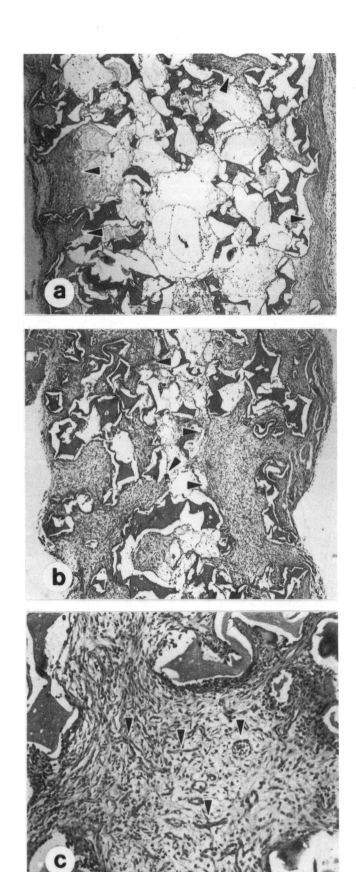

stimulation of collagenase production by cells in the wound area. The source of this difference in collagen response between the earlier and later studies is not clear.

Three other groups have also reported studies on FGF-2 effects in implant-type models in rats. Fiddes *et al.* (1991) utilized the sponge implant model as described by Broadley *et al.* (1988, 1989b) and Davidson and Broadley (1991) to test the activity of a recombinant form of human FGF-2 they had produced; histological examination of the resulting sponges confirmed an increased level of granulation tissue infiltration and angiogenesis in response to the FGF-2 treatment. Staining of the histological sections with Masson's trichrome suggested in addition that an increase in collagen content had occurred in response to FGF-2, but this observation was not quantified with biochemical analyses. Lazarou *et al.* (1989) also utilized sponge implants to test the activity of a form of rhFGF-2, but in this case the FGF-2 was applied to the sponges at the time of implantation. Unexpectedly, analysis of the sponges at 10 days postimplantation showed a significant decrease in tissue infiltration as well as decreased collagen content in the sponges treated with FGF-2. One possible explanation for the difference in granulation tissue formation seen in this study versus the studies reported by Davidson and colleagues and Fiddes and co-workers is that, with this model, the time of acute application of the growth factor after wounding may be critical to obtaining consistent effects. Sprugel *et al.* (1987) emulsified bovine pituitary FGF-2 in a bovine dermal collagen suspension, placed the suspension in porous polytetrafluoroethylene tubes, and then implanted the tubes under the abdominal skin of rats. Analysis of the tube contents 10 days later indicated that the response to FGF-2 was quite variable in this model, but reflected on average an increase in DNA content and vascularization relative to controls. Sprugel *et al.* (1987) also noted increased evidence for remodeling of the collagen matrix in the FGF-2-containing tubes, consistent with the idea that FGF-2 may trigger the production of collagenase in the wound area.

4.1.2. Rabbit Ear Dermal Ulcer Model

While the implant models of deep dermal healing produce responses in granulation tissue deposition that are accessible to biochemical as well as histological analyses, these models suffer from two key drawbacks (Davidson *et al.*, 1988): (1) the implants stimulate a mild foreign body reaction that could generate artifacts; and, (2) more importantly, these models do not examine reepithelialization and in fact exclude possible epithelial influences on the process of dermal repair. To avoid these drawbacks, Mustoe and colleagues (Mustoe *et al.*, 1991; Pierce *et al.*, 1992) developed a new model in which full-thickness circular wounds are made in rabbit ears by excision of punch biopsies down to bare cartilage. Due to an avascular cartilage base, these

---

Figure 3. Formation of granulation tissue in implanted sponges in rats. Sponges were harvested 9 days after implantation, and fixed sections were stained with hematoxylin-eosin. (A) Sponge injected on day 6 post-implantation with vehicle (magnification: × 40). (B) Sponge injected on day 6 postimplantation with FGF-2 (magnification × 40). (C) Higher-magnification (× 100) view of an FGF-2-injected sponge. Arrowheads in A and B indicate the extent of granulation tissue infiltration into the sponges and in C indicate capillaries within the granulation tissue. (Reproduced from Davidson *et al.*, 1985, with permission.)

wounds are essentially noncontracting and heal by migration of new granulation tissue and epithelium from the wound periphery.

In a study reported by Pierce *et al.* (1992), FGF-2 was applied to the ear wounds once at the time of surgery in an aqueous buffer (2 μg rhFGF-2 per wound; 8 μg/cm^2 of wound surface). A mutant form of rhFGF-2 was used in which two of the four cysteines were changed to serines, giving the molecule more stability (Seno *et al.*, 1988). When the wounds were analyzed after 7 days, the FGF-2-treated wounds contained significantly greater amounts of new granulation tissue relative to controls. The glycosaminoglycan (GAG) and fibronectin content in the wound bed was also significantly higher in the FGF-2-treated wounds, as was the amount of neovascularization (in fact, specific lectin staining indicated that the cells within the FGF-2-induced matrix were almost entirely endothelial cells). However, the FGF-2-treated wounds showed significantly less new collagen accumulation over time than controls, and Sirius red staining of histological sections revealed approximately 50% greater erosion of mature collagen bundles in the unwounded dermis around the borders of the FGF-2-treated ear ulcers relative to controls, suggesting FGF-2-induced collagenolysis. Surprisingly, Pierce *et al.* (1992) found that the accumulation of GAG- and fibronectin-rich matrix in the FGF-2-treated wounds continued even after the wounds were fully closed, with the result that a palpable mass had formed by 10 to 14 days postwounding. Since no such phenomena were seen with control, PDGF-treated, or TGF-β1-treated wounds, the investigators proposed that excess FGF-2 may directly or indirectly interrupt the normal signals involved in the resolution of the healing response in this model. Whether this effect is related to the use of the mutant form of FGF-2 remains to be determined.

The study reported by Mustoe *et al.* (1991) was similar in design to that of Pierce *et al.* (1992), except that wild-type rhFGF-2 was used (rather than the mutant form) and the rhFGF-2 was applied in a bovine collagen suspension. Presumably as a result of one or both of these differences in experimental design, the observations reported by Mustoe *et al.* were quite different from those of Pierce *et al.*: while FGF-2 treatment in the Mustoe *et al.* study significantly stimulated the rate of wound reepithelialization and appeared to increase neovascularization, there was no significant stimulation of granulation tissue formation relative to vehicle-treated wounds. Pierce *et al.* (1992) suggested that binding of the FGF-2 to its collagen vehicle in the Mustoe *et al.* study may have adversely affected its ability to interact with cells in the wound.

As part of a study on ischemic versus nonischemic wound repair in the presence of growth factors (see Section 4.2.5.), Mustoe *et al.* (1994) have recently repeated the approach of Pierce *et al.* (1992) of applying the mutant form of rhFGF-2 to the full-thickness rabbit ear wounds (the vehicle used in this new study was not described). In agreement with Pierce *et al.* (1992), Mustoe *et al.* (1994) observed significantly greater angiogenesis and new granulation tissue in the FGF-2-treated, nonischemic wounds relative to vehicle-treated controls.

4.1.3. Rodent Full-Thickness Excisional Wound Models

The effects of FGF-2 on the healing of full-thickness excisional wounds have also been examined in a number of studies using normal mice (Tsuboi and Rifkin, 1990;

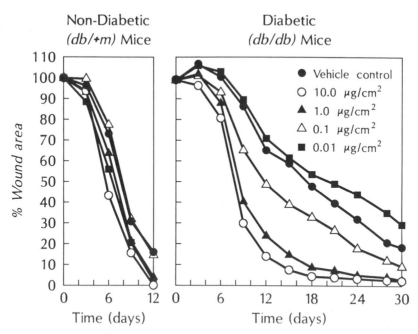

Figure 4. Effect of various doses of rhFGF-2 on the closure of full-thickness excisional wounds in nondiabetic (*db*/+*m*) and diabetic (*db*/*db*) mice. The rhFGF-2 or vehicle was applied once to the wounds on day 0. The open, nonepithelialized area of each wound was traced and quantitated with a digitizer on day 0 (approximate starting wound area: 2 cm^2), and was retraced every third day thereafter. The number of animals in each treatment group ranged from 7–10. (Reprinted from Klingbeil *et al.*, 1991, with permission.)

Greenhalgh *et al.*, 1990; Klingbeil *et al.*, 1991; Uhl *et al.*, 1993) and rats (Stenberg *et al.*, 1989, 1991). In contrast to the rabbit ear results of Pierce *et al.* (1992) and Mustoe *et al.* (1994) described above, all of these investigators reported little or no effect on the rate of wound closure or granulation tissue formation when up to 27 μg/cm^2 (in mice) or 66.7 μg/cm^2 (in rats) of rhFGF-2 was applied to the wounds in a variety of vehicles (see Table II and Fig. 4). The source of the difference in results between the rabbit and rodent studies is not clear, but possible explanations include the species difference and the use of a mutant form of rhFGF-2 in the two rabbit studies.

In the studies by Tsuboi and Rifkin (1990), Greenhalgh *et al.* (1990), Klingbeil *et al.* (1991), and Stenberg *et al.* (1989, 1991), the excisional wounds were made in the backs or sides of the animals, rather than on the ears as in the rabbit studies. Thus, the differential effects of FGF-2 could also be due to differences in wound location, especially given the fact that the back and side wounds of the rodents were healing mainly by contraction, while the rabbit ear wounds were essentially noncontracting. However, in a more recent study, Uhl *et al.* (1993) made full-thickness excisional wounds in the ears of mice and then examined the effects of rhFGF-2 treatment on the rate at which the wound surface area decreased (reepithelialized). The rhFGF-2 was injected around the edge of the wounds on 3 consecutive days postwounding. In agreement with the other rodent studies on full-thickness excisional wounds, the mouse ear wounds did not show an increase in the rate of wound closure in response to FGF-2 (although there was a slight but significant decrease in the average time to complete reepithelialization in the FGF-2-treated wounds).

In addition to measuring the rate of wound closure in their rat study, Stenberg *et al.* (1989, 1991) also excised the wound at the time of closure and measured the breaking strength of the wound scar. Interestingly, despite the lack of an effect on how fast the wounds closed in FGF-2-treated versus control animals, the FGF-2-treated wounds formed scars with significantly less strength than the scars of the control wounds, suggesting that the application of FGF-2 led to the formation of less cross-linked collagen by the time of wound closure.

4.1.4. Incisional Healing Models

Another wound model that incorporates both dermal and epidermal repair events is the healing of full-thickness incisions ("surgical wounds"). To test the effects of FGF-2 in this type of model, McGee *et al.* (1988) made dorsal incisions in rats, closed the wounds with sutures, and then 3 days later injected the wounds with either 0.4 μg rhFGF-2 or saline vehicle. The wounds were analyzed for the degree of repair on days 5, 6, and 7 postwounding. Measurements of the tensile strength of the wounds indicated an increase in strength of about 35–40% in the FGF-2-treated incisions relative to controls, but the increase did not reach statistical significance unless the data from all three analysis days were pooled (54 measurements). Histological examination of the wounds indicated that there was better collagen organization and a more mature epidermal layering in the FGF-2-treated incisions than in controls. These histological results, coupled with the observation that only slight differences in wound collagen content were observed with FGF-2 treatment, led the investigators to suggest that FGF-2 was influencing tensile strength in the incision by accelerating the cross-linking of collagen within the wounds [in contrast to Stenberg *et al.* (1989, 1991), who hypothesized that less collagen cross-linking occurred in their rat excisional wounds after FGF-2 treatment].

Slavin *et al.* (1992) also tested the effects of rhFGF-2 on incisional wound healing in rats, but in this case the rhFGF-2 (or vehicle) was applied on the day of wounding. In initial experiments, the rhFGF-2 was applied in either collagen or saline, and doses from 0.5 μg to 50 μg were used. At day 7 postwounding, no increases in wound strength were observed in FGF-2-treated versus control incisions. In a similar set of experiments, Phillips *et al.* (1993) applied 10 μg of rhFGF-2 to incisions in rats on the day of wounding and also observed no significant change in wound strength versus controls as the incisions healed. Thus, as with the sponge implant model (see Table II), treating the incisional wounds on day 0 appears to give a different result than if the FGF-2 is applied on day 3 [as was done by McGee *et al.* (1988)]. In the Slavin *et al.* (1992) study, wounds treated with 50 μg rhFGF-2 actually showed a significant decrease in strength [reminiscent of Stenberg *et al.* (1989, 1991)], even though histological analysis indicated that the FGF-2 treatment caused a significant increase in wound vascularization and cellularity.

Given the lack of accelerated healing in their initial experiments, Slavin *et al.* (1992) tried delaying the release of the FGF-2 by encapsulating 5 μg of the growth factor in red blood cell ghosts prior to application to the incisions on day 0. This approach resulted in incisions that were now approximately 50% stronger on day 7 relative to wounds treated with ghosts alone, equivalent to the effect seen when 2 μg TGF-β was applied at day 0 in a collagen suspension. These results are thus consistent with the hypothesis that FGF-2 may only be able to accelerate the incisional repair process if it is present after the initial wave of wound-healing events. Interestingly, the FGF-2 wounds treated with the red blood cell ghosts were actually significantly less cellular at day 7 postwounding than control wounds.

Wu and Mustoe (1995) have recently developed an incisional repair model in the rabbit ear, with full-thickness wounds made down to the cartilage. After growth factor or vehicle treatment on day 0, the wounds are sutured, allowed to heal for 10 days, and then excised for measurement of wound breaking strength. Using this model, Wu and Mustoe have tested the incision-healing effects of the mutant form of rhFGF-2 in which two cysteines were changed to serines. As in the other incisional-healing studies described above where the FGF-2 was applied on day 0 (Slavin *et al.*, 1992; Phillips *et al.*, 1993), Wu and Mustoe (1995) observed no differences in wound strength when 20 μg of mutant rhFGF-2 was tested in this manner.

The results of Tsuboi and Rifkin (1990) provide the one exception to the idea that wound strength is not increased if incisions in normal animals are treated once with FGF-2 at the time of wounding. These investigators applied 5 μg of rhFGF-2 to incisions in *db/+m* mice and then measured the breaking strength on day 9 postwounding. The FGF-2-treated incisions showed a small (24%) but significant enhancement of breaking strength when compared to vehicle-treated wounds. The cause of this difference relative to the other incision studies described here remains to be determined, but possibilities include the different species used, the particular mouse strain chosen, or the use of a 1.5% carboxymethylcellulose vehicle.

4.1.5. Partial-Thickness Skin Wound Models

The ability of FGF-2 to stimulate the growth of keratinocytes in culture (O'Keefe *et al.*, 1988; Shipley *et al.*, 1989) suggested that FGF-2, in addition to its effects on granulation tissue formation, might also serve to accelerate epidermal healing in wounds. To test this possibility, Hebda and colleagues (Hebda *et al.*, 1990a; Fiddes *et al.*, 1991) applied rhFGF-2 to 1 cm² partial-thickness wounds made in the paravertebral and thoracic areas of pigs. The rhFGF-2 was applied once to the wounds in a saline vehicle at the time of wounding. Subsets of the wounds were then excised on days 3 to 7 postwounding and examined to determine the number that were completely reepithelialized. Relative to vehicle-treated controls, a significantly greater percentage of wounds treated with 10 μg of rhFGF-2 were found to be fully reepithelialized on days 5, 6, and 7. A significant acceleration in the rate of epithelialization was also seen with the application of 1 μg of rhFGF-2; however, 0.1 μg rhFGF-2 appeared ineffective. Application of 10 μg of growth factor on days 0, 1, and 2 showed no greater effect than a single application on day 0. Histological examinations of the excised wounds indicated that FGF-2 treatment resulted in the earlier appearance of epidermal migration and collagen deposition, as well as an increased neovascularization in the wound bed. Overall, the study indicated that healing in the FGF-2-treated wounds was proceeding about 1 day (~20%) faster than in the control wounds.

Lynch *et al.* (1989) also tested the ability of FGF-2 to stimulate partial-thickness wound healing in pigs, but they observed no significant effects and reported only a trend toward increased epidermal thickness and hydroxyproline content in the wounds in response to FGF-2. It is worth noting, however, that Lynch *et al.* (1989) did not analyze the wounds until day 7 postwounding, past the point at which Hebda *et al.* (1990a) saw the greatest difference between FGF-2- and vehicle-treated wounds. Lynch *et al.* (1989) also applied only 0.5 μg (0.33 μg/cm^2) of FGF-2 to the wounds, less than the minimum effective dose (1 μg/cm^2) in the Hebda *et al.* study. Other possible explanations for the difference in results include the difference in vehicle used (phosphate-buffered saline in the case of Hebda *et al.*; 3% methylcellulose gel in the case of Lynch *et al.*), and the fact that Hebda *et al.* left the wounds uncovered while Lynch *et al.* utilized a semiocclusive dressing.

Consistent with the results of Lynch *et al.* (1989), LeGrand *et al.* (1993) only saw non-statistically significant trends toward accelerated healing when they applied rhFGF-2 to partial-thickness wounds in a rodent (guinea pigs). In this study, a dressing material was saturated with an FGF-2 solution (0.1–30 μg total rhFGF-2) and then applied to the 3 cm^2 wounds on day 0; fresh FGF-2-saturated dressings were placed on the wounds on days 1, 2, 3, and 4. No changes were seen in the degree of wound reepithelialization at day 3 or day 7 relative to vehicle-treated wounds, but histological analysis of the wounds at day 7 postwounding showed a trend toward a greater depth in the new granulation tissue that had formed below the new epidermis.

Eriksson *et al.* (1989) made partial-thickness burn wounds in pigs, removed the damaged epidermis, and then covered each wound with a liquid-tight, closed vinyl chamber. The chambers were filled with either saline vehicle or vehicle containing 10 ng/ml rhFGF-2; the fluids in the chambers were then removed daily and replaced with fresh vehicle or growth factor solution, respectively. Although no statistical analyses were presented in this preliminary report, the investigators found by histological examination of the wounds that the FGF-2-treated burns were 98% reepithelialized by day 6 versus 85% reepithelialization in the vehicle-treated burns.

While the results of Hebda *et al.* (1990a) and Eriksson *et al.* (1989) suggested that application of FGF-2 could be beneficial for the healing of partial-thickness clinical wounds, a study by Cooper *et al.* (1991) indicated that in some cases exogenous FGF-2 could be detrimental to the epithelialization process. The wound-healing model utilized by this group involved the placement of meshed, split-thickness human skin onto full-thickness wounds in athymic (BALB/c *nu/nu*) mice. A polyacrylamide gel dressing with or without incorporated growth factor was placed over the graft to provide slow release delivery of the rhFGF-2 (about 100 ng/cm^2 total dose per dressing application). The wounds were examined every 2 days for the degree of epithelialization in the interstices of the graft, and fresh dressings were applied. The grafts were then excised at day 8 and examined histologically. While a "beefy" granulation tissue bed of vascularized tissue was found in the FGF-2-treated wounds, there were significantly fewer grafts showing reepithelialization in these wounds compared to wounds treated with the polyacrylamide gel dressing alone.

4.2. FGF-2: Models of Impaired Healing

Stimulation of repair in poorly healing or nonhealing wounds in humans has often been proposed as a key clinical problem that might be helped by the exogenous application of growth factors. Given this likely clinical target, a number of laboratories have tested the ability of FGF-2 to modulate repair in impaired healing models in animals (Table III).

4.2.1. Wounds in Diabetic Animals

The first study of FGF-2 in impaired wounds was reported by Broadley *et al.* (1988, 1989b), who examined granulation tissue deposition in rats that had been made acutely diabetic by treatment with streptozotocin (an antibiotic that causes destruction of pancreatic beta cells). Polyvinyl alcohol sponge disks were implanted beneath the ventral panniculus carnosus of the rats, and 5 μg of rhFGF-2 or vehicle alone was injected into the sponges 3 days later. The sponges were removed and analyzed on day 7 or day 9. When compared with vehicle-treated sponges removed from normal rats, the vehicle-treated sponges from the diabetic rats showed significantly less collagen content, protein content, and (at day 7) DNA content, indicating an impaired rate of granulation tissue accumulation. Application of rhFGF-2 resulted in a partial reversal of the healing defect in the diabetic rats, in that the FGF-2-treated sponges from these rats contained almost three times as much DNA by day 9 as did the vehicle-treated sponges and also contained a significantly greater amount of protein. However, consistent with the results they (Broadley *et al.*, 1988, 1989b) and others (Pierce *et al.*, 1992) have reported in normal animals (see Table II), Broadley and colleagues found that the application of FGF-2 in the diabetic rats actually resulted in a further decrease in the collagen level in the sponges.

Phillips *et al.* (1990, 1993) also examined FGF-2 effects on healing in streptozotocin-treated rats, but focused on breaking strength of full-thickness incisions rather than on granulation tissue deposition in sponge implants. In the Phillips *et al.* studies, the rhFGF-2 or aqueous vehicle alone was applied at the time of surgery and was injected into the wound edges (to reduce loss due to leakage). The wounds were sutured and breaking strength was then measured at days 7, 10, and 14 postwounding. On all three harvest days, the vehicle-treated wounds from the diabetic rats showed significantly less strength than similar wounds taken from normal rats. Application of FGF-2 produced no significant changes in the day 7 or day 10 measurements; however, by day 14, the FGF-2-treated wounds from the diabetic rats showed a complete reversal of the healing defect and were as strong as vehicle-treated wounds from the normal rats. In contrast, as indicated above (see Table II), Phillips *et al.* (1993) saw no difference in breaking strength at any time point when FGF-2 was used to treat incisional wounds in normal rats.

The C57BL/KsJ *db/db* mouse strain carries a mutation that results in the development of a number of symptoms reminiscent of human adult-onset diabetes, including severe hyperglycemia, obesity, and insulin resistance. Tsuboi and Rifkin (1990) examined the healing of full-thickness incisions in these animals and found that, on day 9 postwounding, the wounds from the *db/db* mice had about half the breaking strength of wounds in normal, heterozygous (*db/+m*) mice. Similar to the results with the streptozotocin-treated rats (Phillips *et al.*, 1993), application of rhFGF-2 (5 μg on day 0) was found to significantly increase the day 9 breaking strength of the wounds from the *db/db* mice, so that they showed on average about 80% of the strength of wounds taken from *db/+m* mice. However, in contrast to the results of Phillips *et al.* (1993), Tsuboi and Rifkin (1990) also reported an approximately 24% increase in breaking strength when incisions in the nondiabetic (*db/+m*) mice were treated with FGF-2 (Table II).

Several groups (Tsuboi and Rifkin, 1990; Greenhalgh *et al.*, 1990; Klingbeil *et al.*, 1991; Fiddes *et al.*, 1991; Tsuboi *et al.*, 1992) have also used the C57BL/KsJ *db/db* mouse strain to examine the effects of FGF-2 on the healing of full-thickness dermal excisions in the presence of a diabetes-like state. All of these groups reported a profound impairment in the healing of vehicle-treated excisional wounds in the *db/db* animals relative to *db/+m* littermates, with significant delays seen in the entry of inflammatory cells, in the formation of granulation tissue, and in the rate of wound closure. Tsuboi and Rifkin (1990) found that 0.5 μg or 5 μg (1.8 or 18 μg/cm²) rhFGF-2, applied either once on day 0 or once a day for 5 days, almost completely reversed the impairment in granulation tissue formation in the *db/db* wounds when analyzed on day 8 postwounding; however, they saw no significant change in the rate of wound closure after FGF-2 treatment. Greenhalgh *et al.* (1990) examined wounds on day 10 postsurgery, after having treated the wounds with 1 μg (0.44 μg/cm²) of rhFGF-2 per day on days 0–4; these investigators also reported a clear increase in granulation tissue formation with no statistically significant improvement in the degree of wound closure [although increased wound closure after FGF-2 treatment was evident when the wounds were examined at day 21 postwounding (Greenhalgh *et al.*, 1990; Albertson *et al.*, 1993)]. In contrast, Klingbeil *et al.* (1991) noted a rapid increase in wound closure in diabetic wounds treated on day 0 with either 2 μg or 20 μg (1 or 10 μg/cm²) of rhFGF-2 (Fig. 4), and this increase was statistically significant on days 6–30 postwounding (see also Fiddes *et al.*, 1991). The rate of wound closure in the diabetic wounds treated with 20 μg of rhFGF-2 was nearly equivalent to that seen in vehicle-treated nondiabetic (*db/+m*) wounds (Fig. 4).

A possible explanation for the wound closure differences seen between the three *db/db* studies described above may lie with the way the wounds were dressed. Tsuboi and Rifkin (1990) left the wounds open, and speculated that the large scabs that formed may have inhibited keratinocyte migration. Greenhalgh *et al.* (1990) covered the wounds with Opsite, a semipermeable dressing that has itself recently been shown to delay wound closure in *db/db* mice (Lasa *et al.*, 1993). Klingbeil *et al.* (1991) covered the wounds with an occlusive dressing (Bioclusive).

Interestingly, Tsuboi and Rifkin (1990) and Tsuboi *et al.* (1992) noted that the effects of FGF-2 application on granulation tissue formation, wound cellularity, and neovascularization in the *db/db* wounds peaked at around 8–12 days. The amount of granulation tissue, cells, and capillaries in the wounds then began to decrease, suggesting to the investigators that the wound-healing effects of FGF-2 are not unlimited, but are rather subject to normal events of wound-healing resolution and maturation. These results are thus in contrast to those reported by Pierce *et al.* (1992), who noted excessive accumulation of granulation tissue in full-thickness dermal ear ulcers in rabbits that had been treated with a stabilized (mutant) form of rhFGF-2 (see Table II).

Another mouse strain that at least transiently displays diabetes-like metabolic defects is the obese (*ob/ob*) mouse. Klingbeil *et al.* (1991) examined full-thickness excisional wound healing in this strain and, as with the *db/db* mice, found a severe decrease in the rate of wound closure relative to nonobese (*ob/+*) littermates. Treatment of the *ob/ob* mouse wounds with as little as 0.2 μg (0.1 μg/cm^2) of rhFGF-2 on the day of wounding significantly accelerated the closure rate; treatment with 20 μg (10 μg/cm^2) of rhFGF-2 appeared to completely reverse the impairment in the rate of wound closure.

4.2.2. Wounds in Glucocorticoid-Treated Animals

As an additional test of the ability of FGF-2 to stimulate full-thickness wound healing in impaired situations, Klingbeil *et al.* (1991) utilized hairless (*hr/hr*) mice that had been injected with an anti-inflammatory glucocorticoid (prednisolone acetate) 5 hr before excisional wounds were made in the animals. While wounds in normal *hr/hr* mice closed at approximately the same rate as wounds in *db/+* mice, the wounds in the prednisolone-treated mice closed significantly less rapidly. This impairment in wound closure rate was completely reversed by the application of 2 μg (1 μg/cm^2) or 20 μg (10 μg/cm^2) of rhFGF-2 on the day of wounding. A dose of 0.2 μg was also found to significantly accelerate the rate of wound closure.

4.2.3. Wounds in Protein-Malnourished Animals

Another condition that is often associated with clinically relevant wound-healing impairment is malnutrition. To test whether the application of FGF-2 could serve to reverse this type of healing defect, Albertson *et al.* (1993) placed C57BL/KsJ *db/+m* mice on a 1% protein diet and then examined the rate of closure of Opsite-covered full-thickness excisional wounds made in the backs of the mice. Vehicle-treated wounds in these mice showed a small but significant decrease in the extent of wound closure on day 10 postwounding relative to wounds on mice fed regular chow, as expected with the malnourishment. Treating the wounds with rhFGF-2 (0.44 μg/cm^2/day for 5 days), however, did not enhance the extent of wound closure in either the normal or malnourished animals.

Albertson *et al.* (1993) then extended their study to genetically diabetic (C57BL/KsJ *db/db*) mice. As with the *db/+m* mice, the *db/db* mice, when placed on a 1% protein diet, also showed a significantly decreased ability to close full-thickness wounds, relative to mice of the same strain fed normal chow. Histological comparison of the wounds at day 21 from malnourished and nonmalnourished *db/db* mice showed a marked decrease in the cellularity and amount of granulation tissue deposition in the malnourished animals. These malnutrition-associated delays in wound closure, cell accumulation, and granulation tissue formation all appeared to be completely reversed if the wounds in the malnourished animals were treated with 1 μg (0.44 μg/cm^2) rhFGF-2 for 5 days. The investigators speculated that the ability of FGF-2 to overcome the malnutrition healing impairment in the *db/db* but not the *db/+m* mice on the 1% diet may be related to the observations of Greenhalgh *et al.* (1990), who showed that Opsite-covered, full-thickness wounds in *db/+m* mice close mainly by contraction, while the Opsite-dressed *db/db* wounds heal mainly by granulation tissue formation and reepithelialization. FGF-2 may thus be able to stimulate chemotaxis, cellular proliferation, and angiogenesis in the face of malnutrition, but may be unable to promote wound contraction in this model.

4.2.4. Wounds with Bacterial Contamination

Robson and colleagues (Stenberg *et al.*, 1989, 1991; Hayward *et al.*, 1992; Fiddes *et al.*, 1991) have tested the ability of FGF-2 to stimulate healing in wounds inoculated with *Escherichia coli*. Stenberg *et al.* (1989, 1991) used a model in which *E. coli* cells were added to full-thickness excisional wounds in rats, producing a significant increase in the time needed for full closure of the wounds. In this model, one application of 1, 10, or 100 μg (0.67, 6.7, or 67 μg/cm^2, respectively) of rhFGF-2 to the infected wounds was found to completely reverse the healing delay and cause the wounds to close as fast or faster than uninfected wounds. Hayward *et al.* (1992) extended this observation by studying chronic granulating wounds, formed by seeding *E. coli* on day 0 into full-thickness dorsal burns in rats. The wounds were debrided on day 5, and 100 μg/cm^2 of rhFGF-2 was applied. In contrast to the results of Stenberg *et al.* (1989, 1991), Hayward *et al.* detected no improvement in wound closure rate with this single dose of rhFGF-2. However, if 100 μg/cm^2 of rhFGF-2 was applied on days 5, 9, 12, 15, and 18 postburn, the rate of wound closure was markedly accelerated to the point where it was indistinguishable from the wound closure rate in uninfected animals. Doses of 1 μg/cm^2 or 10 μg/cm^2 applied on days 5, 9, 12, 15, and 18 were not effective.

The difference in results obtained by Stenberg *et al.* (1989, 1991) and Hayward *et al.* (1992) when a single dose or lower doses of FGF-2 was used could be due to a number of factors, but it is worth noting that the wound area in the Stenberg *et al.* studies was 1.5 cm^2 versus 30 cm^2 in the Hayward *et al.* study. In addition, the bacterial counts in the wound areas during the healing process were markedly higher in the latter study (10^6–10^8 organisms per gram tissue versus 10^4–10^5 in the Stenberg *et al.* experiments).

4.2.5. Ischemic Wounds

Since a frequent characteristic of chronic nonhealing human wounds is an association with local tissue hypoxia, Mustoe and colleagues (Ahn and Mustoe, 1990; Mustoe *et al.*, 1994; Wu and Mustoe, 1995; Wu *et al.*, 1995) have recently focused on wound-healing models where sustained ischemia is present in the wound area. In these models, rabbit ears are made ischemic by cutting and cauterizing two of the three arteries found at the base of the ear. Wu and Mustoe (1995) then made full-thickness incisions in the ears, treated the incisions with either rhFGF-2 (double cysteine-mutant form) or vehicle, and let the incisions heal for 10 days before testing the wound-breaking strength. A single dose (20 μg) of the mutant rhFGF-2 was found to have no effect on the wound strength under these conditions (or on the strength of incisions in ears that had not been made ischemic) (see Table II).

Mustoe and Wu and their colleagues (Mustoe *et al.*, 1994; Wu *et al.*, 1995) have also made full-thickness excisions in the ischemic ears and then examined the ability of the mutant rhFGF-2 to stimulate new granulation tissue formation and reepithelialization in these dermal ulcers. As in the case of the ischemic incisions, treatment of the ischemic dermal ulcers with a single dose of rhFGF-2 (5–30 μg) failed to produce any significant effects on granulation tissue or epithelium formation, even though increased neovascularization was noted. Interestingly, however, when the rabbits were subjected to hyperbaric oxygen therapy in addition to being treated with the mutant rhFGF-2, the FGF-2 triggered an increase of over 200% in the amount of new granulation tissue formed by day 7 postwounding (relative to ischemic wounds in the contralateral ear, which was treated with vehicle alone). The FGF-2-treated wounds now also showed almost twice as much new epithelium at day 7. Wu *et al.* (1995) therefore suggest that the ability of FGF-2 to stimulate at least some types of wound-healing events may be critically dependent on tissue oxygen tension in the wound area. These investigators have also proposed that inhibition of the actions of endogenous FGF-2 may actually contribute to the wound-healing deficit seen in the presence of ischemia.

Quite different conclusions about the activity of FGF-2 in the presence of ischemia were drawn by Uhl *et al.* (1993), who also examined the healing of full-thickness excisions in ischemic ears. Mice were used rather than rabbits in these experiments, but the ears were again made ischemic by blocking two of the three vascular bundles entering the ear. FGF-2 (recombinant human, nonmutant) was applied on three consecutive days to the full-thickness excisions and was injected around the edge of the wounds rather than applied topically. Under these conditions, FGF-2 treatment resulted in a significant acceleration in the rate at which the wounds closed over time. Histological analyses also revealed a significant increase in neovascularization and granulation tissue formation as a result of FGF-2 application. Given the various differences in the protocols used by these investigators and Wu *et al.* (1995), it is not clear why Uhl *et al.* (1993) saw significant FGF-2 effects in the presence of ischemia while Wu *et al.* (1995) did not. It is worth noting, however, that the FGF-2-treated wounds in the Uhl *et al.* study showed evidence of an inflammatory reaction that persisted through day 13 postwounding, which may have contributed to the granulation tissue and epithelialization effects seen.

4.3. FGF-1

Although FGF-1 was purified to homogeneity and cloned at about the same time as FGF-2, it is at present far less well characterized with regard to *in vivo* wound-healing activity. The effects of FGF-1 have only been reported for a limited number of animal models of dermal tissue repair (Table IV), including: (1) full-thickness excisions in normal mice and rats (Mellin *et al.*, 1992; Matuszewska *et al.*, 1994); (2) full-thickness incisions in normal rats (Mellin *et al.*, 1992); and (3) full-thickness excisions in diabetic (*db/db*) mice (Matuszewska *et al.*, 1994; Mellin *et al.*, 1995).

Mellin *et al.* (1992) utilized recombinant versions of both bovine and human FGF-1 to examine the effects of this factor on tissue repair in rodents under unimpaired conditions. Since the presence of heparin was required for full biological activity of the recombinant FGF-1 *in vitro* (Linemeyer *et al.*, 1987; Jaye *et al.*, 1987), Mellin *et al.* (1992) also included heparin in all treatments with the FGF-1 *in vivo*. In an initial study, full-thickness excisional wounds were made by punch biopsy in the backs of mice, and the wounds were then treated twice a day with 0.25 μg (0.9 μg/cm^2) of recombinant bovine FGF-1. No dressings were applied to the wounds. Planimetry measurements of the wound area remaining open over time indicated that the FGF-1-treated excisions showed a significantly greater amount of closure than did vehicle-treated controls on days 2, 4, and 6 postwounding.

Given the positive results seen in mice, Mellin *et al.* (1992) then turned to full-thickness excisions in rats. In these experiments, recombinant human FGF-1 (rhFGF-1) was used rather than bovine sequence material. As in the mouse study, wounds made with a punch biopsy were treated twice a day with the growth factor and were left undressed. Applications of 0.5 μg (1.8 μg/cm^2) rhFGF-1 in this model were found to significantly increase the rate of wound closure, and the FGF-1-treated excisions on average reached full closure 3–4 days sooner than the vehicle controls. Histological examination of the wounds on day 8 postsurgery indicated that the rhFGF-1 treatments resulted in significantly greater granulation tissue formation and neovascularization than in controls, but did not appear to affect the degree of wound contraction. Interestingly, as had been noted earlier for FGF-2 application in normal rats (Lazarou *et al.*, 1989; Stenberg *et al.*, 1989, 1991; Slavin *et al.*, 1992; Phillips *et al.*, 1993) (see Table II), Mellin *et al.* (1992) found that if they altered the protocol and treated the wounds with rhFGF-1 only on the day of wounding, no significant acceleration of healing was detected.

In a final test of FGF-1's effects on normal healing in rodents, Mellin *et al.* (1992) applied rhFGF-1 to full-thickness dorsal incisions in rats. The sutured wounds were treated daily with 2 μg of rhFGF-1; part of the rhFGF-1 was applied topically while the rest was injected at intervals at the base of the incisions. Wound breaking strength measurements showed that the FGF-1 treatments resulted in wounds that were significantly stronger at day 10 postsurgery than vehicle-treated incisions. Histological analyses also indicated that daily application of rhFGF-1 stimulated collagen deposition and the accumulation of cells in the wound area.

To test the effects of FGF-1 in a model of impaired healing, Matuszewska *et al.* (1994) and K. Thomas and colleagues (Mellin *et al.*, 1995) have applied rhFGF-1 to full-thickness excisions made in the backs of diabetic (*db/db*) mice. In these studies, the rhFGF-1 was generally formulated with heparin and a viscosity-increasing agent (hydroxyethyl cellulose); of the two additives, however, only the heparin was shown to be important for the *in vivo* activity of the FGF-1. As in the experiments of Klingbeil *et al.* (1991) on FGF-2 effects in *db/db* mice, the wounds in these FGF-1 studies were covered with a Bioclusive dressing. Both Matuszewska *et al.* and Mellin *et al.* have found that application of 6 μg (3 μg/cm²) of the rhFGF-1 on days 0, 3, and 7 after surgery resulted in an increase in the rate of wound closure relative to vehicle controls, with the increase becoming statistically significant starting at days 7–10 postwounding. The difference between FGF-1- versus vehicle-treated wounds was particularly apparent as the excisions approached full closure: Mellin *et al.* (1995) calculated that the mean time required to completely heal the FGF-1-treated wounds was 30 days less than the mean time to complete healing in the control wounds. A dose as low as 1.2 μg (0.6 μg/cm²) given on days 0, 3, and 7 was found to be as effective as the 6-μg dose given on these days. Matuszewska *et al.* (1994) also reported significant effects if the rhFGF-1 was applied only once on day 0, but the acceleration in wound closure in this case was not as pronounced as when doses were applied on days 0, 3, and 7. In contrast, similar to the results of Mellin *et al.* (1992) with normal rats, Matuszewska *et al.* (1994) found no effects on the rate of wound closure when the rhFGF-1 was applied once on day 0 to normal, nondiabetic mice (*db/+m*).

4.4. FGF-4

As with FGF-1, only limited information is so far available on the ability of exogenous FGF-4 to modulate dermal wound repair (Table V). Much of that information has been generated by Wu and Mustoe and their collaborators (Wu and Mustoe, 1995; Wu *et al.,* 1995). As described above in the review of FGF-2 animal wound-healing studies, these investigators have developed models in which full-thickness incisions or full-thickness dermal excisions are made down to the cartilage in rabbit ears. By cutting and cauterizing the rostral and central arteries at the base of the ear to be wounded, these investigators can also create an ischemic, impaired wound-healing situation in the ears (Ahn and Mustoe, 1990; Mustoe *et al.,* 1994).

Recombinant human FGF-4 (rhFGF-4) has been tested by Mustoe and Wu and their colleagues in both the incision and excision injury models, under both ischemic and nonischemic conditions. In each set of experiments, a single dose of either rhFGF-4 or the vehicle alone was applied to the wounds on the day of wounding. In the case of the incision model, the sutured wounds were then allowed to heal for 10 days before being tested for wound breaking strength. Surprisingly, while a 20-μg dose of rhFGF-2 had produced no significant effects in this model, an identical dose of rhFGF-4 resulted in a small but significant increase in wound breaking strength under both normal and ischemic conditions (Wu and Mustoe, 1995). Similarly, histological examination of full-thickness excisions treated with 5 μg (20 μg/cm²) of rhFGF-4 revealed significant increases in the amount of new granulation tissue and epithelium formed by day 7 postwounding under both ischemic and nonischemic conditions, relative to vehicle-treated controls (Wu *et al.,* 1995). Particularly striking results were observed in FGF-4-treated excisions under the hypoxic conditions, where a dose of 15 μg (60 μg/cm²) of rhFGF-4 resulted in the formation of twice as much new epithelium and four times as much granulation tissue as in vehicle-treated wounds. In contrast, ischemic ear dermal excisions treated with an identical dose of rhFGF-2 were indistinguishable from vehicle-treated wounds (Table III). Taken together, the results of Wu and Mustoe (1995) and Wu *et al.* (1995) thus suggest that FGF-4 is better able to produce an effect *in vivo* than is FGF-2 under hypoxic conditions, and might therefore be a better candidate therapeutic agent than FGF-2 for the treatment of ischemic nonhealing human wounds. Wu *et al.* (1995) have proposed that the differential effect of hypoxia on the activity of FGF-2 versus FGF-4 might be accomplished through differential regulation of the forms of the FGF receptors expressed in the wound environment.

Hebda *et al.* (1990b; personal communication) have also examined the effects of FGF-4 in their model of partial-thickness dermal healing in pigs. The rhFGF-4 (0.1, 1, or 10 μg) or aqueous vehicle was applied once to the 1 cm^2 wounds at the time of surgery, and the excisions were then examined daily for the percent of the wounds in each treatment group that had become completely reepithelialized. Similar to the results seen with 10 μg of rhFGF-2 (Hebda *et al.*, 1990a; Fiddes *et al.*, 1991) (see Table II), the application of 10 μg of rhFGF-4 was found to produce an acceleration of about 20% in the rate of epidermal healing. Single treatments with 1 μg or 0.1 μg rhFGF-4 also produced significant (although lesser) effects on reepithelialization.

4.5. KGF/FGF-7

Given the proposed role of KGF/FGF-7 as a paracrine mediator of keratinocyte growth and differentiation (Rubin *et al.*, 1989; Finch *et al.*, 1989; Marchese *et al.*, 1990), several groups have conducted studies aimed at determining the effects of exogenous KGF/FGF-7 on epithelial regeneration after wounding (Table VI). Staiano-Coico *et al.* (1993) made partial-thickness excisions in pigs and treated the wounds once on the day of surgery with 1 μg (0.08 μg/cm^2) of recombinant human KGF/FGF-7 (rhKGF/FGF-7). Hebda *et al.* (1993; personal communication) also applied rhKGF/FGF-7 to porcine partial-thickness wounds, utilizing a protocol essentially identical to that used previously by this group to test the effects of rhFGF-2 and rhFGF-4 (Hebda *et al.*, 1990a,b); in this protocol, the wounds were considerably smaller than those made by Staiano-Coico *et al.* (1 cm^2 versus 13 cm^2), and a larger dose of rhKGF/FGF-7 was applied (2.8–28 μg/cm^2). More recently, Pierce *et al.* (1994) modified the rabbit ear dermal ulcer model (Mustoe *et al.*, 1991; Pierce *et al.*, 1992) so that the cartilage at the wound base was removed along with the dermal and epidermal tissue in the wound area. Recombinant hKGF/FGF-7 (1–10 μg; 4–40 μg/cm^2) was then added to determine its ability to stimulate epithelial growth over the newly exposed dermis that had been lying beneath the cartilage on the other side of the ear. In all three of these studies (Staiano-Coico *et al.*, 1993; Hebda *et al.*, 1993; P. Hebda, personal communication; Pierce *et al.*, 1994), the KGF/FGF-7-treated wounds were found to display a significantly increased rate of reepithelialization relative to vehicle-treated wounds, and the new epithelium was noted in each case to be significantly thicker than in the controls. Hebda *et al.* (1993; personal communication) concluded that the acceleration of reepithelialization seen with KGF/FGF-7 in their porcine model was very similar to that seen in their previous studies with FGF-2 and FGF-4.

Interestingly, Pierce *et al.* (1994) noted that the size and number of hair follicles and sebaceous glands were increased in the wounds treated with rhKGF/FGF-7. Bromodeoxyuridine labeling of cells in S-phase indicated that application of rhKGF/FGF-7 had resulted in increased cell proliferation within the follicles and sebaceous glands, and staining of histological sections with oil red o showed enhanced numbers of differentiated, sebum-producing cells in the sebaceous glands. Thus, in addition to accelerating the regeneration of the epithelium in the rabbit ear "deep partial-thickness" wounds, KGF/FGF-7 application in this model resulted in significant effects on dermal adnexal structures.

In the studies reported by Staiano-Coico *et al.* (1993), rhKGF/FGF-7 (1 μg; 0.25 μg/cm^2) was also used to treat full-thickness porcine wounds. As with the partial-thickness excision model results, KGF/FGF-7 treatment of the full-thickness wounds was found to result in the formation of a significantly thicker neoepidermis. The most striking result from these experiments, however, was the observation that the neo-epidermis of the KGF/FGF-7-treated wounds developed a pronounced, sustained and deep rete ridge pattern. In contrast, similar wounds treated with vehicle or rhFGF-2 formed a more flattened epidermis with few if any rete ridges. Histological and ultra-structural analyses of wounds at 2 weeks postsurgery indicated that a more mature basal keratinocyte was present in the KGF/FGF-7-treated wounds. Greater numbers of mature collagen fibers were also present in the superficial dermis of the growth factor-treated wounds at 2 weeks, indicating that KGF/FGF-7 application produced indirect effects that extended beyond the epithelium. Taken together, these results suggested to Staiano-Coico *et al.* (1993) that treatment of full-thickness excisions with KGF/FGF-7 may result in healed wounds that are stronger and more durable, with better attachment of epidermis to dermis.

Unexpectedly, application of rhKGF/FGF-7 to "normal" full-thickness dermal excisions in rabbit ears (leaving the bare cartilage intact) was not observed to accelerate reepithelialization (Pierce *et al.,* 1994), even though such an acceleration was reported after application of rhFGF-2 in this model (Mustoe *et al.,* 1991). If the rabbit ears were made ischemic, however, through partial disruption of the ear blood supply, rhKGF/FGF-7 (5 or 30 μg; 20 or 120 μg/cm^2) was found to significantly accelerate the formation of new epithelium relative to vehicle-treated ischemic controls (Wu *et al.,* 1993). At a dose of 40 μg (160 μg/cm^2) in this ischemic model, the rhKGF/FGF-7 generated a significant increase in new granulation tissue formation, again presumably through the induction of indirect effects (Wu *et al.,* 1993). KGF/FGF-7, like FGF-4, can thus be distinguished from FGF-2 by its ability to be active in an ischemic environment in the rabbit ear model. Consistent with all the reports described above of KGF/FGF-7 effects in tissue-repair models, Wu *et al.* (1993) also noted that the new epithelium on the KGF/FGF-7-treated wounds was thicker than that seen in vehicle-treated excisions.

5. Clinical Trials of FGFs for Treatment of Dermal Wounds

The results summarized above for animal models of wound healing provided strong (although not unanimous) support for the hypothesis that exogenous application of at least some of the FGF family members might be of benefit for accelerating the healing of human wounds. A number of clinical studies therefore have been undertaken in recent years to test this hypothesis. To date, however, the details of only a few of the trials have been published. In all of these published studies, the agent being tested for efficacy (and safety) was rhFGF-2.

Robson *et al.* (1992) reported results from a randomized, blinded, placebo-controlled Phase I/II study in which rhFGF-2 was used to treat chronic pressure sores. The study enrolled 50 patients, all of whom had pressure sores that extended from bone to the subcutaneous tissue (grades III/IV). Only sores with an initial volume of be-tween 10 and 200 cm^3 were considered for treatment. The patients were hospitalized for the first 30 days of the study and were then discharged with follow-up examinations over the next 5 months. Three different doses of rhFGF-2 (1, 5, and 10 μg/cm^2 of ulcer surface area) were evaluated; in each case, the growth factor was applied in an aqueous vehicle by spraying on the wound surface. In part due to the results of Hayward *et al.* (1992), which indicated that multiple doses of rhFGF-2 were necessary to stimulate healing in a chronic granulating wound in rats, Robson *et al.* (1992) also chose to evaluate several different multiday dosing schedules, in which rhFGF-2 (or vehicle control) was applied on (1) the first day of the study (day 1), and again on day 13; (2) days 1, 4, 7, 10, and 13; (3) days 1, 4, 7, 10, 13, 16, 19, and 22; (4) days 1–5, and then days 7, 14, and 21; and (5) days 1–21. The wounds were checked at the start of the study and then at periodic intervals for remaining wound volume, visual appearance, and histological features (in biopsies).

Given the number of dosing regimens in this study, only three to seven patients were enrolled in any one subgroup, and statistically significant effects of FGF-2 ver-sus placebo were not seen when the individual subgroups were compared. However, if the data from all 35 patients treated with FGF-2 were combined and compared with the combined data from the vehicle-treated patients, a significantly larger proportion (60%) of the FGF-2-treated wounds were found to have achieved greater than 70% reduction in wound volume in 30 days (only 29% of the placebo-treated wounds reached this degree of closure by day 30). On average, the wounds that had received FGF-2 showed a 69% reduction in volume during the 30-day hospital stay, while the vehicle-treated wounds declined in volume by 59% in this time period. When the data were graphed as the initial wound volume versus absolute decrease in volume after 30 days, the slope of the regression curve for the FGF-2-treated wounds was signifi-cantly steeper than the slope for the placebo wounds, again indicating that FGF-2 enhanced the degree of wound closure. Histologically, the FGF-2-treated wounds showed greater numbers of capillaries and fibroblasts, consistent with numerous ob-servations from animal studies on FGF-2 (see Tables II and III). No FGF-2-related adverse events were seen in the study. Overall, then, the results of this initial trial in pressure sores were promising, but indicated that the effects of exogenous FGF-2 in the protocols used were not so dramatic that clear efficacy could be shown with only a small number of patients.

In contrast to Robson *et al.* (1992), Mazué *et al.* (1991) chose to utilize a much milder and more uniform wound type for their first human studies with rhFGF-2: epidermal suction blisters created by the investigators on the forearms of 24 healthy volunteers. In this double-blind, placebo-controlled study, the 0.5-cm^2 blisters were made on both forearms of each volunteer. The blisters on one arm were treated daily for 6 days with 0.5 μg (1 μg/cm^2) rhFGF-2. Blisters on the other forearm received vehicle alone on those days. Reepithelialization of the blisters was followed using evaporimetry to periodically measure transepidermal water loss. Similar to the results observed by Hebda *et al.* (1990a) after rhFGF-2 application in a pig partial-thickness wound model (see Table II), Mazué *et al.* (1991) found that a small but significant increase in the rate of reepithelialization was apparent in the human blisters treated with FGF-2. Mazué *et al.* (1991) also reported a greater amount of cellularity, neo-vascularization, and collagen maturation in histological sections taken from the FGF-2-treated versus vehicle-treated wounds 3 days after complete reepithelialization was achieved.

Greenhalgh and Rieman (1994) also examined the effects of exogenously applied rhFGF-2 on the healing of a clinically generated wound, although in this case the wounds were partial-thickness skin graft donor sites in burned children. The donor sites ranged in size from 64 to 200 cm^2. Each of the 11 evaluable patients in the double-blinded study served as his or her own control, with one donor site receiving 5 μg/cm^2 rhFGF-2 while another received the aqueous vehicle (by spray applicator). The donor sites were treated with growth factor or vehicle on the day of graft harvest and then daily for the next 4 days thereafter. The sites were evaluated by visual inspection every 4–5 days for degree of healing. Unlike Mazué *et al.* (1991), Greenhalgh and Rieman (1994) could detect no difference in the rate of reepithelialization or time to complete closure in the FGF-2-treated versus vehicle-treated sites. No adverse events related to FGF-2 were seen in the study, and the FGF-2-treated sites remained indistinguishable cosmetically from the vehicle-treated wounds up to 1 year after healing. Greenhalgh and Rieman (1994) point out that these partial-thickness wounds in children may heal too rapidly under normal circumstances to allow much acceleration of the process by exogenous growth factors. The investigators suggest that a better target for FGF-2 therapy might be graft donor sites in patients who have suffered larger burns, since in these cases donor site healing appears to be more impaired.

Perhaps the largest clinical trial conducted to date with FGF-2 was a Phase III study on patients with diabetic leg ulcers (D. Carmichael, Synergen, Inc., personal communication). The ulcers were evaluated for time to complete closure as well as rate of closure, but no further details of the study have been reported. The study was stopped prior to completion after an interim analysis of the data from one half of the total planned enrollment showed no significant enhancement in the healing of rhFGF-2-treated versus vehicle-treated ulcers. Since ischemic wounds in rabbit ears have shown no acceleration in healing in response to exogenous rhFGF-2 (Mustoe *et al.,* 1994; Wu *et al.,* 1995; Wu and Mustoe, 1995), Mustoe and colleagues have suggested that the lack of effect seen in the diabetic leg ulcer clinical trial may be a reflection of local tissue hypoxia in these wounds creating an environment in which FGF-2 is not active (Mustoe *et al.,* 1994). The results of Wu *et al.* (1995) in the ischemic rabbit ear model (see Table III) also suggest that better results might be obtained in clinical trials if the patients with diabetic leg ulcers were given hyperbaric oxygen therapy in addition to rhFGF-2.

References

Abraham, J. A., Mergia, A., Whang, J. L., Tumolo, A., Friedman, J., Hjerrild, K. A., Gospodarowicz, D., and Fiddes, J. C., 1986, Nucleotide sequence of a bovine clone encoding the angiogenic protein, basic fibroblast growth factor, *Science* **233**:545–548.

Ahn, S. T., and Mustoe, T. A., 1990, Effects of ischemia on ulcer wound healing: A new model in the rabbit ear, *Ann. Plast. Surg.* **24**:17–23.

Albertson, S., Hummel, R. P., Breeden, M., and Greenhalgh, D. G., 1993, PDGF and FGF reverse the healing impairment in protein-malnourished diabetic mice, *Surgery* **114**:368–373.

Avivi, A., Zimmer, Y., Yayon, A., Yarden, Y., and Givol, D., 1991, Flg-2, a new member of the family of fibroblast growth factor receptors, *Oncogene* **6**:1089–1092.

Avivi, A., Yayon, A., and Givol, D., 1993, A novel form of FGF receptor-3 using an alternative exon in the immunoglobulin domain III, *FEBS Lett.* **330**:249–252.

Baird, A., and Böhlen, P., 1990, Fibroblast growth factors, in: *Handbook of Experimental Pharmacology, Volume 95: Peptide Growth Factors and Their Receptors I* (M. B. Sporn and A. B. Roberts, eds.), pp. 369–418, Springer-Verlag, Berlin.

Baird, A., and Klagsbrun, M., 1991, Nomenclature meeting report and recommendations, *Ann. NY Acad. Sci.* **638**:xiii–xvi.

Baird, A., and Ling, N., 1987, Fibroblast growth factors are present in the extracellular matrix produced by endothelial cells *in vitro:* Implication for a role of heparinase-like enzymes in the neovascular response, *Biochem. Biophys. Res. Commun.* **142**:428–435.

Baird, A., Mormede, P., and Böhlen, P., 1985, Immunoreactive fibroblast growth factor in cells of peritoneal exudate suggests its identity with macrophage-derived growth factor, *Biochem. Biophys. Res. Commun.* **126**:358–364.

Bashkin, P., Doctrow, S., Klagsbrun, M., Svahn, C. M., Folkman, J., and Vlodavsky, I., 1989, Basic fibroblast growth factor binds to subendothelial extracellular matrix and is released by heparitinase and heparin-like molecules, *Biochemistry* **28**:1737–1743.

Basilico, C., Newman, K. M., Curatola, A. M., Talarico, D., Mansukhani, A., Velchich, A., and Delli-Bovi, P., 1989, Expression and activation of the K-*fgf* oncogene, *Ann. NY Acad. Sci.* **567**:95–103.

Bates, B., Hardin, J., Zhan, X., Drickamer, K., and Goldfarb, M., 1991, Biosynthesis of human fibroblast growth factor-5, *Mol. Cell. Biol.* **11**:1840–1845.

Bellosta, P., Talarico, D., Rogers, D., and Basilico, C., 1993, Cleavage of K-FGF produces a truncated molecule with increased biological activity and receptor binding affinity, *J. Cell Biol.* **121**:705–713.

Bernfield, M., Kokenyesi, R., Kato, M., Hinkes, M. T., Spring, J., Gallo, R. L., and Lose, E. J., 1994, Biology of the syndecans, *Annu. Rev. Cell Biol.* **8**:1–39.

Blotnick, S., Peoples, G. E., Freeman, M. R., Eberlein, T. J., and Klagsbrun, M., 1994, T lymphocytes synthesize and export heparin-binding epidermal growth factor-like growth factor and basic fibroblast growth factor, mitogens for vascular cells and fibroblasts: Differential production and release by CD4[+] and CD8[+] T cells, *Proc. Natl. Acad. Sci. USA* **91**:2890–2894.

Bottaro, D. P., Rubin, J. S., Ron, D., Finch, P. W., Florio, C., and Aaronson, S. A., 1990, Characterization of the receptor for keratinocyte growth factor: Evidence for multiple fibroblast growth factor receptors, *J. Biol. Chem.* **265**:12767–12770.

Brem, H., and Klagsbrun, M., 1993, The role of fibroblast growth factors and related oncogenes in tumor growth, in: *Oncogenes and Tumor Suppressor Genes in Human Malignancies* (C. C. Benz and E. T. Liu, eds.), pp. 211–231, Kluwer Academic Publishers, Norwell, Massachusetts.

Broadley, K. N., Aquino, A. M., Hicks, B., Ditesheim, J. A., McGee, G. S., Demetriou, A. A., Woodward, S. C., and Davidson, J. M., 1988, Growth factors bFGF and TGFβ accelerate the rate of wound repair in normal and in diabetic rats, *Int. J. Tissue React.* **X**:345–353.

Broadley, K. N., Aquino, A. M., Woodward, S. C., Buckley-Sturrock, A., Sato, Y., Rifkin, D. B., and Davidson, J. M., 1989a, Monospecific antibodies implicate basic fibroblast growth factor in normal wound repair, *Lab. Invest.* **61**:571–575.

Broadley, K. N., Aquino, A. M., Hicks, B., Ditesheim, J. A., McGee, G. S., Demetriou, A. A., Woodward, S. C., and Davidson, J. M., 1989b, The diabetic rat as an impaired wound healing model: Stimulatory effects of transforming growth factor-beta and basic fibroblast growth factor, *Biotechnol. Ther.* **1**:55–68.

Brüstle, O., Aguzzi, A., Talarico, D., Basilico, C., and Kleihues, P., 1992, Angiogenic activity of the K-*fgf/hst* oncogene in neural transplants, *Oncogene* **7**:1177–1183.

Buckley-Sturrock, A., Woodward, S. C., Senior, R. M., Griffin, G. L., Klagsbrun, M., and Davidson, J. M., 1989, Differential stimulation of collagenase and chemotactic activity in fibroblasts derived from rat wound repair tissue and human skin by growth factors, *J. Cell. Physiol.* **138**:70–78.

Buntrock, P., Jentzsch, K. D., and Heder, G., 1982a, Stimulation of wound healing, using brain extract with fibroblast growth factor (FGF) activity. I. Quantitative and biochemical studies into formation of granulation tissue, *Exp. Pathol.* **21**:46–53.

Buntrock, P., Jentzsch, K. D., and Heder, G., 1982b, Stimulation of wound healing, using brain extract with fibroblast growth factor (FGF) activity. II. Histological and morphometric examination of cells and capillaries, *Exp. Pathol.* **21**:62–67.

Buntrock, P., Buntrock, M., Marx, I., Kranz, D., Jentzsch, K. D., and Heder, G., 1984, Stimulation of wound healing, using brain extract with fibroblast growth factor (FGF) activity. III. Electron microscopy, autoradiography, and ultrastructural autoradiography of granulation tissue, *Exp. Pathol.* **26**:247–254.

Burgess, W., and Maciag, T., 1989, The heparin-binding (fibroblast) growth factor family of proteins, *Annu. Rev. Biochem.* **58**:575–606.

Burrus, L. W., Zuber, M. E., Lueddecke, B. A., and Olwin, B. B., 1992, Identification of a cysteine-rich receptor for fibroblast growth factors, *Mol. Cell. Biol.* **12**:5600–5609.

Chedid, M., Rubin, J. S., Csaky, K. G., and Aaronson, S. A., 1994, Regulation of keratinocyte growth factor gene expression by interleukin 1, *J. Biol. Chem.* **269**:10753–10757.

Chen, W. Y. J., Rogers, A. A., and Lydon, M. J., 1992, Characterization of biologic properties of wound fluid collected during early stages of wound healing, *J. Invest. Dermatol.* **99**:559–564.

Cooper, D. M., Yu, E. Z., Hennessey, P., Ko, F., and Robson, M. C., 1994, Determination of endogenous cytokines in chronic wounds, *Ann. Surg.* **219**:688–692.

Cooper, M. L., Hansbrough, J. F., Foreman, T. J., Sakabu, S. A., and Laxer, J. A., 1991, The effects of epidermal growth factor and basic fibroblast growth factor on epithelialization of meshed skin graft interstices, in: *Clinical and Experimental Approaches to Dermal and Epidermal Repair: Normal and Chronic Wounds* (A. Barbul, M. Caldwell, W. Eaglstein, T. Hunt, D. Marshall, E. Pines, and G. Skover, eds.), pp. 429–442, Alan R. Liss, New York.

Coulier, F., Batoz, M., Marics, I., deLapeyriere, O., and Birnbaum, D., 1991, Putative structure of the FGF6 protein and role of the signal peptide, *Oncogene* **6**:1437–1444.

Coulier, F., Pizette, S., Ollendorff, V., deLapeyriere, O., and Birnbaum, D., 1994, The human and mouse fibroblast growth factor 6 (FGF6) genes and their products: Possible implication in muscle development, *Prog. Growth Factor Res.* **5**:1–14.

Davidson, J. M., and Broadley, K. N., 1991, Manipulation of the wound-healing process with basic fibroblast growth factor, *Ann. NY Acad. Sci.* **638**:306–315.

Davidson, J. M., Klagsbrun, M., Hill, K. E., Buckley, A., Sullivan, R., Brewer, P. S., and Woodward, S. C., 1985, Accelerated wound repair, cell proliferation, and collagen accumulation are produced by a cartilage-derived growth factor, *J. Cell Biol.* **100**:1219–1227.

Davidson, J., Buckley, A., Woodward, S., Nichols, W., McGee, G., and Demetriou, A., 1988, Mechanisms of accelerated wound repair using epidermal growth factor and basic fibroblast growth factor, in: *Growth Factors and Other Aspects of Wound Healing: Biological and Clinical Implications* (A. Barbul, E. Pines, M. Caldwell, and T. K. Hunt, eds.), pp. 63–75, Alan R. Liss, New York.

de Lapeyriere, O., Rosnet, O., Benharroch, D., Raybaud, F., Marchetto, S., Planche, J., Galland, F., Mattei, M.-G., Copeland, N. G., Jenkins, N. A., Coulier, F., and Birnbaum, D., 1990, Structure, chromosome mapping and expression of the murine FGF-6 gene, *Oncogene* **5**:823–831.

Delli-Bovi, P., Curatola, A. M., Kern, F. G., Greco, A., Ittman, M., and Basilico, C., 1987, An oncogene isolated by transfection of Kaposi's sarcoma DNA encodes a growth factor that is a member of the FGF family, *Cell* **50**:729–737.

Delli-Bovi, P., Curatola, A. M., Newman, K. M., Sato, Y., Moscatelli, D., Hewick, R. M., Rifkin, D. B., and Basilico, C., 1988, Processing, secretion, and biological properties of a novel growth factor of the fibroblast growth family with oncogenic potential, *Mol. Cell. Biol.* **8**:2933–2941.

Delli-Bovi, P., Mansukhani, A., Ziff, E. B., and Basilico, C., 1989, Expression of the K-*fgf* protooncogene is repressed during differentiation of F9 cells, *Oncogene Res.* **5**:31–37.

Dickson, C., and Peters, G., 1987, Potential oncogene product related to growth factors, *Nature* **326**:833.

Dickson, C., Acland, P., Smith, R., Dixon, M., Deed, R., MacAllan, D., Walther, W., Fuller-Pace, F., Kiefer, P., and Peters, G., 1990, Characterization of *int-2:* A member of the fibroblast growth factor family, *J. Cell Sci. (Suppl.)* **13**:87–96.

Dionne, C. A., Crumley, G., Bellot, F., Kaplow, J. M., Searfoss, G., Ruta, M., Burgess, W. H., Jaye, M., and Schlessinger, J., 1990, Cloning and expression of two distinct high-affinity receptors cross-reacting with acidic and basic fibroblast growth factors, *EMBO J.* **9**:2685–2692.

Dixon, M., Deed, R., Acland, P., Moore, R., Whyte, A., Peters, G., and Dickson, C., 1989, Detection and characterization of the fibroblast growth factor-related oncoprotein INT-2, *Mol. Cell. Biol.* **9**:4896–4902.

Elenius, K., Vainio, S., Laato, M., Salmivirta, M., Thesleff, R., and Jalkanen, M., 1991, Induced expression of syndecan in healing wounds, *J. Cell Biol.* **114**:585–595.

Elenius, K., Maatta, A., Salmivirta, M., and Jalkanen, M., 1992, Growth factors induce 3T3 cells to express bFGF binding syndecan, *J. Biol. Chem.* **9**:6435–6441.

Eriksson, E., Breuing, K., Johansen, L. B., and Miller, D. R., 1989, Growth factor solutions for wound treatment in pigs, *Surg. Forum* **40**:618–620.

Esch, F., Baird, A., Ling, N., Ueno, N., Hill, F., Denoroy, L., Klepper, R., Gospodarowicz, D., Böhlen, P., and Guillemin, R., 1985, Primary structure of bovine pituitary basic fibroblast growth factor (FGF) and comparison with the amino-terminal sequence of bovine brain acidic FGF, *Proc. Natl. Acad. Sci. USA* **82**:6507–6511.

Fiddes, J. C., Hebda, P. A., Hayward, P., Robson, M. C., Abraham, J. A., and Klingbeil, C. K., 1991, Preclinical wound-healing studies with recombinant human basic fibroblast growth factor, *Ann. NY Acad. Sci.* **638**:316–328.

Finch, P. W., Rubin, J. S., Miki, T., Ron, D., and Aaronson, S. A., 1989, Human KGF is FGF-related with properties of a paracrine effector of epithelial cell growth, *Science* **245**:752–755.

Flaumenhaft, R., and Rifkin, D. B., 1991, Extracellular matrix regulation of growth factor and protease activity, *Curr. Opin. Cell Biol.* **3**:817–823.

Florkiewicz, R. Z., and Sommer, A., 1989, Human basic fibroblast growth factor gene encodes four polypeptides: Three initiate translation from non-AUG codons, *Proc. Natl. Acad. Sci. USA* **86**:3978–3981.

Folkman, J., and Klagsbrun, M., 1987, Angiogenic factors, *Science* **235**:442–447.

Folkman, J., Klagsbrun, M., Sasse, J., Wadzinski, M., Ingber, D., and Vlodavsky, I., 1988, A heparin-binding and angiogenic protein—basic fibroblast growth factor—is stored within basement membrane, *Am. J. Pathol.* **130**:393–400.

Gallo, R. L., Ono, M., Povsic, T., Page, C., Eriksson, E., Klagsbrun, M., and Bernfield, M., 1994, Syndecans, cell surface heparan sulfate proteoglycans are induced by a proline-rich antimicrobial peptide from wounds, *Proc. Natl. Acad. Sci. USA* **91**:11035–11039.

Gibran, N. S., Isik, F. F., Heimbach, D. M., and Gordon, D., 1994, Basic fibroblast growth factor in the early human burn wound, *J. Surg. Res.* **56**:226–234.

Gimenez-Gallego, G., Rodkey, K., Bennett, C., Rios-Candelore, M., DiSalvo, J., and Thomas, K. A., 1985, Brain-derived acidic fibroblast growth factor: Complete amino acid sequence and homologies, *Science* **230**:1385–1388.

Gospodarowicz, D., and Cheng, J., 1986, Heparin protects basic and acidic FGF from inactivation, *J. Cell. Physiol.* **128**:475–484.

Grayson, L. S., Hansbrough, J. F., Zapata-Sirvent, R. L., Dore, C. A., Morgan, J. L., and Nicolson, M. A., 1993, Quantitation of cytokine levels in skin graft donor site wound fluid, *Burns* **19**:401–405.

Greenhalgh, D. G., and Rieman, M., 1994, Effects of basic fibroblast growth factor on the healing of partial-thickness donor sites: A prospective, randomized, double-blind trial, *Wound Repair Regen.* **2**:113–121.

Greenhalgh, D. G., Sprugel, K. H., Murray, M. J., and Ross, R., 1990, PDGF and FGF stimulate wound healing in the genetically diabetic mouse, *Am. J. Pathol.* **136**:1235–1246.

Gross, J. L., Moscatelli, D., and Rifkin, D. B., 1983, Increased capillary endothelial cell protease activity in response to angiogenic stimuli *in vitro, Proc. Natl. Acad. Sci. USA* **80**:2623–2627.

Haub, O., Drucker, B., and Goldfarb, M., 1990, Expression of the murine fibroblast growth factor 5 gene in the adult central nervous system, *Proc. Natl. Acad. Sci. USA* **87**:8022–8026.

Hayward, P., Hokanson, J., Heggars, J., Fiddes, J., Klingbeil, C., Goeger, M., and Robson, M., 1992, Fibroblast growth factor reverses the bacterial retardation of wound contraction, *Am. J. Surg.* **163**:288–293.

Hebda, P. A., Klingbeil, C. K., Abraham, J. A., and Fiddes, J. C., 1990a, Basic fibroblast growth factor stimulation of epidermal wound healing in pigs, *J. Invest. Dermatol.* **95**:626–631.

Hebda, P. A., Brady, E. P., Wolfman, N., Stoudemire, J., and Rogers, D., 1990b, Stimulation of epidermal and dermal wound healing by Kaposi sarcoma-derived fibroblast growth factor, *J. Invest. Dermatol.* **94**:534.

Hebda, P. A., Colaiacovo, L., Kruse, S. A., Rodgers, R., Morris, C. F., and Pierce, G. F., 1993, Keratinocyte growth factor: Stimulation of epidermal regeneration in partial thickness wounds in pig skin, *J. Invest. Dermatol.* **100**:557.

Hébert, J. M., Basilico, C., Goldfarb, M., Haub, O., and Martin, G. R., 1990, Isolation of cDNAs encoding four mouse FGF family members and characterization of their expression patterns during embryogenesis, *Dev. Biol.* **138**:454–463.

Hébert, J. M., Rosenquist, T., Götz, J., and Martin, G. R., 1994, FGF5 as a regulator of the hair growth cycle: Evidence from targeted and spontaneous mutations, *Cell* **78**:1017–1025.

Houssaint, E., Blanquet, P. R., Champion-Arnaud, P., Gesnel, M. C., Torriglia, A., Courtois, Y., and Breathnach, R., 1990, Related fibroblast growth factor receptor genes exist in the human genome, *Proc. Natl. Acad. Sci. USA* **87**:8180–8184.

Huang, Y. Q., Li, J. J., Moscatelli, D., Basilico, C., Nicolaides, A., Zhang, W. G., Polesz, B. J., and Friedman-Kien, A. E., 1993, Expression of INT-2 oncogene in Kaposi's sarcoma lesions, *J. Clin. Invest.* **91**:1191–1197.

Hughes, R. A., Sendtner, M., Goldfarb, M., Lindholm, D., and Thoenen, H., 1993, Evidence that fibroblast growth factor 5 is a major muscle-derived survival factor for cultured spinal motoneurons, *Neuron* **10**:369–377.

Jackson, A., Friedman, S., Zhan, X., Engleka, K. A., Forough, R., and Maciag, T., 1992, Heat shock induces the release of fibroblast growth factor 1 from NIH 3T3 cells, *Proc. Natl. Acad. Sci. USA* **89**:10691–10695.

Jaye, M., Howk, R., Burgess, W., Ricca, G. A., Chiu, I. M., Ravara, M. W., O'Brien, S. J., Modi, W. S., Maciag, T., and Drohan, W. N., 1986, Human endothelial cell growth factor: Cloning, nucleotide sequence, and chromosome localization, *Science* **233**:541–545.

Jaye, M., Burgess, W. H., Shaw, A. B., and Drohan, W. N., 1987, Biological equivalence of natural bovine and recombinant human α-endothelial cell growth factors, *J. Biol. Chem.* **262**:16612–16617.

Johnson, D. E., Lee, P. L., Lu, J., and Williams, L. T., 1990, Diverse forms of a receptor for acidic and basic fibroblast growth factors, *Mol. Cell. Biol.* **10**:4728–4736.

Johnson, D. E., Lu, J., Chen, H., Werner, S., and Williams, L. T., 1991, The human fibroblast growth factor receptor genes: A common structural arrangement underlies the mechanism for generating receptor forms that differ in their third immunoglobulin domain, *Mol. Cell. Biol.* **11**:4627–4634.

Kan, M., Wang, F., Xu, J., Crabb, J. W., Hou, J., and McKeehan, W. L., 1993, An essential heparin-binding domain in the fibroblast growth factor receptor kinase, *Science* **259**:1918–1921.

Kandel, J., Bossy-Wetzel, E., Radvany, F., Klagsbrun, M., Folkman, J., and Hanahan, D., 1991, Neovascularization is associated with a switch to the export of bFGF in the multistep development of fibrosarcoma, *Cell* **66**:1095–1104.

Keegan, K., Johnson, D. E., Williams, L. T., and Hayman, M. J., 1991, Isolation of an additional member of the fibroblast growth factor receptor family, FGFR-3, *Proc. Natl. Acad. Sci. USA* **88**:1095–1099.

Kiefer, P., Mathieu, M., Close, M. J., Peters, G., and Dickson, C., 1993, FGF3 from *Xenopus laevis*, *EMBO J.* **12**:4159–4168.

Klagsbrun, M., 1989, The fibroblast growth factor family: Structural and biological properties, *Prog. Growth Factor Res.* **1**:207–235.

Klagsbrun, M., 1990, The affinity of fibroblast growth factors (FGF's) for heparin: FGF–heparan sulfate interactions in cells and extracellular matrix, *Curr. Opin. Cell Biol.* **2**:857–863.

Klagsbrun, M., and Baird, A., 1991, A dual receptor system is required for basic fibroblast growth factor activity, *Cell* **67**:1–20.

Klagsbrun, M., and D'Amore, P. A., 1991, Regulators of angiogenesis, *Annu. Rev. Physiol.* **53:**217–239.

Klagsbrun, M., and Folkman, J., 1990, Angiogenesis, in: *Handbook of Experimental Pharmacology, Volume 95: Peptide Growth Factors and Their Receptors II* (M. B. Sporn and A. B. Roberts, eds.), pp. 549–586, Springer-Verlag, Berlin.

Klagsbrun, M., and Shing, Y., 1985, Heparin affinity of anionic and cationic capillary endothelial cell growth factors: Analysis of hypothalamus-derived growth factors and fibroblast growth factors, *Proc. Natl. Acad. Sci. USA* **82:**805–809.

Klingbeil, C. K., Cesar, L. B., and Fiddes, J. C., 1991, Basic fibroblast growth factor accelerates tissue repair in models of impaired wound healing, in: *Clinical and Experimental Approaches to Dermal and Epidermal Repair: Normal and Chronic Wounds* (A. Barbul, M. Caldwell, W. Eaglstein, T. Hunt, D. Marshall, E. Pines, and G. Skover, eds.), pp. 443–458, Alan R. Liss, New York.

Kornbluth, S., Paulson, K. E., and Hanafusa, H., 1988, Novel tyrosine kinase identified by phosphotyrosine antibody screening of cDNA libraries, *Mol. Cell. Biol.* **8:**5541–5544.

Kuchler, K., and Thorner, J., 1992, Secretion of peptides and proteins lacking hydrophobic signal sequences: The role of adenosine triphosphate-driven membrane translocators, *Endocr. Rev.* **13:**499–514.

Kurita, Y., Tsuboi, R., Ueki, R., Rifkin, D. B., and Ogawa, H., 1992, Immunohistochemical localization of basic fibroblast growth factor in wound healing sites of mouse skin, *Arch. Dermatol. Res.* **284:**193–197.

Lasa, C. I., Kidd, R. R., Nunez, H. A., and Drohan, W. N., 1993, Effect of fibrin glue and Opsite on open wounds in *db/db* mice, *J. Surg. Res.* **54:**202–206.

Lazarou, S. A., Efron, J. E., Shaw, T., Wasserkrug, H. L., and Barbul, A., 1989, Fibroblast growth factor inhibits wound collagen synthesis, *Surg. Forum* **40:**627–629.

Lee, P. L., Johnson, D. E., Cousens, L. S., Fried, V. A., and Williams, L. T., 1989, Purification and complementary DNA cloning of a receptor for basic fibroblast growth factor, *Science* **245:**57–60.

LeGrand, E. K., Burke, J. F., Costa, D. E., and Kiorpes, T. C., 1993, Dose responsive effects of PDGF-BB, PDGF-AA, EGF, and bFGF on granulation tissue in a guinea pig partial thickness skin excision model, *Growth Factors* **8:**307–314.

Linemeyer, D. L., Kelly, L. J., Menke, J. G., Gimenez-Gallego, G., DiSalvo, J., and Thomas, K. A., 1987, Expression in *Escherichia coli* of a chemically synthesized gene for biologically active bovine acidic fibroblast growth factor, *Bio/Technology* **5:**960–965.

Lobb, R. R., Alderman, E. M., and Fett, J. W., 1985, Induction of angiogenesis by bovine brain derived class I heparin binding growth factor, *Biochemistry* **24:**4969–4973.

Lynch, S. E., Colvin, R. B., and Antoniades, H. N., 1989, Growth factors in wound healing: Single and synergistic effects on partial thickness porcine skin wounds, *J. Clin. Invest.* **84:**640–646.

Maciag, T., Mehlman, T., Friesel, R., and Schreiber, A. B., 1984, Heparin binds endothelial cell growth factor, the principal endothelial cell mitogen in bovine brain, *Science* **225:**932–935.

Mansukhani, A., Moscatelli, D., Talarico, D., Levytska, V., and Basilico, C., 1990, A murine fibroblast growth factor (FGF) receptor expressed in CHO cells is activated by basic FGF and Kaposi FGF, *Proc. Natl. Acad. Sci. USA* **87:**4378–4382.

Marchese, C., Rubin, J., Ron, D., Faggioni, A., Torrisi, M. R., Messina, A., Frati, L., and Aaronson, S. A., 1990, Human keratinocyte growth factor activity on proliferation and differentiation of human keratinocytes: Differentiation response distinguishes KGF from EGF family, *J. Cell. Physiol.* **144:** 326–332.

Marics, I., Adelaide, J., Raybaud, F., Mattei, M.-G., Coulier, F., Planche, J., de Lapeyriere, O., and Birnbaum, D., 1989, Characterization of the HST-related FGF.6 gene, a new member of the fibroblast growth factor gene family, *Oncogene* **4:**335–340.

Matuszewska, B., Keogan, M., Fisher, D. M., Soper, K. A., Hoe, C.-M., Huber, A. C., and Bondi, J. V., 1994, Acidic fibroblast growth factor: Evaluation of topical formulations in a diabetic mouse wound healing model, *Pharm. Res.* **11:**65–71.

Mazué, G., Bertolero, F., Jacob, C., Sarmientos, P., and Boncucci, R., 1991, Preclinical and clinical studies with recombinant human basic fibroblast growth factor, *Ann. NY Acad. Sci.* **638:**329–340.

McGee, G. S., Davidson, J. M., Buckley, A., Sommer, A., Woodward, S. C., Aquino, A. M., Barbour, R., and Demetriou, A. A., 1988, Recombinant basic fibroblast growth factor accelerates wound healing, *J. Surg. Res.* **45:**145–153.

McNeil, P. L., 1993, Cellular and molecular adaptations to injurious mechanical stress, *Trends Cell Biol.* **3:**302–307.

McNeil, P. L., Muthukrishnan, L., Warder, E., and D'Amore, P., 1989, Growth factors are released by mechanically wounded endothelial cells, *J. Cell Biol.* **109**:811–822.

Mellin, T. N., Mennie, R. J., Cashen, D. E., Ronan, J. J., Capparella, J., James, M. L., Di Salvo, J., Frank, J., Linemeyer, D., Gimenez-Gallego, G., and Thomas, K. A., 1992, Acidic fibroblast growth factor accelerates dermal wound healing, *Growth Factors* **7**:1–14.

Mellin, T. N., Cashen, D. E., Ronan, J. J., Murphy, B.S., Di Salvo, J., and Thomas, K. A., 1995, Acidic fibroblast growth factor accelerates dermal wound healing in diabetic mice, *J. Invest. Dermatol.* **104**:850–855.

Meyers, S. L., O'Brien, M. T., Smith, T., and Dudley, J. P., 1990, Analysis of the *int-1, int-2, c-myc,* and *neu* oncogenes in human breast carcinomas, *Cancer Res.* **50**:5911–5918.

Mignatti, P., Tsuboi, R., Robbins, W., and Rifkin, D. B., 1989, *In vitro* angiogenesis on the human amniotic membrane: Requirement for basic fibroblast growth factor-induced proteinases, *J. Cell Biol.* **108**: 671–682.

Mignatti, P., Morimoto, T., and Rifkin, D. B., 1992, Basic fibroblast growth factor, a protein devoid of a secretory signal sequence, is released by cells via a pathway independent of the endoplasmic reticulum-Golgi complex, *J. Cell. Physiol.* **151**:81–93.

Miki, T., Fleming, T. P., Bottaro, D. P., Rubin, J. S., Ron, D., and Aaronson, S. A., 1991, Expression cDNA cloning of the KGF receptor by creation of a transforming autocrine loop, *Science* **251**:72–75.

Miki, T., Bottaro, D. P., Fleming, T. P., Smith, C. L., Burgess, W. H., Chan, A. M.-L., and Aaronson, S. A., 1992, Determination of ligand-binding specificity by alternative splicing: Two distinct growth factor receptors encoded by a single gene, *Proc. Natl. Acad. Sci. USA* **89**:246–250.

Miyamoto, M., Naruo, K.-I., Seko, C., Matsumoto, S., Kondo, T., and Kurokawa, T., 1993, Molecular cloning of a novel cytokine cDNA encoding the ninth member of the fibroblast growth factor family, which has a unique secretion property, *Mol. Cell. Biol.* **13**:4251–4259.

Moscatelli, D., 1987, High and low affinity binding sites for basic fibroblast growth factor on cultured cells: Absence of a role for low affinity binding in the stimulation of plasminogen activator production by bovine capillary endothelial cells, *J. Cell. Physiol.* **131**:123–130.

Moscatelli, D., and Quarto, N., 1989, Transformation of NIH 3T3 cells with basic fibroblast growth factor or the hst/K-*fgf* oncogene causes down-regulation of the fibroblast growth factor receptor: Reversal of morphological transformation and restoration of receptor number by suramin, *J. Cell Biol.* **109**:2519–2527.

Muller, W. J., Lee, F. S., Dickson, C., Peters, G., Pattengale, P., and Leder, P., 1990, The *int-2* gene product acts as an epithelial growth factor in transgenic mice, *EMBO J.* **9**:907–913.

Mustoe, T. A., Pierce, G. F., Morishima, C., and Deuel, T. F., 1991, Growth factor-induced acceleration of tissue repair through direct and inductive activities in a rabbit dermal ulcer model, *J. Clin. Invest.* **87**:694–703.

Mustoe, T. A., Ahn, S. T., Tarpley, J. E., and Pierce, G. F., 1994, Role of hypoxia in growth factor responses: Differential effects of basic fibroblast growth factor and platelet-derived growth factor in an ischemic wound model, *Wound Repair Regen.* **2**:277–283.

Muthukrishnan, L., Warder, E., and McNeil, P. L., 1991, Basic fibroblast growth factor is efficiently released from a cytosolic storage site through plasma membrane disruptions of endothelial cells, *J. Cell. Physiol.* **148**:1–16.

Nabel, E. G., Yang, Z.-Y., Plautz, G., Forough, R., Zhan, X., Haudenschild, C. C., Maciag, T., and Nabel, G. J., 1993, Recombinant fibroblast growth factor-1 promotes intimal hyperplasia and angiogenesis in arteries *in vivo, Nature* **362**:844–846.

Niswander, L., and Martin, G. R., 1992, FGF-4 expression during gastrulation, myogenesis, limb and tooth development in the mouse, *Development* **114**:755–768.

Nurcombe, V., Ford, D. M., Wildschut, J. A., and Bartlett, P. F., 1993, Developmental regulation of neural response to FGF-1 and FGF-2 by heparan sulfate proteoglycan, *Science* **260**:103–106.

O'Keefe, E. J., Chiu, M. L., and Payne, R. E., 1988, Stimulation of growth of keratinocytes by basic fibroblast growth factor, *J. Invest. Dermatol.* **90**:767–769.

Olwin, B. B., Hannon, K., and Kudla, A. J., 1994, Are fibroblast growth factors regulators of myogenesis in vivo? *Prog. Growth Factor Res.* **5**:145–158.

Ornitz, D. M., Moreadith, R. W., and Leder, P., 1991, Binary system for regulating transgene expression in mice: Targeting *int-2* gene expression with yeast *GAL4/UAS* control elements, *Proc. Natl. Acad. Sci. USA* **88**:698–702.

Ornitz, D. M., Yayon, A., Flanagan, J. G., Svahn, C. M., Levi, E., and Leder, P., 1992, Heparin is required for cell-free binding of basic fibroblast growth factor to a soluble receptor and for mitogenesis in whole cells, *Mol. Cell. Biol.* **12**:240–247.

Partanen, J., Makela, T. P., Eerola, E., Korhonen, J., Hirvonen, H., Claesson-Welsh, L., and Alitalo, K., 1991, FGFR-4, a novel acidic fibroblast growth factor receptor with a distinct expression pattern, *EMBO J.* **10**:1347–1354.

Partanen, J., Vainikka, S., Korhonen, J., Armstrong, E., and Alitalo, K., 1992, Diverse receptors for fibroblast growth factors, *Prog. Growth Factor Res.* **4**:69–83.

Pasquale, E. B., 1990, A distinctive family of embryonic protein-tyrosine kinase receptors, *Proc. Natl. Acad. Sci. USA* **87**:5812–5816.

Phillips, L. G., Geldner, P., Brou, J., Dobbins, S., Hokanson, J., and Robson, M. C., 1990, Correction of diabetic incisional healing impairment with basic fibroblast growth factor, *Surg. Forum* **41**:602–603.

Phillips, L. G., Abdullah, K. M., Geldner, P. D., Dobbins, S., Ko, F., Linares, H. A., Broemeling, L. D., and Robson, M. C., 1993, Application of basic fibroblast growth factor may reverse diabetic wound healing impairment, *Ann. Plast. Surg.* **31**:331–334.

Pierce, G. F., Tarpley, J. E., Yanagihara, D., Mustoe, T. A., Fox, G. M., and Thomason, A., 1992, Platelet-derived growth factor (BB homodimer), transforming growth factor-β1, and basic fibroblast growth factor in dermal wound healing: Neovessel and matrix formation and cessation of repair, *Am. J. Pathol.* **140**:1375–1388.

Pierce, G. F., Yanagihara, D., Klopchin, K., Danilenko, D. M., Hsu, E., Kenney, W. C., and Morris, C. F., 1994, Stimulation of all epithelial elements during skin regeneration by keratinocyte growth factor, *J. Exp. Med.* **179**:831–840.

Prats, H., Kaghad, M., Prats, A. C., Klagsbrun, M., Lelias, J. M., Liauzun, P., Chalon, P., Tauber, J. P., Amalric, F., Smith, J. A., and Caput, D., 1989, High molecular mass forms of basic fibroblast growth factor are initiated by alternative CUG codons, *Proc. Natl. Acad. Sci. USA* **86**:1836–1840.

Rapraeger, A. C., Krufka, A., and Olwin, B. B., 1991, Requirement of heparan sulfate for bFGF mediated fibroblast growth and myoblast differentiation, *Science* **252**:1705–1708.

Reiland, J., and Rapraeger, A. C., 1993, Heparan sulfate proteoglycan and FGF receptor target basic FGF to different intracellular destinations, *J. Cell Sci.* **105**:1085–1093.

Rifkin, D. B., and Moscatelli, D., 1989, Recent developments in the cell biology of basic fibroblast growth factor, *J. Cell Biol.* **109**:1–6.

Robson, M. C., Phillips, L. G., Lawrence, W. T., Bishop, J. B., Youngerman, J. S., Hayward, P. G., Broemeling, L. D., and Heggers, J. P., 1992, The safety and effect of topically applied recombinant basic fibroblast growth factor on the healing of chronic pressure sores, *Ann. Surg.* **216**:401–408.

Roghani, M., Mansukhani, A., Dell'Era, P., Bellosta, P., Basilico, C., Rifkin, D. B., and Moscatelli, D., 1994, Heparin increases the affinity of basic fibroblast growth factor for its receptor but is not required for binding, *J. Biol. Chem.* **269**:3976–3984.

Rosengart, T. K., Johnson, W. V., Friesel, R., Clark, R., and Maciag, T., 1988, Heparin protects heparin-binding growth factor-1 from proteolytic inactivation *in vitro, Biochem. Biophys. Res. Commun.* **152**:432–440.

Ross, R., 1993, The pathogenesis of atherosclerosis: A perspective for the 1990s, *Nature* **362**:801–809.

Rubin, J. S., Osada, H., Finch, P. W., Taylor, W. G., Rudikoff, S., and Aaronson, S. A., 1989, Purification and characterization of a newly identified growth factor specific for epithelial cells, *Proc. Natl. Acad. Sci. USA* **86**:802–806.

Ruta, M., Burgess, W., Givol, D., Epstein, J., Neiger, N., Kaplow, J., Crumley, G., Dionne, C., Jaye, M., and Schlessinger, J., 1989, Receptor for acidic fibroblast growth factor is related to the tyrosine kinase encoded by the *fms*-like gene (FLG), *Proc. Natl. Acad. Sci. USA* **86**:8722–8726.

Sakaguchi, K., Yanagashita, M., Takeuchi, Y., and Aurbach, G. D., 1991, Identification of heparan sulfate proteoglycan as a high affinity receptor for acidic fibroblast growth factor (aFGF) in a parathyroid cell line, *J. Biol. Chem.* **266**:7270–7278.

Saksela, O., Moscatelli, D., Sommer, A., and Rifkin, D. B., 1988, Endothelial-derived heparan sulfate binds basic fibroblast growth factor and protects it from proteolytic degradation, *J. Cell Biol.* **107**:743–751.

Sato, Y., and Rifkin, D. B., 1988, Autocrine activities of basic fibroblast growth factor: Regulation of endothelial cell movement, plasminogen activator synthesis and DNA analysis, *J. Cell Biol.* **107**:1199–1205.

Schreier, T., Degen, E., and Baschong, W., 1993, Fibroblast migration and proliferation during *in vitro* wound healing, *Res. Exp. Med.* **193**:195–205.

Schweigerer, L., Neufeld, G., Friedman, J., Abraham, J. A., Fiddes, J. C., and Gospodarowicz, D., 1987, Capillary endothelial cells express basic fibroblast growth factor, a mitogen that promotes their own growth, *Nature* **325**:257–259.

Seno, M., Sasada, R., Iwane, M., Sudo, K., Kurokawa, T., Ito, K., and Igarashi, K., 1988, Stabilizing basic fibroblast growth factor using protein engineering, *Biochem. Biophys. Res. Commun.* **151**:701–708.

Shing, Y., Folkman, J., Sullivan, R., Butterfield, C., Murray, J., and Klagsbrun, M., 1984, Heparin affinity: Purification of a tumor-derived capillary endothelial cell growth factor, *Science* **223**:1296–1299.

Shing, Y., Folkman, J., Haudenschild, C., Lund, D., Crum, R., and Klagsbrun, M., 1985, Angiogenesis is stimulated by a tumor-derived capillary endothelial cell growth factor, *J. Cell. Biochem.* **29**:275–287.

Shipley, G. D., Keeble, W. W., Hendrickson, J. E., Coffey, R. J., and Pittelkow, M. R., 1989, Growth of normal human keratinocytes and fibroblasts in serum-free medium is stimulated by acidic and basic fibroblast growth factor, *J. Cell. Physiol.* **138**:511–518.

Slavin, J., Hunt, J. A., Nash, J. R., Williams, D. F., and Kingsnorth, A. N., 1992, Recombinant basic fibroblast growth factor in red blood cell ghosts accelerates incisional wound healing, *Br. J. Surg.* **79**:918–921.

Somers, K. D., Cartwright, S. L., and Schechter, G. L., 1990, Amplification of the *int-2* gene in human head and neck squamous cell carcinomas, *Oncogene* **5**:915–920.

Sommer, A., and Rifkin, D. B., 1989, Interaction of heparin with human basic fibroblast growth factor: Protection of the angiogenic protein from proteolytic degradation by a glycosaminoglycan, *J. Cell. Physiol.* **138**:215–220.

Sprugel, K. H., McPherson, J. M., Clowes, A. W., and Ross, R., 1987, Effects of growth factors *in vivo*. I. Cell ingrowth into porous subcutaneous chambers, *Am. J. Pathol.* **129**:601–613.

Staiano-Coico, L., Krueger, J. G., Rubin, J. S., D'limi, S., Vallat, V. P., Valentino, L., Fahey, T., Hawes, A., Kingston, G., Madden, M. R., Mathwich, M., Gottlieb, A. B., and Aaronson, S. A., 1993, Human keratinocyte growth factor effects in a porcine model of epidermal wound healing, *J. Exp. Med.* **178**:865–878.

Stenberg, B. D., Phillips, L. G., Hokanson, J. A., Heggars, J. P., and Robson, M. C., 1989, Effect of bFGF on the inhibition of contraction caused by bacterial contamination, *Surg. Forum* **40**:629–631.

Stenberg, B. D., Phillips, L. G., Hokanson, J. A., Heggers, J. P., and Robson, M. C., 1991, Effect of bFGF on the inhibition of contraction caused by bacteria, *J. Surg. Res.* **50**:47–50.

Talarico, D., Ittmann, M. M., Bronson, R., and Basilico, C., 1993, A retrovirus carrying the K-*fgf* oncogene induces diffuse meningeal tumors and soft-tissue fibrosarcomas, *Mol. Cell. Biol.* **13**:1998–2010.

Tanaka, A., Miyamoto, K., Minamino, N., Takeda, M., Sato, B., Matsuo, H., and Matsumoto, K., 1992, Cloning and characterization of an androgen-induced growth factor essential for the androgen-dependent growth of mouse mammary carcinoma cells, *Proc. Natl. Acad. Sci. USA* **89**:8928–8932.

Thompson, J. A., Haudenschild, C., Anderson, K. D., DiPietro, J. M., Anderson, W. F., and Maciag, T., 1989, Heparin-binding growth factor 1 induces the formation of organoid neovascular structures *in vivo*, *Proc. Natl. Acad. Sci. USA* **86**:7928–7932.

Tsuboi, R., and Rifkin, D. B., 1990, Recombinant basic fibroblast growth factor stimulates wound healing in healing-impaired *db/db* mice, *J. Exp. Med.* **172**:245–251.

Tsuboi, R., Shi, C.-M., Rifkin, D. B., and Ogawa, H., 1992, A wound healing model using healing-impaired diabetic mice, *J. Dermatol.* **19**:673–675.

Tsuboi, R., Sato, C., Kurita, Y., Ron, D., Rubin, J. S., and Ogawa, H., 1993, Keratinocyte growth factor (FGF-7) stimulates migration and plasminogen activator activity of normal human keratinocytes, *J. Invest. Dermatol.* **101**:49–53.

Uhl, E., Barker, J. H., Bondàr, I., Galla, T. J., Leiderer, R., Lehr, H.-A., and Messmer, K., 1993, Basic fibroblast growth factor accelerates wound healing in chronically ischaemic tissue, *Br. J. Surg.* **80**:977–980.

Vlodavsky, I., Friedman, R., Sullivan, R., Sasse, J., and Klagsbrun, M., 1987a, Aortic endothelial cells synthesize basic fibroblast growth factor which remains cell-associated and platelet-derived growth factor-like protein which is secreted, *J. Cell. Physiol.* **131**:402–408.

Vlodavsky, I., Folkman, J., Sullivan, R., Friedman, R., Ishai-Michaeli, R., Sasse, J., and Klagsbrun, M., 1987b, Endothelial cell-derived basic fibroblast growth factor: Synthesis and deposition into subendothelial extracellular matrix, *Proc. Natl. Acad. Sci. USA* **84:**2292–2296.

Vlodavsky, I., Fuks, Z., Ishai-Michaeli, R., Bashkin, P., Levi, E., Korner, G., Bar-Shavit, R., and Klagsbrun, M., 1991, Extracellular matrix-resident basic fibroblast growth factor: Implication for the control of angiogenesis, *J. Cell. Biochem.* **45:**167–176.

Werner, S., Roth, W. K., Bates, B., Goldfarb, M., and Hans, P., 1991, Fibroblast growth factor 5 proto-oncogene is expressed in normal human fibroblasts and induced by serum growth factors, *Oncogene* **6:**2137–2144.

Werner, S., Peters, K. G., Longaker, M. T., Fuller-Pace, F., Banda, M. J., and Williams, L. T., 1992a, Large induction of keratinocyte growth factor expression in the dermis during wound healing, *Proc. Natl. Acad. Sci. USA* **89:**6896–6900.

Werner, S., Duan, D.-S. R., deVries, C., Peters, K. G., Johnson, D. E., and Williams, L. T., 1992b, Differential splicing in the extracellular region of fibroblast growth factor receptor-1 generates receptor variants with different ligand-binding specificities, *Mol. Cell. Biol.* **12:**82–88.

Werner, S., Breeden, M., Hübner, G., Greenhalgh, D. G., and Longaker, M. T., 1994a, Induction of keratinocyte growth factor expression is reduced and delayed during wound healing in the genetically diabetic mouse, *J. Invest. Dermatol.* **103:**469–473.

Werner, S., Smola, H., Liao, X., Longaker, M. T., Krieg, T., Hofschneider, P. H., and Williams, L. T., 1994b, The function of KGF in morphogenesis of epithelium and reepithelialization of wounds, *Science* **266:**819–822.

Whitby, D. J., and Ferguson, M. W. J., 1991a, Immunohistochemical localization of growth factors in fetal wound healing, *Dev. Biol.* **147:**207–215.

Whitby, D. J., and Ferguson, M. W. J., 1991b, The extracellular matrix of lip wounds in fetal, neonatal, and adult mice, *Development* **112:**651–668.

Wilkinson, D. G., Peters, G., Dickson, C., and McMahon, A. P., 1988, Expression of the FGF-related proto-oncogene *int-2* during gastrulation and neurulation in the mouse, *EMBO J* **7:**691–695.

Wilkinson, D. G., Bhatt, S., and McMahon, A. P., 1989, Expression pattern of the FGF-related proto-oncogene *int-2* suggests multiple roles in fetal development, *Development* **105:**131–136.

Wu, L., and Mustoe, T. A., 1995, Effect of ischemia upon growth factor enhancement of incisional wound healing, *Surgery,* **117:**570–576.

Wu, L., Pierce, G. F., Ladin, D. A., Zhao, L. L., and Mustoe, T. A., 1993, KGF accelerates ischemic dermal ulcer healing in the rabbit ear, *Surg. Forum* **44:**704–706.

Wu, L., Pierce, G. F., Ladin, D. A., Zhao, L. L., Rogers, D., and Mustoe, T. A., 1995, Effects of oxygen on wound responses to growth factors: Kaposi's FGF, but not basic FGF, stimulates repair in ischemic wounds, *Growth Factors,* in press.

Yanagisawa-Miwa, A., Uchida, Y., Nakamura, F., Tomaru, T., Kido, H., Kaimjo, T., Sugimoto, T., Kaji, K., Utsuyama, M., Kurashima, C., and Ito, H., 1992, Salvage of infected myocardium by angiogenic action of basic fibroblast growth factor, *Science* **257:**1401–1403.

Yayon, A., and Klagsbrun, M., 1990, Autocrine transformation by chimeric signal peptide-basic fibroblast growth factor: Reversal by suramin, *Proc. Natl. Acad. Sci. USA* **87:**5346–5350.

Yayon, A., Klagsbrun, M., Esko, J. D., Leder, P., and Ornitz, D. M., 1991, Cell surface, heparin-like molecules are required for binding of basic fibroblast growth factor to its high affinity receptor, *Cell* **64:**841–848.

Yoshida, T., Miyagawa, K., Odagiri, H., Sakamoto, H., Little, P. F. R., Terada, M., and Sugimura, T., 1987, Genomic sequence of *hst*, a transforming gene encoding a protein homologous to fibroblast growth factors and the *int-2* encoded protein, *Proc. Natl. Acad. Sci. USA* **84:**7305–7309.

Zhan, X., Bates, B., Hu, X., and Goldfarb, M., 1988, The human FGF-5 oncogene encodes a novel protein related to fibroblast growth factors, *Mol. Cell. Biol.* **8:**3487–3495.

Zhou, D. J., Casey, G., and Cline, M. J., 1988, Amplification of human *int-2* in breast cancers and squamous carcinomas, *Oncogene* **2:**279–282.

Zhu, X., Komiya, H., Chirino, A., Faham, S., Hsu, B. T., and Rees, D. C., 1991, Three-dimensional structures of acidic and basic fibroblast growth factors, *Science* **251:**90–93.

Chapter 7

Role of Platelet-Derived Growth Factor *in Vivo*

CARL-HENRIK HELDIN and BENGT WESTERMARK

1. Introduction

Platelet-derived growth factor (PDGF) was originally identified as a mitogen for fibroblasts, smooth muscle cells, and glial cells (Kohler and Lipton, 1974; Ross *et al.,* 1974; Westermark and Wasteson, 1976). PDGF was subsequently purified from human platelets (Antoniades *et al.,* 1979; Deuel *et al.,* 1981; Heldin *et al.,* 1979; Raines and Ross, 1982). More recent studies have shown that PDGF is synthesized by a number of cell types and also acts on many different cell types (for reviews on PDGF, see Heldin and Westermark, 1990; Raines *et al.,* 1990).

PDGF stimulates not only the growth of cells, but also chemotaxis, i.e., the migration of cells toward a concentration gradient of PDGF. The cellular effects of PDGF and its specific expression during the embryonal development suggest that PDGF has an important role in the control of this process. Moreover, PDGF is present at the sites of wounds and has potent effects on growth, chemotaxis, and matrix production in the healing process. PDGF also has been implicated in certain pathological conditions. The *sis* oncogene of simian sarcoma virus is structurally related to PDGF, and transformation by this virus involves autocrine stimulation by a PDGF-like growth factor. Similarly, the development of certain human tumors may involve autocrine and paracrine stimulation by PDGF. Overactivity of PDGF may also be part of the development of certain nonmalignant disorders involving excessive cell proliferation, such as atherosclerosis, rheumatoid arthritis, glomerulonephritis, and fibrotic conditions.

The present chapter will focus on the normal function of PDGF in wound healing

CARL-HENRIK HELDIN • Ludwig Institute for Cancer Research, Biomedical Center, S-751 24 Uppsala, Sweden. BENGT WESTERMARK • Department of Pathology, University Hospital, S-751 85 Uppsala, Sweden.

The Molecular and Cellular Biology of Wound Repair (Second Edition), edited by Richard A. F. Clark. Plenum Press, New York, 1996

and the possible clinical use of PDGF to stimulate healing, as well as on the possible *in vivo* consequences of overactivity of PDGF and the design and potential use of PDGF antagonists.

2. Structure of PDGF

PDGF is a family of isoforms consisting of disulfide-bonded homo- or hetero-dimers of products of two genes, the PDGF A-chain gene and PDGF B-chain gene. The heterodimer, PDGF-AB, is the most common isoform in preparations of PDGF from human platelets, but homodimers also occur in platelets (Hammacher *et al.,* 1988; Hart *et al.,* 1990; Soma *et al.,* 1992). Among other cell types, there are examples of cells making only the A or the B chain, which thus assemble as homodimers, and of cells making both PDGF chains, which assemble into all three possible combinations. The different isoforms of PDGF have overlapping but distinct biological effects, since they interact with different affinities with two different receptors (see Fig. 1).

The human genes for the A and B chains of PDGF are localized on chromosomes 7 and 22, respectively, and are organized in a similar way with 7 exons. In each case the signal sequence is encoded by exon 1, the N-terminal prosequence that is removed after synthesis by exons 2 and 3, and most of the mature protein by exons 4 and 5 (Bonthron *et al.,* 1988; Johnsson *et al.,* 1984; Rorsman *et al.,* 1988). Exon 7 is mainly noncoding, and exon 6 of the B-chain gene encodes a C-terminal sequence that may be removed during processing. The A chain occurs as two splice variants, with and without exon 6; the form without exon 6 is the most common variant and encodes a slightly shorter product. Interestingly, exon 6 both in the A-chain and the B-chain gene encodes a 10-amino-acid motif of basic amino acids that has been shown to mediate interactions with other molecules intracellularly, at the cell surface, and in the cell matrix (LaRochelle *et al.,* 1991; Raines and Ross, 1992; Östman *et al.,* 1991a). The presence or absence of this retention motif will thus have a potentially important effect on the compartmental-ization of the factor. Some of the PDGF B chain and the long splice form of the A chain are retained intracellularly and will ultimately be degraded (Östman *et al.,* 1992); the functional importance of this pool of PDGF is unknown. Of the material that is released from the cells, most will remain at the cell surface or in the matrix immediately adjacent to the cell; it is possible that this pool of PDGF efficiently stimulates autocrine cellular events. In contrast, the short form of the PDGF A chain is secreted normally, and thus may also stimulate paracrine events at a distance from the producer cells.

Some insight into the structure of PDGF has recently been obtained following the assignment of the inter- and intrachain disulfide bonds of the molecule (Andersson *et al.,* 1992; Haniu *et al.,* 1993; Jaumann *et al.,* 1991; Östman *et al.,* 1993), and the elucidation of the three-dimensional configuration of PDGF-BB at 3.0 Å resolution (Oefner *et al.,* 1992). The two subunits of the molecule are arranged in an antiparallel manner, with the second and fourth cysteine residues forming the interchain disulfide bonds. The three intrachain disulfide bonds are present in a tight knotlike structure in one end of the molecule. The major part of the molecule consists of two twisted β sheets that end in two loops pointing in the same direction (loops 1 and 3); in the other

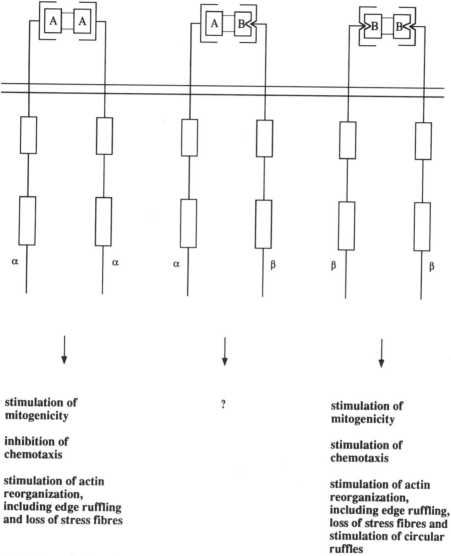

Figure 1. Schematic illustration of the interaction between PDGF isoforms and PDGF α and β receptors. Ligand binding induces receptor dimerization. Responses induced via αα receptor dimers and ββ receptor dimers are indicated. Note that the effect on chemotaxis is cell-type dependent. It is possible that unique signals are transduced via αβ receptor heterodimers (Rupp *et al.*, 1994), but the precise nature of such signals remains to be characterized.

direction, a short region (loop 2) connects the two β sheets. Thus, loop 1 and 3 of one of the subunits in the dimer will be close to loop 2 of the other subunit. Mutational analyses have shown that amino acid residues in loops 1 and 3 are important for receptor binding (Clements *et al.*, 1991; LaRochelle *et al.*, 1991; Östman *et al.*, 1991b). Moreover, a peptide comprising sequences from these two regions was found to have a fairly good ability to compete with the binding of ^{125}I-PDGF to receptors (Engström *et al.,* 1992). It is possible that loop 2 also contributes to the receptor binding region (LaRochelle *et al.,* 1992), but since such an involvement was not observed using other less sensitive assays (Östman *et al.,* 1991b), this loop is likely to be of less importance compared to loops 1 and 3.

3. Signal Transduction via PDGF Receptors

The PDGF isoforms exert their effects on cells by interacting with two structurally similar protein tyrosine kinase receptors. The α receptor binds both A and B chains, whereas the β receptor binds only the B chain. Thus, the different isoforms induce different dimeric receptor complexes: PDGF-AA only αα receptor complexes; PDGF-AB, αα and αβ receptor complexes; and PDGF-BB, αα, αβ, and ββ receptor complexes (Fig. 1) (Heldin and Westermark, 1990; Raines *et al.,* 1990). Both α and β receptors have been shown to induce mitogenic signals. However, whereas the β receptor mediates stimulation of chemotaxis, the α receptor actually inhibits chemotaxis, at least in some cell types. There is also a difference between the two receptors with regard to reorganization of actin filaments; both receptors stimulate the formation of edge ruffles and the loss of stress fibers, but only the β receptor stimulates the formation of circular ruffles on the dorsal surface of the cell (Eriksson *et al.,* 1992).

The dimerization of PDGF receptors induces autophosphorylation of the receptors, which occurs in *trans* between the molecules in the dimer. The autophosphorylation is a key step in the signal transduction. Most information regarding the specific localization of the phosphorylated tyrosine residues is available for the β receptor (reviewed by Claesson-Welsh, 1994). One autophosphorylation site has been localized to a conserved tyrosine residue inside the kinase domain, and may be involved in the regulation of the catalytic activity of the receptor kinase. A total of eight autophosphorylation sites have been localized in the noncatalytic parts of the intracellular domain of the receptor (Fig. 2). These phosphorylated regions form binding sites for signal transduction molecules with SH2 (Src homology 2) domains (reviewed by Cantley *et al.,* 1991; Koch *et al.,* 1991). The SH2 domain is a motif of about 100 amino acids that folds in such a way that a pocket is formed into which a phosphorylated tyrosine residue fits. The immediately adjacent amino acid residues, particularly those three that are localized C-terminal of the phosphorylated tyrosine, determine the specificity in the interactions between individual SH2 domains and different tyrosine-phosphorylated regions.

There are two different categories of SH2 domain containing signal transduction molecules: those with intrinsic catalytic activities, and those without catalytic domains that serve as adaptor molecules between the PDGF receptor and other catalytic molecules. At present, eight SH2-domain-containing signal transduction molecules are known to bind to different autophosphorylation sites in the PDGF β receptor, i.e., Src and other members of this family of tyrosine kinases; phosphatidylinositol-3′-kinase (PI-3-kinase); GTPase-activating protein (GAP) of Ras; proteintyrosine phosphatase-

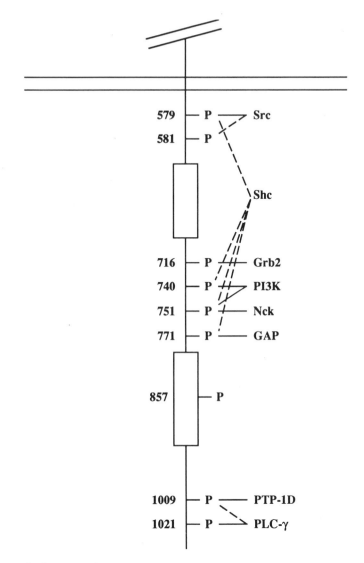

Figure 2. Interaction between activated PDGF β receptor and signal transduction molecules. Known auto-phosphorylation sites are indicated (P) and the tyrosine residues involved numbered. For references, see Arvidsson *et al.* (1994); Claesson-Welsh (1994); Yokote *et al.* (1994).

1D (PTP-1D); phospholipase C-γ (PLC-γ); and the adaptor molecules, Grb2, Nck, and Shc (Claesson-Welsh, 1994) (Fig. 2).

The binding of individual signal transduction molecules to the PDGF receptors initiates several signal transduction pathways. The PI-3-kinase has been shown to be of importance for PDGF-stimulated motility responses, such as chemotaxis and actin reorganization (Wennström *et al.*, 1994a,b). The product of the enzyme, phosphatidyl-3,4,5-trisphosphate, thus appears to be a second messenger regulating motility responses in the cell; its downstream targets, however, remain to be identified. In certain cell types, PI-3-kinase also affects mitogenic signaling, but in most cells mitogenicity is not absolutely dependent on this enzyme. On the other hand, there is much evidence that activation of the mitogen-activated protein (MAP) kinase cascade is important for mitogenic signaling (reviewed by Blumer and Johnson, 1994). Activation is brought about through direct interaction between the first serine/threonine kinase in the cascade, Raf-1, and activated Ras. Ras in turn is activated by the action of a nucleotide exchange protein, Sos, which occurs in a complex with the adaptor molecule Grb2 (reviewed by Schlessinger, 1993). The signal transduction pathway is initiated by the direct or indirect binding of Grb2 to the activated PDGF receptor. PLC-γ (Valius and Kazlauskas, 1993) and Src tyrosine kinases (Twamley-Stein *et al.*, 1993) have also been shown to mediate mitogenic signals; it thus appears that multiple signal transduction pathways are involved in PDGF-stimulated cell growth.

4. *In Vivo* Function of PDGF

4.1. Embryogenesis

The specific spatial and temporal expression of PDGF and PDGF receptor in the embryo (Morrison-Graham *et al.*, 1992; Orr-Urtreger *et al.*, 1992; Schatteman *et al.*, 1992) and in the placenta (Goustin *et al.*, 1985) suggests that PDGF has an important regulatory role during the development. PDGF may have a role during the very early phases of embryonal growth since *Xenopus* oocytes have been shown to contain maternally derived mRNA for PDGF and PDGF receptors (Mercola *et al.*, 1988). Moreover, PDGF stimulates the growth of bovine embryos at the eight-cell stage (Larson *et al.*, 1992; Thibodeaux *et al.*, 1993).

Mouse embryos at the two-cell and blastocyst stages express the PDGF A chain and α receptor. In early postimplantation embryos, PDGF A-chain expression is restricted to the ectoderm, whereas α-receptor expression is found in the mesoderm (Palmieri *et al.*, 1992). These findings suggest that an autocrine PDGF α-receptor stimulation occurs in preimplantation embryos, whereas the activation of the receptor in the mesoderm in later stages is mediated by a paracrine growth factor. Expression of the PDGF B chain and β receptor seems to be absent or very low in early mouse development (Mercola *et al.*, 1990), but seems to be induced in specific tissues, e.g., nervous tissues, at later stages of development (Sasahara *et al.*, 1991; Smits *et al.*, 1991; Yeh *et al.*, 1991).

A specific function for PDGF has been elucidated in the control of the differentiation of glial cells in the optic nerve (Noble *et al.*, 1988; Richardson *et al.*, 1988). O-2A progenitor cells express PDGF α receptors and are prevented from premature differentiation into oligodendrocytes by PDGF-AA added exogenously or provided as a paracrine growth factor by type 2 astrocytes. Interestingly, the addition of fibroblast growth factor (FGF) along with PDGF maintains the O-2A progenitor cells in a proliferative, undifferentiated compartment, allowing for serial propagation of these cells (Bögler *et al.*, 1990). *In vivo* studies have shown that another important function of PDGF may be to prevent cells of the oligodendrocyte lineage from apoptosis (Barres *et al.*, 1992). The presence of functional PDGF β receptors in Schwann cells suggests that PDGF may also be important for growth and regeneration of glial cells of the peripheral nerve system (Eccleston *et al.*, 1990; Weinmaster and Lemke, 1990).

4.2. Wound Healing

PDGF is one of several factors that have been found to stimulate the healing of soft tissues (reviewed in Deuel *et al.*, 1991; Pierce *et al.*, 1991). The knowledge about the role of PDGF in wound healing has come from three lines of research, i.e., studies of the effects of PDGF *in vitro* on different cell types of importance for wound healing, investigations of the expression of PDGF and PDGF receptors during the healing of wounds, and studies of the effect of exogenously added PDGF on wound healing.

Studies *in vitro* have revealed that PDGF is a potent mitogen for connective tissue cells, and that it in addition stimulates chemotaxis of fibroblasts (Seppä *et al.*, 1982), smooth muscle cells (Grotendorst *et al.*, 1981), neutrophils, and macrophages (Deuel *et al.*, 1982; Senior *et al.*, 1983; Siegbahn *et al.*, 1990). Of additional importance is the ability of PDGF to activate macrophages to produce and secrete other growth factors of importance for various aspects of the healing process. In addition to the chemotaxis and proliferation of cells, the production of matrix proteins and proteoglycans is important for the efficient healing of a wound. Although other factors, in particular transforming growth factor-β (TGF-β), have more potent stimulating effects on matrix production, PDGF has been shown to stimulate the production of fibronectin (Blatti *et al.*, 1988) and hyaluronic acid (Heldin *et al.*, 1989) by fibroblasts. Another potentially important effect of PDGF in the healing process is its ability to stimulate contraction of collagen matrices *in vitro* (Clark *et al.*, 1989; Gullberg *et al.*, 1990), implicating an effect on wound contraction *in vivo*. It is also possible that PDGF is important in the later remodeling phase of wound healing, since it has been shown to stimulate the production and secretion of collagenase in fibroblasts (Bauer *et al.*, 1985).

A prerequisite for a role of PDGF in normal wound healing is that PDGF is present at the site of the wound and that cells in the wounded area have PDGF receptors. PDGF is present in large amounts in the platelets that release their contents at the site of the wound and is also secreted by activated macrophages (Shimokado *et al.*, 1985), thrombin-stimulated endothelial cells (Harlan *et al.*, 1986), smooth muscle cells of damaged arteries (Walker *et al.*, 1986), activated fibroblasts (Paulsson *et al.*, 1987), as well as by epidermal keratinocytes (Ansel *et al.*, 1993); thus, PDGF is likely to be present in large quantities in the wound. Consistent with this possibility, PDGF-like factors have been found to be present in wound fluid (Matsuoka and Grotendorst, 1989). Regarding the ability to respond to PDGF, staining for PDGF receptors has revealed that fibroblasts and smooth muscle cells of resting tissues contain low levels of receptors. However, the PDGF β receptor is up-regulated in conjunction with inflammation, for example, thereby making cells responsive to PDGF action (Reuterdahl *et al.*, 1993; Rubin *et al.*, 1988a; Terracio *et al.*, 1988). In addition to expression of PDGF β receptors on connective tissue cells after cutaneous injury, expression has also been noticed on epithelial cells (Antoniades *et al.*, 1991).

An important aspect of tissue repair is the formation of new blood vessels. PDGF has recently been shown to have a weak angiogenic activity (Risau *et al.*, 1992), and PDGF receptors, which are absent from endothelial cells of large vessels, were demonstrated on capillary endothelial cells (Bar *et al.*, 1989; Smits *et al.*, 1989) and on microvascular pericytes (Sundberg *et al.*, 1993). PDGF may also stimulate angiogenesis in an indirect way, by inducing the secretion of endothelial cell growth factors by myofibroblasts (Sato *et al.*, 1993). Clearly, however, the angiogenic effect of PDGF is weaker than that of other growth factors, e.g., of the FGF family (Folkman and Klagsbrun, 1987).

Direct application of PDGF in chambers implanted into rats was found to induce increased formation of granulation tissue (Grotendorst *et al.*, 1985; Sprugel *et al.*, 1987); PDGF-BB was found to be significantly more effective than PDGF-AA (Lepistö *et al.*, 1992). A series of studies by Pierce and collaborators involving local applications of PDGF to incisional wounds in rat skin (Pierce *et al.*, 1988) and to excisional wounds in rabbit ear (Mustoe *et al.*, 1991) have given further support for the notion that PDGF augments wound healing. A single application of PDGF-BB to incisional wounds increased the wound-breaking strength to 150–170% of control wounds, and thus decreased the time for healing (Pierce *et al.*, 1988, 1989). The PDGF-BB-treated wounds were characterized by an increased inflammatory response with larger numbers of neutrophils, monocytes, and fibroblasts during the acute phase of the healing. In the excisional wound model, in which 6-mm-wide pieces of dermis were removed from the rabbit ear down to the level of the cartilage, a single application of PDGF led to an increase of granulation tissue rich in fibroblasts and glycosaminoglycans to 200% of the control wounds after 7 days (Mustoe *et al.*, 1991; Pierce *et al.*, 1991). PDGF also increased the rate of reepithelialization and of neovascularization. An important conclusion from these studies is that PDGF-BB does not alter the normal sequence of repair, but increases its rate.

PDGF in combination with insulin-like growth factor-I (IGF-I) has also been shown to enhance repair of excisional wounds in porcine skin (Lynch *et al.*, 1987, 1989) and in guinea pigs (Hill *et al.*, 1991). Moreover, PDGF has been found capable of improving healing in situations of deficient repair, e.g., in diabetic mice (Greenhalg *et al.*, 1990) or after irradiation (Grotendorst *et al.*, 1985; Mustoe *et al.*, 1989).

Positive effects of local administration of PDGF-BB to decubitus ulcers of patients have also been reported (Robson *et al.*, 1992). Patients treated with 100 μg/ml PDGF-BB, i.e., a rather high dose, for 28 days had a pronounced healing response compared to controls.

PDGF has also been shown to potentiate regeneration of the peridontium in naturally occurring peridontitis in dogs (Lynch *et al.*, 1991), as well as in experimental peridontitis in monkeys (Rutherford *et al.*, 1992); the effect was enhanced if PDGF was combined with IGF-I (Lynch *et al.*, 1991) or dexamethasone (Rutherford *et al.*, 1993).

However, in a rat craniotomy bone regeneration model, PDGF was found to inhibit the bone regeneration induced by osteogenin; rather, PDGF induced a soft tissue repair wound phenotype and response (Marden *et al.,* 1993). Thus, whereas PDGF efficiently stimulates healing of soft tissues, it appears not to have any positive effect on fracture healing.

Comparison of the effect of high concentrations of PDGF on the healing of excisional wounds with that of high concentrations of other factors revealed interesting differences (Pierce *et al.,* 1992). PDGF-BB accelerated the deposition of provisional wound matrix containing in particular glycosaminoglycans and fibronectin. TGF-β1 induced an enhanced synthesis and maturation of collagen. In contrast, basic FGF induced predominantly an angiogenic response and an increased collagenolytic activity that delayed wound maturation. Thus, different growth factors affect the different phases in wound healing differently. Growth factors have been shown to induce the synthesis of themselves in positive autocrine feedback loops, as well as of other growth factors. Thus, a plethora of factors is likely to be present at the site of the wound, which assures efficient enhancement of the different phases of wound healing. The finding that wounds heal more rapidly if the wound is covered by an occlusive dressing (Katz *et al.,* 1991) is consistent with the possibility that locally produced factors are important during wound healing.

4.3. Platelet Aggregability

In addition to the long-known roles of PDGF in embryogenesis and wound healing, PDGF was recently shown to have yet another potentially important effect, i.e., to inhibit platelet aggregation (Bryckaert *et al.,* 1989). Thrombin induces platelet aggregation and release of the platelet granule content; these effects can be inhibited by addition of PDGF. Human platelets have PDGF α receptors but not β receptors (Vassbotn *et al.,* 1994). The finding that thrombin-induced platelet aggregation is accompanied by activation of PDGF α receptors, and that this effect can be blocked by PDGF antibodies, indicates that the PDGF released from the platelets serves an autocrine feedback role in the control of platelet aggregation.

5. PDGF in Disease

5.1. Fibrosis

As discussed above, PDGF has as an important function *in vivo* to stimulate the formation of connective tissue. Consistent with such a function, there are several observations that support the notion that overactivity of PDGF is involved in the development of in various fibrotic conditions.

There is circumstantial evidence that megakaryocytes and PDGF are involved in the pathogenesis of bone marrow fibrosis (Castro-Malaspina *et al.*, 1981; Groopman, 1980). Thus, the levels of PDGF in circulating plasma and urine are significantly elevated in patients with myelofibrosis (Gersuk *et al.*, 1989), and the concentration of PDGF in platelets is correspondingly decreased (Katoh *et al.*, 1988).

The fibrosis that is observed in chronic liver disease is characterized by a dedifferentiation of fat-storing cells (Ito cells) to myofibroblast-like cells and by a proliferation of these cells. As these cells respond to PDGF (P. Heldin *et al.*, 1991; Pinzani *et al.*, 1989, 1991) and activated macrophages from patients with liver disease have been found to secrete large amounts of PDGF (Peterson and Isbrucker, 1992), it is possible that PDGF has a role in the development of liver cirrhosis.

Scleroderma (systemic sclerosis) is a disease characterized by progressive fibrosis in the skin and in a number of visceral organs. A role for PDGF in the progression of this disease is suggested by the observations that PDGF β receptors, which are absent from fibroblasts of normal skin, were found to be expressed in cells of skin from scleroderma patients (Klareskog *et al.*, 1990), and that PDGF immunoreactivity was also demonstrated in scleroderma skin (Gay *et al.*, 1989).

Idiopathic pulmonary fibrosis is characterized by inflammation and fibrosis. Evidence for a role of PDGF in this process was presented by Martinet *et al.* (1986), who showed that alveolar macrophages from patients with idiopathic pulmonary fibrosis produced significantly higher amount of PDGF than those of healthy subjects. The macrophages have been given a central role in this disease as producers of PDGF and other cytokines (Vignaud *et al.*, 1991), but also the alveolar epithelium of patients with idiopathic pulmonary fibrosis has been shown to produce PDGF B chain and other cytokines (Antoniades *et al.*, 1990). Rat lung fibroblasts have been shown to have more PDGF β receptors than α receptors, and thus they respond well to PDGF-BB but only poorly to PDGF-AA, with regard to growth (Bonner *et al.*, 1991; Caniggia *et al.*, 1993), as well as glycosaminoglycan synthesis (Cannigia and Post, 1992); however, exposure to crysolite asbestos *in vitro* leads to an increased responsiveness to A-chain-containing PDGF isoforms (Bonner *et al.*, 1993). Also, other fibrotic conditions in the lung, like those following hypoxid pulmonary hypertension (Katayose *et al.*, 1993), breathing of high concentrations of oxygen (Han *et al.*, 1992; Powell *et al.*, 1992), and obliterative bronchiolitis after lung transplantation (Hertz *et al.*, 1992), involve overexpression of PDGF. The increased PDGF production may result from injury to the pulmonary tissue, since bronchoalveolar lavage fluids from patients with acute diffuse lung injury (Snyder *et al.*, 1991), as well as from rats having received tracheal instillations of bleomycin (Walsh *et al.*, 1993), were found to contain high concentrations of PDGF-like peptides.

In situations of chronic synovial inflammation, such as in the joints of patients with rheumatoid arthritis, the expression of PDGF β receptors is up-regulated (Reuterdahl *et al.*, 1991; Rubin *et al.*, 1988b). Moreover, PDGF is present in high amounts in inflamed joints (Sano *et al.*, 1993). It is thus possible that PDGF plays a role in the stimulation of mesenchymal cell proliferation that often accompanies chronic inflammatory diseases.

5.2. Atherosclerosis

The atherosclerotic process is characterized by an excessive inflammatory–fibroproliferative response to different forms of insults to the endothelium and smooth muscle of the artery wall (reviewed in Ross, 1993). Injury to the endothelium leads to migration of macrophages to the subendothelial space, where they accumulate lipids and become foam cells. Such foam cells, together with T cells, form fatty streaks, which progress to fibrous plaques. Advanced plaques also contain smooth muscle cells that have migrated into the intima layer of the vessel and proliferated there, as well as platelet-containing thrombi. PDGF and other growth factors are likely to have important roles in this process by stimulation of chemotaxis and proliferation of the different cells in the lesion. PDGF has been shown to be synthesized by many of the cell types in the lesion, such as activated macrophages (Shimokado *et al.,* 1985), smooth muscle cells (Nilsson *et al.,* 1985; Seifert *et al.,* 1984), and endothelial cells (DiCorleto and Bowen-Pope, 1983), and is also released from the platelets in thrombi. The expression of PDGF by smooth muscle cells and macrophages have been shown to be increased in atherosclerotic lesions compared to controls (Ross *et al.,* 1990; Wilcox *et al.,* 1988). Moreover, intimal smooth muscle cells in the lesions express increased amounts of PDGF β receptors, and are therefore more responsive to PDGF (Rubin *et al.,* 1988a).

Further evidence for the involvement of PDGF in the atherosclerotic process comes from *in vivo* studies in which neutralizing PDGF antibodies were given to rats subjected to balloon angioplasty of carotid arteries; the intimal thickening that follows this treatment was inhibited by the PDGF antibodies (Ferns *et al.,* 1991). On the other hand, infusion of PDGF-BB into rats after carotid injury caused an increase in the intimal thickening and in the migration of smooth muscle cells from the media of the vessel to the intima (Jawien *et al.,* 1992). In conclusion, there is accumulating evidence that PDGF, acting in concert with other growth factors, has important roles in the development of atherosclerosis.

5.3. Glomerulonephritis

Acute and chronic glomerular inflammation occurs in response to different types of injury and involves remodeling and repair of the damaged glomerular tissue. Besides endothelial cells of the glomeruli, mesangial cells have an important role in this process. Proliferation of mesangial cells and production of extracellular matrix proteins by these cells are associated with the development of many types of glomerulonephritides (reviewed by Abboud, 1993; Johnson *et al.,* 1993; Sterzel *et al.,* 1993). Results obtained during the recent years support the notion that PDGF is involved in the stimulation of proliferation of mesangial cells and TGF-β in the stimulation of matrix production.

Mesangial cells both produce PDGF and respond to PDGF *in vitro* (Shultz *et al.,* 1988; Silver *et al.,* 1989). They express more β receptors than α receptors, and thus respond best to B-chain-containing PDGF isoforms (Floege *et al.,* 1991).

During mesangial proliferative nephritis, the expression of PDGF increases both in patients and in experimental animal models (Gesualdo *et al.*, 1991; Iida *et al.*, 1991). Much work has been performed using the anti-Thy-1 model for glomerulonephritis in rat, in which injection of complement-fixing antibodies against an antigen on the surface of mesangial cells leads to mesangiolysis, which is followed by a infiltration by platelets and monocytes/macrophages, and 2 to 6 days later a massive proliferation of mesangial cells (Johnson *et al.*, 1991). After a phase characterized by extracellular matrix deposition, the lesion is ultimately healed and glomerular structure restored. This model resembles somewhat immunoglobulin A nephropathy, the most common form of glomerulonephritis in humans. The expression of PDGF and PDGF β receptor correlated with the mesangial proliferation in the anti-Thy-1 model (Iida *et al.*, 1991), suggesting a role for PDGF in the process. Apart from the mesangial cells themselves, platelets and macrophages can be the source of PDGF in the diseased kidney, and it is thus possible that PDGF stimulates mesangial cell proliferation through autocrine as well as paracrine mechanisms. A causative role of PDGF in the anti-Thy-1 model is furthermore supported by the finding that administration of neutralizing PDGF antibodies slows down the proliferation of mesangial cells 4 days after injury (Johnson *et al.*, 1992). However, the PDGF antibodies had no effect at day 2, indicating that the early phase of proliferation is driven by other factors. In addition, infusion of PDGF-BB to normal rats induced mild mesangial proliferation and a massive proliferation in rats given subnephritic doses of anti-Thy-1 sera (Floege *et al.*, 1993).

The observations mentioned above provide strong indications that PDGF is involved in the development of glomerulonephritis. Clearly, however, other growth factors and cytokines may also be involved, e.g., TGF-β has been given a major role in the stimulation of matrix production in the disease. There is also evidence that inhibition of TGF-β by infusion of the TGF-β-binding proteoglycan decorin inhibits the development of glomerulonephritis in the anti-Thy-1 model (Border *et al.*, 1990). Consistent with the possibility that PDGF and TGF-β have different effects in the development of glomerulonephritis, Isaka *et al.* (1993) found, using an *in vivo* transfection technique, that introduction of either PDGF B chain or TGF-β into rat kidney induced glomerulosclerosis; significantly, however, TGF-β affected primarily extracellular matrix accumulation and PDGF affected primarily cell proliferation.

5.4. Malignancies

The findings that the normal counterpart of the *sis* oncogene product is the B chain of PDGF (Doolittle *et al.*, 1983; Waterfield *et al.*, 1983) and that *sis*-transformation occurs by autocrine stimulation of PDGF receptors (reviewed by Westermark *et al.*, 1987) boosted analyses of PDGF and PDGF receptor expression in human malignancies. Such studies have revealed that coexpression of PDGF and PDGF receptors are common in certain types of tumors (reviewed in Heldin and Westermark, 1991; Raines *et al.*, 1990).

One example of a type of tumor in which PDGF is often overexpressed is glioblastoma (Fleming *et al.*, 1992; Maxwell *et al.*, 1990; Nistér *et al.*, 1988). Investigations of the expression pattern of PDGF and PDGF receptors in human glioblastoma tumors led to several interesting conclusions (Hermanson *et al.,* 1988, 1992; Plate *et al.,* 1992). Evidence for two autocrine loops was obtained, involving PDGF-A/α receptors in the tumor cells and PDGF-B/β receptors in the stroma compartment of the tumor. Whereas the expression of α receptors on tumor cells and B chain in vessels occurred both in low- and high-grade tumors, the expression of PDGF A chain and β receptor were higher in the more malignant tumors. These observations suggest that autocrine and paracrine mechanisms involving different isoforms of PDGF are of importance for the balanced growth of different cell types in glioblastoma tumors.

Since PDGF is primarily known as a connective tissue mitogen, it was of interest to investigate the expression of PDGF and PDGF receptors on fibromas and fibrosarcomas of different degrees of malignancies. Also, these tumors were found to express PDGF and PDGF receptors in a malignancy-dependent manner (Alman *et al.,* 1992; Smits *et al.,* 1992), suggesting that PDGF overactivity is associated with tumor progression.

Also, tumors originating from epithelial tissue that normally does not respond to PDGF have been shown to produce PDGF (Raines *et al.,* 1990). In certain cases, an autocrine loop may be established by aberrant expression of PDGF receptors, as has been observed, e.g., in the cases of certain thyroid carcinoma cell lines (N.-E. Heldin *et al.,* 1988, 1991), gastric carcinoma cells (Chung and Antoniades, 1992), and lung cancer cell lines (Antoniades *et al.,* 1992). In other cases, PDGF production in carcinomas may affect stroma formation (Chaudhry *et al.,* 1992; Lindmark *et al.,* 1993). Direct evidence for a role of PDGF in stroma formation has been obtained by Forsberg *et al.* (1993), who analyzed the histology of tumors formed in nude mice by human melanoma cells and compared them with tumors formed by cells stably transfected with PDGF B-chain cDNA. Whereas the tumors that arose from the PDGF-producing melanoma cells contained a network of connective tissue with an abundance of blood vessels, the tumors derived from control cells showed less stroma and blood vessels and large areas of necroses. It is possible that also in spontaneous tumors, PDGF or other factors derived from the tumor cells stimulate the formation of a supportive stroma.

Another example that subversion of the mitogenic pathway of PDGF can lead to malignant transformation is the finding that acute myelomonocytic leukemia is often associated with a chromosomal rearrangement that leads to a fusion of the intracellular effector domain and the transmembrane domain of the PDGF β receptor with an Ets-like transcription factor (Tel) (Golub *et al.,* 1994). Presumably, the resulting fusion protein is constitutively active and initiates the signal transduction in a ligand-independent manner.

6. PDGF Antagonists

Since PDGF appears to be involved in many severe disorders, specific PDGF antagonists would be of potential clinical utility. In the design and administration of such antagonists, care must be taken not to interfere with important normal functions of PDGF. Thus, inhibition of PDGF could negatively affect the wound-healing process. However, since there appear to be many different growth factors involved in wound healing with partially overlapping effects, the inhibition of PDGF may slow down wound healing only marginally. Another potentially more serious complication could be the effect on platelet aggregability. Inhibition of PDGF's feedback control of platelets could lead to an increased risk for thrombosis. Since platelets have only α receptors, whereas in many disorders the PDGF effects are exerted mainly via the β receptor, which is found in higher amounts on, e.g., smooth muscle cells, fibroblasts, and mesangial cells, a possible strategy would be to develop antagonists that block PDGF action via β receptor without affecting the α receptor.

Regarding the possibility of interfering with the intracellular pathways regulated by PDGF, a potential problem is specificity, since such pathways are likely to be shared by several factors. One possibility would be to specifically inhibit the receptor kinase, and certain low-molecular-weight inhibitors, denoted tyrphostins, which discriminate to a certain extent between different tyrosine kinases, have been developed (Fig. 3) (Bryckaert *et al.*, 1992).

Another possibility is to interfere with the extracellular interactions between ligand and receptor. There are certain low-molecular-weight substances that have been found to displace PDGF from its receptor, and thus act as antagonists. One example is suramin (Williams *et al.*, 1984). However, suramin is not specific for PDGF, but interferes with a number of other ligand–receptor interactions as well (Betsholtz *et al.*, 1986). Neomycin has also been shown to inhibit the binding of PDGF to its receptor (Vassbotn *et al.*, 1992). Interestingly, the inhibition occurs in a receptor-specific manner; the binding of PDGF-BB to the α receptor is blocked, but not the binding to the β receptor. The binding of PDGF-AA to α receptors is somewhat affected but less than the binding of PDGF-BB. Even if the high concentrations of neomycin needed for any appreciable effect makes this compound unsuitable as a PDGF inhibitor for clinical use, these observations indicate that the α and β receptors recognize slightly different epitopes on PDGF, and that it therefore should be possible to find receptor-specific antagonists.

Due to the conservation of PDGF through the evolution, it has been difficult to obtain high-affinity antibodies against PDGF. However, useful neutralizing sera have been obtained in rabbits and goats that have been shown to inhibit autocrine stimulation in *sis*-transformed cells (Huang *et al.*, 1984; Johnsson *et al.*, 1985), as well as the atherosclerotic process that occurs after deendothelialization of carotid arteries in rats (Ferns *et al.*, 1991) and mesangial proliferation in the development of glomerulonephritis (Johnson *et al.*, 1992). In addition, a soluble form of the PDGF β receptor has been shown to bind PDGF and thereby prevent it from binding to signaling receptors (Duan *et al.*, 1991). Whereas isoform-specific antibodies could be clinically useful as PDGF antagonists, the use of soluble receptors would be complicated by the fact that extracellular domains of both α and β receptor bind PDGF-AA, which is not desirable because of the risk for increased platelet aggregability, as discussed above.

Since receptor dimerization is a key event in signal transduction, another strategy to achieve PDGF antagonism would be to prevent receptor dimerization (Fig. 3). If the

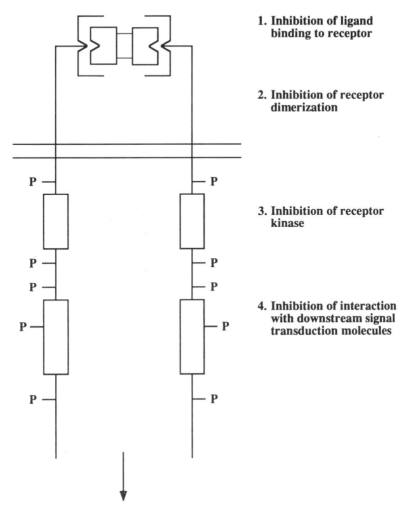

1. **Inhibition of ligand binding to receptor**

2. **Inhibition of receptor dimerization**

3. **Inhibition of receptor kinase**

4. **Inhibition of interaction with downstream signal transduction molecules**

Figure 3. Possible levels of specific inhibition of PDGF signals. Inhibition is also possible further downstream, but in such cases is not expected to be specific since intracellular signal transduction pathways are shared by many factors.

effect of the dimeric PDGF molecule is to bring together two receptors in a dimer, monomeric PDGF may be an antagonist. However, a PDGF mutant, in which the cysteines involved in interchain disulfide bonding were mutated to serine residues, was found to stimulate receptor dimerization and activation (Andersson *et al.*, 1992), probably because the PDGF molecule even in the absence of disulfide bonds occurs as a dimer (Kenney *et al.*, 1994). A more efficient way to prevent receptor dimerization could be to combine a wild-type PDGF chain with a mutant chain, which is unable to bind to receptors and actively prevents receptor dimerization.

7. Future Perspectives

The results of animal experiments and the first clinical trials support the notion that topical application of PDGF stimulates wound healing. It is therefore possible that PDGF will be an important future clinical tool, particularly for the stimulation of soft tissue repair in patients with impaired capacity for wound healing. Much work remains, however, to optimize, e.g., dose, ways of administration, and choice of PDGF isoform. Moreover, critical comparisons with other growth factors should be performed in order to select the best factor, or combinations of factors, for different types of wounds.

Accumulating evidence strongly suggests that overactivity of PDGF is a causative factor in the development of several serious disorders. Specific and efficient PDGF antagonists are therefore highly desirable. As discussed above, it will probably be necessary to design antagonists that discriminate between the two receptors for PDGF. Moreover, it remains to be solved how to administer antagonists to the appropriate localizations in the different diseases. Despite these difficulties, it is likely that suitable PDGF antagonists will be produced and find their way to the clinic in a not too distant future.

ACKNOWLEDGMENTS. We thank Ingegärd Schiller for valuable help in the preparation of this chapter.

References

Abboud, H. E., 1993, Growth factors in glomerulonephritis, *Kidney Int.* **43:**252–267.

Alman, B. A., Goldberg, M. J., Naber, S. P., Galanopoulous, T., Antoniades, H. N., and Wolfe, H. J., 1992, Aggressive fibromatosis, *J. Pediatr. Orthop.* **12:**1–10.

Andersson, M., Östman, A., Bäckström, G., Hellman, U., George-Nascimento, C., Westermark, B., and Heldin, C.-H., 1992, Assignment of interchain disulfide bonds in platelet-derived growth factor (PDGF) and evidence for agonist activity of monomeric PDGF, *J. Biol. Chem.* **267:**11260–11266.

Ansel, J. C., Tiesman, J. P., Olerud, J. E., Krueger, J. G., Krane, J. F., Tara, D. C., Shipley, G. D., Gilbertson, D., Usui, M. L., and Hart, C. E., 1993, Human keratinocytes are a major source of cutaneous platelet-derived growth factor, *J. Clin. Invest.* **92:**671–678.

Antoniades, H. N., Scher, C. D., and Stiles, C. D., 1979, Purification of human platelet-derived growth factor, *Proc. Natl. Acad. Sci. USA* **76:**1809–1812.

Antoniades, H. N., Bravo, M. A., Avila, R. E., Galanopoulos, T., Neville-Golden, J., Maxwell, M., and Selman, M., 1990, Platelet-derived growth factor in idiopathic pulmonary fibrosis, *J. Clin. Invest.* **86:**1055–1064.

Antoniades, H. N., Galanopoulos, T., Neville-Golden, J., Kiritsy, C. P., and Lynch, S. E., 1991, Injury induces *in vivo* expression of platelet-derived growth factor (PDGF) and PDGF receptor mRNAs in skin epithelial cells and PDGF mRNA in connective tissue fibroblasts, *Proc. Natl. Acad. Sci. USA* **88:**565–569.

Antoniades, H. N., Galanopoulos, T., Neville-Golden, J., and O'Hara, C. J., 1992, Malignant epithelial cells in primary human lung carcinomas coexpress *in vivo* platelet-derived growth factor (PDGF) and PDGF receptor mRNAs and their protein products, *Proc. Natl. Acad. Sci. USA* **89:**3942–3946.

Arvidsson, A.-K., Rupp, E., Nånberg, E., Downward, J., Rönnstrand, L., Wennström, S., Schlessinger, J., Heldin, C.-H., and Claesson-Welsh, L., 1994, Tyr716 in the PDGF β-receptor kinase insert is involved in GRB2-binding and Ras activation, *Mol. Cell. Biol.* **14:**6715–6726.

Bar, R. S., Boes, M., Booth, B. A., Dake, B. L., Henley, S., and Hart, M. N., 1989, The effects of platelet-derived growth factor in cultured microvessel endothelial cells, *Endocrinology* **124:**1841–1848.

Barres, B. A., Hart, I. K., Coles, H. S. R., Burne, J. F., Voyvodic, J. T., Richardson, W. D., and Raff, M. C., 1992, Cell death and control of cell survival in the oligodendrocyte lineage, *Cell* **70:**31–46.

Bauer, E. A., Cooper, T. W., Huang, J. S., Altman, J., and Deuel, T. F., 1985, Stimulation of *in vitro* human skin collagenase expression by platelet-derived growth factor, *Proc. Natl. Acad. Sci. USA* **82:**4132–4136.

Betsholtz, C., Johnsson, A., Heldin, C.-H., and Westermark, B., 1986, Efficient reversion of simian sarcoma virus-transformation and inhibition of growth factor-induced mitogenesis by suramin, *Proc. Natl. Acad. Sci. USA* **83:**6440–6444.

Blatti, S. P., Foster, D. N., Ranganathan, G., Moses, H. L., and Getz, M. J., 1988, Induction of fibronectin gene transcription and mRNA is a primary response to growth-factor stimulation of AKR-2B cells, *Proc. Natl. Acad. Sci. USA* **85:**1119–1123.

Blumer, K. J., and Johnson, G. L., 1994, Diversity in function and regulation of MAP kinase pathways, *Trends Biochem. Sci.* **19:**236–240.

Bögler, O., Wren, D., Barnett, S. C., Land, H., and Noble, M., 1990, Cooperation between two growth factors promotes extended self-renewal and inhibits differentiation of oligodendrocyte-type-2 astrocyte (O-2A) progenitor cells, *Proc. Natl. Acad. Sci. USA* **87:**6368–6372.

Bonner, J. C., Osornio-Vargas, A. R., Badgett, A., and Brody, A. R., 1991, Differential proliferation of rat lung fibroblasts induced by the platelet-derived growth factor-AA, -AB, and -BB isoforms secreted by rat alveolar macrophages, *Am. J. Respir. Cell Mol. Biol.* **5:**539–547.

Bonner, J. C., Goodell, A. L., Coin, P. G., and Brody, A. R., 1993, Chrysotile asbestos up-regulates gene expression and production of α-receptors for platelet-derived growth factor (PDGF-AA) on rat lung fibroblasts, *J. Clin. Invest.* **92:**425–430.

Bonthron, D. T., Morton, C. C., Orkin, S. H., and Collins, T., 1988, Platelet-derived growth factor A chain: Gene structure, chromosomal location, and basis for alternative mRNA splicing, *Proc. Natl. Acad. Sci. USA* **85:**1492–1496.

Border, W. A., Okuda, S., Languino, L. R., Sporn, M. B., and Ruoslahti, E., 1990, Suppression of experimental glomerulonephritis by antiserum against transforming growth factor β1, *Nature* **346:**371–374.

Bryckaert, M. C., Rendu, F., Tobelem, G., and Wasteson, Å., 1989, Collagen-induced binding to human platelets of platelet-derived growth factor leading to inhibition of P43 and P20 phosphorylation, *J. Biol. Chem.* **264:**4336–4341.

Bryckaert, M. C., Eldor, A., Fontenay, M., Gazit, A., Osherov, N., Gilon, C., Levitzki, A., and Tobelem, G., 1992, Inhibition of platelet-derived growth factor-induced mitogenesis and tyrosine kinase activity in cultured bone marrow fibroblasts by tyrphostins, *Exp. Cell Res.* **199:**255–261.

Cannigia, I., and Post, M., 1992, Differential effect of platelet-derived growth factor on glycosaminoglycan synthesis by fetal rat lung cells, *Am. J. Physiol.* **263:**L495–L500.

Caniggia, I., Liu, J., Han, R., Buch, S., Funa, K., Tanswell, K., and Post, M., 1993, Fetal lung epithelial cells express receptors for platelet-derived growth factor, *Am. J. Res. Cell Mol. Biol.* **9:**54–63.

Cantley, L. C., Auger, K. R., Carpenter, C., Duckworth, B., Graziani, A., Kapeller, R., and Soltoff, S., 1991, Oncogenes and signal transduction, *Cell* **64:**281–302.

Castro-Malaspina, H., Rabellino, E., Yen, A., Nackman, R., and Moore, M., 1981, Human megakaryocyte stimulation of proliferation of bone marrow fibroblasts, *Blood* **4:**781–787.

Chaudhry, A., Papanicolaou, V., Öberg, K., Heldin, C.-H., and Funa, K., 1992, Expression of platelet-derived growth factor and its receptors in neuroendocrine tumors of the digestive system, *Cancer Res.* **52:**1006–1012.

Chung, C. K., and Antoniades, H. N., 1992, Expression of c-*sis*/platelet-derived growth factor B, insulin-like growth factor I, and transforming growth factor α messenger RNAs and their respective receptor messenger RNAs in primary human gastric carcinomas: *In vivo* studies with *in situ* hybridization and immunocytochemistry, *Cancer Res.* **52:**3453–3459.

Claesson-Welsh, L., 1994, Signal transduction by the PDGF receptors, *Prog. Growth Factor Res.* **5:**37–54.

Clark, R. A. F., Folkvord, J. M., Hart, C. E., Murray, M. J., and McPherson, J. M., 1989, Platelet isoforms of platelet-derived growth factor stimulate fibroblasts to contract collagen matrices, *J. Clin. Invest.* **84:**1036–1040.

Clements, J. M., Bawden, L. J., Bloxidge, R. E., Catlin, G., Cook, A. L., Craig, S., Drummond, A. H., Edwards, R. M., Fallon, A., Green, D. R., Hellewell, P. G., Kirwin, P. M., Nayee, P. D., Richardson, S. J., Brown, D., Chahwala, S. B., Snarey, M., and Winslow, D., 1991, Two PDGF-B chain residues, arginine 27 and isoleucine 30, mediate receptor binding and activation, *EMBO J.* **10:**4113–4120.

Deuel, T. F., Huang, J. S., Proffitt, R. T., Baenziger, J. U., Chang, D., and Kennedy, B. B., 1981, Human platelet-derived growth factor: Purification and resolution into two active protein fractions, *J. Biol. Chem.* **256:**8896–8899.

Deuel, T. F., Senior, R. M., Huang, J. S., and Griffin, G. L., 1982, Chemotaxis of monocytes and neutrophils to platelet-derived growth factor, *J. Clin. Invest.* **69:**1046–1049.

Deuel, T. F., Kawahara, R. S., Mustoe, T. A., and Pierce, G. F., 1991, Growth factors and wound healing: Platelet-derived growth factor as a model cytokine, *Annu. Rev. Med.* **42:**567–584.

DiCorleto, P. E., and Bowen-Pope, D. F., 1983, Cultured endothelial cells produce a platelet-derived growth factor-like protein, *Proc. Natl. Acad. Sci. USA* **80:**1919–1923.

Doolittle, R. F., Hunkapiller, M. W., Hood, L. E., Devare, S. G., Robbins, K. C., Aaronson, S. A., and Antoniades, H. N., 1983, Simian sarcoma virus *onc* gene, v-*sis,* is derived from the gene (or genes) encoding a platelet-derived growth factor, *Science* **221:**275–277.

Duan, D.-S. R., Pazin, M. J., Fretto, L. J., and Williams, L. T., 1991, A functional soluble extracellular region of the platelet-derived growth factor (PDGF) β-receptor antagonizes PDGF-stimulated responses, *J. Biol. Chem.* **266:**413–418.

Eccleston, P. A., Collarini, E. J., Jessen, K. R., Mirsky, R., and Richardson, W. D., 1990, Schwann cells secrete a PDGF-like factor: Evidence for an autocrine growth mechanism involving PDGF, *Eur. J. Neurosci.* **2:**985–992.

Engström, U., Engström, Å., Ernlund, A., Westermark, B., and Heldin, C.-H., 1992, Identification of a peptide antagonist for platelet-derived growth factor, *J. Biol. Chem.* **267:**16581–16587.

Eriksson, A., Siegbahn, A., Westermark, B., Heldin, C.-H., and Claesson-Welsh, L., 1992, PDGF α- and β-receptors activate unique and common signal transduction pathways, *EMBO J.* **11:**543–550.

Ferns, G. A. A., Raines, E. W., Sprugel, K. H., Motani, A. S., Reidy, M. A., and Ross, R., 1991, Inhibition of neointimal smooth muscle accumulation after angioplasty by an antibody to PDGF, *Science* **253:**1129–1132.

Fleming, T. P., Matsui, T., Heidaran, M. A., Molloy, C. J., Artrip, J., and Aaronson, S. A., 1992, Demonstration of an activated platelet-derived growth factor autocrine pathway and its role in human tumor cell proliferation in vitro, *Oncogene* **7:**1355–1359.

Floege, J., Topley, N., Hoppe, J., Barrett, T. B., and Resch, K., 1991, Mitogenic effect of platelet-derived growth factor in human glomerular mesangial cells: Modulation and/or suppression by inflammatory cytokines, *Clin. Exp. Immunol.* **86:**334–341.

Floege, J., Eng, E., Young, B. A., Alpers, C. E., Barrett, T. B., Bowen-Pope, D. F., and Johnson, R. J., 1993, Infusion of platelet-derived growth factor or basic fibroblast growth factor induces selective glomerular mesangial cell proliferation and matrix accumulation in rats, *J. Clin. Invest.* **92:**2952–2962.

Folkman, J., and Klagsbrun, M., 1987, Angiogenic factors, *Science* **235:**442–447.

Forsberg, K., Valyi-Nagy, I., Heldin, C.-H., Herlyn, M., and Westermark, B., 1993, Platelet-derived growth factor (PDGF) in oncogenesis: Development of a vascular connective tissue stroma in xenotransplanted human melanoma producing PDGF-BB, *Proc. Natl. Acad. Sci. USA* **90:**393–397.

Gay, S., Jones, R. E. J., Huang, G. Q., and Gay, R. E., 1989, Immunohistologic demonstration of platelet-derived growth factor (PDGF) and *sis*-oncogene expression in scleroderma, *J. Invest. Dermatol.* **92:**301–303.

Gersuk, G. M., Carmel, R., and Pattengale, P. K., 1989, Platelet-derived growth factor concentrations in platelet-poor plasma and urine from patients with myeloproliferative disorders, *Blood* **74:**2330–2334.

Gesualdo, L., Pinzani, M., Floriano, J. J., Hassan, M. O., Nagy, N. U., Schena, F. P., Emancipator, S. N., and Abboud, H. E., 1991, Platelet-derived growth factor expression in mesangial proliferative glomerulonephritis, *Lab. Invest.* **65:**160–167.

Golub, T. R., Barker, G. F., Lovett, M., and Gilliland, D. G., 1994, Fusion of PDGF receptor β to a novel *ets*-like gene, *tel,* in chronic myelomonocytic leukemia with t(5;12) chromosomal translocation, *Cell* **77:**307–316.

Goustin, A. S., Betsholtz, C., Pfeifer-Ohlsson, S., Persson, H., Rydnert, J., Bywater, M., Holmgren, G., Heldin, C.-H., Westermark, B., and Ohlsson, R., 1985, Coexpression of the *sis* and *myc* proto-oncogenes in developing human placenta suggests autocrine control of trophoblast growth, *Cell* **41**:301–312.

Greenhalg, D. G., Sprugel, K. H., Murray, M. J., and Ross, R., 1990, PDGF and FGF stimulate wound healing in the genetically diabetic mouse, *Am. J. Pathol.* **136**:1235–1246.

Groopman, J. E., 1980, The pathogenesis of myelofibrosis in myeloproliferative disorders, *Ann. Intern. Med.* **92**:857–858.

Grotendorst, G., Seppä, H. E. J., Kleiman, H. K., and Martin, G. R., 1981, Attachment of smooth muscle cells to collagen and their migration toward platelet-derived growth factor, *Proc. Natl. Acad. Sci. USA* **78**:3669–3672.

Grotendorst, G. R., Martin, G. R., Pancev, D., Sodek, J., and Harvey, A. K., 1985, Stimulation of granulation tissue formation by platelet-derived growth factor in normal and diabetic rats, *J. Clin. Invest.* **76**:2323–2339.

Gullberg, D., Tingström, A., Thuresson, A.-C., Olsson, L., Terracio, L., Borg, T. K., and Rubin, K., 1990, β_1 Integrin-mediated collagen gel contraction is stimulated by PDGF, *Exp. Cell Res.* **186**:264–272.

Hammacher, A., Hellman, U., Johnsson, A., Östman, A., Gunnarson, K., Westermark, B., Wasteson, Å., and Heldin, C.-H., 1988, A major part of platelet-derived growth factor purified from human platelets is a heterodimer of one A and one B chain, *J. Biol. Chem.* **263**:16493–16498.

Han, R. N., Buch, S., Freeman, B. A., Post, M., and Tanswell, A. K., 1992, Platelet-derived growth factor and growth-related genes in rat lung. II. Effect of exposure to 85% O_2, *Am. J. Physiol.* **262**:L140–L146.

Haniu, M., Rohde, M. F., and Kenney, W. C., 1993, Disulfide bonds in recombinant human platelet-derived growth factor BB dimer: Characterization of intermolecular and intramolecular disulfide linkages, *Biochemistry* **32**:2431–2437.

Harlan, J. M., Thompson, P. J., Ross, R., and Bowen-Pope, D. F., 1986, α-Thrombin induces release of PDGF-like molecule(s) by cultured human endothelial cells, *J. Cell Biol.* **103**:1129–1133.

Hart, C. E., Bailey, M., Curtis, D. A., Osborn, S., Raines, E., Ross, R., and Forstrom, J. W., 1990, Purification of PDGF-AB and PDGF-BB from human platelet extracts and identification of all three PDGF dimers in human platelets, *Biochemistry* **29**:166–172.

Heldin, C.-H., and Westermark, B., 1990, Platelet-derived growth factor: Mechanism of action and possible *in vivo* function, *Cell Regul.* **1**:555–566.

Heldin, C.-H., and Westermark, B., 1991, Platelet-derived growth factor and autocrine mechanisms of oncogenic processes, *CRC Crit. Rev. Oncog.* **2**:109–124.

Heldin, C.-H., Westermark, B., and Wasteson, Å., 1979, Platelet-derived growth factor: Purification and partial characterization, *Proc. Natl. Acad. Sci. USA* **76**:3722–3726.

Heldin, N.-E., Gustavsson, B., Claesson-Welsh, L., Hammacher, A., Mark, J., Heldin, C.-H., and Westermark, B., 1988, Aberrant expression of receptors for platelet-derived growth factor in an anaplastic thyroid carcinoma cell line, *Proc. Natl. Acad. Sci. USA* **85**:9302–9306.

Heldin, N.-E., Cvejic, D., Smeds, S., and Westermark, B., 1991, Coexpression of functionally active receptors for thyrotropin and platelet-derived growth factor in human thyroid carcinoma cells, *Endocrinology* **129**:2187–2193.

Heldin, P., Laurent, T. C., and Heldin, C.-H., 1989, Effect of growth factors on hyaluronan synthesis in cultured human fibroblasts, *Biochem. J.* **258**:919–922.

Heldin, P., Pertoft, H., Nordlinder, H., Heldin, C.-H., and Laurent, T. C., 1991, Differential expression of platelet-derived growth factor α- and β-receptors on fat-storing cells and endothelial cells of rat liver, *Exp. Cell Res.* **193**:364–369.

Hermanson, M., Nistér, M., Betsholtz, C., Heldin, C.-H., Westermark, B., and Funa, K., 1988, Endothelial cell hyperplasia in human glioblastoma: Coexpression of mRNA for platelet-derived growth factor (PDGF) B chain and PDGF receptor suggests autocrine growth stimulation, *Proc. Natl. Acad. Sci. USA* **85**:7748–7752.

Hermanson, M., Funa, K., Hartman, M., Claesson-Welsh, L., Heldin, C.-H., Westermark, B., and Nistér, M., 1992, Platelet-derived growth factor and its receptors in human glioma tissue: Expression of messenger RNA and protein suggests the presence of autocrine and paracrine loops, *Cancer Res.* **52**:3213–3219.

Hertz, M. I., Henke, C. A., Nakhleh, R. E., Harmon, K. R., Marinelli, W. A., Fox, J. M. K., Kubo, S. H.,

Shumway, S. J., Bolman III, R. M., and Bitterman, P. B., 1992, Obliterative bronchiolitis after lung transplantation: A fibroproliferative disorder associated with platelet-derived growth factor, *Proc. Natl. Acad. Sci. USA* **89:**10385–10389.

Hill, E., Turner-Beatty, M., Grotewiel, M., Fosha-Thomas, S., Cox, C., Turman, C., Drees, D., Baird, L., Maratea, D., Tucker, R., and Counts, D., 1991, The effect of PDGF on the healing of full thickness wounds in hairless guinea pigs, *Comp. Biochem. Physiol.* **100A:**365–370.

Huang, J. S., Huang, S. S., and Deuel, T. F., 1984, Transforming protein of simian sarcoma virus stimulates autocrine growth of SSV-transformed cells through PDGF cell-surface receptors, *Cell* **39:**79–87.

Iida, H., Seifert, R., Alpers, C. E., Gronwald, R. G. K., Phillips, P. E., Pritzl, P., Gordon, K., Gown, A. M., Ross, R., Bowen-Pope, D. F., and Johnson, R. J., 1991, Platelet-derived growth factor (PDGF) and PDGF receptor are induced in mesangial proliferative nephritis in the rat, *Proc. Natl. Acad. Sci. USA* **88:**6560–6564.

Isaka, Y., Fujiwara, Y., Ueda, N., Kaneda, Y., Kamada, T., and Imai, E., 1993, Glomerulosclerosis induced by *in vivo* transfection of transforming growth factor-β or platelet-derived growth factor gene into the rat kidney, *J. Clin. Invest.* **92:**2597–2601.

Jaumann, M., Hoppe, V., Tatje, D., Eichner, W., and Hoppe, J., 1991, On the structure of platelet-derived growth factor AA: C-terminal processing, epitopes, and characterization of cysteine residues, *Biochemistry* **30:**3303–3309.

Jawien, A., Bowen-Pope, D. F., Lindner, V., Schwartz, S. M., and Clowes, A. W., 1992, Platelet-derived growth factor promotes smooth muscle migration and intimal thickening in a rat model of balloon angioplasty, *J. Clin. Invest.* **89:**507–511.

Johnson, R. J., Iida, H., Alpers, C. E., Majesky, M. W., Schwartz, S. M., Pritzl, P., Gordon, K., and Gown, A. M., 1991, Expression of smooth muscle cell phenotype by rat mesangial cells in immune complex nephritis, *J. Clin. Invest.* **87:**847–858.

Johnson, R. J., Raines, E. W., Floege, J., Yoshimura, A., Pritzl, P., Alpers, C., and Ross, R., 1992, Inhibition of mesangial cell proliferation and matrix expansion in glomerulonephritis in the rat by antibody to platelet-derived growth factor, *J. Exp. Med.* **175:**1413–1416.

Johnson, R. J., Floege, J., Couser, W. G., and Alpers, C. E., 1993, Role of platelet-derived growth factor in glomerular disease, *J. Am. Soc. Nephrol.* **4:**119–128.

Johnsson, A., Heldin, C.-H., Wasteson, Å., Westermark, B., Deuel, T. F., Huang, J. S., Seeburg, P. H., Gray, A., Ullrich, A., Scrace, G., Stroobant, P., and Waterfield, M. D., 1984, The c-*sis* gene encodes a precursor of the B chain of platelet-derived growth factor, *EMBO J.* **3:**921–928.

Johnsson, A., Betsholtz, C., Heldin, C.-H., and Westermark, B., 1985, Antibodies against platelet-derived growth factor inhibit acute transformation by simian sarcoma virus, *Nature* **317:**438–440.

Katayose, D., Ohe, M., Yamauchi, K., Ogata, M., Shirato, K., Fujita, H., Shibahara, S., and Takishima, T., 1993, Increased expression of PDGF A- and B-chain genes in rat lungs with hypoxic pulmonary hypertension, *Am. J. Physiol.* **264:**L100–L106.

Katoh, O., Kimura, A., and Kuramoto, A., 1988, Platelet-derived growth factor is decreased in patients with myeloproliferative disorders, *Am. J. Hematol.* **27:**276–280.

Katz, M. H., Alvarez, A. F., Kirsner, R. S., Eaglstein, W. H., and Falanga, V., 1991, Human wound fluid from acute wounds stimulates fibroblast and endothelial cell growth, *J. Am. Acad. Dermatol.* **25:**1054–1058.

Kenney, W. C., Haniu, M., Herman, A. C., Arakawa, T., Costigan, V. J., Lary, J., Yphantis, D. A., and Thomason, A. R., 1994, Formation of mitogenically active PDGF-B dimer does not require interchain disulfide bonds, *J. Biol. Chem.* **269:**12351–12359.

Klareskog, L., Gustafsson, R., Scheynius, A., and Hällgren, R., 1990, Increased expression of platelet-derived growth factor type B receptors in the skin of patients with systemic sclerosis, *Arthritis Rheum.* **33:**1534–1541.

Koch, C. A., Anderson, D., Moran, M. F., Ellis, C., and Pawson, T., 1991, SH2 and SH3 domains: Elements that control interactions of cytoplasmic signaling proteins, *Science* **252:**668–674.

Kohler, N., and Lipton, A., 1974, Platelet as a source of fibroblast growth-promoting activity, *Exp. Cell Res.* **87:**297–301.

LaRochelle, W. J., May-Siroff, M., Robbins, K. C., and Aaronson, S. A., 1991, A novel mechanism regulating growth factor association with the cell surface: Identification of a PDGF retention domain, *Genes Dev.* **5:**1191–1199.

LaRochelle, W. J., Pierce, J. H., May-Siroff, M., Giese, N., and Aaronson, S. A., 1992, Five PDGF B amino acid substitutions convert PDGF A to a PDGF B-like transforming molecule, *J. Biol. Chem.* **267:** 17074–17077.

Larson, R. C., Ignotz, G. G., and Currie, W. B., 1992, Platelet derived growth factor (PDGF) stimulates development of bovine embryos during the fourth cell cycle, *Development* **115:**821–826.

Lepistö, J., Laato, M., Niinikoski, J., Lundberg, C., Gerdin, B., and Heldin, C.-H., 1992, Effects of homodimeric isoforms of platelet-derived growth factor (PDGF-AA and PDGF-BB) on wound healing in rat, *J. Surg. Res.* **53:**596–601.

Lindmark, G., Sundberg, C., Glimelius, B., Påhlman, L., Rubin, K., and Gerdin, B., 1993, Stromal expression of platelet-derived growth factor β-receptor and platelet-derived growth factor B-chain in colorectal cancer, *Lab. Invest.* **69:**682–689.

Lynch, S. E., Nixon, J. C., Colvin, R. B., and Antoniades, H. N., 1987, Role of platelet-derived growth factor in wound healing: Synergistic effects with other growth factors, *Proc. Natl. Acad. Sci. USA* **84:**7696–7700.

Lynch, S. E., Colvin, R. B., and Antoniades, H. N., 1989, Growth factors in wound healing. Single and synergistic effects on partial thickness porcine skin wounds, *J. Clin. Invest.* **84:**640–646.

Lynch, S. E., de Castilla, G. R., Williams, R. C., Kiritsy, C. P., Howell, T. H., Reddy, M. S., and Antoniades, H. N., 1991, The effects of short-term application of a combination of platelet-derived and insulin-like growth factors on periodontal wound healing, *J. Periodontol.* **62:**458–467.

Marden, L. J., Fan, R. S. P., Pierce, G. F., Reddi, A. H., and Hollinger, J. O., 1993, Platelet-derived growth factor inhibits bone regeneration induced by osteogenin, a bone morphogenetic protein, in rat craniotomy defects, *J. Clin. Invest.* **92:**2897–2905.

Martinet, Y., Bitterman, P. B., Mornex, J.-F., Grotendorst, G. R., Martin, G. R., and Crystal, R. G., 1986, Activated human monocytes express the c-*sis* proto-oncogene and release a mediator showing PDGF-like activity, *Nature* **319:**158–160.

Matsuoka, J., and Grotendorst, G. R., 1989, Two peptides related to platelet-derived growth factor are present in human wound fluid, *Proc. Natl. Acad. Sci. USA* **86:**4416–4420.

Maxwell, M., Naber, S. P., Wolfe, H. J., Galanopoulos, T., Hedley-Whyte, E. T., Black, P. M., and Antoniades, H. N., 1990, Coexpression of platelet-derived growth factor (PDGF) and PDGF-receptor genes by primary human astrocytomas may contribute to their development and maintenance, *J. Clin. Invest.* **86:**131–140.

Mercola, M., Melton, D. A., and Stiles, C. D., 1988, Platelet-derived growth factor A chain is maternally encoded in *Xenopus* embryos, *Science* **241:**1223–1225.

Mercola, M., Wang, C., Kelly, J., Brownlee, C., Jackson-Grusby, L., Stiles, C., and Bowen-Pope, D., 1990, Selective expression of PDGF A and its receptor during early mouse embryogenesis, *Dev. Biol.* **138:** 114–122.

Morrison-Graham, K., Schatteman, G. C., Bork, T., Bowen-Pope, D. F., and Weston, J. A., 1992, A PDGF receptor mutant in the mouse (*Patch*) perturbs the development of a non-neuronal subset of neural crest-derived cells, *Development* **115:**133–142.

Mustoe, T. A., Purdy, J., Gramates, P., Deuel, T. F., Thomason, A., and Pierce, G. F., 1989, Reversal of impaired wound healing in irradiated rats by platelet derived growth factor-BB, *Am. J. Surg.* **158:** 345–350.

Mustoe, T. A., Pierce, G. F., Morishima, C., and Deuel, T. F., 1991, Growth factor-induced acceleration of tissue repair through direct and inductive activities in a rabbit dermal ulcer model, *J. Clin. Invest.* **87:**694–703.

Nilsson, J., Sjölund, M., Palmberg, L., Thyberg, J., and Heldin, C.-H., 1985, Arterial smooth muscle cells in primary culture produce a platelet-derived growth factor-like protein, *Proc. Natl. Acad. Sci. USA* **82:**4418–4422.

Nistér, M., Libermann, T. A., Betsholtz, C., Pettersson, M., Claesson-Welsh, L., Heldin, C.-H., Schlessinger, J., and Westermark, B., 1988, Expression of messenger RNAs for platelet-derived growth factor and transforming growth factor-α and their receptors in human malignant glioma cell lines, *Cancer Res.* **48:**3910–3918.

Noble, M., Murray, K., Stroobant, P., Waterfield, M. D., and Riddle, P., 1988, Platelet-derived growth factor promotes division and motility and inhibits premature differentiation of the oligodendrocyte/type-2 astrocyte progenitor cell, *Nature* **333:**560–562.

Oefner, C., D'Arcy, A., Winkler, F. K., Eggimann, B., and Hosang, M., 1992, Crystal structure of human platelet-derived growth factor BB, *EMBO J.* **11:**3921–3926.

Orr-Urtreger, A., Bedford, M. T., Do, M.-S., Eisenbach, L., and Lonai, P., 1992, Developmental expression of the α receptor for platelet-derived growth factor, which is deleted in the embryonic lethal *Patch* mutation, *Development* **115:**289–303.

Östman, A., Andersson, M., Betsholtz, C., Westermark, B., and Heldin, C.-H., 1991a, Identification of a cell retention signal in the B-chain of platelet-derived growth factor and in the long splice version of the A-chain, *Cell Regul.* **2:**503–512.

Östman, A., Andersson, M., Hellman, U., and Heldin, C.-H., 1991b, Identification of three amino acids in the platelet-derived growth factor (PDGF) B-chain that are important for binding to the PDGF β-receptor, *J. Biol. Chem.* **266:**10073–10077.

Östman, A., Thyberg, J., Westermark, B., and Heldin, C.-H., 1992, PDGF-AA and PDGF-BB biosynthesis: Proprotein processing in the Golgi complex and lysosomal degradation of PDGF-BB retained intracellularly, *J. Cell Biol.* **118:**509–519.

Östman, A., Andersson, M., Bäckström, G., and Heldin, C.-H., 1993, Assignment of intrachain disulfide bonds in platelet-derived growth factor B-chain, *J. Biol. Chem.* **268:**13372–13377.

Palmieri, S. L., Payne, J., Stiles, C. D., Biggers, J. D., and Mercola, M., 1992, Expression of mouse PDGF-A and PDGF alpha-receptor genes during pre- and postimplantation development: Evidence for a developmental shift from an autocrine to a paracrine mode of action, *Mech. Dev.* **39:**181–191.

Paulsson, Y., Hammacher, A., Heldin, C.-H., and Westermark, B., 1987, Possible positive autocrine feedback in the prereplicative phase of human fibroblasts, *Nature* **328:**715–717.

Peterson, T. C., and Isbrucker, R. A., 1992, Fibroproliferation in liver disease: Role of monocyte factors, *Hepatology* **15:**191–197.

Pierce, G. F., Mustoe, T. A., Senior, R. M., Reed, J., Griffin, G. L., Thomason, A., and Deuel, T. F., 1988, *In vivo* incisional wound healing augmented by platelet-derived growth factor and recombinant c-*sis* gene homodimeric proteins, *J. Exp. Med.* **167:**974–987.

Pierce, G. F., Mustoe, T. A., Lingelbach, J., Masakowski, V. R., Griffin, G. L., Senior, R. M., and Deuel, T. F., 1989, Platelet-derived growth factor and transforming growth factor-β enhance tissue repair activities by unique mechanisms, *J. Cell Biol.* **109:**429–440.

Pierce, G. F., Berg, J. V., Rudolph, R., Tarpley, J., and Mustoe, T. A., 1991, Platelet-derived growth factor-BB and transforming growth factor beta1 selectively modulate glycosaminoglycans, collagen, and myofibroblasts in excisional wounds, *Am. J. Pathol.* **138:**629–646.

Pierce, G. F., Tarpley, J. E., Yanagihara, D., Mustoe, T. A., Fox, G. M., and Thomason, A., 1992, Platelet-derived growth factor (BB homodimer), transforming growth factor-β1, and basic fibroblast growth factor in dermal wound healing. Neovessel and matrix formation and cessation of repair, *Am. J. Pathol.* **140:**1375–1388.

Pinzani, M., Gesualdo, L., Sabbah, G. M., and Abboud, H. E., 1989, Effects of platelet-derived growth factor and other polypeptide mitogens on DNA synthesis and growth of cultured rat liver fat-storing cells, *J. Clin. Invest.* **84:**1786–1793.

Pinzani, M., Knauss, T. C., Pierce, G. F., Hsieh, P., Kenney, W., Dubyak, G. R., and Abboud, H. E., 1991, Mitogenic signals for platelet-derived growth factor isoforms in liver fat-storing cells, *Am. J. Physiol.* **260:**C485–C491.

Plate, K. H., Breier, G., Farrell, C. L., and Risau, W., 1992, Platelet-derived growth factor receptor-β is induced during tumor development and up-regulated during tumor progression in endothelial cells in human gliomas, *Lab. Invest.* **67:**529–534.

Powell, P. P., Wang, C. C., and Jones, R., 1992, Differential regulation of the genes encoding platelet-derived growth factor receptor and its ligand in rat lung during microvascular and alveolar wall remodeling in hyperoxia, *Am. J. Respir. Cell Mol. Biol.* **7:**278–285.

Raines, E. W., and Ross, R., 1982, Platelet-derived growth factor. I. High yield purification and evidence for multiple forms, *J. Biol. Chem.* **257:**5154–5160.

Raines, E. W., and Ross, R., 1992, Compartmentalization of PDGF on extracellular binding sites dependent on exon-6-encoded sequences, *J. Cell Biol.* **116:**533–543.

Raines, E. W., Bowen-Pope, D. F., and Ross, R., 1990, Platelet-derived growth factor, in: *Handbook of Experimental Pharmacology. Peptide Growth Factors and Their Receptors,* Vol. 95, Part I (M. B. Sporn and A. B. Roberts, eds.), pp. 173–262, Springer-Verlag, Heidelberg.

Reuterdahl, C., Tingström, A., Terracio, L., Funa, K., Heldin, C.-H., and Rubin, K., 1991, Characterization of platelet-derived growth factor β-receptor expressing cells in the vasculature of human rheumatoid synovium, *Lab. Invest.* **64:**321–329.

Reuterdahl, C., Sundberg, C., Rubin, K., Funa, K., and Gerdin, B., 1993, Tissue localization of β receptors for platelet-derived growth factor and platelet-derived growth factor B chain during wound repair in humans, *J. Clin. Invest.* **91:**2065–2075.

Richardson, W. D., Pringle, N., Mosley, M. J., Westermark, B., and Dubois-Dalcq, M., 1988, A role for platelet-derived growth factor in normal gliogenesis in the central nervous system, *Cell* **53:**309–319.

Risau, W., Drexler, H., Mironov, V., Smits, A., Siegbahn, A., Funa, K., and Heldin, C.-H., 1992, Platelet-derived growth factor is angiogenic *in vivo, Growth Factors* **7:**261–266.

Robson, M. C., Phillips, L. G., Thomason, A., Robson, L. E., and Pierce, G. F., 1992, Platelet-derived growth factor BB for the treatment of chronic pressure ulcers, *Lancet* **339:**23–25.

Rorsman, F., Bywater, M., Knott, T. J., Scott, J., and Betsholtz, C., 1988, Structural characterization of the human platelet-derived growth factor A-chain cDNA and gene: Alternative exon usage predicts two different precursor proteins, *Mol. Cell. Biol.* **8:**571–577.

Ross, R., 1993, The pathogenesis of atherosclerosis: A perspective for the 1990s, *Nature* **362:**801–809.

Ross, R., Glomset, J., Kariya, B., and Harker, L., 1974, A platelet-dependent serum factor that stimulates the proliferation of arterial smooth muscle cells *in vitro, Proc. Natl. Acad. Sci. USA* **71:**1207–1210.

Ross, R., Masuda, J., Raines, E. W., Gown, A. M., Katsuda, S., Sasahara, M., Malden, L. T., Masuko, H., and Sato, H., 1990, Localization of PDGF-B protein in macrophages in all phases of atherogenesis, *Science* **248:**1009–1012.

Rubin, K., Tingström, A., Hansson, G. K., Larsson, E., Rönnstrand, L., Klareskog, L., Claesson-Welsh, L., Heldin, C.-H., Fellström, B., and Terracio, L., 1988a, Induction of B-type receptors for platelet-derived growth factor in vascular inflammation: Possible implications for development of vascular proliferative lesions, *Lancet* **1:**1353–1356.

Rubin, K., Terracio, L., Rönnstrand, L., Heldin, C.-H., and Klareskog, L., 1988b, Expression of platelet-derived growth factor receptors is induced on connective tissue cells during chronic synovial inflammation, *Scand. J. Immunol.* **27:**285–294.

Rupp, E., Siegbahn, A., Rönnstrand, L., Wernstedt, C., Claesson-Welsh, L., and Heldin, C.-H., 1994, A unique autophosphorylation site in the PDGF α-receptor from a heterodimeric receptor complex, *Eur. J. Biochem.* **225:**29–41.

Rutherford, R. B., Niekrash, C. E., Kennedy, J. E., and Charette, M. F., 1992, Platelet-derived and insulin-like growth factors stimulate regeneration of periodontal attachment in monkeys, *J. Periodontal Res.* **27:**285–290.

Rutherford, R. B., Ryan, M. E., Kennedy, J. E., Tucker, M. M., and Charette, M. F., 1993, Platelet-derived growth factor and dexamethasone combined with a collagen matrix induce regeneration of the periodontium in monkeys, *J. Clin. Periodontol.* **20:**537–544.

Sano, H., Engleka, K., Mathern, P., Hla, T., Crofford, L. J., Remmers, E. F., Jelsema, C. L., Goldmuntz, E., Maciag, T., and Wilder, R. L., 1993, Coexpression of phosphotyrosine-containing proteins, platelet-derived growth factor-B, and fibroblast growth factor-1 *in situ* in synovial tissues of patients with rheumatoid arthritis and Lewis rats with adjuvant or streptococcal cell wall arthritis, *J. Clin. Invest.* **91:**553–565.

Sasahara, M., Fries, J. W. U., Raines, E. W., Gown, A. M., Westrum, L. E., Frosch, M. P., Bonthron, D. T., Ross, R., and Collins, T., 1991, PDGF B-chain in neurons of the central nervous system, posterior pituitary, and in a transgenic model, *Cell* **64:**217–227.

Sato, N., Beitz, J. G., Kato, J., Yamamoto, M., Clark, J. W., Calabresi, P., and Frackelton, Jr., A. R., 1993, Platelet-derived growth factor indirectly stimulates angiogenesis *in vitro, Am. J. Pathol.* **142:**1119–1130.

Schatteman, G. C., Morrison-Graham, K., van Koppen, A., Weston, J. A., and bowen-Pope, D. F., 1992, Regulation and role of PDGF receptor α-subunit expression during embryogenesis, *Development* **115:**123–131.

Schlessinger, J., 1993, How receptor tyrosine kinases activate Ras, *Trends Biochem. Sci.* **18:**273–275.

Seifert, R. A., Schwartz, S. M., and Bowen-Pope, D. F., 1984, Developmentally regulated production of platelet-derived growth factor-like molecules, *Nature* **311:**669–671.

Senior, R. M., Griffin, G. L., Huang, J. S., Walz, D. A., and Deuel, T. F., 1983, Chemotactic activity of platelet alpha granule proteins for fibroblasts, *J. Cell Biol.* **96:**382–385.

Seppä, H. E. J., Grotendorst, G. R., Seppä, S. I., Schiffman, E., and Martin, G. R., 1982, Platelet-derived growth factor is chemotactic for fibroblasts, *J. Cell Biol.* **92:**584–588.

Shimokado, K., Raines, E. W., Madtes, D. K., Barrett, T. B., Benditt, E. P., and Ross, R., 1985, A significant part of macrophage-derived growth factor consists of at least two forms of PDGF, *Cell* **43:**277–286.

Shultz, P. J., DeCorleto, P. E., Silver, B. J., and Abboud, H. E., 1988, Mesangial cells express PDGF mRNAs and proliferate in response to PDGF, *Am. J. Physiol.* **255:**F674–F684.

Siegbahn, A., Hammacher, A., Westermark, B., and Heldin, C.-H., 1990, Differential effects of the various isoforms of platelet-derived growth factor on chemotaxis of fibroblasts, monocytes, and granulocytes, *J. Clin. Invest.* **85:**916–920.

Silver, B. J., Jaffer, F. E., and Abboud, H. E., 1989, Platelet-derived growth factor synthesis in mesangial cells: Induction by multiple peptide mitogens, *Proc. Natl. Acad. Sci. USA* **86:**1056–1060.

Smits, A., Hermanson, M., Nistér, M., Karnushina, I., Heldin, C.-H., Westermark, B., and Funa, K., 1989, Rat brain capillary endothelial cells express functional PDGF B-type receptors, *Growth Factors* **2:**1–8.

Smits, A., Kato, M., Westermark, B., Nistér, M., Heldin, C.-H., and Funa, K., 1991, Neurotrophic activity of platelet-derived growth factor (PDGF): Rat neuronal cells possess functional PDGF β-type receptors and respond to PDGF, *Proc. Natl. Acad. Sci. USA* **88:**8159–8163.

Smits, A., Funa, K., Vassbotn, F. S., Beausang-Linder, M., af Ekenstam, F., Heldin, C.-H., Westermark, B., and Nistér, M., 1992, Expression of platelet-derived growth factor and its receptors in proliferative disorders of fibroblastic origin, *Am. J. Pathol.* **140:**639–648.

Snyder, L. S., Hertz, M. I., Peterson, M. S., Harmon, K. R., Marinelli, W. A., Henke, C. A., Greenheck, J. R., Chen, B., and Bitterman, P. B., 1991, Acute lung injury. Pathogenesis of intraalveolar fibrosis, *J. Clin. Invest.* **88:**663–673.

Soma, Y., Dvonch, V., and Grotendorst, G. R., 1992, Platelet-derived growth factor AA homodimer is the predominant isoform in human platelets and acute human wound fluid, *FASEB J.* **6:**2996–3001.

Sprugel, K. H., McPherson, J. M., Clowes, A. W., and Ross, R., 1987, Effects of growth factors *in vivo*. I. Cell ingrowth into porous subcutaneous chambers, *Am. J. Pathol.* **129:**601–613.

Sterzel, R. B., Schulze-Lohoff, E., and Marx, M., 1993, Cytokines and mesangial cells, *Kidney Int.* **43**(Suppl. 39):S26–S31.

Sundberg, C., Ljungström, M., Lindmark, G., Gerdin, B., and Rubin, K., 1993, Microvascular pericytes express platelet-derived growth factor-β receptors in human healing wounds and colorectal adenocarcinoma, *Am. J. Pathol.* **143:**1377–1388.

Terracio, L., Rönnstrand, L., Tingström, A., Rubin, K., Claesson-Welsh, L., Funa, K., and Heldin, C.-H., 1988, Induction of platelet-derived growth factor receptor expression in smooth muscle cells and fibroblasts upon tissue culturing, *J. Cell Biol.* **107:**1947–1957.

Thibodeaux, J. K., Del Vecchio, R. P., and Hansel, W., 1993, Role of platelet-derived growth factor in development of *in vitro* matured and *in vitro* fertilized bovine embryos, *J. Reprod. Fertil.* **98:**61–66.

Twamley-Stein, G. M., Pepperkok, R., Ansorge, W., and Courtneidge, S. A., 1993, The Src family tyrosine kinases are required for platelet-derived growth factor-mediated signal transduction in NIH 3T3 cells, *Proc. Natl. Acad. Sci. USA* **90:**7696–700.

Valius, M., and Kazlauskas, A., 1993, Phospholipase C-γ1 and phosphatidylinositol 3 kinase are the downstream mediators of the PDGF receptor's mitogenic signal, *Cell* **73:**321–334.

Vassbotn, F. S., Östman, A., Siegbahn, A., Holmsen, H., and Heldin, C.-H., 1992, Neomycin is a platelet-derived growth factor (PDGF) antagonist that allows discrimination of PDGF α- and β-receptor signals in cells expressing both receptor types, *J. Biol. Chem.* **267:**15635–15641.

Vassbotn, F. S., Havnen, O. K., Heldin, C.-H., and Holmsen, H., 1994, Negative feedback regulation of human platelets via autocrine activation of the platelet-derived growth factor α-receptor, *J. Biol. Chem.* **269**:13874–13879.

Vignaud, J.-M., Allam, M., Martinet, N., Pech, M., Plenat, F., and Martinet, Y., 1991, Presence of platelet-derived growth factor in normal and fibrotic lung is specifically associated with interstitial macrophages, while both interstitial macrophages and alveolar epithelial cells express the c-*sis* proto-oncogene, *Am. J. Respir. Cell Mol. Biol.* **5**:531–538.

Walker, L. N., Bowen-Pope, D. F., Ross, R., and Reidy, M. A., 1986, Production of PDGF-like molecules by cultured arterial smooth muscle cells accompanies proliferation after arterial injury, *Proc. Natl. Acad. Sci. USA* **83**:7311–7315.

Walsh, J., Absher, M., and Kelley, J., 1993, Variable expression of platelet-derived growth factor family proteins in acute lung injury, *Am. J. Respir. Cell Mol. Biol.* **9**:637–644.

Waterfield, M. D., Scrace, G. T., Whittle, N., Stroobant, P., Johnsson, A., Wasteson, Å., Westermark, B., Heldin, C.-H., Huang, J. S., and Deuel, T. F., 1983, Platelet-derived growth factor is structurally related to the putative transforming protein p28*sis* of simian sarcoma virus, *Nature* **304**:35–39.

Weinmaster, G., and Lemke, G., 1990, Cell-specific cyclic AMP-mediated induction of the PDGF receptor, *EMBO J.* **9**:915–920.

Wennström, S., Hawkins, P., Cooke, F., Hara, K., Yonezawa, K., Kasuga, M., Jackson, T., Claesson-Welsh, L., and Stephens, L., 1994a, Activation of phosphoinositide 3-kinase is required for PDGF-stimulated membrane ruffling, *Curr. Biol.* **4**:385–393.

Wennström, S., Siegbahn, A., Yokote, K., Arvidsson, A.-K., Heldin, C.-H., Mori, S., and Claesson-Welsh, L., 1994b, Membrane ruffling and chemotaxis transduced by the PDGF β-receptor require the binding site for phosphatidylinositol 3′ kinase, *Oncogene* **9**:651–660.

Westermark, B., and Wasteson, Å., 1976, A platelet factor stimulating human normal glial cells, *Exp. Cell Res.* **98**:170–174.

Westermark, B., Betsholtz, C., Johnsson, A., and Heldin, C.-H., 1987, Acute transformation by simian sarcoma virus is mediated by an externalized PDGF-like growth factor, in: *Viral Carcinogenesis* (N. O. Kjeldgaard and J. Forchhammer, eds.), pp. 445–457, Munksgaard, Copenhagen.

Wilcox, J. N., Smith, K. M., Williams, L. T., Schwartz, S. M., and Gordon, D., 1988, Platelet-derived growth factor mRNA detection in human atherosclerotic plaques by in situ hybridization, *J. Clin. Invest.* **82**:1134–1143.

Williams, L. T., Tremble, P. M., Lavin, M. F., and Sunday, M. E., 1984, Platelet-derived growth factor receptors form a high affinity state in membrane preparations, *J. Biol. Chem.* **259**:5287–5294.

Yeh, H.-J., Ruit, K. G., Wang, Y.-X., Parks, W. C., Snider, W. D., and Deuel, T. F., 1991, PDGF A-chain gene is expressed by mammalian neurons during development and in maturity, *Cell* **64**:209–216.

Yokote, K., Mori, S., Hansen, K., McGlade, J., Pawson, T., Heldin, C.-H., and Claesson-Welsh, L., 1994, Direct interaction between Shc and the PDGF β-receptor, *J. Biol. Chem.* **269**:15337–15343.

Chapter 8

Transforming Growth Factor-β

ANITA B. ROBERTS and MICHAEL B. SPORN

1. Introduction

Transforming growth factor-β (TGF-β) is generally acknowledged to be the cytokine with the broadest range of activities in repair of injured tissue, based both on the variety of cell types that produce and/or respond to it and on the spectrum of its cellular responses (Roberts and Sporn, 1990). TGF-β is released from degranulating platelets and secreted by all of the major cell types participating in the repair process, including lymphocytes, macrophages, endothelial cells, smooth muscle cells, epithelial cells, and fibroblasts (see Fig. 1). A unique feature of this molecule is that its autoinduction results in sustained expression at the site of a wound and extends the effectiveness of both the initial burst of endogenous TGF-β released upon injury and exogenous TGF-β that might be applied to a wound. The ability of TGF-β to improve and/or accelerate tissue repair has been studied extensively in a variety of animal models of both normal and impaired healing. A limited number of clinical trials are in progress, but it is anticipated that many new applications for TGF-β will ultimately be found, once problems with appropriate timing and formulation can be solved.

In this chapter, we will give a brief overview of the basic biochemistry and biology of TGF-β, review results from a variety of animal models of wound healing, and discuss potential clinical applications of TGF-β to repair of tissue injury. We will also briefly summarize the extensive literature concerning the roles of TGF-β in fibrotic disease and in carcinogenesis, both of which, in many respects, can be viewed as aberrations of the normal repair process. We will focus principally on articles published in the past 4 years.

1.1. Multiple Forms of TGF-β

The term *TGF-β* refers generically to any of the TGF-β isoforms, each of which are highly homologous and often interchangeable in a variety of biological assays, but

ANITA B. ROBERTS and MICHAEL B. SPORN • Laboratory of Chemoprevention, National Cancer Institute, Bethesda, Maryland 20892-5055.

The Molecular and Cellular Biology of Wound Repair (Second Edition), edited by Richard A. F. Clark. Plenum Press, New York, 1996

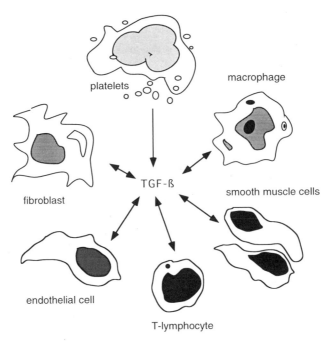

platelets

macrophage

TGF-ß

fibroblast

smooth muscle cells

endothelial cell

T-lymphocyte

Figure 1. Many different cell types participating in tissue repair processes are capable of both responding to and producing TGF-β. TGF-β released from degranulating platelets or from matrix stores contributes to inductive and autoregulatory processes that result in enhanced expression of TGF-β1 by responsive cells involved in tissue repair. We thank Dr. John Letterio for the artistic presentation.

are encoded by distinct genes and have unique promoters (Roberts and Sporn, 1990). Mammals express three isoforms of TGF-β designated TGF-β1, -2, and -3; TGF-β1 is the most abundant isoform in all tissues, and in human platelets it is the *only* isoform of the peptide. Certain cells such as retinal pigment epithelial cells secrete predominantly TGF-β2 (Connor *et al.*, 1989), and certain body fluids such as the aqueous and vitreous of the eye (Connor *et al.*, 1989; Jampel *et al.*, 1990), amniotic fluid (Altman *et al.*, 1990), saliva, and breast milk (Jin *et al.*, 1991) contain principally TGF-β2. It is noteworthy that >85% of the TGF-β in adult wound fluid is the type 1 isoform, whereas in wound fluid from second trimester fetuses, which do not scar, the type 2 isoform predominates (Longaker *et al.*, 1994). TGF-β3 is the least studied of the TGF-β isoforms. It has been isolated from human umbilical cord (ten Dijke *et al.*, 1988) and is secreted from certain cells, including myoblast cell lines (Lafyatis *et al.*, 1991); however, it is usually less abundant than either TGF-β1 or -2 in both tissue and cell extracts.

Burt and Law (1994) have suggested that TGF-β4 in chickens and TGF-β5 in *Xenopus* are the homologues of mammalian TGF-β1. If this hypothesis is correct, it suggests that TGF-β1 is the most evolutionarily diverged isoform, since TGF-β4 and -5 are each only 82% identical to TGF-β1. In contrast, chicken and *Xenopus* TGF-β2 are 99 and 95% identical to mammalian TGF-β2, respectively. Based on this hypothesis, TGF-β4 and -5 should play dominant roles in wound healing in avian and amphibian species, respectively.

The various isoforms of TGF-β are usually interchangeable *in vitro* or when applied exogenously to a wound (Schmid *et al.,* 1993b; Ksander *et al.,* 1993). As an example, TGF-β1, -2, -3, and -5 have similar activity in stimulating the chemotaxis of human macrophages and in inhibiting the growth of human B lymphocytes (Roberts *et al.,* 1990). However, in assay medium containing serum, it can appear as if TGF-β1 or -3 are more potent than TGF-β2, since α_2-macroglobulin, which is abundant in serum, binds TGF-β2 with about tenfold higher affinity than TGF-β1 (Danielpour and Sporn, 1990). In wound healing, exogenous application of TGF-β1 and -2 appear to have equivalent outcomes (Ksander *et al.,* 1993), whereas application of TGF-β3 has resulted in less scarring of incisional wounds (Shah *et al.,* 1995).

1.2. Transcriptional Regulation of TGF-β1, -2, and -3

The three mammalian isoforms of TGF-β each have distinct promoter regions and 5' and 3' untranslated regions that regulate their transcription and translation, respectively (Roberts and Sporn, 1992a; Kim *et al.,* 1992). Because of this, a particular stimulus such as wounding will affect the expression of each isoform differently. Thus, TGF-β1 expression is selectively induced in response to a variety of stimuli following wounding, ischemia, or anoxia, and in carcinogenesis and fibrogenesis. In contrast, expression of TGF-β2 and -3 is regulated primarily in response to hormonal and developmental signals.

The TGF-β1 promoter lacks classic TATA or CAAT boxes, but includes several response elements important in wounding, such as AP-1 and Egr-1 sites (Kim *et al.,* 1989; Dey *et al.,* 1994) (see Fig. 2). Expression of TGF-β1 is induced in response to Egr-1; to products of the oncogenes *jun, fos, src, abl,* and *ras;* to the transactivator proteins of viruses including human T-cell leukemia virus type 1, human cytomegalovirus, and hepatitis B virus; and, perhaps most importantly, to TGF-β itself, by a mechanism termed *autoinduction* (Kim *et al.,* 1989). Autoinduction is of central

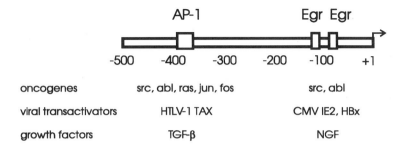

Figure 2. The TGF-β1 promoter contains both AP-1 and Egr sites, which are responsible for its selective induction and autoinduction following injury. These sites, which are not found in either the TGF-β2 or -3 promoters, are activated by oncogenes, viral transactivators, as well as growth factors (see Section 1.2).

importance in wound healing since it results in sustained expression of TGF-β1, which persists long after the initiating stimulus. It is mediated through the AP-1 sites in the TGF-β1 promoter (Kim *et al.*, 1989). These sites bind the transcription factor complex formed between the oncogene products Jun and Fos, each of which is rapidly induced in response to both TGF-β and wounding (Martin and Nobes, 1992; Wang and Johnson, 1994). Thus the sustained elevation of the expression of TGF-β1 seen following injury or in a variety of disease states is due, at least in part, to the wide variety of "distress signals" that activate transcription of the type 1 isoform selectively and do not affect the transcription of TGF-β2 and -3 (Roberts and Sporn, 1992a).

In contrast, the TGF-β2 and -3 promoters each contain a TATA box with a functional cyclic AMP-responsive element/activating transcription factor (CRE/ATF) site just 5′ of the TATA box (Lafyatis *et al.*, 1990; O'Reilly *et al.*, 1992). Although in some models, expression of TGF-β2 and -3 has been shown to be enhanced following wounding (Schmid *et al.*, 1993a,b), the mechanistic basis of this increase has not been studied, nor has it been determined whether increased synthesis or activation of existing stores of peptide is involved.

Although the long 5′ and 3′ untranslated regions of all 3 TGF-β isoforms probably play a role in posttranscriptional regulation of TGF-β expression (Romeo *et al.*, 1993; Kim *et al.*, 1992), thus far, it has not been demonstrated how or whether these elements might be modulated to affect posttranscriptional control.

1.3. Activation of Latent TGF-β

TGF-β is secreted from cells and released from platelets in a form that is unable to bind and activate its signaling receptors directly, and therefore is called "latent" TGF-β (for reviews, see Miyazono *et al.*, 1993; Harpel *et al.*, 1992). Two forms of latent TGF-β have been described and are termed the small and large latent complexes (see Fig. 3). The small latent complex is derived from a single gene product and consists of a noncovalent complex between the mature C-terminal 112-amino-acid TGF-β and the N-terminal remainder of its own pre-pro-domain (residues 30–278), called the latency-associated protein (LAP). In the large latent complex, LAP is covalently linked to another protein called the latent TGF-β-binding protein (LTBP). LTBP has been cloned and contains 16 epidermal growth factorlike repeats, eight cysteine repeats, as well as an RGD (Arg-Gly-Asp) sequence and the cellular binding domain of the laminin B2 chain; the platelet form appears to have been proteolytically processed and is 125–160 kDa. Although the exact role of LTBP in activation of the latent complex is not fully understood, it appears to increase the efficiency of secretion of TGF-β and may promote binding of TGF-β to matrix and facilitate its activation, analogous to fibrillin to which it shows some homology (Taipale *et al.*, 1994; Flaumenhaft *et al.*, 1993a). Certain cells, such as osteoblasts, have been shown to secrete TGF-β in the form of the small latent complex (Dallas *et al.*, 1994); the implications of this for activation are not presently understood. However, recent observations on prostate cancer cells have shown that failure of cells to secrete LTBP is associated with a poor prognosis (Eklov *et al.*, 1993).

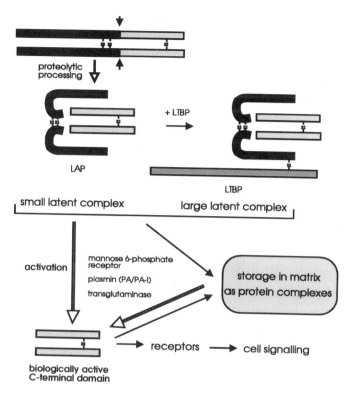

Figure 3. The formation and activation of latent complexes of TGF-β are highly regulated processes. Latent TGF-β consists of a noncovalent complex between LAP and mature TGF-β; in the large latent complex, LAP is covalently bound to LTBP. These complexes can be either sequestered by matrix or activated by a cooperative proteolytic process involving the mannose-6-phosphate receptor and surface-bound protease. Only the mature C-terminal domain of TGF-β binds to its signaling receptors, although certain larger complexes, such as that between TGF-β and thrombospondin, are also biologically active (see Section 1.3 for details).

In addition to complexes formed with LAP or LAP/LTBP, active TGF-β can also associate with a variety of matrix proteins including biglycan, decorin, type IV collagen, fibronectin, and thrombospondin (Noble *et al.,* 1992). TGF-β can also form both covalent and noncovalent complexes with α_2-macroglobulin; it has been proposed that α_2-macroglobulin might function either to scavenge TGF-β2 or to deliver it to particular tissues such as the liver (O'Connor-McCourt and Wakefield, 1987).

Understanding of the elements involved in regulation of the activation of latent TGF-β is likely to be critical in providing insight into its roles in wound healing, fibrotic disease, and carcinogenesis. The ability of TGF-β to be stored extracellularly in an inactive form provides a reservoir of the peptide, which can be readily made available following activation, even in the absence of transcriptional induction. Studies involving coculture of endothelial cells and smooth muscle cells have shown that activation of the latent complex is a multifactorial process depending, in part, on the mannose-6-phosphate receptor, plasmin, and transglutaminase (Harpel *et al.*, 1992). This has led to a model of cell-surface-mediated activation of the latent complex in which both latent TGF-β and plasmin must be surface-bound. The anchoring of latent TGF-β to activation sites on the cell surface or extracellular matrix is proposed to occur via LTBP (Taipale *et al.*, 1994) and/or via binding of mannose-6-phosphate residues on LAP to the mannose-6-phosphate receptor (Harpel *et al.*, 1992). Transglutaminase is thought to mediate the cross-linking of plasminogen to the cell surface or to extracellular matrix.

Other mechanisms have also been described for activation of latent TGF-β. Analogous to the use of low pH to activate latent TGF-β *in vitro*, it is thought that the acidic local environment around osteoclasts might serve to activate latent TGF-β in resorbing bone (Brown *et al.*, 1990; Bonewald *et al.*, 1991). Binding of TGF-β to thrombospondin has also been reported to result in its activation, suggesting that certain matrix proteins are able to bind TGF-β in a conformation that leaves the receptor-binding epitope exposed or that constrains the molecule so that this epitope is optimally configured for binding (Schultz-Cherry and Murphy-Ullrich, 1993). Since thrombospondin is released in large quantities from platelets and secreted from cells, and since its synthesis is induced by growth factors including TGF-β, it is likely that it plays an important in the activation of TGF-β in wound healing (Bornstein, 1992), possibly by transfer of TGF-β from an inactive to an active matrix protein complex. Treatment of cells with ligands of the steroid receptor superfamily such as retinoic acid, 1,25-dihydroxy vitamin D_3, tamoxifen, or gestodene also results in activation of TGF-β (Roberts and Sporn, 1992b), though, in certain cases, this has been demonstrated to result from increased protease secretion from cells treated with these agents (Kojima and Rifkin, 1993). Some of the reported vulnerary qualities of retinoids might be based, in part, on their ability both to enhance secretion and to activate TGF-β (Glick *et al.*, 1991; Hunt *et al.*, 1969; Weinzweig *et al.*, 1990; Varani *et al.*, 1991).

1.4. TGF-β Receptors

The TGF-β receptors have recently been cloned and shown to belong to a growing family of receptors with intrinsic serine–threonine kinase activity (for reviews, see Massagué *et al.*, 1994; ten Dijke *et al.*, 1994a; Lin and Moustakas, 1994). The most widely accepted model for the functional signaling unit of these receptors is a heterodimer or higher-order multimer composed of both type I (approximately 55 kDa) and II (approximately 80 kDa) receptors, but many variations on this model are under active investigation (Lin and Moustakis, 1994; Yamashita *et al.*, 1994b). Receptors in these two classes currently implicated in binding and signaling of the TGF-β include a single type II receptor (Lin *et al.*, 1992) and two distinct type I receptors: either Tsk-7L/ALK-2/R1, which can also function as a type I activin receptor (He *et al.*, 1993; Ebner *et al.*, 1993; ten Dijke *et al.*, 1994b), or ALK-5/R4, which appears to be specific for the TGF-β (Bassing *et al.*, 1994). Recent data based on use of kinase-defective receptors suggest a novel signaling mechanism dependent on phosphorylation of specific sites of the type I receptor by the type II receptor kinase (Wrana *et al.*, 1994).

Two other cell surface proteins, the type III receptor (also called betaglycan) and endoglin, which are about 70% homologous in their transmembrane and cytoplasmic domains, also bind TGF-β with high affinity; these proteins have only short cytoplasmic tails and lack a kinase domain (Massagué *et al.*, 1994). The binding site for TGF-β in betaglycan has been shown to reside in the proximal third of the extracellular domain, near the transmembrane domain (Fukushima *et al.*, 1993; Pepin *et al.*, 1994); this region has only very short stretches of homology to endoglin, suggesting that their binding sites may be distinct. It has been postulated that both of these proteins facilitate binding of TGF-β to its signaling receptors by "presenting" it in some appropriate conformation (Yamashita *et al.*, 1994a). A high-molecular-weight type V TGF-β receptor exhibiting serine–threonine kinase activity has also been described; however, its role in TGF-β signaling is not yet clear (Q. Liu *et al.*, 1994).

2. Biological Effects of TGF-β Relevant to Tissue Repair

TGF-β affects nearly every aspect of tissue repair. It is released in the form of a large latent complex from the α granules of platelets, when they degranulate by exposure to thrombin. Once activated, TGF-β attracts cells to the wound site by chemotaxis and then stimulates the formation of granulation tissue. In this section, we will discuss the individual steps as they might be influenced by TGF-β.

2.1. Chemotaxis

TGF-β is perhaps the most potent stimulator of chemotaxis known, stimulating the migration of monocytes, lymphocytes, neutrophils, and fibroblasts in the femtomolar concentration range (Wahl *et al.*, 1987; Postlethwaite *et al.*, 1987; Adams *et al.*, 1991; Brandes *et al.*, 1991). In all of the cell types studied, the dose–response to TGF-β is in the form of a bell-shaped curve with concentrations either less than or greater than the active range having no effect. Since chemotaxis is stimulated only at very low concentrations of TGF-β and not at higher concentrations that are required to activate gene transcription in responsive cells, the TGF-β-directed migration of these cells into a wound site must occur either prior to activation of significant amounts of latent TGF-β released from platelets or at sites quite distant from the major source of TGF-β, where the concentrations of active TGF-β would be quite low.

TGF-β also modulates the migration of cells, although the direction of its effects appear to be both cell-specific and contextual. Thus TGF-β stimulates the migration of an intestinal epithelial cell line, IEC-6 (Ciacci *et al.*, 1993), but inhibits the migration of smooth muscle cells after wounding *in vitro* in monolayer (Sato and Rifkin, 1989); it also increases the migration of keratinocytes both *in vitro* and in organ culture (Hebda, 1988; Nickoloff *et al.*, 1988). Although TGF-β has a chemotactic effect on fibroblasts in Boyden chamber assays, it inhibits their migration into three-dimensional collagen gel matrices (Ellis *et al.*, 1992).

2.2. Angiogenesis

The specific manner in which TGF-β stimulates angiogenesis *in vivo* is still being debated. The first evidence that TGF-β had angiogenic properties came from injection of TGF-β into Hunt-Schilling chambers implanted subcutaneously into the back of rats (Sporn *et al.,* 1983) and from subcutaneous injection of TGF-β into an uninjured area of skin in newborn mice (Roberts *et al.,* 1986). In each of these cases, TGF-β induced the formation of highly vascularized granulation tissue at the site of injection. In the rabbit corneal model, it has been clearly demonstrated that TGF-β induces angiogenesis indirectly by recruiting inflammatory cells which then stimulate angiogenesis by secreting endothelial cell mitogens (Phillips *et al.,* 1992).

In direct contrast, TGF-β has been shown to be a potent inhibitor of the growth of endothelial cells in monolayer. In coculture of vascular endothelial cells and either smooth muscle cells or pericytes, but not in homotypic cultures, latent TGF-β secreted by the cells is activated and growth of the cells is inhibited (Flaumenhaft *et al.,* 1993b). This has led to the suggestion that local activation of TGF-β at the edge of the pericyte cover of growing capillaries contributes to cessation of growth of the endothelial cells. In *in vitro* models of angiogenesis, the ability of TGF-β either to stimulate or to inhibit the formation of endothelial tubes has been shown to be contextual, depending on its concentration, the presence of other cytokines such as basic fibroblast growth factor, and the pericellular environment of the responding endothelial cell (Pepper *et al.,* 1993; Merwin *et al.,* 1990; Gajdusek *et al.,* 1993). The effects of TGF-β in *in vitro* angiogenesis have been shown to be mediated, in part, by its effects on the "proteolytic balance," which are important in the interplay between invasion of the extracellular matrix and formation of a lumen in endothelial cell cords (Pepper *et al.,* 1993).

2.3. Production and Remodeling of Extracellular Matrix

The effects of TGF-β on extracellular matrix are more complex and more profound than those of any other growth factor and are central to its effects on increasing the maturation and strength of wounds, as well as on pathological matrix accumulation characteristic of fibrotic disease (Roberts and Sporn, 1992a; Noble *et al.,* 1992). In a dermal site, the target cell is the wound fibroblast, which is first stimulated to migrate chemotactically in response to femtomolar concentrations of TGF-β at the periphery of the wounded area and then is activated transcriptionally by higher concentrations of TGF-β within the wound site. Matrix production and/or processing by keratinocytes (Vollberg *et al.,* 1991; Wikner *et al.,* 1990; Keski-Oja and Koli, 1992), osteoblasts (Fawthrop *et al.,* 1992; Sodek and Overall, 1992), renal mesangial cells (Border *et al.,* 1990a), endothelial cells, and smooth muscle cells (Basson *et al.,* 1992) are also regulated by TGF-β. TGF-β regulates the transcription of a wide spectrum of matrix proteins including collagen, fibronectin, and glycosaminoglycans, of matrix-degrading proteases and their inhibitors, and of integrin receptors, increasing production of matrix proteins while decreasing their proteolysis and modulating their interactions with cells as mediated by integrin receptors. The specific target genes affected are cell- and context-specific. Evidence suggests that effects of TGF-β on matrix may be mediated through a receptor complex or signaling pathway different from that which mediates its effects on growth, since certain cells that have lost responsivity to the growth inhibitory effects of TGF-β are still sensitive to its effects on induction of fibronectin and plasminogen activator inhibitor synthesis (Geiser *et al.,* 1992). Understanding of the basis of these differences may eventually permit selective manipulation of one or the other of these pathways.

Recent experiments have shown that the interaction of TGF-β and matrix is bidirectional (Nathan and Sporn, 1991). In mammary epithelial cells, the level of TGF-β1 expression is transcriptionally down-regulated when the cells are cultured on reconstituted basement membrane compared to the same cells cultured on plastic (Streuli *et al.,* 1993). Expression of TGF-β2 was unchanged in the two culture conditions, in agreement with the very different nature of the promoters of TGF-β1 and -2 and underscoring their different roles in tissue repair and disease (Roberts and Sporn, 1992a). These results have important implications for wound healing since they suggest (1) that disruption of a normal basement membrane may be one of the signals contributing to increased and prolonged expression of TGF-β1, and (2) that the re-formed basement membrane, possibly signaling through integrin receptors, may play an active role in terminating expression of one of its inducers, TGF-β1. They also suggest that pathological accumulation of matrix may result not only from excessive production of TGF-β1, but also from defects in the signaling pathways emanating from matrix that would ordinarily function to terminate or suppress expression of TGF-β1.

3. Animal Models of Tissue Repair

Animal models of both basal and impaired tissue repair in rats, guinea pigs, pigs, and rabbits with experimentally induced incisional, excisional, and burn wounds or with implants of wire mesh Hunt-Schilling chambers or polyvinyl alcohol sponges have been utilized to evaluate the role of endogenous TGF-β and the efficacy of exogenous TGF-β in the repair process (for reviews, see Cromack *et al.,* 1991; Martin *et al.,* 1992; Roberts and Sporn, 1993; Lawrence and Diegelmann, 1994). Endpoints involve evaluation of the quality and rate of healing as assessed by physical measurements of the tensile strength of linear incisional wounds, measurement of rates of wound closure, histomorphometric analyses, biochemical measurements of synthesis of DNA and matrix proteins, and assessment of scarring.

At the time of injury, latent TGF-β1 is released from degranulating platelets into the wound bed as a bolus. Subsequently, injury-induced expression of immediate-early genes contributes to the transcriptional activation and autoinductive pathways of TGF-β1 expression (Martin and Nobes, 1992; Wang and Johnson, 1994), resulting in elevated levels of expression of endogenous TGF-β1 that persist over a protracted period. Since large stores of latent TGF-β are localized to pericellular matrix, in part via LTBP binding (Taipale *et al.,* 1992, 1994), the proteolytic environment characteristic of the early stages of wound healing might also serve to release TGF-β locally from its matrix stores (Falcone *et al.*, 1993; Kane *et al.*, 1991), prior to or independent of transcriptional activation of the TGF-β1 gene. Thus, the levels of TGF-β in wound fluid remain elevated for up to 14 days, with peak levels on days 7 to 9 following implantation of wire mesh Hunt-Schilling chambers in the back of rats, at the time of maximum fibroblast proliferation and collagen synthesis (Cromack *et al.*, 1987).

Immunohistochemical studies have shown that each of the three TGF-β isoforms are expressed in unique patterns following wounding (Kane *et al.*, 1991; Levine *et al.*, 1993). However, it has been shown that regardless of which TGF-β isoform is applied exogenously to a wound, expression of TGF-β1, and not TGF-β2 and -3, is induced endogenously. Thus, topical application of either TGF-β1, -2, or -3 resulted in approximately equivalent induction of TGF-β1 mRNA in the wound bed of thermal injuries in mice (Schmid *et al.*, 1993b); TGF-β2 or -3 mRNAs were not detected. Similarly, in a model of impaired healing in mice, exogenous application of TGF-β2 resulted in expression of endogenous TGF-β1, as determined by immunohistochemistry (Ksander *et al.*, 1993). These results and those showing autoinduction of TGF-β1 mRNA and protein following its exogenous application in two different wound models in pigs (Quaglino *et al.*, 1990, 1991) again underscore the critical role of the type 1 isoform of TGF-β in tissue repair.

Study of the expression patterns of the TGF-β isoforms in uninjured human skin and in wounds suggests that there may be differences compared to the animal models. Whereas little expression of TGF-β2 and -3 was seen in mice, TGF-β3 mRNA and protein are prominantly expressed in keratinocytes of intact human dermis (Schmid *et al.*, 1993a); TGF-β1 mRNA expression was seen at the reepithelialization front of acute wounds. The TGF-β type II receptor was also expressed in the epidermis with stronger expression in the upper, more differentiated layers (Schmid *et al.*, 1993a). Now that tools are available for detection of the TGF-β receptors, it will be important to examine the spacial and temporal patterns of their expression in tissue repair and in fibrogenesis to determine whether any regulation of the TGF-β response occurs at the receptor level. Recent studies in carcinogenesis demonstrate that cells that are no longer sensitive to regulation by TGF-β have either impaired expression of the type II receptor or impaired receptor signaling, suggesting that it may be important to examine receptor expression in chronic wounds (Park *et al.*, 1994). A caveat in extrapolating functional significance from *in situ* expression of receptor mRNA alone is illustrated by recent studies of receptor expression in hepatic lipocytes, which are suggested to serve as a paradigm for mesenchymal cells. These studies show that receptors may have to become activated, possibly by formation of multimeric complexes, before they can bind ligand, and, because of this, that expression of receptor mRNA does not always correspond with TGF-β binding activity.

3.1. Unimpaired Wound Models

3.1.1. Delivery Vehicle

TGF-β is typically delivered to the wound in either a saline vehicle (Pierce *et al.*, 1992), a 3% methycellulose suspension (Beck *et al.*, 1990a), an emulsified bovine collagen vehicle (Zyderm II) (Mustoe *et al.,* 1991), or a sponge made of bovine collagen either with or without heparin (Ksander *et al.,* 1990a,b). In the only study in which delivery vehicles were actually compared, it was shown that TGF-β applied topically as a single dose to an incisional wound in a rat in saline was ineffective compared to the same dose delivered in a collagen suspension, suggesting that prolonged local exposure of the wound to TGF-β was beneficial (Mustoe *et al.,* 1987).

3.1.2. Dose Levels and Application Schedule

The design of most wound-healing studies has consisted of a single application of TGF-β applied topically at the time of wounding. In studies addressing the question of dosing schedule, it was shown in a rabbit ear dermal ulcer model that a single application of TGF-β at the time of wounding had an effect equal to that of multiple doses, whereas application 24 hr after wounding had no beneficial effect (Beck *et al.,* 1990a). Timing of application TGF-β is therefore of critical importance and suggests that there is a particular cell population present in the early stages of tissue repair that is optimally targeted by TGF-β. However, in glucocorticoid-impaired healing of excisional wounds in the rabbit ear, where healing was delayed sufficiently to evaluate the extended repair process, repeated application of TGF-β, spaced several days apart, was of greater benefit than a single treatment (Beck *et al.,* 1991).

Important, and conceptually novel, is the demonstration that systemic administration of TGF-β, even up to 24 hour before wounding, is effective in enhancing repair (Sporn and Roberts, 1993). Following injection of a single intravenous dose of TGF-β1 (100 or 500 μg/kg), Beck *et al.* (1993b) showed that the tensile strength of incisional wounds in either old or glucocorticoid-impaired rats was increased to the level observed in normal untreated young rats; marginal effects were observed using the lowest dose of 10 μg/kg. Systemic TGF-β1 was effective if administered as early as 24 hr before wounding or up to 4 hr after wounding, and significantly reduced if administered 48 hr prior to wounding. These results suggest that TGF-β can "prime" cells for increased responsiveness to factors released at the wound site, and that such signals can persist for as long as 24 hr. Circulating monocytes and possibly also tissue fibroblasts are potential targets of systemic TGF-β; *in vitro* experiments have shown that monocytes exposed to TGF-β have enhanced responses to secondary stimuli (Mustoe *et al.,* 1987). The ability of TGF-β to autoregulate its own production may also contribute; this would result in amplification of the initial "endocrinelike" action of the systemic TGF-β by subsequent paracrine and autocrine mechanisms.

Doses of TGF-β used to treat wounds topically have varied over a 1000-fold concentration range from as little as a few nanograms (Shah *et al.,* 1994a) to several micrograms (Ksander *et al.,* 1990a; Beck *et al.,* 1990b, 1993b). Interestingly, in the rabbit ear ulcer model, two independent studies varied by a factor of 10 in terms of optimal dose of TGF-β, although each showed a biphasic dose-dependency. Thus, Beck *et al.* (1990a) showed positive effects of TGF-β1 on wound healing, which increased at doses of 5–100 ng TGF-β1 and decreased at higher doses, while Mustoe *et al.* (1991) showed optimal effects at 1μg with decreasing effects at lower or higher amounts of TGF-β1. In these two studies, TGF-β1 was applied in a methylcellulose or collagen suspension, respectively; it is not known how this might have affected the determination of dose range. However, the response curves clearly indicate that dose range studies will be important and that optimal levels will vary depending on the mode of delivery and wound type.

3.1.3. Effects of TGF-β on the Tensile Strength of Incisional Wounds

One way to assess healing is to measure the strength of an incisional wound; this is commonly done by testing the breaking strength of formalin-fixed or freshly excised incisions. Such measurements consistently show that topical application of TGF-β increases the breaking strength in incisional wounds in rats (Mustoe *et al.*, 1987; Brown *et al.*, 1988; Ammann *et al.*, 1990; Pierce *et al.*, 1989b; Beck *et al.*, 1993b) and guinea pigs (Ksander *et al.*, 1990a) with maximum effects at early time points and negligible effects at later time points when healing was complete in both treated and control wounds. Tensile strength of punch wounds in pigs (Beck *et al.*, 1990b) and guinea pig wounds healing by secondary intent was also increased (Ksander *et al.*, 1990b). A newly developed *in vivo* method that does not require excision of the wound site confirmed that TGF-β1 (2 µg/rat) significantly increased wound strength of abdominal linear incisional wounds in rats at day 5 (Perry *et al.*, 1993).

These results derive, in part, from the profound and unique effects of TGF-β on synthesis of collagen and other extracellular matrix proteins. Injection of labeled proline showed that TGF-β enhanced collagen synthesis in a dose-dependent manner in the rabbit ear ulcer model (Beck *et al.*, 1990a; Chen *et al.*, 1992) as had been shown previously in wire mesh chambers (Sporn *et al.*, 1983) and in a model for granulation tissue formation in uninjured skin (Roberts *et al.*, 1986). Others have used morphological analysis with either Masson trichrome or reticulin stain (Mustoe *et al.*, 1987), specific antibodies to procollagen type I (Pierce *et al.*, 1991a), or picosirius red (Shah *et al.*, 1994a; Pierce *et al.*, 1992) to show increases in newly synthesized collagen in wounds treated with TGF-β. The effect of TGF-β on collagen synthesis is transient and parallels the increase in tensile strength; subsequent collagen cross-linking, on the other hand, is not a major determinant of TGF-β-dependent improvement in repair (Pierce *et al.*, 1991a).

Comparisons of the effects of TGF-β with those of platelet-derived growth factor (PDGF-BB) and basic fibroblast growth factor (bFGF) in both the rabbit ear dermal ulcer model (Pierce *et al.*, 1991b, 1992; Mustoe *et al.*, 1991) and an incisional model in rats (Pierce *et al.*, 1991a) demonstrate that individual growth factors may regulate different aspects of repair. Specifically, with respect to their effects on collagen synthesis, it was shown that ulcer wounds treated with TGF-β1 appeared to bypass the inflammatory phase of wound repair, with early deposition and maturation of collagen into large bundles at the leading edge of the wound, as assessed by picosirius red staining with polarization optics (Pierce *et al.*, 1992). In contrast, PDGF-BB stimulated wound closure by augmenting deposition of provisional matrix composed of glycosaminoglycans and fibronectin at the edge of new granulation tissue and did not stimulate new collagen synthesis until rather late in the repair process. Wounds treated with bFGF contained little to no collagen even after complete closure (Pierce *et al.*, 1992). These results are consistent with the action of these growth factors on fibroblasts *in vitro*. In contrast to dominant effects of TGF-β on the wound fibroblast phenotype, PDGF-BB had stronger effects on inflammatory cell influx and bFGF on angiogenesis than did TGF-β1. Nonetheless, TGF-β1, injected subcutaneously into uninjured skin can, by itself, induce all the features of granulation tissue, possibly, in part, by induction of other growth factor activities (Roberts *et al.*, 1986). Not unexpectedly, synergistic effects are seen using combinations of growth factors to treat partial-thickness porcine skin wounds (Lynch *et al.*, 1989).

3.1.4. Effects of TGF-β on Reepithelialization

Because it inhibits the growth of keratinocytes *in vitro*, it had been predicted that TGF-β would also inhibit reepithelialization of open wounds. However, Hebda (1988) showed that TGF-β increased the rate of outgrowth of keratinocytes from porcine skin explants, again emphasizing that results from cells cultured on plastic are often not predictive for behavior of cells in the context of their natural *in vivo* environment. In wound models, the effects depend on the dose of TGF-β used and the particular model. Thus, in the rabbit ear ulcer model, which precludes contraction and is thus considered a useful model for the healing of human ulcer wounds that also do not heal by contraction, high doses of TGF-β1 (1 μg/wound), which were optimal in terms of formation of new granulation tissue, had no effect on reepithelialization; however, 5 μg/wound significantly inhibited reepithelialization (Mustoe *et al.*, 1991). In contrast, in a similar model, treatment with 100 ng TGF-β actually enhanced the migration of epithelial cells from the wound margin toward the center without affecting their proliferation (Chen *et al.*, 1992). Similarly, in dermal wounds in pigs, the extent of surface covered by epithelium was again unaffected by TGF-β over a tenfold high dose range from 1 to 10 μg/cm² (Ksander *et al.*, 1990a), whereas application of lower doses of 70 or 700 ng/day TGF-β1 for 5 days resulted in increases in the area of regenerated epidermis and increased cross-sectional depths of regenerated dermis, characterized by granulation tissue, compared to wounds treated with saline vehicle alone (Jones *et al.*, 1991). In glucocorticoid-impaired healing of excisional wounds in pigs, TGF-β1 caused a substantial reduction in the degree of epithelialization after 5 days' treatment with doses >500 ng/wound per day, applied daily; however, all wounds were fully epithelialized at 10 days regardless of the treatment (Quaglino *et al.*, 1990). Together, these data suggest that it is unlikely that TGF-β, used in an appropriate dose range, will have adverse effects on reepithelialization in clinical treatment of wounds.

3.2. Healing-Impaired Models

Much of the need for clinical management of wounds is in a patient population with compromised healing, often as a complication of other basal metabolic derangements such as diabetes or poor vascular circulation, or of treatment protocols such as steroids, chemotherapy, or radiation that often complicate the postoperative management of surgery for malignancy. Studies of animal models of impaired healing suggest that the effects of growth factor treatment are substantial, often increasing the healing rates to a level comparable to that of unimpaired, control animals.

3.2.1. Corticosteroids

Pharmacological levels of corticosteroids reduce inflammatory cell influx and angiogenesis and inhibit fibroblastic collagen synthesis, resulting in a marked healing deficit (for review, see Wahl, 1989). In rats treated with methylprednisolone with deficits in wound strength of 50% at 7 days, application of TGF-β1 0.25–1 μg/incision (Pierce *et al.*, 1989a) or 2–8 μg/incision (Beck *et al.*, 1991) restored the deficit in a dose-dependent fashion. It is significant that PDGF-BB, used in the same study design, failed to restore breaking strength (Pierce *et al.*, 1989a). Since macrophages were notably absent in wounds of the glucocorticoid-treated rats, the data suggest that these cells normally mediate the effects of PDGF on induction of procollagen synthesis in fibroblasts, whereas TGF-β is able to act directly on the fibroblasts. Indeed, while both TGF-β and PDGF increased the number of fibroblasts in wounds of treated rats, only TGF-β treatment stimulated their synthesis of type I procollagen (Pierce *et al.*, 1989a). As discussed in Section 3.1.2, systemic administration of TGF-β1 was also highly effective in restoring the deficit in wound-breaking strength in steroid-treated rats (Beck *et al.*, 1993b).

TGF-β1 (100 or 500 ng/wound) also improved the healing of ulcer wounds in the ears of rabbits treated with methylprednisolone to suppress healing, increasing the amount of granulation tissue and increasing reepithelialization (Beck *et al.*, 1991). Moreover, in this model, the delayed healing permitted evaluation of multiple applications of TGF-β; two applications of TGF-β spaced several days apart improved the healing compared to a single application (Beck *et al.*, 1991). In glucocorticoid-impaired healing of excisional wounds in pigs, TGF-β1 caused a transient reduction in the degree of epithelialization after 5 days' treatment with doses >500 ng/wound per day, applied daily; however, epithelialization was complete at later times (Quaglino *et al.*, 1990).

3.2.2. Chemotherapeutic Agents

Adriamycin (doxorubicin) is an antineoplastic agent in widespread clinical use that impairs wound healing when used in large doses; the mechanisms of its action on wound healing are not understood. However, some insight is provided by the observation that mRNA levels of both TGF-β1 and type 1 collagen are decreased in wounds of rats treated with adriamycin compared to untreated controls; topical application of TGF-β1 (2 μg) to these wounds restored the levels of type 1 collagen mRNA to normal (Salomon *et al.*, 1990). In other experiments in which adriamycin pretreatment of rats decreased wound tear strength and energy by about 50% at 10 days following wounding, treatment with TGF-β1 (2 μg) resulted in a transient reversal in the wound-healing impairment (Curtsinger *et al.*, 1989). Together, these studies show that decreased expression of endogenous TGF-β1 may underlie the impaired healing associated with certain chemotherapeutic regimens and that its topical replacement may correct the defect.

3.2.3. Radiotherapy

Impaired or delayed healing of irradiated tissue is a significant clinical problem. As in other models of impaired healing, TGF-β has been shown to reverse the radiation-induced deficit. In guinea pigs, irradiation of skin flaps 2 days prior to incisional wounding resulted in a 72% reduction in wound-bursting strength measured 7 days postwounding. TGF-β1 at 1 or 5 μg/wound nearly doubled the strength of radiation-treated wounds; treatment with higher doses (20 μg) decreased the wound strength (Bernstein *et al.*, 1991). The low doses of TGF-β, but not the high dose, were also shown to increase expression of pro-α1(I) collagen mRNA at the wound site. Interestingly, enhanced and sustained expression of TGF-β1 mRNA and protein are also thought to be involved in radiation-induced fibrosis (M. Martin *et al.*, 1993).

Comparison of the effects of total body irradiation, which reduces levels of circulating monocytes without having significant local effects on skin, with that of megavolt electron beam surface irradiation, which impairs the function of dermal fibroblasts, has provided interesting insight into the mechanisms of TGF-β action in wound healing. Using a rat linear skin incision model, Cromack *et al.* (1993) have shown that TGF-β1 (2 μg/wound) significantly accelerated repair and wound-breaking strength in animals receiving total body irradiation, but not in those receiving surface irradiation. These results are consistent with those from glucocorticoid-treated rats that showed that the effects of TGF-β are mediated through its action on dermal fibroblasts and are less dependent on bone-marrow-derived monocytes/macrophages. As anticipated, PDGF-BB, which is dependent on tissue macrophages for its effects on healing, was effective in the surface irradiation model (Mustoe *et al.*, 1989). Appreciation of these mechanistic differences in the action of TGF-β and PDGF will be important in optimal management of clinical wounds.

3.2.4. Others

Other models of impaired healing involve use of aged or diabetic animals or covering wounds with occlusive dressings; in every case TGF-β has been shown to improve the outcome and restore the healing deficit toward normal. Thus, TGF-β1 injected directly into the wound site day 3 postwounding increased both the accumulation of granulation tissue and the tensile strength of implanted polyvinyl alcohol sponges and incisional wounds, respectively, at day 7 in streptozotocin-induced diabetic rats (Broadley *et al.*, 1989). Age-dependent deficits in wound strength in rats were also reversed both by topical (1–4 μg/wound) and systemic (100–500 μg/kg) TGF-β1 (Beck *et al.*, 1993b); TGF-β2 was also shown to effective (Cox *et al.*, 1992). Decreased expression of TGF-β1 was observed in wounds covered with occlusive dressings, which resulted in impairment of granulation tissue formation; this could be partially reversed by treatment with either TGF-β1 or -2 (Ksander *et al.*, 1993).

3.3. Healing of Nondermal Wounds

Although much of the focus on the ability of TGF-β to stimulate tissue repair has been on dermal wounds, in part because such wounds are easily treated topically and their progress monitored visually, many of the same cellular mechanisms underlie repair in other sites of the body. The efficacy of TGF-β in the limited number of animal models studied and described below suggests that it could improve the outcome in a variety of clinical settings. Moreover, in treatment of bony defects or fractures, gastrointestinal anastomoses, oral lesions, or in the eye, TGF-β can be applied locally, so that problems of targeting or systemic toxicity need not be of major concern.

3.3.1. Bone

Repeated injections of TGF-β subperiosteally into uninjured rat femurs stimulates formation of cartilage that subsequently mineralizes to form bone (Joyce et al., 1990), analogous to the observation that repeated injections of TGF-β subcutaneously into uninjured skin induces local formation of granulation tissue (Roberts et al., 1986). The osteoinductive properties of TGF-β have also been demonstrated in a skull defect in rabbits where application of TGF-β (0.4–5 μg) at the time of surgery resulted in a dose-dependent closure and bony bridging of the defects, suggesting that TGF-β might be applicable to therapy of nonhealing bony defects (Beck et al., 1993a). Once the dosing, timing, and delivery of TGF-β can be optimized, it is expected that it will find application in fracture healing as well (Joyce et al., 1991). Progress has been made in this regard in the design of a biodegradable controlled release system for TGF-β consisting of poly(DL-lactic-co-glycolic acid) and demineralized bone matrix; the TGF-β is released over a 20-day period, depending on the loading and the percent bone matrix in the device (Gombotz et al., 1993).

Mechanistically, the role of TGF-β in bone repair involves many of the same processes involved in soft tissue repair: it is initiated by platelet degranulation, cell recruitment by chemotaxis, angiogenesis, and formation of matrix (Joyce et al., 1991). In transgenic mice, synthesis and secretion of TGF-β1 in osteogenic zones coincided with areas of expression of the α_2(I) collagen promoter (D'Souza et al., 1993). TGF-β stimulates the chemotaxis of osteoblastlike cells (Pfeilschifter et al., 1990) as well as differentiation of mesenchymal cells into chondrocytes, resulting in intramembranous or endochondral ossification in the parietal bones of neonatal and adult rats, respectively (Taniguchi et al., 1993).

3.3.2. Gastrointestinal

Anastomotic dehiscence resulting from loss of wound integrity is a major cause of morbidity and mortality in gastrointestinal surgery. A rabbit model of a gastric incision was developed to test whether TGF-β might promote closure (Mustoe et al., 1990). Despite the fact that this wound site is populated principally by smooth muscle cells with relatively few fibroblasts, TGF-β1 (0.1–10 μg) applied topically at the time of surgery increased both wound-breaking strength and energy at days 7–11 postsurgery.

Optimal effects were seen at a dose of 2 μg, with approximately a 4-day acceleration of healing (Mustoe et al., 1990). Human intestinal smooth muscle cells respond to TGF-β in vitro by increasing synthesis of collagen (Graham et al., 1990), suggesting that the animal model might translate to clinical application in anastomotic surgery. It has subsequently been shown that TGF-β (5 μg/wound) also increased the breaking strength of longitudinal ileal wounds in pigs, partially reversing the steroid-induced impairment of healing of these wounds (Slavin et al., 1992). In a different model of intestinal injury, TGF-β promoted healing of perforated rat mesentery when injected into the peritoneal cavity for either 2 or 4 days postwounding; this effect was presumably mediated directly by mesenchymal cells, as no recruitment of peritoneal macrophages was observed (Franzén and Schultz, 1993). Not surprisingly, caution must be exercised in administering repeated doses of TGF-β intraperitoneally, as it has been shown that daily injections for 5 days can significantly increase the severity of postoperative intestinal adhesions in rats; adhesions did not result from injection of TGF-β into the peritoneum of uninjured rats (Williams et al., 1992).

3.3.3. Oral Mucosal

One of the proposed clinical applications of TGF-β is for treatment of oral mucositis, or ulcerative stomatitis, a painful condition that affects approximately one third of patients undergoing high-dose chemotherapy for treatment of cancer (Dreizen et al., 1981); the frequency is even higher in pediatric patients. Administration of 5-fluorouracil intraperitoneally to Syrian golden hamsters produces an ulcerative mucositis that has been used as a model for this condition. Surprisingly, topical administration of TGF-β during the period of active mucositis impaired healing of the oral mucosa (Sonis et al., 1994). However, topical application of four doses of 20 μg TGF-β3/dose to the cheek pouch over a 24-hr period prior to injection of the 5-fluorouracil resulted in a transient decrease in proliferation of the oral mucosa and reduced the severity and duration of the resulting mucositis. Clearly, this outcome is based not on the ability of TGF-β to promote repair, but rather on its ability to inhibit the proliferation of epithelial cells, in this case protecting the cells from the cytotoxic effects of the chemotherapy.

3.3.4. Eye

Both animal models and recent clinical data demonstrate a role for TGF-β in repair of injury in the eye. In models of corneal injury, it has been shown that TGF-β is chemotactic for corneal cells at femtomolar concentrations (Grant et al., 1992) and appears to stimulate matrix synthesis by corneal fibroblasts (Schultz et al., 1992). It has also been shown that direct intraocular application of TGF-β1 to the site of a retinal tear in a rabbit eye induces a chorioretinal adhesion (Smiddy et al., 1989); doses either tenfold less than or greater than the optimal 700 ng/eye were ineffective. Remarkably, these data have translated directly to the clinical application of TGF-β for treatment of macular holes (Smiddy et al., 1993; Lansing et al., 1993; Glaser et al., 1992; see also Section 6). Overproduction of TGF-β is implicated in cataractlike changes in rat lens epithelial explants (J. Liu *et al.*, 1994) and in the vitreous of human eyes with intraocular fibrosis resulting from proliferative vitreoretinopathy (Connor *et al.*, 1989), substantiating the close link between the mechanisms involved in fibrogenesis and healing (see Section 4).

3.3.5. Central Nervous System

Functional recovery from injury to the brain or spinal cord is poor, and it is not yet known how to prevent gliosis and promote neuronal growth. Wound closure occurs rapidly following a penetrating injury in the rat brain with reformation of the blood–brain barrier; however, the formation of a fibrous scar inhibits reconnection of neural pathways (reviewed in Logan and Berry, 1993). The cellular events following injury to the brain mimic to some extent that seen in peripheral dermal wounds in that they are initiated by platelet degranulation and followed by invasion of cells of the monocyte/macrophage lineage; reactive microglia, astrocytes, and neurons are also involved. After a penetrating brain injury in the rat, TGF-β1 mRNA is up-regulated predominantly in cells with an astrocyte phenotype (Lindholm *et al.*, 1992; Logan *et al.*, 1992). Whether or not the expressed TGF-β mediates repair is not known, but it has been suggested that TGF-β has a critical role in organizing the response of the brain to neurodegeneration and to injury, particularly by protecting it against invading lymphocytes and by suppressing microglial activity (Finch *et al.*, 1993).

Addition of TGF-β1 to the site of a brain injury results in increased influx of macrophages, increased deposition of fibronectin, and subsequent scarring; conversely, antibodies to TGF-β1 block those effects and reduce levels of scarring below that of control wounds (Logan and Berry, 1993). However, even when scarring was blocked by anti-TGF-β1 or by other treatments such as infusion of decorin, neuronal regeneration was not seen, suggesting that the physical barrier of scar tissue is not the limiting factor and that additional strategies, possibly involving neurotropic factors, will have to be developed.

3.3.6. Others

Since TGF-β acts on most cells and since mechanisms of repair in all tissues follow similar patterns of cell recruitment, cell proliferation, and deposition of matrix, it follows that TGF-β will likely mediate repair of many different tissues. A role of repair in the myocardium is strongly suggested, although the mechanisms involved are still not understood (Roberts *et al.*, 1993). In other tissues including lung, kidney, and liver, as well as the eye, excessive or prolonged expression of TGF-β has been implicated in pathological accumulation of matrix (see Section 4). However, since fibrosis can be viewed as an aberration of the repair process, the implication would be that controlled expression of TGF-β in response to local injury would mediate repair in these same tissues.

Intravenous administration of TGF-β1 has been shown to reduce the cellular damage associated with ischemia/reperfusion injury, be that in the splanchnic circulation, where TGF-β attenuates the development of vascular resistance in a rabbit model

(Thomas and Thibodaux, 1992); the heart, where it also protects against reperfusion injury and maintains myocyte integrity in rats as well as cats (Lefer *et al.,* 1990, 1993); or in the brain, where it reduces ischemic injury in mice (Prehn *et al.,* 1993) and piglets (Armstead *et al.,* 1993) and reduces infarct size in a rabbit model of thromboembolic stroke (Gross *et al.,* 1993). The mechanisms involved appear to be indirect and center on the ability of systemic TGF-β1 to reduce adherence of circulating neutrophils to endothelium, thereby preventing their extravasation into tissue and blocking the subsequent elaboration by these cells of inflammatory cytokines such as tumor necrosis factor-α and interleukin-1β, which mediate cellular injury (Lefer *et al.,* 1993; Roberts *et al.,* 1993). TGF-β also suppresses the synthesis of hydrogen peroxide (Tsunawaki *et al.,* 1988) and nitric oxide (Vodovotz *et al.,* 1993) by activated macrophages. However, data suggest that direct effects of TGF-β on the target tissues are also important. Thus, in the heart, expression of TGF-β is increased at the margins of infarcted tissue and it maintains the contractility of cultured cardiac myocytes and inhibits their synthesis and secretion of the inducible form of nitric oxide synthase (Roberts *et al.,* 1993). And although investigations are only beginning, local expression of TGF-β in the brain following injury may also be mediating repair (Wiessner *et al.,* 1993; Finch *et al.,* 1993); an indication of this is the ability of TGF-β to prevent degeneration of cultured rat neurons after treatment with toxic levels of glutamate (Prehn *et al.,* 1993).

3.4. Scarring versus Nonscarring

Formation of scar tissue represents a significant clinical problem in that it results in functional impairment and disfiguration. It has been known for some time that fetal dermal wounds heal without scarring (for reviews, see Mast *et al.,* 1992; Adzick and Lorenz, 1994), and surgery on fetuses is now in limited use to repair defects with otherwise lethal outcomes (Adzick and Harrison, 1994). In fetal wounds, collagen is deposited in a reticular pattern indistinguishable from that of the surrounding normal dermis, whereas in adult wounds, it is deposited in large, parallel bundles perpendicular to the wound surface (Whitby and Ferguson, 1991a; Longaker *et al.,* 1990). Since fetal wounds typically contain less TGF-β1 than their adult counterparts (Whitby and Ferguson, 1991b), experiments have been designed to examine the effect of exogenous TGF-β on fetal wounds. Complementing these studies are those addressed at reduction of scarring by blocking endogenous TGF-β activity in adult wounds.

3.4.1. Scarring in Fetal Wounds

Fetal wounds do not involve formation of a fibrin/platelet clot, do not have a significant inflammatory reaction, do not neovascularize, and do not scar. Platelets and macrophages, major sources of TGF-β1 in the adult wound, are significantly reduced or absent in embryonic and fetal wounds (Longaker *et al.,* 1994). The extracellular matrix deposited in fetal wounds is similar in composition to that of adult wounds, but differs in spacial and temporal organization. Immunohistochemical localization of TGF-β in wounds made in mouse embryos has shown that levels of TGF-β isoforms are significantly reduced (Whitby and Ferguson, 1991b), but, probably most significantly, that TGF-β1 mRNA and protein are expressed only transiently following wounding, being detectable as early as 1 hr postwounding but returning to background levels by 18 hr (P. Martin *et al.*, 1993). These data suggest either that autoinductive cascades are in some unknown manner impaired or repressed or that they are simply not sustained in the fetus in the absence of external sources of TGF-β from macrophages and lymphocytes. Wounding of mouse embryos in which the TGF-β1 gene has been knocked out (Kulkarni *et al.*, 1993) demonstrate that TGF-β1 plays a critical role in contraction of mesenchymal cells (P. Martin, personal communication), much as it has been shown to enhance contractile capacity of wound fibroblasts embedded in a collagen gel (Montesano and Orci, 1988).

One other aspect that has not been examined thoroughly is the ratio of the TGF-β isoforms in wounds. Analysis of the relative levels of TGF-β1 and -2 in wound fluid of sheep into which wire mesh Hunt-Schilling chambers had been implanted subcutaneously showed that levels of TGF-β1 were approximately equal in 100-day-old fetuses compared to adults, whereas levels of TGF-β2 were about 14-fold higher in the fetuses (Longaker *et al.*, 1994). The significance of this pattern of endogenous expression is not presently understood. However, TGF-β1, -2, or -3 produce approximately equivalent results when applied topically (Schmid *et al.*, 1993b; Ksander *et al.*, 1993).

The question of whether the limited expression of TGF-β1 in fetal wounds is critical for the absence of scarring has been addressed by adding TGF-β1 to fetal wounds (Fig. 4). Subcutaneous implantation of polyvinyl alcohol sponges into fetal rabbits showed that addition of TGF-β1 resulted in a fibrotic response accompanied by a significant increase in the cellularity of the sponges, especially as reflected in an increase in the number of infiltrating T lymphocytes (Adolph *et al.*, 1993). Moreover, in fetal rat limbs cultured in serum-free medium, addition of TGF-β1 to day 14 limbs that do not scar resulted in scarring, whereas addition of TGF-β antibodies to day 18 limbs that heal with scarring reduced the degree of scarring, as determined by quantitation of collagen and assessment of fiber organization using picosirius red (Houghton *et al.*, 1995). These data clearly show that differences in the levels and the kinetics of TGF-β expression in the fetal compared to the adult wound must, in part, be responsible for the scarring pattern observed.

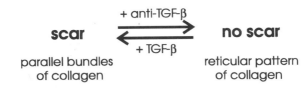

Figure 4. Alteration of the levels of TGF-β at a site of tissue injury can modify the degree of scarring as assessed by the pattern of collagen organization in the healed wound. Fetal wounds that do not scar can be induced to scar by addition of TGF-β. Conversely, the scarring of adult wounds is reduced by addition of antibodies to TGF-β. Higher levels of TGF-β also result in increased numbers of inflammatory cells in the wound bed, whereas antibodies to TGF-β reduce the influx of these cells (see Section 3.4).

3.4.2. Scarring in Adult Wounds

Enhanced expression of mRNAs for TGF-β1 and types I, III, or VI collagen has been observed in tissues of patients with systemic sclerosis (Kulozik *et al.*, 1990; Peltonen *et al.*, 1990), in postburn hypertrophic scar tissue (Ghahary *et al.*, 1993), in diffuse fasciitis (Peltonen *et al.*, 1991b), and in keloids (Peltonen *et al.*, 1991a). Elucidation of the mechanisms underlying scarring will require that the process be studied in its early rather than mature stages. In rat incisional wounds, application of neutralizing antibodies to TGF-β1 and -2, applied at the time of wounding, reduced cutaneous scarring as assessed by the architecture of the neoepidermis (Shah *et al.*, 1994a). Analogous to effects in brain wounds (Logan and Berry, 1993), use of anti-TGF-β1 and -2 antibodies reduced the inflammatory and angiogenic responses as well as extracellular matrix deposition in the early stages of wound healing. The tensile strength of the wounds treated with antibody was equal to that of control, untreated wounds at day 7, whereas wounds treated with a single application of 10 ng TGF-β had twice the strength at that time. However, by days 14 and 70 postwounding, there were no differences in tensile strength between any of the groups (Shah *et al.*, 1992).

Preliminary observations indicate that mannose-6-phosphate, which has been shown to reduce activation of latent TGF-β (Miyazono and Heldin, 1989; Harpel *et al.*, 1992), can also reduce scarring in incisional wounds in rats (McCallion *et al.*, 1995). More surprisingly, TGF-β3, used in the same rat model, reduces the level of immunohistochemical staining for TGF-β1 and -2 and has effects on neodermal architecture similar to that of neutralizing antibodies to TGF-β1 and -2 (Shah *et al.*, 1993, 1994b). These observations must be expanded to other animal models and ultimately tested in human wounds; but, minimally, they suggest that varying the TGF-β profile of a healing wound can have significant effects on the cellularity and neodermal architecture.

4. Fibroproliferative Disorders

Tissue fibrosis, like carcinogenesis discussed in Section 5, is an aberration of the wound-healing process. The process is initiated by injury, often of a chronic nature, and many elements in the process are similar to those in wound healing. However, in a simplistic sense, the stop signals are missing, leading to uncontrolled matrix deposition or scarring (for a review, see Border and Noble, 1994). Just as TGF-β plays a key role in wound healing, so is it thought to be central to development of pathological tissue fibrosis, acting principally on mesenchymal cells to stimulate synthesis of matrix and reduce its proteolysis (Noble *et al.*, 1992). Target tissues in which TGF-β has been implicated in fibrogenesis include kidney, lung, liver, and skin.

In a model of pulmonary fibrosis induced by bleomycin in rats, the temporal relationships between expression of TGF-β1 and type I collagen mRNA and protein strongly suggest that expression of active TGF-β1 by alveolar macrophages is causative in the accumulation of matrix (Khalil and Greenberg, 1991; Khalil *et al.*, 1993). Increased expression of TGF-β1 and extracellular matrix proteins has also been reported in nonnecrotizing granulomas of pulmonary sarcoidosis (Limper *et al.,* 1994) and very early after radiation injury resulting in chronic radiation fibrosis (Finkelstein *et al.,* 1994).

Similarly, in experimental glomerulonephritis in rats, enhanced expression of TGF-β1 correlates with the pathological accumulation of matrix (reviewed in Border and Noble, 1994). *In vitro* studies show that mesangial cells respond to TGF-β treatment with increased secretion of proteoglycans, whereas glomerular epithelial cells secrete increased fibronectin, type IV collagen, and laminin (Nakamura *et al.,* 1992). Kidney biopsies from patients with immunoglobulin A nephropathy, a disease histologically similar to glomerulonephritis, also show elevated glomerular staining for TGF-β (Yoshioka *et al.,* 1993). Elevated levels of expression have also been observed in both a rat model of diabetic nephropathy and in human disease (Yamamoto *et al.,* 1993).

TGF-β has also been implicated in pathogenesis of liver fibrosis (Castilla *et al.,* 1991) and in experimental models of liver disease in the rat, including carbon tetrachloride and galactosamine hepatoxicity and radiation hepatitis (Czaja *et al.,* 1989; Anscher *et al.,* 1990; Armendariz-Borunda *et al.,* 1990).

Current thinking is that development of effective TGF-β antagonists will ultimately provide a new treatment paradigm for these diseases. Injection of antiserum to TGF-β1 has reduced the accumulation of matrix in experimental glomerulonephritis (Border *et al.,* 1990b), in bleomycin-induced pulmonary disease (Giri *et al.,* 1993), as well as in the injury models previously described in brain (Logan and Berry, 1993) and skin (Shah *et al.,* 1994a). Other approaches to interfere with the action of TGF-β in these diseases are based on development of receptor antagonists, soluble receptors or binding proteins including LAP, or agents that would block the pathways of TGF-β activation.

5. Carcinogenesis

The relationship between wound healing and carcinogenesis has been appreciated for a long time. Haddow (1972) proposed that, "the wound may be regarded as a tumor which heals itself," whereas Dvorak (1986) put it another way, stating that, "tumors are wounds that do not heal" and that "tumors appear to the host as an unending series of wounds that continually initiate healing but never heal completely." Each understood that tumor stroma contains many of the same elements as granulation tissue of a healing wound, even though the two processes are initiated by quite distinct mechanisms. Some insight into the molecular basis of this relationship is provided by the discoveries that wounding of chickens infected with Rous sarcoma virus (Sieweke *et al.,* 1990) or of transgenic mice carrying either the *v-jun* oncogene under control of a widely expressed promoter (Schuh *et al.,* 1990) or the bovine papilloma virus type 1 genome (Sippola-Theile *et al.,* 1989) leads to tumor formation. In the chicken study, it was shown that injection of TGF-β, but not other growth factors expressed in wounding, could substitute completely for wounding in tumor induction (Sieweke *et al.,*

1990). These data suggest that wounding-induced autoinduction of TGF-β1, as amplified and sustained by expression of these viral proteins or oncogene products, promotes tumorigenesis possibly in part by enhancing formation of tumor stroma.

6. Clinical Applications

While animal models are obviously not ideal substitutes for human studies, they have nonetheless demonstrated a strong rationale for clinical evaluation of TGF-β in treatment of patients with both acute and chronic wounds. The safety of topical application of TGF-β is demonstrated by pharmacokinetic studies that have shown that high-dose dermal application of TGF-β resulted in local effects at wound sites without systemic toxicity; chronic systemic administration, on the other hand, produced a spectrum of lesions in multiple target tissues, especially liver and kidney (Terrell et al., 1993). Other studies using radiolabeled TGF-β1 have shown that 35% of a topical dose applied to excisional wounds in rats can be recovered intact after 24 hr, again with little or low systemic absorption (Zioncheck et al., 1994).

The first successful clinical application of TGF-β has been in the eye for treatment of full-thickness macular holes that cause a significant reduction in visual acuity. Clinical observations that spontaneous improvement in visual acuity in patients with macular hole was always accompanied by flattening of the rim of the hole, and that TGF-β acted as a chorioretinal glue in treatment of tears in rabbit eyes (Smiddy et al., 1989), suggested that local application of TGF-β might induce a chorioretinal adhesion along the edge of the macular hole. Indeed, application of TGF-β2 to the edge of the hole resulted in both flattening of the detachment and significant improvement in visual acuity (Glaser et al., 1992; Smiddy et al., 1993); the latter was observed most commonly in eyes receiving the highest dose of TGF-β2 (1330 ng/eye). Only the type 2 isoform of TGF-β has been tested, though its use is consistent with the predominance of this isoform in the vitreous (Connor et al., 1989). The intraocular treatment with TGF-β2 simplifies the vitrectomy surgery, eliminating the need for epiretinal membrane peeling (Lansing et al., 1993). Immunohistochemical localization of TGF-β by electron microscopy shows distinct localization of the different isoforms of TGF-β in the retina: TGF-β1 and TGF-β3 are localized to Mueller's glia and retinal ganglion cells; TGF-β2 to the cytoplasm of retinal pigment epithelial cells; TGF-β1 and -2 to the photoreceptor outer segment; and TGF-β3 in the mitochondria of the photoreceptor cell inner segments (Anderson et al., 1995). These findings suggest that TGF-β may play a role in restoration of photoreceptor cell function and not only in reapposition of the rim of detached retinal tissue to the underlying choroid (Glaser et al., 1992; Smiddy et al., 1993).

Its ability to improve or accelerate healing in animal models of injury suggests that TGF-β will be effective in treatment of chronic, nonhealing ulcers in patients. Wound fluid from chronic wounds has been shown to have abnormal growth factor profiles and to have an altered proteolytic balance in favor of proteolysis, compared to "acute" mastectomy wound fluid (Wysocki et al., 1993). Since TGF-β shifts the proteolytic balance in the opposite direction, it might be particularly effective in this environment. Clinical trials have recently begun for evaluation of TGF-β2 in treat-ment of venous stasis ulcers. An open label trial utilizing TGF-β2 applied to the wound in a collagen sponge at 0.5 μg/cm² three times per week for 6 weeks showed improvement in the rate of closure of the ulcer. A subsequent three-arm prospective blind trial using 2.5 μg/cm² TGF-β2 applied in the same manner with the same treatment schedule also showed improvement; there was no indication of recurrence in patients whose ulcers had closed completely (Robson *et al.*, 1995). As in animal models, there was some suggestion that lower doses might be more effective than higher doses. Further studies of the effects of TGF-β on healing of other types of chronic, nonhealing wounds as well as surgical wounds in healing-impaired patients are now critically needed.

7. Perspectives

Clearly, these studies are only a beginning. The multiplicity of effects of TGF-β on cells involved in tissue repair and its proven effectiveness in a wide variety of animal models of tissue injury provide a compelling argument for its clinical applica-tion. Although most attention has been focused on use of TGF-β for healing of cutaneous wounds, evidence is rapidly accumulating that it will also promote repair of injury to many different organs and tissues by both direct and indirect mechanisms. Its clinical application is limited only by our ability to devise appropriate delivery modes and treatment schedules, be they local or systemic, single dose or repeated application; toxicity does not pose a problem for either topical application or for acute administra-tion systemically. The clinical need for an agent that will promote healing is enormous, both for treatment of chronic, nonhealing wounds such as venous stasis, decubitus, or diabetic ulcers, and as a prophylactic treatment for surgical patients, especially those predicted to have impaired healing responses (Sporn and Roberts, 1993). As develop-ment of clinical applications of all three isoforms of TGF-β by several companies in the pharmaceutical industry proceeds, it is hoped that the enormous promise of this molecule will soon be realized.

References

Adams, D. H., Hathaway, M., Shaw, J., Burnett, D., Elias, E., and Strain, A. J., 1991, Transforming growth factor-beta induces human T lymphocyte migration *in vitro, J. Immunol.* **147:**609–612.

Adolph, V. R., DiSanto, S. K., Bleacher, J. C., Dillon, P. W., and Krummel, T. M., 1993, The potential role of the lymphocyte in fetal wound healing, *J. Pediatr. Surg.* **28:**1316–1320.

Adzick, N. S., and Harrison, M. R., 1994, Fetal surgical therapy, *Lancet* **343:**897–902.

Adzick, N. S., and Lorenz, H. P., 1994, Cells, matrix, growth factors, and the surgeon. The biology of scarless fetal wound repair, *Ann. Surg.* **220:**10–18.

Altman, D. J., Schneider, S. L., Thompson, D. A., Cheng, H. L., and Tomasi, T. B., 1990, A transforming growth factor beta 2 (TGF-beta 2)-like immunosuppressive factor in amniotic fluid and localization of TGF-beta 2 mRNA in the pregnant uterus, *J. Exp. Med.* **172:**1391–1401.

Ammann, A. J., Beck, L. S., DeGuzman, L., Hirabayashi, S. E., Lee, W. P., McFatridge, L., Nguyen, T., Xu,

Y., and Mustoe, T. A., 1990, Transforming growth factor-beta. Effect on soft tissue repair, *Ann. NY Acad. Sci.* **593**:124–134.

Anderson, D. H., Guerin, C. J., Hageman, G. S., Pfeffer, B. A., and Flanders, K. C., 1995, The distribution of TGF-β isoforms in the mammalian retina, retinal pigment epithelium, and choroid, *J. Neurosci. Res.* **42**:63–79.

Anscher, M. S., Crocker, I. R., and Jirtle, R. L., 1990, Transforming growth factor-beta 1 expression in irradiated liver, *Radiat. Res.* **122**:77–85.

Armendariz-Borunda, J., Seyer, J. M., Kang, A. H., and Raghow, R., 1990, Regulation of TGF beta gene expression in rat liver intoxicated with carbon tetrachloride, *FASEB J.* **4**:215–221.

Armstead, W. M., Mirro, R., Zuckerman, S. L., Shibata, M., and Leffler, C. W., 1993, Transforming growth factor-beta attenuates ischemia-induced alterations in cerebrovascular responses, *Am. J. Physiol.* **264**:H381–H385.

Bassing, C. H., Yingling, J. M., Howe, D. J., Wang, T., He, W. W., Gustafson, M. L., Shah, P., Donahoe, P. K., and Wang, X. F., 1994, A transforming growth factor beta type I receptor that signals to activate gene expression, *Science* **263**:87–89.

Basson, C. T., Kocher, O., Basson, M. D., Asis, A., and Madri, J. A., 1992, Differential modulation of vascular cell integrin and extracellular matrix expression *in vitro* by TGF-beta 1 correlates with reciprocal effects on cell migration, *J. Cell. Physiol.* **153**:118–128.

Beck, L. S., Chen, T. L., Hirabayashi, S. E., DeGuzman, L., Lee, W. P., McFatridge, L. L., Xu, Y., Bates, R. L., and Ammann, A. J., 1990a, Accelerated healing of ulcer wounds in the rabbit ear by recombinant human transforming growth factor-beta 1, *Growth Factors* **2**:273–282.

Beck, L. S., Chen, T. L., Mikalauski, P., and Ammann, A. J., 1990b, Recombinant human transforming growth factor-beta 1 (rhTGF-beta 1) enhances healing and strength of granulation skin wounds, *Growth Factors* **3**:267–275.

Beck, L. S., DeGuzman, L., Lee, W. P., Xu, Y., McFatridge, L. A., and Amento, E. P., 1991, TGF-beta 1 accelerates wound healing: Reversal of steroid-impaired healing in rats and rabbits, *Growth Factors* **5**:295–304.

Beck, L. S., Amento, E. P., Xu, Y., DeGuzman, L., Lee, W. P., Nguyen, T., and Gillett, N. A., 1993a, TGF-beta 1 induces bone closure of skull defects: Temporal dynamics of bone formation in defects exposed to rhTGF-beta 1, *J. Bone Miner. Res.* **8**:753–761.

Beck, L. S., DeGuzman, L., Lee, W. P., Xu, Y., Siegel, M. W., and Amento, E. P., 1993b, One systemic administration of transforming growth factor-beta 1 reverses age- or glucocorticoid-impaired wound healing, *J. Clin. Invest.* **92**:2841–2849.

Bernstein, E. F., Harisiadis, L., Salomon, G., Norton, J., Sollberg, S., Uitto, J., Glatstein, E., Glass, J., Talbot, T., Russo, A., and Mitchell, J. B., 1991, Transforming growth factor-beta improves healing of radiation-impaired wounds, *J. Invest. Dermatol.* **97**:430–434.

Bonewald, L. F., Wakefield, L., Oreffo, R. O. C., Escobedo, A., Twardzik, D. R., and Mundy, G. R., 1991, Latent forms of transforming growth factor β (TGF-β) derived from bone cultures: Identification of a naturally occurring 100 kDa complex with similarity to recombinant latent TGF-β, *Mol. Endocrinol.* **5**:741–751.

Border, W. A., and Noble, N. A., 1994, Transforming growth factor-β in tissue fibrosis: Understanding the molecular basis of fibrotic disease, *N. Engl. J. Med.* **331**:1286–1292.

Border, W. A., Okuda, S., Languino, L. R., and Ruoslahti, E., 1990a, Transforming growth factor-beta regulates production of proteoglycans by mesangial cells, *Kidney Int.* **37**:689–695.

Border, W. A., Okuda, S., Languino, L. R., Sporn, M. B., and Ruoslahti, E., 1990b, Suppression of experimental glomerulonephritis by antiserum against transforming growth factor beta 1, *Nature* **346**: 371–374.

Bornstein, P., 1992, Thrombospondins—structure and regulation of expression, *FASEB J.* **6**:3290–3299.

Brandes, M. E., Mai, U. E., Ohura, K., and Wahl, S. M., 1991, Type I transforming growth factor-beta receptors on neutrophils mediate chemotaxis to transforming growth factor-beta, *J. Immunol.* **147**: 1600–1606.

Broadley, K. N., Aquino, A. M., Hicks, B., Ditesheim, J. A., McGee, G. S., Demetriou, A. A., Woodward, S. C., and Davidson, J. M., 1989, The diabetic rat as an impaired wound healing model: Stimulatory effects of transforming growth factor-beta and basic fibroblast growth factor, *Biotechnol. Ther.* **1**:55–68.

Brown, G. L., Curtsinger, L. J., White, M., Mitchell, R. O., Pietsch, J., Nordquist, R., von Fraunhofer, A., and Schultz, G. S., 1988, Acceleration of tensile strength of incisions treated with EGF and TGF-beta, *Ann. Surg.* **208**:788–794.

Brown, P. D., Wakefield, L. M., Levinson, A. D., and Sporn, M. B., 1990, Physicochemical activation of recombinant latent transforming growth factor-beta's 1, 2, and 3, *Growth Factors* **3**:35–43.

Burt, D. W., and Law, A. S., 1994, Evolution of the transforming growth factor-beta superfamily, *Prog. Growth Factor Res.* **5**:99–118.

Castilla, A., Prieto, J., and Fausto, N., 1991, Transforming growth factors beta 1 and alpha in chronic liver disease. Effects of interferon alfa therapy, *N. Engl. J. Med.* **324**:933–940.

Chen, T. L., Bates, R. L., Xu, Y., Ammann, A. J., and Beck, L. S., 1992, Human recombinant transforming growth factor-beta 1 modulation of biochemical and cellular events in healing of ulcer wounds, *J. Invest. Dermatol.* **98**:428–435.

Ciacci, C., Lind, S. E., and Podolsky, D. K., 1993, Transforming growth factor beta regulation of migration in wounded rat intestinal epithelial monolayers, *Gastroenterology* **105**:93–101.

Connor, Jr., T. B., Roberts, A. B., Sporn, M. B., Danielpour, D., Dart, L. L., Michels, R. G., de Bustros, S., Enger, C., Kato, H., Lansing, M., Hayashi, H., and Glaser, B. M., 1989, Correlation of fibrosis and transforming growth factor-beta type 2 levels in the eye, *J. Clin. Invest.* **83**:1661–1666.

Cox, D. A., Kunz, S., Cerletti, N., McMaster, G. K., and Burk, R. R., 1992, Wound healing in aged animals— Effects of locally applied transforming growth factor beta 2 in different model systems, *EXS* **61**:287–295.

Cromack, D. T., Sporn, M. B., Roberts, A. B., Merino, M. J., Dart, L. L., and Norton, J. A., 1987, Transforming growth factor beta levels in rat wound chambers, *J. Surg. Res.* **42**:622–628.

Cromack, D. T., Pierce, G. F., and Mustoe, T. A., 1991, TGF-beta and PDGF mediated tissue repair: Identifying mechanisms of action using impaired and normal models of wound healing, *Prog. Clin. Biol. Res.* **365**:359–373.

Cromack, D. T., Porras-Reyes, B., Purdy, J. A., Pierce, G. F., and Mustoe, T. A., 1993, Acceleration of tissue repair by transforming growth factor beta 1: Identification of *in vivo* mechanism of action with radiotherapy-induced specific healing deficits, *Surgery* **113**:36–42.

Curtsinger, L. J., Pietsch, J. D., Brown, G. L., von Fraunhofer, A., Ackerman, D., Polk, Jr., H. C., and Schultz, G. S., 1989, Reversal of adriamycin-impaired wound healing by transforming growth factor-beta, *Surg. Gynecol. Obstet.* **168**:517–522.

Czaja, M. J., Flanders, K. C., Biempica, L., Klein, C., Zern, M. A., and Weiner, F. R., 1989, Expression of tumor necrosis factor-alpha and transforming growth factor-beta 1 in acute liver injury, *Growth Factors* **1**:219–226.

Dallas, S. L., Park-Snyder, S., Miyazono, K., Twardzik, D., Mundy, G. R., and Bonewald, L. F., 1994, Characterization and autoregulation of latent transforming growth factor beta (TGF beta) complexes in osteoblast-like cell lines. Production of a latent complex lacking the latent TGF beta-binding protein, *J. Biol. Chem.* **269**:6815–6821.

Danielpour, D., and Sporn, M. B., 1990, Differential inhibition of transforming growth factor beta 1 and beta 2 activity by alpha 2-macroglobulin, *J. Biol. Chem.* **265**:6973–6977.

Dey, B. R., Sukhatme, V. P., Roberts, A. B., Sporn, M. B., Rauscher III, F. J., and Kim, S.-J., 1994, Repression of the transforming growth factor-β1 gene by the Wilms' tumor suppressor WT1 gene product, *Mol. Endocrinol.* **8**:595–602.

Dreizen, S., McCredie, K. B., and Keating, M. J., 1981, Chemotherapy induced mucositis in adult leukemia, *Postgrad. Med.* **69**:103–108.

D'Souza, R. N., Niederreither, K., and de Crombrugghe, B., 1993, Osteoblast-specific expression of the alpha 2(I) collagen promoter in transgenic mice: Correlation with the distribution of TGF-beta 1, *J Bone Miner. Res.* **8**:1127–1136.

Dvorak, H. F., 1986, Tumors: Wounds that do not heal, *N. Engl. J. Med.* **315**:1650–1659.

Ebner, R., Chen, R. H., Lawler, S., Zioncheck, T., and Derynck, R., 1993, Determination of type I receptor specificity by the type II receptors for TGF-beta or activin, *Science* **262**:900–902.

Eklov, S., Funa, K., Nordgren, H., Olofsson, A., Kanzaki, T., Miyazono, K., and Nilsson, S., 1993, Lack of the latent transforming growth factor beta binding protein in malignant, but not benign prostatic tissue, *Cancer Res.* **53**:3193–3197.

Ellis, I., Grey, A. M., Schor, A. M., and Schor, S. L., 1992, Antagonistic effects of TGF-beta 1 and MSF on fibroblast migration and hyaluronic acid synthesis. Possible implications for dermal wound healing, *J Cell Sci.* **102**:447–456.

Falcone, D. J., McCaffrey, T. A., Haimovitz-Friedman, A., Vergilio, J. A., and Nicholson, A. C., 1993, Macrophage and foam cell release of matrix-bound growth factors. Role of plasminogen activation, *J. Biol. Chem.* **268**:11951–11958.

Fawthrop, F. W., Oyajobi, B. O., Bunning, R. A., and Russell, R. G., 1992, The effect of transforming growth factor beta on the plasminogen activator activity of normal human osteoblast-like cells and a human osteosarcoma cell line MG-63, *J Bone Miner. Res.* **7**:1363–1371.

Finch, C. E., Laping, N. J., Morgan, T. E., Nichols, N. R., and Pasinetti, G. M., 1993, TGF-beta 1 is an organizer of responses to neurodegeneration, *J. Cell. Biochem.* **53**:314–322.

Finkelstein, J. N., Johnston, C. J., Baggs, R., and Rubin, P., 1994, Early alterations in extracellular matrix and transforming growth factor beta gene expression in mouse lung indicative of late radiation fibrosis, *Int. J Radiat. Oncol. Biol. Phys.* **28**:621–631.

Flaumenhaft, R., Abe, M., Sato, Y., Miyazono, K., Harpel, J., Heldin, C. H., and Rifkin, D. B., 1993a, Role of the latent TGF-beta binding protein in the activation of latent TGF-beta by co-cultures of endothelial and smooth muscle cells, *J. Cell Biol.* **120**:995–1002.

Flaumenhaft, R., Kojima, S., Abe, M., and Rifkin, D. B., 1993b, Activation of latent transforming growth factor beta, *Adv. Pharmacol.* **24**:51–76.

Franzén, L. E., and Schultz, G. S., 1993, Transforming growth factor-β enhances connective tissue repair in perforated rat mesentery but not peritoneal macrophage chemotaxis, *Wound Rep. Reg.* **1**:149–55.

Fukushima, D., Butzow, R., Hildebrand, A., and Rouslahti, E., 1993, Localization of transforming growth factor-β binding site in betaglycan, *J. Biol. Chem.* **268**:22710–22715.

Gajdusek, C. M., Luo, Z., and Mayberg, M. R., 1993, Basic fibroblast growth factor and transforming growth factor beta-1: Synergistic mediators of angiogenesis *in vitro, J. Cell. Physiol.* **157**:133–144.

Geiser, A. G., Burmester, J. K., Webbink, R., Roberts, A. B., and Sporn, M. B., 1992, Inhibition of growth by transforming growth factor-beta following fusion of two nonresponsive human carcinoma cell lines. Implication of the type II receptor in growth inhibitory responses, *J. Biol. Chem.* **267**:2588–2593.

Ghahary, A., Shen, Y. J., Scott, P. G., Gong, Y., and Tredget, E. E., 1993, Enhanced expression of mRNA for transforming growth factor-beta, type I and type III procollagen in human post-burn hypertrophic scar tissues, *J. Lab. Clin. Med.* **122**:465–473.

Giri, S. N., Hyde, D. M., and Hollinger, M. A., 1993, Effect of antibody to transforming growth factor beta on bleomycin induced accumulation of lung collagen in mice, *Thorax* **48**:959–966.

Glaser, B. M., Michels, R. G., Kuppermann, B. D., Sjaarda, R. N., and Pena, R. A., 1992, Transforming growth factor-beta 2 for the treatment of full-thickness macular holes. A prospective randomized study, *Ophthalmology* **99**:1162–1172.

Glick, A. B., Abdulkarem, N., Flanders, K. C., Lumadue, J. A., Smith, J. M., and Sporn, M. B., 1991, Complex regulation of TGFβ expression by retinoic acid in the vitamin A- deficient rat, *Development* **111**:1081–1086.

Gombotz, W. R., Pankey, S. C., Bouchard, L. S., Ranchalis, J., and Puolakkainen, P., 1993, Controlled release of TGF-beta 1 from a biodegradable matrix for bone regeneration, *J. Biomater. Sci. Polym. Ed.* **5**:49–63.

Graham, M. F., Bryson, G. R., and Diegelmann, R. F., 1990, Transforming growth factor beta 1 selectively augments collagen synthesis by human intestinal smooth muscle cells, *Gastroenterology* **99**:447–453.

Grant, M. B., Khaw, P. T., Schultz, G. S., Adams, J. L., and Shimizu, R. W., 1992, Effects of epidermal growth factor, fibroblast growth factor, and transforming growth factor-beta on corneal cell chemotaxis, *Invest. Ophthalmol. Vis. Sci.* **33**:3292–3301.

Gross, C. E., Bednar, M. M., Howard, D. B., and Sporn, M. B., 1993, Transforming growth factor-beta 1 reduces infarct size after experimental cerebral ischemia in a rabbit model, *Stroke* **24**:558–562.

Haddow, A., 1972, Molecular repair, wound healing, and carcinogenesis: Tumor production a possible overhealing? *Adv. Cancer Res.* **16**:181–234.

Harpel, J. G., Metz, C. N., Kojima, S., and Rifkin, D. B., 1992, Control of transforming growth factor-beta activity: Latency vs. activation, *Prog. Growth Factor. Res.* **4**:321–335.

He, W. W., Gustafson, M. L., Hirobe, S., and Donahoe, P. K., 1993, Developmental expression of four novel serine/threonine kinase receptors homologous to the activin/transforming growth factor-β type II receptor family, *Dev. Dyn.* **196:**133–142.

Hebda, P. A., 1988, Stimulatory effects of transforming growth factor-beta and epidermal growth factor on epidermal cell outgrowth from porcine skin explant cultures, *J. Invest. Dermatol.* **91:**440–445.

Houghton, P. E., Keefer, K. A., and Krummel, T. M., 1995, The role of transforming growth factor beta (TGF-β) in the conversion from scarless healing to healing with scar formation, *Wound Rep. Reg.* **3:**229–236.

Hunt, T. K., Ehrlich, H. P., Garcia, J. A., and Dunphy, J. E., 1969, Effect of vitamin A on reversing the inhibitory effect of cortisone on healing of open wounds in animals and man, *Ann. Surg.* **170:**633–641.

Jampel, H. D., Roche, N., Stark, W. J., and Roberts, A. B., 1990, Transforming growth factor-beta in human aqueous humor, *Curr. Eye Res.* **9:**963–969.

Jin, Y., Cox, D. A., Knecht, R., Raschdorf, F., and Cerletti, N., 1991, Separation, purification, and sequence identification of TGF-beta 1 and TGF-beta 2 from bovine milk, *J. Protein Chem.* **10:**565–575.

Jones, S. C., Curtsinger, L. J., Whalen, J. D., Pietsch, J. D., Ackerman, D., Brown, G. L., and Schultz, G. S., 1991, Effect of topical recombinant TGF-beta on healing of partial thickness injuries, *J. Surg. Res.* **51:**344–352.

Joyce, M. E., Roberts, A. B., Sporn, M. B., and Bolander, M. E., 1990, Transforming growth factor-beta and the initiation of chondrogenesis and osteogenesis in the rat femur, *J. Cell Biol.* **110:**2195–2207.

Joyce, M. E., Jingushi, S., Scully, S. P., and Bolander, M. E., 1991, Role of growth factors in fracture healing, *Prog. Clin. Biol. Res* **365:**391–416.

Kane, C. J., Hebda, P. A., Mansbridge, J. N., and Hanawalt, P. C., 1991, Direct evidence for spatial and temporal regulation of transforming growth factor beta 1 expression during cutaneous wound healing, *J. Cell. Physiol.* **148:**157–173.

Keski-Oja, J., and Koli, K., 1992, Enhanced production of plasminogen activator activity in human and murine keratinocytes by transforming growth factor-beta 1, *J. Invest. Dermatol.* **99:**193–200.

Khalil, N., and Greenberg, A. H., 1991, The role of TGF-beta in pulmonary fibrosis, *Ciba Found. Symp.* **157:**194–207.

Khalil, N., Whitman, C., Zuo, L., Danielpour, D., and Greenberg, A., 1993, Regulation of alveolar macrophage transforming growth factor-beta secretion by corticosteroids in bleomycin-induced pulmonary inflammation in the rat, *J. Clin. Invest.* **92:**1812–1818.

Kim, S. J., Jeang, K. T., Glick, A. B., Sporn, M. B., and Roberts, A. B., 1989, Promoter sequences of the human transforming growth factor-beta 1 gene responsive to transforming growth factor-beta 1 autoinduction, *J. Biol. Chem.* **264:**7041–7045.

Kim, S. J., Park, K., Koeller, D., Kim, K. Y., Wakefield, L. M., Sporn, M. B., and Roberts, A. B., 1992, Posttranscriptional regulation of the human transforming growth factor-beta 1 gene, *J. Biol. Chem.* **267:**13702–13707.

Kojima, S., and Rifkin, D. B., 1993, Mechanism of retinoid-induced activation of latent transforming growth factor-beta in bovine endothelial cells, *J. Cell. Physiol.* **155:**323–332.

Ksander, G. A., Chu, G. H., McMullin, H., Ogawa, Y., Pratt, B. M., Rosenblatt, J. S., and McPherson, J. M., 1990a, Transforming growth factors-beta 1 and beta 2 enhance connective tissue formation in animal models of dermal wound healing by secondary intent, *Ann. NY Acad. Sci.* **593:**135–147.

Ksander, G. A., Ogawa, Y., Chu, G. H., McMullin, H., Rosenblatt, J. S., and McPherson, J. M., 1990b, Exogenous transforming growth factor-beta 2 enhances connective tissue formation and wound strength in guinea pig dermal wounds healing by secondary intent, *Ann. Surg.* **211:**288–294.

Ksander, G. A., Gerhardt, C. O., and Olsen, D. R., 1993, Exogenous transforming growth factor-β2 enhances connective tissue formation in transforming growth factor-β1-deficient, healing-impaired dermal wounds in mice, *Wound Rep. Reg.* **1:**137–148.

Kulkarni, A. B., Huh, C. G., Becker, D., Geiser, A., Lyght, M., Flanders, K. C., Roberts, A. B., Sporn, M. B., Ward, J. M., and Karlsson, S., 1993, Transforming growth factor beta 1 null mutation in mice causes excessive inflammatory response and early death, *Proc. Natl. Acad. Sci. USA* **90:**770–774.

Kulozik, M., Hogg, A., Lankat-Buttgereit, B., and Krieg, T., 1990, Co-localization of transforming growth factor beta 2 with alpha 1(I) procollagen mRNA in tissue sections of patients with systemic sclerosis, *J. Clin. Invest.* **86:**917–922.

Lafyatis, R., Lechleider, R., Kim, S. J., Jakowlew, S., Roberts, A. B., and Sporn, M. B., 1990, Structural and functional characterization of the transforming growth factor beta 3 promoter. A cAMP-responsive element regulates basal and induced transcription, *J. Biol. Chem.* **265:**19128–19136.

Lafyatis, R., Lechleider, R., Roberts, A. B., and Sporn, M. B., 1991, Secretion and transcriptional regulation of transforming growth factor-β3 during myogenesis, *Mol. Cell Biol.* **11.7:**3795–3803.

Lansing, M. B., Glaser, B. M., Liss, H., Hanham, A., Thompson, J. T., Sjaarda, R. N., and Gordon, A. J., 1993, The effect of pars plana vitrectomy and transforming growth factor-beta 2 without epiretinal membrane peeling on full-thickness macular holes, *Ophthalmology* **100:**868–871.

Lawrence, W. T., and Diegelmann, R. F., 1994, Growth factors in wound healing, *Clin. Dermatol.* **12:**157–169.

Lefer, A. M., Tsao, P., Aoki, N., and Palladino, Jr., M. A., 1990, Mediation of cardioprotection by transforming growth factor-beta, *Science* **249:**61–64.

Lefer, A. M., Ma, X. L., Weyrich, A. S., and Scalia, R., 1993, Mechanism of the cardioprotective effect of transforming growth factor beta 1 in feline myocardial ischemia and reperfusion, *Proc. Natl. Acad. Sci. USA* **90:**1018–1022.

Levine, J. H., Moses, H. L., Gold, L. I., and Nanney, L. B., 1993, Spatial and temporal patterns of immunoreactive transforming growth factor beta 1, beta 2, and beta 3 during excisional wound repair, *Am. J. Pathol.* **143:**368–380.

Limper, A. H., Colby, T. V., Sanders, M. S., Asakura, S., Roche, P. C., and DeRemee, R. A., 1994, Immunohistochemical localization of transforming growth factor-beta 1 in the nonnecrotizing granulomas of pulmonary sarcoidosis, *Am. J. Respir. Crit. Care Med.* **149:**197–204.

Lin, H. Y., and Moustakas, A., 1994, TGF-β receptors: Structure and function, *Cell. Mol. Biol.* **40:**337–349.

Lin, H. Y., Wang, X.-F., Ng-Eaton, E., Weinberg, R. A., and Lodish, H. F., 1992, Expression cloning of the TGF-β type II receptor, a functional transmembrane serine/threonine kinase, *Cell* **68:**775–785.

Lindholm, D., Castren, E., Kiefer, R., Zafra, F., and Thoenen, H., 1992, Transforming growth factor-beta 1 in the rat brain: Increase after injury and inhibition of astrocyte proliferation, *J. Cell Biol.* **117:**395–400.

Liu, J., Hales, A. M., Chamberlain, C. G., and McAvoy, J. W., 1994, Induction of cataract-like changes in rat lens epithelial explants by transforming growth factor beta, *Invest. Ophthalmol. Vis. Sci.* **35:**388–401.

Liu, Q., Huang, S. S., and Huang, J. S., 1994, Kinase activity of the type V transforming growth factor beta receptor, *J. Biol. Chem.* **269:**9221–9226.

Logan, A., and Berry, M., 1993, Transforming growth factor-beta 1 and basic fibroblast growth factor in the injured CNS, *Trends Pharmacol. Sci.* **14:**337–342.

Logan, A., Frautschy, S. A., Gonzalez, A. M., Sporn, M. B., and Baird, A., 1992, Enhanced expression of transforming growth factor beta 1 in the rat brain after a localized cerebral injury, *Brain Res.* **587:**216–225.

Longaker, M. T., Whitby, D. J., Adzick, N. S., Crombleholme, T. M., Langer, J. C., Duncan, B. W., Bradley, S. M., Stern, R., Ferguson, M. W. J., and Harrison, M. R., 1990, Studies in fetal wound healing. VI. Second and early third trimester fetal wounds demonstrate rapid collagen deposition without scar formation, *J. Pediatr. Surg.* **25:**63–69.

Longaker, M. T., Bouhana, K. S., Harrison, M. R., Danielpour, D., Roberts, A. B., and Banda, M. J., 1994, Wound healing in the fetus: Possible role for inflammatory macrophages and transforming growth factor-β isoforms, *Wound Rep. Reg.* **2:**104–112.

Lynch, S. E., Colvin, R. B., and Antoniades, H. N., 1989, Growth factors in wound healing. Single and synergistic effects on partial thickness porcine skin wounds, *J. Clin. Invest.* **84:**640–646.

Martin, M., Lefaix, J. L., Pinton, P., Crechet, F., and Daburon, F., 1993, Temporal modulation of TGF-beta 1 and beta-actin gene expression in pig skin and muscular fibrosis after ionizing radiation, *Radiat. Res* **134:**63–70.

Martin, P., and Nobes, C. D., 1992, An early molecular component of the wound healing response in rat embryos—induction of c-fos protein in cells at the epidermal wound margin, *Mech. Dev.* **38:**209–215.

Martin, P., Hopkinson-Woolley, J., and McCluskey, J., 1992, Growth factors and cutaneous wound repair, *Prog. Growth Factor Res.* **4:**25–44.

Martin, P., Dickson, M. C., Millan, F. A., and Akhurst, R. J., 1993, Rapid induction and clearance of TGF beta 1 is an early response to wounding in the mouse embryo, *Dev. Genet.* **14:**225–238.

Massagué, J., Attisano, L., and Wrana, J. L., 1994, The TGF-β family and its composite receptors, *Trends Cell Biol.* **4:**172–178.

Mast, B. A., Diegelmann, R. F., Krummel, T. M., and Cohen, I. K., 1992, Scarless wound healing in the mammalian fetus, *Surg. Gynecol. Obstet.* **174:**441–451.

McCallion, R., Wood, J., Foreman, D., Shah, M., and Ferguson, M. W. J., 1995, Exogenous mannose-6-phosphate prevents scarring in cutaneous wounds, *Laucet,* in press.

Merwin, J. R., Anderson, J. M., Kocher, O., Van Itallie, C. M., and Madri, J. A., 1990, Transforming growth factor beta 1 modulates extracellular matrix organization and cell–cell junctional complex formation during *in vitro* angiogenesis, *J. Cell. Physiol.* **142:**117–128.

Miyazono, K., and Heldin, C. H., 1989, Role for carbohydrate structures in TGF-beta 1 latency, *Nature* **338:**158–160.

Miyazono, K., Ichijo, H., and Heldin, C. H., 1993, Transforming growth factor-beta: Latent forms, binding proteins and receptors, *Growth Factors* **8:**11–22.

Montesano, R., and Orci, L., 1988, Transforming growth factor beta stimulates collagen–matrix contraction by fibroblasts: Implications for wound healing, *Proc. Natl. Acad. Sci. USA* **85:**4894–4897.

Mustoe, T. A., Pierce, G. F., Thomason, A., Gramates, P., Sporn, M. B., and Deuel, T. F., 1987, Accelerated healing of incisional wounds in rats induced by transforming growth factor-beta, *Science* **237:**1333–1336.

Mustoe, T. A., Purdy, J., Gramates, P., Deuel, T. F., Thomason, A., and Pierce, G. F., 1989, Reversal of impaired wound healing in irradiated rats by platelet-derived growth factor, *Am. J. Surg.* **158:**345–350.

Mustoe, T. A., Landes, A., Cromack, D. T., Mistry, D., Griffin, A., Deuel, T. F., and Pierce, G. F., 1990, Differential acceleration of healing of surgical incisions in the rabbit gastrointestinal tract by platelet-derived growth factor and transforming growth factor, type beta, *Surgery* **108:**324–329.

Mustoe, T. A., Pierce, G. F., Morishima, C., and Deuel, T. F., 1991, Growth factor-induced acceleration of tissue repair through direct and inductive activities in a rabbit dermal ulcer model, *J. Clin. Invest.* **87:**694–703.

Nakamura, T., Miller, D., Ruoslahti, E., and Border, W. A., 1992, Production of extracellular matrix by glomerular epithelial cells is regulated by transforming growth factor-beta 1, *Kidney Int.* **41:**1213–1221.

Nathan, C., and Sporn, M. B., 1991, Cytokines in context, *J. Cell Biol.* **113:**981–986.

Nickoloff, B. J., Mitra, R. S., Riser, B. L., Dixit, V. M., and Varani, J., 1988, Modulation of keratinocyte motility. Correlation with production of extracellular matrix molecules in response to growth promoting and antiproliferative factors, *Am. J. Pathol.* **132:**543–551.

Noble, N. A., Harper, J. R., and Border, W. A., 1992, *In vivo* interactions of TGF-beta and extracellular matrix, *Prog. Growth Factor Res.* **4:**369–382.

O'Connor-McCourt, M. D., and Wakefield, L. M., 1987, Latent transforming growth factor-beta in serum. A specific complex with alpha 2-macroglobulin, *J. Biol. Chem.* **262:**14090–14099.

O'Reilly, M. A., Geiser, A. G., Kim, S. J., Bruggeman, L. A., Luu, A. X., Roberts, A. B., and Sporn, M. B., 1992, Identification of an activating transcription factor (ATF) binding site in the human transforming growth factor-beta 2 promoter, *J. Biol. Chem.* **267:**19938–19943.

Park, K., Kim, S.-J., Bang, Y-J., Park, J.-G., Kim, N. K., Roberts, A. B., and Sporn, M. B., 1994, Genetic change in the transforming growth factor-β (TGF-β) type II receptor gene in human gastric cancer cells: Correlation with sensitivity to growth inhibition by TGF-β, *Proc. Natl. Acad. Sci. USA* **91:**8772–8776.

Peltonen, J., Kähäri, L., Jaakkola, S., Kähäri, V. M., Varga, J., Uitto, J., and Jimenez, S. A., 1990, Evaluation of transforming growth factor beta and type I procollagen gene expression in fibrotic skin diseases by *in situ* hybridization, *J. Invest. Dermatol.* **94:**365–371.

Peltonen, J., Hsiao, L. L., Jaakkola, S., Sollberg, S., Aumailley, M., Timpl, R., Chu, M. L., and Uitto, J., 1991a, Activation of collagen gene expression in keloids: Co-localization of type I and VI collagen and transforming growth factor-beta 1 mRNA, *J. Invest. Dermatol.* **97:**240–248.

Peltonen, J., Varga, J., Sollberg, S., Uitto, J., and Jimenez, S. A., 1991b, Elevated expression of the genes for transforming growth factor-beta 1 and type VI collagen in diffuse fasciitis associated with the eosinophilia-myalgia syndrome, *J. Invest. Dermatol.* **96:**20–25.

Pepin, M. C., Beauchemin, M., Plamondon, J., and O'Connor-McCourt, M. D., 1994, Mapping of the ligand binding domain of the transforming growth factor β receptor type III by deletion mutagenesis, *Proc. Natl. Acad. Sci. USA* **91**:6997–7001.

Pepper, M. S., Vassalli, J. D., Orci, L., and Montesano, R., 1993, Biphasic effect of transforming growth factor-beta 1 on *in vitro* angiogenesis, *Exp. Cell Res.* **204**:356–363.

Perry, L. C., Connors, A. W., Matrisian, L. M., Nanney, L. B., Charles, D. P., Reyes, D. P., Kerr, L. D., and Fisher, J., 1993, Role of transforming growth factor-β1 and epidermal growth factor in the wound-healing process: An *in vivo* biomechanical evaluation, *Wound Rep. Reg.* **1**:41–46.

Pfeilschifter, J., Wolf, O., Naumann, A., Minne, H. W., Mundy, G. R., and Ziegler, R., 1990, Chemotactic response of osteoblastlike cells to transforming growth factor beta, *J. Bone Miner. Res.* **5**:825–830.

Phillips, G. D., Whitehead, R. A., and Knighton, D. R., 1992, Inhibition by methylprednisolone acetate suggests an indirect mechanism for TGF-B induced angiogenesis, *Growth Factors* **6**:77–84.

Pierce, G. F., Mustoe, T. A., Lingelbach, J., Masakowski, V. R., Gramates, P., and Deuel, T. F., 1989a, Transforming growth factor beta reverses the glucocorticoid-induced wound-healing deficit in rats: Possible regulation in macrophages by platelet-derived growth factor, *Proc. Natl. Acad. Sci. USA* **86**:2229–2233.

Pierce, G. F., Mustoe, T. A., Lingelbach, J., Masakowski, V. R., Griffin, G. L., Senior, R. M., and Deuel, T. F., 1989b, Platelet-derived growth factor and transforming growth factor-beta enhance tissue repair activities by unique mechanisms, *J. Cell Biol.* **109**:429–440.

Pierce, G. F., Brown, D., and Mustoe, T. A., 1991a, Quantitative analysis of inflammatory cell influx, procollagen type I synthesis, and collagen cross-linking in incisional wounds: Influence of PDGF-BB and TGF-beta 1 therapy, *J. Lab. Clin. Med.* **117**:373–382.

Pierce, G. F., Vande Berg, J., Rudolph, R., Tarpley, J., and Mustoe, T. A., 1991b, Platelet-derived growth factor-BB and transforming growth factor beta 1 selectively modulate glycosaminoglycans, collagen, and myofibroblasts in excisional wounds, *Am. J. Pathol.* **138**:629–646.

Pierce, G. F., Tarpley, J. E., Yanagihara, D., Mustoe, T. A., Fox, G. M., and Thomason, A., 1992, Platelet-derived growth factor (BB homodimer), transforming growth factor-beta 1, and basic fibroblast growth factor in dermal wound healing. Neovessel and matrix formation and cessation of repair, *Am. J. Pathol.* **140**:1375–1388.

Postlethwaite, A. E., Keski-Oja, J., Moses, H. L., and Kang, A. H., 1987, Stimulation of the chemotactic migration of human fibroblasts by transforming growth factor beta, *J. Exp. Med.* **165**:251–256.

Prehn, J. H., Backhauss, C., and Krieglstein, J., 1993, Transforming growth factor-beta 1 prevents glutamate neurotoxicity in rat neocortical cultures and protects mouse neocortex from ischemic injury *in vivo, J. Cereb. Blood Flow Metab.* **13**:521–525.

Quaglino, Jr., D., Nanney, L. B., Kennedy, R., and Davidson, J. M., 1990, Transforming growth factor-beta stimulates wound healing and modulates extracellular matrix gene expression in pig skin. I. Excisional wound model, *Lab. Invest.* **63**:307–319.

Quaglino, Jr., D., Nanney, L. B., Ditesheim, J. A., and Davidson, J. M., 1991, Transforming growth factor-beta stimulates wound healing and modulates extracellular matrix gene expression in pig skin: Incisional wound model, *J. Invest. Dermatol.* **97**:34–42.

Roberts, A. B., and Sporn, M. B., 1990, The transforming growth factor βs, in: *Handbook of Experimental Pharmacology. Peptide Growth Factors and Their Receptors,* 95th ed. (M. B. Sporn and A. B. Roberts, eds.), pp. 419–472, Springer-Verlag, New York.

Roberts, A. B., and Sporn, M. B., 1992a, Differential expression of the TGF-beta isoforms in embryogenesis suggests specific roles in developing and adult tissues, *Mol. Reprod. Dev.* **32**:91–98.

Roberts, A. B., and Sporn, M. B., 1992b, Mechanistic Interrelationships between two superfamilies: The steroid/retinoid receptors and transforming growth factor-β, *Cancer Surv.* **14**:205–219.

Roberts, A. B., and Sporn, M. B., 1993, Physiological actions and clinical applications of transforming growth factor-beta (TGF-beta), *Growth Factors* **8**:1–9.

Roberts, A. B., Sporn, M. B., Assoian, R. K., Smith, J. M., Roche, N. S., Wakefield, L. M., Heine, U. I., Liotta, L. A., Falanga, V., Kehrl, J. H., and Fauci, A. S., 1986, Transforming growth factor type beta: Rapid induction of fibrosis and angiogenesis *in vivo* and stimulation of collagen formation *in vitro, Proc. Natl. Acad. Sci. USA* **83**:4167–4171.

Roberts, A. B., Rosa, F., Roche, N. S., Coligan, J. E., Garfield, M., Rebbert, M. L., Kondaiah, P., Danielpour, D., Kehrl, J. H., Wahl, S. M., David, I. B., and Sporn, M. B., 1990, Isolation and characterization of

TGF-beta 2 and TGF-beta 5 from medium conditioned by *Xenopus* XTC cells, *Growth Factors* **2**: 135–147.

Roberts, A. B., Sporn, M. B., and Lefer, A. M., 1993, Cardioprotective actions of transforming growth factor-β, *Trends Cardiol. Med.* **3**:77–81.

Robson, M. C., Phillips, L. G., Cooper, D. M., Robson, L. E., Hanham, A. F., and Ksander, G. A., 1995, Transforming growth factor-β2 accelerates healing of venous stasis ulcers in an open-label, placebo-controlled clinical study, *Wound Rep. Reg.*, **3**:157–167.

Romeo, D. S., Park, K., Roberts, A. B., Sporn, M. B., and Kim, S. J., 1993, An element of the transforming growth factor-beta 1 5′-untranslated region represses translation and specifically binds a cytosolic factor, *Mol. Endocrinol.* **7**:759–766.

Salomon, G. D., Kasid, A., Bernstein, E., Buresh, C., Director, E., and Norton, J. A., 1990, Gene expression in normal and doxorubicin-impaired wounds: Importance of transforming growth factor-beta, *Surgery* **108**:318–322.

Sato, Y., and Rifkin, D. B., 1989, Inhibition of endothelial cell movement by pericytes and smooth muscle cells: Activation of a latent transforming growth factor-beta 1-like molecule by plasmin during co-culture, *J. Cell Biol.* **109**:309–315.

Schmid, P., Cox, D., Bilbe, G., McMaster, G., Morrison, C., Stahelin, H., Luscher, N., and Seiler, W., 1993a, TGF-beta S and TGF-beta type II receptor in human epidermis: Differential expression in acute and chronic skin wounds, *J. Pathol.* **171**:191–197.

Schmid, P., Kunz, S., Cerletti, N., McMaster, G., and Cox, D., 1993b, Injury induced expression of TGF-beta 1 mRNA is enhanced by exogenously applied TGF-beta S, *Biochem. Biophys. Res Commun.* **194**: 399–406.

Schuh, A. C., Keating, S. J., Monteclaro, F. S., Vogt, P. K., and Breitman, M. L., 1990, Obligatory wounding requirement for tumorigenesis in *v-jun* transgenic mice, *Nature* **346**:756–760.

Schultz, G., Chegini, N., Grant, M., Khaw, P., and MacKay, S., 1992, Effects of growth factors on corneal wound healing, *Acta Ophthalmol. Suppl.* **202**:60–66.

Schultz-Cherry, S., and Murphy-Ulrich, J. E., 1993, Thrombospondin causes activation of latent transforming growth factor-β secreted by endothelial cells by a novel mechanism, *J. Cell Biol.* **122**:923–932.

Shah, M., Foreman, D. M., and Ferguson, M. W., 1992, Control of scarring in adult wounds by neutralising antibody to transforming growth factor beta, *Lancet* **339**:213–214.

Shah, M., Roberts, A. B., Gold, L. I., and Ferguson, M. W. J., 1993, Immunolocalization of TGF-β isoforms in normal and experimentally modulated incisional wounds in adult rodents, *Wound Rep. Reg.* **1**:124.

Shah, M., Foreman, D. M., and Ferguson, W. J., 1994a, Neutralizing antibody to TGF-β1,2 reduces cutaneous scarring in adult rodents, *J. Cell Sci.* **107**:1137–1157.

Shah, M., Foreman, D. M., and Ferguson, W. J., 1995, Neutralization of TGF-β1 and TGF-β2 or exogenous addition of TGF-β3 to cutaneous rat wounds reduces scarring, *J. Cell Sci.*, **108**:985–1002.

Sieweke, M. H., Thompson, N. L., Sporn, M. B., and Bissell, M. J., 1990, Mediation of wound-related Rous sarcoma virus tumorigenesis by TGF-beta, *Science* **248**:1656–1660.

Sippola-Thiele, M., Hanahan, D., and Howley, P. M., 1989, Cell-heritable stages of tumor progression in transgenic mice harboring the bovine papillomavirus type 1 genome, *Mol. Cell. Biol.* **9**:925–934.

Slavin, J., Nash, J. R., and Kingsnorth, A. N., 1992, Effect of transforming growth factor beta and basic fibroblast growth factor on steroid-impaired healing intestinal wounds, *Br. J. Surg.* **79**:69–72.

Smiddy, W. E., Glaser, B. M., Green, W. R., Connor, Jr., T. B., Roberts, A. B., Lucas, R., and Sporn, M. B., 1989, Transforming growth factor beta. A biologic chorioretinal glue, *Arch. Ophthalmol.* **107**:577–580.

Smiddy, W. E., Glaser, B. M., Thompson, J. T., Sjaarda, R. N., Flynn, Jr., H. W., Hanham, A., and Murphy, R. P., 1993, Transforming growth factor-beta 2 significantly enhances the ability to flatten the rim of subretinal fluid surrounding macular holes. Preliminary anatomic results of a multicenter prospective randomized study, *Retina* **13**:296–301.

Sodek, J., and Overall, C. M., 1992, Matrix metalloproteinases in periodontal tissue remodelling, *Matrix Suppl.* **1**:352–362.

Sonis, S. T., Lindquist, L., Van Vugt, A., Stewart, A. A., Stam, K., Qu, G. Y., Iwata, K. K., and Haley, J. D., 1994, Prevention of chemotherapy-induced ulcerative mucositis by transforming growth factor beta 3, *Cancer Res.* **54**:1135–1138.

Sporn, M. B., and Roberts, A. B., 1993, A major advance in the use of growth factors to enhance wound healing, *J. Clin. Invest.* **92**:2565–2566.

Sporn, M. B., Roberts, A. B., Shull, J. H., Smith, J. M., Ward, J. M., and Sodek, J., 1983, Polypeptide transforming growth factors isolated from bovine sources and used for wound healing *in vivo, Science* **219**:1329–1331.

Streuli, C. H., Schmidhauser, C., Kobrin, M., Bissell, M. J., and Derynck, R., 1993, Extracellular matrix regulates expression of the TGF-beta 1 gene, *J. Cell Biol.* **120**:253–260.

Taipale, J., Koli, K., and Keski-Oja, J., 1992, Release of transforming growth factor-beta 1 from the pericellular matrix of cultured fibroblasts and fibrosarcoma cells by plasmin and thrombin, *J. Biol. Chem.* **267**:25378–25384.

Taipale, J., Miyazono, K., Heldin, C. H., and Keski-Oja, J., 1994, Latent transforming growth factor-beta 1 associates to fibroblast extracellular matrix via latent TGF-beta binding protein, *J. Cell Biol.* **124**: 171–181.

Taniguchi, Y., Tanaka, T., Gotoh, K., Satoh, R., and Inazu, M., 1993, Transforming growth factor beta 1-induced cellular heterogeneity in the periosteum of rat parietal bones, *Calcif. Tissue Int.* **53**:122–126.

ten Dijke, P., Hansen, P., Iwata, K. K., Pieler, C., and Foulkes, J. G., 1988, Identification of another member of the transforming growth factor type β gene family, *Proc. Natl. Acad. Sci. USA* **85**:4715–4719.

ten Dijke, P., Franzen, P., Yamashita, H., Ichijo, H., Heldin, C.-H., and Miyazono, K., 1994a, Serine/threonine kinase receptors, *Prog. Growth Factor Res.* **5**:55–72.

ten Dijke, P., Yamashita, H., Ichijo, H., Franzen, P., Laiho, M., Miyazono, K., and Heldin, C. H., 1994b, Characterization of type I receptors for transforming growth factor-beta and activin, *Science* **264**: 101–104.

Terrell, T. G., Working, P. K., Chow, C. P., and Green, J. D., 1993, Pathology of recombinant human transforming growth factor-beta 1 in rats and rabbits, *Int. Rev. Exp. Pathol.* **34**(Pt B):43–67.

Thomas, G. R., and Thibodaux, H., 1992, Transforming growth factor-beta 1 inhibits postischemic increases in splanchnic vascular resistance, *Biotechnol. Ther.* **3**:91–100.

Tsunawaki, S., Sporn, M., Ding, A., and Nathan, C., 1988, Deactivation of macrophages by transforming growth factor-beta, *Nature* **334**:260–262.

Varani, J., Jones, J., Dame, M., Sulavik, C., Gibbs, D. F., and Johnson, K. J., 1991, Effects of all-*trans* retinoic acid on neutrophil-mediated endothelial cell injury *in vitro* and immune complex injury in rats, *Am. J. Pathol.* **139**:901–909.

Vodovotz, Y., Bogdan, C., Paik, J., Xie, Q. W., and Nathan, C., 1993, Mechanisms of suppression of macrophage nitric oxide release by transforming growth factor beta, *J. Exp. Med.* **178**:605–613.

Vollberg, T. M., George, M. D., and Jetten, A. M., 1991, Induction of extracellular matrix gene expression in normal human keratinocytes by transforming growth factor beta is altered by cellular differentiation, *Exp. Cell Res.* **193**:93–100.

Wahl, S. M., 1989, Glucocorticoids and wound healing, in: *Antiinflammatory Steroid Action: Basic and Clinical Aspects* (R. P. Schleimer, H. N. Claman, and A. L. Oronsky, eds.), pp. 280–302, Academic Press, New York.

Wahl, S. M., Hunt, D. A., Wakefield, L. M., McCartney-Francis, N., Wahl, L. M., Roberts, A. B., and Sporn, M. B., 1987, Transforming growth factor type beta induces monocyte chemotaxis and growth factor production, *Proc. Natl. Acad. Sci. USA* **84**:5788–5792.

Wang, J. Y., and Johnson, L. R., 1994, Expression of protooncogenes *c-fos* and *c-myc* in healing of gastric mucosal stress ulcers, *Am. J. Physiol.* **266**:G878–886.

Weinzweig, J., Levenson, S. M., Rettura, G., Weinzweig, N., Mendecki, J., Chang, T. H., and Seifter, E., 1990, Supplemental vitamin A prevents the tumor-induced defect in wound healing, *Ann. Surg.* **211**:269–276.

Whitby, D. J., and Ferguson, M. W. J., 1991a, The extracellular matrix of lip wounds in fetal, neonatal and adult mice, *Development* **112**:651–668.

Whitby, D. J., and Ferguson, M. W., 1991b, Immunohistochemical localization of growth factors in fetal wound healing, *Dev. Biol.* **147**:207–215.

Wiessner, C., Gehrmann, J., Lindholm, D., Topper, R., Kreutzberg, G. W., and Hossmann, K. A., 1993, Expression of transforming growth factor-beta 1 and interleukin-1 beta mRNA in rat brain following transient forebrain ischemia, *Acta Neuropathol. (Berl)* **86**:439–446.

Wikner, N. E., Elder, J. T., Persichitte, K. A., Mink, P., and Clark, R. A., 1990, Transforming growth factor-beta modulates plasminogen activator activity and plasminogen activator inhibitor type-1 expression in human keratinocytes *in vitro, J. Invest. Dermatol.* **95:**607–613.

Williams, R. S., Rossi, A. M., Chegini, N., and Schultz, G., 1992, Effect of transforming growth factor beta on postoperative adhesion formation and intact peritoneum, *J. Surg. Res.* **52:**65–70.

Wysocki, A. B., Staiano-Coico, L., and Grinnell, F., 1993, Wound fluid from chronic leg ulcers contains elevated levels of metalloproteinases MMP-2 and MMP-9, *J. Invest. Dermatol.* **101:**64–68.

Wrana, J. L., Attisano, L., Wieser, R., Ventura, F., and Massagué, J., 1994, Mechanism of activation of the TGF-β receptor, *Nature* **370:**341–347.

Yamamoto, T., Nakamura, T., Noble, N. A., Ruoslahti, E., and Border, W. A., 1993, Expression of transforming growth factor beta is elevated in human and experimental diabetic nephropathy, *Proc. Natl. Acad. Sci. USA* **90:**1814–1818.

Yamashita, H., Ichijo, H., Grimsby, S., Moren, A., ten Dijke, P., and Miyazono, K., 1994a, Endoglin forms a heteromeric complex with the signaling receptors for transforming growth factor-beta, *J. Biol. Chem.* **269:**1995–2001.

Yamashita, H., ten Dijke, P., Franzen, P., Miyazono, K., and Heldin, C.-H., 1994b, Formation of hetero-oligomeric complexes of type I and type II receptors for transforming growth factor-β, *J. Biol. Chem.* **269:**20172–20178.

Yoshioka, K., Takemura, T., Murakami, K., Okada, M., Hino, S., Miyamoto, H., and Maki, S., 1993, Transforming growth factor-beta protein and mRNA in glomeruli in normal and diseased human kidneys, *Lab. Invest.* **68:**154–163.

Zioncheck, T. F., Chen, S. A., Richardson, L., Mora-Worms, M., Lucas, C., Lewis, D., Green, J. D., and Mordenti, J., 1994, Pharmacokinetics and tissue distribution of recombinant transforming growth factor-β1 after topical and intravenous administration in male rats, *Pharm. Res.* **11:**213–220.

Part III

New Tissue Formation:
The Cutaneous Paradigm

Chapter 9

Integrins in Wound Repair

KENNETH M. YAMADA, JAMES GAILIT,
and RICHARD A. F. CLARK

1. Introduction

The integrin family of cell adhesion receptors consists of over 20 members, which mediate cell surface interactions with extracellular matrix or (in some cases) with other cells (Akiyama *et al.*, 1990b; Albelda and Buck, 1990; Clark, 1990; Hemler, 1990; Hogg, 1991; Ruoslahti, 1991; Shattil and Brugge, 1991; Yamada, 1991; Damsky and Werb, 1992; Ginsberg *et al.*, 1992; Hynes, 1992; Akiyama and Yamada, 1993; Gailit and Clark, 1993; Glukhova and Thiery, 1993; Gumbiner, 1993; Juliano and Haskill, 1993; Sastry and Horwitz, 1993; Sonnenberg, 1993; Tuckwell *et al.*, 1993; Zetter, 1993; Springer, 1994). Each integrin is a heterodimer, consisting of one α and one β subunit in a noncovalent complex. As summarized in Fig. 1, only certain combinations of integrins are observed: major groupings include integrins of the $\beta 1$ subfamily and integrins containing the αv subunit. Changes in either the α or the β subunit of integrin heterodimers alter their specificity for ligands, as summarized in Table I.

2. Integrin Structure

Integrins appear to share the distinctive shape shown in Fig. 2. As visualized using a variety of electron microscopic methods (Nermut *et al.*, 1988), integrins consist of a globular or mushroom-shaped head region that is thought to contain substantial contributions from both α and β subunits, and that mediates integrin binding to ligands. Integrins have long, semirigid legs that hold the head approximately 20 nm above the cell surface and penetrate through the plasma membrane via hydrophobic transmembrane domains. Integrins have carboxy-terminal cytoplasmic tails or domains

KENNETH M. YAMADA • Laboratory of Developmental Biology, National Institute of Dental Research, National Institutes of Health, Bethesda, Maryland 20892-4370. JAMES GAILIT and RICHARD A. F. CLARK • Department of Dermatology, Health Sciences Center, State University of New York at Stony Brook, Stony Brook, New York 11794-8165.

The Molecular and Cellular Biology of Wound Repair (Second Edition), edited by Richard A. F. Clark. Plenum Press, New York, 1996

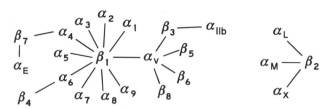

Figure 1. Schematic diagram showing the subunit associations of human integrins. Integrins are heterodimers containing one alpha and one beta subunit. Their patterns of association are specific. The β1, αv, and β2 subunits have multiple possible partners. Many others, including the α1 and β6 subunits, have only one known partner. These specific patterns of noncovalent association create dimers with varying specificities for many different extracellular matrix molecules or cellular targets (see Table I).

protruding into the cytoplasm, which are often short (15–60 amino acids in length), although the β4 tail is more than 1000 amino acids long (Sastry and Horwitz, 1993). This transmembrane organization permits integrins to bind to extracellular ligands and to transmit signals into the cytoplasm. Conversely, however, the intracellular tails interact with cytoplasmic molecules to regulate extracellular ligand-binding affinity. These functions have been termed "outside-in" and "inside-out" signaling, respectively (Ginsberg *et al.,* 1992). As reviewed in Section 3, specificity of integrin function can be attributed to the structures of both the extracellular and intracellular domains of integrins.

A number of integrin subunits can exist as different isoforms due to alternative splicing of their precursor mRNA molecules. Spliced subunits include α3, α4, α6, α7, β1, β3, and β4 (Tamura *et al.,* 1990; Balzac *et al.,* 1993; Collo *et al.,* 1993; Song *et al.,* 1993). Although the functional significance of this alternative splicing is not known in all cases, such alterations in β1 and β3 integrin cytoplasmic domains alter the capacity of these domains to target the movement of receptors to adhesion sites (focal contacts). The originally described β1 sequence (now termed β1A) readily targets receptor localization to these sites, but the β1B isoform remains diffuse in localization in spite of its similar α-subunit associations and capacity to bind fibronectin (Balzac *et al.,* 1993). The alternatively spliced version of the cytoplasmic tail of β3 (the alternative isoform is termed β3B) is also unable to direct a reporter to focal contacts (LaFlamme *et al.,* 1994). The β3B isoform is also unable to mediate signaling via tyrosine phosphorylation of focal adhesion kinase (FAK), in clear contrast with the β1A and β3A cytoplasmic domains (Akiyama *et al.,* 1994). Alternative spicing may therefore help to regulate integrin localization and signal transduction.

Integrin Glycosylation, Subunit Assembly, and Acquisition of Function

The subunits of integrin receptors are glycosylated by asparagine-linked oligosaccharides. The presence and processing of these sugar moieties appears to be crucial for

Table I. The Integrin Family of Adhesion Receptors

β_1* Integrins (VLAs)[a]	β_2 Integrins	β_3* Integrins	Other integrins
α1β1 Laminin, collagen	αLβ2 ICAM-1, ICAM-2, ICAM-3	αIIbβ3 Fg, Fn, vWF, Vn, Tsp	α6β4* Laminins
α2β1 Collagen (laminin)	αMβ2 iC3b, Fg, factor X, ICAM-1	αvβ3 Vn, Fg, Fn, vWF, Tsp	αvβ5 Vitronectin, fibronectin
α3*β1 Ln, Col, Fn, En, Epi	αXβ2 iC3b, Fg		αvβ6 Fibronectin
α4*β1 Fn IIICS, VCAM-1			α4β7 VCAM-1, Fn IIICS, MAdCAM-1
α5β1 Fibronectin			αEβ7 ?
α6*β1 Laminin			αvβ8 Vitronectin
α7*β1 Laminin			
α8β1 Vitronectin, fibronectin			
α9β1 Tenascin			
αvβ1 Vitronectin, fibronectin			

[a]*, Alternatively spliced.
Abbreviations: Ln, laminin; Col, collagen; Fn, fibronectin; En, entactin (nidogen); Epi, epiligrin; Fg, fibrinogen; vWF, von Willebrand factor; Vn, vitronectin; Tsp, thrombospondin.

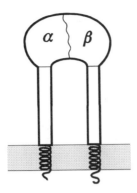

Figure 2. Diagram of the general structure of integrin receptors. Each is composed of an alpha and beta subunit that are joined noncovalently at amino-terminal domains to form the extracellular head region, which binds to ligands. Long extracellular legs appear to allow the integrin to protrude approximately 20 nm above the plasma membrane. Each integrin subunit spans the plasma membrane with a transmembrane segment and each has a unique intracellular domain.

integrin function, as established for the $\alpha5\beta1$ fibronectin receptor. Immature $\beta1$ chains are characterized by the high-mannose form of N-linked oligosaccharide, and they are not normally expressed on the cell surface and cannot bind ligand (Akiyama and Yamada, 1987). Treating cells with inhibitors of N-linked oligosaccharide maturation or with enzymes that remove oligosaccharides results in loss of binding function; moreover, the latter treatment clearly inhibits subunit association (Akiyama *et al.*, 1989; Zheng *et al.*, 1994). Mature carbohydrates thus appear to promote subunit association and secretion of a mature integrin able to bind to the ligand. Some conditions such as malignant transformation or transforming growth factor-β (TGF-β) treatment of human fibroblasts change the ratio of immature to mature $\beta1$ subunits, enhancing the proportion of total $\beta1$ subunits on the cell surface (Heino *et al.*, 1989; Akiyama *et al.*, 1990a).

It is currently thought that integrin $\beta1$ and αv subunits are generally produced in excess, and that the rates of generation of final cell surface heterodimers depend on the production and ratios of their appropriate α or β subunit partners (Cheresh and Spiro, 1987; Heino *et al.*, 1989; Koivisto *et al.*, 1994). In fact, $\beta1$ integrins can be shown to exist initially in the endoplasmic reticulum as a complex with calnexin, a 90-kDa membrane-bound chaperone (Lenter and Vestweber, 1994). Calnexin's chaperone function facilitates the proper assembly of other molecular complexes, including the major histocompatibility (MHC) class I molecules, the T-cell receptor complex, and membrane-associated immunoglobulins. The $\beta1$ integrin–calnexin complex remains in the endoplasmic reticulum until an α chain binds to calnexin and displaces it to form a mature α/β integrin heterodimer complex. It will be interesting to learn the relationship of integrin oligosaccharide maturation to calnexin binding and assembly of α/β complexes.

3. Integrin Function

3.1. Ligand Specificity

Cell behavior is dependent on the ligand-binding specificities of the integrins present on the cell surface. For example, cells lacking the $\alpha4\beta1$ integrin cannot adhere to the alternatively spliced CS1 site in fibronectin. One striking characteristic of the integrins is their overlapping but nonidentical binding specificities. As can be seen in Table I, many integrins can bind to fibronectin or to laminin, and yet each integrin has a distinctive repertoire of binding to other ligands. Binding is based on protein–protein recognition, and in many cases a central element of the molecular recognition event involves a short adhesive peptide sequence (Yamada, 1991). As reviewed in Chapter 2 on provisional matrix, even though the same short peptide sequence is used as a part of the recognition process by a number of integrins, additional specificity arises from separate "synergy" sequences.

The types and concentrations of integrin receptors on cells are important determinants of the specificity of binding to extracellular matrix molecules, although their activation state and spatial distribution (diffuse vs. aggregated) also appear important. During epithelial wound repair, there are increases in levels of the $\alpha5\beta1$ integrin and αv-containing integrins (Grinnell, 1990, 1992; Guo *et al.*, 1990; Kurpakus *et al.*, 1991; Douglass *et al.*, 1992; Hertle *et al.*, 1992; Murakami *et al.*, 1992; Paallysaho *et al.*, 1992; Tervo *et al.*, 1992; Cavani *et al.*, 1993; Gipson *et al.*, 1993; Hergott *et al.*, 1993; Juhasz *et al.*, 1993; Larjava *et al.*, 1993; Stepp *et al.*, 1993; Gailit *et al.*, 1994) including $\alpha v\beta6$ (H. Larjava and R. A. F. Clark, personal communication).

Adult human keratinocytes are initially unable to adhere to fibronectin, even though they have the ability to adhere to collagen and to vitronectin (Guo *et al.*, 1990). Over a span of several days after *in vivo* wounding or *in vitro* culturing in serum-containing media, keratinocytes adhere to fibronectin. This process is accompanied by changes in the glycosylation state of the $\beta1$ subunit and in levels of $\alpha5\beta1$ (Guo *et al.*, 1990; L. T. Kim *et al.*, 1992). Whether alterations in αv integrins contribute to this process remains to be established. Interestingly, amphibian epidermal wound healing does not involve these changes in integrin subunits or such activation of keratinocytes, and it proceeds quite rapidly without scar formation (Donaldson *et al.*, 1994). Whether these differences can be attributed to differences in integrin functional state even before wounding remains to be determined.

3.2. Ligand-Binding Sites

A series of studies has defined the sites on several α and β subunits that bind to ligands (D'Souza *et al.*, 1990; Loftus *et al.*, 1990; Smith and Cheresh, 1990; Bajt *et al.*, 1992; Plow *et al.*, 1992; Takada *et al.*, 1992; Calvete *et al.*, 1994). The major platelet receptor $\alpha IIb\beta3$ has been studied in the most detail: The ligand-binding pocket is thought to consist of contributions from amino acids in the regions of residues 114–128 and 212–219 in the β3 subunit, combined with contributions of residues 294–314 and probably 657–665 in the αIIb subunit (Plow *et al.*, 1992; Calvete *et al.*, 1994). The notion that an integrin heterodimer has multiple polypeptide regions intimately involved in contact with a ligand would be predicted if those ligands had more than one contact site, i.e., a primary peptide recognition site and one or more synergy sites as described for fibronectin (Yamada *et al.*, 1992). Remarkably, a short 11-residue peptide sequence from the αIIb subunit (TDVNGDGRHDL, residues 296–306) by itself can mimic the capacity of the entire integrin to bind to fibrinogen (D'Souza *et al.*, 1991). It will be of interest to learn the extent of ligand-binding specificity of this tiny sequence.

A central element of the ligand recognition process for RGD (Arg-Gly-Asp) peptides may be the formation of a putative divalent cation-binding site by sequences similar to the EF-hand, a high-affinity cation-binding motif characteristic of molecules such as parvalbumin and calmodulin (Loftus *et al.*, 1990; Tuckwell *et al.*, 1992). The integrin EF-hand structure, however, lacks one of the key consensus coordinating residues. This missing residue may be provided by the aspartic acid residue in peptide recognition sequences such as RGD or LDV (Leu-Asp-Vol) (Loftus *et al.*, 1990; Tuckwell *et al.*, 1992). Alternatively, or in addition, the ligand may displace a previously bound divalent cation (D'Souza *et al.*, 1994).

Evidence for two distinct domains involved in integrin binding to a non-RGD ligand (a counterreceptor molecule) appears to exist for the lymphocyte function-associated antigen (LFA-1) molecule (αLβ2). A direct binding site for the transmembrane counterreceptor ICAM-1 (intercellular adhesion molecule-1) is present in the "I" (inserted) domain of LFA-1 (Randi and Hogg, 1994). Additional binding regions correspond to 10- and 20-amino-acid subregions in domains V and VI, respectively, which are associated with predicted divalent-cation-binding motifs (Stanley *et al.*, 1994). It will be interesting to learn whether interaction sites corresponding to each of these integrin-binding sites can be identified on the ligand, e.g., the ICAM-1 molecule.

Another type of ligand-binding mechanism involving three-dimensional structural recognition appears to be required for the binding of native collagen by the α2β1 integrin. Cells are known to bind to native collagen via integrins such as α2β1, and synthetic peptide inhibitors containing RGD or DGEA (Asp-Gly-Glu-Ala) sequences in collagen have been proposed to represent critical sites for binding (Dedhar *et al.*, 1987; Gehlsen *et al.*, 1988; Staatz *et al.*, 1991). Nevertheless, such recognition appears relatively weak or nondemonstrable in other systems (Tuckwell *et al.*, 1994). In fact, even though the α2β1 integrin appears to recognize R and D residues in native type IV collagen, the recognition involves residues on two different chains. Even though these two residues are not adjacent in the primary sequence, they are located side-by-side in the three-dimensional structure of triple-helical collagen (Eble *et al.*, 1993). Studies of type VI collagen show that binding to the native triple-helical molecules is mediated by α1β1 and α2β1 and is resistant to RGD peptide inhibition, while binding to apparently exposed RGD sequences in denatured collagen was mediated by αvβ3 and α5β1 (Pfaff *et al.*, 1993). Recently it has been demonstrated that recombinant α2β1 expressed in

Chinese hamster ovary (CHO) cells binds type I collagen through the I domain of the α2 subunit (Kamata and Takada, 1994). Likewise, α1β1 appears to bind collagen through its I domain (Kern *et al.*, 1994). In contrast, binding to denatured type I collagen (gelatin) occurs either directly through αvβ3 interactions with an unmasked RGD site (Davis, 1992) or indirectly via a fibronectin intermediate, which bridges the gelatin molecule (binding by its collagen-binding site) and a cell surface receptor such as the integrin α5β1 (Klebe, 1974; Tuckwell *et al.*, 1994).

3.3. Integrin Activation

Integrins can potentially exist in at least three states of activation: inactive, partially activated, and highly activated (Ginsberg *et al.*, 1992; Hynes, 1992; Faull *et al.*, 1993; Juliano and Haskill, 1993; Sastry and Horwitz, 1993). Platelet activation switches the major platelet integrin αIIbβ3 from an "inactive" state (inability to bind soluble fibrinogen and fibronectin, although able to bind RGD-containing peptides and substrate-bound fibrinogen) to an activated state. This process involves inside-out signaling involving a conformational change in the external domain of αIIbβ3 (Ginsberg *et al.*, 1992). The altered conformation can be detected by monoclonal antibodies, which bind only to activated αIIbβ3 (Abrams *et al.*, 1994), and by changes in susceptibility to protease digestion (Calvete *et al.*, 1994). According to the latter studies, activation may cause a major reorientation of the extracellular domains at the interface between the α and β subunits, such that a series of peptide residues become buried as the integrin is activated to bind RGD-containing proteins.

The process of platelet activation is dependent on function of the α cytoplasmic domain (O'Toole *et al.*, 1991, 1994), as is phorbol ester-mediated activation of α4β1 (Kassner and Hemler, 1993). Overexpression of a competing β cytoplasmic tail will inhibit the αIIbβ3-signaling process (Y. P. Chen *et al.*, 1994), and an activating anti-β1 monoclonal antibody can activate an α4β1 chimera that is otherwise refractory to activation due to absence of an α cytoplasmic domain. Although the mechanisms of these important activation events are not known in detail, they appear to involve binding of one or more cytoplasmic effectors to the β tail. This process appears to be modulated by the α tail, which produces a conformational change in the extracellular domain that permits the binding of large ligands such as fibrin and fibronectin. Interestingly, binding of a short RGD peptide to the unactivated extracellular domain can apparently produce similar conformational and functional changes that are transient rather than permanent (Du *et al.*, 1991).

Other integrin receptors can also exist in inactive or resting states. Many of the β1 integrins on unactivated lymphocytes, such as the α5β1 fibronectin receptor and the α6β1 laminin receptor, are not active (or are only poorly active) for mediating adhesion unless activated by phorbol esters (Dustin and Springer, 1989; Shimizu *et al.*, 1990; Ginsberg *et al.*, 1992). The α5β1 fibronectin receptor can in fact be shown to exist in three states of activation (Faull *et al.*, 1993).

The β2 integrin LFA-1 shows striking switches in activation state. For example, T-cell receptor activation due to cross-linking can trigger LFA-1 (αLβ2) adhesive function (Dustin and Springer, 1989). The mechanism is not yet known. Interestingly, the related αMβ2 integrin (Mac-1) of monocytes can be activated to bind factor X and fibrinogen by the drug cytochalasin B, which disrupts actin microfilament organization (Elemer and Edgington, 1994). The authors suggest that a transient disassembly of actin filaments may release constraints on this integrin, permitting conformational changes in the receptor that permits it to bind ligands.

3.4. Cell Surface Distribution of Integrin Receptors

Cells can be roughly classified into two general phenotypes: migratory and stationary (Couchman and Rees, 1979; Duband *et al.*, 1988a). The migratory state is found in rapidly migrating embryonic cells, but may also be a characteristic of granulation tissue fibroblasts. Many cells display the "stationary" phenotype, including embryonic cells that have halted migration or granulation tissue fibroblasts that become firmly associated with fibronectin and collagen fibrils. At a molecular level, the migratory phenotype is characterized by a diffuse distribution of β1 integrin receptors, little organized cytoskeleton, and generally poor organization of cell surface fibronectin fibrils. The stationary phenotype is characterized by integrin organization into focal adhesion sites. These focal adhesions are associated extracellularly with fibronectin fibrils (Singer, 1979; Chen and Singer, 1982) and intracellularly with specialized cytoskeletal complexes composed of talin, α-actinin, and paxillin into which actin bundles appear to insert (Geiger, 1989; Turner and Burridge, 1991).

These two distinct cellular phenotypes are likely to have important functions in wound repair. Migration of cells into wounds requires the rapid making and breaking of contacts with extracellular molecules and cytoskeletal plasticity to permit rapid cell movement. In fact, excessive cell adhesion to substrates will inhibit rates of migration (Duband *et al.*, 1991; DiMilla *et al.*, 1993; Schmidt *et al.*, 1993). After cells cease migration and begin to enter the phase of wound contraction, they must bind tightly to extracellular collagen and fibronectin fibrils and organize a contractile cytoskeleton (Welch *et al.*, 1990). Integrin receptors probably play central roles in these processes, forming the link between extracellular matrix fibrils and the contractile cytoskeleton (Singer *et al.*, 1984).

3.5. Integrin–Cytoskeletal Interactions

The cytoskeletal protein talin, which binds to actin and appears in focal contacts, also binds to β1 integrins with relatively modest affinity ($K_d = 10^{-6}$ M) (Horwitz *et al.*, 1986). This interaction may be inhibited by integrin tyrosine phosphorylation (Tapley *et al.*, 1989). Tyrosine phosphorylation, however, does not appear necessary for cell motility (Duband *et al.*, 1988b; Hayashi *et al.*, 1990). α-Actinin, another cytoskeletal protein that is located in focal contacts but also along actin microfilament bundles, binds to β1 integrin tails (Otey *et al.*, 1990, 1993). Interestingly, α-actinin binds with a 100-fold higher affinity to synthetic β1 integrin tails than to intact integrins ($K_d = 2 \times$

10^{-8} M vs. 2×10^{-6} M, respectively) (Otey *et al.,* 1990). This finding suggests that the α tail might possibly suppress access of α-actinin to the β tail to which it binds. Although strong binding of α-actinin to integrins has been difficult to demonstrate in most immunoprecipitation experiments, neutrophils responding to the formyl-Met-Leu-Phe (FMLP) chemotactic peptide show a strong but short-lived binding (10–15 min) of α-actinin to β2 integrins (Pavalko and LaRoche, 1993). Similarly, antibody activation of the T-cell receptor may induce a protein kinase C (PKC)-dependent association of β2 integrins with the actin-containing cytoskeleton (Pardi *et al.,* 1992).

The binding of α-actinin to integrins has been localized to two regions on the β1 cytoplasmic domain and to two other sites on β2 tails (Otey *et al.,* 1993; Pavalko and Otey, 1994). It is conceivable that the differences described in estimated binding affinities of α-actinin point to differences between binding to unoccupied integrins and to activated or exposed β1 tails, as has been postulated to occur after ligand occupancy of integrins (LaFlamme *et al.,* 1992).

Increasing evidence points to a direct link between extracellular ligands, clustering of integrins, and organization of cytoskeletal proteins into submembranous aggregates. Binding of beads coated with ligands such as fibronectin leads to accumulations of cytoskeletal proteins on the inner surface of the plasma membrane (Grinnell and Geiger, 1986; Mueller *et al.,* 1989; Plopper and Ingber, 1993). In fact, clustering of chimeric receptors containing only the β1 tail of integrins (lacking any extracellular or transmembrane domain other than a neutral reporter domain) will also lead to striking accumulations of a variety of cytoskeletal molecules including talin, α-actinin, tensin, vinculin, and F-actin on the inner surface of the membrane (S. Miyamoto, S. E. LaFlamme, and K. M. Yamada, unpublished results). Regions of the β cytoplasmic domain needed for binding and transmembrane accumulation of various cytoskeletal molecules could be characterized by progressive truncation of β_1 integrin tails, which were expressed in α–β integrin heterodimer complexes and aggregated by antibody-coated beads (Lewis and Schwartz, 1995). The capacity of β_1 integrin to induce accumulation of talin and FAK requires the carboxy-terminal 4 amino acids of the cytoplasmic domain, while a site needed for α-actinin localization was up to 9 amino acids more amino-terminal. Even though α-actinin is known to bind to actin filaments directly, binding of α-actinin to one of the mutants in the absence of talin and FAK (and possibly other molecules) does not lead to actin accumulation; integrin-mediated clustering of α-actinin is therefore insufficient to organize actin accumulation (Lewis and Schwartz, 1995). These and other studies point to a direct link between the β1 integrin tail and the cytoskeleton, especially after external ligation and clustering of the receptors. In contrast, the α5 tail appears to lack this function except for weak redistribution of α-actinin, and may instead serve to modulate the activity of the β subunits (LaFlamme *et al.,* 1992; Briesewitz *et al.,* 1993; Ylanne *et al.,* 1993).

Inside-Out Adhesion Site Organization

In addition, information intrinsic to β cytoplasmic domain sequences can also target receptors directly to focal contacts. This intrinsic propensity of certain β tails to target receptors to adhesion sites is most clearly displayed in chimeric receptors, which are targeted to these sites if they contain β1 integrin tails (Geiger *et al.*, 1992; LaFlamme *et al.*, 1992). In fact, certain tails, but not others modified by alternative splicing, appear to play roles in such adhesion site organization (LaFlamme *et al.*, 1992, 1994; Balzac *et al.*, 1993). It appears likely, then, that intracellular targeting information cooperates with extracellular inputs as integrins accumulate along fibrils of extracellular matrix to generate formation of the focal contact or fibronexus.

Chimeric swaps of integrin cytoplasmic domains provide a complementary approach to analyzing integrin cytoplasmic domain functions. The α2 and α5 cytoplasmic tails of integrin α subunits appear to be more effective in supporting collagen gel contraction, while α4 tails appear more effective in promoting cell migration (Schiro *et al.*, 1991; Chan *et al.*, 1992). Functions of α cytoplasmic domains in cell adhesion may have relatively low specificity, since short cytoplasmic stubs of differing sequence were able to substitute for full-length α tails in permitting integrin-mediated cell adhesion, and tails were not needed for ligand binding or fibronectin matrix assembly (Bauer *et al.*, 1993; Kassner *et al.*, 1994).

These studies could indicate that α cytoplasmic domains play significant structural or binding roles in interactions of integrins with cytoplasmic molecules, or they could indicate a function for α tails in regulation of β tail function. Until α cytoplasmic domains can be clearly shown to function in isolation, roles in modulating β tail function appear somewhat more likely, as also appears more likely for integrin inside-out activation.

Swaps of β1 and β5 integrin cytoplasmic domains have provided further evidence for distinct functions of these domains. Replacement of the β1 cytoplasmic tail with the β5 tail resulted in a loss of capacity to localize to focal adhesionlike structures (Pasqualini and Hemler, 1994). Similarly, the β5 integrin localizes less than β1 or β3 receptors to focal contacts (Wayner *et al.*, 1991). This difference, however, may be due to differences in cells or culture conditions (Conforti *et al.*, 1994). Nevertheless, chimeric β1 subunit with a β5 tail was more effective at mediating cell migration in a haptotactic migration assay. In addition, cells expressing the chimera failed to proliferate in response to fibronectin plus an integrin-activating antibody and to phosphorylate a 90-kDa protein that correlates with proliferation (Pasqualini and Hemler, 1994). These studies suggest that certain β integrin tails have distinct functions, at least in the context of chimeras.

3.6. Integrin-Mediated Signaling

Integrins can transduce a series of signals after ligand binding or receptor clustering. For example, integrin-mediated adhesion of cells to a variety of extracellular matrix molecules including fibronectin, vitronectin, laminin, and collagen will lead to enhanced tyrosine phosphorylation of the molecule FAK (Guan *et al.*, 1991; Kornberg *et al.*, 1991, 1992; Guan and Shalloway, 1992). This protein localizes to focal contacts and may be involved in their organization (although its function is not yet clarified). As a model for transmembrane signal transduction via integrins, however, it has been quite informative. Antibody-mediated clustering, at least with certain anti-β3 and anti-β1 monoclonal antibodies, will trigger this phosphorylation process (Kornberg *et al.,* 1991, 1992), and overexpression and especially clustering of even isolated β1 cytoplasmic tails will induce signaling (Akiyama *et al.,* 1994; Lukashev *et al.,* 1994).

This mechanism of signaling by clustering of cytoplasmic domains is strongly reminiscent of hormone- and growth factor-induced signal transduction. In the latter cases, however, the mechanisms often involve bringing tyrosine kinase domains into close proximity, thus inducing autophosphorylation of receptor tails and subsequent binding of cytoplasmic effectors via phosphotyrosine-recognizing SH2-domain molecules (Schlessinger, 1988; de Vos *et al.,* 1992; Metzger, 1992). Integrins, in contrast, have no known enzymatic activities; instead they may cause clustering of associated kinases. The signaling mediated by integrin β tails has specificity. Studies with isolated tails indicate that the β1, β3, and β5 tails all readily signal tyrosine phosphorylation of FAK, whereas the alternatively spliced version of the β3 integrin does not signal (Akiyama *et al.,* 1994). Alternative splicing can thus apparently help regulate integrin signal transduction.

Binding of fibroblast integrins to extracellular matrix ligand can also trigger tyrosine phosphorylation pathways involved in the activation of transcription factor function. The kinases termed extracellular regulated kinases-1 and 2 (ERK1 and ERK2), which are prominent members of the mitogen-activated protein (MAP) kinase pathway, are transiently activated after adhesion of cells to fibronectin (Q. Chen *et al.,* 1994; Morino *et al.,* 1995). This activation is accompanied by translocation into the nucleus, where these molecules may transcriptionally activate as yet unknown fibroblast genes. An independent pathway, termed the Jun kinase or stress-activated protein (SAP) kinase pathway can also be activated by integrin ligands or anti-integrin antibodies, but with different kinetics than ERKs (Miyamoto *et al.,* 1995b). It is remarkable that two major signaling pathways, one specific for cytokines such as platelet-derived growth factor (PDGF) and epidermal growth factor (EGF) (the ERK pathway) and the other for stresses and inflammatory cytokines such as tumor necrosis factor-α (TNF-α) (the Jun kinase pathway), can both be activated by interaction of integrins with an extracellular ligand such as fibronectin.

Integrin-mediated adhesion of cells to fibronectin and cross-linking of integrins can trigger a rise in intracellular calcium (Schwartz, 1993; Schwartz and Denninghoff, 1994). In endothelial cells, this rise is due to an influx of extracellular calcium ion. This influx can be mediated by the αv integrin subunit (Schwartz and Denninghoff, 1994), probably via association with a putative calcium transporter termed integrin-associated protein (IAP) (Lindberg *et al.,* 1993; Schwartz, 1993). The latter protein is ubiquitous, found even on erythrocytes (which appear to lack integrins). Therefore, IAP may have some additional function besides serving as an integrin-triggered calcium channel. In some cells, ligation of integrins can also induce a rise in intracellular pH, changes in inositol lipid metabolism, and increased GTP-bound p21ras, which are signaling pathways often associated with cellular responses to growth factors (Schwartz *et al.,* 1991a, b; Schwartz and Lechene, 1992; Kapron-Bras *et al.,* 1993; McNamee *et al.,* 1993).

Since integrins often accumulate at focal contacts, it is interesting to consider the regulatory proteins known to be present at these sites, which may be candidates for proteins that interact with integrin cytoplasmic tails or associated molecules to help mediate transmembrane signaling. Molecules found at focal contacts include FAK and its truncated form pp41[FRNK], pp60[v-src], pp59[fyn], the alpha (type 3) isoform of protein kinase C, and calpain II (calcium-dependent protease II) (Hyatt *et al.*, 1990; Pavalko and Otey, 1994). It is important to determine which of these proteins are involved in interactions with integrin cytoplasmic domains.

A tantalizing series of studies has established that the KXGFFKR amino acid sequence present in the cytoplasmic domains of all integrin α subunits can bind calreticulin (Rojiani *et al.*, 1991; Dedhar *et al.*, 1994; Leung-Hagesteijn *et al.*, 1994). Calreticulin is a calcium-binding protein that normally resides in the endoplasmic reticulum, but it may have additional functions in the cytoplasm and the nucleus. Antibody-induced redistribution of integrins results in calreticulin redistribution, and a partial decrease in calreticulin levels using antisense oligonucleotides is accompanied by major inhibition of cell adhesion (Leung-Hagesteijn *et al.*, 1994). In addition, calreticulin can bind to retinoic acid and androgen receptors via the same KXRRKR sequence, resulting in inhibition of receptor binding to the respective DNA response elements and inhibition of transcriptional activity (Burns *et al.*, 1994; Dedhar *et al.*, 1994). It has thus been suggested that calreticulin may be involved in transducing integrin-mediated signals to the nucleus and thereby modulating gene expression (Leung-Hagesteijn *et al.*, 1994). Whether these effects are direct, indirect, or merely coincidental remains to be determined.

Other signaling molecules are likely to be identified in close or regulated association with integrin cytoplasmic domains. FAK binds to β1 integrin cytoplasmic domain sequences *in vitro* (Schaller *et al.*, 1995), and can be clustered into intracellular complexes by experimental aggregation of integrins using anti-integrin antibodies bound to beads (Miyamoto *et al.*, 1995a). In fact, certain anti-α5β1 antibodies can induce clustering of only FAK and the cytoskeletal protein tensin, but not seven other cytoskeletal proteins; combining integrin aggregation with ligand occupancy triggers accumulation of the full complement of cytoskeletal proteins (Miyamoto *et al.*, 1995a). These types of study suggest that particularly close associations exist between β1 integrin tails and FAK and tensin, slightly more peripherally with talin and α-actinin, and only after formation of large complexes for actin and other cytoskeletal molecules.

Distinct hierarchies of cytoskeletal and signaling molecules can be identified during formation of the transmembrane aggregates induced by integrins. After aggregation-induced accumulation of tensin and FAK, integrin occupancy triggers accumulation of talin, α-actinin, and vinculin (Miyamoto *et al.*, 1995b). If tyrosine phosphorylation is permitted and actin microfilaments are intact, aggregation and occupancy can also lead to accumulation of F-actin, paxillin, and filamin. Interestingly, aggregation even without ligand occupancy will induce the accumulation of 19 other signal transduction molecules if tyrosine phosphorylation is not inhibited; moreover, under these conditions, ERK and JUN kinase pathways are also activated (Miyamoto *et al.,* 1995b). Thus, hierarchical groupings of signaling and cytoskeletal proteins characterize the responses to integrins. This complexity provides multiple possible points of regulation of integrin responses by both extracellular and intracellular stimuli, as well as a variety of sites for potential pharmacological and therapeutic intervention.

3.7. Integrin Regulation of Gene Expression and Cell Proliferation

For many years, the extracellular matrix has been proposed to provide important regulatory information to cells, helping to control specific gene expression and cell proliferation (Hay, 1991). Integrins are now known to be central mediators of these forms of regulation (Damsky and Werb, 1992; Hynes, 1992; Juliano and Haskill, 1993). In rabbit synovial fibroblasts, adhesion to fragments of fibronectin containing its central cell-adhesive domain or anti-α5 antibodies that are clustered or immobilized on substrates stimulates secretion of proteinases such as stromelysin and interstitial collagenase (Werb *et al.,* 1989). Interestingly, these effects are not produced by intact fibronectin, which appears to provide suppressive information primarily from the alternatively spliced CS1 site of fibronectin, which binds to the $\alpha_4\beta_1$ integrin (Huhtala *et al.,* 1995). This binding apparently results in a signaling event that suppresses activation otherwise provided by the central cell-binding domain. This type of biological activity based on a fragment of fibronectin resembles a previous report of monocyte chemotaxis toward a cell-binding domain of fibronectin but not intact fibronectin (Clark *et al.,* 1988).

In monocytes, adhesion to different extracellular matrix components results in selective patterns of expression of specific genes (Juliano and Haskill, 1993). For example, adhesion of monocytes to fibronectin results in induction of colony-stimulating factor (CSF-1) and monocyte adherence-derived inflammatory gene-15 (MAD-15) genes. Other genes are induced by adhesion to type IV collagen. Genes such as interleukin-1 (IL-1) and IL-8 are induced on all substrates. Roles of transcription factors involved in these regulatory events are being characterized. For example, NF-kB undergoes translocation into the nucleus (Juliano and Haskill, 1993). Mammary epithelial cell interaction with the basement membrane analogue Matrigel via a β1-integrin pathway results in striking morphogenetic changes and enhanced expression of the differentiated phenotype, e.g., induction of casein and whey acidic protein production (Streuli *et al.,* 1991; Howlett and Bissell, 1993; Lin and Bissell, 1993).

Besides effects on specific gene expression, integrins can induce cell proliferation (Symington, 1990, 1992; Guadagno *et al.,* 1993; Han *et al.,* 1993). Matrix regulation of proliferation occurs at a cell cycle control point similar to the yeast cell cycle control point, termed START. Interestingly, its function can be replaced in certain cells by the cytokine transforming growth factor-β (TGF-β) (Han *et al.,* 1993). GRGDS peptide occupancy of α5β1 fibronectin receptors on erythroleukemia cells stimulates kinase activities associated with the cyclin A and cdc2 within 2 hr; appearance of the 110-kDa form of the retinoblastoma protein within 3 hr; and cell cycle progression within 24 hr

(Symington, 1992). These studies provide links between integrin ligand occupancy and intracellular growth regulators.

4. Cell–Matrix Interactions during Wound Repair

4.1. Introduction

Integrin receptors are clearly required for all phases of wound repair. Immediately after injury, platelet $\alpha IIb\beta3$ integrins mediate interactions with extracellular matrix (ECM) molecules, such as fibrin, fibronectin, vitronectin, thrombospondin, and von Willebrand factor (Plow *et al.,* 1992). These platelet–ECM interactions are essential for stable clot formation. Over the next 24 hr, integrins of the $\beta2$ family are obligatory for leukocyte accumulation in the wound (Springer, 1994). These inflammatory cells cleanse the wound of contaminating microorganisms and tissue debris. Once granulation tissue formation begins, integrins are required for neovascularization (Clark *et al.,* 1995b) and presumably for fibroblast ingrowth (Clark *et al.,* 1990). Later, during wound contraction, $\alpha2\beta1$ and $\alpha5\beta1$ integrins probably provide the linkage between fibroblasts and the ECM, allowing the tensile forces generated by fibroblasts to be transmitted over the breadth of the wound (Welch *et al.,* 1990; Schiro *et al.,* 1991). Best studied in the context of wound repair, however, are keratinocyte integrins. Despite the lack of direct evidence that integrins function during reepithelialization, many epidermal wound-healing studies have demonstrated a correlation between integrin appearance on the wound keratinocytes and the appropriate ECM ligand beneath the migrating epidermis, strongly suggesting a functional interaction. In addition, the requirement of integrins for motility of cultured keratinocytes further suggests their importance for *in vivo* epidermal migration. Thus we have chosen to expand on keratinocyte integrins and their probable function in wound repair.

4.2. Keratinocyte Integrins—An Example of Cell–Matrix Interactions during Wound Repair

During skin reepithelialization, keratinocytes at the wound margin dissolve their complex attachment with the underlying basement membrane and migrate over a provisional matrix consisting of fibrin(ogen), vitronectin, tenascin, and type I collagen (Odland and Ross, 1968; Clark *et al.,* 1982; Mackie *et al.,* 1988; Betz *et al.,* 1993; Cavani *et al.,* 1993). Keratinocytes use integrins for attachment and migration on these provisional matrix proteins (Gailit and Clark, 1993). Sometimes the cells utilize more than one integrin to interact with a particular matrix protein. For example, keratinocytes can use both $\alpha3\beta1$ and $\alpha5\beta1$ integrins to bind fibronectin. Of these, however, the $\alpha5\beta1$ integrin is more important for migration even though the cells express more $\alpha3\beta1$ integrin on their surface. When $\alpha5\beta1$ integrin on human keratinocytes was blocked by monospecific antibodies, migration was dramatically inhibited (L. T. Kim *et al.*, 1992).

4.2.1. Keratinocyte α5β1

Although α5β1 is expressed by human keratinocytes *in situ* and in culture, it is not fully expressed nor completely functional *in situ*. When these cells are freshly isolated from epidermis and placed into culture, they exhibit little α5β1 on the cell surface, they do not attach to fibronectin (Toda *et al.*, 1987), and the β1 subunit is not localized into focal adhesions (Grinnell, 1990). Interestingly, freshly isolated keratinocytes attach, but do not spread, on collagen and laminin (Grinnell, 1990). After 7–10 days in culture, keratinocytes attach to fibronectin, and β1 localizes with vinculin to focal adhesions (Toda *et al.*, 1987; Grinnell, 1990). Attachment to fibronectin is inhibited by peptides containing RGD and by polyclonal antibodies against fibronectin receptor. Thus, the expression and function of α5β1 are markedly increased on normal human keratinocytes after a short time in culture (Guo *et al.*, 1990). Such up-regulation of integrin expression and function during cell culture has also been observed with other normal and malignant cells (Albelda *et al.*, 1990; Horton, 1990).

The effect of cell culture on α5β1 expression in keratinocytes appears to be very similar to the effect of TGF-β on integrin expression in other cell types. Cell culture conditions may increase the expression of functional α5β1 by increasing the rate at which the β1 precursor is converted to mature glycoprotein (Guo *et al.*, 1991). Immunoprecipitates from freshly harvested keratinocytes contained only immature β1 subunits associated with α2, α3, and α5. Identical immunoprecipitates from cells cultured 4–7 days contained mostly mature β1. Although the rate of β1 processing in freshly isolated keratinocytes has not been measured, in cultured keratinocytes it is relatively fast: half-maximal maturation occurs in about 3 hr (Larjava, 1991).

Rabbit keratinocytes freshly isolated from ear skin, like freshly isolated human keratinocytes, do not adhere to fibronectin; but keratinocytes taken from wound beds of the same animals do adhere to fibronectin (Grinnell *et al.*, 1988). The induction of fibronectin receptor observed during culture *in vitro*, therefore, has a corollary during wound repair *in vivo*.

The α5β1 fibronectin receptor appears to be involved in keratinocyte motility on fibronectin (J. P. Kim *et al.*, 1992a). It has been known, in fact, for some time that fibronectin, as well as collagen types I and IV, promotes the migration of human keratinocytes (O'Keefe *et al.*, 1985). Interestingly, α5 is found only in keratinocytes capable of migration. When keratinocytes migrating out of skin explants are examined after 7 days, α5 is found predominantly in the migrating cells; β1, α2, and α3 are found both in the skin and in the migrating cells (Guo *et al.*, 1991). Furthermore, freshly isolated keratinocytes are relatively immobile cells (Grinnell, 1990) and they contain little fully processed α5β1; cultured cells are motile and do express functional α5β1 (Guo *et al.*, 1991). These observations indicate a consistent correlation between the presence of functional α5β1 and the ability to migrate. Thus, α5β1 expressed on wound epidermis as it migrates over the fibronectin-rich provisional matrix (Clark, 1990; Cavani *et al.*, 1993; Juhasz *et al.*, 1993; Gailit *et al.*, 1994) probably is required for this migration. Furthermore, the TGF-β1 that colocalizes with α5β1 to the migrating epidermis might be responsible for full expression and function of this fibronectin receptor (Gailit *et al.*, 1994).

4.2.2. Keratinocyte αv Integrins

Vitronectin, another plasma protein, is deposited in wounds along with fibrin(ogen) and fibronectin. Vitronectin was previously called serum spreading factor or epibolin since it is the factor in serum that induces keratinocytes at the cut edge of a human skin explant to migrate and eventually to surround the dermal component of the explant by a process called "epiboly" (Stenn, 1981). Vitronectin is also required for keratinocyte spreading on fibronectin, collagen, and laminin (Stenn, 1987).

Cultured human keratinocytes adhere to vitronectin through αvβ5. Antibodies against αv inhibit keratinocyte adhesion to vitronectin (Adams and Watt, 1991), while antibodies against β1 do not (Larjava *et al.*, 1990; Adams and Watt, 1991). Normal keratinocytes do not express αvβ3, the classic vitronectin receptor, in culture or in skin. Immunoprecipitations from detergent lysates of radiolabeled cells revealed complexes containing αv and β5 (Larjava *et al.*, 1990; Adams and Watt, 1991) but not β3 or β1 (Adams and Watt, 1991; Marchisio *et al.*, 1991). Although keratinocytes do express β1, the complex αvβ1, another potential vitronectin receptor, is not formed in normal cultured keratinocytes cells. The mRNA for β5 is present in keratinocytes (Adams and Watt, 1991; Marchisio *et al.*, 1991) and in other epithelial cell types (Ramaswamy and Hemler, 1990; Suzuki *et al.*, 1990).

It is important to distinguish among the different αv integrins as they seem to bind to different ligands, distribute differently in response to binding, and may well provide distinct signals to the cell. For example, M21 human melanoma cells contain αvβ3 and αvβ5 (Wayner *et al.*, 1991). When these cells were plated on vitronectin, αvβ3 colocalized in focal contacts with vinculin, talin, and the ends of actin filaments; αvβ5 showed a punctate distribution on the ventral surface that was entirely different from the focal contact staining. Similar results have been obtained from a strain of non-differentiating keratinocytes. This strain, called ndk, expresses αvβ1, αvβ3, and αvβ5 (Adams and Watt, 1991). When adherent cultures of these cells were examined, the results again suggested that αvβ3 and αvβ5 had different cellular distribution patterns.

A role for αvβ5 in keratinocyte motility was first suggested by two separate observations. First, in growing keratinocyte cultures, αv has been identified in basal cells at the periphery of expanding colonies (Adams and Watt, 1991; Marchisio *et al.*, 1991). Suprabasal cells, which have differentiated, did not contain αv. It is known that in an expanding colony the cells at the periphery migrate outward as the colony enlarges (Barrandon and Green, 1987). Thus, the expression of αvβ5 is characteristic of a migratory phenotype but not of a differentiated and stationary phenotype. Second, vitronectin promotes many forms of keratinocyte motility. As mentioned, vitronectin stimulates the movement of epithelial sheets, a process called epiboly (Stenn, 1981). Although vitronectin increases local motility of keratinocytes, in an almost circular pattern (Brown *et al.*, 1991), it does not stimulate the same directed, linear motility produced by collagen or fibronectin (O'Keefe *et al.*, 1985). In fact, vitronectin reduces keratinocyte motility on collagen type I (Brown *et al.*, 1991). Recently, it has been shown that αvβ5 is the integrin that mediates this modest motility (Kim *et al.*, 1994). Thus, vitronectin has several unique effects on cell motility, and αv integrins are probably required for keratinocyte interaction with vitronectin. In fact, αv, and more specifically αvβ5, is expressed on epidermal cells as they migrate over a provisional matrix containing vitronectin during epithelialization of cutaneous wounds (Cavani *et al.*, 1993; Gailit *et al.*, 1994). In all cells, except platelets and megakaryocytes, the only integrins that bind to vitronectin all contain αv.

More recently, both *in vitro* and *in vivo* evidence suggests that the migrating epidermis also expresses αvβ6 (Haapasalmi *et al.*, 1995). However, this integrin may be a tenascin receptor rather than a vitronectin receptor. Interestingly, there is a switch in αv integrins expressed during reepithelialization (Clark *et al.*, 1995a; Zambruno *et al.*, 1995). During the first few days of migration, the basal epidermal cells express αvβ5. In contrast, αvβ6 is present around the time of epidermal fusion. This progression of αv receptor expression corresponds to the early presence of vitronectin, a ligand for αvβ5, and the later appearance of tenascin, a ligand for αvβ6.

Clearly, fibronectin, vitronectin, and tenascin stimulate different, complementary types of motility (Brown *et al.*, 1991) and these responses are regulated during wound repair (Cavani *et al.*, 1993; Juhasz *et al.*, 1993; Gailit *et al.*, 1994; Haapasalmi *et al.*, 1995). Keratinocytes in normal unwounded skin are probably not in contact with either of these proteins. Nevertheless, freshly isolated keratinocytes respond immediately to vitronectin by attaching and spreading; they respond to fibronectin, however, only after some time in culture. The significance of this difference, that keratinocytes are constitutively responsive to vitronectin and facultatively responsive to fibronectin, is unclear. These responses are probably mediated by αvβ5 and α5β1.

4.2.3. Keratinocyte α2β1 and α3β1

The integrins α2β1 and α3β1 are prominently expressed by keratinocytes in culture and in skin (Carter *et al.*, 1990b; Adams and Watt, 1991; Marchisio *et al.*, 1991). Unlike keratinocyte adhesion and motility on fibronectin and vitronectin, interactions with collagen are not mediated by RGD even though collagens contain these sequences.

The α2β1 collagen receptor mediates keratinocyte interactions with collagens as demonstrated by the fact that blocking α2β1 with monospecific antibodies inhibits keratinocyte adhesion and migration on both types I and IV collagen (Carter *et al.*, 1990b; Staquet *et al.*, 1990; Adams and Watt, 1991). Moreover, α2β1 is concentrated in the focal adhesion plaques formed when keratinocytes attach to collagen type I, but not in the focal adhesions formed when the cells attach to fibronectin (Carter *et al.*, 1990b). Clearly, α2β1 mediates cultured keratinocyte migration (J. P. Kim *et al.*, 1992a), and it probably also mediates keratinocytes migration over denuded dermal type 1 collagen during cutaneous wound repair. Keratinocyte α2β1 collagen receptors are up-regulated by EGF, and such stimulated cells demonstrate enhanced migration on collagen (Chen *et al.*, 1993). Since increased EGF receptors are expressed on the migrating epidermis during cutaneous wound repair (Wenczak *et al.*, 1992), EGF or

TGF-α may up-regulate expression or function of α2β1 on the migrating keratinocytes. Paradoxically, the expression of α2β1 in the migrating wound epidermis is less, not more, than normal epidermal expression (Cavani *et al.,* 1993). Thus the explicit activity of α2β1 during epidermal repair warrants further study.

There are several possible matrix ligands for α3β1, but it probably serves keratinocytes as a receptor for laminins. In certain situations, α3β1 can bind to collagen and fibronectin, as well as laminin (Elices *et al.,* 1991). The α2β1 and α3β1 may act in concert to mediate cell binding to laminin. Antibodies against α2 (Carter *et al.,* 1990b) and α3 (Carter *et al.,* 1990b; Adams and Watt, 1991) individually inhibit attachment to laminin, but together the antibodies inhibit more effectively (Carter *et al.,* 1990b; Adams and Watt, 1991). However, α3, but not α2, is identified in focal adhesion plaques formed by cells attached to laminin (Carter *et al.,* 1990b). Keratinocyte α3β1 probably does not bind to collagen (Carter *et al.,* 1990b; Staquet *et al.,* 1990; Adams and Watt, 1991). Evidence for keratinocyte α3β1 binding to fibronectin is also contradictory.

The most biologically significant ligand for keratinocyte α3β1 is probably epiligrin, a glycoprotein synthesized by keratinocytes (Carter *et al.,* 1991). Originally identified as a product of keratinocytes in culture, epiligrin has also been identified in the basement membrane of skin and other epithelial tissues and is now recognized as an isoform of laminin (Marinkovich *et al.,* 1993). Now called laminin 5, this isoform is closely associated with anchoring fibrils that traverse the lamina lucida beneath hemidesmosomes (Carter *et al.,* 1991; Rousselle *et al.,* 1991). Like classical laminin, laminin 5 consists of three chains organized in a crosslike structure. The proposal that epiligrin, or laminin 5, is a ligand for α3β1 is supported by several observations (Carter *et al.,* 1991): Keratinocytes, and other cell types expressing α3β1, attach to purified epiligrin; monoclonal antibodies specific for α3 or β1 inhibit attachment to epiligrin; and, furthermore, α3β1 is concentrated in the focal adhesion plaques formed when fibroblasts attach to purified epiligrin. Laminin 5 (Marinkovich *et al.,* 1993), like classical laminin (O'Keefe *et al.,* 1985), inhibits human keratinocyte migration. Thus the reformation of the laminin-rich basement membrane beneath the neoepidermis of a cutaneous wound may stop the lateral movement of keratinocytes (Stanley *et al.,* 1981; Clark *et al.,* 1982).

Besides playing an important function in keratinocyte cell–matrix adhesion, as discussed above, α2β1 and α3β1 interact with each other (Symington *et al.,* 1993), and thus play a role in keratinocyte cell–cell interactions. Studies of human epidermal tissue have demonstrated, using conventional immunohistochemistry, that staining for the β1 subunit is discontinuous near the basement membrane and mostly continuous in areas of cell–cell contact (Larjava *et al.,* 1990). Subsequent studies of cultured keratinocytes confirmed that β1 was localized in areas of cell–cell contact (Carter *et al.,* 1990b; Larjava *et al.,* 1990) and revealed that α2 and α3, but not α5, also accumulated in the same areas when cell aggregation was induced by addition of calcium or serum (Carter *et al.,* 1990b). Furthermore, treating keratinocyte colonies with a monoclonal antibody specific for β1 dissociated the colonies by disrupting cell–cell contacts but apparently without disturbing cell–matrix adhesion (Larjava *et al.,* 1990). Monoclonal antibodies specific for β1 or α3 also inhibited cell–cell adhesion in an assay measuring the attachment of keratinocytes in suspension to a confluent cell layer (Carter *et al.*, 1990b). Finally, a study of cultured human cells demonstrated, using immunoelectron microscopy, that α3β1 was strongly associated with intercellular contact sites (Kaufmann *et al.*, 1989).

Although the role of integrins in keratinocyte cell–cell interactions remains controversial (Tenchini *et al.*, 1993), there is precedent for an integrin to have a dual function. The α4β1 functions in cell–matrix adhesion as a receptor for fibronectin and it functions in cell–cell adhesion as a mediator for lymphocyte–endothelial cell interactions (Cheresh, 1991). The α4β1 binds to vascular cell surface adhesion molecule-1 (VCAM-1) on the surface of endothelial cells activated by cytokines.

4.2.4. Keratinocyte α6β4

The integrin α6β4 is restricted to the basal surface of keratinocytes along the basement membrane zone and is apparently a component of the hemidesmosome, a structure unique to epithelial tissue (Stepp *et al.*, 1990). These adhesion junctions help anchor basal cells to the basement membrane. The α6β4 is a prominent integrin of keratinocytes in culture and in skin (Carter *et al.*, 1990a; De Luca *et al.*, 1990; Marchisio *et al.*, 1991). Although α6 can combine with β1 in some cells and tissues, immunoprecipitations demonstrated that in cultured human keratinocytes α6 and β4 combine exclusively with each other (De Luca *et al.*, 1990; Adams and Watt, 1991). The α6 and β4 were first identified in hemidesmosomes of corneal epithelium using immunoelectron microscopy (Stepp *et al.*, 1990). Subsequent work has confirmed the presence of α6β4 in hemidesmosomes in mucosal epithelium and in a cultured cell line (Jones *et al.*, 1991). The rat bladder carcinoma cell line 804G has the unusual ability to form hemidesmosomes *in vitro*. These cells apparently express α6β4 along their basal surface in a pattern that coincides with the distribution of a hemidesmosomal marker, bullous pemphigoid antigen (Jones *et al.*, 1991). A polyclonal antiserum, directed primarily against the β4 subunit of α6β4, was able to prevent the formation of hemidesmosomes by 804G cells. This antiserum also prevented the assembly of hemidesmosomes by corneal epithelial cells in an *in vitro* model of wound healing, but it did not inhibit their migration (Kurpakus *et al.*, 1991). Cultured human keratinocytes produce stable anchoring contacts similar to hemidesmosomes, and these structures also contain α6β4 and bullous pemphigoid antigens (Carter *et al.*, 1990b). These stable anchoring contacts are associated with keratin intermediate filaments rather than stress fibers, a finding that distinguishes them from focal adhesions.

Although the ligand for α6β4 in a human adenocarcinoma cell line has been identified as laminin (Lee *et al.*, 1992), the ligand for keratinocyte α6β4 has not been established unequivocally. Anchoring filaments in the basement membrane have been suggested as one possibility (Carter *et al.*, 1990a), but laminin seems a more likely candidate. Polyclonal antibodies specific for β4 were able to detach keratinocytes from a laminin substrate and to inhibit cell attachment to laminin (De Luca *et al.*, 1990). Although the binding sites on laminin for other receptors have been characterized, a distinct binding site for α6β4 has not been identified (Sonnenberg *et al.*, 1990; Sonnenberg, 1993). This situation is also complicated by the possibility that cultured human keratinocytes bind to laminin at three different sites (Wilke and Skubitz, 1991). Whatever the ligand for α6β4, the complex structure of the hemidesmosome further stabilizes the attachment of the neoepidermis of a wound to its newly formed basement membrane.

References

Abrams, C., Deng, Y. J., Steiner, B., O'Toole, T., and Shattil, S. J., 1994, Determinants of specificity of a baculovirus-expressed antibody Fab fragment that binds selectively to the activated form of integrin αIIbβ3, *J. Biol. Chem.* **269**:18781–18788.

Adams, J. C., and Watt, F. M., 1991, Expression of β1, β3, β4, and β5 integrins by human epidermal keratinocytes and non-differentiating keratinocytes, *J. Cell Biol.* **115**:829–841.

Akiyama, S. K., and Yamada, K. M., 1987, Biosynthesis and acquisition of biological activity of the fibronectin receptor, *J. Biol. Chem.* **262**:17536–17542.

Akiyama, S. K., and Yamada, K. M., 1993, Introduction: Adhesion molecules in cancer. Part I, *Semin. Cancer Biol.* **4**:215–218.

Akiyama, S. K., Yamada, S. S., and Yamada, K. M., 1989, Analysis of the role of glycosylation of the human fibronectin receptor, *J. Biol. Chem.* **264**:18011–18018.

Akiyama, S. K., Larjava, H., and Yamada, K. M., 1990a, Differences in the biosynthesis and localization of the fibronectin receptor in normal and transformed cultured human cells, *Cancer Res.* **50**:1601–1607.

Akiyama, S. K., Nagata, K., and Yamada, K. M., 1990b, Cell surface receptors for extracellular matrix components, *Biochim. Biophys. Acta* **1031**:91–110.

Akiyama, S. K., Yamada, S. S., Yamada, K. M., and LaFlamme, S. E., 1994, Transmembrane signal transduction by integrin cytoplasmic domains expressed in single-subunit chimeras, *J. Biol. Chem.* **269**:15961–15964.

Albelda, S. M., and Buck, C. A., 1990, Integrins and other cell adhesion molecules, *FASEB J.* **4**:2868–2880.

Albelda, S. M., Mette, S. A., Elder, D. E., Stewart, R., Damjanovich, L., Herlyn, M., and Buck, C. A., 1990, Integrin distribution in malignant melanoma: Association of the β3 subunit with tumor progression, *Cancer Res.* **50**:6757–6764.

Bajt, M. L., Ginsberg, M. H., Frelinger, A., Berndt, M. C., and Loftus, J. C., 1992, A spontaneous mutation of integrin αIIbβ3 (platelet glycoprotein IIb-IIIa) helps define a ligand binding site, *J. Biol. Chem.* **267**:3789–3794.

Balzac, F., Belkin, A. M., Koteliansky, V. E., Balabanov, Y. V., Altruda, F., Silengo, L., and Tarone, G., 1993, Expression and functional analysis of a cytoplasmic domain variant of the β1 integrin subunit, *J. Cell Biol.* **121**:171–178.

Barrandon, Y., and Green, H., 1987, Cell migration is essential for sustained growth of keratinocytes colonies: The roles of transforming growth factor-α and epidermal growth factor, *Cell* **50**:1131–1137.

Bauer, J., Varner, J., Schreiner, C., Kornberg, L., Nicholas, R., and Juliano, R. L., 1993, Functional role of the cytoplasmic domain of the integrin α5 subunit, *J. Cell Biol.* **122**:209–221.

Betz, P., Nerlich, A., Tubel, J., Penning, R., and Eisenmenger, W., 1993, Localization of tenascin in human skin wounds: An immunohistochemical study, *Int. J. Legal. Med.* **105**:325–328.

Briesewitz, R., Kern, A., and Marcantonio, E. E., 1993, Ligand-dependent and -independent integrin focal contact localization: The role of the alpha chain cytoplasmic domain, *Mol. Biol. Cell.* **4**:593–604.

Brown, C., Stenn, K. S., Falk, R. J., Woodley, D. T., and O'Keefe, E. J., 1991, Vitronectin: Effects on keratinocyte motility and inhibition of collagen-induced motility, *J. Invest. Dermatol.* **96**:724–728.

Burns, K., Duggan, B., Atkinson, E. A., Famulski, K. S., Nemer, M., Bleackley, R. C., and Michalak, M., 1994, Modulation of gene expression by calreticulin binding to the glucocorticoid receptor, *Nature* **367**:476–480.

Calvete, J. J., Mann, K., Schafer, W., Fernandez-Lafuente, R., and Guisan, J. M., 1994, Proteolytic degradation of the RGD-binding and non-RGD-binding conformers of human platelet integrin glycoprotein IIb/IIIa: Clues for identification of regions involved in the receptor's activation, *Biochem. J.* **298**:1–7.

Carter, W. G., Kaur, P., Gil, S. G., Gahr, P. J., and Wayner, E. A., 1990a, Distinct functions for integrins α3β1 in focal adhesions and α6β4/bullous pemphigoid antigen in a new stable anchoring contact (SAC) of keratinocytes: Relation to hemidesmosomes, *J. Cell Biol.* **111:**3141–3154.

Carter, W. G., Wayner, E. A., Bouchard, T. S., and Kaur, P., 1990b, The role of integrins α2β1 and α3β1 in cell–cell and cell–substrate adhesion of human epidermal cells, *J. Cell Biol.* **110:**1387–1404.

Carter, W. G., Ryan, M. C., and Gahr, P. A., 1991, Epiligrin, a new cell adhesion ligand for integrin α3β1 in epithelial basement membranes, *Cell* **65:**599–610.

Cavani, A., Zambruno, G., Marconi, A., Manca, V., Marchetti, M., and Giannetti, A., 1993, Distinctive integrin expression in the newly forming epidermis during wound healing in humans, *J. Invest. Dermatol.* **101:**600–604.

Chan, B. M., Kassner, P. D., Schiro, J. A., Byers, H. R., Kuppe, T. S., and Hemler, M. E., 1992, Distinct cellular functions mediated by different VLA integrin α subunit cytoplasmic domains, *Cell* **68:**1051–1060.

Chen, J. D., Kim, J. P., Zhang, K., Sarret, Y., Wynn, K. C., Kramer, R. H., and Woodley, D. T., 1993, Epidermal growth factor (EGF) promotes human keratinocyte locomotion on collagen by increasing the α2 integrin subunit, *Exp. Cell Res.* **209:**216–223.

Chen, Q., Kinch, M. S., Lin, T. H., Burridge, K., and Juliano, R. L., 1994, Integrin-mediated cell adhesion activates mitogen-activated protein kinases. Integrin-mediated cell adhesion activates mitogen-activated protein kinases, *J. Biol. Chem.* **269:**26602–26605.

Chen, W. T., and Singer, S. J., 1982, Immunoelectron microscopic studies of the sites of cell–substratum and cell–cell contacts in cultured fibroblasts, *J. Cell Biol.* **95:**205–222.

Chen, Y. P., O'Toole, T. E., Shipley, T., Forsyth, J., LaFlamme, S. E., Yamada, K. M., Shattil, S. J., and Ginsberg, M. H., 1994, Inside-out signal transduction inhibited by isolated integrin cytoplasmic domains, *J. Biol. Chem.* **269:**18307–18310.

Cheresh, D. A., 1991, Structure, function and biological properties of integrin αvβ3 on human melanoma cells, *Cancer Metastasis Rev.* **10:**3–10.

Cheresh, D. A., and Spiro, R. C., 1987, Biosynthetic and functional properties of an Arg-Gly-Asp-directed receptor involved in human melanoma cell attachment to vitronectin, fibrinogen, and von Willebrand factor, *J. Biol. Chem.* **262:**17703–17711.

Clark, R. A. F., 1990, Fibronectin matrix deposition and fibronectin receptor expression in healing and normal skin, *J. Invest. Dermatol.* **94:**128S–134S.

Clark, R. A. F., Lanigan, J. M., DellaPelle, P., Manseau, E., Dvorak, H. F., and Colvin, R. B., 1982, Fibronectin and fibrin provide a provisional matrix for epidermal cell migration during wound reepithelialization, *J. Invest. Dermatol.* **70:**264–269.

Clark, R. A. F., Wikner, N. E., Doherty, D. E., and Norris, D. A., 1988, Cryptic chemotactic activity of fibronectin for human monocytes resides in the 120-kDa fibroblastic cell-binding fragment, *J. Biol. Chem.* **263:**12115–12123.

Clark, R. A. F., Gailit, J., Pierschbacher, M. D., and Ruoslahti, E., 1990, Expression of fibronectin and vitronectin receptors in wound fibroblasts, *Clin. Res.* **38:**630A.

Clark, R. A. F., Spencer, J., Larjava, H., and Ferguson, M., 1995a, Reepithelialization of normal human excisional wounds is associated with a switch from αvβ5 to αvβ6 integrins, *J. Invest. Dermatol.*

Clark, R. A. F., Tonnesen, M. G., Gailit, J., and Cheresh, D. A., 1995b, Transient functional expression of αvβ3 on vascular cells during wound repair, *Am. J. Path.,* in press.

Collo, G., Starr, L., and Quaranta, V., 1993, A new isoform of the laminin receptor integrin α7β1 is developmentally regulated in skeletal muscle, *J. Biol. Chem.* **268:**19019–19024.

Conforti, G., Calza, M., and Beltran-Nunez, A., 1994, αvβ5 integrin is localized at focal contacts by HT-1080 fibrosarcoma cells and human skin fibroblasts attached to vitronectin, *Cell. Adhesion Commun.* **1:**279–293.

Couchman, J. R., and Rees, D. A., 1979, The behaviour of fibroblasts migrating from chick heart explants: Changes in adhesion, locomotion and growth, and in the distribution of actomyosin and fibronectin, *J. Cell Sci.* **39:**149–165.

Damsky, C. H., and Werb, Z., 1992, Signal transduction by integrin receptors for extracellular matrix: Cooperative processing of extracellular information, *Curr. Opin. Cell Biol.* **4:**772–781.

Davis, E. D., 1992, Affinity of integrins for damaged extracellular matrix: αvβ3 binds to denatured collagen type I through RGD sites, *Biochem. Biophys. Res. Commun.* **182:**1025–1031.

Dedhar, S., Ruoslahti, E., and Pierschbacher, M. D., 1987, A cell surface receptor complex for collagen type I recognizes the Arg-Gly-Asp sequence, *J. Cell Biol.* **104**:585–593.

Dedhar, S., Rennie, P. S., Shago, M., Hagesteijn, C. Y., Yang, H., Filmus, J., Hawley, R. G., Bruchovsky, N., Cheng, H., and Matusik, R. J., 1994, Inhibition of nuclear hormone receptor activity by calreticulin, *Nature* **367**:480–483.

De Luca, M., Tamura, R. N., Kajiji, S., Bondanza, S., Rossino, P., Cancedda, R., Marchisio, P. C., and Quaranta, V., 1990, Polarized integrin mediates human keratinocyte adhesion to basal lamina, *Proc. Natl. Acad. Sci. USA* **87**:6888–6892.

de Vos, A. M., Ultsch, M., and Kossiakoff, A. A., 1992, Human growth hormone and extracellular domain of its receptor: Crystal structure of the complex, *Science* **255**:306–312.

DiMilla, P. A., Stone, J. A., Quinn, J. A., Albelda, S. M., and Lauffenburger, D. A., 1993, Maximal migration of human smooth muscle cells on fibronectin and type IV collagen occurs at an intermediate attachment strength, *J. Cell Biol.* **122**:729–737.

Donaldson, D. J., Mahan, J. T., Hui, Y., and Yamada, K. M., 1994, Integrin and phosphotyrosine expression in normal and migrating newt keratinocytes, *Anat. Rec.* **241**:49–58.

Douglass, G. D., Zhang, K., and Kramer, R. H., 1992, The role of integrin adhesion receptors in gingival wound healing, *J. Calif. Dent. Assoc.* **20**:37–40.

D'Souza, S. E., Ginsberg, M. H., Burke, T. A., and Plow, E. F., 1990, The ligand binding site of the platelet integrin receptor GPIIb–IIIa is proximal to the second calcium binding domain of its α subunit, *J. Biol. Chem.* **265**:3440–3446.

D'Souza, S. E., Ginsberg, M. H., Matsueda, G. R., and Plow, E. F., 1991, A discrete sequence in a platelet integrin is involved in ligand recognition, *Nature* **350**:66–68.

D'Souza, S. E., Haas, T. A., Piotrowicz, R. S., Byers-Ward, V., McGrath, D. E., Soule, H. R., Cierniewski, C., Plow, E. F., and Smith, J. W., 1994, Ligand and cation binding are dual functions of a discrete segment of the integrin β3 subunit: Cation displacement is involved in ligand binding, *Cell* **79**:659–667.

Du, X. P., Plow, E. F., Frelinger, A. L., O'Toole, T. E., Loftus, J. C., and Ginsberg, M. H., 1991, Ligands activate integrin αIIbβ3 (platelet GPIIb–IIIa), *Cell* **65**:409–416.

Duband, J. L., Nuckolls, G. H., Ishihara, A., Hasegawa, T., Yamada, K. M., Thiery, J. P., and Jacobson, K., 1988a, Fibronectin receptor exhibits high lateral mobility in embryonic locomoting cells but is immobile in focal contacts and fibrillar streaks in stationary cells, *J. Cell Biol.* **107**:1385–1396.

Duband, J. L., Dufour, S., Yamada, K. M., and Thiery, J. P., 1988b, The migratory behavior of avian embryonic cells does not require phosphorylation of the fibronectin–receptor complex, *FEBS Lett.* **230**:181–185.

Duband, J. L., Dufour, S., Yamada, S. S., Yamada, K. M., and Thiery, J. P., 1991, Neural crest cell locomotion induced by antibodies to β1 integrins. A tool for studying the roles of substratum molecular avidity and density in migration, *J. Cell Sci.* **98**:517–532.

Dustin, M. L., and Springer, T. A., 1989, T-cell receptor cross-linking transiently stimulates adhesiveness through LFA-1, *Nature* **341**:619–624.

Eble, J. A., Golbik, R., Mann, K., and Kuhn, K., 1993, The α1β1 integrin recognition site of the basement membrane collagen molecule [α1(IV)]^2 α2(IV)], *EMBO J.* **12**:4795–4802.

Elemer, G. S., and Edgington, T. S., 1994, Microfilament reorganization is associated with functional activation of αMβ2 on monocytic cells, *J. Biol. Chem.* **269**:3159–3166.

Elices, M. J., Urry, L. A., and Hemler, M. E., 1991, Receptor functions for the integrin VLA-3: Fibronectin, collagen, and laminin binding are differentially influenced by Arg-Gly-Asp peptide and by divalent cations, *J. Cell Biol.* **112**:169–181.

Faull, R. J., Kovach, N. L., Harlan, J. M., and Ginsberg, M. H., 1993, Affinity modulation of integrin α5β1: Regulation of the functional response by soluble fibronectin, *J. Cell Biol.* **121**:155–162.

Gailit, J., and Clark, R. A. F., 1993, Integrins in the skin, *Adv. Dermatol.* **8**:129–152.

Gailit, J., Welch, M. P., and Clark, R. A. F., 1994, TGF-β 1 stimulates expression of keratinocyte integrins during re-epithelialization of cutaneous wounds, *J. Invest. Dermatol.* **103**:221–227.

Gehlsen, K. R., Argraves, W. S., Pierschbacher, M. D., and Ruoslahti, E., 1988, Inhibition of *in vitro* tumor cell invasion by Arg-Gly-Asp-containing synthetic peptides, *J. Cell Biol.* **106**:925–930.

Geiger, B., 1989, Cytoskeleton-associated cell contacts, *Curr. Opin. Cell Biol.* **1**:103–109.

Geiger, B., Salomon, D., Takeichi, M., and Hynes, R. O., 1992, A chimeric N-cadherin/β1-integrin receptor which localizes to both cell–cell and cell–matrix adhesions, *J. Cell Sci.* **103**:943–951.

Ginsberg, M. H., Du, X., and Plow, E. F., 1992, Inside-out integrin signalling, *Curr. Opin. Cell Biol.* **4**:766–771.

Gipson, I. K., Spurr-Michaud, S., Tisdale, A., Elwell, J., and Stepp, M. A., 1993, Redistribution of the hemidesmosome components α6β4 integrin and bullous pemphigoid antigens during epithelial wound healing, *Exp. Cell Res.* **207**:86–98.

Glukhova, M. A., and Thiery, J. P., 1993, Fibronectin and integrins in development, *Semin. Cancer Biol.* **4**:241–249.

Grinnell, F., 1990, The activated keratinocyte: Up-regulation of cell adhesion and migration during wound healing, *J. Trauma* **30**:S144–S149.

Grinnell, F., 1992, Wound repair, keratinocyte activation and integrin modulation, *J. Cell Sci.* **101**:1–5.

Grinnell, F., and Geiger, B., 1986, Interaction of fibronectin-coated beads with attached and spread fibroblasts. Binding, phagocytosis, and cytoskeletal reorganization, *Exp. Cell Res.* **162**:449–461.

Grinnell, F., Toda, K.-I., and Takashima, A., 1988, Role of fibronectin in epithelialization and wound healing, in: *Growth Factors and Other Aspects of Wound Healing: Biological and Clinical Implications* (A. Barbul, E. Pines, M. Caldwell, and T. K. Hunt, eds.), pp. 259–272, Alan R. Liss, New York.

Guadagno, T. M., Ohtsubo, M., Roberts, J. M., and Assoian, R. K., 1993, A link between cyclin A expression and adhesion-dependent cell cycle progression, *Science* **262**:1572–1575.

Guan, J. L., and Shalloway, D., 1992, Regulation of focal adhesion-associated protein tyrosine kinase by both cellular adhesion and oncogenic transformation, *Nature* **358**:690–692.

Guan, J. L., Trevithick, J. E., and Hynes, R. O., 1991, Fibronectin/integrin interaction induces tyrosine phosphorylation of a 120-kDa protein, *Cell Regul.* **2**:951–964.

Gumbiner, B. M., 1993, Proteins associated with the cytoplasmic surface of adhesion molecules, *Neuron* **11**:551–564.

Guo, M., Toda, K., and Grinnell, F., 1990, Activation of human keratinocyte migration on type I collagen and fibronectin, *J. Cell Sci.* **96**:197–205.

Guo, M., Kim, L. T., Akiyama, S. K., Gralnick, H. R., Yamada, K. M., and Grinnell, F., 1991, Altered processing of integrin receptors during keratinocyte activation, *Exp. Cell Res.* **195**:315–322.

Haapasalmi, K., Zhang, K., Tonnesen, M. G., Olerud, J., Sheppard, D., Kramer, R., Clark, R. A. F., Uitto, V.-J., and Larjava, H., 1995, Keratinocytes in human wounds express αvβ6 integrin, *J. Invest. Dermatol.* submitted.

Han, E. K., Guadagno, T. M., Dalton, S. L., and Assoian, R. K., 1993, A cell cycle and mutational analysis of anchorage-independent growth: Cell adhesion and TGF-β1 control G1/S transit specifically, *J. Cell Biol.* **122**:461–471.

Hay, E., 1991, *Cell Biology of Extracellular Matrix,* Plenum Press, New York.

Hayashi, Y., Haimovich, B., Reszka, A., Boettige, D., and Horwitz, A., 1990, Expression and function of chicken integrin β1 subunit and its cytoplasmic domain mutants in mouse NIH 3T3 cells, *J. Cell Biol.* **110**:175–184.

Heino, J., Ignotz, R. A., Hemler, M. E., Crouse, C., and Massague, J., 1989, Regulation of cell adhesion receptors by transforming growth factor-β. Concomitant regulation of integrins that share a common β1 subunit, *J. Biol. Chem.* **264**:380–388.

Hemler, M. E., 1990, VLA proteins in the integrin family: Structures, functions, and their role on leukocytes, *Annu. Rev. Immunol.* **8**:365–400.

Hergott, G. J., Nagai, H., and Kalnins, V. I., 1993, Inhibition of retinal pigment epithelial cell migration and proliferation with monoclonal antibodies against the β1 integrin subunit during wound healing in organ culture, *Invest. Ophthalmol. Vis. Sci.* **34**:2761–2768.

Hertle, M. D., Kubler, M. D., Leigh, I. M., and Watt, F. M., 1992, Aberrant integrin expression during epidermal wound healing and in psoriatic epidermis, *J. Clin. Invest.* **89**:1892–1901.

Hogg, N., 1991, An integrin overview, *Chem. Immunol.* **50**:1–12.

Horton, M., 1990, Vitronectin receptor: Tissue specific expression or adaptation to culture? *Int. J. Exp. Pathol.* **71**:741–759.

Horwitz, A., Duggan, K., Buck, C., Beckerle, M. C., and Burridge, K., 1986, Interaction of plasma membrane fibronectin receptor with talin—A transmembrane linkage, *Nature* **320**:531–533.

Howlett, A. R., and Bissell, M. J., 1993, The influence of tissue microenvironment (stroma and extracellular matrix) on the development and function of mammary epithelium, *Epithelial Cell. Biol.* **2**:79–89.

Huhtala, P., Humphries, M. J., McCarthy, J. B., Tremble, P. M., Werb, Z., and Damsky, C. H., 1995, Cooperative signaling by α5β1 and α4β1 integrins regulates metalloproteinase gene expression in fibroblasts adhering to fibronectin, *J. Cell Biol.* **129**:867–879.

Hyatt, S. L., Klauck, T., and Jaken, S., 1990, Protein kinase C is localized in focal contacts of normal but not transformed fibroblasts, *Mol. Carcinog.* **3**:45–53.

Hynes, R. O., 1992, Integrins: Versatility, modulation, and signaling in cell adhesion, *Cell* **69**:11–25.

Jones, J. C. R., Kurpakus, M. A., Cooper, H. M., and Quaranta, V., 1991, A function for the integrin α6β4 in the hemidesmosome, *Cell Regul.* **2**:427–438.

Juhasz, I., Murphy, G. F., Yan, H. C., Herlyn, M., and Albelda, S. M., 1993. Regulation of extracellular matrix proteins and integrin cell substratum adhesion receptors on epithelium during cutaneous human wound healing *in vivo, Am. J. Pathol.* **143**(5):1458–1469.

Juliano, R. L., and Haskill, S., 1993, Signal transduction from the extracellular matrix, *J. Cell Biol.* **120**:577–585.

Kamata, T., and Takada, Y., 1994, Direct binding of collagen to the I domain of integrin α2β1 (VLA-2, CD49b/CD29) in a divalent cation-independent manner, *J. Biol. Chem.* **269**(42):26006–26010.

Kapron-Bras, C., Fitz-Gibbon, L., Jeevaratnam, P., Wilkins, J., and Dedhar, S., 1993, Stimulation of tyrosine phosphorylation and accumulation of GTP-bound p21ras upon antibody-mediated α2β1 integrin activation in T-lymphoblastic cells, *J. Biol. Chem.* **268**:20701–20704.

Kassner, P. D., and Hemler, M. E., 1993, Interchangeable α chain cytoplasmic domains play a positive role in control of cell adhesion mediated by VLA-4, a β1 integrin, *J. Exp. Med.* **178**:649–660.

Kassner, P. D., Kawaguchi, S., and Hemler, M. E., 1994, Minimum α chain cytoplasmic tail sequence needed to support integrin-mediated adhesion, *J. Biol. Chem.* **269**:19859–19867.

Kaufmann, R., Frosch, D., Westphal, C., Weber, L., and Klein, C. E., 1989, Integrin VLA-3: Ultrastructural localization at cell-cell contact sites of human cell cultures, *J. Cell Biol.* **109**:1807–1815.

Kern, A., Briesewitz, R., Bank, I., and Marcantonio, E. E., 1994, The role of the I domain in ligand binding of the human integrin α1β1, *J. Biol. Chem.* **269**:22811–22816.

Kim, J. P., Chen, J. D., and Woodley, D. T., 1992a, Mechanism of human keratinocyte migration on fibronectin: Unique roles of RGD site and integrins, *J. Cell. Physiol.* **151**:443–450.

Kim, J. P., Zhang, K., Kramer, R. H., Schall, T. J., and Woodley, D. T., 1992b, Integrin receptors and RGD sequences in human keratinocyte migration: Unique antimigratory function of α3β1, *J. Invest. Dermatol.* **98**:764–770.

Kim, J. P., Zhang, K., Chen, J. D. Kramer, R. H., and Woodley, D. T., 1994, Vitronectin-driven human keratinocyte locomotion is mediated by the αvβ5 integrin receptor, *J. Biol. Chem.* **269**:26926–26932.

Kim, L. T., Ishihara, S., Lee, C. C., Akiyama, S. K., Yamada, K. M., and Grinnell, F., 1992, Altered glycosylation and cell surface expression of β1 integrin receptors during keratinocyte activation, *J. Cell Sci.* **103**:743–753.

Klebe, R. J., 1974, Isolation of a collagen-dependent cell attachment factor, *Nature* **250**:248–251.

Koivisto, L., Heino, J., Hakkinen, L., and Larjava, H., 1994, The size of the intracellular β1-integrin precursor pool regulates maturation of β1-integrin subunit and associated α-subunits, *Biochem. J.* **300**:771–779.

Kornberg, L. J., Earp, H. S., Turner, C. E., Prockop, C., and Juliano, R. L., 1991, Signal transduction by integrins: Increased protein tyrosine phosphorylation caused by clustering of β1 integrins, *Proc. Natl. Acad. Sci. USA* **88**:8392–8396.

Kornberg, L., Earp, H. S., Parsons, J. T., Schaller, M., and Juliano, R. L., 1992, Cell adhesion or integrin clustering increases phosphorylation of a focal adhesion-associated tyrosine kinase, *J. Biol. Chem.* **267**:23439–23442.

Kurpakus, M. A., Quaranta, V., and Jones, J. C., 1991, Surface relocation of α6β4 integrins and assembly of hemidesmosomes in an *in vitro* model of wound healing, *J. Cell Biol.* **115**:1737–1750.

LaFlamme, S. E., Akiyama, S. K., and Yamada, K. M., 1992, Regulation of fibronectin receptor distribution, *J. Cell Biol.* **117**:437–447.

LaFlamme, S. E., Thomas, L. A., Yamada, S. S., and Yamada, K. M., 1994, Single-subunit chimeric integrins as mimics and inhibitors of endogenous integrin functions. *J. Cell Biol.* **126**:1287–1298.

Larjava, H., 1991, Expression of β1 integrins in normal human keratinocytes. *Am. J. Med. Sci.* **301**:63–68.

Larjava, H., Peltonen, J., Akiyama, S. K., Yamada, S. S., Gralnick, H. R., Uitto, J., and Yamada, K. M., 1990, Novel function for β1 integrins in keratinocyte cell–cell interactions, *J. Cell Biol.* **110**:803–815.

Larjava, H., Salo, T., Haapasalmi, K., Kramer, R. H., and Heino, J., 1993, Expression of integrins and basement membrane components by wound keratinocytes, *J. Clin. Invest.* **92**:1425–1435.

Lee, E. C., Lotz, M. M., Stelle, G. D., and Mercurio, A. M., 1992, The integrin α6β4 is a laminin receptor, *J. Cell Biol.* **117**:671–678.

Lenter, M., and Vestweber, D., 1994, The integrin chains β1 and α6 associate with the chaperone calnexin prior to integrin assembly, *J. Biol. Chem.* **269**:12263–12268.

Leung-Hagesteijn, C. Y., Milankov, K., Michalak, M., Wilkins, J., and Dedhar, S., 1994, Cell attachment to extracellular matrix substrates is inhibited upon down-regulation of expression of calreticulin, an intracellular integrin α-subunit-binding protein, *J. Cell Sci.* **107**:589–600.

Lewis, J. M., and Schwartz, M. A., 1995, Mapping *in vivo* associations of cytoplasmic proteins with integrin β1 cytoplasmic domain mutants, *Mol. Biol. Cell* **6**:151–160.

Lin, C. Q., and Bissell, M. J., 1993, Multi-faceted regulation of cell differentiation by extracellular matrix, *FASEB J.* **7**:737–743.

Lindberg, F. P., Gresham, H. D., Schwarz, E., and Brown, E. J., 1993, Molecular cloning of integrin-associated protein: an immunoglobulin family member with multiple membrane-spanning domains implicated in αvβ3-dependent ligand binding, *J. Cell Biol.* **123**:485–496.

Loftus, J. C., O'Toole, T. E., Plow, E. F., Glass, A., Frelinger, A. L., and Ginsberg, M. H., 1990, A β3 integrin mutation abolishes ligand binding and alters divalent cation-dependent conformation, *Science* **249**:915–918.

Lukashev, M. E., Sheppard, D., and Pytela, R., 1994, Disruption of integrin function and induction of tyrosine phosphorylation by the autonomously expressed β1 integrin cytoplasmic domain, *J. Biol. Chem.* **269**:18311–18314.

Mackie, E. J., Halfter, W., and Liverani, D., 1988, Induction of tenascin in healing wounds, *J. Cell Biol.* **107**:2757–2767.

Marchisio, P. C., Bondanza, S., Cremona, O., Cancedda, R., and De Luca, M., 1991, Polarized expression of integrin receptors (α6β4, α2β1, α3β1, and αvβ5) and their relationship with the cytoskeleton and basement membrane matrix in cultured human keratinocytes, *J. Cell Biol.* **112**:761–773.

Marinkovich, M. P., Keene, D. R., Rimberg, C. S., and Burgeson, R. E., 1993, Cellular origin of the dermal–epidermal basement membrane, *Dev. Dynam.* **197**:255–267.

McNamee, H. P., Ingber, D. E., and Schwartz, M. A., 1993, Adhesion to fibronectin stimulates inositol lipid synthesis and enhances PDGF-induced inositol lipid breakdown, *J. Cell Biol.* **121**:673–678.

Metzger, H., 1992, Transmembrane signaling: The joy of aggregation, *J. Immunol.* **149**:1477–1487.

Miyamoto, S., Akiyama, S. K., and Yamada, K. M., 1995a, Synergistic roles for receptor occupancy and aggregation in integrin transmembrane function, *Science* **267**:883–885.

Miyamoto, S., Teramoto, H., Coso, O. A., Gutkind, J. S., Burbelo, P. D., Akiyama, S. K., and Yamada, K. M., 1995b, Integrin function: Molecular hierarchies of cytoskeletal and signaling molecules, *J. Cell Biol.*, **131**:791–805.

Morino, N., Mimura, T., Hamasaki, K., Tobe, K., Ueki, K., Kikuchi, K., Takehara, K., Kadowaki, T., Yazaki, Y., and Nojima, Y., 1995, Matrix/integrin interaction activates the mitogen-activated protein kinase, p44erk-1 and p42erk-2, *J. Biol. Chem.* **270**:269–273.

Mueller, S. C., Kelly, T., Dai, M. Z., Dai, H. N., and Chen, W. T., 1989, Dynamic cytoskeleton–integrin associations induced by cell binding to immobilized fibronectin, *J. Cell Biol.* **109**:3455–3464.

Murakami, J., Nishida, T., and Otori, T., 1992, Coordinated appearance of β1 integrins and fibronectin during corneal wound healing, *J. Lab. Clin. Med.* **120**:86–93.

Nermut, M. V., Green, N. M., Eason, P., Yamada, S. S., and Yamada, K. M., 1988, Electron microscopy and structural model of human fibronectin receptor, *EMBO J.* **7**:4093–4099.

Odland, G., and Ross, R., 1968, Human wound repair. I. Epidermal regeneration, *J. Cell Biol.* **39**:135–157.

O'Keefe, E. J., Payne, Jr., R. E., Russell, N., and Woodley, D. T., 1985, Spreading and enhance motility of human keratinocytes on fibronectin, *J. Invest. Dermatol.* **85**:125–130.

Otey, C. A., Pavalko, F. M., and Burridge, K., 1990, An interaction between α-actinin and the β1 integrin subunit *in vitro*, *J. Cell Biol.* **111**:721–729.

Otey, C. A., Vasquez, G. B., Burridge, K., and Erickson, B. W., 1993, Mapping of the α-actinin binding site within the β1 integrin cytoplasmic domain, *J. Biol. Chem.* **268**:21193–21197.

O'Toole, T. E., Katagiri, Y., Faull, R. J., Peter, K., Tamura, R., Quaranta, V., Loftus, J. C., Shattil, S. J., and Ginsberg, M. H., 1994, Integrin cytoplasmic domains mediate inside-out signal transduction, *J. Cell Biol.* **124**(6):1047–1059.

O'Toole, T. E., Mandelman, D., Forsyth, J., Shattil, S. J., Plow, E. F., and Ginsberg, M. H., 1991, Modulation of the affinity of integrin αIIbβ3 (GPIIb-IIIa) by the cytoplasmic domain of αIIb, *Science* **254**:845–847.

Paallysaho, T., Tervo, T., Virtanen, I., and Tervo, K., 1992, Integrins in the normal and healing corneal epithelium, *Acta Ophthalmol. Suppl.* **202**:22–25.

Pardi, R., Inverardi, L., Rugarli, C., and Bender, J. R., 1992, Antigen–receptor complex stimulation triggers protein kinase C-dependent CD11a/CD18–cytoskeleton association in T lymphocytes, *J. Cell Biol.* **116**:1211–1220.

Pasqualini, R., and Hemler, M. E., 1994, Contrasting roles for integrin β1 and β5 cytoplasmic domains in subcellular localization, cell proliferation, and cell migration, *J. Cell Biol.* **125**:447–460.

Pavalko, F. M., and LaRoche, S. M., 1993, Activation of human neutrophils induces an interaction between the integrin β2-subunit (CD18) and the actin binding protein alpha-actinin, *J. Immunol.* **151**:3795–3807.

Pavalko, F. M., and Otey, C. A., 1994, Role of adhesion molecule cytoplasmic domains in mediating interactions with the cytoskeleton, *Proc. Soc. Exp. Biol. Med.* **205**:282–293.

Pfaff, M., Aumailley, M., Specks, U., Knolle, J., Zerwes, H. G., and Timpl, R., 1993, Integrin and Arg-Gly-Asp dependence of cell adhesion to the native and unfolded triple helix of collagen type VI, *Exp. Cell Res.* **206**(1):167–176.

Plopper, G., and Ingber, D. E., 1993, Rapid induction and isolation of focal adhesion complexes, *Biochem. Biophys. Res. Commun.* **193**:571–578.

Plow, E. F., D'Souza, S. E., and Ginsberg, M. H., 1992, Ligand binding to GPIIb-IIIa: A status report, *Semin. Thromb. Hemost.* **18**:324–332.

Ramaswamy, H., and Hemler, M. E., 1990, Cloning, primary structure and properties of a novel human integrin β subunit, *EMBO J.* **9**:1561–1568.

Randi, A. M., and Hogg, N., 1994, I domain of β2 integrin lymphocyte function-associated antigen-1 contains a binding site for ligand intercellular adhesion molecule-1, *J. Biol. Chem.* **269**:12395–12398.

Rojiani, M. V., Finlay, B. B., Gray, V., and Dedhar, S., 1991, *In vitro* interaction of a polypeptide homologous to human Ro/SS-A antigen (calreticulin) with a highly conserved amino acid sequence in the cytoplasmic domain of integrin α subunits, *Biochemistry* **30**:9859–9866.

Rousselle, P., Lunstrum, G. P., Keene, D. R., and Burgeson, R. E., 1991, Kalinin: An epithelium-specific basement membrane adhesion molecule that is a component of anchoring filaments, *J. Cell Biol.* **114**:567–576.

Ruoslahti, E., 1991, Integrins, *J. Clin. Invest.* **87**:1–5.

Sastry, S. K., and Horwitz, A. F., 1993, Integrin cytoplasmic domains: Mediators of cytoskeletal linkages and extra- and intracellular initiated transmembrane signaling, *Curr. Opin. Cell Biol.* **5**:819–831.

Schaller, M. D., Otey, C. A., Hildebrand, J. D., and Parsons, J. T., 1995, Focal adhesion kinase and paxillin bind to peptides mimicking beta integrin cytoplasmic domains, *J. Cell Biol.* **130**:1181–1187.

Schiro, J. A., Chan, B. M., Roswit, W. T., Kassner, P. D., Pentland, A. P., Hemler, M. E., Eisen, A. Z., and Kupper, T. S., 1991, Integrin α2β1 (VLA-2) mediates reorganization and contraction of collagen matrices by human cells, *Cell* **67**:403–410.

Schlessinger, J., 1988, Signal transduction by allosteric receptor oligomerization, *Trends Biochem. Sci.* **13**:443–447.

Schmidt, C. E., Horwitz, A. F., Lauffenburger, D. A., and Sheetz, M. P., 1993, Integrin–cytoskeletal interactions in migrating fibroblasts are dynamic, asymmetric, and regulated, *J. Cell Biol.* **123**:977–991.

Schwartz, M. A., 1993, Spreading of human endothelial cells on fibronectin or vitronectin triggers elevation of intracellular free calcium, *J. Cell Biol.* **120**:1003–1010.

Schwartz, M. A., and Denninghoff, K., 1994, αv Integrins mediate the rise in intracellular calcium in endothelial cells on fibronectin even though they play a minor role in adhesion, *J. Biol. Chem.* **269**:11133–11137.

Schwartz, M. A., and Lechene, C., 1992, Adhesion is required for protein kinase C-dependent activation of the Na$^+$/H$^+$ antiporter by platelet-derived growth factor, *Proc. Natl. Acad. Sci. USA* **89**:6138–6141.

Schwartz, M. A., Ingber, D. E., Lawrence, M., Springer, T., and Lechene, C., 1991a, Multiple integrins share the ability to induce elevation of intracellular pH, *Exp. Cell Res.* **195**:533–535.

Schwartz, M. A., Lechene, C., and Ingber, D. E., 1991b, Insoluble fibronectin activates the Na/H antiporter by clustering and immobilizing integrin α1β5, independent of cell shape, *Proc. Natl. Acad. Sci. USA* **88**:7849–7853.

Shattil, S. J., and Brugge, J. S., 1991, Protein tyrosine phosphorylation and the adhesive functions of platelets, *Curr. Opin. Cell Biol.* **3**:869–879.

Shimizu, Y., Van Seventer, G. A., Horgan, K. J., and Shaw, S., 1990, Regulated expression and binding of three VLA (β1) integrin receptors on T cells, *Nature* **345**:250–253.

Singer, I. I., 1979, The fibronexus: A transmembrane association of fibronectin-containing fibers and bundles of 5 nm microfilaments in hamster and human fibroblasts, *Cell* **16**:675–685.

Singer, I. I., Kawka, D. W., Kazazis, D. M., and Clark, R. A. F., 1984, *In vivo* codistribution of fibronectin and actin fibers in granulation tissue: Immunofluorescence and electron microscope studies of the fibronexus at the myofibroblast surface, *J. Cell Biol.* **98**:2091–2106.

Smith, J. W., and Cheresh, D. A., 1990, Integrin (αvβ3)–ligand interaction. Identification of a heterodimeric RGD binding site on the vitronectin receptor, *J. Biol. Chem.* **265**:2168–2172.

Song, W. K., Wang, W., Sato, H., Bielser, D. A., and Kaufman, S. J., 1993, Expression of α7 integrin cytoplasmic domains during skeletal muscle development: Alternate forms, conformational change, and homologies with serine/threonine kinases and tyrosine phosphatases, *J. Cell Sci.* **106**:1139–1152.

Sonnenberg, A., 1993, Integrins and their ligands, *Curr. Top. Microbiol. Immunol.* **184**:7–35.

Sonnenberg, A., Linders, C. J. T., Modderman, P. W., Damsky, C. H., Aumailley, M., and Timpl, R., 1990, Integrin recognition of different cell-binding fragments of laminin (P1, E3, E8) and evidence that α6β1 but not α6β4 functions as a major receptor for fragment E8, *J. Cell Biol.* **110**:2145–2155.

Springer, T. A., 1994, Traffic signals for lymphocyte recirculation and leukocyte emigration: The multistep paradigm, *Cell* **76**:301–314.

Staatz, W. D., Fok, K. F., Zutter, M. M., Adams, S. P., Rodriguez, B. A., and Santoro, S. A., 1991, Identification of a tetrapeptide recognition sequence for the α2β1 integrin in collagen, *J. Biol. Chem.* **266**:7363–7367.

Stanley, J. R., Alvarez, O. M., Bere, E. W., Eaglstein, W. H., and Katz, S. I., 1981, Detection of membrane zone antigens during epidermal wound healing in pigs, *J. Invest. Dermatol.* **7**:240–243.

Stanley, P., Bates, P. A., Harvey, J., Bennett, R. I., and Hogg, N., 1994, Integrin LFA-1 alpha subunit contains an ICAM-1 binding site in domains V and VI, *EMBO J.* **13**:1790–1798.

Staquet, M. J., Levarlet, B., Dezutter-Dambuyant, C., Schmitt, D., and Thivolet, J., 1990, Identification of specific human epithelial cell integrin receptors as VLA proteins, *Exp. Cell Res.* **187**:277–283.

Stenn, K. S., 1981, Epibolin: A protein of human plasma that supports epithelial cell movement, *Proc. Natl. Acad. Sci. USA* **78**:6907–6911.

Stenn, K. S., 1987, Coepibolin, the activity of human serum that enhances the cell-spreading properties of epibolin, associates with albumin, *J. Invest. Dermatol.* **89**:59–63.

Stepp, M. A., Spurr-Michaud, S., Tisdale, A., Elwell, J., and Gipson, I. K., 1990, α6β4 Integrin heterodimer is a component of hemidesmosomes, *Proc. Natl. Acad. Sci. USA* **87**:8970–8974.

Stepp, M. A., Spurr-Michaud, S., and Gipson, I. K., 1993, Integrins in the wounded and unwounded stratified squamous epithelium of the cornea, *Invest. Ophthalmol. Vis. Sci.* **34**:1829–1844.

Streuli, C. H., Bailey, N., and Bissell, M. J., 1991, Control of mammary epithelial differentiation: Basement membrane induces tissue-specific gene expression in the absence of cell–cell interaction and morphological polarity, *J. Cell Biol.* **115**:1383–1395.

Suzuki, S., Huang, Z.-S., and Tanihara, H., 1990, Cloning of an integrin β subunit exhibiting high homology with integrin β3 subunit, *Proc. Natl. Acad. Sci. USA* **87**:5354–5358.

Symington, B. E., 1990, Fibronectin receptor overexpression and loss of transformed phenotype in a stable variant of the K562 cell line, *Cell. Regul.* **1**:637–648.

Symington, B. E., 1992, Fibronectin receptor modulates cyclin-dependent kinase activity, *J. Biol. Chem.* **267**:25744–25747.

Symington, B. E., Takada, Y., and Carter, W. G., 1993, Interaction of integrins α3β1 and α2β1: Potential role in keratinocyte intercellular adhesion, *J. Cell Biol.* **120**:523–535.

Takada, Y., Ylanne, J., Mandelman, D., Puzon, W., and Ginsberg, M. H., 1992, A point mutation of integrin β1 subunit blocks binding of α5β1 to fibronectin and invasin but not recruitment to adhesion plaques, *J. Cell Biol.* **119**:913–921.

Tamura, R. N., Rozzo, C., Starr, L., Chambers, J., Reichardt, L. F., Cooper, H. M., and Quaranta, V., 1990, Epithelial integrin α6β4: Complete primary structure of α6 and variant forms of β4, *J. Cell Biol.* **111**:1593–1604.

Tapley, P., Horwitz, A., Buck, C., Duggan, K., and Rohrschneider, L., 1989, Integrins isolated from Rous sarcoma virus-transformed chicken embryo fibroblasts, *Oncogene* **4**:325–333.

Tenchini, M. L., Adams, J. C., Gilbert, C., Steel, J., Hudson, D. L., Malcovati, M., and Watt, F. M., 1993, Evidence against a major role for integrins in calcium-dependent intercellular adhision of epidermal keratinocytes, *Cell Adhesion Commun.* **1**:55–66.

Tervo, T., van Setten, G. B., Paallysaho, T., Tarkkanen, A., and Tervo, K., 1992, Wound healing of the ocular surface, *Ann. Med.* **24**:19–27.

Toda, K.-I., Tuan, T.-L., Brown, P. J., and Grinnell, F., 1987, Fibronectin receptors of human keratinocytes and their expression during cell culture, *J. Cell Biol.* **105**:3097–3104.

Tuckwell, D. S., Brass, A., and Humphries, M. J., 1992, Homology modelling of integrin EF-hands. Evidence for widespread use of a conserved cation-binding site, *Biochem. J.* **285**:325–331.

Tuckwell, D. S., Weston, S. A., and Humphries, M. J., 1993, Integrins: A review of their structure and mechanisms of ligand binding, *Symp. Soc. Exp. Biol.* **47**:107–136.

Tuckwell, D. S., Ayad, S., Grant, M. E., Takigawa, M., and Humphries, M. J., 1994, Conformation dependence of integrin-type II collagen binding. Inability of collagen peptides to support α2β1 binding, and mediation of adhesion to denatured collagen by a novel α5β1-fibronectin bridge, *J. Cell Sci.* **107**(Pt 4):993–1005.

Turner, C. E., and Burridge, K., 1991, Transmembrane molecular assemblies in cell–extracellular matrix interactions, *Curr. Opin. Cell. Biol.* **3**:849–853.

Wayner, E. A., Orlando, R. A., and Cheresh, D. A., 1991, Integrins αvβ3 and αvβ5 contribute to cell attachment to vitronectin but differentially distribute on the cell surface, *J. Cell Biol.* **113**:919–929.

Welch, M. P., Odland, G. F., and Clark, R. A. F., 1990, Temporal relationships of F-actin bundle formation, collagen and fibronectin matrix assembly, and fibronectin receptor expression to wound contraction, *J. Cell Biol.* **110**:133–145.

Wenczak, B. A., Lynch, J. B., and Nanney, L. B., 1992, Epidermal growth factor receptor distribution in burn wounds. Implications for growth factor-mediated repair, *J. Clin. Invest.* **90**:2392–2401.

Werb, Z., Tremble, P. M., Behrendtsen, O., Crowley, E., and Damsky, C. H., 1989, Signal transduction through the fibronectin receptor induces collagenase and stromelysin gene expression, *J. Cell Biol.* **109**:877–889.

Wilke, M. S., and Skubitz, A. P. N., 1991, Human keratinocytes adhere to multiple distinct peptide sequences of laminin, *J. Invest. Dermatol.* **97**:141–146.

Yamada, K. M., 1991, Adhesive recognition sequences, *J. Biol. Chem.* **266**:12809–12812.

Yamada, K. M., Aota, S., Akiyama, S. K., and LaFlamme, S. E., 1992, Mechanisms of fibronectin and integrin function during cell adhesion and migration, *Cold Spring Harb. Symp. Quant. Biol.* **57**:203–212.

Ylanne, J., Chen, Y., O'Toole, T. E., Loftus, J. C., Takada, Y., and Ginsberg, M. H., 1993, Distinct functions of integrin α and β subunit cytoplasmic domains in cell spreading and formation of focal adhesions, *J. Cell Biol.* **122**(1):223–233.

Zambruno, G., Marchisio, P. C., Marconi, A., Vaschieri, C., Melchiori, A., Giannetti, A., and De Luca M., 1995, Transforming growth factor-β1 modulates β1 and β5 integrin receptors and induces the de novo expression of the αvβ6 heterodimer in normal human keratinocytes: Implications for wound healing, *J. Cell Biol.* **129**:853–865.

Zetter, B. R., 1993, Adhesion molecules in tumor metastasis, *Semin. Cancer Biol.* **4**:219–229.

Zheng, M., Fang, H., and Hakomori, S., 1994, Functional role of N-glycosylation in α5β1 integrin receptor. De-N-glycosylation induces dissociation or altered association of α5β1 subunits and concomitant loss of fibronectin binding activity, *J. Biol. Chem.* **269**:12325–12331.

Chapter 10

Reepithelialization

DAVID T. WOODLEY

1. Definition of Reepithelialization

Reepithelialization is the term used in common parlance to indicate the covering of a skin wound with a new epithelium. In clinical practice, this term is truly ill-defined and usually does not take into account the complexity and specialty cells of an unwounded, mature, human epidermal layer. In the examination of a healed or healing wound, the clinician often says that the wound is "reepithelialized" if the moist erythematous vascular granulation bed is covered by a dry film of epithelium. At the clinical level, the physician usually does not take into account other functions of this epithelial membrane such as its immune function directed by epidermal Langerhan's cells, the role of pigment-producing melanocytes, the sensory function of epithelial Merkel's cells, the barrier function of an organized and mature stratum corneum, and the stable epidermal–dermal adherence that occurs by a fully formed neobasement membrane zone between the epidermis and the underlying neodermis. In the future, as we advance our abilities to measure these functions, it is hoped that the definition of reepithelialization on the clinical level will undergo more refinement and discrimination.

In general, true reepithelialization involves multiple processes including the formation of a provisional wound bed matrix that is formed by an insoluble protein exudate, the migration of epidermal keratinocytes from cut edges, the proliferation of keratinocytes that feed the advancing and migrating epithelial tongue, the stratification and differentiation of the neoepithelium, the reformation of an intact basement membrane zone, and the repopulation of specialized cells that direct sensory functions, pigmentation, and immune parameters.

2. The Morphology of Reepitheliazing Wounds

Cell migration is an early and necessary event for reepithelialization to occur. Immediately after a wound is created, there is a short lag period lasting for several

DAVID T. WOODLEY • Department of Dermatology, Northwestern University, Chicago, Illinois 60611-3008.

The Molecular and Cellular Biology of Wound Repair (Second Edition), edited by Richard A. F. Clark. Plenum Press, New York, 1996

hours (Stenn and Depalma, 1988). The keratinocytes that form the cut edge of the wound begin to migrate within 24 to 48 hr. Clark and colleagues (1982) have shown that these migratory keratinocytes use a provisional matrix in the newly created wound bed that consists primarily of fibronectin and fibrin. Initially, the keratinocytes laterally migrate and cover the wound by a process of migration. There is very little contribution early from cellular division. The keratinocytes migrate not only from the wound periphery but from cut epidermal appendages throughout the wound bed. Usually, then, there is a myriad of small epidermal islands sprinkled throughout the wound bed that contribute to the reepithelialization process in addition to the keratinocytes at the periphery of the wound circumferentially.

Very little is known about the process of cellular migration and reepithelialization. Nevertheless, some basic tenets form our conventional wisdom about this process. First, it is generally accepted that if the wound environment is kept moist by occlusion or by some semipermeable membrane, these wounds will reepithelialize faster than wounds that are allowed to desiccate and form a scab (Winter, 1962; Eaglstein *et al.,* 1988, Woodley and Kim, 1992). It has been shown that this effect is somewhat time-limited and that there is a window of opportunity to use moist occlusion of a wound to promote more rapid reepithelialization (Eaglstein *et al.,* 1988). Second, there appears to be a "ying–yang" relationship between keratinocyte migration and keratinocyte cell division. For example, if keratinocytes are exposed to transforming growth factor-β (TGF-β), the proliferative potential of these cells is driven to negligible levels. Despite the inability to proliferate, these cells are quite capable of migrating on connective tissue matrices (Sarret *et al.,* 1992b). A number of investigators have shown that the blockage of cell division appears to have little if any effect on single-cell epithelial motility or the motility of epidermal sheets.

Odland and Ross (1968) examined the morphology of epithelial cells at the initiation of migration after wounding by using the electron microscope. These investigators and others observed that the keratinocyte dramatically changes its shape as it goes from a stationary basal keratinocyte to a migrating cell. In the unwounded stable state where the cell is juxtaposed to its own basement membrane zone, the basal keratinocyte demonstrates polarity. As the cell begins to migrate across the wound bed, it becomes flat and elongated. Long cytoplasmic extensions called lamellipodia are observed, along with ruffling cytoplasmic projections (Stenn and Depalma, 1988; Odland and Ross, 1968). In the unwounded state, the basal keratinocytes are joined to neighboring keratinocytes by cell–cell junctions called desmosomes and to its juxtaposed basement membrane zone by hemidesmosomes. When these structures are viewed by the electron microscope, they appear as electron-dense concretions studded across the plasma membrane. When the cell then becomes migratory and cell–cell junctions and cell matrix junctions are retracted away from the plasma membrane of the migrating cell toward a perinuclear localization. Tonafilament bundles, which usually converge outward from the cell upon desmosomal and hemidesmosomal connections, are retracted when the cell begins migration (Stenn and Depalma, 1988; Odland and Ross, 1968). Bereiter-Hahn and colleagues (1981) demonstrated that the expression of actin and α-actinin becomes manifest in the lamellipodia extended by the migrating epidermal cell. Moreover, microfilament bundles in a peripheral beltlike distribution are observed. The expression of both actinin and myosin in the cell can be detected by using monospecific antibodies against these elements. Although the cell–cell and cell–matrix junctions are retracted in a perinuclear location, gap junctions appear to be more numerous in migrating keratinocytes than stationary keratinocytes (Stenn and Depalma, 1988; Odland and Ross, 1968; Bereiter-Hahn *et al.,* 1981). Gabbiani *et al.* (1978) demonstrated an increase in the proportion of epidermal cell surface consumed by gap junctions along with an overall increase in the density of gap junctions around the cell. This observation appears to be correlated with the expression of contractile proteins, suggesting some sort of synchronism between these two cellular mechanisms.

The keratinocytes at the leading edge of the migrating epithelial tongue are highly phagocytic. They probably ingest wound debris and some of the elements in the provisional wound matrix. Nevertheless, phagocytosis does not seem to directly stimulate keratinocyte migration. Takashima and Grinnell (1984) have coated latex beads with fibronectin and shown that human keratinocytes readily phagocytose these beads to a much greater level than uncoated beads. Using this methodology, however, it was found that stimulating keratinocyte phagocytosis with fibronectin-coated latex beads did not alter human keratinocyte migration on fibronectin or other connective tissue components (Woodley *et al.,* 1988a).

3. Substrate and Epithelial Migration in Wounds

Connective tissue matrix can influence the biological behavior of epidermal cells and vice versa. For example, the type of substratum on which human keratinocytes are apposed can influence the keratinocyte's proliferative potential, ability to bind to extracellular matrix, ability to spread on connective tissue, and the ability to migrate (Woodley *et al.* 1988a, 1990b). The keratinocytes themselves are known to synthesize much of their own basement membrane (Woodley *et al.,* 1980a, b, 1988b), including laminin, type IV collagen, anchoring filament-associated components (laminin 5, also called epiligrin, kalinin, nicein, BM 600), and anchoring fibril (type VII) collagen (Woodley *et al.,* 1985b, 1987). The bullous pemphigoid antigen has been considered part of the basement membrane beneath the basal keratinocytes (Woodley *et al.,* 1980a; Westgate *et al.,* 1985; Regnier *et al.,* 1981; Mutasim *et al.,* 1985; Schaumburg-Lever *et al.,* 1975), but it is now recognized to be closely associated with the hemidesmosome (Westgate *et al.,* 1985; Regnier *et al.,* 1981; Mutasim *et al.,* 1985). Further, there are two bullous pemphigoid antigens (BPA) (Robledo *et al.,* 1990; Stanley *et al.,* 1981, 1988). One, BPAg1, is a 230,000-Da noncollagenous protein that is localized within the keratinocyte (Stanley *et al.,* 1982b, 1988) and is associated with the hemidesmosome dense plaque. The other, BPAg2, is a transmembrane glycoprotein with an extracellular collagen domain containing typical Gly-X-Y sequences that lie outside of the cell in the upper lamina lucida space of the basement membrane zone (Guidice *et al.,* 1991; Diaz *et al.,* 1990; Robledo *et al.,* 1990). It is likely that some of our information about the localization of all of these basement membrane components is artifactual and that the true nature of the basement membrane zone may not be compartmentalized as well as that visualized by electron microscopy.

Clark and colleagues (1982) demonstrated that during wound healing, when the intact basement membrane zone is abbrogated, migrating keratinocytes use a provisional matrix in the wound bed that is rich in fibronectin and fibrin. This fibronectin has several sources. First, much of it comes from the serum coagulum that occurs during clotting and early wound formation. Second, the resident dermal fibroblasts synthesize cell-surface-associated fibronectin. Third, human keratinocytes can themselves synthesize and deposit fibronectin (Kubo *et al.*, 1984; O'Keefe *et al.*, 1984, 1985). Cornelius and colleagues (1986) demonstrated in human wounds that fibronectin is expressed as a wide amorphous material in the wound bed early during wounding. Around days 5–10 after wounding when the migrating epithelial tongue is actively marching across the wound bed, fibronectin becomes more tightly localized in the upper papillary dermis, giving it the appearance of a basement membrane component. In later stages of the healing wound, the fibronectin expression again becomes more dispersed, amorphous, and less intensely expressed (Cornelius *et al.*, 1986).

Fibronectin is a large glycoprotein ($M_r = 440,000$) consisting of two similar chains disulfide-linked at the carboxyl-terminus. A number of biologically functional domains have been identified within the two chains. One functional domain is the cell-binding domain, which contains an Arg-Gly-D-Asp (RGD) acid tripeptide that is used by cells to attach to fibronectin (Ruoslahti, 1981, 1987). Gilchrest and colleagues (1980) demonstrated that human keratinocytes preferentially attach to a fibronectin matrix over other matrices including laminin and type IV (basement membrane) collagen. Others have also demonstrated that epidermal cells adhere and spread on a fibronectin matrix as well or better than laminin and type IV collagen (Clark *et al.*, 1985; Woodley *et al.*, 1990b). Fibronectin has affinity for anchoring fibril (type VII) collagen and dermal collagen (Woodley *et al.*, 1983, 1987). These interactions may play a role in epidermal–dermal adherence.

In an excised newt limb model of wound healing and epithelial migration, Donaldson and Mahan (1983) demonstrated that the RGDS-binding domain of fibronectin was important in the mediation of the epithelium of the newt limb to migrate onto a glass plate inserted into a wound in the limb. Likewise, Kim *et al.*, (1992a) demonstrated that the process of human keratinocyte migration on a fibronectin matrix was mediated by the 120-kDa fibronectin cell-binding site via RGD within this domain. It was further demonstrated that the fibronectin cell-binding domain alone could be immobilized as a matrix and that it supported human keratinocyte migration to the same degree as whole fibronectin. Even when the small RGD site itself was immobilized as a matrix, keratinocytes were able to use this to support motility, but not to the same degree as whole fibronectin or the cell-binding domain.

Human keratinocytes "sense" fibronectin, using a specific integrin (Kim *et al.*, 1992a). Integrins are cell receptors for connective tissue components and have in common two heterdimeric chains. Two integrins that bind to fibronectin have been described in human keratinocytes, $\alpha3\beta1$ and $\alpha5\beta1$. The $\alpha3\beta1$ integrin is more abundant on the cell than the $\alpha5\beta1$ receptor, but the $\alpha3\beta1$ receptor is much less specific for fibronectin. Kim *et al.*, (1992) demonstrated that when the $\alpha5\beta1$ integrin on human keratinocytes was blocked by monospecific antibodies or by the addition of RGD, migration was dramatically inhibited. It was concluded that fibronectin-driven human keratinocyte migration was mediated by the $\alpha5\beta1$ integrin on the cell side and by the RGD sequence within the cell-binding domain of fibronectin.

In addition to this *in vitro* work, there is some *in vivo* work to suggest the importance of fibronectin in reepithelialization and wound healing. For example, there are instances when the exogenous application of intact fibronectin has been helpful in closing human skin and corneal wounds (Nishida *et al.*, 1985; Wysocki *et al.*, 1988; Kono *et al.*, 1985; Scheel *et al.*, 1991). In support of this concept is the observation that in acute healing wounds the fibronectin molecule is found intact within the wound fluid, whereas in chronic nonhealing wounds fibronectin is degraded into small molecular species (Wysocki and Grinnell, 1990). Why fibronectin is degraded in chronic nonhealing wound is unclear, but it is believed that this is due to a different profile of proteases in chronic wounds when compared to acute healing wounds (Wysocki and Grinnell, 1990). Originally, it was thought that metalloproteinases were the salient proteases, but recently it has been demonstrated that serine protease inhibitors are defective in chronic nonhealing wounds (Rao *et al.*, 1995).

Fibronectin is in plasma and serum, and yet in the setting of a cutaneous wound it acts as a provisional matrix (Clark *et al.*, 1982). Vitronectin is another plasma–serum component that can also function as a matrix component in the setting of wound healing, since it too forms part of the provisional matrix. Vitronectin has been also called serum spreading factor and epibolin. Stenn (1981) identified a factor in serum that induced keratinocytes at the cut edge of a human skin explant to migrate and divide and eventually surround the dermal component of the explant by a process called "epiboly." This factor was called "epibolin" and was later shown to be the same as vitronectin. Stenn also noted that the ability of keratinocytes to spread on fibronectin, collagen, and laminin was ineffective without the presence of serum or epibolin/vitronectin (Stenn and Dvoretzky, 1979; Stenn, 1987).

A number of different kinds of cells have been shown to relate to vitronectin by integrin receptors (Haymen *et al.*, 1985). Integrins that relate to vitronectin include $\alpha v\beta1$, $\alpha v\beta3$, $\alpha v\beta5$, and $\alpha v\beta6$. By ^{125}I surface-labeling human keratinocytes and chromatographing the labeled plasma membrane proteins on a vitronectin column, Kim and colleagues (1995b) demonstrated that the salient integrin on human keratinocytes is the $\alpha v\beta5$ receptor. Brown and colleagues (1991) demonstrated that human keratinocyte motility was supported by a vitronectin matrix. Blocking the $\alpha v\beta5$ receptor on keratinocytes by monospecific antibodies or by the RGD tripeptide inhibited human keratinocyte migration on vitronectin (Kim *et al.*, 1994b). Moreover, it has been shown that the presence of TGF-β in reepithelializing wounds induces an increased expression of $\alpha v\beta5$ (Gailit *et al.*, 1994). Therefore, it appears that two plasma–serum components, fibronectin and vitronectin, can serve as matrix components within the provisional matrix in the early wound bed and can support human keratinocyte attachment and migration. It appears that human keratinocytes use integrins to relate to fibronectin and vitronectin, and that migration on these two components is mediated by RGD sequences within these molecules (Kim *et al.*, 1992a, 1994b).

During early wound healing, the keratinocytes at the unwounded margins of the wound leave the basement membrane lying beneath them and migrate over the wound bed. In doing so, they come into contact with a whole host of new connective tissue components, cells, and serum elements (Woodley *et al.,* 1985a). In the unwounded state, they rest on a 35-nm bed of laminin, a large glycoprotein localized within the lamina lucida of the basement membrane. During wounding and cell migration, the keratinocyte plasma membrane now contacts basement membrane (types IV and VII) and interstitial collagens (types I, III, and VI). Even more than vitronectin and fibronectin, certain collagens (types I and IV) dramatically promote human keratinocyte motility (Woodley *et al.,* 1988a). Unlike fibronectin–vitronectin-driven keratinocyte motility, collagen-driven motility is not mediated by RGD, even though collagens contain these sequences (Kim *et al.,* 1992, 1995). Keratinocytes use the collagen receptor $\alpha2\beta1$ to migrate on collagens. Blocking the $\alpha2\beta1$ receptor with monospecific antibody will readily inhibit human keratinocyte migration on a collagen matrix. Further, if the $\alpha2\beta1$ collagen receptor on human keratinocytes is artificially up-regulated, the cells demonstrate enhanced migration on collagen (Chen *et al.,* 1993a). Basement membrane (type IV) collagen contains a globular amino terminus (NC1), a helical middle domain, and a collagenase-resistant amino-terminus called 7S. A small Hep III domain within the helical domain binds to heparan sulfate proteoglycan, serves as a cell attachment site (Wilke and Furcht, 1990), and mediates human keratinocyte migration on type IV collagen. Neither NC1 or 7S can support human keratinocyte motility (Kim *et al.,* 1995).

In a survey of the influences of most of the known matrix molecules on human keratinocyte migration, it was found that laminin uniquely inhibits migration. Laminin is a large, cross-shaped, noncollagenous glycoprotein within the lamina lucida of the basement membrane beneath basal keratinocytes (Woodley *et al.,* 1988a). It is likely that the plasma membrane of these keratinocytes are in intimate contact with laminin. When human keratinocytes are placed in apposition with collagen, they attach and then begin to migrate. If laminin is then added, the collagen-driven migration is inhibited in a concentration-dependent fashion. Like fibronectin, laminin has specific domains that exhibit biological activity. For example, laminin contains specific domains that bind basement membrane collagen, heparan sulfate proteoglycan, and cells. The carboxyl-terminus of the laminin A chain contains a CSIKVAVS peptide sequence that inhibits human keratinocyte migration on collagen (Sarret *et al.,* 1992). Laminin is known to have affinity for other basement membrane components such as type IV collagen and heparan sulfate proteoglycan, and these affinities are mediated by specific laminin subdomains (Woodley *et al.,* 1983). In healing wounded skin, a neobasement membrane must ultimately form beneath the migrating keratinocytes. It has been shown in animal wounds that the expression of these basement membrane components reappear in an orderly and predictable sequence (Clark *et al.,* 1982; Stanley *et al.,* 1981), and that this *in vivo* sequence is very similar to what is seen *in vitro* using human skin organ cultures (Hintner *et al.,* 1980; Woodley *et al.,* 1980a). Further, in the one *in vivo* human skin wound study done by Cornelius *et al.* (1986), it appears that a similar sequence is found with the bullous pemphigoid antigen first observed beneath the reepithelializing tongue of keratinocytes followed by type IV collagen and laminin. Clark *et al.* (1982) have observed in animals that laminin does not appear consistently until the migratory epithelium has almost completed its migration and is becoming stationary. This *in vivo* observation would support the notion that laminin is linked to a nonmigratory state and may be the connective tissue "brake" for keratinocyte motility (Woodley *et al.*, 1988a).

Recently, isoforms of laminin have been found to be localized within the lamina lucida space and closely associated with the anchoring fibrils that traverse the lamina lucida beneath hemidesmosomes (Carter *et al.*, 1991; Verrando *et al.*, 1987; Rousselle *et al.*, 1991; Marinkovitch *et al.*, 1994). These isoforms, like classical laminin, consist of three chains organized in a cruciate arrangement. They have been given names such as epiligrin, kalinin, nicein, and BM600. Recently it has been shown that these are identical or very similar molecules and have been placed under the umbrella term "laminin 5" (Marinkovitch *et al.*, 1994; Ceilley *et al.*, 1993). Laminin 5, like classical laminin, inhibits human keratinocyte migration (Marinkovitch *et al.*, 1994). Laminin 5 strongly promotes keratinocyte attachment and is thought to anchor the keratinocytes to the substratum. The integrin receptor for laminin 5 is the $\alpha3\beta1$. Kim *et al.* (1992) demonstrated that if the $\alpha3\beta1$ (laminin 5 receptor) was blocked by antibody, human keratinocytes exhibited hypermotility on collagen and fibronectin. Likewise, the keratinocytes of patients with junctional epidermolysis bullosa have genetically abnormal expression of laminin 5, and they exhibit hypermotility on collagen and fibronectin (Chen *et al.*, 1993a, b; Woodley *et al.*, 1988b). Taken together, it appears that laminin 5 serves as a major cell adhesion factor for keratinocytes, and perturbations in the receptor for laminin 5 or laminin 5 itself cause the keratinocytes to be less adherent to their substratum and consequently able to increase their migration over these substrata.

Like laminin, it appears that type VII collagen expression and anchoring fibril formation is a very late event in healing human skin wounds (Cornelius *et al.*, 1986). The functional parallel of this late expression is that healed human wounds exhibit poor epidermal–dermal adherence for a long period of time (month to years) after reepithelialization is completed (Woodley *et al.*, 1988b, 1990a).

4. Soluble Factors and Epithelial Migration in Wounds

In human skin organ cultures, Hebda (1988) demonstrated that TGF-β promoted the expansion of the epidermal outgrowth. In animal models, Mustoe *et al.* (1987) also demonstrated that TGF-β promoted reepithelialization of skin wounds and wound closure. In acute human skin wounds, it has been shown that the application of recombinent epidermal growth factor promotes reepithelialization and wound closure (Brown *et al.*, 1989). These intriguing studies suggested that growth factors could accelerate wound closure by a number of mechanisms including the possible enhancement of epithelial migration. Organ cultures, animal models, and human wounds do not allow a dissection of the cellular mechanisms responsible for the effects of growth factors. It was shown in a pure keratinocyte migration assay that TGF-β could inhibit the keratinocyte proliferative potential to zero and yet the keratinocytes could readily migrate on matrix (Sarret *et al.*, 1992b). However, TGF-β itself did not appear to dramatically increase keratinocyte motility. Neither did the addition of nerve growth factor, fibroblast growth factor, or interleukin-8 (Sarret *et al.*, 1992b). In contrast, epidermal growth factor and TGF-α promoted human keratinocyte migration on collagen by increasing the cell surface expression of the collagen integrin α2β1 (Chen *et al.*, 1993a). This observation is an example of how soluble factors can influence keratinocyte motility by enhancing the known cellular mechanisms by which keratinocytes migrate on matrix. Recently, interleukin-1 (IL-1) has been shown to promote keratinocyte migration by a mechanism independent from integrin expression and matrix interactions (Chen *et al.*, 1995). Therefore, it appears that other intracellular mechanisms may be invoked by growth factors that promote cell motility.

Another link between soluble factors and keratinocyte motility is via the expression of collagenase. Collagenases are metalloproteinases that make specific cleavage sites in collagens. By immunofluorescence staining of human skin, mammalian collagenase is predominently localized within the papillary dermis (Bauer *et al.*, 1977). Dermal fibroblasts synthesize collagenase (Bauer, 1977), and IL-1 from keratinocytes induces fibroblasts to synthesize collagenase (Postlethwaite *et al.*, 1982). In addition, it is now well documented that human keratinocytes constitutively synthesize and secrete collagenase (Woodley *et al.*, 1986; Petersen *et al.*, 1989, 1990). Certain types of collagenases [such as those that degrade basement membrane (type IV) collagen] are known to be associated with the ability of malignant cells to cross tissue planes and establish metastatic foci (Liotta *et al.*, 1979). Human keratinocytes migrating on dermal collagen synthesize and deposit collagenases that degrade type I and type IV collagens (Woodley *et al.*, 1986). Moreover, it has been demonstrated that the degree of collagenase synthesis by human keratinocytes is linked to the migratory mode of the cell (Peterson *et al.*, 1989). For example, keratinocytes apposed to type I collagen and induced to migrate express much more type I collagenase than nonmigratory cells apposed to plastic or laminin (Petersen *et al.*, 1989).

Cytokines may play a role in promoting keratinocyte motility. It is known that human keratinocytes synthesize and secrete IL-1, IL-8, tumor necrosis factor, and granulocyte–macrophage colony-stimulating factor (Luger *et al.*, 1981; Sauder *et al.*, 1982, 1988; Kupper *et al.*, 1986). Some of these cytokines, such as IL-1, can serve as autocrines for the cells themselves. IL-1 is known to stimulate collagenase synthesis is some cells (Postlethwaite *et al.*, 1982). Recently, Chen and co-workers (1995) have demonstrated that IL-1 can stimulate human keratinocyte migration on a collagen matrix and that this occurs by an integrin-independent mechanism. It is likely that the increased cellular expression of collagenase driven by the presence of IL-1 plays a role in the observed increase in migration.

There are wounds that somehow spiral into a chronic, nonhealing mode and seem to refuse to reepithelialize. Several methods have been used to convert a chronic, nonhealing wound into a healing mode. Meleney's ulcers were a prime example of a nonhealing wound. Meleney converted these ulcers to a healing mode by the application of zinc peroxide. Some nonhealing leg ulcers can be induced to heal by the application of 20% benzoyl peroxide (Colman and Roenigk, 1978; Lyon and Reynolds, 1929). Nonhealing decubitus ulcers and leg ulcers have been induced to heal by the application of an electric current (Gentzkow *et al.*, 1993; Mertz *et al.*, 1993). Many surgeons believe that aggressive debridement of nonhealing chronic wounds and essentially transforming them into acute wounds is helpful to initiate the healing mode. What do all of these methods have in common? They all invoke considerable trauma on the wound. It is known that trauma induces IL–1 expression in human keratinocytes. One might hypothesize that trauma on wounds induces IL-1 (and perhaps other cytokines) and promotes reepithelialization.

5. Metabolic Requirements and Epithelial Migration

5.1. Protein Synthesis

Protein synthesis is required for epithelial sheets to migrate (Gibbins, 1973; Stenn *et al.*, 1979; Rocha *et al.*, 1986). Certain studies suggest that selected proteins may be more important than others in the support of keratinocyte motility. For example, in mouse skin, the synthesis and secretion of type V collagen may be necessary for mouse keratinocytes to move over dermal collagen (Stenn *et al.*, 1979). In these mouse skin explant studies, the presence of cycloheximide, a protein synthesis inhibitor, or the presence of a proline-rich analogue that selectively inhibited collagen synthesis readily inhibited keratinocyte movement over dermal collagen (Stenn *et al.*, 1979). Likewise, it has been shown that cell surface glycoproteins on epithelial cells have a different profile depending on whether the cells are stationary or migratory (Anderson and Fejerskov, 1974). Migratory cells have more lectin-binding sites than stationary cells and exhibit significantly higher rates of glycoprotein synthesis (Gipson *et al.*, 1982). Blocking asparagine-linked glycoprotein synthesis by the presence of tunicamycin was found to inhibit the migration of corneal epithelial cells (Gipson and Anderson, 1980).

5.2. cAMP

In a newt limb organ culture system in which a cutaneous wound is created and cells are allowed to migrate out on a coverslip inserted into the wound, Donaldson and Mahan (1984) demonstrated that the presence of catecholamines, such as isoproterenol, inhibited the outward migration of the epidermal cells. When this experiment was repeated in the presence of propranolol, a β-antagonist, the inhibitory effect of isoproterenol was nullified. Therefore, it was thought that epithelial migration involved a β_2-adrenergic receptor and a second messenger, such as cAMP. Compared with stationary newt epithelial cells, the levels of intracellular cAMP were found to be elevated (Dunlap, 1980). Dibutyryl cAMP is a known mitogen for human keratinocytes (Falanga *et al.*, 1991). Nevertheless, the proliferative effects of dibutyryl cAMP on human keratinocytes in culture occurs after 48 hr of incubation. Recently, Iwasaki *et al.* (1994) demonstrated that an early effect of dibutyryl cAMP on human keratinocytes in culture is to promote migration at "physiological" concentrations, and that this effect was associated with an enhancement of keratinocyte synthesis of collagenases. It was found that the concentration of dibutyryl cAMP was critical. At high concentrations it was toxic to the cells and at low levels it was ineffective.

5.3. Divalent Cations

Very little is known about the influence of divalent cations on epithelial cell motility. Magnesium, calcium, and maganese support cell adhesion and spreading on matrix (Fritsch *et al.*, 1979; Stenn and Core, 1986), but how these functions relate to cell motility is unclear. It is now very clear that cell adhesion to matrix and cell motility on a substratum are very different processes (Duband *et al.*, 1988). In an *in vitro* human keratinocyte motility assay, it was found that medium with a low calcium concentration (0.1mM) supported cell migration on collagen much better than medium with higher calcium concentrations (0.4 M) (Sarret *et al.*, 1992a). When intracellular free calcium concentrations were studied, it was found that when human keratinocytes are induced to migrate, intracellular calcium concentrations significantly decrease (Sarret *et al.*, 1992a, b). The mechanism of this decrease is unknown, but it could be due to the free calcium being consumed during signal transduction or due to binding to calmodulin.

5.4. Cytoskeleton

When colchicine or colcemid is added to migrating corneal epithelial cells, no inhibition of the migration occurs (Gipson *et al.*, 1982). Likewise, when these agents are added to the newt limb organ culture system of Donaldson (Dunlap and Donaldson, 1978), the epithelial skin cells from the limb continue to migrate out on the inserted glass coverslip to the same degree as controls. These agents block the formation of microtubules. Therefore, since there is no inhibition of epithelial cell migration in the presence of these agents, it has been hypothesized that microtubule formation plays a minimal role in epithelial cell movement (Stenn and Depalma, 1988; DiPasquale, 1975; Dunlap and Donaldson, 1978; Gipson *et al.*, 1982).

5.5. Energy

One can inhibit cellular respiration, the Krebs cycle, oxidative phosphorylation, and gluconeogenesis and demonstrate little perturbation on epithelial cell migration (Stenn and Depalma, 1988; Kuwabara *et al.*, 1976; Gibbins, 1972). Inhibitors of glycolytic and sulfhydryl enzymes block epidermal cell movement (Stenn and Depalma, 1988; Kuwabara *et al.*, 1976; Gibbins, 1972). Glycogen consumption appears to be the energy source for migrating cells, and migrating epithelial cells rely on an intact glycolytic pathway (Stenn and Depalma, 1988).

5.6. Serum Protein

As outlined earlier, fibronectin and vitronectin (also called S protein, serum-spreading factor, and epibolin) are serum–plasma proteins that induce human keratinocyte migration on matrix via the $\alpha v \beta 5$ integrin, and this interaction is RGD-mediated. These components promote motility and also cell spreading (Stenn, 1978; Stenn and Dvoretzky, 1979). Both fibronectin and vitronectin are present in the early wound bed and probably serve as a provisional matrix for keratinocyte migration in early wounds (Clark *et al.*, 1982).

5.7. Hypoxia

Winter (1962) demonstrated that when skin wounds are kept moist, they heal more rapidly than those allowed to dry. This observation was the impetus for the current use of semiocclusive dressings that keep skin wounds moist and promote healing (Gentzkow *et al.*, 1993; Mertz *et al.*, 1993). Why moist wounds heal more rapidly than dry ones still remains an open question, but it is thought that the formation of a dried eschar inhibits the epithelial tongue of migrating keratinocytes from readily covering the wound bed. This would suggest that their main effect is upon reepithelialization rather than other wound processes such as angiogenesis, fibroplasia, or contraction. In a double-bind controlled study in identical human wounds that did not abbrogate the basement membrane zone, it was shown that the use of a polyurethane occlusive dressing promoted reepithelialization over dressings that were nonocclusive (Woodley and Kim, 1992). This study also would suggest that occlusive dressings act more on reepithelialization than deeper processes involving the formation of the wound bed and granulation tissue. Varghese *et al.* (1986) measured the pH and oxygen tension of wounds covered with occlusive dressings. They found that although the wounds healed well, the oxygen tension and pH were very low. This *in vivo* study suggested that a hypoxic environment is not inhibitory to reepithelialization. In accordance with this observation, it has recently been demonstrated in an *in vitro* human keratinocyte migration assay that when the cells are maintained under hypoxic conditions (oxygen tensions of 2 or 0.2%), they exhibit significantly enhanced migration on connective tissue matrices compared to normoxic cells (Peavey *et al.*, 1994). Hypoxia does not alter the integrin profile of the keratinocytes or total protein synthesis. Nevertheless, the hypoxic cells do demonstrate enhanced collagenase expression. In addition, hypoxia induces up-regulation of the lamella podia-associated components ezrin and moesin.

These observations suggest that hypoxia may be one initiating event for human keratinocyte motility. During an acute skin wound, blood rapidly clots in the small dermal blood vessels. The rent fills with a serum coagulum, fibronectin, fibrinogen, fibrin, and vitronectin. The epidermal cells at the edge of the wound are within avascular tissue in the epidermis and separated from the clotted blood vessels by the early wound bed. It is conceivable that the first basal keratinocytes that must migrate over the provisional wound matrix are experiencing extreme hypoxia. This hypoxia may stimulate keratinocyte processes that are helpful for cellular migration over matrices.

6. Summary

We are just beginning to understand some of the cellular mechanisms responsible for human keratinocyte migration on matrix. It is believed that an understanding of keratinocyte motility will have direct relevance to the problem of wound reepithelialization. The work to date demonstrates that soluble factors (vitronectin, IL-1, epidermal growth factor, TGF-α), intracellular elements (cAMP, calcium), and connective tissue components may all influence human keratinocyte motility. The initiating events, the value or weight of importance of each element, the intracellular mechanisms, and the sequential orchestration of these soluble and matrix components are unknown.

ACKNOWLEDGMENTS. This work was supported by NIH grants PO1 AR41045 and RO1 AR33625, and by funding from the Dermatology Foundation.

References

Bauer, E. A., 1977, Cell culture density as a modulator of collagenase expression in normal human fibroblast cultures, *Exp. Cell Res.* **107:**209–276.

Bauer, E. A., Gordon, J. M., Reddick, M. E., and Eisen, A. Z., 1977, Quantitation and immunocytochemical localization of human skin collagenase in basal cell carcinoma, *J. Invest. Dermatol.* **69:**363–367.

Bereiter-Hahn, J., Strohmeier, R., Kunzenbacher, I., Beck, K., and Voth, M., 1981, Locomotion of *Xenopus* epidermis cells in primary culture, *J. Cell. Sci.* **52:**289–311.

Brown, C., Stenn, K. S., Falk, R. J., Woodley, D. T., and O'Keefe, E. J., 1991, Vitronectin: Effects on keratinocyte motility and inhibition of collagen-induced motility, *J. Invest. Dermatol.* **96:**724–728.

Brown, G. L., Nanney, L. B., Griffen, J., Cramer, A. B., Yancey, J. M., Curtsinger, III, L. J., Holtzin, L., Schultz, G. S., Jurkiewicz, M. J., and Lynch, J. B., 1989, Enhancement of wound healing by topical treatment with epidermal growth factor, *N. Engl. J. Med.* **321:**76–79.

Carter, W. G., Ryan, M. C., and Gahn, P. J., 1991, Epiligrin, a new cell adhesion ligand for integrin α3β1 in epithelial basement membranes, *Cell* **65:**599–610.

Ceilley, E., Watanabe, N., Shapiro, D., Verrando, P., Bauer, E. A., Burgeson, R., Briggaman, R. A., and Woodley, D. T., 1993, Labeling of fractured human skin with antibodies to BM 600/nicein, epiligrin, kalinin and other matrix components, *J. Dermatol. Sci.* **5:**97–103.

Chen, J. D., Kim, J. P., Zhang, K., Sarret, Y., Wynn, K. C., Kramer, R. H., and Woodley, D. T., 1993a, Epidermal growth factor (EGF) promotes human keratinocyte locomotion on collagen by increasing the α2 integrin subunit, *Exp. Cell Res.* **209:**216–223.

Chen, J. D., Langhofer, M., Iwasaki, T., Kim, Y. H., Jones, J. C. R., Krueger, J. G., Carter, D. M., and Woodley, D. T., 1993b, Junctional epidermolysis bullosa (JEB) keratinocytes fail to secrete hemidesmosome (HD)-associated matrix elements and demonstrate enhanced locomotion, *J. Invest. Dermatol.* **11(4):**170 (Abstract).

Chen, J. D., Lapierre, J.-C., Sauder, D., Peavey, C., and Woodley, D. T., 1995, Interleukin-1 alpha stimulates keratinocyte migration through an EGF/TGF-alpha independent pathway, *J. Invest. Dermatol.* **104:**729–733.

Clark, R. A. F., Lanigan, J. M., DellaPelle, P., Manseau, E., Dvorak, H. F., and Colvin, R. B., 1982, Fibronectin and fibrin provide a provisional matrix for epidermal cell migration during wound reepithelialization, *J. Invest. Dermatol.* **79:**264–269.

Clark, R. A. F., Folkvord, J. M., and Wertz, R. L., 1985, Fibronectin, as well as other extracellular matrix proteins, mediate human keratinocyte adherence, *J. Invest. Dermatol.* **85:**368–383.

Colman, G. J., and Roenigk, H. H., 1978, Topical therapy of leg ulcers with 20 percent benzoyl peroxide lotion, *Cutis* **21:**491–494.

Cornelius, L. A., Woodley, D. T., Cronce, D. J., and Briggaman, R. A., 1986, Dermal–epidermal junction reformation following human skin wounding studied by correlative ultrastructural and immunochemical techniques, *J. Invest. Dermatol.* **86:**469 (Abstract).

Diaz, L. A., Ratrie, H., Saunders, W. S., Futamura, S., Squiquera, H. R., Anhalt, G. J., and Guidice, G. J., 1990, Isolation of a human epidermal cDNA corresponding to the 180 kD autoantigen recognized by bullous pemphigoid and herpes gestationis sera. Immunolocaliazation of this protein to the hemidesmosome, *J. Clin. Invest.* **86:**1088–1094.

DiPasquale, A., 1975, Locomotion of epithelial cells, *Exp. Cell Res.* **95:**425–439.

Donaldson, D. J., and Mahan, J. T., 1983, Fibrinogen and fibronectin on substrates from epidermal cell migration during wound closure, *J. Cell Sci.* **62:**117–123.

Donaldson, D. J., and Mahan, J. T., 1984, Influence of catecholamines on epidermal cell migration during wound closure in adult newts, *Comp. Biochem. Physiol.* **78C:**267–270.

Duband, J. L., Nuckolls, G. H., Ishihara, A., Hasegawa, T., Yamada, K. M., Thiery, J. P., and Jacobson, K., 1988, Fibronectin receptor exhibits high lateral motility in embryonic locomoting cells but is immobile in focal contacts and fibrillar streaks in stationary cells, *J. Cell Biol.* **107:**1385–1396.

Dunlap, M. K., 1980, Cyclic AMP levels in migrating and non-migrating newt epidermal cells, *J. Cell. Physiol.* **104:**367–373.

Dunlap, M. K., and Donaldson, D. J., 1978, Inability of colchicine to inhibit newt epidermal cell migration or prevent concanavalin A-mediated inhibition of migration studies *in vivo, Exp. Cell Res.* **116:**15–19.

Eaglstein, W. H., Davis, S. C., Mehle, A. L., and Mertz, P. M., 1988, Optimal use of an occlusive dressing to enhance healing, *Arch. Dermatol.* **124:**392–395.

Falanga, V., Katz, M. H., and Alvarez, A. F., 1991, Dibutryl cyclic AMP by itself or in combination with growth factors can stimulate or inhibit growth of human keratinocytes or dermal fibroblasts, *Wounds* **3:**70–78.

Fritsch, P., Tappeiner, G., and Huspek, G., 1979, Keratinocyte substrate adhesion is magnesium-dependent and calcium independent, *Cell Biol. Int. Rep.* **3:**593–598.

Gabbiani, G., Chaponnier, C., and Huttner, I., 1978, Cytoplasmic filament and gap functions in epithelial cells and myofibroblasts during wound healing, *J. Cell Biol.* **76:**561–568.

Gailit, J., Welch, M. P., and Clark, R. A. F., 1994, TGF-β1 stimulates expression of keratinocyte integrins during re-epithelialization of cutaneous wounds, *Invest. Dermatol.* **103:**221–227.

Gentzkow, G. D., Alon, G., Taler, G., Eltorai, I., and Montroy, R., 1993, Healing of refractory stage III and IV pressure ulcers by a new electrical stimulation device, *Wounds* **5**(3):160–172.

Gibbins, J. R., 1972, Metabolic requirements for epithelial migration as defined by the use of metabolic inhibitors in organ culture, *Exp. Cell Res.* **71:**329–337.

Gibbins, J. R., 1973, Epithelial migration in organ culture. Role of protein synthesis as determined by metabolic inhibitors, *Exp. Cell Res.* **80:**281–290.

Gilchrest, B. A., Nemore, R. E., and Maciag, T., 1980, Growth of human keratinocytes on fibronectin-coated plates, *Cell Biol. Int. Rep.* **4:**1009–1016.

Gipson, I. K., and Anderson, R. A., 1980, Effect of lectin on migration of the corneal epithelium, *Invest. Ophthalmol. Vis. Sci.* **19:**341–349.

Gipson, I. K., and Kiorpes, T. C., 1982, Epithelial sheet movement: Protein and glycoprotein synthesis, *Dev. Biol.* **92:**259–262.

Gipson, I. K., Westcott, M. J., and Brooksby, N. G., 1982, Effects of cytochalasins B and D and colchicine on migration of the corneal epithelium, *Invest. Ophthal. Vis. Sci.* **22:**633–642.

Guidice, G., Squiquera, H. L., Elias, P. M., and Diaz, L. A., 1991, Identification of two collagen domains within the bullous pemphigoid autoantigen, BP180, *J. Clin. Invest.* **87:**734–738.

Haymen, E. G., Pierschbacher, M. D., Suzuki, S., and Ruoslahti, E., 1985, Vitronectin: A major cell attachment-promoting protein in filal bound serum, *Exp. Cell. Res.* **160:**245–258.

Hebda, P. A., 1988, Stimulatory effects of transforming growth factor beta and epidermal growth factor on epidermal cell outgrowth from porcine skin explant cultures, *J. Invest. Dermatol.* **91:**440–445.

Hebda, P. A., Klingbeil C., Abraham J., and Fiddes, J. C., 1988, Acceleration of epidermal wound healing by human basic fibroblast growth factor, *J. Invest. Dermatol.* **90:**568a.

Hintner, H., Fritsch, P. O., Foidart, T. M., Stingl, G., Schuler, G., and Katz, S. I., 1980, Expression of basement membrane zone antigens at the dermo-epibolic junction in organ cultures of human skin, *J. Invest. Dermatol.* **74:**200–204.

Iwasaki, T., Kim, J. P., Wynn, K. C., and Woodley, D. T., 1994, Dibutryl cyclic AMP modulates keratinocyte locomotion, *J. Invest. Dermatol.* **102**:891–897.

Kim, J. P., Chen, J. D., and Woodley, D. T., 1992a, Mechanism of human keratinocyte migration on fibronectin: Unique roles of RGD site and integrins, *J. Cell. Physiol.* **151**:443–450.

Kim, J. P., Zhang, K., Kramer, R. H., Schall, T. J., and Woodley, D. T., 1992b, Integrin receptors and RGD sequences in human keratinocyte migration: Unique anti-migratory function of α3β1, *J. Clin. Invest.* **98**:764–770.

Kim, Y. H., Kim, J. P., Chen, J. D., Iwasaki, T., Hernandez, G., Saraf, P., Bauer, E. A., and Woodley, D. T., 1993, Biologic characteristics of recessive dystrophic epidermolysis bullosa (RDEB) keratinocytes, *J. Invest. Dermatol.* **11**(4):551 (Abstract).

Kim, J. P., Schall, T. J., Kleinman, H. K., and Woodley, D. T., 1994a, Human keratinocyte migration on type IV collagen: Unique roles of heparin binding site and integrins, *Lab. Invest.* **71**:401–408.

Kim, J. P., Zhang, K., Chen, J. D., Kramer, R. H., and Woodley, D. T., 1994b, Vitronectin-driven human keratinocyte locomotion is mediated the αvβ5 integrin receptor, *J. Biol. Chem.* **43**:26926–26932.

Kono, I., Matsumoto, Y., Kano, K., Yasuhisa, I., Narushima, K., Kabashima, T., Yamane, K., Sakurai, T., and Kashiwagi, H., 1985, Beneficial effect of topical fibronectin in patients with keratoconjunctivitis sicca of Sjorgren's syndrome, *J. Rheumatol.* **12**:487–489.

Kubo, M., Norris, D. A., Howell, S. E., and Clark, R. A. F., 1984, Human keratinocytes synthesize, secrete and deposit fibronectin in the pericellular matrix, *J. Invest. Dermatol.* **82**:580–586.

Kupper, T. S., Ballard, D. W., Chua, A. O., McGuire, J. S., Flood, P. M., Horowitz, M. C., Langdon, L., and Gubler, V., 1986, Expression of mRNA homologous to interleukin-1 in human epidermal cells, *J. Exp. Med.* **64**:2095–2098.

Kuwabara, T., Perkins, D. G., and Cogan, D. G., 1976, Sliding of the epithelium in experimental corneal wounds, *Invest. Ophthalmol.* **15**:4–14.

Liotta, L. A., Siegeto, A., Gebron-Robey, P., and Martin, A. K., 1979, Preferential digestion of basement membrane collagen by an enzyme derived from a metastatic tumor, *Proc. Natl. Acad. Sci. USA* **76**:2268–2272.

Luger, T. A., Stadler, B. M., Katz, S. I., and Oppenheimer, J. J., 1981, Epidermal cell derived thymocyte activating factor (ETAF), *J. Immunol.* **127**:1493–1498.

Lyon, R. A., and Reynolds, T. E., 1929, Promotion of healing by benzoyl peroxide and other agents, *Proc. Soc. Exp. Biol. Med.* **27**:122–151.

Marinkovich, M. P., Peavey, C. L., Burgeson, R. E., and Woodley, D. T., 1994, Kalinin inhibits collagen-driven human keratinocyte migration, *J. Clin. Invest.* **102**(4):157 (Abstract).

Mertz, P., Davis, C., Cazzaniga, A., Cheng, K., Reich, J., and Eaglstein, W., 1993, Electrical stimulation: Acceleration of soft tissue repair by varying the polarity, *Wounds* **5**(3):153–159.

Mustoe, T. A., Pierce, G. F., Thomason, A., Sporn, M., Gramates, P. H., and Deuel, T. F., 1987, Accelerated healing of incisional wounds in rats induced by transforming growth factor β, *Science* **237**:1333–1335.

Mutasim, D. F., Takahashi, Y., Ramzy, L. S., Anhalt, G. J., Patel, H. P., and Diaz, L. A., 1985, A pool of bullous pemphigoid antigen(s) is intracellular and associated with the basal cell cytoskeleton–hemidesmosome complex, *J. Invest. Dermatol.* **84**:47–53.

Nishida, T., Nakagawa, S., and Manabe, R., 1985, Clinical evaluation of fibronectin eye drops on epithelial disorders after herpetic keratitis, *Ophthalmology* **92**:213–216.

Odland, G., and Ross, R., 1968, Human wound repair. I Epidermal regeneration, *J. Cell Biol.* **39**:135–151.

O'Keefe, E. J., Woodley, D., Castillo, G., Russell, N., and Payne, R. E., 1984, Production of soluble and cell associated fibronectin by cultured keratinocytes, *J. Invest. Dermatol.* **82**:150–155.

O'Keefe, E. J., Payne, R. E., Russell, N., and Woodley, D. T., 1985, Spreading and enhanced motility of human keratinocytes on fibronectin, *J. Invest. Dermatol.* **85**:125–130.

O'Keefe, E. J., Chiu, M. L., and Payne, R. E., 1988, Stimulation of growth of keratinocytes by basic fibroblast growth factor, *J. Invest. Dermatol.* **90**:767–769.

Peavey, C. L., Ladin, D. A., Mustoe, T. A., and Woodley, D. T., 1994, Hypoxia stimulates human keratinocyte migration on interstitial collagen, *J. Clin. Invest.* **102**(4):699 (Abstract).

Petersen, M. J., Woodley, D. T., Stricklin, G. P., and O'Keefe, E. J., 1989, Constitutive production of procollagenase and collagenase inhibitor by human keratinocytes in culture, *J. Invest. Dermatol.* **92**:156–159.

Petersen, M. J., Woodley, D. T., Stricklin, G. P., and O'Keefe, E. J., 1990, Enhanced synthesis of collagenase by human keratinocytes cultured on type I or type IV collagen, *J. Invest. Dermatol.* **94:**341–346.

Postlethwaite, A. E., Lachman, L. B., Mainardi, C. L., and Kang, A. H., 1982, Interleukin I stimulation of collagenase production by cultured fibroblasts, *J. Exp. Cell Biol.* **157:**801–806.

Rao, C. N., Ladine, D., Liu, Y., Hou, Z., Chilukuri, K., and Woodley, D. T., 1995, Alpha 1 antitypsin is degraded and non-functional in chronic wounds: The inhibitor protects fibronectin from degradation by chronic wound fluid enzymes, *J. Invest. Dermatol.* in press.

Regnier, M., Prunieras, M., and Woodley, D., 1981, Growth and differentiation of adult human epidermal cells on dermal substrate, *Front. Matrix Biol.* **9:**4–32.

Robledo, M. A., Kim, S.-C., Korman, N. J., Stanley, J. R., Labib, R. S., Futamura, S., and Anhalt, G. J., 1990, Studies of the relationship of the 230 kD and 180 kD bullous pemphigoid antigens, *J. Invest. Dermatol.* **94:**793–797.

Rocha, V., Hom, Y. K., and Marinkovich, M. P., 1986, Basal lamina inhibition suppresses synthesis of calcium-dependent proteins associated with mammary epithelial cell spreading, *Exp. Cell Res.* **165:**450–460.

Rousselle, P., Lunstrum, G. P., Keene, D. R., and Burgeson, R. E., 1991, Kalinin: An epithelium-specific basement membrane adhesion molecule that is a component of anchoring filaments, *J. Cell Biol.* **114:**567–576.

Ruoslahti, E., and Pierschbacher, M. D., 1987, New perspectives in cell adhesion: RGD and integrins, *Science* **238:**491–497.

Ruoslahti, E., Engvall, E., and Hayman, E. G., 1981, Fibronectin: Current concepts of its structure and function, *Coll. Res.* **1:**95–128.

Sarret, Y., Kleinman, H. K., and Woodley, D. T., 1991, The peptide (CSIKVAVS-NH$_2$) near the amino terminus of the laminin A chain markedly inhibits human keratinocyte locomotion, *Clin. Res.* **39**(2)**:**514A (Abstract).

Sarret, Y., Raftery, K., and Woodley, D. T., 1992a, Intracellular and extracellular calcium levels dramatically alter human keratinocyte migration, *J. Invest. Dermatol.* **98**(4)**:**572 (Abstract).

Sarret, Y., Woodley, D. T., Grigsby, K., Wynn, K. C., and O'Keefe, E. J., 1992b, Human keratinocyte locomotion: The effect of selected cytokines, *J. Invest. Dermatol.* **98:**12–16.

Sauder, D. N., Carter, C., Katz, S. I., and Oppenheim, J. J., 1982, Epidermal cell production of thymocyte activating factor (ETAF), *J. Invest. Dermatol.* **79:**34–39.

Sauder, D. N., Stanulis-Prager, B. M., and Gilchrist, B. A., 1988, Autocrine growth stimulation of human keratinocytes by epidermal cell derived thymocyte activating factor, *Arch. Dermatol. Res.* **280:**71–78.

Schaumburg-Lever, G., Rule, R. A., Schmidt-Ullrich, B., and Lever, W. F., 1975, Ultrastructural localization of *in vivo* bound immunoglobulins in bullous pemphigoid: A preliminary report, *J. Invest. Dermatol.* **64:**47–49.

Scheel, G., Rahsoth, B., Franke, J., and Grau, P., 1991, Acceleration of wound healing by local application of fibronectin, *Arch. Orthop. Trauma Surg.* **110:**284–287.

Stanley, J. R., Alvarez, O. M., Bere, E. W., Eaglstein, W. H., and Katz, S. I., 1981, Detection of membrane zone antigens during epidermal wound healing in pigs, *J. Invest. Dermatol.* **7:**240–243.

Stanley, J. R., Woodley, D. T., Katz, S. I., and Martin, G. R., 1982a, Structure and function of basement membrane, *J. Invest. Dermatol.* **79:**69s–72s.

Stanley, J. R., Hawley-Nelson, P., Yaar, M., Martin, G. R., and Katz, S. I., 1982b, Laminin and bullous pemphigoid antigen are distinct basement membrane proteins synthesized by epidermal cells, *J. Invest. Dermatol.* **78:**456–459.

Stanley, J. R., Tanaka, T., Mueller, S., Klaus-Kouan, V., and Roop, D., 1988, Isolation of complementary DNA for bullous pemphigoid antigen by use of patients' autoantibodies, *J. Clin. Invest.* **82:**1864–1870.

Stenn, K. S., 1978, The role of serum in the epithelial outgrowth of mouse skin explants, *Br. J. Dermatol.* **98:**411–416.

Stenn, K. S., 1981, Epibolin: A protein of human plasma which supports epithelial cell movement, *Proc. Natl. Acad. Sci. USA* **78:**6907–6911.

Stenn, K. S., 1987, Coephibolin, the activity of human serum that enhances the cell-spreading properties of epibolin, associates with albumin, *J. Invest. Dermatol.* **89:**59–63.

Stenn, K. S., and Core, N. G., 1986, Calton dependence of guinea pig epidermal cell spreading, *In Vitro Cell. Dev. Biol.* **22:**217–222.

Stenn, K. S., and Depalma, L., 1988, Re-epithelialization, in: *The Molecular and Cellualr Bilolgy of Wound Repair,* 1st ed. (R. A. F. Clark and P. M. Hensen, eds.), pp. 321–325, Plenum Press, New York.

Stenn, K. S., and Dvoretzky, I., 1979, Human serum and epithelial spread in tissue culture, *Arch. Dermatol. Res.* **246:**3–15.

Stenn, K. S., Madri, J. A., and Roll, F. J., 1979, Migrating epidermis produces AB2 collagen and requires continual collagen synthesis for movement, *Nature* **277:**229–232.

Takashima, A., and Grinnell, F., 1984, Human keratinocyte adhesion and phagocytosis promoted by fibronectin, *J. Invest. Dermatol.* **83:**352–358.

Varghese, M. C., Balin, A. K., Carter, M., and Caldwell, D., 1986, Local environment of chronic wounds under synthetic dressings, *Arch. Dermatol.* **122:**52–56.

Verrando, P., Hsi, B. L., Yeh, C.-J., Pisani, A., Serieys, N., and Ortonne, J.-P., 1987, Monoclonal antibody GB3, a new probe in the study of human basement membranes and hemidesmosomes, *Exp. Cell Res.* **170:**116–128.

Westgate, G. E., Weaver, A. C., and Couchman, J. R., 1985, Bullous pemphigoid antigen localization suggests an intracellular association with hemidesmosomes, *J. Invest. Dermatol.* **84:**218–224.

Wilke, M. S., and Furcht, L. T., 1990, Human keratinocytes adhere to a unique heparin-binding peptide sequence within the triple helical domain of type IV collagen, *J. Invest. Dermatol.* **95:**264–270.

Winter, G. D., 1962, Formation of the scab and the rate of epithelialization of superficial wounds in the skin of the young domestic pig, *Nature* **193:**293–294.

Woodley, D. T., and Kim, Y. H., 1992, A double-blind comparison of wound dressings using uniform suction blister wounds, *Arch. Dermatol.* **128:**1354–1357.

Woodley, D. T., Didierjean, L., Regnier, M., Saurat, J., and Prunieras, M., 1980a, Bullous pemphigoid antigen synthesized *in vitro* by human epidermal cells, *J. Invest. Dermatol.* **75:**148–151.

Woodley, D. T., Regnier, M., and Prunieras, M., 1980b, *In vitro* basal lamina formations may require nonepidermal cell living substrate, *Br. J. Dermatol.* **103:**397–404.

Woodley, D. T., Rao, C. N., Hassell, J. R., Liotta, L. A., Martin, G. R., and Kleinman, H. K., 1983, Interactions of basement membrane components, *Biochim. Biophys. Acta* **761:**278–283.

Woodley, D. T., O'Keefe, E. J., and Prunieras, M., 1985a, Cutaneous wound healing: A model for cell–matrix interactions, *J. Am. Acad. Dermatol.* **12:**420–433.

Woodley, D. T., Briggaman, R. A., Gammon, W. R., and O'Keefe, E. J., 1985b, Epidermolysis bullosa acquisita antigen is synthesized by human keratinocytes cultured in serum-free medium, *Biochem. Biophys. Res. Commun.* **130:**1267–1272.

Woodley, D. T., Kelebec, T., Banes, A. J., Link, W., Prunieras, M., and Liotta, L. A., 1986, Adult human keratinocytes migrating over nonviable dermal collagen produce collagenolytic enzymes that degrade type I and type IV collagen, *J. Invest. Dermatol.* **86:**418–423, 1986.

Woodley, D. T., O'Keefe, E. J., McDonald, J. A., 1987, Specific affinity between fibronectin and the epidermolysis bullosa acquisita antigen, *J. Clin. Invest.* **179:**1826–1830.

Woodley, D. T., Bachmann, P. M., and O'Keefe, E. J., 1988a, Laminin inhibits human keratinocyte migration, *J. Cell. Physiol.* **136:**140–146.

Woodley, D. T., Peterson, H. D., Herzog, S. R., Stricklin, G. P., Burgeson, R. E., Briggaman, R. A., Cronce, D. J., and O'Keefe, E. J., 1988b, Burn wounds resurfaced by cultured epidermal autografts show abnormal reconstitution of anchoring fibrils, *J. Am. Med. Assoc.* **259:**2566–2571.

Woodley, D. T., Briggaman, R. A., Herzog, S., Meyers, A., Peterson, H. D., and O'Keefe, E. J., 1990a, Characterization of neo-dermis formation beneath cultured human epidermal autografts transplanted on muscle fascia, *J. Invest. Dermatol.* **95:**20–26.

Woodley, D. T., Wynn, K. C., and O'Keefe, E. J., 1990b, Type IV collagen and fibronectin enhance human keratinocyte thymidine incorporation, *J. Invest. Dermatol.* **94:**139–143.

Wysocki, A., Baxter, C. R., Bergstresser, P. R., Grinnell, F., Horowitz, M. S., and Horowitz, B., 1988, Topical fibronectin therapy for treatment of a patient with chronic status ulcers, *Arch. Dermatol.* **124:**175–177.

Wysocki, A. B., and Grinnell, F., 1990, Fibronectin profiles in normal and chronic wound fluid, *Lab. Invest.* **63:**825–831.

Chapter 11

Angiogenesis

JOSEPH A. MADRI, SABITA SANKAR, and ANNE M. ROMANIC

1. Introduction

Angiogenesis, the formation of new vessels during development, in response to injury and tumor angiogenic factors is a dynamic process that is controlled by many diverse, sometimes complex factors acting together in a local environment. The principal cell type involved in the process of angiogenesis is the microvascular endothelial cell. This cell type is quite distinct from the endothelia lining the larger vessels of the circulatory system in its normal physiological functions and in its response to injury (Madri *et al.,* 1991, 1992a, b). Following denudation injury (angioplasty, endarterectomy, synthetic and autologous bypass grafting), large-vessel endothelial cells undergo sheet migration that is modulated by both existing and newly synthesized extracellular matrix components and soluble factors (Madri *et al.,* 1988b, 1991, 1992a, b; Madri and Bell, 1992). In contrast, following injury, microvascular endothelial cells initiate an angiogenic process also modulated by both existing and newly synthesized extracellular matrix components and soluble factors, consisting of local disruption of their investing basement membrane, migration into the local interstitial stroma, cell proliferation, new vessel formation, stabilization, and eventually involution of the newly formed vascular bed (Madri and Pratt, 1988; Madri and Marx, 1992; Madri *et al.,* 1992a, b; Marx *et al.,* 1994). Distinct behavioral patterns exhibited by these two different endothelial cell populations have led to the development of the hypothesis that large-vessel endothelial cells exhibit "dysfunctional" behavior in response to injury-induced changes in the extracellular matrix and soluble factor environments, favoring the development of arteriosclerosis. In contrast, microvascular endothelial cells display a "plastic" phenotype in response to injury-induced changes in the local extracellular matrix and soluble factor environments, displaying a variety of phenotypes normally observed in the microvasculature during angiogenesis (Madri *et al.,* 1991; Madri and Marx, 1992). Since publication of the first edition of this book, a great amount of information has

JOSEPH A. MADRI, SABITA SANKAR, and ANNE M. ROMANIC • Department of Pathology, Yale University School of Medicine, New Haven, Connecticut 06510.

The Molecular and Cellular Biology of Wound Repair (Second Edition), edited by Richard A. F. Clark. Plenum Press, New York, 1996

been accrued regarding vascular cell biology in general and the process of angiogenesis in particular. Of particular importance is the application of molecular biological and biophysical methods to existing and more-recently developed *in vivo* and *in vitro* models of angiogenesis.

This chapter, similar to the one published in the previous edition of this book (Madri and Pratt, 1988), does not attempt an exhaustive general review of endothelial cell biology, but instead considers selected aspects of cell–cell, cell–extracellular matrix, and cell–soluble factor interactions that are thought to be important in the modulation of angiogenesis. As before, we have divided our discussions into severjal main areas: stimulation, migration, proliferation, tube formation, stabilization–differentiation, regression–remodeling–involution, and current and future therapeutic issues.

2. Activation

Since the publication of the first edition of this book there has been significant advances in our understanding of endothelial cell activation–stimulation that occurs early, initiating the angiogenic response. Several investigators have demonstrated the induction of selected proteases and protease inhibitors following activation–stimulation with particular growth factors [basic fibroblast growth factor (bFGF) and transforming growth factor-beta (TGF-β)] known to be present following injury and during the process of angiogenesis (Pepper and Montesano, 1990; Pepper *et al.*, 1992, 1993; Saksela *et al.*, 1987; Mignatti *et al.*, 1992; Tsuboi *et al.*, 1990). In addition to these studies, other groups utilizing monoclonal antibody and cloning methodologies have demonstrated the presence of inducible adhesion molecules on the endothelial cell surface following stimulation with a variety of cytokines (Pober, 1988; Pober and Cotran, 1990; Springer, 1990; Albelda and Buck, 1990).

Just as it has been demonstrated that controlled protease–protease inhibitor systems modulate, in part, the migration of large-vessel endothelial cell and vascular smooth muscle cell migration following injury (Bell and Madri, 1990; Bell *et al.*, 1992; Madri and Bell, 1992; Saksela *et al.*, 1987), investigators have shown that tightly regulated proteolysis is necessary for angiogenesis to proceed and capillaries to develop and mature (Pepper and Montesano, 1990; Bacharach *et al.*, 1992; Blasi, 1993). Specifically, it has been demonstrated that bFGF, phorbol myristate acetate (PMA), and vanadate induce microvascular endothelial cell invasion and tube formation in fibrin gels (Pepper and Montesano, 1990). This behavior has been correlated with increased urokinase activity and mRNA levels. More recently, investigators have shown that levels of plasminogen activator inhibitor (PAI-1) protein and mRNA are also elevated in response to these three agents, suggesting that a tight local control of proteolysis is necessary for normal angiogenesis to occur (Pepper *et al.*, 1992, 1993). Other soluble factors present following injury have also been shown to elicit induction of proteases and protease inhibitors in microvascular endothelial cells. The relative inductions of urokinase-type plasminogen activator (uPA) and PAI-1 vary depending on the factors present and the state of the endothelial cells. Investigators have demonstrated differences in the induction of uPA and PAI-1 by bFGF and TGF-β1 (Pepper and Montesano, 1990). They found that bFGF induced a large increase in proteolysis, while TGF-β1 elicited a net antiproteolysis. Cultures treated with bFGF developed widely patent lumina, while cultures treated with bFGF and TGF-β1 exhibited only superficial invasion of the gel in the form of solid cords. In additional studies, Montesano *et al.* (1990) have also illustrated the importance of the tight control of proteolysis in the process of angiogenesis. In these studies endothelial cells transduced with the middle T (mT) oncogene from polyoma virus (which induces hemangiomas in mice) formed large, ectatic saclike structures resembling hemangiomas when cultured in fibrin gels. These cells were found to express increased uPA activity and decreased PAI-1 activity compared to normal endothelial cells. When these cells were cultured in fibrin gels in the presence of the plasmin inhibitors Trasylol (aprotinin) or ε-amino caproic acid (ε-ACA), the cells formed branching networks of capillarylike tubes with patent lumina similar to normal endothelial cells (Pepper and Montesano, 1990; Montesano *et al.*, 1990). These studies have suggested that factor-driven control of proteolysis mediates tube formation during angiogenesis, with bFGF-induced proteolysis being necessary for lumen formation and TGF-β1-induced antiproteolysis being necessary for controlling the diameters and stabilizing the newly formed vessels.

In addition to this direct factor-mediated induction of proteases during angiogenesis, the endothelial cells in an area of injury are also induced to express uPA following adhesion by mononuclear cells. During the course of inflammation, local endothelial cells are exposed to a variety of cytokines that induce a series of endothelial surface adhesion molecules in a temporal fashion (Pober, 1988; Pober and Cotran, 1990; Hauser *et al.*, 1993a). One of these inducible molecules, vascular cell surface adhesion molecule-1 (VCAM-1), a member of the immunoglobulin (Ig) supergene family, is the counterreceptor for very late antigen-4 (VLA-4), a surface protein present on lymphocytes and monocytes. Romanic and Madri (1994a) have shown that the adhesion of lymphocytes to endothelial cells via a VCAM-1–VLA-4 interaction induces the increased expression of uPA and a decreased expression of PAI-1 in the endothelial cells. These changes in uPA and PAI-1 could directly elicit basement membrane proteolysis and/or activate plasmin and basement membrane and interstitial collagenases that would also contribute to proteolysis of the endothelial basement membrane (Fig. 1).

In addition to these two mechanisms of endothelial cell activation and basement membrane proteolysis, the mononuclear cells that adhere to and transmigrate through endothelial cell monolayers at sites of injury (Hauser *et al.*, 1993a, b) are also likely contributors to local basement membrane proteolysis, which would function to release local endothelial cells from the constraints of their investing basement membranes. Recently, we have demonstrated that T lymphocytes that adhere to endothelial cells via VCAM-1–VLA-4 interactions and transmigrate through these endothelial cell monolayers exhibit induction of 72-kDa gelatinase (basement membrane collagenase) mRNA, protein, and activity (Romanic and Madri, 1994b) (Fig. 1).

Thus, endothelial cell activation can be triggered by a variety of soluble factors present at sites of injury. These factors [bFGF, TGF-β1, interleukin-1 (IL-1), tumor

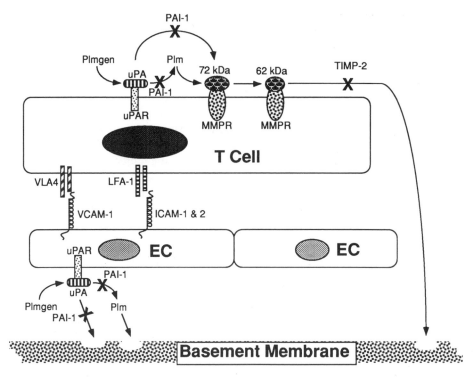

Figure 1. Scheme depicting endothelial cell–leukocyte interactions and modulation of selected proteinase and proteinase–proteinase inhibitor systems by this interaction. Leukocytes (monocytes and T cells) are known to adhere to endothelial cells via VLA-4/VCAM-1 and LFA-1/ICAM-1 interactions. We propose that the modulation of endothelial cell and leukocyte proteinase and proteinase–proteinase inhibitor systems is mediated, in part, by engagement of these ligand pairs. The cell labeled "T cell" represents a T lymphocyte. The smaller rectangular-shaped cells labeled "EC" represent endothelial cells. Shaded ovals within each cell represent the cell nuclei. Abbreviations: uPA, urokinase plasminogen activator; PAI-1, plasminogen activator inhibitor-1; Plmgen, plasminogen; Plm, plasmin; 72 kDa, procollagenase (MMP-2); 62 kDa, active collagenase (MMP2); thick stipled line, basement membrane; VLA-4, very late antigen 4 ($\alpha 4\beta 1$ integrin); VCAM-1, vascular cell adhesion molecule 1; LFA-1, leukocyte function-associated antigen 1; ICAM-1, intercellular adhesion molecule 1; ICAM-2, intercellular adhesion molecule 2; uPAR, EC surface uPA receptor; MMPR, putative 72 kDa/62 kDa EC surface receptor; TIMP-2, tissue inhibitor of metalloproteinase-2. PA can convert PlmG to Plm, ProC'ase to C'ase, and also it can degrade Fn, Ln, and IV. Plm can degrade Fn, Ln, and IV and can convert ProC'ase to C'ase; C'ase can degrade collagen types I, III, IV, and V. PAI binds to and inactivates PA in a reversible manner. (Data used in generating this figure were taken from Romanic and Madri, 1994a, b.)

necrosis factor (TNF), etc.] are capable of eliciting differing endothelial cell responses depending on the state of the endothelial cells as modulated by the local extracellular matrix composition and organization, type, and quality of cell–cell interactions, the presence or absence of other cell types, and the differential expression of specific surface receptors (Madri *et al.*, 1991; Madri and Marx, 1992; Marx *et al.*, 1994). The differential expression of microvascular endothelial cell surface receptors can be mediated by the composition and organization of the surrounding and newly synthesized

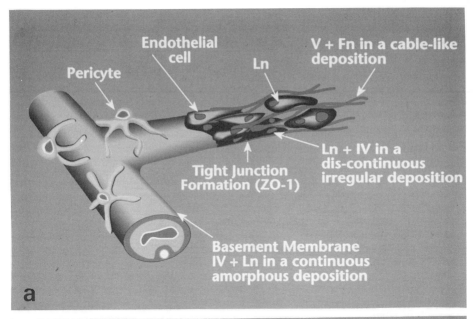

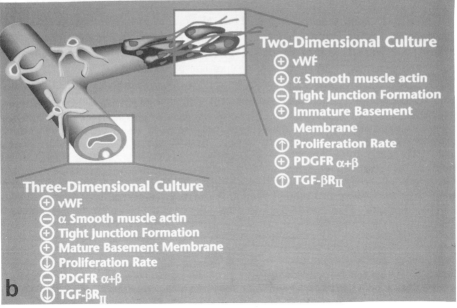

Figure 2. (a) Schematic representation of the matrix modulation of microvascular endothelial cell (EC) phenotype during angiogenesis. Cells at the distal tip of the angiogenic sprout deposit and interact with an extracellular matrix different in composition and organization compared to cells nearer the parent vessel. Cells at the distal tip of an angiogenic sprout express features of undifferentiated EC, having a high proliferative rate, PDGF-α and -β chain receptors, α smooth muscle actin, and a high TGF-β receptor type II to type I ratio. Cells nearer the parent vessel deposit and interact with a mature basement membrane and express features of differentiated EC, having a low proliferative rate, tight junction formation, loss of both PDGF-α and -β chain receptors and α smooth muscle actin, and a low TGF-β receptor type II to type I ratio. (b) Specific culture conditions mimic either undifferentiated endothelial cells at the distal tip of an angiogenic sprout (two-dimensional culture) or differentiated endothelial cells nearer the parent vessel (three-dimensional culture). (Data used in generating this figure were taken from Form *et al.*, 1986; Nicosia and Madri, 1987; Kocher and Madri, 1989; Madri *et al.*, 1988a, 1992a; Merwin *et al.*, 1990, 1991; Marx and Madri, 1992; Madri and Marx, 1992; Marx *et al.*, 1994.)

extracellular matrix. For example, Marx *et al.* (1994) have shown that the surface expression of platelet-derived growth factor (PDGF) receptor α and β chains (PDGFR-α and -β) on microvascular endothelial cells and the cells' responsiveness to PDGF isoforms is modulated by the organization of the local extracellular matrix (Marx *et al.*, 1994). In this study it was found that microvascular endothelial cells would express PDGFR-α and -β when grown in traditional two-dimensional culture, a culture condition that mimics the migrating, proliferating endothelial cell population at or near the tip of an angiogenic sprout (Madri and Marx, 1992). However, when these same cells were cultured in a three-dimensional collagen gel (a culture condition that mimics the quiescent, differentiating endothelial cell population distal to the tip of the angiogenic sprout and closer to the parent vessel), the cells lost their expression of PDGFR-α and -β (Fig. 2a, b) (Kocher and Madri, 1989; Madri *et al.*, 1988a, Merwin *et al.*, 1990; Nicosia and Madri, 1987; Madri and Marx, 1992; Marx *et al.*, 1994).

3. Migration

Migration of microvascular endothelial cells into an area of injury is a necessary component of the process of angiogenesis. Migration of the activated endothelial cells now released from the constraints of their investing basement membranes is modulated by (1) the responsiveness of the cells to a variety of soluble factors including bFGF, PDGF, and TGF-β; (2) interactions of the cells with the various extracellular matrix components comprising the local environment mediated by the repertoire of substrate adhesion molecules present on the cell surfaces; and (3) endothelial cell–endothelial cell interactions (Madri *et al.*, 1989; Bell and Madri, 1989; Basson *et al.*, 1990, 1992).

Following injury, bFGF is released from injured cells and the local extracellular matrix (Mignatti *et al.*, 1992; Vlodavsky *et al.*, 1987) and has been associated with changes in the migratory phenotype of local endothelial cells (Mignatti *et al.*, 1992). bFGF has been shown to stimulate migration of large-vessel and microvascular endothelial cells and has been associated with the induction of urokinase-type plasminogen activator (uPA) and its specific high-affinity receptor (uPAr) (Pepper and Montesano, 1990; Pepper *et al.*, 1992, 1993; Saksela *et al.*, 1987; Tsuboi *et al.*, 1990). Inhibition of endogenous bFGF with antibodies directed against bFGF inhibits migration and uPA and uPAr induction. In contrast, factors such as TGF-β1 and angiotensin 2, which inhibit endothelial cell uPA and increase PAI-1 levels, have been shown to inhibit migration (Bell and Madri, 1990; Bell *et al.*, 1992). In the instance of angiotensin 2, modulation of uPA activity correlates inversely with endogenous c-src levels, and overexpression of c-src in large-vessel and microvascular endothelial cells is associated with increases in c-src protein and activity and faster migration rates, suggesting that c-src participates in the uPA regulatory pathway (Bell *et al.*, 1992; Madri and Bell, 1992). Similar to the observations made following endothelial cell activation, PAI-1 expression is also induced in migrating endothelial cells (Pepper *et al.*, 1992), suggesting a tightly regulated expression of proteases and protease inhibitors resulting in local, controlled extracellular matrix degradation at the cell surfaces of the cells at or near the tips of the angiogenic sprouts and no proteolysis in quiescent, differentiating cells distal to the migrating tips.

In addition to the important role of controlled proteolysis in regulating endothelial cell migration, endothelial cell–extracellular matrix interactions also play pivotal roles in regulating this important process. At the time of publication of the first edition of this text it was known that endothelial cell migration was modulated by extracellular matrix composition and organization and that cellular interactions with the matrix was likely mediated via integrins (Madri *et al.,* 1988c; Basson *et al.,* 1990, 1992). It is now known that specific integrin surface expression and engagement can modulate cell attachment, spreading, migration, and matrix assembly, as well as activate intracellular signaling pathways and protein phosphorylation (Hynes, 1992; Juliano and Haskill, 1993; Wang *et al.,* 1993). Using Chinese hamster ovary (CHO) and 3T3 cells, investigators have demonstrated that decreased expression of $\alpha5\beta1$ integrins is associated with decreased migration of CHO cell variants, and that transfection of the $\alpha5\beta1$ integrin-deficient variants with $\alpha5$ subunit cDNA resulted in normal levels of $\alpha5\beta1$ surface expression and restoration of motility of the cells (Bauer *et al.,* 1992). In studies utilizing endothelial cells, investigators found that migration was inhibited when $\beta1$ integrins were inhibited with functional antibodies directed against the $\beta1$ integrin chain or RGD (Arg-Gly-Asp)-containing peptides (Basson *et al.,* 1990). In recent studies, investigators found that particular extracellular matrix components associated with different levels of endothelial cell migration elicited different integrin organizational patterns (Basson *et al.,* 1992), while soluble factors that are known to modulate endothelial cell migration were found to alter specific integrin surface expression (Basson *et al.,* 1992). These data suggest an important role for integrins in modulating endothelial cell migration during the angiogenic response and illustrate the dynamic interactions among soluble factors, extracellular matrix, and substrate adhesion molecules in this process (Madri and Marx, 1992, Madri *et al.,* 1988a, 1992a).

In addition to the above-mentioned modulation of endothelial migration by soluble factors, extracellular matrix, and substrate adhesion molecules, vascular cell migration is thought to be regulated, in part, by cell–cell interactions. The importance of cell–cell interaction in cell migration has been demonstrated in neural development. During neural development, specific cell adhesion molecules [neural cell adhesion molecule (NCAM) and neuron-glia cell adhesion molecule (NgCAM)] appear to regulate neural outgrowth (Doherty *et al.,* 1991). Recently, endothelial cells were found to express cell adhesion molecules on their surfaces, and the roles of these molecules in cell–cell adhesion, migration, and signaling are being actively investigated (Albelda and Buck, 1990; Albelda *et al.,* 1991). One such molecule, platelet endothelial cell adhesion molecule-1 (PECAM-1), a 130-kDa integral membrane glycoprotein that is a member of the Ig superfamily has been found constitutively on the surfaces of endothelial cells (Albelda *et al.,* 1991; Schimmenti *et al.,* 1992). PECAM-1 exhibits a lateral localization, being present at the lateral borders of cells in contact with each other. When endothelial cells are induced to migrate, their PECAM-1 localization pattern changes from a lateral one in the confluent state to a diffuse pattern covering the entire cell in cells at or near the migrating front, suggesting a role of PECAM-1 as a regulator of endothelial cell migration (Schimmenti *et al.,* 1992). Agents that inhibit endothelial cell migration (such as TGF-β1) appear to stabilize the lateral localization of PECAM-1, while agents that increase cell migration appear to disrupt PECAM-1 lateral localization and promote a diffuse cellular localization. Further, when cells normally not expressing PECAM-1 (specifically 3T3 cells) are stably transfected with PECAM-1 cDNA and overexpress the molecule, they exhibit greater cell aggregation and a decreased migration rate (Schimmenti et al., 1992). These studies support the concept that cell adhesion molecules play important roles as modulators of endothelial cell migration during angiogenesis.

4. Proliferation

Endothelial cell proliferation during angiogenesis is a tightly controlled process limited to specific local cell populations in the angiogenic sprout (Madri and Pratt, 1988; Ausprunk and Folkman, 1977; Burger and Klintworth, 1981). Specifically, the endothelial cells near the leading migratory cells at the tip of the angiogenic sprouts make up this proliferating population. The proliferative response is mediated by diverse, interrelated stimuli, including local soluble autocrine and paracrine factors, extracellular matrix composition and organization, and cell–cell interactions. This localized proliferative response can be explained by local differences in the presence and/or absence of soluble growth factors/wound hormones; local differences in the composition and organization of the surrounding and newly synthesized extracellular matrix; and/or local differences in cell–cell interactions.

It is known that the extracellular matrix synthesized and deposited during the process of angiogenesis differs along the length of an angiogenic sprout as one proceeds from the tips of angiogenic sprouts to more distal regions of the sprouts (Fig. 2a) (Form et al., 1986, Nicosia and Madri, 1987; Langdon et al., 1988). The presence of particular matrix components correlates with the observed proliferation rates in vivo and those obtained in vitro with microvascular endothelial cells plated on particular matrix components (Form et al., 1986, Nicosia and Madri, 1987). The mechanism(s) involved in this phenomenon have been difficult to elucidate. However, the advent of several culture systems designed for microvascular endothelial cells have shed some light on a potential explanation for these observations. Previously, investigators found that culturing microvascular endothelial cells in traditional two-dimensional culture results in a cell phenotype resembling an undifferentiated endothelial cell having a high proliferative rate. In contrast, when microvascular endothelial cells are cultured in a three-dimensional environment, they acquire the phenotype of differentiated endothelial cells having a very low mitotic rate (Fig. 2b) (Madri and Marx, 1992; Marx et al., 1994; Madri et al., 1992a; Merwin et al., 1990, 1991).

Recently, we observed that microvascular endothelial cells cultured in a two-dimensional environment exhibit a proliferative response to PDGF-BB and -AB, while the same cells grown in a three-dimensional environment do not respond to these PDGF isoforms (Marx et al., 1994). It has also been shown that these cells exhibit a dramatic inhibition of proliferation in response to TGF-β1 when in two-dimensional culture, but display no change in proliferative behavior in response to TGF-β1 in three-dimensional culture (Madri *et al.*, 1988a; Merwin *et al.*, 1990, 1991). When these cells were examined for their expression of PDGFR-α and -β chains, they were noted to express both chains when in two-dimensional culture, but did not express either chain when in three-dimensional culture (Marx *et al.*, 1994). In similar studies, we have demonstrated that these cells express greatly reduced levels of type II TGF-β receptor when cultured in a three-dimensional environment compared to the same cells cultured in a two-dimensional environment (S. Sankar, N. Mahooti-Brooks, and J. A. Madri, submitted) (Fig. 2a, b). These studies suggest that changes in the extracellular matrix organization are informational, possibly mediating signaling through mechanorecep-tors on the cell surface (Wang *et al.*, 1993) and effecting changes in surface receptor expression that leads to changes in the cellular responses to particular soluble factors. This can, in part, explain the differences observed in proliferative rates in local endothelial cell populations along the angiogenic sprout. Further experimentation in this area is necessary to elucidate the signaling pathways involved in this dynamic process.

5. Tube Formation/Stabilization

Tube formation during the angiogenic process is a complex process that involves complex cell–cell and cell–matrix interactions. Although our understanding of this aspect of angiogenesis is still incomplete, there have been significant advances in our understanding of this area since publication of the first edition of this volume. During the process of tube formation, microvascular endothelial cells must recognize each other and form stable cell–cell interactions if a capillary network is to form. In order for this to occur, the cells must express a variety of surface proteins that have as their functions the abilities to interact with a changing repertoire of surrounding extracellular matrix components and to effect cell–cell interactions. Investigators have demonstrated the presence of particular cell adhesion molecules (PECAM-1) (Albelda and Buck, 1990) and substrate adhesion molecules (several β1 and β3 integrins) (Albelda and Buck, 1990) on the surfaces of microvascular endothelial cells and have begun to elucidate their roles in the process of tube formation. During tube formation (Fig. 2a), the local populations of microvascular endothelial cells along the length of the sprout interact with an extracellular matrix that displays changes in its composition and organization along its length (Form *et al.*, 1986; Nicosia and Madri, 1987). These changes are thought to modulate several aspects of endothelial cell behavior, including cell proliferation, migration, extracellular matrix synthesis and deposition, and cell–cell interactions (Form *et al.*, 1986; Nicosia and Madri, 1987; Madri *et al.*, 1988a, 1992a; Merwin *et al.*, 1990; Madri and Marx, 1992; Ment *et al.*, 1991, 1992; Wang *et al.*, 1993).

Tube formation has been modeled successfully *in vitro* using a variety of tissue culture systems (Madri and Williams, 1983; Madri *et al.*, 1988a; Pepper and Montesano, 1990; Nicosia and Madri, 1987; Gamble *et al.*, 1993). In the presence of angiogenic factors, three-dimensional cultures of microvascular endothelial cells form branching networks of tubes having lumina and abluminal basal laminae (Madri *et al.,* 1988a; Merwin *et al.,* 1990; Madri and Marx, 1992; Gamble *et al.,* 1993). Cells dispersed in these three-dimensional connective tissue gels utilize integrins to interact with the matrix components present in the processes of invasion into the matrix, migration, and initiation and maintenance of cell contact. Initiation of cell–cell contact is likely mediated by cell adhesion molecules, several of which have now been identified on the surfaces of endothelial cells (Albelda and Buck, 1990). One such molecule, PECAM-1, has been observed to play a role in mediating early cell–cell contact during tube formation. Early aspects of tube formation (initial cell–cell contact) appears to be mediated by PECAM–PECAM interactions, since functional antibodies directed against PECAM-1 and soluble purified PECAM-1 can inhibit tube formation *in vitro* (Madri *et al.,* 1992a). Later aspects of tube formation (stabilization of the newly formed cell–cell contacts by organization of $\beta1$ integrins in these areas and the formation of tight junctions) are not affected by these reagents. In contrast, functional antibodies directed against $\beta1$ integrins and RGD-containing peptides are effective inhibitors of both the early and later aspects of tube formation, presumably by interfering with critical cell–matrix interactions that are necessary for formation of initial cell–cell contacts, stabilization of formed cell–cell contacts, and tight junction formation.

Other studies (Gamble *et al.,* 1993) have shown that inhibition of specific integrin heterodimeric pairs (anti-$\alpha2\beta1$ in cells cultured in type I collagen gels and anti-$\alpha v\beta3$ in cells cultured in fibrin gels) resulted in enhanced tube formation, suggesting that disengagement of particular integrin heterodimeric pairs during the process of tube formation modulates microvascular endothelial cell phenotype from proliferation to differentiation. Another possible interpretation of these results stems from recent studies of tumor cell transmigration and invasion. Investigators have demonstrated that melanoma cells express $\alpha v\beta3$ on their surfaces, and when it is engaged the cells are induced to express 72-kDa gelatinase on their surfaces, leading to increased motility (Seftor *et al.,* 1992). Thus, engagement of microvascular endothelial cell $\alpha2\beta1$ and/or $\alpha v\beta3$ might induce the expression of 72-kDa gelatinase, resulting in increased migration and proliferation. Inhibition of engagement of these integrin pairs by functional antibodies might prevent protease induction and promote differentiation by stabilizing cell–matrix interactions mediated by other integrin pairs, thus allowing for formation of stable cell–cell interactions leading to tight junction formation (Fig. 3). Thus, during the process of tube formation a complex hierarchical, multistep process involving the assembly and surface organization of selected cell adhesion molecules (CAMs), substrate adhesion molecules (SAMs), and junction-associated molecules (JAMs) occurs in a three-dimensional soluble factor-rich environment.

As mentioned in Section 4, the extracellular matrix composition and organization of an angiogenic sprout differs along its length (Fig. 2a) (Form *et al.,* 1986; Nicosia and Madri, 1987). The synthesis and abluminal deposition of a morphologically identifiable basement membrane in the areas of angiogenic sprouts distal to the migrating tips serves to stabilize the vessel, and the eventual presence of such a structure investing all the newly formed vessels following an angiogenic stimulus is necessary for the continued stability of the newly formed vascular bed. Failure or loss of investiture by a basement membrane can have profound and sometimes catastrophic effects. Investiga-

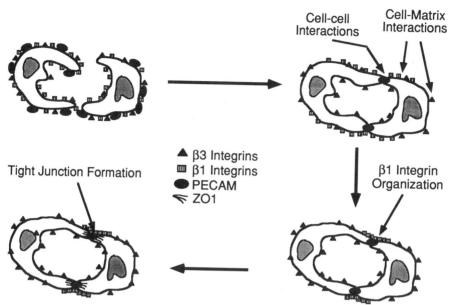

Figure 3. Model for cell adhesion molecule, substrate adhesion molecule, and junction-associated molecule hierarchical expression and organization during angiogenic factor-induced *in vitro* angiogenesis. This figure illustrates putative temporospatial relationships of selected CAMs (PECAM-1), SAMs (β1 integrins), and JAMs (ZO-1) in three-dimensional cultures of microvascular endothelial cells. TGF-β1 induces cell–cell contact initially via PECAM-1 homotypic interactions and integrin–extracellular matrix interactions. Following this stage of the *in vitro* angiogenesis process, β1 integrins exhibit abluminal organization in areas of cell–cell contact, which likely serves to stabilize the newly formed tubelike structures. Following stabilization of the tubelike structures by the β1 integrins, tight junction formation occurs, which correlates with ZO-1 assembly and organization. (Based on data from Merwin *et al.*, 1990, 1991; Madri *et al.*, 1992a.)

tors have begun to investigate the roles of extracellular matrix synthesis, deposition, and degradation during vascular remodeling in the CNS, specifically in the germinal matrix of premature infants (Ment *et al.*, 1991, 1992). Intraventricular hemorrhage or hemorrhage into the remodeling germinal matrix during the period immediately following premature birth from 24 to 28 weeks of gestation is a common problem that results in neurodevelopmental handicaps. This hemorrhage is thought to be due, in part, to the fragile basal laminae of the germinal matrix microvessels that undergo extensive remodeling and involution in the immediate postnatal period following premature delivery. Incidence of hemorrhage decreases to normal over the first 5 postnatal days, and this correlates with increases in basal laminae deposition of the extracellular matrix proteins laminin and type V collagen. Furthermore, treatment with indomethacin, which increases basal laminae deposition of laminin and type V collagen around germinal matrix microvessels compared to nontreated subjects, dramatically decreases the incidence of hemorrhage during this period. These data (Ment *et al.*, 1991, 1992) support the concept that extracellular matrix deposition, organization, and maintenance are critical factors in stabilizing microvessels.

6. Regression/Remodeling/Involution

During development and late in the healing process, there is significant vascular remodeling and involution. Since 1988, several reports have been published that have demonstrated the amelioration of disease coincident with involution of a particular vascular bed. In one study (White *et al.,* 1989), investigators treated a patient suffering from pulmonary hemangiomatosis with interferon-α and noted a dramatic reduction of the hemangioma and a concomitant improvement in the pulmonary function and over-all health of the patient. These reports underscore the potential value in down-regulating angiogenesis and eliciting involution of vascular beds in a variety of disease states, including tumor angiogenesis, vascularization following corneal injury, and possibly vascularization of keloids. Approaches that have been taken include the use of angiostatic steroids, first described by Crum *et al.* (1985), and more recently fungal metabolites (Brem *et al.,* 1993; Kusaka *et al.,* 1994). Despite our advances in this area, our understanding of the mechanisms that initiate and control this aspect of the angiogenic process is still incomplete.

During vascular involution, apoptosis (programed cell death) occurs in a tightly controlled spatiotemporal fashion, resulting in the regression of the neovascular bed with maintenance of the parent vessels that gave rise to the new, transient vascular bed. One such example of this process is the maturation of the germinal matrix during gestational weeks 24 to 32 in human development (Ment *et al.,* 1991, 1992). Normally, the microvessel density is halved during this period of time, with the surviving vessels becoming the lenticulostriate vessels. However, it has been well documented that premature delivery triggers this involution/stabilization process precipitously during the first 5 days following delivery (Ment *et al.,* 1991, 1992). During this period, the vessel density of the germinal matrix is halved. The surviving vessels deposit increased amounts of basal lamina components, develop a mature, morphologically identifible basement membrane and tight junctions, and complete investment by glial end feet, while the involuting vessel segments undergo dissolution of their basement membranes and endothelial apoptosis. The control of this apoptotic event is as yet undefined. Possibilities include specific spatiotemporally distinct signals from surrounding glia mediated by direct contact via glial end feet or by soluble factors secreted by particular glial cells. Alternatively, the process of selective endothelial apoptosis may be triggered by the loss of specific (integrin-mediated) interactions with the underlying basement membrane (Meredith *et al.,* 1993; Frisch and Francis, 1994; Bates *et al.,* 1994). Possible mechanisms might involve the modulation of proteinases and proteinase inhibitors (uPA, PAI-1, and PAI-2), matrix metalloproteinases (MMPs), and tissue inhibitors of metalloproteinases (TIMPs), since their presence and activity have been documented in the brain and they are likely to be important components in the processes of extracellular matrix metabolism in the brain (Romanic and Madri, 1994b). Since there are several good animal models of this process and endothelial cells derived from the germinal matrix region of these animals have been isolated and cultured as have glial cells, the process of controlled endothelial apoptosis and the interaction of endothelial cells, glial cells, and the extracellular matrix in this process can be explored systematically in depth *in vitro.*

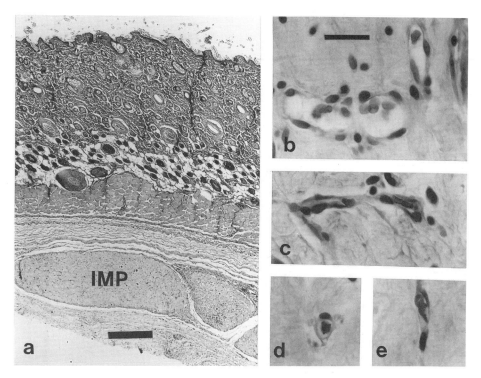

Figure 4. (a) Low-power micrograph of rat skin and subcutaneous tissue illustrating a gel composed of type I collagen (5.0 mg/ml) containing a dense network of anastomosing rat microvessels formed *in vitro* and implanted subcutaneously 2 weeks prior to removal from the host. IMP, gel implant containing vascular networks formed *in vitro*. Stained with hematoxylin and eosin. Scale bar = 400 μm. Original magnification ×2. (Based on data from Kennedy *et al.*, 1994.) (b–d) High-power micrographs of representative microvessels formed *in vitro* in a collagen gel, now containing circulating blood elements of the host. Stained with hematoxylin and eosin. Scale bar = 20 μm. Original magnification ×40.

7. Current and Future Therapeutic Issues

A major therapeutic issue in the angiogenesis field continues to be the development and implementation of strategies and agents that will modulate the up-regulation, maintenance, and/or regression of new vessel formation specifically. Such reagents would enable the medical community to optimize vascularization at cutaneous graft sites or eliminate vascularization following corneal injury and during tumor growth and metastasis (Madri and Pratt, 1988; Langdon *et al.*, 1988; Merwin *et al.*, 1992; Brem *et al.*, 1993; Kusaka *et al.*, 1994; Brooks *et al.*, 1994).

In addition, since the publication of the first edition of this text, there has been a tremendous advancement in our ability to genetically engineer cells and to use such cells in the treatment of a variety of disease states (Culliton, 1989a–c; Dichek *et al.*,

1989; Nabel *et al.*, 1989; Wilson *et al.*, 1989; Zwiebel *et al.*, 1989). One approach considered for the treatment of deficiency diseases is the use of genetically engineered endothelial cells cultured in collagen-coated Gore-Tex fibers (Culliton, 1989b). Such a system would consist of isolating and culturing a patient's endothelial cells from a superficial vein segment or adipose tissue, introducing the gene of choice, and stably expressing the gene product in the endothelial cells. These cells could then be expanded and placed in the Gore-Tex organoid and placed in the patient's circulation. Alternatively, microvascular endothelial cells dispersed in collagen gels at high density have been shown to form branching vascular networks composed of capillarylike vessels with lumina and abluminal basal lamina (Madri *et al.*, 1988a; Merwin *et al.*, 1990). When these gel pellets are placed subcutaneously into animals (rats), the vessels in the collagen gels functionally link up with host vessels by the process of inosculation, and blood flow can be demonstrated in the implants (Kennedy *et al.*, 1994) (Fig. 4). Such "bioreactors" would provide needed proteins directly and continuously to the vascular compartment.

Thus, our increasing knowledge base in the area of angiogenesis should allow continued progress in our ability to beneficially modulate this critical aspect of the wound-healing process.

References

Albelda, A. M., and Buck, C. A., 1990, Integrins and other cell adhesion molecules, *FASEB J.* **4:**2868–2880.

Albelda, S. M., Muller, W. A., Buck, C. A., and Newman, P. J., 1991, Molecular and cellular properties of PECAM-1 (endoCAM/CD31): A novel vascular cell–cell adhesion molecule, *J. Cell Biol.* **114:**1059–1068.

Ausprunk, D. H., and Folkman, J., 1977, Migration and proliferation of endothelial cells in preformed and newly formed blood vessels during tumor angiogenesis, *Microvasc. Res.* **14:**53–65.

Bacharach, E., Itin, A., and Keshet, E., 1992, *In vivo* patterns of expression of urokinse and its inhibitor PAI-1 suggest a concerted role in regulating physiological angiogenesis, *Proc. Natl. Acad. Sci. USA* **89:**10686–10690.

Basson, C. T., Knowles, W. J., Abelda, S., Bell, L., Castronovo, V., Liotta, L. A., and Madri, J. A., 1990, Spatiotemporal segregation of endothelial cell integrin and non-integrin extracellular matrix binding proteins during adhesion events, *J. Cell Biol.* **110:**789–802.

Basson, C. T., Kocher, O., Basson, M. D., Asis, A., and Madri, J. A., 1992, Differential modulation of vascular cell integrin and extracellular matrix expression *in vitro* by TGF-β1 correlates with reciprocal effects on cell migration, *J. Cell. Physiol.* **153:**118–128.

Bates, R. C., Buret, A., van Helden, D. F., Horton, M. A., and Burns, G. F., 1994, Apoptosis induced by inhibition of intercellular contact, *J. Cell Biol.* **125:**403–415.

Bauer, J. S., Schreiner, C. L., Giancotti, F. G., Ruoslahti, E., and Juliano, R. L., 1992, Motility of fibronectin receptor-deficient cells on fibronectin and vitronectin: Collaborative interactions among integrins, *J. Cell Biol.* **116:**477–487.

Bell, L., and Madri, J. A., 1989, The effects of soluble platelet factors on bovine aortic enothelial and smooth muscle cell migration, *Circ. Res.* **65:**1057–1065.

Bell, L., and Madri, J. A., 1990, Influence of the angiotensin system on endothelial and smooth muscle cell migration *in vitro, Am. J. Pathol.* **137:**7–12.

Bell, L., Luthringer, D. J., Madri, J. A., and Warren, S. L., 1992, Autocrine angiotensin system regulation of endothelial cell behavior involves modulation of pp60[c-src] expression, *J. Clin. Invest.* **89:**315–320.

Blasi, F., 1993, Urokinase and urokinase receptor: A paracrine/autocrine system regulating cell migration and invasiveness, *Bioessays* **15**:105–111.

Brem, H., Gresser, I., Grosfeld, J., and Folkman, J., 1993, The combination of antiangiogenic agents to inhibit primary tumor growth and metastasis, *J. Pediatr. Surg.* **28**:1253–1257.

Brooks, P. C., Clark, R. A. F., and Cheresh, D. A., 1994, Requirement of vascular integrin $\omega\beta3$ for angiogenesis, *Science* **264**:569–571.

Burger, P. C., and Klintworth, G. K., 1981, Autoradiographic study of corneal neovascularization induced by chemical cautery, *Lab. Invest.* **45**:328–335.

Crum, R., Szabo, S., and Folkman, J., 1985, A new class of steriods inhibits angiogenesis in the presence of heparin or a heparin fragment, *Science* **230**:1375–1378.

Culliton, B. J., 1989a, Designing cells to deliver drugs, *Science* **246**:746.

Culliton, B. J., 1989b, Gore Tex organoids and genetic drugs, *Science* **246**:747–749.

Culliton, B. J., 1989c, A genetic shield to prevent emphysema? *Science* **246**:750–751.

Dichek, D. A., Neville, R. F., Zwiebel, J. A., Freeman, S. M., Leon, M. B., and Anderson, W. F., 1989, Seeding of intravascular stents with genetically engineered endothelial cells, *Circulation* **80**:1347–1353.

Doherty, P., Rowett, L. H., Moore, S. E., Mann, D. A., and Walsh, F. S., 1991, Neurite outgrowth in response to transfected N-CAM and N-cadherin reveals fundamental differences in neuronal responsiveness to CAMs, *Neuron* **6**:247–258.

Form, D. M., Pratt, B. M., and Madri, J. A., 1986, Endothelial cell proliferation during angiogenesis: *In vitro* modulation by basement membrane components, *Lab. Invest.* **55**:521–530.

Frisch, S. M., and Francis, H., 1994, Disruption of epithelial cell–matrix interaction induces apoptosis, *J. Cell Biol.* **124**:619–626.

Gamble, J. R., Matthias, L. J., Meyer, G., Kaur, P., Russ, G., Faull, R., Berndt, M. C., and Vadas, M. A., 1993, Regulation of *in vitro* capillary tube formation by anti-integrin antibodies, *J. Cell Biol.* **121**:931–943.

Hauser, I., Johnson, D. R., and Madri, J. A., 1993a, Differential induction of VCAM-1 on human iliac venous and arterial endothelial cells and its role in adhesion, *J. Immunol.* **151**:1–14.

Hauser, I., Setter, E., Bell, L., and Madri, J. A., 1993b, Fibronectin expression correlates with U937 cell adhesion to migrating bovine aortic endothelial cells *in vitro, Am. J. Pathol.* **143**:173–180.

Hynes, R. O., 1992, Integrins: Versatility, modulation and signaling in cell adhesion, *Cell* **69**:11–25.

Juliano, R. L., and Haskill, S., 1993, Signal transduction from the extracellular matrix, *J. Cell Biol.* **120**:577–585.

Kennedy, S. P., Smith, J. D., Burton, W. V., Springhorn, J. P., Squinto, S. P., Madri, J. A., and Zavoico, G. B., 1994, Novel gene product delivery system utilizing capillary endothelial cells in a three-dimensional matrix, *J. Cell. Biochem.* **18**(abstr.):301.

Kocher, O., and Madri, J. A., 1989, Modulation of actin mRNAs in cultured capillary endothelial and aortic endothelial and smooth muscle cells by matrix components and TGF-$\beta1$, *In Vitro* **25**:424–434.

Kusaka, M., Sudo, K., Matsutani, E., Kozai, Y., Marui, S., Fujita, T., Ingber, D., and Folkman, J., 1994, Cytostatic inhibition of endothelial cell growth by the angiogenesis inhibitor TNP-470 (AGM-1470), *Br. J. Cancer* **69**:212–216.

Langdon, R., Cuono, C., Birchall, N., Madri, J. A., Kuklinska, E., McGuire, J., and Moellmann, G., 1988, Reconstitution of dermoepidermal and endothelial basement membrane zones in composite skin grafts derived from autologous cultured keratinocytes and cryopreserved allogeneic dermis, *J. Invest. Dermatol.* **91**:478–485.

Madri, J. A., and Bell, L., 1992, Vascular cell responses to injury: Modulation by extracellular matrix and soluble factors, in: *Ultrastructure, Membranes and Cell Interactions in Atherosclerosis* (H. Robenek and N. Severs, eds.), pp. 167–181, CRC Press, Boca Raton, Florida.

Madri, J. A., and Marx, M., 1992, Matrix composition, organization and soluble factors: Modulators of microvascular cell differentiation in vitro, *Kidney Int.* **41**:560–565.

Madri, J. A., and Pratt, B. M., 1988, Angiogenesis, in: *The Molecular and Cellular Biology of Wound Healing,* 1st ed. (R. F. Clark and P. Henson, eds.), pp. 337–358, Plenum Press, New York.

Madri, J. A., and Williams, S. K., 1983, Capillary endothelial cell cultures: Phenotypic modulation by matrix components, *J. Cell Biol.* **97**:153–165.

Madri, J. A., Pratt, B. M., and Tucker, A. M., 1988a, Phenotypic modulation of endothelial cells by transforming growth factor-β depends upon the composition and organization of the extracellular matrix, *J. Cell Biol.* **106:**1375–1384.

Madri, J. A., Pratt, B. M., and Yannariello-Brown, J., 1988b, Endothelial cell extracellular matrix interactions: Matrix as a modulator of cell function, in: *Endothelial Cell Biology in Health and Disease* (N. Simionescu and M. Simionescu, eds.), pp. 167–188, Plenum Press, New York.

Madri, J. A., Pratt, B. M., and Yannariello-Brown, J., 1988c, Matrix driven cell size changes modulate aortic endothelial cell proliferation and sheet migration, *Am. J. Pathol.* **132:**18–27.

Madri, J. A., Reidy, M., Kocher, O., and Bell, L., 1989, Endothelial cell behavior following denudation injury is modulated by TGF-β and fibronectin, *Lab. Invest.* **60:**755–765.

Madri, J. A., Bell, L., Marx, M., Merwin, J. R., Basson, C. T., and Prinz, C., 1991, The effects of soluble factors and extracellular matrix components on vascular cell behavior *in vitro* and *in vivo:* Models of de-endothelialization and repair, *J. Cell. Biochem.* **45:**1–8.

Madri, J. A., Bell, L., and Merwin, J. R., 1992a, Modulation of vascular cell behavior by transforming growth factors beta, *Mol. Reprod. Dev.* **32:**121–126.

Madri, J. A., Merwin, J. R., Bell, L., Basson, C. T., Kocher, O., Perlmutter, R., and Prinz, C., 1992b, Interactions of matrix components and soluble factors in vascular cell responses to injury: Modulation of cell phenotype, in: *Endothelial Cell Dysfunction* (N. Simionescu and M. Simionescu, eds.), pp. 11–30, Plenum Press, New York.

Marx, M., Perlmutter, R., and Madri, J. A., 1994, Modulation of PDGF-receptor expression in microvascular endothelial cells during *in vitro* angiogenesis, *J. Clin. Invest.* **93:**131–139.

Ment, L. R., Stewart, W. B., Ardito, T. A., and Madri, J. A., 1991, Vascular basement membrane re-modeling during germinal matrix maturation in the neonate: Associations with interventricular hemorrhage in the beagle pup model, *Stroke* **22:**390–395.

Ment, L. R., Stewart, W. B., Ardito, T. A., Huang, E., and Madri, J. A., 1992, Indomethacin promotes germinal matrix microvasculature maturation in the newborn pup, *Stroke* **23:**1132–1137.

Meredith, J. E., Fazeli, B., and Schwartz, M. A., 1993, The extracellular matrix as a cell survival factor, *Mol. Biol. Cell* **4:**953–961.

Merwin, J. R., Anderson, J., Kocher, O., van Itallie, C., and Madri, J. A., 1990, Transforming growth factor β1 modulates extracellular matrix organization and cell–cell junctional complex formation during *in vitro* angiogenesis, *J. Cell Physiol.* **142:**117–128.

Merwin, J. R., Newman, W., Beall, D., Tucker, A., and Madri, J. A., 1991, Vascular cells respond differentially to transforming growth factors-beta₁ and beta₂, *Am. J. Pathol.* **138:**37–51.

Merwin, J. R., Lynch, M. J., Madri, J. A., Pastan, I., and Seigall, C. B., 1992, Acidic FGF-*Pseudomonas* exotoxin (aFGF-PE) chimeric protein elicits anti-angiogenic effects on endothelial cells, *Cancer Res.* **52:**4995–5001.

Mignatti, P., Morimoto, T., and Rifkin, D. B., 1992, Basic fibroblast growth factor released by single isolated cells stimulates their migration in an autocrine manner, *Proc. Natl. Acad. Sci. USA* **88:**11007–11011.

Montesano, R., Pepper, M. S., Mohle-Steinlen, U., Risau, W., Wagner, E. F., and Orci, L., 1990, Increased proteolytic activity is responsible for the aberrant morphogenetic behavior of endothelial cells expressing middle T oncogene, *Cell* **62:**435–445.

Nabel, E. G., Plautz, G., Boyce, F. M., Stanley, J. C., and Nabel, G. J., 1989, Recombinant gene expression *in vivo* within endothelial cells of the artery wall, *Science* **244:**1342–1344.

Nicosia, R. F., and Madri, J. A., 1987, The microvascular extracellular matrix: Developmental changes during angiogenesis in the aortic ring-plasma clot model, *Am. J. Pathol.* **128:**78–90.

Pepper, M. S., and Montesano, R., 1990, Proteolytic balance and capillary morphogenesis, *Cell Differ. Dev.* **32:**319–328.

Pepper, M. S., Sappino, A. P., Montesano, R., Orci, L., and Vassalli, J.-D., 1992, Plasminogen activator inhibitor-1 is induced in migrating endothelial cells, *J. Cell. Physiol.* **153:**129–139.

Pepper, M. S., Sappino, A.-P., Stocklin, R., Montesano, R., Orci, L., and Vassalli, J.-D., 1993, Up-regulation of urokinase receptor expression on migrating endothelial cells, *J. Cell Biol.* **122:**673–684.

Pober, J. S., 1988, Cytokine-mediated activation of vascular endothelium, *Am. J. Pathol.* **133:**426–433.

Pober, J. S., and Cotran, R. S., 1990, Cytokines and endothelial cell biology, *Physiol. Rev.* **70:**427–451.

Romanic, A. M., and Madri, J. A., 1994a, The induction of 72 kDa gelatinase in T cells upon adhesion to endothelial cells is VCAM-1 dependent, *J. Cell Biol.* **125:**1165–1178.

Romanic, A. M., and Madri, J. A., 1994b, Extracellular matrix-degrading proteinases in the nervous system, *Brain Pathol.* **4:**145–156.

Saksela, O., Moscatelli, D., and Rifkin, D. B., 1987, The opposing effects of basic fibroblast growth factor and transforming growth factor beta on the regulation of plasminogen activator activity in capillary endothelial cells, *J. Cell Biol.* **105:**957–963.

Schimmenti, Yan, H-C., Madri, J. A., and Albelda, S., 1992, Cell adhesion molecule PECAM-1 modulates cell migration, *J. Cell. Physiol.* **153:**417–428.

Seftor, R. E. B., Seftor, E. A., Gehlsen, K. R., Stetler-Stevenson, W. G., Brown, P. D., Rouslahti, E., and Hendrix, M. J. C., 1992, Role of the αvβ3 integrin in human melanoma cell invasion, *Proc. Natl. Acad. Sci. USA* **89:**1557–1561.

Springer, T. A., 1990, Adhesion receptors of the immune system, *Nature* **346:**425–432.

Tsuboi, R., Sato, Y., and Rifkin, D. B., 1990, Correlation of cell migration, cell invasion, receptor number, proteinase production and basic fibroblast growth factor levels in endothelial cells, *J. Cell Biol.* **110:**511–517.

Vlodavsky, I., Folkman, J., Sullivan, R., Fridman, R., Ishai-Michaeli, R., Sasse, J., and Klagsbrun, M., 1987, Endothelial cell derived basic fibroblast growth factor: Synthesis and deposition into subendothelial extracellular matrix, *Proc. Natl. Acad. Sci. USA* **84:**2292–2296.

Wang, N., Butler, J. P., and Ingber, D. E., 1993, Mechanotransduction across the cell surface and through the cytoskeleton, *Science* **260:**1124–1127.

White, C. W., Sondheimer, H. M., Crouch, E. C., Wilson, H., and Fan, L. L., 1989, Treatment of pulmonary hemangiomatosis with recombinant interferon alfa-2a, *New Engl. J. Med.* **320:**1197–1200.

Wilson, J. M., Birinyi, L. K., Salomon, R. N., Libby, P., Callow, A. D., and Mulligan, R. C., 1989, Implantation of vascular grafts lined with genetically modified endothelial cells, *Science* **244:**1344–1346.

Zwiebel, J. A., Freeman, S. M., Kantoff, P. W., Cornetta, K., Ryan, U. S., and Anderson, W. F., 1989, High-level recombinant gene expression in rabbit endothelial cells transduced by retroviral vectors, *Science* **243:**220–222.

Chapter 12

Mechanisms of Parenchymal Cell Migration into Wounds

JAMES B. McCARTHY, JOJI IIDA, and LEO T. FURCHT

1. Motility Is Regulated by Many Aspects of a Changing Wound Environment

Cellular motility is a fundamental consideration in the successful healing of wounds. While much effort has appropriately been recently placed on understanding the molecular basis of adhesion and signaling mechanisms involved in cell motility, it is important to keep in mind that these processes occur in the context of a complex and changing wound environment, which includes soluble and insoluble (i.e., density) gradients of cell motility-promoting components. A number of cell types must enter the wound in a relatively coordinated fashion, and this is controlled, in part, by modulating both the increased random and directed migration of cells.

Increased random and directed cell migration can be stimulated by certain soluble factors such as certain cytokines or bioactive lipids that are produced in the local wound environment as a result of tissue damage, cellular activation, or the limited action of certain enzymes. Additionally, the deposition of extracellular matrix (ECM) components within the wound can have profound influence on cell motility within the local environment. In this regard, it is important to take into account the changing composition of the ECM within wounds at various stages of healing (Gailit and Clark, 1994), since the ECM can modulate signal transduction, cytoskeletal organization, and transcriptional/translational control of genes. Early wounds are rich in plasma-derived substances such as fibrin, fibronectin, and thrombospondin. As the wound heals, the composition of the granulation tissue changes until new ECM components are deposited, such as hyaluronan, types I and III collagens, ECM-associated proteoglycans, SPARC (secreted protein acidic and rich in cysteine), and tenascin, among others. All of these individual components have been implicated in modulating cell motility, some by virtue of the fact that they can directly promote cell adhesion (e.g., collagens,

JAMES B. McCARTHY, JOJI IIDA, and LEO T. FURCHT • Department of Laboratory Medicine and Pathology, Biomedical Engineering Center, University of Minnesota, Minneapolis, Minnesota 55455.

The Molecular and Cellular Biology of Wound Repair (Second Edition), edited by Richard A. F. Clark. Plenum Press, New York, 1996

fibronectin) and some by virtue of their ability to discourage cell adhesion (e.g., tenascin, ECM-associated chondroitin sulfate proteoglycans). In addition to considering the changing composition of the ECM, it is also important to emphasize that the structural organization of the ECM has very important influences on cell migration. Thus, the motility of cells into wounds can be influenced by many discrete and overlapping mechanisms.

2. Mechanisms of Directing Cell Motility

Various mechanisms have been advanced to explain the phenomenon of directed cell motility, a property of cells that is basic not only to wound healing but to many aspects of development and metastatic spread of malignant neoplasms as well (Trinkaus, 1984). Clearly, motility per se is a phenomenon of cell behavior subject to the control of highly interrelated and complex biochemical–biophysical processes; obviously, a detailed treatment of this topic is beyond the scope of this chapter. Although distinctions may be made *in vitro* concerning the different mechanisms of directing motility, they all share several qualities. Regardless of the mechanism, cells must become asymmetric in order to move, forming thrusting protrusions, and must establish an active dominant leading edge. The leading edge must in turn establish adherence to the substratum, and ultimately, through contractile forces, pull the rear edge of the cell forward. Directional influences *in vivo* must persist long enough to establish a net direction for the motility of entire cell populations. The stimulus for directional movement can be a soluble attractant (chemotaxis), a substratum-bound gradient of a particular matrix constituent (haptotaxis), or the three-dimensional array of ECM within the tissue (contact guidance). The directional movement of epithelial cell sheets presents a special biological problem. For cells within the sheet, directionality results from marginal cells extending lamellapodia away from the sheet. This phenomenon has been termed the free edge effect. Mechanisms must also exist *in vivo* acting to open up spaces within the ECM as cells migrate into wounds. With these principles in mind, the following account highlights the major mechanisms that coordinate the ordered migration of specific cell populations that debride and heal wounds.

2.1. Chemotaxis

Chemotaxis (defined as the directed migration of cells in response to a concentration gradient of a soluble attractant) and chemokinesis (accelerated random migration) are complex processes, involving several discrete but overlapping steps. These steps include modulating cell–cell or cell–ECM adhesion and modulating the integrity of the cytoskeleton (Trinkaus, 1976; Zigmond, 1989; Erickson, 1990). Chemotactic and chemokinetic locomotion require a functional cytoskeletal system, including actin, possibly myosin, tubulin, and intermediate filaments (Zigmond, 1989). As might be expected, a complex process such as cell motility must be regulated by several signal transduction pathways. Analysis of the action of several chemotactic factors indicates that such factors can modulate calcium fluxes, activation of the Na^+/H^+ antiporter with subsequent alkalinization of the cytoplasm, activation of phosphoinositide turnover, and protein phosphorylation (Caterina and Devreotes, 1991). Understanding the precise relationships of these signaling pathways in the context of cytoskeletal rearrangement and the regulation of cell adhesion receptor function is an important topic in contemporary cell biology.

The biological effect of a chemoattractant is regulated in part by diffusion of the attractant from its source into an attractant-poor environment. Much of the knowledge about chemotaxis in higher organisms stems from extensive research on the migration of neutrophils and monocytes in response to attractants generated during inflammation. Leukocyte migration has been studied *in vitro* using several assay systems, including in Boyden chambers (Zigmond and Hirsch, 1973), under agarose (Nelson *et al.*, 1978), and in special orientation chambers (Zigmond, 1977). In studies using Boyden chambers, stimulated neutrophil migration was demonstrated to be composed of two components: (1) directional migration, in which cells moved up a positive concentration gradient of attractant called chemotaxis, and (2) increased random migration, in which cells moved independent of a concentration gradient termed chemokinesis (Zigmond and Hirsch, 1973). Later studies by Zigmond (1977) used a special orientation chamber in order to visualize cells directly that are put into a concentration gradient of attractant.

Zigmond observed that cells structurally orient in response to gradients of chemotactic substances, with a leading lamellapodium extending toward the source of the attractant and with a trailing uropod. Cells could orient in gradients as small as 1% across the cell surface (Zigmond, 1977). Other investigators have shown that this cytological polarization can be accompanied by asymmetry in the distribution of receptors and membrane activities. Concanavalin A (Con A) and Fc receptors, coated pits, and pinocytotic vesicles have been shown to concentrate in the trailing uropod of an oriented cell (Davis *et al.*, 1982). These studies, as well as others, suggest that new receptors are first inserted anteriorly during migration and are then swept to the trailing edge as the cell moves forward. It has been hypothesized that the anterior–posterior sweeping of receptor–ligand complexes on the cell surface could serve to amplify even relatively shallow attractant gradients across the cell (Zigmond *et al.*, 1981; Davis *et al.*, 1982), helping maintain directional motility during chemotaxis.

2.2. Haptotaxis

A second mechanism of promoting directional single-cell migration is along an adhesion gradient; this is termed haptotaxis. On the basis of time lapse and other studies (Trinkaus, 1984), haptotaxis may be distinguished *in vitro* from chemotaxis. The directional information during migration by haptotaxis comes from the substratum as opposed to fluids surrounding the cell. Directional haptotactic motility does not result from the stimulation of protrusions at the leading edge of the cell, as occurs in chemotaxis. By contrast, cells migrating by haptotaxis extend lamellapodia more or less randomly, and each of these protruding lamellapodia competes for a finite amount of membrane, such that when one lamellapodium adheres, spreads, and becomes dominant, tension on the remainder of the cell inhibits further protrusive activity. Thus, the increasing adhesion gradients on the substratum appear to influence directional movement by favoring stabilization of a leading edge on one side of a motile cell. After the cell becomes asymmetric, with a trailing and leading edge, the cell advances and the trailing edge is pulled forward, forming retraction fibers, which are created by the residual adherence of mature focal adhesion plaques. As the cell lunges forward and breaks old adhesions, the excess membrane at the trailing edge becomes available for incorporation at the leading edge; hence motility continues. Chen (1981) observed that mechanically lifting the trailing edge of a migrating cell off the substratum accelerated the advance of the leading edge of the cell, a phenomenon that he termed retraction-induced spreading. Again, detachment of the trailing edge probably provides excess membrane for the leading lamellapodium, consistent with the observations of Harris (1973). There is also evidence that new membrane from the Golgi apparatus may be inserted selectively at the leading edge of the cell (Bergmann *et al.*, 1983).

The concept of haptotaxis for regulating directional cell movement was originally suggested by Carter (1967a, b). In these early studies, substrata of varying adhesiveness were used to manipulate directional cell migration *in vitro*. Briefly, hydrophilic palladium was evaporated in a gradient fashion across a hydrophobic cellulose acetate substratum. L cells plated onto such substrata accumulated in areas of greater adhesivity (i.e., hydrophilic palladium). Similar results could be demonstrated using haptotactic islands, created by shadowing palladium onto cellulose acetate strips protected by electron microscope grids (Carter, 1967a). As with gradient shadowing, cells localized preferentially to the areas of increasing adhesion (i.e., the squares of palladium). Harris (1973) extended these findings by demonstrating that the direction of adhesion-mediated migration was related to the relative hydrophilicity of substrates used. For example, chicken fibroblasts would localize on palladium squares when coated onto a nonwettable background, such as underivatized polystyrene; however, the same cells would localize on the highly hydrophilic polystyrene derivatized for tissue culture when it was used as a background for palladium shadowing. Cells can therefore discriminate between two substrata of differing hydrophilicity and segregate according to this hierarchy toward more hydrophilic substrata. Although such artificial patterns have proven useful in understanding general aspects of cell behavior in the context of haptotactic migration, applications of these principles *in vivo* will rely on understanding the way in which biologically relevant adhesive substrata can modulate haptotaxis. In this regard, specific patterns of the ECM proteins laminin and fibronectin have been used to direct patterns of neurite extension (Hammarback *et al.*, 1988), suggesting that knowledge of haptotaxis could prove useful to biomedical engineers in designing and controlling the regeneration of specific tissues.

2.3. Contact Guidance

Contact guidance, initially proposed by Weiss (1945, 1985) as a mechanism for directing cell movements, is probably closely related to haptotaxis. Briefly, contact guidance simply refers to the tendency of cells to align along discontinuities in substrata to which they are attached. As an example, cells migrating on scratched substrata *in vitro* tend to align with and move along the scratches. In addition, cells implanted at two discrete foci in a plasma clot soon begin to orient and migrate toward each other, a phenomenon that Weiss termed the two-center effect (Trinkaus, 1984). The mutually opposing contractile forces created by the cells exert tension on the clot matrix, reorganizing random fibrin strands into collinear fibrils along which the cells then migrate. Presumably, the reason for orientation in contact guidance is similar to that proposed for haptotaxis. In contact guidance, however, competition of lamellapodia for a finite amount of membrane is subject to the added influence of the three-dimensional orientation (as well as adhesive quality) of the substratum, accentuating orientation. This mechanism of orientation may in fact account for the highly oriented appearance of infiltrating fibroblasts as these cells enter the granulation tissue. The orientation of fibrils within resolving granulation tissue can also have dramatic effects on wound contraction, hence wound closure. The contractile forces of the myofibroblasts, which are enriched in more mature granulation tissue, can effectively operate along the lines of stress that contact guidance forces have set up. As the wound heals, lines of stress and patterns of orientation develop along axes parallel to the wound surface (Repesh *et al.*, 1982). Wound contraction along these stress lines efficiently closes the wound and can reduce the diameter of the original defect by 75% (Peacock and Van Winkle, 1976). Previous work has demonstrated a role for cell motility in organizing the architecture of connective tissue not only during wound healing but also in certain phases of development (Stopak and Harris, 1982; Stopak *et al.*, 1985).

3. Extracellular Matrices

The extracellular space in both normal and wounded tissue is occupied by ECM and plasma transudate. Certain components of the ECM can serve to promote the adhesion and migration of cells directly, while other constituents might actually impede cell adhesion. Many ECM components have been extensively characterized for the ability to support cell adhesion, motility, invasion, and growth *in vitro* (Akiyama *et al.*, 1990; Damsky and Werb, 1992; Faassen *et al.*, 1992a; Hynes, 1992; Yamada *et al.*, 1992; Juliano and Haskill, 1993; Sastry and Horwitz, 1993; Gailit and Clark, 1994). These studies suggest that complex ECMs, such as exist *in vivo*, contain informational arrays that are recognized by cells in very complex ways. Such molecular recognition events have localized effects on the plasma membrane of cells, effecting receptor distribution and activation of cytoskeletal reorganization. These events are also of fundamental importance in regulating gene transcription and translation. The importance of the ECM to wound healing dictates that a brief description of the better-understood constituents that have been studied with regard to cell movement within granulation tissue be mentioned.

Normal human skin is composed of a stratified squamous epithelium and an underlying connective tissue that constitute the epidermis and dermis, respectively. As with all surface linings, the epithelium is separated from the underlying issue by a basal lamina and reticular lamina, representing the dermal–epidermal junction. The dermis, which for the most part constitutes dense irregular connective tissue, is enriched in types I and III collagen, dermatan–chondroitin, heparan sulfate, and keratan sulfate proteoglycans (Pringle *et al.*, 1985). The proteoglycan core proteins present within dermis include versican, decorin, and biglycan (Lennon *et al.*, 1991; Willen *et al.*, 1991; Yeo *et al.*, 1991; Schonherr *et al.*, 1993; Yamagata *et al.*, 1993; Scholzen *et al.*, 1994; Zimmermann *et al.*, 1994). Fibronectin in the dermis is associated either with collagen fibers or it exists as independent fibrils within the ground substance.

When normal dermis is wounded, the composition of the ECM changes drastically (Grinnnell *et al.*, 1981; Repesh *et al.*, 1982; Gailit and Clark, 1994). As a result of the vascular response following injury, a transudate of plasma escapes into the extravascular space and coagulates within the lesion, forming a clot. The fibrin within the clot is decorated with fibronectin, vitronectin, thrombospondin, and other cell-adhesion-promoting proteins originating from plasma. As the wound matures and granulation tissue is formed, the composition of the ECM changes. Fibroblast infiltration into the region coincides with an increase in the appearance of reticular (type III) collagen fibers and hyaluronan. Hyaluronan is a glycosaminoglycan that is elevated in tissues that are undergoing extensive cellular infiltration and remodeling, such as granulation tissue, or stroma that are immediately adjacent to invading neoplasms (see Section 6). Tenascin, an ECM protein that can bind chondroitin sulfate proteoglycan and facilitate cell de-adhesion (Erickson and Bourdon, 1989), is also present within granulation tissue. Fibronectin is also a prominent component in granulation tissue, and experiments have demonstrated that this fibronectin is also largely derived from the infiltrating fibroblasts. Keratinocytes migrating in from the margins can synthesize fibronectin as well (Clark *et al.*, 1983, 1985). As the granulation tissue resolves, fibrillar (type I) collagen begins to predominate, and the wound undergoes remodeling. It is against this backdrop of a changing environment that the complex regulation of directional cell movement must occur into wounds.

4. Cytoskeletal Rearrangement and Intracellular Signals Are Important for Cell Motility

The ability of the cell to modulate the polymerized state of actin at sites of lamellapodial extension is a fundamental consideration in cell motility. It is beyond the scope of this chapter to treat this topic in detail; however, the reader is referred to several excellent reviews (Devrotes and Zigmond, 1988; Stossel, 1989, 1990; Condeelis *et al.*, 1992; Divecha and Irvine, 1995). In most cells, actin constitutes almost 10% of the total protein of the cell, with approximately half of this actin existing in the F (polymerized) state. All else being equal, it has been calculated that the amount of G-actin within cells is sufficient as to cause a spontaneous polymerization, such that almost all of the actin within cells would exist as F-actin. However, various actin-binding proteins, which interact either with sites on growing F-actin polymers or with individual G-actin monomers, act to regulate the amount of F- and G-actin at any given time or location within cells (Stossel, 1989, 1990). These proteins act to allow the cells to regulate the state of actin polymerization, and among other things are critical for the control of cell motility. Proof for this comes in part from several experiments in which actin-binding proteins have been genetically manipulated (Cunningham *et al.*, 1991) or manipulated by the use of specific toxins (Cooper, 1991), with a concomitant effect on cell motility.

In order for cells to initiate motility, there must be mechanisms in place to allow the cell to extend the boundaries of its plasma membrane beyond previously established locations. One mechanism by which such a process can occur is by regulating the local osmotic pressure at specific sites within the cytoplasm (Stossel, 1990). According to this model, agonists that stimulate cell motility would be predicted to cause localized depolymerization of actin, creating a localized increase in osmotic pressure. This increase in osmotic pressure could cause the formation of a pseudopod, which is in turn stabilized by the repolymerization of G-actin into F-actin. Such pulses of actin polymerization–depolymerization have indeed been observed in neutrophils that are responding to chemotactic stimuli such as f-Met-Leu-Phe (Harvath, 1990). There are many cytoskeletal proteins that can interact with actin and modulate cell motility, including proteins such as filamin, vinculin, paxillin, and talin (Stossel, 1989; Cooper, 1991; Kellie *et al.*, 1991). Furthermore, there are proteins that bind to G-actin [profilin (Stossel, 1990)] and F-actin [gelsolin (Cunningham *et al.*, 1991)] that collectively act to depolymerize actin. Gelsolin will break preformed actin networks (Stossel *et al.*, 1985), whereas profilin, by binding to G-actin, acts to inhibit the localized addition of actin onto available ends of growing actin filaments (Stossel, 1989). The regulation of these two actin-binding proteins, therefore, can have fundamental consequences on cell motility. Indeed, cells that overexpress gelsolin exhibit increased rates of cell migration (Cunningham *et al.*, 1991). Clearly, there are additional signaling pathways that regulate cytoskeletal organization, membrane ruffling, and pseudopodal formation, such as the rho/rac pathway (Takaishi *et al.*, 1994; Ridley *et al.*, 1995). A brief focus on profilin and gelsolin, however, can offer one example of the way multiple-signaling pathways can impact on actin polymerization and cell motility.

Products of phosphatidylinositol metabolism, specifically PtdIns $(4,5)P_2$, apparently play an important role in modulating the activity of actin-binding proteins such as gelsolin and profilin (McCarthy and Turley, 1993; Divecha and Irvine, 1995). In addition to being an important component of plasma membranes, this phosphoinositide also had been shown previously to bind both gelsolin and profilin (Lassing and Lindberg, 1985; Jamney and Stossel, 1987). When bound to profilin, PtdIns $(4,5)P_2$ inhibits the ability of the profilin to bind to actin monomers, thus favoring the formation of actin filaments. Enzymatic hydrolysis of PtdIns $(4,5)P_2$ bound to profilin occurs via the action of phospholipase C (Goldschmidt-Clermont *et al.*, 1991). This process is normally inhibited, unless the phospholipase C is first tyrosine-phosphorylated (Goldschmidt-Clermont *et al.*, 1991). Thus, tyrosine phosphorylation of phospholipase C could be an important first step in releasing PtdIns $(4,5)P_2$ from profilin, allowing profilin to interact with actin monomers and prevent F-actin formation. Furthermore, the localized production of Ins $(1,4,5)P_3$ and diacylglycerol could have important effects on Ca^{2+} release and the activation of protein kinase C, respectively (Divecha and Irvine, 1995). These events can also regulate the activity of gelsolin, further contributing to a localized breakdown of filamentous actin, creating a transient increase in the osmotic pressure in the region of actin depolymerization (McCarthy and Turley, 1993; Divecha and Irvine, 1995). The cycle could be completed when PtdIns $(4,5)P_2$ is locally regenerated, perhaps as a result of the translocation of PtdIns(4)P 5-kinase to the cytoskeleton (Payrastre et al., 1991). This regeneration of PtdIns $(4,5)P_2$ would then favor the reestablishment of a filamentous actin network, contributing to the stability of the newly forming protrusion.

Such a model, although far from complete, is useful to partially explain how the mechanism by agonists that stimulate tyrosine phosphorylation can modulate actin polymerization, pseudopod formation, and cell motility. From this type of model, it should also be clear that factors that contribute to stabilizing filamentous actin are also important for cell motility. The ability of specific adhesion receptors to cause the nucleation of actin monomers into F-actin undoubtedly plays an important role in stabilizing cell adhesion and in the formation of signaling complexes within the context of the cytoskeleton (see Section 5 and Chapter 9, this volume). Furthermore, disruption of the gene for filamin, an actin-binding protein that cross-links F-actin into orthogonal networks, also inhibits cell locomotion (Cunningham et al., 1991). Thus, it is important that a cell not only be able to disrupt the cytoskeleton, but also that it be able to selectively stabilize its polymerized state, in order to effectively control motility.

5. Integrins Are the Major Adhesion Receptor for ECM

It is clear that cell adhesion is intimately associated with the process of cell motility. As a result, understanding the molecular basis of cell adhesion can contribute significantly to our understanding of some of the mechanism(s) involved in cell motility. The primary adhesion receptors for extracellular matrices are members of the integrin group of cell surface proteins. These heterodimeric receptors are the best understood of the cell surface molecules that participate in recognition of the ECM. These receptors recognize the ECM and transmit signals into the interior of the cell that cause cytoskeletal reorganization, changes in cell shape, and changes in the transcription, translation, and secretion of proteins (Akiyama et al., 1990; Damsky and Werb, 1992; Faassen et al., 1992; Ginsberg et al., 1992; Hynes, 1992; Yamada et al., 1992; Juliano and Haskill, 1993; Sastry and Horwitz, 1993; Gailit and Clark, 1994). As might be expected, the signals transmitted into cells (so-called outside-in signals) as a result of integrin function are complex and involve tyrosine phosphorylation, changes in calcium influx, and activation of phospholipid metabolism, to name a few. Importantly, integrin function can also be influenced by signals generated inside the cell (so-called inside-out signals), and thus cytokines or accessory recognition molecules that transmit their own signals to the cell interior (Woods and Couchman, 1992; Iida et al., 1994) can influence the activity of integrins. While a detailed treatment of integrin structure function can be found elsewhere (see reviews cited above and Chapter 9, this volume), several aspects of integrin function will be highlighted here.

There is ample evidence to suggest that the cytoplasmic tails of both α and β subunits in a particular integrin heterodimer can influence inside-out or outside-in signaling events characteristic of integrins. Although the cytoplasmic tails of β integrins can directly enter into focal contacts (Sastry and Horwitz, 1993; Chapter 9, this volume), integrins containing the same α subunit but different β subunits will localize to different sites on the plasma membrane. For example, while αvβ3 integrins can be localized to focal adhesions, αvβ5 integrins are predominantly excluded from such structures (Wayner et al., 1991). While a direct role for interacting with cytoskeletal components has not been demonstrated for α integrin cytoplasmic domains, it has been hypothesized that the α integrin subunit tail may regulate the binding activities of the β integrin tail, and in this way modulate the activity of the integrin heterodimer. Furthermore, there is evidence that the α integrin subunit tail can modulate the binding affinity of integrins for specific ligands (Kassner and Hemler, 1993; O'Toole et al., 1994) by an inside-out mechanism. The cytoplasmic tails of β1 integrins have been shown to bind numerous cytoplasmic proteins, including talin, α-actinin, and pp125[FAK] (reviewed in Chapter 9, this volume; Schaller and Parsons, 1994). While direct linkages between the α integrin cytoplasmic tails and cytoskeletal components have not been shown, there have been recent studies to suggest that the α integrin subunit may bind calreticulin, an endoplasmic reticulum protein that binds calcium and has been implicated in transcriptional control via binding to retinoic acid and androgen receptors (Dedhar et al., 1994). This binding is thought to be mediated by the KXGFFKR sequence contained within juxtamembrane regions of α integrin subunit cytoplasmic tails.

The discrete distribution of integrins within adherent cells argues that each heterodimer can participate in discrete signaling events within the cell. One outside-in signaling event that has received much attention recently is the ability of integrins to stimulate the tyrosine phosphorylation of pp125[FAK], a cytoplasmic tyrosine kinase that localizes to focal adhesions and regulates discrete signaling pathways within the cell, such as the activation of ras and MAP kinase (Schlaepfer et al., 1994). While ligand-induced or antibody-induced clustering of many integrins will stimulate the tyrosine phosphorylation of pp125[FAK], the mechanisms by which this occurs as well as the significance of this event are not yet understood. Importantly, while clustering of β integrin subunits has been implicated in the tyrosine phosphorylation of pp125[FAK], certain alternatively spliced β integrin tails (e.g., alternatively spliced β3 integrin) do not stimulate tyrosine phosphorylation of pp125[FAK], again emphasizing the importance of the cytoplasmic tails in transmitting signals (Akiyama et al., 1994). While tyrosine phosphorylation of pp125[FAK] has not been directly linked to cell motility, it may be representative of additional intracellular tyrosine phosphorylation events that are linked to outside-in integrin signaling. Such integrin-mediated events could modulate specific cytoskeletal rearrangements important for pseudopod formation or stabilization.

Additionally, ligating certain integrins can stimulate the flux of divalent calcium ions from the exterior to the interior of the cell. In particular, ligating the αv subunit of integrins is implicated in activating calcium flux, possibly as a result of the cooperative action of integrin-associated protein (Lindberg et al., 1993; Schwartz, 1993). This integrin subunit has been associated with the motility of normal and transformed cells, and it is possible that localized calcium influxes that occur as a result of the cross-linking of αv integrin may be partly responsible for the activity of heterodimers containing this subunit.

The mechanism by which integrins transmit signals has recently been shown to depend not only the engagement of the receptor by the ligand, but also by the ability of the ligand to cross-link the receptor in the plane of the plasma membrane. Partial engagement of α5β1 integrin with specific anti-integrin monoclonal antibodies stimulates the localization of pp125^{FAK} and the cytoskeletal protein tensin (Miyamoto *et al.*, 1995), whereas other antibodies can be used to stimulate the localized accumulation of multiple cytoskeletal components, including a-actinin, vinculin, talin, and others. Thus, it is important to consider both ligand occupancy of the receptor, as well as the clustered status of the integrin within the plasma membrane, when considering mechanisms by which integrins can modulate the polymerized state of the cytoskeleton. Such interactions are critical in order for cells to control their motile phenotype.

Although the mechanisms by which integrins modulate motility are far from understood, there are some reports that begin to shed light on certain structural properties of integrin subunits that are important in this process. For example, cells expressing a β1 integrin chimera containing a β5 integrin cytoplasmic tail were more motile than cells that expressed only the β1 integrin subunit, implicating the β integrin subunit tail in regulating some aspects of cell motility (Pasqualini and Hemler, 1994). Although a less compelling case has been made for the α subunit in terms of directly interacting with cytoskeletal components, a chimeric approach has been used to demonstrate the importance for the cytoplasmic tail of α4 integrin in modulating cell motility on α4β1 integrin-dependent ligands (Chan *et al.*, 1992).

While these studies are instructive with respect to the importance of integrins in modulating cell adhesion and motility, not all ligands that bind integrins will stimulate motility (Faassen *et al.*, 1992a). Specifically, while melanoma cells expressing α4β1 integrin will migrate on fragments of fibronectin that contain both proteoglycan binding and α4β1 integrin binding sites, the cells adhere, but do not migrate on several cell adhesion promoting synthetic peptides from within these fragments that bind either integrin or cell surface proteoglycan (Iida *et al.*, 1992, 1994). Removal of the cell surface proteoglycan from these cells using enzymatic and/or pharmacological approaches tends to inhibit cell motility and cell adhesion on the fragments, suggesting that cell motility on such ligands results from the contribution of cell surface proteoglycans and integrins (S. Meijne, J. B. McCarthy, J. Iida, L. T. Furcht, unpublished observations). While the exact explanation for these observations is not yet known, it appears that the cell surface proteoglycan core protein can signal α4β1 integrin by modifying intracellular signaling pathways (Iida *et al.*, 1995) as well as by affect the affinity state of the integrin by virtue of the influence of the glycosaminoglycan (S. Meijne, J. B. McCarthy, J. Iida, L. T. Furcht, unpublished observations). Furthermore, the binding of α4β1 integrin to this region of fibronectin involves multiple interaction sites (reviewed in Iida *et al.*, 1994), reminiscent of previous findings demonstrating that integrin interaction with ECM ligands can involve multiple sites on the ligand (termed synergy sites, as described in chapter 9, this volume). Such multisite interactions for integrins, as well as the ligand-(or cell-) induced coordinate engagement of integrins with other cell surface receptors for the ECM (such as cell surface proteoglycans) (Iida *et al.*, 1994; see also chapter 15 this volume; section 6) can modulate the signals transmitted to the cell. Understanding the complex factors that regulate molecular recognition of the ECM is essential if we are to understand how the ECM controls cell motility as well as other aspects of cellular phenotype.

6. ECM Receptors Specifically Associated with Cell Motility within Granulation Tissue

Despite the well-recognized importance of integrins with respect to cell adhesion and motility, there are additional receptors that have been described that have very specific influences on cell motility. These include cell surface lectins (Barondes, 1988) and glycosyl transferases (Shur, 1989), the latter of which have been implicated in mediating cell motility on laminin. Although these are clearly important in the general context of cell motility, space limitations prevent a detailed discussion of these molecules. For the purposes of the current discussion, we will focus on CD44 and RHAMM (receptor for hyaluronan mediated motility), which are two ECM receptors implicated in modulating the motility of normal and transformed cells. Both of these receptors can interact with the ECM component hyaluronan, which is enriched in granulation tissue and has been implicated in modulating the motility of cells into tissues. Hyaluronan (HA) appears to modulate cell invasion into tissues by its ability to interact with specific cell surface receptors such as CD44 and RHAMM, as well by its ability to become hydrated and thus cause the localized swelling/disruption of ECM (Laurent and Fraser, 1992; Turley, 1992; Lesley *et al.*, 1993a; McCarthy and Turley, 1993; Sherman *et al.*, 1994). CD44 can also interact with other ECM components, such as fibronectin and collagen, due to the fact that it can be expressed as a part-time cell surface proteoglycan, and thus interact with glycosaminoglycan-binding domains of these ECM components. Although the mechanism(s) by which these two molecules modulate the motility of normal and transformed cells is not understood, there is clear evidence indicating their importance in different model systems.

While CD44 and RHAMM both contain a common structural motif that binds HA (Yang *et al.*, 1994), the products of these two different genes are very distinct in primary structure. Both genes are alternatively spliced, suggesting that there is some transcriptional control over cell type-specific localization or function of the various isoforms of these two ECM receptors. Additionally, CD44 can also be rather extensively N- and O-glycosylated (Lokeshwar and Borguignon, 1991; Lesley *et al.*, 1993a; Iida *et al.*, 1994; Sherman *et al.*, 1994), and when expressed as chondroitin sulfate proteoglycan, CD44 has an important influence on the motility and invasion of tumor cells (Faassen *et al.*, 1992b, 1993).

CD44 promotes cell adhesion to HA-coated substrata and it promotes the motility of certain cell types on HA (Lesley and Hyman, 1992; Sherman *et al.*, 1994). As is the case with integrins (see Chapter 9, this volume), the expression of CD44 on the surface of cells does not always correlate with the functional capacity of the CD44 to bind HA. In some cases, the ability of CD44 to bind HA can be up-regulated by signal transduction activators such as phorbol esters (Liao *et al.*, 1993). The cytoplasmic tail of CD44 binds the cytoskeletal protein ankyrin (Lokeshwar *et al.*, 1994) and it is also a substrate for protein kinase C (Wayner and Carter, 1988). Although the cytoplasmic tail is not necessary for HA binding (Lesley and Hyman, 1992), it is apparently required to support cell adhesion to HA-coated substrata (Lokeshwar *et al.*, 1994).

Certain antibodies against CD44 have been identified that have the property of activating cell surface CD44 to bind HA, which is similar to what has been observed using activating antibodies to stimulate integrin function (see Chapter 9, this volume, and other reviews). However, in contrast to what has been observed with integrins, such CD44-activating antibodies are thought to act by clustering the CD44 on the surface of the plasma membrane (Lesley *et al.*, 1993b), whereas those that activate integrins are thought to do so by inducing a conformational change in the integrin (reviewed in Chapter 9, this volume). Most, if not all, alternatively spliced variants of CD44 can bind HA (Lesley *et al.*, 1993a; Sherman *et al.*, 1994). Interestingly, certain CD44 alternatively spliced isoforms, which arise by insertion of exons into a specific point in the ectodomain of the protein, appear to influence tumor progression (Günthert *et al.*, 1991; Rudy *et al.*, 1993), consistent with a role for CD44 in cellular invasion/motility. Whether such isoforms are up-regulated during cellular invasion of granulation tissue remains an open and potentially interesting question in the area of wound healing.

CD44 is also expressed as a cell surface proteoglycan, which can contain either heparan sulfate (HS) or chondroitin sulfate (CS) glycosaminoglycans (see Chapter 15, this volume). This was originally described in keratinocytes, where CD44/HS was identified as a type III collagen receptor, and in lymphocytes, where CD44/CS has been shown to bind ECM components such as fibronectin (Jalkanen and Jalkanen, 1992). Interestingly, while the CD44/HS expressed by keratinocytes binds to type I collagen, it is not a primary adhesion receptor for type I collagen (Wayner and Carter, 1987). Recently, this form of CD44 has been shown to bind fibroblast growth factor, suggesting that it might somehow be involved in modulating the activity of this, and perhaps other, heparin-binding growth factors (Bennett *et al.*, 1995). This may represent an important mechanism by which CD44/HS can modulate the cyokine–growth-factor-dependent adhesion, migration, or proliferation of normal and transformed cells.

Other cells (e.g., melanoma cells) express CD44 as a chondroitin sulfate proteoglycan (Faassen *et al.*, 1992b, 1993). In the case of melanoma-associated CD44/CS, the expression and function of the molecule has been shown to be important in mediating the motility and invasion of cells within type I collagen three-dimensional gels (Faassen *et al.*, 1992b, 1993) or through reconstituted basement membranes (Knutson and McCarthy, submitted). Furthermore, isolated CD44/CS binds to type I collagen (Faassen *et al.*, 1992b) and type IV collagen (Knutson and McCarthy, submitted) affinity columns. This binding is at least in part due to the CS, since alkaline borohydride-released CS also binds to the collagen affinity columns, although at a slightly reduced affinity compared to the intact proteoglycan (Faassen *et al.*, 1992b). Furthermore, treatment of melanoma cells with transforming growth factor-beta (TGF-β) stimulates tumor invasion and motility and increases the production of CD44/CS in melanoma cells (Faassen *et al.*, 1993). As has been observed in the case of TGF-β-mediated increases in syndecan I expression (Rapraeger, 1989), the effect of TGF-β on CD44/CS is at the level of the added glycosaminoglycan chain. CS isolated from TGF-β-treated melanoma cells is 1.5 to 2 times longer than CS isolated from untreated controls, but the relative levels of CD44 core protein remain essentially constant in both cell populations (Faassen *et al.*, 1993). Inhibition of CS addition in either case inhibits melanoma cell motility and invasion in type I collagen gels, indicating that CD44/CS is required in either case. Again, it may prove instructive to evaluate CD44/CS proteoglycan expression by cells as they enter granulation tissue, since increases in CS proteoglycan expression have been noted as cells infiltrate granulation tissue. If a role for CD44/CS proteoglycan in cell migration into wounds could be established, then this might lead to strategies that could be used to control this process.

As was observed for the keratinocyte CD44/HS, the expression of CD44/CS by melanoma cells does not appear to influence adhesion to collagen, as least as assessed in relatively short-term adhesion assays (Faassen *et al.*, 1992b, 1993). Time-lapse analysis of melanoma cells migrating on the surface of and into collagen gels has been used as one approach for evaluating the mechanism by which CD44/CS might act to modulate cell motility. When CD44/CS expression is compromised, the cells still attempt to extend leading lamellapodia; however, the structures fail to stabilize and they retract back into the cell body. Taking these observations into account with other data demonstrating that CD44 can be expressed on microspikes extending from the leading lamellapodia of cells (Brown *et al.*, 1991), it would appear that CD44/CS may act, in part, at the leading edge of cells to stabilize newly forming protrusions (Faassen *et al.*, 1993). It is possible that the ability of CD44 to bind ankyrin may somehow impact on the ability of this receptor to facilitate F-actin generation and stabilize newly forming pseudopodia. In this way, CD44/CS proteoglycan, by binding to ECM components at the leading edge of migrating cells, may accentuate the asymmetry between the leading and trailing edges of cells as they migrate. It is also possible that CD44/CS may alter integrin function as has been previously shown in other experimental systems (Iida *et al.*, 1994).

RHAMM is another well studied receptor that interacts with HA (McCarthy and Turley, 1993; Sherman *et al.*, 1994). In contrast to CD44, RHAMM has not been shown to be expressed as a cell surface proteoglycan, and the only ECM ligand to date for RHAMM is HA. The RHAMM receptor was originally identified on H-ras-transfected cells, which exhibited elevated motility in conjunction with elevated HA production (Turley *et al.*, 1991). Blocking antibodies against HA-binding proteins from these cells inhibited motility and were used to identify and clone RHAMM (Hardwick *et al.*, 1992). The protein is alternatively spliced and has an unusual primary structure in that it contains neither a leader sequence nor a putative transmembrane domain. Although a direct association of RHAMM with plasma membrane of cells has not been shown, it has been proposed that at least certain isoforms of RHAMM can interact with the cell surface via a docking protein (Turley, 1992). RHAMM is expressed on a number of cell types, including fibroblasts, macrophages, and lymphocytes, and it has been shown to regulate the motility of these cell types.

As has been observed for CD44/CS, the level of RHAMM increases on cell surfaces following TGF-β stimulation, and blocking this stimulation can inhibit the increased motility induced by TGF-β (Samuel *et al.*, 1993). While the mechanism by which RHAMM transmits signals is not well understood, several observations have been made that suggest that it has a complex mechanism of action. Recently it has been reported that stimulation of RHAMM by HA causes a transient increase in the tyrosine phosphorylation of several cytoplasmic proteins, including pp125^{FAK}, which is then followed by a decrease in the tyrosine phosphorylation pattern and a subsequent dissolution of focal contacts or adhesions as the cells begin to migrate (Hall *et al.,* 1994). This leads to the intriguing possibility that the dephosphorylation of pp125^{FAK} (and perhaps other signaling molecules as well) might be important for RHAMM stimulation of cell motility.

7. Concluding Remarks

Understanding the complex mechanisms by which cells can regulate their motility will have important consequences, not only for understanding the process of wound repair/closure, but also for understanding pathologies associated with cell motility, including chronic inflammation and tumor invasion and metastasis. Identification of the receptors involved and their associated signal transduction mechanism will contribute significantly toward understanding this complex process. A better understanding of the pathways associated with the activation of individual receptors, in conjunction with the interplay that must occur between different groups of receptors, will be required to understand the molecular basis of cell motility. Ultimately, the ability of these receptor-initiated pathways to modulate transcription, protein synthesis, and secretion will have to be fully understood in order to rationally design strategies and approaches to control some of the pathologies associated with cell motility.

References

Akiyama, S. K., Nagata, K., and Yamada, K., 1990, Cell surface receptors for extracellular matrix components, *Biochim Biophys Acta* **1031**:91–110.

Akiyama, S. K., Yamada, S. S., Yamada, K. M., and LaFlamme, S. E., 1994, Transmembrane signal transduction by integrin cytoplasmic domains expressed in single-subunit chimeras, *J. Biol. Chem.* **269**:15961–15964.

Barondes, S. H., 1988, Bifunctional properties of lectins: lectins redefined. *Trends in Biochem. Sci.* **13**:480–482.

Bennett, K. L., Jackson, D. G., Simon, J. C., Tanczos, E., Peach, R., Modrell, B., Stamenkovic, I., Plowman, G., and Aruffo, A., 1995, CD44 isoforms contain exon V3 are responsible for presentation of heparin-binding growth factor, *J. Cell Biol.* **128**:687–698.

Bergmann, J. E., Kupfer, A., and Singer, S. J., 1983, Membrane insertion at the leading edge of motile fibroblasts, *Proc. Natl. Acad. Sci. USA* **80**:1367–1371.

Brown, T. A., Bouchard, T., St. John, T., Wayner, E., and Carter, W. G., 1991, Human keratinocytes express a new CD44 core protein (CD44E) as a heparan-sulfate intrinsic membrane proteoglycan with additional exons, *J. Cell Biol.* **113**:207–221.

Carter, S. B., 1967, Haptotactic Islands. A method of confining single cells to study individual cell reactions and clone fermentation, *Exp. Cell Res.* **48**:188–193.

Carter, S. B., 1967, Haptotaxis and the mechanism of cell motility, *Nature* **213**:256–260.

Caterina, M. J., and Devreotes, P. N., 1991, Molecular insights into eukaryotic chemotaxis, *FASEB J.* **5**:3078–3085.

Chan, B. M., Kassner, P. D., Schiro, J. A., Byers, H. R., Kuppe, T. S., and Hemler, M. E., 1992, Distinct cellular functions mediated by different VLA integrin a subunit cytoplasmic domains, *Cell* **68:**1051–1060.

Chen, W.-T., 1981, Mechanism of retraction of the trailing edge during fibroblast movement, *J. Cell Biol.* **90:**187–200.

Clark, R. A. F., Winn, H. J., Dvorak, H. F., and Colvin, R. B., 1983, FIbronectin beneath reepithelializing epidermis in vivo: Sources and significance, *J. Invest. Dermatol.* **80(Suppl):**26s–30s.

Clark, R. A. F., Nielsen, L. D., Howell, S. E., and Folkvord, J. M., 1985, Human keratinocytes that have not terminally differentiated synthesize laminin and fibronectin by deposit on fibronectin in the pericellular matrix, *J. Cell. Biochem.* **28:**127–141.

Condeelis, J., Jones, J., and Segall, J. E., 1992, Chemotaxis of metastatic tumor cells: clues to mechanisms from dictyostelium paradigm, *Cancer Metastasis Rev.* **11:**55–68.

Cooper, J. A., 1991, The role of actin polymerization in cell motility, *Annual Review of Physiology* **53:**585–605.

Cunningham, C. C., Stossel, T. P., and Kwiatkowski, D. J., 1991, Enhanced motility of NIH 3T3 fibroblasts that overexpress gelsolin, *Science* **251:**1233–1236.

Damsky, C. H., and Werb, Z., 1992, Signal transduction by integrin receptors for extracellular matrix: coooperative processing of extracellular information, *Cur. Opin. Cell Biol.* **4:**772–781.

Davis, B. H., Walter, R. J., Pearson, C. B., Becker, E. L., and Oliver, J. M., 1982, Membrane activity and topography of f-MET-Leu-Phe-treated polymorphonuclear leukocytes, *Am. J. Pathol.* **108:**206–213.

Dedhar, S., Rennie, P. S., Shago, M., Hagesteijn, C. Y., Yang, H., Filmus, J., Hawley, R. G., Bruchovsky, N., Cheng, H., and Matusik, R. J., 1994, Inhibition of nuclear hormone receptor activity by calreticulin, *Nature* **367:**480–483.

Devrotes, P., and Zigmond, S., 1988, Chemotaxis in eukaryotic cells, *Annu. Rev. Cell Biol.* **4:**649–686.

Divecha, N., and Irvine, R. F., 1995, Phospholipid signaling, *Cell* **80:**269–278.

Erickson, C. A., 1990, Cell migration in the embryo and adult organism, *Curr. Opin. Cell Biol.* **2:**67–74.

Erickson, H. P., and Bourdon, M. A., 1989, Tenascin: An extracellular matrix protein prominent on specialized embryonic tissues and tumors, *Annu. Rev. Cell Biol.* **5:**71–92.

Faassen, A. E., Drake, S. L., Iida, J., Knutson, J. R., and McCarthy, J. B., 1992, Mechanisms of normal cell adhesion to the extracellular matrix and alterations associated with tumor invasion and metastasis, *Adv. Pathol. Lab. Med.* **5:**229–259.

Faassen, A. E., Schrager, J. A., Klein, D. J., Oegema, T. R., Couchman, J. R., and McCarthy, J. B., 1992, A cell surface chondroitin sulfate proteoglycan, immunologically related to CD44, is involved in type I collagen mediated melanoma cell motility and invasion, *J. Cell Biol.* **116:**521–531.

Faassen, A. E., Mooradian, D. L., Tranquillo, R. T., Dickinson, R. B., Letourneau, P. C., Oegema, T. R., and McCarthy, J. B., 1993, Cell surface CD-44 related chondroitin sulfate proteoglycan is required for transforming growth factor β-stimulated mouse melanoma cell motility and invasive behavior on type I collagen, *J. Cell Science* **105:**501–511.

Gailit, J., and Clark, R. A. F., 1994, Wound repair in the context of the extracellular matrix, *Curr. Opin. Cell Biol.* **6(5):**717–725.

Ginsberg, M. H., Du, X., and Plow, E. F., 1992, Inside-out integrin signaling, *Curr. Opin. Cell Biol.* **4:**766–771.

Goldschmidt-Clermont, P. J., Kim, J. W., Machesky, L. M., Rhee, S.-G., and Pollard, T. D., 1991, Regulation of phospholipase C-g1 by profilin and tyrosine phosphorylation, *Science* **251:**1231–1233.

Grinnnell, F., Billingham, R. E., and Burgess, L., 1981, Distribution of fibronectin during wound healing in vivo, *J. Invest. Dermatol.* **76:**181–189.

Günthert, U., Hofmann, M., Rudy, W., Reber, S., Zöller, M., Haußmann, I., Matzku, S., Wenzel, A., Ponta, H., and Herrlich, P., 1991, A new variant of glycoprotein CD44 confers metastatic potential to rat carcinoma cells, *Cell* **65:**13–24.

Hall, C. L., Wang, C., Lange, L. A., and Turley, E. A., 1994, Hyaluronan and the hyaluronan receptor RHAMM promote focal adhesion turnover and transient tyrosine kinase activity, *J. Cell Biol.* **126:**575–588.

Hammarback, J. A., McCarthy, J. B., Palm, S. L., Furcht, L. T., and Letourneau, P. C., 1988, Growth cone guidance by substrate-bound laminin pathways is correlated with neuron-to-pathway adhesivity, *Dev. Biol.* **126(1):**29–39.

Hardwick, C., Hoare, K., Owens, R., Hohn, H. P., Höök, M., Moore, M., Cripps, V., Austen, L., Nance, D. M., and Turley, E. A., 1992, Molecular cloning of a novel hyaluronan receptor that mediated tumor cell motility, *J. Cell Biol.* **117:**1343–1350.

Harris, A. K., 1973, The behavior of cultured cells on substrata of various adhesiveness, *Exp. Cell Res.* **77:**285–297.

Harvath, L., 1990, Regulation of neutrophil chemotaxis: correlations with actin polymerization, *Cancer Invest.* **8:**651–654.

Hynes, R. O., 1992, Integrins: versatility, modulation, and signaling in cell adhesion, *Cell* **69:**11–25.

Iida, J., Milius, R. P., Oegema, T. R., Furcht, L. T., and McCarthy, J. B., 1994, Role of cell surface proteoglycans in tumor cell recognition of fibronectin, *Trends Glycosc. Glycotechnol.* **6:**1–16.

Jalkanen, S., and Jalkanen, M., 1992, Lymphocyte CD44 binds the COOH-terminal heparin binding domain of fibronectin, *J. Cell Biol.* **116:**817–825.

Jamney, P. A., and Stossel, T. P., 1987, Modulation of gelsolin function by phosphatidylinositol 4,5-bisphosphate, *Nature* **325:**362–364.

Juliano, R. L., and Haskill, S., 1993, Signal transduction from the extracellular matrix, *J. Cell Biol.* **120:**577–585.

Kassner, P. D., and Hemler, M. E., 1993, Interchangeable a chain cytoplasmic domains play a positive role in control of cell adhesion mediated by VLA-4, a $\beta 1$ integrin, *J. Exp. Med.* **178:**649–660.

Kellie, S., Horvath, A. R., and Elmore, M. A., 1991, Cytoskeletal targets for oncogenic tyrosine kinases, *J. Cell Sci.* **99:**207–211.

Knutson, J. R., Iida, J., Fields, G. B., and McCarthy, J. B., 1995, CD44/Chondroitin sulfate proteoglycan and $\alpha 2 \beta 1$ integrin mediate human melanoma cell migration on type IV collagen and invasion of basement membranes, *Mol. Biol. Cell,* in press.

Lassing, I., and Lindberg, V., 1985, Specific interaction between phosphatidylinositol 4,5-bisphosphate and profilin, *Nature* **314:**472–474.

Laurent, T. C., and Fraser, J. R., 1992, Hyaluronan, *FASEB J.* **6:**2397–2404.

Lennon, D. P., Carrino, D. A., Baber, M. A., and Caplan, A. I., 1991, Generation of a monoclonal antibody against avian small dermatan sulfate proteoglycan: immunolocalization and tissue distribution of PG-II (decorin) in embryonic tissues, *Matrix* **11**(6):412–427.

Lesley, J., and Hyman, R., 1992, CD44 can be activated to function as an hyaluronic acid receptor in normal murine T cells, *J. Immunol.* **22:**2719–2723.

Lesley, J., Hyman, R., and Kincade, P. W., 1993a, CD44 and its interaction with the cellular matrix, *Adv. Immunol.* **54:**271–335.

Lesley, J., Kincade, P. W., and Hyman, R., 1993b, Antibody-induced activation of the hyaluronan receptor function in CD44 requires multivalent binding by antibody, *Eur. J. Immunol.* **23:**1902–1909.

Liao, H. X., Levesque, M. C., Patton, K., Bergamo, B., Jones, D., Moody, M. A., Telen, M. J., and Haynes, B. F., 1993, Regulation of CD44H and CD44E isoform binding both hyaluronan by phorbol myristate acetate and anti-CD44 monoclonal and polyclonal antibodies, *J. Immunol.* **151:**6490–6499.

Lindberg, F. P., Gresham, H. D., Schwartz, E., and Brown, E. J., 1993, Molecular cloning of integrin-associated protein: an immunoglobulin family member with multiple membrane spanning domains implicated in av$\beta 3$ dependent ligand binding, *J. Cell Biol.* **123:**485–496.

Lokeshwar, V. B., and Borguignon, L. Y. W., 1991, Post-translational protein modification and expression of ankyrin-binding sites(s) in GP85 (Pgp-1/CD44) and its biosynthetic precursors during T-lymphoma membrane biosynthesis, *J. Biol. Chem.* **266:**17983–17989.

Lokeshwar, V. B., Fregien, N., and Bourguignon, L. Y. W., 1994, Ankyrin-binding domain of CD44 (GP85) is required for the expression of hyaluronic acid-mediated adhesion function, *J. Cell Biol.* **126**(4):1099–1109.

McCarthy, J., and Turley, E. A., 1993, Effects of extracellular matrix components on cell locomotion, *Crit. Rev. Oral Biol. Med.* **4:**619–637.

Miyamoto, S., Akiyama, S. K., and Yamada, K. M., 1995, Synergistic roles for receptor occupancy and aggregation in integrin transmembrane function, *Science* **267:**883–885.

Nelson, R. D., McCormack, R. T., and Fiegel, V. D., 1978, Chemotaxis of human leukocytes under agarose, in: *Leukocyte Chemotaxis* (J. I. Gallin and P. C. Quie, eds.), pp. 25–42, Raven Press, New York.

O'Toole, T. E., Katagiri, Y., Faull, R. J., Peter, K., Quranta, V., Loftus, J. C., Shattil, S. J., and Ginsberg, M. H., 1994, Integrin cytoplasmic domains mediate inside-out signal transduction, *J. Cell Biol.* **124:**1047–1059.

Pasqualini, R., and Hemler, M. E., 1994, Contrasting roles for integrin β1 and β5 cytoplasmic domains in subcellular localization, proliferation and cell migration, *J. Cell Biol.* **125:**447–460.

Payrastre, B., Van Bergen en Henegouwen, P. M. P., Breton, M., den Hartigh, J. C., Plantavid, M., Verkleij, A. J., and Boonstra, J., 1991, Phosphoinositide kinase diacylglycerol kinase and phospholipase C activities associated to the cytoskeleton: effect of epidermal growth factor, *J. Cell Biol.* **115:**121–128.

Peacock, E. E., and Van Winkle, W., 1976, *Wound Repair.* W. B. Saunders, Philadelphia.

Pringle, G. A., Dodd, C. M., Osborn, J. W., Pearson, C. H., and Mosmann, T. R., 1985, Production and characterization of monoclonal antibodies to bovine proteodermatan sulfate, *Collagen Related Res.* **5:**23–29.

Rapraeger, A., 1989, Transforming growth factor (type β) promotes the addition of chondroitin sulfate chains to the cell surface proteoglycan (syndecan) of mouse mammary epithelia, *J. Cell Biol.* **109:**2509–2518.

Repesh, L. A., Fitzgerald, T. J., and Furcht, L. T., 1982, Fibronectin involvement in granulation tissue and wound healing in rabbits, *J. Histochem. Cytochem.* **30:**351–358.

Ridley, A. J., Comoglio, P. M., and Hall, A., 1995, Regulation of scatter factor/hepatocyte growth factor responses by Ras, Rac, and Rho in MDCK cells, *Mol. Cell. Biol.* **15**(2):1110–1122.

Rudy, W., Hoffman, M., Schwartz-Albiez, R., Zöller, M., Heider, K.-H., Ponta, H., and Herrlich, P., 1993, The two major CD44 proteins expressed on a metastatic rat tumor line are derived from different splice variants: each one individually suffices to confer metastatic behavior, *Cancer Res.* **53:**1262–1268.

Samuel, S. K., Hurta, R. A., Spearman, M. A., Wright, J. A., Turley, E. A., and Greenberg, A. H., 1993, TGF-Beta 1 stimulation of cell locomotion utilizes the hyaluronan receptor RHAMM and hyaluronan, *J. Cell Biol.* **123:**749–758.

Sastry, S. K., and Horwitz, A. F., 1993, Integrin cytoplasmic domains: mediators of cytoskeletal linkages and extra- and intracellular initiated transmembrane signaling, *Curr. Opin. Cell Biol.* **5:**819–831.

Schaller, M. D., and Parsons, J. T., 1994, Focal Adhesion Kinase and Associated Proteins, *Curr. Opin. Cell Biol.* **6**(5):705–710.

Schlaepfer, D. D., Hanks, S. K., Hunter, T., and van der Geer, P., 1994, Integrin-mediated signal transduction linked to Ras pathway by GRB2 binding to focal adhesion kinase, *Nature* **372:**786–791.

Scholzen, T., Solursh, M., Suzuki, S., Reiter, R., Morgan, J. L., Buchberg, A. M., Siracusa, L. D., and Iozzo, R. V., 1994, The murine decorin. Complete cDNA cloning, genomic organization, chromosomal assignment, and expression during organogenesis and tissue differentiation, *J. Biol. Chem.* **269**(45):28270–28281.

Schonherr, E., Beavan, L. A., Hausser, H., Kresse, H., and Culp, L. A., 1993, Differences in decorin expression by papillary and reticular fibroblasts in vivo and in vitro, *Biochem. J.* **290**(Pt 3):893–899.

Schwartz, M. A., 1993, Spreading of human endothelial cells on fibronectin or vitronectin triggers elevation of intracellular free calcium, *J. Cell Biol.* **120:**1003–1010.

Sherman, L., Sleeman, J., Herrlich, P., and Ponta, H., 1994, Hyaluronate receptors: key players in growth, differentiation, migration and tumor progression, *Curr. Opin. Cell Biol.* **6**(5):726–733.

Shur, B. D., 1989, Glycoconjugates as mediators of cellular interactions during development, *Curr. Opin. Cell Biol.* **1:**905–912.

Stopak, D., and Harris, A. K., 1982, Connective tissue morphogenisis by fibroblast traction, *Dev. Biol.* **90:**383–392.

Stopak, D., Wessells, N. K., and Harris, A. K., 1985, Morphogenetic rerrangement of injected collagen in developing chicken limb buds, *Proc. Natl. Acad. Sci. USA* **82:**2804–2808.

Stossel, T. P., 1989, From signal to pseudopod formation. How cells control cytoplasmic actin assembly, *J. Biol. Chem.* **264:**18261–18264.

Stossel, T. P., 1990, How cells crawl: with the discovery that the cellular motor contains muscle proteins, we can begin to describe cell motility in molecular detail, *Sci. Amer.* **78:**408–423.

Stossel, T. P., Chaponnier, C., Ezzell, R. M., Hartwig, J. H., Jamney, P. A., Kwiartkowski, D. J., and Lind, S. E., 1985, Non muscle actin binding proteins, *Ann. Rev. Cell Biol.* **1:**353–402.

Takaishi, K., Sasaki, T., Kato, M., Yamochi, W., Kuroda, S., Nakamura, T., Takeichi, M., and Takai, Y., 1994,

Involvement of Rho p21 small GTP-binding protein and its regulator in the HGF-induced cell motility, *Oncogene* **9**(1):273–279.

Trinkaus, J. P., 1976, On the mechanism of metazoan cell movements, in: *The Cell Surface in Animal Embryogenisis and Development* (G. Poste and G. J. Nicolson, eds.), pp. 225–329, North Holland, Amsterdam.

Trinkaus, J. P., 1984, *Cell into Organs. The Forces That Shape the Embryo.* Prentice-Hall, Englewood Cliffs, N.J.

Turley, E. A., 1992, Hyaluronan and cell locomotion, *Cancer Metastasis Rev.* **11**:21–30.

Turley, E. A., Austen, L., Vandeligt, K., and Clary, C., 1991, Hyaluronan and a cell-associated binding protein regulate the locomotion of ras-transformed cells, *J. Cell Biol.* **112**:1041–1047.

Wayner, E. A., and Carter, W. G., 1987, Identification of multiple cell adhesion receptors for collagen and fibronectin in human fibrosarcoma cells possessing unique α and common β subunits, *J. Cell Biol.* **105**:1873–1884.

Wayner, E. A., and Carter, W. G., 1988, Characterization of the class III collagen receptor, a phosphorylated, transmembrane glycoprotein expressed in nucleated human cells, *J. Biol. Chem.* **263**:4193–4201.

Wayner, E. A., Orlando, R. A., and Cheresh, D. A., 1991, Integrins avβ3 and avβ5 contribute to cell attachment to vitronectin but differentially distribute on the cell surface, *J. Cell Biol.* **113**:919–929.

Weiss, P., 1945, The problem of specificity in growth and development, *Yale J. Biol. Med.* **19**:239–278.

Weiss, P., 1985, Cell contact, *Int. Rev. Cytol.* **7**:391–423.

Willen, M. D., Sorrell, J. M., Lekan, C. C., Davis, B. R., and Caplan, A. I., 1991, Patterns of glycosaminogly-can/proteoglycan immunostaining in human skin during aging, *J. Invest. Dermatol.* **96**(6):968–974.

Woods, A., and Couchman, J. R., 1992, Protein kinase C involvement in focal adhesion formation, *J. Cell Sci.* **101**:277–290.

Yamada, K. M., Aota, S., Akiyama, S. K., and LaFlamme, S. E., 1992, Mechanisms of fibronectin and integrin function during cell adhesion and migration, *Cold Spring Harbor Symp. Quant. Biol.* **57**:203–212.

Yamagata, M., Shinomura, T., and Kimata, K., 1993, Tissue variation of two large chondroitin sulfate proteoglycans (PG-M/versican and PG-H/aggrecan) in chick embryos, *Anat. Embryol.* **187**(5):433–444.

Yang, B., Yang, B. L., Savani, R. C., and Turley, E. A., 1994, Identification of a common hyaluronan binding motif in the hyaluronan binding proteins RHAMM, CD44 and Link Protein, *EMBO J.* **13**:286–296.

Yeo, T. K., Brown, L., and Dvorak, H. F., 1991, Alterations in proteoglycan synthesis common to healing wounds and tumors, *Am. J. Pathol.* **138**(6):1437–1450.

Zigmond, S. H., 1977, Ability of polymorphonuclear leukocytes to orient in gradients of chemotactic factors, *J. Cell Biol.* **75**:606–616.

Zigmond, S. H., 1989, Cell locomotion and chemotaxis, *Curr. Opin. Cell Biol.* **1**:80–86.

Zigmond, S. H., and Hirsch, J. G., 1973, Ability of polymorphonuclear leukocytes to orient in gradients of chemotactic factors, *J. Exp. Med.* **137**:387–410.

Zigmond, S. H., Levitsky, H. I., and Kreel, B. J., 1981, Cell polarity: An examination of its behavior expression and its consequences for polymorphonuclear leukocyte chemotaxis, *J. Cell Biol.* **89**:585–592.

Zimmermann, D. R., Dours-Zimmermann, M. T., Schubert, M., and Bruckner-Tuderman, L., 1994, Versican is expressed in the proliferating zone in the epidermis and in association with the elastic network of the dermis, *J. Cell Biol.* **124**(5):817–825.

Chapter 13

The Role of the Myofibroblast in Wound Healing and Fibrocontractive Diseases

ALEXIS DESMOULIÈRE and GIULIO GABBIANI

1. Historical Remarks and Definition of the Myofibroblast

Following tissue injury, tissue repair takes place in well-characterized steps. After clot formation, inflammatory cells, essentially mononuclear cells and granulocytes, invade the injured tissue; then, fibroblasts migrate, proliferate, and synthesize extracellular matrix components, participating in the formation of granulation tissue. Finally, following reepithelialization and wound closure, tissue remodeling implicates extracellular matrix degradation, decrease of cellularity, and constitution of the scar.

This sequence of histological changes has been well known for many years, but the factors regulating the modulation of various cellular activities still remain poorly known. Moreover, the process of wound contraction, which was defined in modern terms by Carrel in 1922, is not well understood. In 1971, a peculiar phenotype of granulation tissue fibroblastic cells was described (Gabbiani *et al.,* 1971). While normal fibroblasts contain a well-developed rough endoplasmic reticulum with dilated cisternae and an oval nucleus, these modified granulation tissue fibroblasts (Fig. 1) still show an abundant rough endoplasmic reticulum but in addition express bundles of microfilaments with dense bodies similar to those found in smooth muscle (SM) cells; this feature suggested that these cells, called myofibroblasts, are responsible for the production of the force determining wound contraction. Furthermore, the nucleus of myofibroblasts shows indentations, an ultrastructural feature that has been correlated with cellular contraction in several systems (Franke and Schinko, 1969; Majno *et al.,* 1969). Myofibroblasts are interconnected by gap junctions and are connected to the extracellular matrix by structures called fibronexus (plural fibronexi), which are transmembrane complexes of intracellular microfilaments in apparent continuity with extra-

ALEXIS DESMOULIÈRE • Department of Pathology, University of Geneva, 1211 Geneva 4, Switzerland, and CNRS-URA 1459, Institut Pasteur de Lyon, 69365 Lyon cedex 7, France.   GIULIO GABBIANI • Department of Pathology, University of Geneva, 1211 Geneva 4, Switzerland.

The Molecular and Cellular Biology of Wound Repair (Second Edition), edited by Richard A. F. Clark. Plenum Press, New York, 1996.

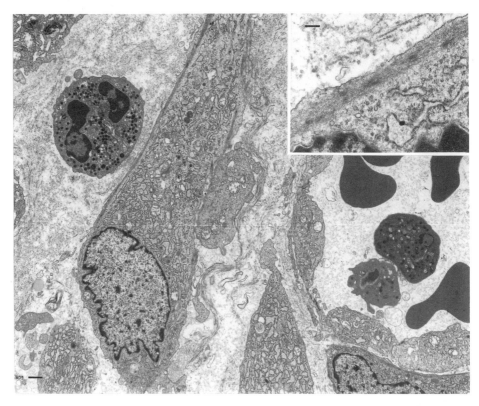

Figure 1. Transmission electron micrograph showing the classical organization of a granulation tissue 12 days after experimental wounding in rats. A small vessel is surrounded by some inflammatory cells (a granulocyte is visible), several typical myofibroblasts with microfilament bundles (one of which is illustrated in the inset), and abundant extracellular matrix. Bar = 1 μm; inset = 0.25 μm.

cellular fibronectin fibres (Singer, 1979; Singer *et al.*, 1984). In this chapter, we shall discuss the mechanisms of myofibroblast appearance and regression and the factors influencing their phenotype.

2. The Myofibroblast *in Vivo*

To characterize myofibroblast phenotypic features, cytoskeletal proteins are of particular interest since they display multiple variants, which are encoded by multigene families or are the result of differential mRNA splicing, that represent reliable differentiation markers. Furthermore, it is well accepted that cytoskeletal proteins play a key role during the process of cell contraction. Using cytoskeletal markers, the presence of four main myofibroblastic phenotypes has been described (for review, see Sappino *et al.*, 1990b; Schürch *et al.*, 1992). In addition to cytoplasmic actin isoforms, they coexpress: (1) vimentin (V-type), (2) vimentin and desmin (VD-type), (3) vimentin and

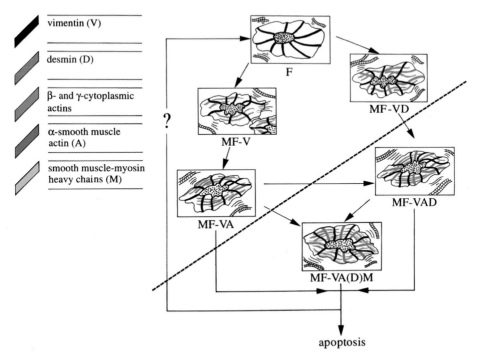

Figure 2. Schematic representation of myofibroblast evolution. The fibroblast (F) generally contains vimentin ß- and γ-cytoplasmic actin isoforms that are organized in a network and never in bundles. In several normal conditions (Table I) and during pathological situations (Table II), cytoplasmic actins organize into microfilament bundles; this modified fibroblast is now called myofibroblast (MF-V). MF-V may then acquire the expression of other cytoskeletal proteins, representing further steps toward smooth muscle differentiation. These proteins are: desmin (MF-VD), a-muscle actin (MF-VA), and smooth muscle myosin heavy chains (MF-VADM). Brackets in MF-VA(D)M indicate that desmin may be absent in myofibroblasts expressing smooth muscle myosin heavy chains. Myofibroblast phenotypes below the dotted line are observed only in pathological conditions.

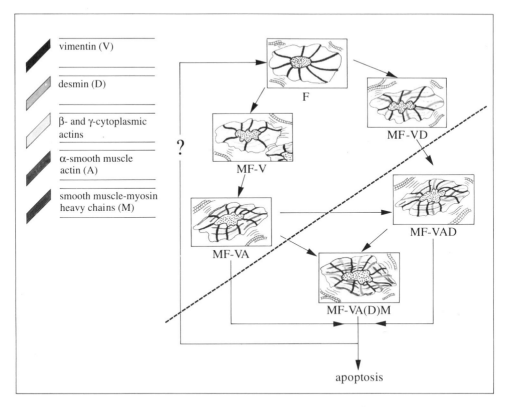

Figure 2. Schematic representation of myofibroblast evolution. The fibroblast (F) generally contains vimentin and β- and γ-cytoplasmic actin isoforms that are organized in a network and never in bundles. In several normal conditions (Table I) and during pathological situations (Table II), cytoplasmic actins organize into microfilament bundles; this modified fibroblast is now called myofibroblast (MF-V). MF-V may then acquire the expression of other cytoskeletal proteins, representing further steps toward smooth muscle differentiation. These proteins are: desmin (MF-VD), α-smooth muscle actin (MF-VA), and smooth muscle-myosin heavy chains (MF-VADM). Brackets in MF-VA(D)M indicate that desmin may be absent in myofibroblasts expressing smooth-muscle myosin heavy chains. Myofibroblast phenotypes below the dotted line are observed only in pathological situations. (Color version of this figure follows page 392.)

α-SM actin (VA-type), and (4) vimentin, desmin, and α-SM actin (VAD-type). α-SM actin is the actin isoform typical of contractile vascular SM cells (Gabbiani *et al.*, 1981) and is also expressed by practically all myofibroblastic populations *in vivo*.

Other new and important markers of myofibroblastic differentiation are the SM myosin heavy chains (MHC). Recent studies suggest that MHC isoforms are useful in precisely defining different well-differentiated myofibroblastic phenotypes (Buoro *et al.*, 1993; Chiavegato *et al.*, 1995). Figure 2 summarizes the different myofibroblastic phenotypes that can be characterized *in vivo* and *in vitro*. It is noteworthy that the spectrum of myofibroblastic differentiation includes cells with phenotypic features similar to those of classical fibroblasts and of classical SM cells, suggesting that, at least under certain conditions, fibroblasts can evolve into bona fide SM cells.

2.1. Distribution

It has been suggested that different organs contain fibroblasts with specific features (for review, see Sappino *et al.*, 1990b; Shimizu and Yoshizato, 1992). Komuro (1990) has proposed classifying fibroblasts into subtypes depending on their main functions: (1) fibrogenesis, (2) tissue skeleton or barrier, (3) intercellular communication system, (4) gentle contractile machinery, (5) endocrine activity, and (6) vitamin A storing. Other specific functions or features such as growth factor or cytokine production (Aggarwal and Pocsik, 1992; Bennett and Schultz, 1993), interaction with the immune system (Phipps *et al.*, 1990), and determination of epithelial differentiation (Cunha *et al.*, 1991; Hayashi *et al.*, 1993) could be added to this list, indicating the complexity of fibroblast biology.

2.1.1. Normal Tissues

Typical myofibroblasts have been found in a variety of organs (for review, see Sappino *et al.*, 1990b; Schmitt-Gräff *et al.*, 1994) (Table I). Immunohistochemical studies have shown that they may express proteins typical of contractile cells such as desmin, SM-MHC, and α-SM actin, suggesting that these cells participate in visceral contraction and/or organ remodeling. This view is supported by the observation that generally myofibroblasts are present in organs in which the capacities of remodeling are important (Schmitt-Gräff *et al.*, 1994). We do not know whether the proportion of SM cell markers (i.e., α-SM actin, desmin, and SM-MHC) in different fibroblast populations reflects precise functional activities or whether specific properties related to the expression of different cytoskeletal proteins remains to be defined.

2.1.2. Wound Healing and Fibrotic Conditions

During wound repair, fibroblasts participate in the formation of granulation tissue and modulate into myofibroblasts (Gabbiani *et al.*, 1971; Darby *et al.*, 1990). Myofibroblasts are poorly developed in early granulation tissue and are most numerous in the phase of wound contraction. At wound contraction, they are well organized in the architecture of the tissue in the form of several almost continuous layers parallel to the tissue surface, whereas small vessels are disposed perpendicularly to the fibroblastic layers and the wound surface (MacSween and Whaley, 1992); myofibroblasts and small vessels progressively disappear in the scar (Darby *et al.*, 1990). It is conceivable that either the myofibroblastic phenotype reverts to a quiescent form when the wound is closed or myofibroblasts disappear selectively through apoptosis (Darby *et al.*, 1990; Clark, 1993) (Fig. 2; see also Section 8). This last possibility has recently been demonstrated (Desmoulière et al., 1995).

A heterogeneity of cytoskeletal protein expression has been observed in myofibroblasts during experimental and human wound healing as well as in fibrocontractive diseases. During wound healing, the V-type is present in early granulation tissue and is replaced by the VA-type during the period of active retraction (Darby *et al.*, 1990). In general, granulation tissue myofibroblasts are devoid of desmin and SM-

Table I. Fibroblastic Cells of Normal Organs Displaying Ultrastructural and/or Immunochemical Features of SM Differentiation

Localization	Stress fibers	Desmin	α-SM actin	SM-MHC	References
Uterine submucosa	−	+	−	−	Glasser and Julian (1986)
Reticular cells of lymph nodes and spleen	+	+	+	+	Toccanier-Pelte et al. (1987)
Intestinal pericryptal cells	+	+	+	Not known	Sappino et al. (1989)
Intestinal villous core	+	+	+	+	Kaye et al. (1968)
Testicular stoma	+	+	+	+	Skalli et al. (1986)
Theca externa of the ovary	+	Not known	+	Not known	Czernobilsky et al. (1989)
Periodontal ligament	+	Not known	Not known	Not known	Beertsen et al. (1974)
Adrenal gland capsule	+	Not known	Not known	Not known	Bressler (1973)
Hepatic perisinusoidal cells	−	+	−	−	Yokoi et al. (1984)
Lung septa	+	+	−	−	Kapanci et al. (1992)
Bone marrow stroma	+	Not known	+	Not known	Charbord et al. (1990)

MHC, while these proteins are expressed more permanently in hypertrophic scar, fibromatosis, and stroma reaction to tumors (Skalli *et al.*, 1989; Schürch *et al.*, 1992; Chiavegato *et al.*, 1995) (Table II).

Pathological settings in which myofibroblasts represent the main cellular component may be classified into three groups (Table II): response to injury and repair phenomena (hypertrophic scars and burn contracture, organ fibrosis), quasineoplastic proliferative conditions (fibromatosis), and stromal response to neoplasia (for review, see Schürch *et al.*, 1992; Schmitt-Gräff *et al.*, 1994).

Recently, we investigated collagen organization and the possible presence of α-SM-actin-expressing myofibroblasts in two types of excessive scarring: keloid and hypertrophic scar (Ehrlich *et al.*, 1994). Contrary to hypertrophic scars, keloids do not regress with time, are difficult to revise surgically, and do not provoke scar contractures. Keloids contained large, thick collagen fibers. In contrast, in nodular structures of hypertrophic scars, fine, randomly organized collagen fibers were present. Furthermore, only nodules of hypertrophic scars contained α-SM-actin-expressing myofibroblasts (Table II). The presence in hypertrophic scar myofibroblasts of α-SM actin may represent an important element in the pathogenesis of contraction. When placed in culture, fibroblasts from hypertrophic scars and keloids express similar amounts of α-SM actin, suggesting that microenvironmental factors influence *in vivo* the expression of this protein. These results illustrate the complex mechanisms by which growth factors and extracellular matrix components regulate cell activities and myofibroblastic differentiation (see Section 5).

2.2. Cellular Origin of the Myofibroblast

It is clear now that, as suggested by earlier observations (MacDonald, 1959; Grillo, 1963; Ross *et al.*, 1970), granulation tissue fibroblasts arise from quiescent connective tissue cells. Theoretically, however, myofibroblasts can derive from at least three different mesenchymal cell types: fibroblasts, pericytes, and SM cells. Shum and McFarlane (1988) have proposed on the basis of morphological observations that myofibroblasts derive from vascular SM cells. An intimate relationship between myofibroblasts and blood vessel wall has been described at the electron microscopic level, particularly during the initial steps of granulation tissue formation (Janssen, 1902; Larsen and Posch, 1958; Kisher *et al.*, 1982). It seems likely that in a majority of situations, myofibroblasts derive from preexisting fibroblasts, but there are cases in which they may derive from pericytes and/or SM cells (Grimaud and Borojevic, 1977).

The analysis of cytoskeletal proteins does not allow a perfect definition of the origin of a mesenchymal cell. Fibroblasts, SM cells, and pericytes can express α-SM actin, SM myosin, and desmin. In experimental granulation tissue, myofibroblasts derived from local fibroblasts temporarily acquire markers of SM differentiation, such as α-SM actin, which disappear when the wound is closed (Darby *et al.*, 1990). In this case, it seems clear that some local stimuli, probably distinct from those producing proliferation, induce SM differentiation markers in resident fibroblasts. The factors

Table II. Fibroblastic Cells Displaying Ultrastructural and/or Immunochemical Features of SM Differentiation in Pathological Tissues

Localization and/or situations	Stress fibers	Desmin	α-SM actin	SM-MHC	References
Normally healing granulation tissue	+	−	+[a]	−	Darby et al. (1990)
Response to injury and repair phenomena					
Hypertrophic scars	+	+[a]	+	+[b]	Baur et al. (1975)
Keloids	+	−	−	−	Ehrlich et al. (1994)
Asthma (bronchial mucosa)	+	Not known	+	Not known	Roche (1991)
Anterior capsular cataract	+	Not known	+	Not known	Schmitt-Gräff et al. (1990)
Liver cirrhosis	+	+[a]	+	+[a]	Rockey and Friedman (1992)
Lung fibrosis	+	+[a]	+	+[a]	Kapanci et al. (1990)
Kidney fibrosis	+	−	+	Not known	Johnson et al. (1991)
Fibromatosis					
Dupuytren's disease (nodule)	+	+[a]	+	+[b]	Skalli et al. (1989)
Scleroderma (dermis and esophageal submucosa)	+	+[a]	+	−	Sappino et al. (1990a)
Stroma reaction to tumors	+	+	+	+[a]	Schürch et al. (1992)

[a] Only in a proportion of cells.
[b] Only in a proportion of cases.

eliciting α-SM actin expression can be produced by neighboring epithelial or mesenchymal cells, underlying the role of mesenchymal–epithelial interactions in these phenomena. Whether the distinct heterogeneity in the cytoskeletal phenotype of myofibroblasts is attributable to differentiation from a common cell type or from different cell types remains uncertain. It is conceivable that a common ancestor cell gives rise to myofibroblasts, pericytes, and SM cells, which would then represent examples of cellular isoforms (Caplan *et al.,* 1983). Moreover, myofibroblasts during pathological situations may also originate from more specialized cells such as interstitial cells in the lung septa (Mitchell *et al.,* 1989; Kapanci *et al.,* 1990; Kuhn and McDonald, 1991; Vyalov *et al.,* 1993), mesangial cells in the glomerulus (Johnson *et al.,* 1991), and perisinusoidal cells in the liver (Ballardini *et al.,* 1988; Ramadori, 1991; Schmitt-Gräff *et al.,* 1991; Friedman, 1993).

Recent work (Pascolini *et al.,* 1992) has reported the presence of an α-SM actinlike protein in planaria (*Dugesia lugubris* s.l.). When a planaria was induced to regenerate by head amputation, immunostaining for the α-SM actinlike molecule labeled the basal portion of the epidermal cells, the undifferentiated mesenchymelike cells, and the myoblasts. These results show that α-SM actin can be acquired under pathological conditions by poorly differentiated cells that play a specific role during regeneration or tissue repair. In higher vertebrates, myofibroblasts could derive from fibroblasts that have conserved the properties of these undifferentiated mesenchymal cells (common ancestor cell) present in planaria.

3. Fetal Wound Healing

The presence of myofibroblasts in normal healing and in the pathological sequelae of healing, such as scarring and contracture, is well established. However, it is unclear what contributions this cell makes to tissue repair *in utero,* which appears to evolve differently compared to postnatal tissue repair. In many species, the fetus possesses the unique ability to heal skin wounds without scar formation (Burrington, 1971; Goss, 1977; Adzick *et al.,* 1985; Rowsell, 1986; Krummel *et al.,* 1987; Adzick and Longaker, 1992). Longaker *et al.* (1991) have previously shown that fetal excisional wounds in lambs at 100 days' gestation contract and contain myofibroblasts, based on morphological and immunohistochemical criteria. Recently, we have examined the development and ultrastructural features of myofibroblasts in fetal lamb wounds made at 75 through 120 days' gestation (term = 145 days) (Estes *et al.,* 1994). Our purpose was to investigate possible differences between a wound that heals without scar formation (e.g., at 75 days' gestation) and one that heals with scarring (e.g., at 100 and 120 days' gestation). α-SM actin was practically absent in wounds made at 75 days' gestation, but it was present in progressively greater amounts in wounds made at 100 and 120 days' gestation (Estes *et al.,* 1994). These experiments show for the first time that in the fetus the transition from scarless tissue repair to healing with scar formation coincides with the expression of α-SM actin by myofibroblasts, and thus supports the possibility that α-SM actin is not only a marker of granulation tissue contraction but participates actively in this phenomenon.

It is noteworthy that the lamellipodia generally observed in migrating epithelial cells of adult wounds are absent in embryonic wounds. In embryonic wounds, contraction of an actin cable present in basal cells at the free edge of the epidermis allows wound closure (Martin and Lewis, 1992). Recently, Martin *et al.* (1994) suggested that c-*fos* and transforming growth factor-β1 (TGF-β1), which are rapidly induced at the fetal wound margin, participate in the activation of mesenchymal contraction mechanisms.

4. Cultured Fibroblasts

When grown *in vitro,* fibroblasts derived from normal tissues acquire several phenotypic features of myofibroblasts such as a system of microfilaments, called stress fibers. Microfilaments are mainly composed of actin, as shown by immunofluorescence or immunoelectron microscopy with specific antibodies (Goldman *et al.,* 1975; Willingham *et al.,* 1981). Several studies have shown that stress fibers contain actin-associated proteins such as myosin, tropomyosin, α-actinin, and filamin (for review, see Skalli and Gabbiani, 1988). In addition to stress fibers, cultured fibroblasts develop gap junctions (Bellows *et al.,* 1981), similar to what was observed in myofibroblasts *in vivo* (Gabbiani *et al.,* 1978).

The presence of α-SM actin in primary and passaged fibroblastic populations has been reported by several laboratories (Vandekerckhove and Weber, 1981; Leavitt *et al.,* 1985; Skalli *et al.,* 1986); it has also been shown that the expression of this protein is decreased after viral transformation (Leavitt *et al.,* 1985; Okomoto-Inoue *et al.,* 1990). However, it has always been controversial whether these α-SM-actin-expressing cells derive from SM cells and/or pericytes present in the tissue from which cultures have been produced or whether they represent a true feature of fibroblastic cultures. We have observed that α-SM actin is always present in a variable proportion of cells in rat, mouse, and human fibroblastic cultures (Desmoulière *et al.,* 1992a). Furthermore, to evaluate if a subpopulation of bona fide fibroblasts has the potential to express α-SM actin, we have cloned and subcloned fibroblastic populations. Even after cloning and subcloning, a certain percentage of cells were positive for α-SM actin. It is noteworthy that α-SM actin can be expressed by a proportion of cells in a population cultured from a single α-SM-actin-positive or α-SM-actin-negative cell. We believe that α-SM actin expression in cultured fibroblastic populations is a feature of fibroblastic cultures themselves, which may be related to functions exerted by fibroblasts under particular environmental conditions *in vivo.* This assumption is corroborated by the finding that α-SM actin is expressed by fibroblasts cultured from organs where *in situ* such cells have been verified not to contain this protein. This has been described in lens cells (Schmitt-Gräff *et al.,* 1990), mammary gland stroma (Ronnov-Jessen *et al.,* 1990), perisinusoidal cells of the liver (Rockey and Friedman, 1992), and mesangial cells

(Elger *et al.,* 1993), and has been correlated with the observation that under pathological conditions *in vivo* lens cells, mammary gland fibroblasts, and perisinusoidal and
mesangial cells express α-SM actin (for review, see Schmitt-Gräff *et al.,* 1994). Presently, however, the genetic and environmental factors regulating α-SM actin expression in fibroblasts are poorly known. The microenvironmental factors described in
Sections 5.1 and 5.2 may be important in producing the selection of α-SM-actin-
positive fibroblasts in a given population. Using different markers (prostaglandin E2,
glycosaminoglycan, collagen or collagenase synthesis, proliferative rate, response to
mononuclear cell-derived mediators, etc.) clonal heterogeneity has been reported in
morphologically homogeneous fibroblastic populations (for representative references
in this field, see Desmoulière *et al.,* 1992a). Cultured fibroblasts may also express
different phenotypic features (for review, see Macieira-Coelho, 1988). A whole spectrum of differentiation steps has been described for fibroblastic cells *in vitro*
(Bayreuther *et al.,* 1988, 1991). *In vivo,* the concept of fibroblast heterogeneity is now
well accepted (for review, see Schmitt-Gräff *et al.,* 1994).

Cytoskeletal proteins such as desmin and SM-MHC are also differentially expressed by cultured fibroblast derived from different organs or pathological situations.
However, the expression of desmin and SM-MHC is generally low; in several populations, no desmin or SM-MHC positive cells are found (Desmoulière *et al.,* 1992a).
During subculture, contrary to the expression of α-SM actin, which increases, desmin
and SM-MHC expression generally decreases.

5. Factors Influencing Fibroblast Phenotype

Little is known concerning the mechanisms leading to the development of cytoskeletal features similar to those of SM cells in fibroblastic cells and producing the
persistence of these features in pathological situations. Factors influencing myofibroblast differentiation have been previously reviewed (Schmitt-Gräff *et al.,* 1994).
As with SM cells, cytokines and extracellular matrix components are good candidates
for modulating fibroblast phenotype and cytoskeletal protein expression. Furthermore,
it is necessary to keep in mind that complex interactions between cytokines and
extracellular matrix components modify their reciprocal activities (for review, see
Mauviel and Uitto, 1993). Table III shows the effect of some cytokines and extracellular matrix components on granulation tissue formation and myofibroblast differentiation.

5.1. Cytokines

It is now clear that these mediators are produced by a large range of cells not
normally considered part of the immune system, such as epithelial and endothelial
cells, SM cells, and fibroblasts. Cytokines now include interferons, cytotoxic factors,
interleukins, hematopoietic colony-stimulating factors, inflammatory cytokines, and

Table III. Action of Cytokines and Extracellular Matrix Components
on Cytoskeletal Features of Fibroblasts

Local application of	Stress fibers	Desmin	α-SM actin	SM-MHC	References
NaCl	—	—	—	—	
PDGF	↗	—	—	—	Desmoulière *et al.* (1992b)
Interleukin-1α	↗	—	—	Not known	Rubbia-Brandt *et al.* (1991)
Interleukin-2	↗	—	—	Not known	Rubbia-Brandt *et al.* (1991)
γ-IFN	↘	—	↘	Not known	Desmoulière *et al.* (1992a)
GM-CSF	↗	—	↗	—	Vyalov *et al.* (1993)
TNF-α	↗	—	—	—	Desmoulière *et al.* (1992b)
Heparin	—	—	—	—	Desmoulière *et al.* (1992b)
TNF-α + heparin	↗	↗	↗	—	Desmoulière *et al.* (1992b)
TGF-β1	↗	—	↗	—	Desmoulière *et al.* (1993)

growth factors (Aggarwal and Pocsik, 1992). During the last few years, the role of cytokines in cell differentiation during development and in cell–cell interactions in adult animals has been well documented (Blau and Baltimore, 1991). Furthermore, these mediators are implicated in pathological phenomena, particularly in the cascade of events observed during tissue repair (Kovacs, 1991; Robson, 1991; Martin *et al.*, 1992; Bennett and Schultz, 1993). A plethora of cytokines is released by platelets and other cells during the initial events after injury and later during granulation tissue development and scar formation. Each cytokine may play different roles and cytokines generally act in a coordinated sequence. In this section, we will focus on the main biological effect of each cytokine and discuss cytokines known to interfere with proliferation, migration, and phenotypic modulation of fibroblastic cells *in vitro* and/or *in vivo*.

5.1.1. Proliferation

After injury, platelet degranulation induces the release within and around the wound clot of abundant amounts (for review, see Bennett and Schultz, 1993) of platelet-derived growth factor (PDGF), TGF-β, epidermal growth factor (EGF), and insulinlike growth factor-I (IGF-I). Tumor necrosis factor-α (TNF-α) secreted mainly by inflammatory cells (Aggarwal and Pocsik, 1992), connective tissue growth factor (CTGF) secreted by endothelial cells (Bradham *et al.*, 1991), and fibroblast growth factor (FGF) produced by macrophages and endothelial cells (Baird *et al.*, 1985; Muthukrishnan *et al.*, 1991) are also present. All these cytokines are potent mitogens for fibroblastic cells. However, contradictory results concerning the role of TGF-β in the control of the cell cycle have been published (Thornton *et al.*, 1990; Kovacs, 1991). This discrepancy may be explained by the differences in doses and in target cells used. In any event, TGF-β effects on cell proliferation may be less important than its chemotactic capacity, as well as its activity in increasing extracellular matrix deposition. As shown by Antoniades *et al.* (1991), injury induces *in vivo* an expression of PDGF and PDGF receptor mRNAs in epidermal cells and of PDGF mRNA in fibroblasts. It seems that application of pharmacological doses of PDGF does not modify significantly the healing of an experimental wound, although contradictory observations have been described in different models or situations (Lynch *et al.*, 1987; Pierce *et al.*, 1991). As we shall discuss and as suggested by Lynch *et al.*, (1987), PDGF may have synergistic effects with other growth factors.

Basic FGF increases the accumulation of granulation tissue in subcutaneously implanted sponges by inducing fibroblast proliferation and collagen accumulation (Davidson *et al.*, 1985), but, as we have said for PDGF, the results concerning the exogenous application of FGFs on wounds depend on the model (Fredj-Reygrobellet *et al.*, 1986; Greenhalgh *et al.*, 1990; Fina *et al.*, 1991), and further studies are needed to evaluate the role of PDGF and FGF in combination with other growth factors or extracellular matrix components. The influence of TNF-α and FGF on connective tissue repair process appears prominent, particularly through their potent angiogenic activity. It has been shown that various components of the blood-clotting system, including thrombin and fibrin, increase growth of fibroblasts and wound healing; fibrinogen, however, had no effect (Pohl *et al.*, 1979; Carney *et al.*, 1992).

5.1.2. Migration

In response to injury, resident fibroblasts proliferate and migrate into the wounded site (Clark, 1993). The specific factors inducing fibroblast migration are poorly known. However, it is well accepted that PDGF (Seppa *et al.*, 1982) and basic FGF (Robson, 1991) are potent chemotactic agents for fibroblastic cells. CTGF is both mitogenic and chemotactic for SM cells and fibroblasts. TGF-β is able to induce selectively CTGF production and secretion, while PDGF, EGF, and FGF do not modify levels of the peptide (Soma and Grotendorst, 1989). Recently, Igarashi *et al.* (1993) have studied the regulation of CTGF gene expression in human skin fibroblasts and during wound healing. They confirm *in vitro* and *in vivo* the role of TGF-β in the production of CTGF, and their results illustrate the cascade process during tissue repair. Pierce *et al.* (1989) have observed that TGF-β1 is chemotactic for both monocytes and fibroblasts. IGF-I has been shown to promote migration of endothelial cells into the wound area, resulting in increased neovascularization (Bennett and Schultz, 1993).

5.1.3. Fibroblast Phenotypic Modulation

In this section, we shall discuss the cytokines that are known to modify the cytoskeletal features of fibroblastic cells. The modifications of extracellular matrix induced by fibroblast activation will be examined in Part 4 of this volume.

5.1.3a. PDGF, FGF, and TNF-α. PDGF appears to decrease the expression of α-SM actin mRNA and protein in cultured SM cells (Corjay *et al.*, 1990) and fibroblasts (A. Desmoulière and G. Gabbiani, unpublished observations). Subcutaneous delivery of the heterodimeric form of PDGF induced the formation of an important granulation tissue, albeit devoid of α-SM-actin-positive myofibroblats (Rubbia-Brandt *et al.,* 1991). Further investigations are needed to evaluate the role of PGDF in combination with others factors on myofibroblast phenotypic modulation and tissue repair. Preliminary results have shown that basic FGF decreases the expression of α-SM actin in cultured fibroblasts (A. Desmoulière and G. Gabbiani, unpublished observations). TNF-α treatment does not induce α-SM actin in granulation tissue fibroblasts and does not modify significantly the expression of α-SM actin in cultured fibroblastic cells (Rubbia-Brandt *et al.,* 1991; Desmoulière *et al.,* 1992b).

5.1.3b. Granulocyte–Macrophage-Colony Stimulating Factor (GM-CSF). GM-CSF has been shown to induce the proliferation of hematopoietic progenitor cells in the myeloid and erythroid lineages and the stimulation of mature monocytes–macrophages and neutrophils (Gasson, 1991). This effect on monocyte–macrophage activities is responsible for recently developed clinical applications (Jones, 1993). Furthermore, some extrahematopoietic activities of GM-CSF have been described (Dedhar *et al.,* 1988; Bussolino *et al.,* 1989). We have observed that the application of GM-CSF to the rat subcutaneous tissue induces the formation of an important granulation tissue rich in α-SM-actin-positive myofibroblasts (Rubbia-Brandt *et al.,* 1991). *In vitro* experiments have shown that GM-CSF does not directly stimulate α-SM actin expression when added to the culture medium of rat or human fibroblasts. We will discuss (see Section 6.1) the mechanisms of this *in vivo* action of GM-CSF. In a chronic granulating wound model, GM-CSF was shown to reestablish normal wound closure rates (Hayward *et al.,* 1990).

5.1.3c. TGF-β1. Among factors secreted by activated macrophages and able to modulate the expression of α-SM actin, TGF-β1 is probably the most efficient. In human arterial SM cells, TGF-β1 induced a growth inhibition and increased the expression of α-SM actin (Björkerud, 1991). Recently, we have shown that TGF-β1 stimulates the expression of α-SM actin in granulation tissue myofibroblasts (Desmoulière *et al.,* 1993). As discussed above, other cytokines and growth factors, such as PDGF, FGF, and TNF-α, despite their profibrotic activity, do not induce α-SM actin expression in myofibroblasts. Furthermore, the expression of α-SM actin protein and mRNA by TGF-β1 is induced in both growing and quiescent cultured fibroblastic populations. The expression of α-SM actin observed in fibroblasts cultured in the presence of fetal calf serum is partly inhibited by the addition of antibodies against TGF-β1. Thus TGF-β1 could represent one of the main regulators of α-SM actin expression in fibroblasts. It has been shown that α-SM actin expression is induced by TGF-β1 in quiescent human breast fibroblasts (Ronnov-Jessen and Petersen, 1993). This result illustrates the key role that TGF-β1 may play in the context of tumor cell–fibroblast interactions (see Section 6.3) during breast neoplasia. The action of TGF-β1 on α-SM actin expression confirms and extends the notion that TGF-β plays an important role in both fibroblast differentiation and fibrosis formation. It is well known that TGF-β increases the accumulation of extracellular matrix compounds leading to the development of fibrosis (Border and Ruoslahti, 1992). Recently, Beck *et al.* (1993) have shown that systemic administration of TGF-β1 is effective in enhancing healing in age- or glucocorticoid-impaired wound healing when it is given at the time of wounding, 4 hr after wounding, or, most surprisingly, even 24 hr before wounding.

5.1.3d. Endothelin. In a croton oil-induced granulation tissue model in rats, endothelin-1 induced a reversible concentration-dependent contraction of the granulation tissue (Appleton *et al.*, 1992). The contractile activity is correlated with the development of myofibroblastic features. Endothelin-1 has mitogenic activity on cultured rat fibroblasts (MacNulty *et al.*, 1990) and acts synergistically with polypeptide growth factors such as TGF-β. Endothelin-1 is able to induce the expression of α-SM actin in cultured vascular SM cells (Hahn *et al.*, 1992). It would be interesting to know whether endothelin-1 is able to act directly on myofibroblastic differentiation by inducing α-SM actin expression. Recent reports concerning the action of a vasopressor substance on myofibroblast differentiation support the role of this cytokine in the processes leading to wound contraction (Thiemermann and Corder, 1992).

5.1.3e. γ-Interferon (γ-IFN). We have observed that in cultured fibroblasts, γ-IFN, a cytokine produced by T-helper lymphocytes, decreases α-SM actin protein and mRNA expression as well as cell proliferation (Desmoulière *et al.*, 1992a). γ-IFN has been shown with some exceptions to decrease proliferative activity and collagen production in fibroblastic cells (Duncan and Berman, 1985). The properties of this cytokine make it a good candidate to exert antifibrotic activity *in vivo,* as already suggested by Grandstein *et al.* (1990) and Larrabee *et al.* (1990). Preliminary results (Pittet *et al.*, 1994) have shown that γ-IFN treatment decreases the clinical manifestations and the size of hypertrophic scars and Dupuytren's nodules. In hypertrophic scars, immunofluorescence examination showed that α-SM actin expression was decreased in myofibroblasts of treated lesions. These results suggest that γ-IFN could represent a useful adjunct to the nonsurgical therapy of hypertrophic scars and Dupuytren's disease.

5.2. Extracellular Matrix Components

It is well accepted that the extracellular matrix represents a structural support for cellular constituents, but evidence exists showing that the matrix plays a central role as a source of signals influencing growth and differentiation of different cell types, including fibroblasts (for review, see Juliano and Haskill, 1993). Furthermore, the fibroblast is the main cell type implicated in the production of extracellular matrix components during the repair process. Recently, with a combined immunohistochemical and *in situ* hybridization study, it has been shown that α-SM-actin-expressing myofibroblasts are the main producer of collagen during pulmonary fibrosis (Zhang *et al.*, 1994).

Among extracellular matrix components, different types of collagen, glycoproteins, and proteoglycans are involved in fibroblastic differentiation. *In vitro*, adhesion, proliferation, and migration of fibroblasts are modulated by extracellular matrix components. Recently, Streuli *et al.* (1993) have shown that extracellular matrix regulates the expression of the TGF-β1 gene. The authors propose that there is a feedback loop whereby TGF-β1-induced synthesis of basement membrane is repressed once a functional basement membrane is present. This result illustrates the complex mechanisms by which growth factors and extracellular matrix components regulate cell activities. The components of the basal lamina, i.e., collagen type IV, laminin, and heparan sulfate, are known to maintain SM cells in a differentiated state. These components may also induce differentiation in undifferentiated cells (i.e., fetal cells) or partially differentiated cells such as quiescent fibroblasts.

It is well known that heparin decreases the proliferation of SM cells *in vivo* (Clowes and Karnovsky, 1977) and *in vitro* (Hoover *et al.,* 1980). Furthermore, heparin has also been shown to inhibit SM cell modulation from a contractile to a synthetic phenotype (Chamley-Campbell and Campbell, 1981; Majack and Bornstein, 1984), as well as the switch in actin isoform expression observed in SM cells after a balloon catheter-induced endothelial injury (Clowes *et al.,* 1988). Heparin is able to induce α-SM actin expression in cultured SM cells (Desmoulière *et al.,* 1991). In cultured fibroblasts, heparin increases the expression of both α-SM actin protein and mRNA (Desmoulière *et al.,* 1992b). This induction is observed in many different fibroblastic populations. We assume (see Section 5.2) that heparin facilitates the presentation to cell receptors of differentiation or maturation factors present in serum. The analysis of [^{3}H]thymidine incorporation in synchronized cells suggests that heparin produces a selection of α-SM-actin-expressing fibroblasts, since after heparin treatment the proportion of cells entering into the cell cycle is higher among the α-SM-actin-positive cells and lower among the α-SM-actin-negative cells compared to controls. For *in vivo* studies, osmotic minipumps filled with saline solution, nonanticoagulant heparin derivatives, or recombinant murine TNF-α without or with nonanticoagulant heparin derivatives were implanted subcutaneously in rats (Desmoulière *et al.,* 1992b). After 14 days, the newly formed connective tissue around the perfusion pumps was collected. In control animals and in animals treated with heparin derivatives, the capsule was thin and α-SM actin staining was never detected. TNF-α produced a significant fibroblast accumulation, but the fibroblastic cells were positive for α-SM actin only when TNF-α was combined with heparin derivatives. Thus proliferation and α-SM actin synthesis by fibroblasts appear to be distinct phenomena during the formation and progression of granulation tissue. Furthermore, heparin and heparan sulfate proteoglycans are known to bind many growth factors (Flaumenhaft and Rifkin, 1991; Ruoslahti and Yamaguchi, 1991) such as basic FGF (Ornitz *et al.,* 1992), TNF-α (Lantz *et al.,* 1991), and TGF-β (McCaffrey *et al.,* 1989, 1992), and these interactions may be essential to their activity. In conclusion, heparin is likely to participate in the complex modulation mechanisms during different connective tissue reactions.

6. Interactions between Fibroblasts and Other Granulation Tissue Components

As we have seen, different cells can produce cytokines and extracellular matrix components capable of modifying the myofibroblastic phenotype. During all steps of the wound-healing process, i.e., inflammation, angiogenesis, and reepithelialization, fibroblasts interact with different cell types and the extracellular matrix environment is modified (Fig. 1). We will focus our discussion on inflammatory cells, vascular cells, and epithelial cells. We will also discuss the modifications of cell adhesion molecule pattern during fibroblast phenotypic changes.

6.1. Macrophages and Inflammatory Cells

In vitro experiments have shown that GM-CSF does not directly stimulate α-SM actin expression when added to the culture medium of rat or human fibroblasts. To clarify this point, we have studied chronologically the formation of granulation tissue induced by GM-CSF treatment. After GM-CSF local treatment, the appearance of α-SM-actin-rich myofibroblasts (Rubbia-Brandt *et al.*, 1991) is preceded by a characteristic clusterlike accumulation of macrophages (Vyalov *et al.*, 1993), suggesting that such macrophages are important in the stimulation of α-SM actin synthesis by myofibroblasts; macrophages may produce one or more α-SM actin expression-inducing factors. Among factors secreted by activated macrophages and capable of modulating the expression of α-SM actin, TGF-β1 appears to be the most efficient (Desmoulière *et al.*, 1993). Moreover, in transgenic mice expressing GM-CSF, fibrotic nodules developed in areas where macrophages accumulate (Lang *et al.*, 1987). In lung fibrosis, factors secreted by macrophages are implicated in the genesis and the maintenance of the disease. In human pulmonary fibrosis, Broekelmann *et al.* (1991) have observed that macrophage and TGF-β localization coincide with extracellular matrix deposition. Khalil *et al.* (1989) have shown that in pulmonary fibrosis induced by intratracheal instillation of bleomycin, an accumulation of α-SM-actin-expressing myofibroblasts is observed around clustered macrophages exhibiting a high expression of TGF-β.

It is an old clinical observation that mast cell proliferation is accompanied by fibrotic changes (Claman, 1985). Functional features of mast cells are regulated by fibroblasts (Levi-Schaffer *et al.*, 1985). Our results (Desmoulière *et al.*, 1992b) suggest that mast cells in turn exert a regulatory influence on fibroblast activities through their products, such as heparin.

6.2. Vascular Cells (Endothelial Cells and Pericytes)

Many studies have examined *in vitro* the relationships between endothelial cells and SM cells or pericytes. In normal conditions, endothelial cells participate in the homeostasis of the arterial wall by secreting substances such as heparan sulfate proteoglycans or nitric oxide. It seems probable that endothelial cells and pericytes influence the development of the myofibroblastic phenotype. In turn, fibroblastic cells play an important role during angiogenesis. Capillary morphogenesis by microvascular endothelial cells cultured in collagen or fibrin gels is modulated by cytokines such as vascular endothelial growth factor, a potent mitogen acting selectively on endothelial cells, basic FGF, or TGF-β (Pepper *et al.*, 1993). Swiss 3T3 fibroblasts are able to stimulate formation of capillarylike tubules by cloned bovine microvascular endo-thelial cells (Montesano *et al.*, 1993).

Angiogenesis plays an important role in many pathological processes (Arnold and West, 1991). During the transformation of a granulation tissue in an irreversibly fibro-tic tissue, some important factors presumably reach the lesion via newly formed micro-vessels. The presence of rounded endothelial cells reducing the lumen of new vessels of hypertrophic scar and keloid has been reported (Kischer *et al.*, 1982). Fibroblast proliferation may be induced by local hypoxia with associated release of oxygen free radicals (Murrel *et al.*, 1987, 1989). Furthermore, it has been shown that hypoxia is a stimulus for fibroblasts in culture (Hunt *et al.*, 1985). Rounded endothelial cells pro-jecting into the lumen of vessels may have a critical role in the development and maintenance of hypertrophic scar and keloid and may represent a common pathway in the development of fibrotic lesions.

6.3. Epithelial Cells

Interactions between epithelium and mesenchyme play an important role during normal development and for the maintenance of differentiation in the adult. It has been demonstrated that the growth and differentiation of normal epithelium is regulated inductively or permissively by neighboring mesenchymal components.

TGF-β suppresses the appearance of the transformed phenotype in cocultures of *ras*-transformed keratinocytes and normal dermal fibroblasts (Missero *et al.*, 1991). Loréal *et al.* (1993) have shown that a high level of matrix protein synthesis by liver cells *in vitro* is not sufficient to induce extracellular matrix deposition; cooperation between hepatocyte and perisinusoidal cells (Ito cells) is necessary for this process.

In an *in vitro* system in which Madin-Darby canine kidney (MDCK) epithelial cells are cocultured in collagen gels with fibroblasts, Montesano *et al.* (1991) have observed that hepatocyte growth factor, a fibroblast-derived molecule, plays a crucial role in the control of epithelial morphogenesis. Among stromal mediators of epithelial cell proliferation, keratinocyte growth factor, unlike other molecules in the FGF family, does not appear to cause mesenchymal stromal cell proliferation (Finch *et al.*, 1989; Ulich *et al.*, 1994).

During wound healing, keratinocytes change from a sedentary to a migratory phenotype (Grinnell, 1992). It is conceivable that fibroblasts and myofibroblasts pre-sent in granulation tissue contribute to keratinocyte phenotypic modulation by secret-ing specific extracellular matrix components. In turn, we can assume that keratinocytes are able to act on myofibroblast differentiation. Coculture results in stimulation of cell growth and total protein synthesis in both keratinocytes and dermal fibroblasts when compared to monocultured controls. Coculture with fibroblasts also facilitates ker-atinocyte terminal differentiation, while both cell types secrete paracrine factors affect-ing cell growth, motility, and protein expression (for review, see Birchmeier and Birchmeier, 1993). The observation that in human cervix intraepithelial neoplasia α-SM actin develops in stromal fibroblasts prior to invasive phenomena (Cintorino *et al.*, 1991) supports the possibility that epithelial changes influence the fibroblastic phenotype through an apparently intact basement membrane.

Stromal influences on epithelial neoplasia have also been documented using different models *in vivo*. For example, transformed fibroblasts coinoculated with epithelial cells accelerated the growth and shortened the latency period of human epithelial tumors in athymic mice (Camps *et al.*, 1990). Fabra *et al.* (1992) have shown that specific fibroblasts influence the invasive potential of highly metastatic human colon carcinoma cells. By using low- and high-metastatic murine lung carcinoma cell lines, Nakanishi *et al.* (1994) have demonstrated that the ability of tumor cells to induce a stromal response of the host tissue was inversely correlated with their metastatic potential. However, the relationship between the stromal response of the host tissue and tumor invasion and metastasis remains to be defined. Recent work showing that the steroid analogue tamoxifen induces the formation of TGF-β in the stroma of human breast tumors (Butta *et al.*, 1992) furnishes an intriguing new interpretation of the action of this substance and may allow the study of the role of cytokines as mediators in cell–cell interaction mechanisms.

6.4. Cell Adhesion Molecules

Adhesive interactions between cells and collagen are required for contraction and are mediated by α2β1 integrins (Klein *et al.*, 1991; Schiro *et al.*, 1991). Fibronectin α5β1 receptors appear on granulation tissue fibroblasts on day 7 after injury, when fibroblasts become resident within the wound (Clark, 1993); at this time they produce important amounts of type I collagen and fibronectin (Welch *et al.*, 1990). Furthermore, fibronectin receptors play a key role for fibronectin and collagen matrix assembly (McDonald *et al.*, 1987). As we have emphasized, myofibroblasts form specific junctions (fibronexi) with the surrounding fibronectin-rich granulation tissue (Singer *et al.*, 1984). Interestingly, the presence of numerous fibronexi has been observed in Dupuytren's disease, a lesion characterized by a high contraction activity (Tomasek and Haaksma, 1991). Recently, the close relationship between α-SM actin expression in fibroblasts and collagen remodeling has been underlined by Arora and McCulloch (1994); their results indicate that gel contraction is dependent on α-SM actin expression, and that α-SM actin is a functional marker for a fibroblast subtype involved in extracellular matrix remodeling. Fibroblasts, SM cells, and endothelial cells develop interactions with elastic fibers (Perdomo *et al.*, 1994). This type of interaction can play an important role during wound healing and fibrotic process, since some tissues such as skin and lung are very rich in elastin.

7. Mechanisms of Granulation Tissue Contraction and *in Vitro* Wound Contraction Models

While the relationship between myofibroblast and granulation tissue contraction appears well established, the mechanisms regulating this process remain at present mysterious. Numerous physiological and pharmacological agents known to act on SM cell have been reported to influence the contraction of granulation tissue strips *in vitro* (Majno *et al.*, 1971; Ryan *et al.*, 1974; for review, see Skalli and Gabbiani, 1988). It is well accepted that α-SM actin is a marker for cells with contractile features. Epidermal cells of the planaria that express α-SM actin may help to contract the wound edges to create the first protective epidermal covering (Pascolini *et al.*, 1992). These planarian epidermal cells may exert functions similar to those of myoepithelial cells. Human lens cells of anterior capsular cataract and cultured bovine lens-forming cells (Greenburg and Hay, 1986; Schmitt-Gräff *et al.*, 1990), despite their embryonic ectodermal origin, express α-SM actin under some pathological or experimental conditions and may be modulated into mesenchymal-like cells. In this context, they can play a contractile role in which α-SM actin is implicated. Furthermore, the presence of fibronexus (Singer *et al.*, 1984) illustrates the tight connections that develop between myofibroblasts and the surrounding extracellular matrix, allowing granulation tissue contraction.

The force generated by cultured fibroblasts can distort a sheet of silicon on which they are grown (Harris *et al.*, 1981; Boswell *et al.*, 1992). Stress fibers are thought to be the force-generating element involved in wound contraction. Moreover, as suggested by Boswell *et al.* (1992) and Streuli *et al.* (1993), normal fibroblasts develop *in vitro* "differentiated" properties that resemble those exhibited in a "wounded" environment, presumably because cell culture conditions resemble basically those of an open wound. However, good models to evaluate fibroblast contraction in culture are not presently available.

To study wound contraction *in vitro*, different models using fibroblasts cultured in collagen or fibrin matrices have been developed. In a floating collagen matrix, contraction occurs in a mechanically relaxed tissue, while in an anchored collagen matrix, contraction occurs in a stressed tissue. As suggested by Grinnell (1994), free matrices resemble dermis and anchored matrices resemble granulation tissue. Fibroblasts in anchored matrices develop myofibroblastic features (Bell *et al.*, 1979) in contrast to fibroblasts in floating collagen matrices (Tomasek *et al.*, 1992). Fibroblasts in floating versus anchored matrices show important differences in cell proliferation (Nishiyama *et al.*, 1989) and in collagen biosynthesis (Nusgens *et al.*, 1984). Furthermore, the tension forces developed in a restrained collagen lattice by myofibroblasts obtained from human skin granulation tissue explants are more intense than those exerted by human dermis fibroblasts (Delvoye *et al.*, 1991). These results suggest that mechanical forces within the tissue can regulate cell proliferation, cell differentiation, and extracellular matrix organization.

8. Senescence and Apoptosis

Fibroblasts grown *in vitro* can be stimulated to divide as well as to leave the cell cycle and "differentiate" (i.e., express α-SM actin) as fibroblasts do *in vivo* in granulation tissue. Fibroblastic cells repeatedly pressed to divide *in vitro* attain a state in which they are refractory to further mitotic stimulation and to subculturing. Cultured human fibroblasts have a finite *in vitro* life span that correlates with the age of the donor, and it has been proposed that this loss of proliferative capacity serves as a model of cellular aging (Hayflick, 1965). However, we can assume, as suggested by previous works (Martin *et al.*, 1974; Bell *et al.*, 1978; Macieira-Coelho and Taboury, 1982), that loss of division potential *in vitro* represents differentiation instead of aging. It seems clear that the loss of proliferation capacity observed *in vivo* during aging and *in vitro* after several passages represent different features induced by mechanisms that cannot be always correlated (Bruce, 1991). The accelerated growth rate of human fetal cells results primarily from developmental events intrinsic to the cells and is associated with high responsiveness to the mitogenic action of peptide growth factors (Fant, 1991). Human newborn skin fibroblasts senesce *in vitro* without acquiring growth factor requirements typical of fibroblasts cultured from adults (Wharton, 1984). Furthermore, cloning and subcloning experiments show that features of the cultured parental population are representative of a subpopulation present *in vivo* and furnish evidence for selection events in mass culture (Zavala *et al.*, 1978).

It is well established that the evolution of granulation tissue into scar tissue implies a massive decrease in cellularity, with the most involved cells being fibroblasts, endothelial cells, and pericytes. When granulation tissue cells are not eliminated, there is development of hypertrophic scar or keloid, both characterized by a high degree of cellularity. By means of *in situ* end-labeling (ISEL) of fragmented DNA (Table IV) and of morphometry at the electron microscopic level, we have shown that the proportion

Table IV. Evaluation of Cells Showing ISEL of Fragmented DNA[a,b]

Days after wounding	Granulation tissue	
	Fibroblasts	Vascular cells
7	−	−
8	−	+/−
10	−	+/−
12	+/−	+
16	++	++
20	+++	++
25	++	++
30	+/−	+
60	−	−

[a]The staining with ISEL method was observed blindly and independently by two researchers and classified as: −, no staining; +/−, staining in less than 3% of cells; +, staining in 3–6% of cells; ++, staining in 6–9% of cells; and +++, staining in 9–12% of cells.
[b]From Desmoulière *et al.* (1995), with permission.

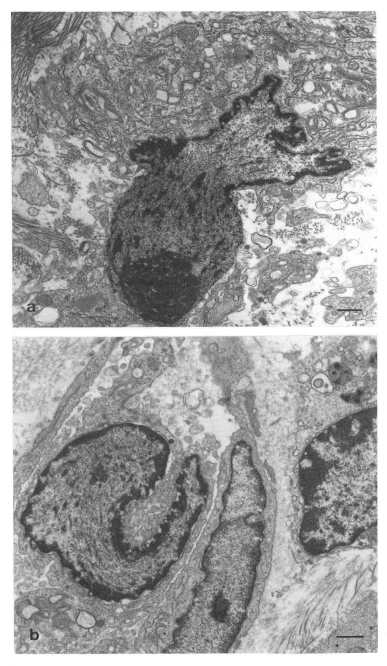

Figure 3. Transmission electron micrographs showing early apoptotic features in granulation tissue 20 days after wounding. (a) The nucleus of an apoptotic fibroblast shows chromatin condensation and appears in part extruded from the cytoplasm. (b) A macrophage with phagolysosomes (right) and an endothelial cell with initial apoptotic changes are observed. Bar = 1 μm.

of myofibroblastic and vascular cells undergoing apoptosis dramatically increases as the wound closes (Desmoulière *et al.*, 1995). Apoptotic fibroblasts showed condensation and margination of chromatin, cytoplasmic vacuoles, and convoluted cell surface (Fig. 3a). In addition, capillaries with endothelial cells undergoing apoptosis were observed (Fig. 3b). Macrophages containing phagolysosomes were seen near fibroblastic apoptotic cells, suggesting that this is the major route of removal of apoptotic bodies (Fig. 3b).

Our data indicate that apoptosis of granulation tissue cells takes place essentially after wound closure and affects target cells consecutively rather than producing a single wave of cell disappearance. These observations are in line with the gradual resorption of granulation tissue after wound closure.

The question that remains to be answered is what is the stimulus leading to apoptosis during wound healing. In the last years, gene products regulating cell death have been identified (for review, see Evans, 1993). In fibroblasts, the c-myc protein (Evan *et al.*, 1992) and interleukin-1β-converting enzyme, the mammalian homologue of the *Caenorhabditis elegans* cell death gene *ced-3* (Miura *et al.*, 1993), has been shown to induce apoptosis. A possible mechanism for apoptosis induction could be via direct action and/or withdrawal of cytokines or growth factors (Robaye *et al.*, 1991; Jürgensmeier *et al.*, 1994; Moulton *et al.*, 1994). Several cytokines have been shown to be present in the normally healing wound, released mainly by platelets and inflammatory cells. It is probable that as the wound resolves, there is a decrease in the level of these factors. A possible explanation for the death of at least a subpopulation of myofibroblasts and vascular cells would be that they are growth factor-dependent. Alternatively, factors selectively causing the death of fibroblastic and vascular cells could be liberated after epithelialization has been completed. Recently, Jürgensmeier *et al.* (1994) have shown that TGF-β-treated fibroblasts eliminate transformed fibroblasts by induction of apoptosis. This mechanism could be potentially acting during wound healing. Myofibroblasts could be induced to die by a subpopulation of fibroblasts that is not implicated in myofibroblastic differentiation but participates in the control of cellularity. Studies along these lines will be useful for the understanding of normal and pathological wound healing and possibly for modifying the evolution of keloid and hypertrophic scar.

Interactions between cells and extracellular matrix components have been suggested to play a key role in fibroblast apoptosis (Ruoslahti and Reed, 1994). When their anchorage is modified, cells can undergo apoptosis, and this process implicates integrin-mediated signaling as the controlling factor. Up to now this type of apoptotic response appears to be limited to epithelial and endothelial cells (Meredith *et al.*, 1993; Frisch and Francis, 1994), but we can suppose that a similar mechanism could be implicated in fibroblastic cell apoptosis.

Our *in vivo* and *in vitro* results suggest that under normal conditions, the process of myofibroblast differentiation ends with the death of these cells. Thus, myofibroblasts could be considered to be terminally differentiated cells. As suggested by McCulloch and Bordin (1991), we can assume that the development of fibroblast subpopulations is organized similarly to that of the hematopoietic system.

9. Conclusions and Perspectives

Although many biological features of myofibroblasts remain a mystery, it is now accepted that these cells arise from preexisting fibroblasts during cutaneous wound healing (Darby *et al.,* 1990). In organs other than the skin, myofibroblasts can develop from specialized cells such as SM cells, glomerular mesangial cells, and hepatic perisinusoidal cells (for review, see Schmitt-Gräff *et al.,* 1994). During their development, myofibroblasts express to various extents proteins normally present in SM cells, suggesting that they assume contractile features (Skalli and Gabbiani, 1988). Moreover, they organize architecturally in a well-defined pattern that is typically observed in the granulation tissue of a cutaneous open wound and consists of arrays of cells aligned in layers parallel to the wound surface, with small vessels disposed perpendicularly. It has been shown that cultured fibroblasts, which have a cytoskeletal organization similar to that of *in vivo* myofibroblasts, exert an important tension on their substratum (Harris *et al.,* 1981). It is conceivable that myofibroblasts exert a similar tension or isometric contraction on the surrounding tissues. It has been proposed that tissue shape in general is regulated by a cytoskeleton-dependent tensionally integrated structure (tensegrity); this view implies that architecture is more important than individual molecules or even simple mechanics (Ingber, 1993). In this context, the well-established architectural organization of granulation tissue during wound healing correlates well with the appearance of wound contraction. Loss of myofibroblasts by apoptosis (Desmoulière *et al.,* 1995) would explain the decreased tension after wound epithelialization.

The factors regulating myofibroblast differentiation and modulation are becoming better known (for review, see Desmoulière and Gabbiani, 1994). In this context, the observation that γ-IFN can reduce the size of hypertrophic scars and Dupuytren's nodules (Pittet *et al.,* 1994) awaits further clinical confirmation. Similarly, recent reports that experimental scarring as well as fibrotic lesions (Border *et al.,* 1990; Shah *et al.,* 1992; Beck *et al.,* 1993) are reduced by pretreatment or treatment with antibodies against TGF-β1 appear promising.

The acquisition of SM cell cytoskeletal features, especially of α-SM actin expression, by myofibroblasts not only represents a good marker of the contractile and/or tensile activity of these cells, but probably also plays a role in the actual production of contractile or tensile forces. The recent observation that α-SM actin expression can be selectively inhibited in cultured SM cells microinjected with the acetylated N-terminal tetrapeptide of α-SM actin (Chaponnier *et al.,* 1995) will probably furnish a tool to study the role of this actin isoform in myofibroblast differentiation and activity.

In conclusion, during the last years, many biological features of the myofibroblast have been clarified. In the future we should learn more about the role of these cells in normal and pathological wound healing and we shall probably become able to interfere efficiently with their activity during pathological situations.

ACKNOWLEDGMENTS. This work has been partially supported by the Swiss National Science Foundation (grant No. 31–40372–94). We thank the American Journal of Pathology for allowing reproduction of Table IV and Mrs. G. Gillioz for typing the manuscript.

References

Adzick, N. S., and Longaker, M. T., 1992, Scarless fetal wound healing: Therapeutic implications, *Ann. Surg.* **215**:3–7.

Adzick, N. S., Harrison, M. R., Glick, P. L., Beckstaed, J. L., Villa, R. L., Scheuenstuhl, H., and Goodson, W. H., 1985, Comparison of fetal, newborn, and adult wound healing by histologic, enzyme-histochemical, and hydroxyproline determinations, *J. Pediatr. Surg.* **20**:315–319.

Aggarwal, B. B., and Pocsik, E., 1992, Cytokines: From clone to clinic, *Arch. Biochem. Biophys.* **292**:335–359.

Antoniades, H. N., Galanopoulos, T., Neville-Golden, J., Kiritsy, C. P., and Lynch, S. E., 1991, Injury induces *in vivo* expression of platelet-derived growth factor (PDGF) and PDGF receptor mRNAs in skin epidermal cells and PDGF mRNA in connective tissue fibroblasts, *Proc. Natl. Acad. Sci. USA* **88**:565–569.

Appleton, I., Tomlinson, A., Chander, C. L., and Willoughby, D. A., 1992, Effect of endothelin-1 on croton oil-induced granulation tissue in the rat, *Lab. Invest.* **67**:703–710.

Arnold, F., and West, D. C., 1991, Angiogenesis in wound healing, *Pharm. Ther.* **52**:407–422.

Arora, P. D., and McCulloch, C. A. G., 1994, Dependence of collagen remodelling on α-smooth muscle actin expression by fibroblasts, *J. Cell. Physiol.* **159**:161–175.

Baird, A., Mormede, P., and Bohlen, P., 1985, Immunoreactive fibroblast growth factor in cells of peritoneal exudate suggests its identity with macrophage-derived growth factor, *Biochem. Biophys. Res. Commun.* **126**:358–364.

Ballardini, G., Fallani, M., Biagini, G., Bianchi, F. B., and Pisi, E., 1988, Desmin and actin in the identification of Ito cells and in monitoring their evolution to myofibroblasts in experimental liver fibrosis, *Virchows Arch. [B] Cell Pathol.* **56**:45–49.

Baur, P. S., Larson, D. L., and Stacey, T. R., 1975, The observation of myofibroblasts in hypertrophic scars, *Surg. Gynecol. Obstet.* **141**:22–26.

Bayreuther, K., Rodemann, H. P., Hommel, R., Dittmann, K., Albiez, M., and Francz, P. I., 1988, Human skin fibroblasts *in vitro* differentiate along a terminal cell lineage, *Proc. Natl. Acad. Sci. USA* **85**:5112–5116.

Bayreuther, K., Francz, P. I., Gogol, J., Hapke, C., Maier, M., and Meinrath, H. G., 1991, Differentiation of primary and secondary fibroblasts in cell culture systems, *Mutat. Res.* **256**:233–242.

Beck, L. S., DeGuzman, L., Lee, W. P., Xu, Y., Siegel, M. W., and Amento, E. P., 1993, One systemic administration of transforming growth factor-β1 reverses age- or glucocorticoid-impaired wound healing, *J. Clin. Invest.* **92**:2841–2849.

Beertsen, W., Events, V., and van den Hoof, A., 1974, Fine structure of fibroblasts in the periodontal ligament of the rat incisor and their possible role in tooth eruption, *Arch. Oral Biol.* **19**:1097–1098.

Bell, E., Marek, L. F., Levinstone, D. S., Merrill, C., Sher, S., Young, I. T., and Eden, M., 1978, Loss of division potential *in vitro:* Aging or differentiation? Departure of cells from cycle may not be a sign of aging, but a sign of differentiation, *Science* **202**:1158–1163.

Bell, E., Ivarsson, B., and Merrill, C., 1979, Production of a tissue-like structure by contraction of collagen lattices by human fibroblasts of different proliferative potential *in vitro, Proc. Natl. Acad. Sci. USA* **76**:1274–1278.

Bellows, C. G., Melcher, A. H., and Aubin, J. E., 1981, Contraction and organization of collagen gels by cell cultured from periodontal ligament, gingiva and bone suggest functional differences between cell types, *J. Cell Sci.* **211**:1052–1054.

Bennett, N. T., and Schultz, G. S., 1993, Growth factors and wound healing: Biochemical properties of growth factors and their receptors, *Am. J. Surg.* **165**:728–737.

Birchmeier, C., and Birchmeier, W., 1993, Molecular aspects of mesenchymal–epithelial interactions, *Annu. Rev. Cell Biol.* **9**:511–540.

Björkerud S., 1991, Effects of transforming growth factor-β1 on human arterial smooth muscle cells *in vitro, Arterioscler. Thromb.* **11**:892–902.

Blau, H. M., and Baltimore, D., 1991, Differentiation requires continuous regulation, *J. Cell Biol.* **112**:781–783.

Border, W. A., and Ruoslahti, E., 1992, Transforming growth factor-β in disease: The dark side of tissue repair, *J. Clin. Invest.* **90**:1–7.

Border, W. A., Okuda, S., Languino, L. R., Sporn, M. B., and Ruoslahti, E., 1990, Suppression of experimental glomerulonephritis by antiserum against transforming growth factor beta 1, *Nature* **346**:371–374.

Boswell, C. A., Majno, G., Joris, I., and Ostrom, K. A., 1992, Acute endothelial cell contraction *in vitro:* A comparison with vascular smooth muscle cells and fibroblasts, *Microsvasc. Res.* **43**:178–191.

Bradham, D. M., Igarashi, A., Potter, R. L., and Grotendorst, G. R., 1991, Connective tissue growth factor: A cysteine-rich mitogen secreted by human vascular endothelial cells is related to the SRC induced immediate early gene product CEF-10, *J. Cell Biol.* **114**:1285–1294.

Bressler, R. S., 1973, Myoid cells in the capsule of the adrenal gland and in monolayers derived from cultured adrenal capsules, *Anat. Rec.* **177**:525–531.

Broekelmamm, T. J., Limper, A. H., Colby, T. V., and McDonald, J. A., 1991, Transforming growth factor $\beta1$ is present at sites of extracellular matrix gene expression in human pulmonary fibrosis, *Proc. Natl. Acad. Sci. USA* **88**:6642–6646.

Bruce, S. A., 1991, Ultrastructure of dermal fibroblasts during development and aging: Relationship to *in vitro* senescence of dermal fibroblasts, *Exp. Gerontol.* **26**:3–16.

Buoro, S., Ferrarese, P., Chiavegato, A., Roelofs, M., Scatena, M., Pauletto, P., Passerini-Glazel, G., Pagano, F., and Sartore, S., 1993, Myofibroblast-derived smooth muscle cells during remodelling of rabbit urinary bladder wall induced by partial outflow obstruction, *Lab. Invest.* **69**:589–602.

Bussolino, F., Wang, J. M., Defilippi, P., Turrini, F., Sanavio, F., Edgell, C. J. S., Aglietta, M., Arese, P., and Mantovani, A., 1989, Granulocyte- and granulocyte–macrophage-colony stimulating factors induce human endothelial cells to migrate and proliferate, *Nature* **337**:471–473.

Burrington, J. D., 1971, Wound healing in the fetal lamb, *J. Pediatr. Surg.* **6**:523–527.

Butta, A., MacLennan, K., Flanders, K. C., Sacks, N. P. M., Smith, I., McKinna, A., Dowsett, M., Wakefield, L. M., Sporn, M. B., Baum, M., and Coletta, A. A., 1992, Induction of transforming growth factor $\beta1$ in human breast cancer *in vivo* following tamoxifen treatment, *Cancer Res.* **52**:4261–4264.

Camps, J. L., Chang, S. M., Hsu, T. C., Freeman, M. R., Hong, S. J., Zhau, H. E., von Eschenbach, A. C., and Chung, L. W. K., 1990, Fibroblast-mediated acceleration of human epithelial tumor growth *in vivo*, *Proc. Natl. Acad. Sci. USA* **87**:75–79.

Caplan, A. I., Fiszman, M. Y., and Eppenberger, H. M., 1983, Molecular and cell isoforms during development, *Science* **221**:921–927.

Carney, D. H., Mann, R., Redin, W. R., Pernia, S. D., Berry, D., Heggers, J. P., Hayward, P. G., Robson, M. C., Christie, J., Annable, C., Fenton II, J. W., and Glenn, K. C., 1992, Enhancement of incisional wound healing and neovascularization in normal rats by thrombin and synthetic thrombin receptor-activating peptides, *J. Clin. Invest.* **89**:1469–1477.

Carrel, A., 1922, Growth-promoting function of leucocytes, *J. Exp. Med.* **36**:385–391.

Chamley-Campbell, J. H., and Campbell, G. R., 1981, What controls smooth muscle phenotype? *Atherosclerosis* **40**:347–357.

Chaponnier, C., Goethals, M., Janmey, P. A., Gabbiani, F., Gabbiani, G., and Vandekerckhove, J., 1995, The specific NH2-terminal sequence Ac-EEED of α-smooth muscle actin plays a role in polymerization in vitro and in vivo, *J. Cell Biol.* **130**:887–895.

Charbord, P., Lerat, H., Newton, I., Tamayo, E., Gown, A. M., Singer, J. W., and Herve, P., 1990, The cytoskeleton of stromal cells from human bone marrow cultures resembles that of cultured smooth muscle cells, *Exp. Hematol.* **18**:276–282.

Chiavegato, A., Bochaton-Piallat, M. L., D'Amore, E., Sartore, S., and Gabbiani, G., 1995, Expression of myosin heavy chain isoforms in mammary epithelial cells and in myofibroblasts from different fibrotic settings during neoplasia, *Virchows Arch.* **426**:77–86.

Cintorino, M., Bellizi de Marco, E., Leoncini, P., Tripodi, S. A., Ramaekers, F. C., Sappino, A. P., Schmitt-Gräff, A., and Gabbiani, G., 1991, Expression of α-smooth-muscle actin in stromal cells of the uterine cervix during epithelial neoplastic changes, *Int. J. Cancer* **47**:843–846.

Claman, H. N., 1985, Mast cells, T cells and abnormal fibrosis, *Immunol. Today* **6**:192–195.

Clark, R. A. F., 1993, Regulation of fibroplasia in cutaneous wound repair, *Am. J. Med. Sci.* **306**:42–48.

Clowes, A. W., and Karnovsky, M. J., 1977, Suppression by heparin of smooth muscle cell proliferation in injured arteries, *Nature* **265**:625–626.

Clowes, A. W., Clowes, M. M., Kocher, O., Ropraz, P., Chaponnier, C., and Gabbiani, G., 1988, Arterial smooth muscle cells *in vivo:* Relationship between actin isoform expression and mitogenesis and their modulation by heparin, *J. Cell Biol.* **107:**1939–1945.

Corjay, M. H., Blank, R. S., and Owens, G. K., 1990, Platelet-derived growth factor-induced destabilization of smooth muscle alpha-actin mRNA, *J. Cell. Physiol.* **145:**391–397.

Cuhna, G. R., Hayashi, N., and Wong, Y. C., 1991, Regulation of differentiation and growth of normal adult and neoplastic epithelia by inductive mesenchyme, *Cancer Surv.* **11:**73–90.

Czernobilsky, B., Shezen, E., Lifschitz-Mercer, B., Fogel, M., Luzon, A., Jacob, N., Skalli, O., and Gabbiani, G., 1989, Alpha smooth muscle actin (α-SM actin) in normal human ovaries, in ovarian stromal hyperplasia and in ovarian neoplasma, *Virchows Arch. [B] Cell. Pathol.* **57:**55–61.

Darby, I., Skalli, O., and Gabbiani, G., 1990, α-Smooth muscle actin is transiently expressed by myofibroblasts during experimental wound healing, *Lab. Invest.* **63:**21–29.

Davidson, J. M., Klagsbrun, M., Hill, K. E., Buckley, A., Sullivan, R., Brewer, P. S., and Woodward, S. C., 1985, Accelerated wound repair, cell proliferation, and collagen accumulation are produced by a cartilage-derived growth factor, *J. Cell Biol.* **100:**1219–1227.

Dedhar, S., Gaboury, L., Galloway, P., and Eaves, C., 1988, Human granulocyte–macrophage colony-stimulating factor is a growth factor active on a variety of cell types of nonhemopoietic origin, *Proc. Natl. Acad. Sci. USA* **85:**9253–9257.

Delvoye, P., Wiliquet, R., Leveque, J. L., Nusgens, B. V., and Lapière, C. M., 1991, Measurement of mechanical forces generated by skin fibroblasts embedded in the three-dimensional collagen gel, *J. Invest. Dermatol.* **97:**898–902.

Desmoulière, A., and Gabbiani, G., 1994, Modulation of fibroblastic cytoskeletal features during pathological situations: The role of extracellular matrix and cytokines, *Cell Motil. Cytoskeleton* **29:**195–203.

Desmoulière, A., Rubbia-Brandt, L., and Gabbiani, G., 1991, Modulation of actin isoform expression in cultured arterial smooth muscle cells by heparin and culture conditions, *Arterioscler. Thromb.* **11:**244–253.

Desmoulière, A., Rubbia-Brandt, L., Abdiu, A., Walz, T., Macieira-Coelho, A., and Gabbiani, G., 1992a, α-Smooth muscle actin is expressed in a subpopulation of cultured and cloned fibroblasts and is modulated by γ-interferon, *Exp. Cell Res.* **201:**64–73.

Desmoulière, A., Rubbia-Brandt, L., Grau, G., and Gabbiani, G., 1992b, Heparin induces α-smooth muscle actin expression in cultured fibroblasts and in granulation tissue myofibroblasts, *Lab. Invest.* **67:**716–726.

Desmoulière, A., Geinoz, A., Gabbiani, F., and Gabbiani, G., 1993, Transforming growth factor-β1 induces α-smooth muscle actin expression in granulation tissue myofibroblasts and in quiescent and growing cultured fibroblasts, *J. Cell Biol.* **122:**103–111.

Desmoulière, A., Redard, M., Darby, I., and Gabbiani, G., 1995, Apoptosis mediates the decrease in cellularity during the transition between granulation tissue and scar, *Am. J. Pathol.* **146:**56–66.

Duncan, M. R., and Berman, B., 1985, γ Interferon is the lymphokine and β interferon the monokine responsible for inhibition of fibroblast collagen production and late but not early fibroblast proliferation, *J. Exp. Med.* **162:**516–527.

Ehrlich, H. P., Desmoulière, A., Diegelmann, R. F., Cohen, I. K., Compton, C. C., Garner, W. L., Kapanci, Y., and Gabbiani, G., 1994, Morphological and immunochemical differences between keloid and hypertrophic scar, *Am. J. Pathol.* **145:**105–113.

Elger, M., Drenckhahn, D., Nobiling, R., Mundel, P., and Kriz, W., 1993, Cultured rat mesangial cells contain smooth muscle α-actin not found in vivo, *Am. J. Pathol.* **142:**497–509.

Estes, J. M., Vande Berg, J. S., Adzick, N. S., MacGillivray, T. E., Desmoulière, A., and Gabbiani, G., 1994, Phenotypic and functional features of myofibroblasts in sheep fetal wounds, *Differentiation* **56:**173–181.

Evan, G. I., Wyllie, A. H., Gilbert, C. S., Littlewood, T. D., Land, H., Brooks, M., Waters, C. M., Penn, L. Z., and Hancock, D. C., 1992, Induction of apoptosis in fibroblasts by c-myc protein, *Cell* **69:**119–128.

Evans, V. G., 1993, Multiple pathways to apoptosis, *Cell Biol. Int.* **17:**461–476.

Fabra, A., Nakajima, M., Bucana, C. D., and Fidler, I. J., 1992, Modulation of the invasive phenotype of human colon carcinoma cells by organ specific fibroblasts of nude mice, *Differentiation* **52:**101–110.

Fant, M. E., 1991, *In vitro* growth rate of placental fibroblasts is developmentally regulated, *J. Clin. Invest.* **88:**1697–1702.

Fina, M., Bresnick, S., Baird, A., and Ryan, A., 1991, Improved healing of tympanic membrane perforations with basic fibroblast growth factor, *Growth Factors* **5**:265–272.

Finch, P. W., Rubin, J. S., Miki, T., Ron, D., and Aaronson, S. A., 1989, Human KGF is FGF-related with properties of a paracrine effector of epithelial cell growth, *Science* **245**:752–755.

Flaumenhaft, R., and Rifkin, D. B., 1991, Extracellular matrix regulation of growth factor and protease activity, *Curr. Opin. Cell Biol.* **3**:817–823.

Franke, W. W., and Schinko, W., 1969, Nuclear shape in muscle cells, *J. Cell Biol.* **42**:326–331.

Fredj-Reygrobellet, D., Plouet, J., Delayre, T., Baudouin, C., Bourret, F., and Lapalus, P., 1986, Effects of aFGF and bFGF on wound healing in rabbit corneas, *Curr. Eye Res.* **6**:1205–1209.

Friedman, S. L., 1993, The cellular basis of hepatic fibrosis. Mechanisms and treatment strategies, *N. Engl. J. Med.* **25**:1828–1835.

Frisch, S. M., and Francis, H., 1994, Disruption of epithelial cell-matrix interactions induces apoptosis, *J. Cell Biol.* **124**:619–626.

Gabbiani, G., Ryan, G. B., and Majno, G., 1971, Presence of modified fibroblasts in granulation tissue and their possible role in wound contraction, *Experientia* **27**:549–550.

Gabbiani, G., Chaponnier, C., and Hüttner, I., 1978, Cytoplasmic filaments and gap junctions in epithelial cells and myofibroblasts during wound healing, *J. Cell Biol.* **76**:561–568.

Gabbiani, G., Schmid, E., Winter, S., Chaponnier, C., de Chastonay, C., Vanderkerckhove, J., Weber, K., and Franke, W. W., 1981, Vascular smooth muscle cells differ from other smooth muscle cells: Predominance of vimentin filaments and a specific α-type actin, *Proc. Natl. Acad. Sci. USA* **78**:298–302.

Gasson, J. C., 1991, Molecular physiology of granulocyte–macrophage colony-stimulating factor, *Blood* **77**:1131–1145.

Glasser, S. R., and Julian, J., 1986, Intermediate filament protein as a marker of uterine stromal cell decidualization, *Biol. Reprod.* **35**:463–474.

Goldman, R. D., Lazarides, E., Pollack, R., and Weber, K., 1975, The distribution of actin in nonmuscle cells. The use of actin antibody in the localization of actin within the microfilament bundles of mouse 3T3 cells, *Exp. Cell Res.* **90**:333–344.

Goss, A. N., 1977, Intrauterine healing of fetal rat oral mucosa, skin and cartilage, *J. Oral Pathol.* **6**:35–38.

Grandstein, R. D., Rook, A., Flotte, T. J., Hass, A., Gallo, R. L., Jaffe, H. S., and Amento, E. P., 1990, A controlled trial of intralesional recombinant interferon-γ in the treatment of keloidal scarring, *Arch. Dermatol.* **126**:1295–1302.

Greenburg, G., and Hay, E. D., 1986, Cytodifferentiation and tissue phenotype change during transformation of embryonic lens epithelium to mesenchyme-like cells *in vitro*, *Dev. Biol.* **115**:363–379.

Greenhalgh, D. G., Sprugel, K. H., Murray, M. J., and Ross, R., 1990, PDGF and FGF stimulate wound healing in the genetically diabetic mouse, *Am. J. Pathol.* **136**:1235–1246.

Grimaud, J. A., and Borojevic, R., 1977, Myofibroblasts in hepatic schistosomal fibrosis, *Experientia* **33**:890–892.

Grillo, H. C., 1963, Origin of fibroblasts in wound healing: An autoradiographic study of inhibition of cellular proliferation by local x-irradiation, *Ann. Surg.* **157**:453–467.

Grinnell, F., 1992, Wound repair, keratinocyte activation and integrin modulation, *J. Cell Sci.* **101**:1–5.

Grinnell, F., 1994, Fibroblasts, myofibroblasts, and wound contraction, *J. Cell Biol.* **124**:401–404.

Hahn, A. W. A., Resink, T. J., Kern, F., and Bühler, F. R., 1992, Effects of endothelin-1 on vascular smooth muscle cell phenotypic differentiation, *J. Cardiovasc. Pharmacol.* **20**(Suppl. 12):533–536.

Harris, A. K., Stopack, D., and Wild, P., 1981, Fibroblast traction as a mechanism for collagen morphogenesis, *Nature* **290**:249–251.

Hayashi, N., Cuhna, G. R., and Parker, M., 1993, Permissive and instructive induction of adult rodent prostatic epithelium by heterotypic urogenital sinus mesenchyme, *Epithelial Cell Biol.* **2**:66–78.

Hayflick, L., 1965, The limited *in vitro* lifetime of human diploid cell strains, *Exp. Cell Res.* **37**:614–636.

Hayward, P. G., Geldner, P., Altrock, B., Pierce, G., and Robson, M. C., 1990, Granulocyte–macrophage colony stimulating factor in open wound healing, *Surg. Forum* **41**:621–623.

Hébert, L., Pandey, S., and Wang, E., 1994, Commitment to cell death is signaled by the appearance of a terminin protein of 30 kDa, *Exp. Cell Res.* **210**:10–18.

Hoover, R. L., Rosenberg, R., Hearing, W., and Karnovsky, M. J., 1980, Inhibition of rat arterial smooth muscle cell proliferation by heparin, *Circ. Res.* **47**:578–583.

Hunt, T. K., Banda, M. J., and Silver, I. A., 1985, Cell interactions in posttraumatic fibrosis, in: *Fibrosis,* Ciba Foundation Symposium 114 (A. Bailey, ed.), pp. 127–149, Pitman, London.

Igarashi, A., Okochi, H., Bradham, D. M., and Grotendorst, G. R., 1993, Regulation of connective tissue growth factor gene expression in human skin fibroblasts and during wound repair, *Mol. Biol. Cell* **4:**637–645.

Ingber, D. E., 1993, Cellular tensegrity: Defining new rules of biological design that govern the cytoskeleton, *J. Cell Sci.* **104:**613–627.

Janssen, P., 1902, Zur Lehre von der Dupuytren'schen Fingerkontraktur mit besonderer Berücksichtigung der operativen Beseitigung und der pathologischen Anatomie des Leidens, *Arch. Klin. Chir.* [*Am.*] **67:**761–789.

Johnson, R. J., Iida, H., Alpers, C. E., Majesky, M. W., Schwartz, S. M., Pritzl, P., Gordon, K., and Gown, A. M., 1991, Expression of smooth muscle cell phenotype by rat mesangial cells in immune complex nephritis. α-Smooth muscle actin is a marker of mesangial cell proliferation, *J. Clin. Invest.* **87:**847–858.

Jones, T. C., 1993, The effects of rhGM-CSF on macrophage function, *Eur. J. Cancer* **29A:**S10–S13.

Juliano, R. L., and Haskill, S., 1993, Signal transduction from the extracellular matrix, *J. Cell Biol.* **120:**577–585.

Jürgensmeier, J. M., Schmitt, C. P., Viesel, E., Höfler, P., and Bauer, G., 1994, Transforming growth factor β-treated normal fibroblasts eliminate transformed fibroblasts by induction of apoptosis, *Cancer Res.* **54:**393–398.

Kapanci, Y., Burgan, S., Pietra, G. G., Conne, B., and Gabbiani, G., 1990, Modulation of actin isoform expression in alveolar myofibroblasts (contractile interstitial cells) during pulmonary hypertension, *Am. J. Pathol.* **136:**881–889.

Kapanci, Y., Ribaux, C., Chaponnier, C., and Gabbiani, G., 1992, Cytoskeletal features of alveolar myofibroblasts and pericytes in normal human and rat lung, *J. Histochem. Cytochem.* **40:**1955–1963.

Kaye, G. I., Lane, N., and Pascal, P. R., 1968, Colonic pericryptal fibroblast sheet: Replication, migration and cytodifferentiation of a mesenchymal cell-system in adult tissue. II. Fine structural aspects of normal rabbit and human colon, *Gastroenterology* **54:**852–865.

Khalil, N., Bereznay, O., Sporn, M., and Greenberg, A. H., 1989, Macrophage production of transforming growth factor β and fibroblast collagen synthesis in chronic pulmonary inflammation, *J. Exp. Med.* **170:**727–737.

Kischer, C. W., Thies, A. C., and Chvapil, M., 1982, Perivascular myofibroblasts and microvascular occlusion in hypertrophic scars and keloids, *Hum. Pathol.* **13:**819–824.

Klein, C. E., Dressel, D., Steinmayer, T., Mauch, C., Eckes, B., Krieg, T., Bankert, R. B., and Weber, L., 1991, Integrin α2β1 is up-regulated in fibroblasts and highly aggressive melanoma cells in three-dimensional collagen lattices and mediates the reorganization of collagen I fibrils, *J. Cell Biol.* **115:**1427–1436.

Komuro, T., 1990, Re-evaluation of fibroblasts and fibroblast-like cells, *Anat. Embryol.* **182:**103–112.

Kovacs, E. J., 1991, Fibrogenic cytokines: The role of immune mediators in the development of scar tissue, *Immunol. Today* **12:**17–23.

Krummel, T. M., Nelson, J. M., Diegelmann, R. F., Lindblad, W. J., Salzberg, A. M., Greenfield, L. J., and Cohen, I. K., 1987, Fetal response to injury in the rabbit, *J. Pediatr. Surg.* **22:**640–644.

Kuhn, C., and McDonald, J. A., 1991, The roles of the myofibroblast in idiopathic pulmonary fibrosis, *Am. J. Pathol.* **138:**1257–1265.

Lang, R. A., Metcalf, D., Cuthbertson, R. A., Lyons, I., Stanley, E., Kelso, A., Kannourakis, G., Williamson, D. J., Klintworth, G. K., Gonda, T. J., and Dunn, AR., 1987, Transgenic mice expressing a hemopoietic growth factor gene (GM-CSF) develop accumulations of macrophages, blindness, and a fatal syndrome of tissue damage, *Cell* **51:**675–686.

Lantz, M., Thysell, H., Nilsson, E., and Olsson, I., 1991, On the binding of tumor necrosis factor (TNF) to heparin and the release *in vivo* of the TNF-binding protein I by heparin, *J. Clin. Invest.* **88:**2026–2031.

Larrabee, Jr., W. F., East, C. A., Jaffe, H. S., Stephenson, C., and Peterson, K. E., 1990, Intralesional interferon gamma treatment for keloids and hypertrophic scars, *Arch. Otolaryngol. Head Neck Surg.* **116:**1159–1162.

Larsen, R. D., and Posch, J. L., 1958, Dupuytren's contracture. With special reference to pathology, *J. Bone Joint Surg. Am.* **40A:**773–792.

Leavitt, J., Gunning, P., Kedes, L., and Jariwalla, R., 1985, Smooth muscle α-actin is a transformation-sensitive marker for mouse NIH 3T3 and rat-2 cells, *Nature* **316:**840–842.

Levi-Schaffer, F., Austen, K. F., Caulfield, J. P., Hein, A., Bloes, W. F., and Stevens, R. L., 1985, Fibroblasts maintain the phenotype and viability of the rat heparin-containing mast cells *in vitro, J. Immunol.* **135:**3454–3462.

Longaker, M. T., Burd, D. A. R., Gown, A. M., Yen T. S. B., Jennings, R. W., Duncan, B. W., Harrison, M. R., and Adzick, N. S., 1991, Midgestational excisional fetal lamb wounds contract *in utero, J. Pediatr. Surg.* **26:**942–948.

Loréal, O., Levavasseur, F., Fromaget, C., Gros, D., Guillouzo, A., and Clément, B., 1993, Cooperation of Ito cells and hepatocytes in the deposition of an extracellular matrix *in vitro, Am. J. Pathol.* **143:**538–544.

Lynch, S. E., Nixon, J. C., Colvin, R. B., and Antoniades, H. N., 1987, Role of platelet-derived growth factor in wound healing: Synergistic effects with other growth factors, *Proc. Natl. Acad. Sci. USA* **84:**7696–7700.

MacDonald, R. A., 1959, Origin of fibroblasts in experimental healing wounds: Autoradiographic studies using tritiated thymidin, *Surgery* **46:**376–382.

Macieira-Coelho, A., 1988, Biology of normal proliferating cells *in vitro.* Relevance for *in vivo* aging, in: *Interdisciplinary Topics in Gerontology,* Vol. 23 (H. P. von Hahn, ed.), pp. 1–218, Karger, Basel.

Macieira-Coelho, A., and Taboury, F., 1982, A re-evaluation of the changes in proliferation in human fibroblasts during ageing *in vitro, Cell Tissue Kinet.* **15:**213–224.

MacNulty, E. E., Plevin, R., and Wakelam, M. J. O., 1990, Stimulation of the hydrolysis of phosphatidylinositol 4,5-biphosphate and phosphatidylcholine by endothelin, a complete mitogen for Rat-1 fibroblasts, *Biochem. J.* **272:**761–766.

MacSween, R. N. M., and Whaley, K., 1992, Inflammation, healing and repair, in: *Muir's Textbook of Pathology,* 13th ed., pp. 112–165, Edward Arnold, London.

Majack, R. A., and Bornstein, P., 1984, Heparin and related glycosaminoglycans modulate the secretory phenotype of vascular smooth muscle cells, *J. Cell Biol.* **99:**1688–1695.

Majno, G., Shea, S. M., and Leventhal, M., 1969, Endothelial contraction induced by histamine-like mediators. An electron microscopic study, *J. Cell Biol.* **42:**647–672.

Majno, G., Gabbiani, G., Hirschel, B. J., Ryan, G. B., and Statkov, P. R., 1971, Contraction of granulation tissue *in vitro:* Similarity to smooth muscle, *Science* **173:**548–550.

Martin, G. M., Sprague, C. A., Norwood, T. H., and Pendergrass, W. R., 1974, Clonal selection, attenuation and differentiation in an *in vitro* model of hyperplasia, *Am. J. Pathol.* **74:**137–154.

Martin, P., and Lewis, J., 1992, Actin cables and epidermal movement in embryonic wound healing, *Nature* **360:**179–182.

Martin, P., Hopkinson-Woolley, J., and McCluskey, J., 1992, Growth factors and cutaneous wound repair, *Prog. Growth Factor Res.* **4:**25–44.

Martin, P., Nobes, C., McCluskey, J., and Lewis, J., 1994, Repair of exisional wounds in the embryo, *Eye* **8:**155–160.

Mauviel, A., and Uitto, J., 1993, The extracellular matrix in wound healing: Role of the cytokine network, *Wounds* **5:**137–152.

McCaffrey, T. A., Falcone, D. J., Brayton, C. F., Agarwal, L. A., Welt, F. G. P., and Weksler, B. B., 1989, Transforming growth factor-β activity is potentiated by heparin via dissociation of the transforming growth factor-β/α$_2$-macroglobulin inactive complex, *J. Cell Biol.* **109:**441–448.

McCaffrey, T. A., Falcone, D. J., and Du, B., 1992, Transforming growth factor-β1 is a heparin-binding protein: Identification of putative heparin-binding regions and isolation of heparins with varying affinity for TGF-β1, *J. Cell. Physiol.* **152:**430–440.

McCulloch, C. A. G., and Bordin, S., 1991, Role of fibroblast subpopulations in periodontal physiology and pathology, *J. Periodont. Res.* **26:**144–154.

McDonald, J. A., Quade, B. J., Broekelmann, T. J., LaChane, R., Forsman, K., Hasegawa, E., and Akiyama, S., 1987, Fibronectin's cell-adhesive domain and an amino-terminal matrix assembly domain participate in the assembly into fibroblast pericellular matrix, *J. Biol. Chem.* **262:**2957–2967.

Meredith, Jr., J. E., Fazeli, B., and Schwartz, M. A., 1993, The extracellular matrix as a cell survival factor, *Mol. Biol. Cell* **4**:953–961.

Missero, C., Ramon y Cajal, S., and Dotto, G. P., 1991, Escape from transforming growth factor β control and oncogene cooperation in skin tumor development, *Proc. Natl. Acad. Sci. USA* **88**:9613–9617.

Mitchell, J. J., Woodcock-Mitchell, J., Reynolds, S., Low, R., Leslie, K., Adler, K., Gabbiani, G., and Skalli, O., 1989, α-Smooth muscle actin in parenchymal cells of bleomycin-injured rat lung, *Lab. Invest.* **60**:643–650.

Miura, M., Zhu, H., Rotello, R., Hartwieg, E. A., and Yuan, J., 1993, Induction of apoptosis in fibroblasts by IL-1β-converting enzyme, a mammalian homolog of the *C. elegans* cell death gene *ced*-3, *Cell* **75**:653–660.

Montesano, R., Matsumoto, K., Nakamura, T., and Orci, L., 1991, Identification of a fibroblast-derived epithelial morphogen as hepatocyte growth factor, *Cell* **67**:901–908.

Montesano, R., Pepper, M. S., and Orci, L., 1993, Paracrine induction of angiogenesis *in vitro* by Swiss 3T3 fibroblasts, *J. Cell Sci.* **105**:1013–1024.

Moulton, B. C., 1994, Transforming growth factor-β stimulates endometrial stromal apoptosis *in vitro*, *Endocrinology* **134**:1055–1060.

Murrel, G. A. C., Francis, M. J. O., and Bromley, L., 1987, Free radicals and Dupuytren's contracture, *Br. Med. J.* **295**:1373–1375.

Murrel, G. A. C., Francis, M. J. O., and Howlett, C. R., 1989, Dupuytren's contracture. Fine structure in relation to aetiology, *J. Bone Joint Surg.* **71B**:367–373.

Muthukrishnan, L., Warder, E., and McNeil, P. L., 1991, Basic fibroblast growth factor is efficiently released from a cytosolic storage site through plasma membrane disruptions of endothelial cells, *J. Cell. Physiol.* **148**:1–16.

Nakanishi, H., Oguri, K., Takenaga, K., Hosoda, S., and Okayama, M., 1994, Differential fibrotic stromal responses of host tissue to low- and high-metastatic cloned Lewis lung carcinoma cells, *Lab. Invest.* **70**:324–332.

Nishiyama, T., Tsunenaga, M., Nakayama, Y., Adachi, E., and Hayashi, T., 1989, Growth rate of human fibroblasts is repressed by the culture within reconstituted collagen matrix but not by the culture on the matrix, *Matrix* **9**:193–199.

Nusgens, B., Merrill, C., Lapière, C., and Bell, E., 1984, Collagen biosynthesis by cells in a tissue equivalent matrix *in vitro*, *Collagen Relat. Res.* **4**:351–363.

Okomoto-Inoue, M., Taniguchi, S., Sadano, H., Kawano, T., Kimura, G., Gabbiani, G., and Baba, T., 1990, Alteration in expression of smooth muscle α-actin associated with transformation of rat 3Y1 cells, *J. Cell Sci.* **96**:631–637.

Ornitz, D. M., Yayon, A., Flanagan, J. G., Svahn, C. M., Levi, E., and Leder, P., 1992, Heparin is required for cell-free binding of basic fibroblast growth factor to a soluble receptor and for mitogenesis in whole cells, *Mol. Cell. Biol.* **12**:240–247.

Pascolini, R., Di Rosa, I., Fagotti, A., Panara, F., and Gabbiani, G., 1992, The mammalian anti-α-smooth muscle actin monoclonal antibody recognizes an α-actin-like protein in planaria (*Dugesia lugubris* s.l.), *Differentiation* **51**:177–186.

Pepper, M. S., Vassalli, J. D., Orci, L., and Montesano, R., 1993, Biphasic effect of transforming growth factor-β1 on *in vitro* angiogenesis, *Exp. Cell Res.* **204**:356–363.

Perdomo, J. J., Gounon, P., Schaeverbeke, M., Schaeverbeke, J., Groult, V., Jacob, M. P., and Robert, L., 1994, Interaction between cells and elastin fibers: An ultrastructural and immunocytochemical study, *J. Cell. Physiol.* **158**:451–458.

Phipps, R. P., Penney, D. P., Keng, P., Silvera, M., Harkins, S., and Derdak, S., 1990, Immune functions of subpopulations of lung fibroblasts, *Immunol. Res.* **9**:275–286.

Pierce, G. F., Mustoe, T. A., Lingelbach, J., Masakowski, V. R., Gramates, P., and Deuel, T. F., 1989, Transforming growth factor β reverses the glucocorticoid-induced wound-healing deficit in rats: Possible regulation in macrophages by platelet-derived growth factor, *Proc. Natl. Acad. Sci. USA* **86**:2229–2233.

Pierce, G. F., Mustoe, T. A., Altrock, B. W., Deuel, T. F., and Thomasson, A., 1991, Role of platelet-derived growth factor in wound healing, *J. Cell. Biochem.* **45**:319–326.

Pittet, B., Rubbia-Brandt, L., Desmoulière, A., Sappino, A. P., Roggero, P., Guerret, S., Grimaud, J. A.,

Lacher, R., Montandon, D., and Gabbiani, G., 1994, Action of γ-interferon on the clinical and biologic evolution of hypertrophic scars and Dupuytren's disease: An open pilot study, *Plast. Reconstr. Surg.* **93:**1224–1235.

Pohl, J., Bruhn, H. D., and Christophers, E., 1979, Thrombin and fibrin-induced growth of fibroblasts: Role in wound repair and thrombus organization, *Klin. Wochenschr.* **57:**273–277.

Ramadori, G., 1991, The stellate cell (Ito-cell, fat-storing cell, lipocyte, perisinusoidal cell) of the liver, *Virchows Arch. [B] Cell Pathol.* **61:**147–158.

Robaye, B., Mosselmans, R., Fiers, W., Dumont, J. E., and Galand, P., 1991, Tumor necrosis factor induces apoptosis (programmed cell death) in normal endothelial cells *in vitro, Am. J. Pathol.* **138:**447–453.

Robson, M. C., 1991, Growth factors as wound healing agents, *Curr. Opin. Biotech.* **2:**863–867.

Roche, W. R., 1991, Fibroblasts and asthma, *Clin. Exp. Allergy* **21:**545–548.

Rockey, D. C., and Friedman, S. L., 1992, Cytoskeleton of liver perisinusoidal cells (lipocytes) in normal and pathological conditions, *Cell Motil. Cytoskeleton* **22:**227–234.

Ronnov-Jessen, L., van Deurs, B., Celis, J. E., and Petersen, O. W., 1990, Smooth muscle differentiation in cultured human breast gland stromal cells, *Lab. Invest.* **63:**532–543.

Ronnov-Jessen, L., and Petersen, O. W., 1993, Induction of α-smooth muscle actin by transforming growth factor-β1 in quiescent human breast gland fibroblasts. Implications for myofibroblast generation in breast neoplasia, *Lab. Invest.* **68:**696–707.

Ross, R., Everett, N. B., and Tyler, R., 1970, Wound healing and collagen formation. VI. The origin of the wound fibroblast studied in parabiosis, *J. Cell Biol.* **44:**645–654.

Rowsell, A. R., 1986, The intra-uterine healing of fetal muscle wounds: Experimental study in the rat, *Br. J. Plast. Surg.* **37:**635–642.

Rubbia-Brandt, L., Sappino, A. P., and Gabbiani, G., 1991, Locally applied GM-CSF induces the accumulation of α-smooth muscle actin containing myofibroblasts, *Virchows Arch. [B] Cell. Pathol.* **60:**73–82.

Ruoslahti, E., and Yamaguchi, Y., 1991, Proteoglycans as modulators of growth factor activities, *Cell* **64:**867–869.

Ruoslahti, E., and Reed, J. C., 1994, Anchorage dependence, integrins, and apoptosis, *Cell* **77:**477–478.

Ryan, G. B., Cliff, W. J., Gabbiani, G., Irle, C., Montandon, D., Statkov, P. R., and Majno, G., 1974, Myofibroblasts in human granulation tissue, *Hum. Pathol.* **5:**55–67.

Sappino, A. P., Dietrich, P. Y., Widgren, S., and Gabbiani, G., 1989, Colonic pericryptal fibroblasts. Differentiation pattern in embryogenesis and phenotypic modulation in epithelial proliferative lesions, *Virchows Arch. [A] Pathol. Anat.* **415:**551–557.

Sappino, A. P., Masouyé, I., Saurat, J. H., and Gabbiani, G., 1990a, Smooth muscle differentiation in scleroderma fibroblastic cells, *Am J. Pathol.* **137:**585–591.

Sappino, A. P., Schürch, W., and Gabbiani, G., 1990b, Differentiation repertoire of fibroblastic cells: Expression of cytoskeletal proteins as marker of phenotypic modulations, *Lab. Invest.* **63:**144–161.

Schiro, J. A., Chan, B. M. C., Roswit, W. R., Kassner, P. D., Pentland, A. P., Hemler, M. E., Eisen, A. Z., and Kupper, T. S., 1991, Integrin α2β1 (VLA-2) mediates reorganization and contraction of collagen matrices by human cells, *Cell* **67:**403–410.

Schmitt-Gräff, A., Pau, H., Spahr, R., Piper, H. M., Skalli, O., and Gabbiani, G., 1990, Appearance of alpha-smooth muscle actin in human eye lens cells of anterior capsular cataract and in cultured bovine lens-forming cells, *Differentiation* **43:**115–122.

Schmitt-Gräff, A., Krüger, S., Bochard, F., Gabbiani, G., and Denk, H., 1991, Modulation of alpha smooth muscle actin and desmin expression in perisinusoidal cells of normal and diseased human livers, *Am. J. Pathol.* **138:**1233–1242.

Schmitt-Gräff, A., Desmoulière, A., and Gabbiani, G., 1994, Heterogeneity of myofibroblast phenotype features: An example of fibroblastic cell plasticity, *Virchows Arch.* **425:**3–24.

Schürch, W., Seemayer, T. A., and Gabbiani, G., 1992, Myofibroblast, in: *Histology for Pathologists* (S. S. Sternberg, ed.), pp. 109–144, Raven Press, New York.

Seppa, H. E. J., Grotendorst, G. R., Seppa, S. I., Schiffmann, E., and Martin, G. R., 1982, Platelet-derived growth factor is chemotactic for fibroblasts, *J. Cell Biol.* **92:**584–588.

Shah, M., Foreman, D. M., and Fergusson, M. W., 1992, Control of scarring in adult wounds by neutralising antibody to transforming growth factor beta, *Lancet* **339:**213–214.

Shimizu, K., and Yoshizato, K., 1992, Organ-dependent expression of differentiated states in fibroblasts cultured *in vitro, Dev. Growth Differ.* **34**:43–50.

Shum, D. T., and McFarlane, R. M., 1988, Histogenesis of Dupuytren's disease: An immunohistochemical study of 30 cases, *J. Hand Surg.* **13A**:61–67.

Singer, I. I., 1979, The fibronexus: A transmembrane association of fibronectin-containing fibers and bundles of 5 nm microfilaments in hamster and human fibroblasts, *Cell* **16**:675–685.

Singer, I. I., Kawka, D. W., Kazazis, D. M., and Clark, R. A. F., 1984, *In vivo* codistribution of fibronectin and actin fibers in granulation tissue: Immunofluorescence and electron microscope studies of the fibronexus at the myofibroblast surface, *J. Cell Biol.* **98**:2091–2106.

Skalli, O., and Gabbiani, G., 1988, The biology of the myofibroblast. Relationship to wound contraction and fibrocontractive diseases, in: *The Molecular and Cellular Biology of Wound Repair* (R. A. F. Clark, and P. M. Henson, eds.), pp. 373–402, Plenum Press, New York.

Skalli, O., Ropraz, P., Trzeciak, A., Benzonana, G., Gillessen, D., and Gabbiani, G., 1986, A monoclonal antibody against α-smooth muscle actin: A new probe for smooth muscle differentiation, *J. Cell Biol.* **103**:2787–2796.

Skalli, O., Schürch, W., Seemayer, T. A., Lagacé, R., Montandon, D., Pittet, B., and Gabbiani, G., 1989, Myofibroblasts from diverse pathological settings are heterogeneous in their content of actin isoforms and intermediate filament proteins, *Lab. Invest.* **60**:275–285.

Soma, Y., and Grotendorst, G. R., 1989, TGFβ stimulates primary human skin fibroblasts DNA synthesis via an autocrine production of PDGF-related peptides, *J. Cell. Physiol.* **140**:246–253.

Streuli, C. H., Schmidhauser, C., Kobrin, M., Bissell, M. J., and Derynck, R., 1993, Extracellular matrix regulates expression of the TGF-β1 gene, *J. Cell Biol.* **120**:253–260.

Thiemermann, C., and Corder, R., 1992, Is endothelin-1 the regulator of myofibroblast contraction during wound healing? *Lab. Invest.* **67**:677–679.

Thornton, S. C., Por, S. B., Walsh, B. J., Penny, R., and Breit, S. N., 1990, Interaction of immune and connective tissue cells: I. The effect of lymphokines and monokines on fibroblast growth, *J. Leucocyte Biol.* **47**:312–320.

Toccanier-Pelte, M. F., Skalli, O., Kapanci, Y., and Gabbiani, G., 1987, Characterization of stromal cells with myoid features in lymph nodes and spleen in normal and pathologic conditions, *Am. J. Pathol.* **129**:109–118.

Tomasek, J. J., and Haaksma, C. J., 1991, Fibronectin filaments and actin microfilaments are organized into a fibronexus in Dupuytren's diseased tissue, *Anat. Rec.* **230**:175–182.

Tomasek, J. J., Haaksma, C. J., Eddy, R. J., and Vaughan, M. B., 1992, Fibroblast contraction occurs on release of tension in attached collagen lattices: Dependency on an organized actin cytoskeleton and serum, *Anat. Rec.* **232**:358–368.

Ulich, T. R., Yi, E. S., Cardiff, R., Yin, S., Bikhazi, N., Biltz, R., Morris, C. F., and Pierce, G. F., 1994, Keratinocyte growth factor is a growth factor for mammary epithelium *in vivo*. The mammary epithelium of lactating rats is resistant to the proliferative action of keratinocyte growth factor, *Am. J. Pathol.* **144**:862–868.

Vandekerckhove, J., and Weber, K., 1981, Actin typing on total cellular extracts. A highly sensitive protein chemical procedure able to distinguish different actins, *Eur. J. Biochem.* **113**:595–603.

Vyalov, S., Desmoulière, A., and Gabbiani, G., 1993, GM-CSF-induced granulation tissue formation: Relationships between macrophage and myofibroblast accumulation, *Virchows Arch. [B] Cell. Pathol.* **63**:231–239.

Welch, M. P., Odland, G. F., and Clark, R. A. F., 1990, Temporal relationships of F-actin bundle formation, collagen and fibronectin matrix assembly, and fibronectin receptor expression to wound contraction, *J. Cell Biol.* **110**:133–145.

Wharton, W., 1984, Newborn human skin fibroblast senesce *in vitro* without acquiring adult growth factor requirements, *Exp. Cell Res.* **154**:310–314.

Willingham, M. C., Yamada, S. S., Davies, P. J. A., Rutherford, A. V., Gallo, M. G., and Pastan, I., 1981, Intracellular localization of actin in cultured fibroblasts by electron microscopic immunochemistry, *J. Histochem. Cytochem.* **29**:17–37.

Yokoi, Y., Namihisa, T., Kuroda, H., Komatsu, I., Miyazaki, A., Watanabe, S., and Usui, K., 1984, Immunocytochemical detection of desmin in fat-storing cells (Ito cells), *Hepatology* **4**:709–714.

Zavala, C., Herner, G., and Fialkow, P. J., 1978, Evidence for selection in cultured diploid fibroblast strains, *Exp. Cell Res.* **117:**137–144.

Zhang, K., Rekhter, M. D., Gordon, D., and Phan, S. H., 1994, Myofibroblasts and their role in lung collagen gene expression during pulmonary fibrosis. A combined immunohistochemical and *in situ* hybridization study, *Am. J. Pathol.* **145:**114–125.

Part IV

Essentials of Tissue Remodeling

Chapter 14

Proteinases and Tissue Remodeling

PAOLO MIGNATTI, DANIEL B. RIFKIN, HOWARD G. WELGUS, and WILLIAM C. PARKS

1. Introduction

The term *tissue remodeling* describes transient or permanent changes in tissue architecture that involve breaching of histological barriers such as basement membranes, basal laminae, and interstitial stroma [extracellular matrix (ECM)]. Tissue remodeling is important to several stages of wound repair, such as inflammation and granulation tissue formation, and in a variety of other physiological or pathological states. These include ovulation, spermatogenesis, trophoblast implantation, mammary involution following lactation, uterine involution, nerve regeneration, rheumatoid arthritis, tumor invasion, and metastasis formation. A common feature of tissue remodeling involves the production of high levels of extracellular proteolytic activities by parenchymal and/or connective tissue cells. The ECM is organized into highly complex structures, each of which consists of different components including various collagen types, glycoproteins such as fibronectin and laminin, elastin, glycosaminoglycans (GAGs), and proteoglycans. Because these ECM components have distinct hydrolytic requirements for their degradation, remodeling of the ECM involves the action of an array of degradative enzymes.

The ECM-degrading proteinases produced by most cells can be subdivided into three main classes: (1) serine proteinases, (2) metalloproteinases, and (3) cysteine proteinases (cathepsins) (Mignatti and Rifkin, 1993). In addition, a number of endo- and exoglycosidases selectively degrade GAGs and the amino sugar moieties of proteoglycans. A consistent body of experimental evidence has shown that serine proteinases and metalloproteinases play major roles in most physiological and pathological states involving tissue remodeling. These enzymes will be discussed in this section.

PAOLO MIGNATTI • Department of Genetics and Microbiology, University of Pavia, 27100 Pavia, Italy.   DANIEL B. RIFKIN • Department of Cell Biology and Kaplan Cancer Center, New York University Medical Center, New York, New York 10016–6402.   HOWARD G. WELGUS and WILLIAM C. PARKS • Division of Dermatology, Department of Medicine, Washington University School of Medicine at the Jewish Hospital, St. Louis, Missouri 63110.

The Molecular and Cellular Biology of Wound Repair (Second Edition), edited by Richard A. F. Clark. Plenum Press, New York, 1996

2. Serine Proteinases: The Plasminogen Activators

The major members of the family of the serine proteinases include the plasminogen activators (PAs), leukocyte elastase, and cathepsin G. Of these, the former are the best characterized, and a considerable body of experimental evidence indicates their involvement in many conditions involving tissue remodeling (for reviews, see Mullins and Rohrlich, 1983; Danø et al., 1985; Moscatelli and Rifkin, 1988; Saksela and Rifkin, 1988; Mignatti and Rifkin, 1993). PAs convert the zymogen plasminogen, a plasma protein ubiquitous in the body, to plasmin. Originally known as fibrinolysin because of its ability to degrade the fibrin clot, plasmin has a broad, trypsinlike substrate specificity and can degrade several ECM components including fibronectins, laminin, and the protein core of proteoglycans (Werb et al., 1980). It does not degrade elastin and native collagens, but it can degrade gelatins, the partially degraded or denatured forms of the collagens. In addition, plasmin also activates certain prometalloproteinases (see Section 4.3) (Werb et al., 1977; Matrisian, 1990; Murphy et al., 1992b), as well as latent elastase (Chapman and Stone, 1984).

A number of vertebrate proteins, including plasma kallikrein, the blood coagulation factors XI and XII, and the bacterial protein streptokinase can activate plasminogen (Colman, 1969; Mandle and Kaplan, 1977; Bouma and Griffin, 1978; Summaria et al., 1982). However, the term PA is currently restricted to two enzymes—the urokinase-type PA (uPA) and the tissue-type PA (tPA)—which are kinetically very efficient activators of plasminogen.

2.1. Domain Structure of PAs

Urokinase (55 kDa) and tPA (70 kDa) are the products of two distinct genes that have evolved from a common ancestor gene by duplication and subsequent mutations (Edlund et al., 1983; Pennica et al., 1983; Verde et al., 1984). The amino acid and nucleotide sequences of uPA and tPA reveal interesting features. Both PAs are typical "mosaic" proteins consisting of distinct domains that have structural and/or functional homology to functional units of ECM structural proteins or growth factors. These modular structures result from "exon shuffling" or "exon insertion," a process by which structural or functional domains are exchanged between proteins during evolution. tPA was the first mosaic protein to be described. Three functional "modules" are present both in tPA and uPA (Fig. 1): (1) a C-terminal proteinase domain homologous to the proteinase domain of trypsin and other trypsinlike proteinases, and containing the His, Asp, and Ser residues typical of the charge relay active site of all serine proteinases; (2) the "kringle" module, a cysteine-rich sequence folded by three internal disulfide bridges to form a structure that resembles the Danish cake bearing this name; and (3) the N-terminal "growth factor" domain, a cysteine-rich sequence with high homology to epidermal growth factor (EGF) (Li and Graur, 1991). Whereas uPA has one kringle, tPA has two. In addition, the 43 N-terminal residues of tPA have no counterpart in uPA. This sequence forms a fingerlike structure homologous to the "finger" domains of

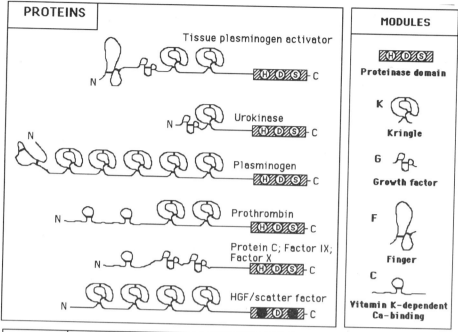

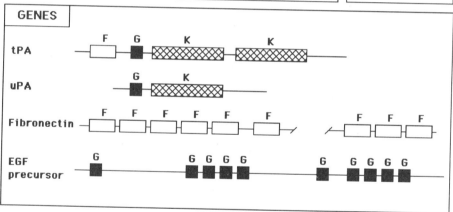

Figure 1. Modular structures of urokinase and tissue-type plasminogen activators. The protein and gene structures of uPA and tPA are shown in comparison to those of other mosaic proteins. In the proteinase domain of hepatocyte growth factor, mutations of the His and Ser residues into Gln and Tyr, respectively, result in loss of catalytic activity.

fibronectin that confer fibrin affinity on this ECM protein. tPA, unlike uPA, has a strong affinity for fibrin. Because plasminogen also binds to fibrin, the positioning of plasminogen and tPA on the common ligand results in a 60-fold decrease of the K_M (increase in the affinity) for plasminogen activation. Whereas tPA is poorly active in the absence of fibrin, its activity is strongly enhanced by fibrin. On the basis of these features, it has been proposed that the two PAs may have different physiological roles, tPA being primarily involved in clot lysis and uPA mediating tissue-remodeling processes.

Kringle domains are also present in other serine proteinases: plasminogen has five kringles and prothrombin has two (Li and Graur, 1991). In addition, apolipoprotein "a" possesses a number of kringles with high homology to plasminogen but is devoid of an active proteinase domain (Ichinose, 1992). Interestingly, the recently identified hepatocyte growth factor/scatter factor (HGF/SF) also has structural and sequence homology to plasminogen (Ponting *et al.*, 1992). HGF/SF has four kringles and a proteinase domain that lacks catalytic activity because of a mutation in the active site (Tashiro *et al.*, 1990; Nakamura, 1991; Gherardi *et al.*, 1993; Mizuno and Nakamura, 1993). It is noteworthy that, like plasminogen, HGF/SF is also efficiently activated by uPA or other serine proteinases (Naldini *et al.*, 1992). In addition, HGF/SF stimulates uPA expression in epithelial and endothelial cells (Bussolino *et al.*, 1992; Pepper *et al.*, 1992a,b). "Growth factor" domains are also present in the blood coagulation factors IX and X, and in protein C (Li and Graur, 1991) (Fig. 1).

In their catalytically active forms, uPA and tPA consist of two polypeptide chains (A and B). The B chains of both enzymes, which include the C-terminal portion of the proteins, are similar and contain the active site. The A chains also show a high degree of homology but differ considerably in size, the difference being accounted for by the second kringle and the finger domain of tPA. The amino acid sequence encompassing residues 13–30 of the N-terminal, EGF-like domain of the noncatalytic A chain of uPA mediates binding to a specific cell membrane receptor (uPA receptor) (see Section 2.2.1) (Appella *et al.*, 1987).

2.2. Regulation of PA Activity

The expression of PA activity is regulated by complex control mechanisms that act both transcriptionally and posttranscriptionally. Similar mechanisms also control matrix metalloproteinase (MMP) activities (Fig. 2). In a variety of cells, PA gene transcription is modulated by a number of agents, including tumor promoters, oncogenes, growth factors, cyclic AMP, retinoids, prostaglandins, and UV light (Table I) (Danø *et al.*, 1985). Posttranscriptional control of enzyme activity occurs at different levels. These include: (1) proenzyme activation, (2) enzyme focalization on the cell membrane, and (3) interaction with specific tissue inhibitors (Fig. 2).

As is the case for all extracellular serine proteinases and MMPs (see Section 4), PAs are secreted in the form of inactive, single-chain zymogens (pro-uPA or sc-uPA, pro-tPA or sc-tPA) that are converted to the active, two-chain form by limited proteolysis (Fig. 3) (Danø *et al.*, 1985; Petersen *et al.*, 1988). One of the more important features of the PA–plasmin modulatory system is the amplification loop resulting from plasminogen and pro-uPA activation. Trace amounts of plasmin activate pro-uPA (Cubellis *et al.*, 1986; Petersen *et al.*, 1988), thus generating a self-maintained feedback mechanism of pro-uPA and plasminogen activation. The concentration of circulating plasminogen is relatively high (approx. 2 μM, or 200 μg/ml) and in humans approx-

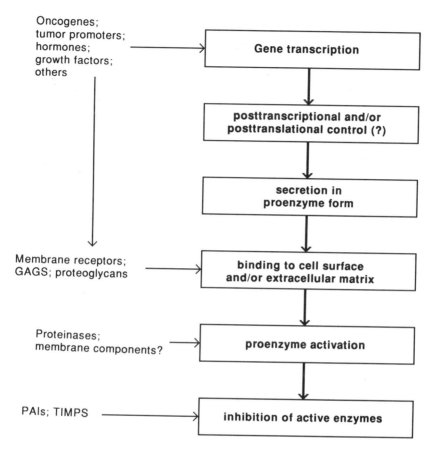

Figure 2. Transcriptional and posttranscriptional mechanisms that control the activity of serine proteinases and metalloproteinases.

Table I. Transcriptional Modulators of uPA and MMPs

uPA	MMPs
Steroid hormones	Steroid hormones
Peptide hormones (gonadotropins)	Extracellular matrix
Prostaglandins	Prostaglandins
UV light	UV light
cAMP	Lipopolysaccaride
Cholera toxin	Retinoic acid
Retinoic acid	Growth factors
Growth factors	Tumor promoters
Tumor promoters	Oncogenes
Oncogenes	Neoplastic transformation
Neoplastic transformation	

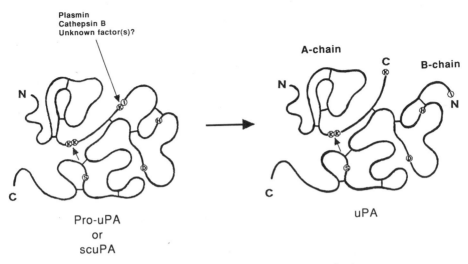

Figure 3. Molecular mechanism of prourokinase activation.

imately 40% of the plasminogen is located in extravascular sites, including the basal layer of epidermis (Robbins and Summaria, 1970; Isseroff and Rifkin, 1983). Thus, the production of small amounts of PA can result in high local concentrations of plasmin.

Plasmin-mediated activation is probably not the only way by which sc-uPA can be converted to active tc-uPA *in vivo*. In chicken fibroblasts, sc-uPA activation is associated with a membrane-bound arginine-specific serine proteinase. However, the expression of this enzyme appears to be restricted to Rous sarcoma virus (RSV)-transformed chicken fibroblasts (Berkenpas and Quigley, 1991). Cathepsin B can activate both the soluble and receptor-bound forms of pro-uPA *in vitro* (Kobayashi *et al.*, 1991). However, the presence of relatively high concentrations of plasminogen in virtually all vertebrate tissues implicate plasmin as the most important activator of pro-uPA.

2.2.1. The uPA Receptor

The amplification loop achieved by plasmin activation of pro-uPA is further modulated by the high-affinity interaction (K_d = 50–150 pM) of uPA with a specific plasma membrane binding protein. The uPA receptor (uPAR) is a highly glycosylated, 55- to 60-kDa protein linked to the plasma membrane by a glycosyl-phosphatidyl inositol (GPI) anchor (Nielsen *et al.*, 1988; Estreicher *et al.*, 1989; Behrendt *et al.*, 1990, 1991; Roldan *et al.*, 1990; Ploug *et al.*, 1991). The nonglycosylated polypeptide has a molecular weight of 35 kDa, as predicted by its deduced amino acid sequence (Roldan *et al.*, 1990). The high degree of glycosylation probably has an important role in determining uPAR affinity for its ligand (Møller *et al.*, 1993). The protein consists of three internal repeats (domains 1 to 3), each of which contains approximately 90 amino acid residues. The N-terminal 87 amino acid residues constitute the ligand-binding domain (domain 1) (Behrendt *et al.*, 1990; Rønne *et al.*, 1991; Pöllanen, 1993). On the surface of human U937 monocytelike cells, uPAR is present in two forms. One form has ligand-binding properties and consists of three domains; the second is a two-domain form that lacks domain 1 [uPAR(2+3)] and is devoid of uPA-binding activity. The two-domain form results from uPA or plasmin cleavage of the three-domain molecule (Høyer-Hansen *et al.*, 1992). The cleavage of uPAR by uPA or plasmin may represent a control mechanism by which these proteinases modulate their own activity on the cell surface.

uPAR has been found on the membranes of a variety of cells, including fibroblasts, endothelial cells, macrophages, and keratinocytes (Ellis *et al.*, 1993; Behrendt *et al.*, 1993). As is the case for uPA, uPAR expression is also modulated by several agents, including cytokines, hormones, and tumor promoters (Stoppelli *et al.*, 1986; Kirchheimer *et al.*, 1988; Lu *et al.*, 1988; Estreicher *et al.*, 1989; Picone *et al.*, 1989; Lund *et al.*, 1991a,b; Mignatti *et al.*, 1991). In the human monocytelike U937 cells, the tumor promoter phorbol 12-myristate 13-acetate (PMA), which induces macrophage differentiation, strongly increases uPAR gene transcription, as well as the cells' uPA-binding capacity (Picone *et al.*, 1989; Lund *et al.*, 1991b). Interestingly, monocytes induced to migrate in a chemotactic gradient rapidly polarize their uPAR to the leading front of the cell (Estreicher *et al.*, 1990). This effect provides an additional mechanism for the modulation of uPA activity and imparts versatility to the role of uPA and uPAR in cell migration and tissue remodeling.

Following secretion, uPA binds to uPAR in its inactive pro-uPA form through a specific N-terminal sequence of its noncatalytic A chain (Appella *et al.*, 1987). The bound zymogen is then activated by proteolytic cleavage (Cubellis *et al.*, 1986; Petersen *et al.*, 1988). Unlike most plasma membrane receptors, uPAR is not phosphorylated after binding the ligand, and formation of uPA–uPAR complexes does not result in internalization or receptor down-regulation. The interaction of uPA with uPAR on the plasma membrane has three important consequences: (1) the localization of enzyme activity at focal contact sites (Hébert and Baker, 1988; Pöllanen *et al.*, 1988); (2) a dramatic lowering (40-fold) of the K_M for plasminogen activation (Ellis *et al.*, 1991; Lee *et al.*, 1994); and (3) the internalization and rapid degradation of uPAR-bound uPA after complex formation with the type 1 plasminogen activator inhibitor (PAI-1) (Cubellis *et al.*, 1990).

2.2.2. Binding Sites for tPA and Plasminogen

Binding sites for tPA and/or tPA-inhibitor (PAI-1) complexes (see Section 2.2.3) have also been described on the membrane of certain cells including fibroblasts, endothelial cells, monocytes–macrophages, and melanoma cells (Hajjar and Hamel, 1990; Morton *et al.*, 1990; Bizik *et al.*, 1993; Carroll *et al.*, 1993; Felez *et al.*, 1993). A high-affinity binding site for tPA on the membrane of human vascular endothelial cells is associated with a 40-kDa protein that also binds plasminogen (Hajjar and Hamel, 1990; Felez *et al.*, 1993; Hajjar, 1991, 1993). However, PAI-1 associated with the surface of endothelial cells also appears to be a major binding site for tPA (Wittwer and Sanzo, 1990; Ramakrishnan *et al.*, 1990; Russell *et al.*, 1990). Unlike uPA-uPAR interactions, binding of tPA–PAI-1 complexes requires elements of the PAI-1 moiety and/or regions of the protease domain of tPA (Morton *et al.*, 1990).

Although high-affinity plasma membrane binding sites for plasminogen have not yet been characterized, an alpha-enolase-related molecule and the Hyman nephritis autoantigen (gp330) have been implicated as candidate receptors (Kanalas and Makker, 1991; Miles *et al.*, 1991). In general, cell surface proteins with C-terminal lysyl residues appear to function as plasminogen binding sites; alpha-enolase is a prominent representative of this class of receptors (Miles *et al.*, 1991). This molecule, as well as the putative endothelial cell tPA receptor, also interacts with tPA (Miles *et al.*, 1991; Felez *et al.*, 1993). These findings suggest that several high-affinity binding sites for tPA may be shared with plasminogen. In addition, low-affinity, high-capacity binding sites for plasminogen appear to be present in the chondroitin sulfate proteoglycans of the ECM and of the cell surface (Hajjar *et al.*, 1986; Miles and Plow, 1985; Plow *et al.*, 1986; Plow and Miles, 1990). Interestingly, uPA has a significant affinity for heparin and heparan sulfate proteoglycans (Andrade-Gordon and Strickland, 1986). Binding sites for plasminogen and tPA are also present on fibronectin and laminin (Moser *et al.*, 1993). Fibronectin binds both plasminogen and tPA via a 55-kDa N-terminal fragment (Moser *et al.*, 1993). Unlike fibrin, intact fibronectin does not enhance the rate of tPA-catalyzed plasminogen activation; however, a mixture of proteolytically degraded fibronectin fragments stimulates the activation reaction, resulting in an 11-fold increase in the k_{cat}/K_M (Stack and Pizzo, 1993). Therefore, both plasminogen and PAs are colocalized either on the cell surface and/or in the ECM (Plow *et al.*, 1986). The activation of the PA–plasmin system and the resulting plasmin activity appear not to occur in the soluble phase but on insoluble substrates, a feature that presents interesting similarities with reactions of the coagulation cascade (Moscatelli and Rifkin, 1988).

2.2.3. The PA Inhibitors

The third mechanism for the extracellular control of PA activity is mediated by specific protein inhibitors present in most tissues. PA-producing cells often also express PA inhibitors. The expression of these inhibitors, like PA and uPAR synthesis, can be modulated by a number of biological agents, including tumor promoters and growth factors. Although PAs can form complexes with several members of the serine proteinase inhibitor (serpin) superfamily (Carrel and Travis, 1985), only three inhibitors have a sufficiently high affinity to be effective *in vivo*. The first of these, the type 1 PA inhibitor (PAI-1), is a 45-kDa protein produced by a variety of cell types and present in platelets and plasma (Loskutoff and Edgington, 1977; Hekman and Loskutoff, 1985). The second inhibitor, the type 2 PA inhibitor (PAI-2), is a 46.6-kDa protein expressed most notably by cells of the monocyte–macrophage lineage (Kawano *et al.*, 1970; Astedt *et al.*, 1985; Kruithof *et al.*, 1986). The third inhibitor, protease nexin I (PN-I), is a 45-kDa protein originally purified from cultured fibroblasts, but also produced by several other cell types (Baker *et al.*, 1980; Eaton *et al.*, 1984). A fourth, less-characterized inhibitor, called PAI-3, has been isolated from human urine and is identical to the protein C inactivator, but it is considerably less efficient than the other inhibitors (Heeb *et al.*, 1987). Both PAI-1 and PAI-2 bind the active two-chain forms of uPA and tPA, rapidly forming 1:1 molar complexes, but they have a poor affinity for the one-chain zymogen forms of both PAs. The association rate constant of PAI-1 for tc-uPA and tc-tPA ($K_a = 10^7-10^8$ $M^{-1}s^{-1}$) is higher than that of PAI-2 ($K_a = 10^5-10^6$ $M^{-1}s^{-1}$). PN-I is less specific for PA than PAI-1 and PAI-2. It inhibits tc-uPA effectively ($K_a = 10^5$ $M^{-1}s^{-1}$) but has virtually no effect on sc-uPA and sc- and tc-tPA. In contrast, PN-I is an extremely rapid inhibitor of thrombin and also inactivates trypsin and plasmin (Saksela and Rifkin, 1988).

PA inhibitors appear to participate in the turnover of PAs. Urokinase–PN-I complexes bind to the membrane of fibroblasts and are internalized and degraded (Baker *et al.*, 1980; Low *et al.*, 1981). When uPA bound to uPAR reacts with PAI-1, uPAR–uPA–PAI-1 complexes are rapidly internalized and degraded (Cubellis *et al.*, 1990). Thus, the interaction of PA inhibitors with uPA results in a rapid blockade of the pro-uPA–uPA–plasmin loop and causes inactivated enzyme molecules to be cleared from the cell surface. Recent findings have elucidated the mechanism for clearance from the extracellular space of uPA, tPA, and PAI-1. PA–PAI-1 complexes and uncomplexed tPA bind to α_2-macroglobulin receptor/low-density lipoprotein receptor-related protein (LRP or α_2-MR), a multifunctional receptor shared by a variety of ligands including α_2-macroglobulin, apoprotein E-enriched beta-very-low-density lipoprotein, tPA, and *Pseudomonas* exotoxin A. This receptor mediates endocytosis and degradation of the uPAR–uPA–PAI-1 complex on cell surfaces, and participates, in cooperation with other receptors, in the hepatic clearance of PA–PAI-1 complexes and uncomplexed tPA from blood plasma (Herz *et al.*, 1992; Bu *et al.*, 1993; Kounnas *et al.*, 1993; Grobmyer *et al.*, 1993; Iadonato *et al.*, 1993; Andreasen *et al.*, 1994). This mechanism represents a novel type of molecular recognition of serine proteinases and serpins by their cellular binding sites.

2.3. General Features of the PA–Plasmin System

The most relevant feature of the PA–plasmin system is the amplification achieved by the conversion of plasminogen to plasmin. Because of the high concentration of plasminogen in virtually all tissues, the production of small amounts of PA can result in high local concentrations of plasmin. Plasmin degrades several ECM components and at the same time activates interstitial procollagenase and prostromelysins (see Section 4.3), as well as the latent form of elastase. Although the physiological mechanisms of activation of the type IV procollagenases are not yet understood, some MMPs can activate these enzymes *in vitro* (see Section 4.3). Thus, the production of even small amounts of PA results in the generation of high local concentrations of activated serine proteinases and MMPs. This cascade can be blocked at different levels by specific inhibitors. The blockade of plasminogen activation by PAIs will inhibit all subsequent events; the blockade of plasmin formation will result in the repression of MMP and elastase activation. In contrast, inhibition of MMPs by their specific tissue inhibitors (TIMPs) (see Section 4.5) will block these enzymes but leave plasmin unaffected.

Several important features of PAs have become apparent in recent years. First, extracellular enzyme activation and proteolysis occur not in the soluble phase but on insoluble substrates, such as the plasma membrane or the ECM. This feature and the cascade of proteolytic activation reactions involving PAs and MMPs show interesting similarities with the activation of the proteinases involved in the blood coagulation cascade. As is the case for blood clotting, most activation reactions involved in ECM remodeling are believed to occur on cell surfaces with high-affinity binding sites serving to localize zymogens. The localization of zymogens at privileged sites results in enhanced rates of activation, as well as in the protection of active species from inactivation by inhibitors. This concept has received considerable support from the characterization of uPAR and the findings of its involvement in tissue remodeling. Recently, high-affinity binding sites have also been described for the 72-kDa form of type IV collagenase (MMP-2 or gelatinase A) (Emonard *et al.*, 1992; Monsky *et al.*, 1993). These binding sites are associated with specific regions of the cell membrane (*invadopodia* or *podosomes*) involved in cell migration and invasion (Monsky *et al.*, 1993). An integral plasma membrane component also appears to be required for pro-MMP-2 activation (Brown *et al.*, 1993). However, these molecules have not yet been purified and characterized. Second, tissue remodeling results not only from increased proteinase production but also from a decrease in proteinase inhibitors. This observation raises an interesting point as to the optimal level of proteolytic activity for ECM turnover and tissue remodeling. Finally, besides being involved in ECM degradation, PAs also participate in controlling tissue remodeling through two parallel mechanisms: (1) the mobilization of growth factors, such as the fibroblast growth factors (FGFs) that are associated with ECM components, and (2) the activation of cytokines such as HGF/SF and transforming growth factor-β (TGF-β). These growth factors are potent angiogenesis inducers and can also modulate proteinase expression in a number of cell types.

2.4. PA–Cytokine Interactions

Basic FGF has a high affinity for the heparan sulfate proteoglycans of the ECM (Saksela *et al.*, 1988; Bashkin *et al.*, 1989; Saksela and Rifkin, 1990). The growth factor is found associated with the ECM *in vitro* and with basement membranes *in vivo* (Bashkin *et al.*, 1989; DiMario *et al.*, 1989) where it retains its biological activity and is protected from proteolytic degradation (Rogelj *et al.*, 1989; Saksela *et al.*, 1988). Also, uPA and plasminogen have significant affinities for heparan sulfate and chondroitin sulfate (Miles and Plow, 1985; Andrade-Gordon and Strickland, 1986; Hajjar *et al.*, 1986; Plow *et al.*, 1986). Thus, both growth factors and components of the PA–plasmin system are located in the ECM in an insoluble phase.

Several lines of experimental evidence have indicated that basic fibroblast growth factor (bFGF) release from the cell surface and ECM can be mediated by plasmin. When plasmin activity is increased, the release of bFGF–proteoglycan complexes is heightened. On the contrary, when plasmin formation is inhibited, the release of bFGF–proteoglycan complexes is suppressed (Saksela and Rifkin, 1990). This process of growth factor mobilization is of particular importance for the biological activity of bFGF because it permits bFGF to partition into the aqueous phase, rather than into the insoluble matrix, and to diffuse in the tissue and interact with its plasma membrane receptors (Flaumenhaft *et al.*, 1990). The mobilization of bFGF from the ECM can be achieved not only through the action of plasmin but also by other degradative enzymes, including heparitinases (Vlodavski *et al.*, 1983). However, the relative abundance of plasminogen in virtually all tissues and the efficient amplification mechanism achieved by PA-mediated plasminogen activation implicate plasmin as the most important proteinase in this process.

The mobilization of bFGF from the ECM contributes to the control of the extracellular proteolysis mediated by vascular endothelial cells during angiogenesis. The bFGF released from the ECM stimulates PA, uPAR, and collagenase expression by microvascular endothelial cells (Moscatelli *et al.*, 1986; Mignatti *et al.*, 1991). These proteinases are required to degrade the vascular basement membrane and permit vascular endothelial cell invasion during blood vessel formation (Mignatti *et al.*, 1986, 1989; Montesano *et al.*, 1986). This highly controlled invasive process is also modulated by PAs through the activation of latent TGF-β1 in the ECM. TGF-β1, a potent inhibitor of cell proliferation, migration, and proteinase production in vascular endothelial cells (Heimark *et al.*, 1986; Muller *et al.*, 1987; Saksela *et al.*, 1987), is secreted constitutively by a variety of cell types, including endothelial cells, as part of a high-molecular-weight complex. The growth factor (25 kDa) is noncovalently linked to a 75-kDa latency-associated peptide (LAP) that is disulfide-linked to a 125- to 190-kDa binding protein (Miyazono *et al.*, 1988). Mature TGF-β1 must be released from this complex to interact with its plasma membrane receptor and elicit a biological response. Plasmin is an efficient activator of latent TGF-β1 both *in vitro* and in cell cultures (Lyons *et al.*, 1988, 1990; Sato and Rifkin, 1989; Sato *et al.*, 1990). The active TGF-β1 formed in the ECM by the action of plasmin counteracts the stimulatory effects of bFGF on vascular endothelial cells by down-regulating uPA and collagenase gene expression and by stimulating PAI-1 and TIMP synthesis (Saksela *et al.*, 1987; Pepper *et al.*, 1990). As a consequence, plasmin formation is blocked, active MMPs are inhibited, and pro-MMPs are no longer activated. However, this blockade of extracellular proteolysis also turns off the plasmin-mediated activation of TGF-β1. When no more active TGF-β1 is present, the effects of bFGF on endothelial cells again become prevalent and proteinase production increases again (Flaumenhaft *et al.*, 1992).

This self-regulatory mechanism has profound implications in angiogenesis. Electron microscopic analysis of capillary formation has shown that the proteolytic degradation of the vessels' basal lamina and endothelial cell migration are temporally followed by a stage in which the newly forming capillaries synthesize and organize a new lamina propria. During this process, extracellular proteolysis must be locally inhibited to permit the deposition and assembly of ECM components. After a capillary loop is formed, degradation of the newly formed basement membrane occurs at the tip of the loop. Endothelial cells invade from this location, and a new capillary sprout is formed (Ausprunk and Folkman, 1977). Thus, from a biochemical point of view the process of capillary formation during angiogenesis can be thought of as resulting from alternate cycles of activation and inhibition of extracellular proteolysis.

3. Role of PAs in Wound Healing

Several lines of evidence implicate PAs in inflammation, granulation tissue formation, matrix formation, and reepithelialization, the different stages of wound healing. PAs are produced by different cell types in the skin. These can be classified as follows: (1) epidermal cells: keratinocytes; (2) dermal cells: fibroblasts and capillary endothelial cells; and (3) cells that have migrated into the skin and are active during the different stages of wound repair: granulocytes and macrophages. The PAs secreted by these cells have concerted roles in tissue remodeling. These include dissolution of the fibrin clot, ECM degradation, growth factor mobilization and activation, cell migration into the wound area, angiogenesis, and reepithelialization.

3.1. Inflammation

During the early and late inflammatory stages of wound repair, the damaged area is supplied with high levels of proteolytic activity. From the onset of tissue injury, blood vessel disruption results in the extravasation of plasma constituents. Plasminogen and PAs are released into the open tissue. Some of the same factors that take part in the coagulation cascade and support the initial inflammatory process can also directly activate plasminogen or convert proactivators into active PAs. The blood coagulation factors XI and XII, as well as serum kallikrein, activate plasminogen, although with low efficiency (Colman, 1969; Mandle and Kaplan, 1977; Bouma and Griffin, 1978). Tissue kallikrein and factor Xa activate tPA (Andreasen *et al.*, 1984; Ichinose *et al.*, 1984). Trace amounts of plasmin activate the pro-uPA also brought into the wound region with blood plasma extravasation (Skriver *et al.*, 1982; Eaton *et al.*, 1984; Cubellis *et al.*, 1986; Petersen *et al.*, 1988), and thus trigger the efficient amplification loop of pro-uPA–plasminogen activation.

The formation of the fibrin clot provides an early, primitive form of ECM, which potentiates the migration of inflammatory cells into the lesion. Granulocytes and macrophages, the first cells that infiltrate into an area of injury and inflammation, secrete high amounts of uPA (Gordon *et al.*, 1974; Unkeless *et al.*, 1974; Granelli-Piperno *et al.*, 1977; Vassalli *et al.*, 1984). The high level of plasmin formed by the action of the PAs produced by these cells, together with the tPA derived from blood vessel disruption and plasma extravasation, lead to the ultimate dissolution of the fibrin clot. Because of its affinity for fibrin, the PA form mainly responsible for clot lysis is believed to be tPA (Danø *et al.*, 1985).

However, there is emerging evidence in PA-deficient transgenic mice (Carmeliet *et al.*, 1994) that nonplasmin pathways may also be important in fibrinolysis. Given the proximity of fibrin and monocytes at sites of vascular injury, it has been proposed that these cells may use an alternative mechanism for fibrin degradation and clearance. Monocytes have recently been shown to possess a plasmin-independent fibrinolytic pathway that uses the integrin Mac-1. Binding by this adhesion molecule results in fibrinogen–fibrin internalization and lysosomal degradation (Simon *et al.*, 1993).

The role played by PAs is not limited to fibrin clot degradation. The high level of proteolytic activity produced by monocytes has been associated with macrophage activation (Gordon *et al.*, 1974; Unkeless *et al.*, 1974). During their differentiation into macrophages, monocytes also express high levels of uPAR (Stoppelli *et al.*, 1985; Picone *et al.*, 1989). Interestingly, migrating monocytes polarize their uPAR to the leading front of the cell (Estreicher *et al.*, 1990), indicating an important role for uPA–uPAR interactions in cell migration and tissue invasion. Monocytes also produce PAI-1 and PAI-2 (Kruithof *et al.*, 1986; Antalis and Dickinson, 1992; Hamilton *et al.*, 1993a,b). The expression of these inhibitors is up-regulated by several cytokines, including macrophage colony-stimulating factor (M-CSF), granulocyte–macrophage CSF (GM-CSF), and TGF-β. However, the synthesis of each PAI appears to be independently regulated. PAI-1 expression is up-regulated by TGF-β and down-regulated by lipopolysaccharide, particularly in the presence of TGF-β. In contrast, LPS up-regulates PAI-2 levels, whereas TGF-β reduces both the basal levels and LPS-induced levels of PAI-2. The glucocorticoid dexamethasone up-modulates PAI-1; interleukin-4 (IL-4) is ineffective (Hamilton *et al.*, 1993a,b). These findings indicate that both PAI-1 and PAI-2 are involved in the control of PA activity in monocyte–macrophages at sites of inflammation and tissue remodeling.

The high level of local proteolytic activity produced by macrophages is an essential component of the process of phagocytosis that leads to the debridement of the injured region. In addition, the plasmin produced by macrophages degrades glycoprotein components of the ECM and activates the pro-MMPs and latent elastase produced by the same cells (Werb *et al.*, 1977, 1980; Edlund *et al.*, 1983; Chapman and Stone, 1984; Matrisian, 1990). Whereas this process constitutes a first step in tissue remodeling, the degradation of fibrin and ECM components triggers the recruitment of other inflammatory cells through the generation of fibrin, collagen, elastin, fibronectin, and laminin degradation products, all of which are chemotactic for granulocytes and macrophages, as well as for endothelial cells (Stecher and Sorkin, 1972; Postlethwaite and Kang, 1976; Fernandez *et al.*, 1978; Senior *et al.*, 1980; Bowersox and Sorgente, 1982; Norris *et al.*, 1982).

3.2. Granulation Tissue Formation

In the stage of granulation tissue formation, fibroblasts and endothelial cells move into the wound space. This phenomenon has been studied in simplified *in vitro* models in which a monolayer of cultured cells is wounded with a razor blade or a rubber policeman, and cell migration into the denuded area is measured. Early experiments clearly associated PA activity with the migration of fibroblasts (Lipton *et al.*, 1971; Bürk, 1973; Ossowski *et al.*, 1973). Recent studies have shown the involvement of several components of the plasminogen–plasmin system in endothelial cell migration and angiogenesis. Vascular endothelial cells produce PAs both *in vitro* and *in vivo*. Immunohistochemical surveys of tissue sections have shown that human endothelial cells produce tPA but no detectable uPA (Kristensen *et al.*, 1984). Primary cultures of human endothelial cells secrete tPA and at later passages both tPA and uPA (Philips *et al.*, 1984). Long-term or immortalized cultures of endothelial cells primarily express only uPA (Tsuboi *et al.,* 1990; Peverali *et al.,* 1994). Although the PA produced by vascular endothelial cells *in vivo* is tPA, forming capillaries appear to express uPA (Bacharach *et al.,* 1992). In wound-healing experiments *in vitro,* endothelial cells have been shown to up-regulate uPA, uPAR, and PAI-1 expression (Pepper *et al.,* 1992c, 1993). The synthesis of these proteins is modulated in vascular endothelial cells by a number of cytokines that induce angiogenesis both *in vitro* and *in vivo* (angiogenic factors). These include bFGF, TGF-β, HGF/SF, and vascular endothelial growth factor (VEGF) (Basilico and Moscatelli, 1992; Grant *et al.,* 1993; Roberts *et al.,* 1986; Yang and Moses, 1990; Ferrara *et al.,* 1991; K. J. Kim *et al.,* 1992; Pepper *et al.,* 1991). Basic FGF and TGF-β have opposing effects on the PA activity of endothelial cells. The former is a potent inducer of uPA expression and has a relatively modest effect on PAI-1 synthesis, whereas the latter strongly down-regulates uPA and up-regulates PAI-1 expression (Saksela *et al.,* 1987; Pepper *et al.,* 1990). In contrast, bFGF and VEGF have a potent synergistic effect on endothelial cell PA production and angiogenesis (Pepper *et al.,* 1992a). HGF/SF, which has structural homology to plasminogen and is activated by uPA (Naldini *et al.,* 1992; Mizuno and Nakamura, 1993), also up-regulates uPA expression in epithelial and endothelial cells (Bussolino *et al.,* 1992; Pepper *et al.,* 1992a), indicating the existence of an amplification mechanism of uPA production and HGF/SF activation. Interestingly, bFGF applied locally to open wounds of healing-impaired, diabetic (*db/db*) mice restores normal wound repair (Tsuboi and Rifkin, 1990).

The concerted action of several angiogenic factors on endothelial cell PA activity indicates that plasmin formation must be finely modulated during angiogenesis. This view is also supported by the abnormal behavior of endothelial cells that express high levels of uPA. Endothelioma cells expressing the polyoma virus middle T (mT) oncogene (End cells) produce high amounts of uPA and low levels of PA inhibitors. When End cells are grown within fibrin gels, they invade the substrate and form large hemangiomalike cystic structures. Neutralization of excess proteolytic activity by exogenous serine proteinase inhibitors corrects the aberrant behavior of the cells and results in the formation of capillarylike tubular structures (Montesano *et al.,* 1990).

Whereas these findings implicate PAs and PAIs in cell migration and angiogenesis, the mechanism(s) of action of PA in cell movement is not yet understood. Plasmin has been shown to disrupt actin cables in rat embryo cells (Pollack and Rifkin, 1975), suggesting that PA or plasmin may be involved in the continuous rearrangement of cytoskeletal components that occurs during cell migration. However, more recent evidence has shown that the uPA-mediated stimulation of cell motility requires interaction of the N-terminal peptide with uPAR and is independent of the enzyme's catalytic activity (Fibbi *et al.,* 1988; Odekon *et al.,* 1992). Because uPAR is a GPI-anchored protein and has no cytoplasmic domain, the mechanism by which intracellular signaling for movement can be generated remains unclear.

The importance of PAs in angiogenesis is also supported by other findings *in vitro* and *in vivo.* It has long been known that plasma-derived fibrinolytic agents interact with ECM components and play important roles in maintaining vascular integrity. Treatment of ECM with plasmin or trypsin stimulates *in vitro* endothelial cell organization into capillarylike structures (Maciag, 1984). This finding has generated the pothesis that ECM degradation initiates a cascade of events that modulate angiogenesis. It must be considered that *in vivo* PAs and other ECM-degrading proteinases originate from a variety of cell types, including granulocytes, macrophages, and fibroblasts present in the wound area well before angiogenesis actually begins. A scenario of cell cooperation may be established that is mediated by proteolytic enzymes. Fibrin, fibronectin, and collagens and their respective degradation products are chemotactic for endothelial cells and potent angiogenesis inducers *in vivo* (Bowersox and Sorgente, 1982; Alessandri *et al.*, 1983). ECM degradation products are produced very early after tissue injury. As soon as clot lysis is initiated by the PAs in the wound area, fibrin and fibronectin peptides are released in the lesion and may promote angiogenesis by attracting endothelial cells from adjacent capillaries and/or venules and by favoring their organization into new capillaries.

3.3. Matrix Formation and Reepithelialization

In the remodeling process that follows granulation tissue formation, the elimination of fibronectin from the early ECM is accomplished by the secretion of proteinases into the wound area. The involvement of the PA–plasmin system at this stage is suggested by its similarity with other physiological processes in which PA production has been associated with tissue remodeling. Examples are mammary involution following lactation, ovulation, and spermatogenesis, trophoblast implantation, and uterine involution postpartum (for a review, see Danø *et al.*, 1985). In these processes, as well as in wound healing, the ability of plasmin to activate certain pro-MMPs (see Section 4) may represent a key feature in the control of connective tissue turnover.

Several lines of evidence also suggest a role for uPA in the process of reepithelialization. uPA is produced by cultured human epidermal cells (Hashimoto *et al.*, 1983). In cultures of differentiating mouse keratinocytes, the amount of cell-associated uPA increases with advanced differentiation. The enzyme levels decrease shortly after squame production, and the highest PA levels are found in squames that have detached from the culture surface. Because plasminogen is present in the basal layers of epidermis (Isseroff and Rifkin, 1983), these findings have suggested that the plasmin generated by endogenous keratinocyte PA may facilitate the terminal differentiative events of nuclear dissolution that characterize keratinocyte differentiation and be responsible for squame detachment (Risch *et al.*, 1980; Isseroff *et al.*, 1983). Interestingly, uPA and uPAR expression is up-regulated by several cytokines that also modulate migration and invasion of epithelial cells. These include EGF, HGF/SF, and keratinocyte growth factor (KGF), a member of the fibroblast growth factor family (FGF-7) (Estreicher *et al.*, 1989; Pepper *et al.*, 1992b; Tsuboi *et al.*, 1993). Interestingly, as is the case for macrophages, the localization pattern of uPAR in keratinocytes suggests that this molecule may be coupled to cell migration during cutaneous wounding (McNeill and Jensen, 1990).

Recent findings also implicate components of the PA–plasmin system in peripheral nerve and liver regeneration. The regenerative ability of the peripheral nervous system is dependent, at least in part, on Schwann cell properties. Rat dorsal root ganglion neurons, after a wounding stimulus *in vivo* or when they are cultured *in vitro*, synthesize TGF-β. This growth factor elicits multiple Schwann cell responses, including proliferation and morphological changes, decreased tPA, and augmented PAI-1 secretion (Rogister *et al.*, 1993). This generates conditions that are thought to favor successful neuritic regrowth. During partial hepatectomy, a transient increase in PAI-1 mRNA is apparent within 1 to 2 hr after surgery. PAI-1 expression is much higher in the region adjacent to the tissue injury than in more distal regions, and is induced primarily in hepatocytes in the transition zone between viable and necrotic tissue. Capsular mesothelial cells, subcapsular hepatocytes, and venous endothelial cells bordering the area also express PAI-1 mRNA, whereas a much weaker synthesis is evident in hepatocytes dispersed throughout the remaining intact lobes. These findings suggest that PAI-1 may be of importance in the local tissue remodeling that accompanies liver regeneration (Schneiderman *et al.*, 1993).

4. Matrix Metalloproteinases

The MMPs comprise a gene family of enzymes that share important common properties. Their catalytic mechanism requires an active site Zn^{2+}; they are secreted as inactive zymogens; they can degrade various components of the extracellular matrix; and their proteolytic activity is inhibited by tissue-derived inhibitors (TIMPs). In addition, the various metalloproteinases share a high degree of structural similarity reflected by about a 40% amino acid homology among all members of the family. To date, 11 different MMPs representing 11 distinct gene products have been characterized, and it is likely that additional members of this gene family will be identified in the future. Additional information can be found in other reviews (Birkedal-Hansen *et al.*, 1993; Matrisian, 1992; Woessner, 1991).

4.1. Classification of MMPs

MMPs are categorized by their capacity to degrade various extracellular matrix substrates (Table II), properties that are conferred by constituent structural domains. The best-characterized, and historically oldest, subgroup of MMPs are the interstitial collagenases, which possess the unique ability to cleave the triple helix of native types I, II, and III collagens. Three interstitial collagenases have thus far been identified (Freije *et al.*, 1994; Hasty *et al.*, 1990; Stricklin *et al.*, 1977), all of which cleave native type I collagen at a single locus (Gly^{775}-Ile^{776} in the α1 chain; Gly^{775}-Leu^{776} in α2), which is located about three fourths the distance from the N-terminus of the collagen molecule. At physiological temperature (37 °C), the three-fourth- and one-fourth-length fragments of collagenase digestion denature spontaneously into randomly coiled gelatin peptides and can be further attacked by a variety of enzymes, including the gelatinases. However, the single cleavage of the collagen triple helix catalyzed by interstitial collagenases is the rate-limiting step of collagen degradation (Welgus *et al.*,

Table II. Substrate Specificity of Matrix Metalloproteinases

	Collagenases	Gelatinases	Stromelysins	Matrilysin	Metalloelastase
Laminin			+.	+	
Entactin		+		+	
Type IV		+	+	+	+
Type VII		+	+		
Fibronectin			+	+	
Proteoglycans			+		
Collagens I, II, III	+				
Gelatin		+	+	+	
Type V		+	+		
Elastin		+		+	+

1981), and these enzymes are believed to be uniquely capable of initiating type I collagen degradation *in vivo* at neutral pH. Collagenase-1 is produced in the human by a variety of epithelial and mesenchymal cell types including keratinocytes, fibroblasts, macrophages, chondrocytes, and smooth muscle cells. Collagenase-2 (Hasty *et al.*, 1990) is a product only of the polymorphonuclear leukocyte and is contained within neutrophil granules, in distinction to all other MMPs, which are rapidly secreted without significant intracellular stores. Collagenase-3 (Freije *et al.*, 1994) is a newly described enzyme found in breast cancer, but may represent the predominant interstitial collagenase in certain rodent species, such as the rat and mouse (Quinn *et al.*, 1990).

The stromelysins are so designated because of their broad substrate specificity. Three stromelysins have been characterized, and two of these (stromelysin-1 and stromelysin-2) possess very similar catalytic activity, but exhibit quite different gene regulation. Stromelysins-1 and -2 are strong proteoglycanases and can also degrade basement membranes, laminin, fibronectin, and the nonhelical telopeptides of some collagens (e.g., types IV and IX) (Murphy *et al.*, 1991). Stromelysin-3 appears to exhibit only weak proteolytic activity (Murphy *et al.*, 1993).

There are two metallogelatinases, of molecular weight 72,000 and 92,000 kDa, which possess virtually identical substrate specificity, but are expressed in different tissues and are subject to distinct regulation. The 72-kDa gelatinase (gelatinase A) is produced constitutively *in vitro* by most cell types, including fibroblasts, osteoblasts, and smooth muscle cells; but, unlike other metalloproteinases, its expression is not regulated by cytokines, growth factors, or hormones, with the exception of TGF-β (Overall *et al.*, 1991). This paucity of regulatory modification most likely reflects the lack of an AP-1 sequence in its promoter, a unique property of 72-kDa gelatinase as compared to all other MMPs. The 92-kDa gelatinase (gelatinase B) is actively expressed by eosinophils (Ståhle-Bäckdahl *et al.*, 1994; Ståhle-Bäckdahl and Parks, 1993), monocytes–macrophages, and epithelial-derived cells (e.g., keratinocytes) and is stored by neutrophils (Hasty *et al.*, 1990; Ståhle-Bäckdahl and Parks, 1993). The 92-kDa gelatinase promoter contains two AP-1 sites and its expression is subject to modification by a variety of physiological signals (Huhtala *et al.*, 1991). Both the 92-kDa and 72-kDa metallogelatinases efficiently degrade denatured collagens (i.e., gelatins) of all genetic types, and these enzymes also attack basement membranes, fibronectin, and insoluble elastin (Murphy *et al.,* 1991).

Matrilysin is the smallest MMP (molecular weight 28,000 kDa) but possesses broad and potent catalytic activity against ECM substrates. Matrilysin is a stronger proteoglycanase than stromelysin and also degrades basement membranes, insoluble elastin, laminin, fibronectin, gelatin, and entactin (Murphy *et al.,* 1991; Sires *et al.,* 1993). Matrilysin appears to be produced only by a select population of cell types, most prominent of which are the glandular epithelia of a variety of tissues, including the endometrium, breast, prostate, pancreas, and parotid, and sweat glands of the skin (Rodgers *et al.,* 1993; Saarialho-Kere *et al.,* 1995).

Two MMPs have been recently characterized that do not belong to any of the subgroups described above: macrophage metalloelastase (Shapiro *et al.,* 1992, 1993b) and a transmembrane metalloproteinase (Sato *et al.,* 1994). Macrophage metalloelastase possesses high specific activity against insoluble elastin and can also degrade type IV collagen. The remainder of its potential substrates and whether nonmacrophage cell types can express this protease have not been determined. Finally, Sato *et al.* (1994) have recently identified a novel MMP with a transmembrane domain that directs a cell surface localization and can activate secreted 72-kDa progelatinase.

4.2. Domain Structure of MMPs

The MMPs are organized into structural domains (Fig. 4) that impart their specific biological functions. All MMPs have a catalytic domain of 21 kDa that binds the active site Zn^{2+}. The Zn-coordinating region is highly conserved, containing the sequence HEXGHXXGLXH, in which the three His residues represent three of four Zn-interactive ligands (Goldberg *et al.,* 1986). Single amino acid mutations introduced into this sequence render the enzyme catalytically inactive (Woessner, 1991). This catalytic domain is also the site of TIMP interaction with active metalloenzyme and contributes to determining the substrate specificity of MMPs (Murphy *et al.,* 1992a,b).

All MMPs also contain an N-terminal prodomain of 8 kDa that is responsible for maintaining the enzymes is an inactive, or zymogen, state. This propeptide is characterized by the invariant sequence PRCGVPD, with the Cys residue representing the fourth interactive ligand of the active site Zn^{2+}. Enzyme latency is conferred by this Cys–Zn interaction (Van Wart and Birkedal-Hansen, 1990). MMP zymogens can be activated by exposure to proteases, chaotropes, or sulfhydryl chelators, all of which disrupt the Cys–Zn association, causing the enzymes to cleave themselves by an intramolecular process at the start of their catalytic domain (see Section 4.3). Matrilysin, the smallest, and perhaps primordial MMP, contains only the "pro" and catalytic domains.

Other than matrilysin, all MMPs contain a 22-kDa C-terminal domain that has structural homology to the heme-binding protein, hemopexin. This domain has been implicated, at least partially, in two major MMP functions: determining their substrate specificity and modulating their interactions with TIMPs. The contribution of the hemopexinlike domain to substrate specificity is variable, but is most clearly illustrated

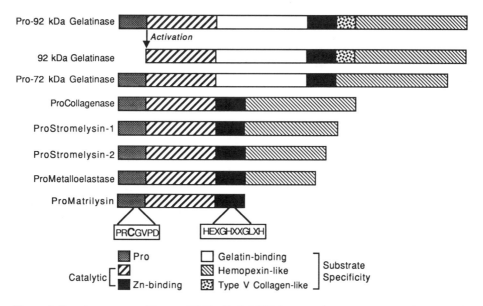

Figure 4. Domain structure of human MMPs. Each MMP is secreted as a proenzyme whose state is maintained by the interaction of a conserved cysteine residue in the prodomain with sequences in the Zn^{2+}-binding domain (see Fig. 5). The prodomain is cleaved in the extracellular space, producing an active proteinase, and catalytic activity is conferred by the Zn^{2+}-binding domain. Substrate specificity is thought to be conferred by sequences within the hemopexinlike, collagenlike, and gelatin-binding domains. TIMPs interact with both the catalytic and hemopexinlike domains.

in the case of interstitial collagenase. Here, truncation of the hemopexinlike domain yields an enzyme that can nonspecifically attack casein, but is incapable of degrading type I collagen (Murphy *et al.,* 1992a). Furthermore, substitution of stromelysin's hemopexinlike domain onto the collagenase catalytic domain fails to restore collagenolytic activity. Yet, for other enzymes, such as stromelysin (Murphy *et al.,* 1992a) and the metallogelatinases (Murphy *et al.,* 1992c; O'Connell *et al.,* 1994), C-terminal truncation does not appear to alter substrate specificity. The metallogelatinases are unique among the MMPs in the capacity of their zymogen forms to bind TIMPs. This TIMP-binding capacity of the metallogelatinase zymogens is conferred entirely by their hemopexinlike domains (Murphy *et al.,* 1992c; O'Connell *et al.,* 1994). Furthermore, this domain also participates, in addition to the catalytic domain, in the binding of TIMP to all *active* MMPs (Baragi *et al.,* 1994). The collagenases, stromelysins, and macrophage metalloelastase are comprised of catalytic, "pro," and hemopexinlike domains.

The metallogelatinases (72 kDa and 92 kDa) both contain three head-to-tail repeats of the Fn-2 domain of the cell adhesion protein, fibronectin. These Fn-2 repeats divide the catalytic domain of both gelatinases and are specifically responsible for these enzymes' strong gelatin-binding affinity (Collier *et al.,* 1992; Murphy *et al.,* 1994). The 92-kDa gelatinase, in addition, contains a unique collagenlike domain whose function is unknown. The most recently characterized MMP, a cell surface-

-associated enzyme capable of activating 72-kDa progelatinase, contains a 24-amino-acid insert in its hemopexinlike domain that is highly hydrophobic and likely serves a transmembrane-integrating function (Sato *et al.,* 1994). This transmembrane domain is essential for the enzyme's capacity to activate 72-kDa progelatinase at the cell surface.

4.3. Proenzyme Activation

As discussed earlier, all MMPs are secreted from cells in a catalytically inactive, or zymogen, form. Retention of latency depends on the interaction of a Cys residue in the proenzyme domain with the active site Zn^{2+} (Van Wart and Birkedal-Hansen, 1990). Dissociation of this Cys–Zn interaction by molecular perturbations, including exposure to chaotropic agents (e.g., thiocyanates), Cys chelators (e.g., organomercurials), or protease cleavage of the "pro"-domain near the Cys residue, all destabilize this association, resulting in autocatalytic cleavage of the enzyme and removal of the 8 kDa "pro"-domain (Fig. 5). Several serine proteinases, including plasmin, trypsin, neutrophil elastase, and mast cell chymase, can activate a variety, but not all, of the MMPs (Matrisian, 1992; Saarinen *et al.,* 1994; Woessner, 1991). Furthermore, some of the active MMPs can activate other MMP zymogens, although the biological significance of this is unknown. Plasmin activation of MMPs (He *et al.,* 1989) and cell surface activation of 72-kDa progelatinase (Sato *et al.,* 1994) appear to be particularly relevant physiologically. Nevertheless, precisely how MMPs are activated *in vivo,* during normal tissue homeostasis, inflammation, or tumor invasion, remains unknown.

4.4. Regulation of MMP Expression

In general, MMPs are not expressed constitutively *in vivo,* but rather are induced in response to cytokines, growth factors, hormones, oncogenes, and cell contact with extracellular matrix or other cell types (Table I). Regulation of MMPs is cell type-specific, and certain agents, such as TGF-β, can even have opposite effects on different cell types. This cytokine stimulates MMP production in keratinocytes (Salo *et al.,* 1991) but inhibits expression by fibroblasts (Edwards *et al.,* 1987). The major inducers of fibroblast MMP expression seem to be proinflammatory mediators, particularly IL-1 and tumor necrosis factor-alpha (TNF-α) (Mauviel, 1993). Keratinocytes are stimulated to express MMPs in response to EGF, TGF-α, and cell contact with type I collagen (Saarialho-Kere *et al.,* 1993a; Woodley *et al.,* 1986). Macrophage production of MMPs is induced by lipopolysaccharide and the ingestion of particulate matter (Welgus *et al.,* 1990). The glucocorticoid and retinoid hormones are potent suppressors of MMP production in a variety of cell types (Shapiro *et al.,* 1991). The lymphokines gamma interferon (IFN-γ), IL-4, and IL-10 all inhibit MMP expression in macrophages, but have little or no effect on enzyme production in other cell types (Lacraz *et al.,* 1992, 1994; Shapiro *et al.,* 1990). Interestingly, a novel mechanism for the induction of metalloproteinase expression in fibroblasts and monocytes has recently been reported to be mediated by cell surface glycoproteins via direct cell–cell contact of these target cells with activated T lymphocytes (Lacraz *et al.,* 1994).

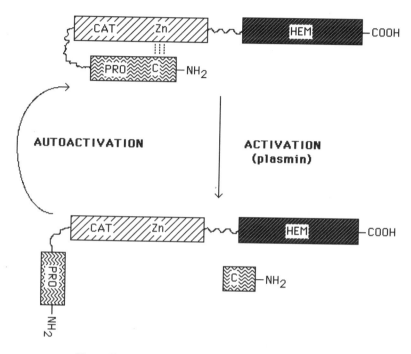

Figure 5. Molecular mechanisms of MMP activation.

The induction of MMPs in several cell types by phorbol esters, IL-1, and TNF-α is dependent on the binding of transcription factors to activator protein-1 (AP-1) and polyomavirus enhancer A-binding protein (PEA-3) elements located in the MMP promoters (Angel and Karin, 1992; Buttice and Kurkinen, 1994; Buttice *et al.*, 1991; Gaire *et al.*, 1994; Vincenti *et al.*, 1994). Both protein kinase C and tyrosine kinase signal transduction pathways have been implicated in metalloproteinase induction (Sudbeck *et al.*, 1994). Prostaglandins, in particular PGE_2, control metalloproteinase expression in macrophages, but not in other cell types (Corcoran *et al.*, 1992; Mauviel *et al.*, 1994; Shapiro *et al.*, 1993a; Sudbeck *et al.*, 1994).

4.5. Tissue Inhibitors of Metalloproteinases

The catalytic activity of MMPs is controlled, at least in part, by a gene family of tissue-derived metalloproteinase inhibitors called TIMPs. To date, three such TIMPs have been characterized, receiving the designations TIMP (or TIMP-1), TIMP-2, and TIMP-3. TIMPs block the catalytic activity of MMPs but have no efficacy against serine, sulfhydryl, or acid proteases. In addition to their metalloproteinase inhibitory effects, the TIMPs have been reported to possess growth-promoting activities for a wide range of cells (Hayakawa *et al.*, 1992).

TIMP is a 28-kDa glycoprotein with a protein core of 21 kDa. It inhibits all MMPs of mammalian origin, but will not block the activity of other classes of metalloenzymes, for example, thermolysin or bacterial-derived collagenase. TIMP inhibits the activity of MMPs by forming a high-affinity ($K_i \sim 10^{-10}$ M), yet noncovalent, enzyme-inhibitor complex. TIMP inhibition of active MMPs requires interactions primarily with their catalytic domain but secondarily with their hemopexinlike domain, which does contribute to the high affinity of binding (Baragi *et al.*, 1994; Murphy *et al.*, 1992c). TIMP is conformationally restrained by six disulfide bonds, which provide the molecule with high resistance to extremes of pH and temperature (Carmichael *et al.*, 1986). However, this tight secondary structure is required for enzyme inhibition; reduction and alkylation renders the molecule inactive. The three largest of the six disulfide loops are required to bring essential amino acids into spatial proximity to interact with MMPs (O'Shea *et al.*, 1992). In agreement with this, various peptide regions within these loops inhibit collagenase activity (Bodden *et al.*, 1994), and these could be potentially used as therapeutic agents. TIMP binds only active metalloproteinases and not the MMP zymogens, with the exception of 92-kDa progelatinase. Here, the C-terminal, hemopexinlike domain by itself binds TIMP in the zymogen state, and 92-kDa progelatinase is secreted from cells already in complex with TIMP (O'Connell *et al.*, 1994). Upon activation, the formation of an active 92-kDa gelatinase–TIMP complex is dependent on interactions with both the catalytic and hemopexinlike domains. In addition to inhibiting the matrix-degrading activity of MMPs, TIMP also inhibits their intramolecular activation from a zymogen to an activated state.

TIMP-2 appears designed specifically to interact with the 72-kDa gelatinase (Boone *et al.*, 1990; Howard *et al.*, 1991). A protein of 21 kDa, TIMP-2 contains 12 Cys residues precisely conserved in location to those of TIMP, and which fold the molecule into the same six disulfide loops. Unlike TIMP, TIMP-2 is not glycosylated. In addition, TIMP-2 contains nine C-terminal amino acids that appear to be important in directing this enzyme's specificity toward 72-kDa gelatinase. TIMP-2 binds to the hemopexinlike domain of 72-kDa progelatinase, and this enzyme is secreted from cells as a 72-kDa progelatinase–TIMP-2 complex (Murphy *et al.*, 1992b,c). TIMP-3 is a newly described MMP inhibitor of 24 kDa, containing the six disulfide loop structure of the other TIMPs, but which exhibits a predominantly extracellular matrix-associated localization (Leco *et al.*, 1994).

5. Role of MMPs in Wound Repair

5.1. Interstitial Collagenase Production by Basal Keratinocytes

Wound healing is an orderly process that involves inflammation, reepithelialization, matrix deposition, and tissue remodeling. In most injuries, especially chronic wounds, healing is accompanied by inflammation, angiogenesis, and the formation of granulation tissue. Degradation of extracellular matrix is required to remove damaged tissue and provisional matrices and to permit vessel formation and cell migration, and these remodeling processes involve various proteinases. Clearly, numerous cell types contribute proteinases of various classes and types that affect tissue restructuring during healing, and metalloproteinases, in particular interstitial collagenase, seemingly play a critical role in various stages of cutaneous repair.

Many studies have shown that collagenase is present in the wound environment (Ågren *et al.,* 1992; Buckley-Sturrock *et al.,* 1989), and it has often been thought that the enzyme is produced by fibroblasts, macrophages, and other cells within the granulation tissue (Porras-Reyez *et al.,* 1991). By *in situ* hybridization and immunohistochemistry, active collagenase expression is indeed detected in fibroblasts and macrophages in samples of human burns (Stricklin *et al.,* 1993) and in some samples of wounded skin and necrobiotic disorders (Saarialho-Kere *et al.,* 1992, 1993a,b; Stricklin *et al.,* 1993). However, in a thorough examination of the role of metalloproteinases in cutaneous wounds, expression of collagenase in the dermis was present in less than 50% of the greater than 100 samples representing chronic wounds, including pyogenic granuloma, decubitus ulcers, sepsis ulcers, and nonspecific ulcers. When detected in these samples, the expression was typically low and confined to a few cells (Saarialho-Kere *et al.,* 1992, 1993a). Furthermore, collagenase mRNA is not detected in dermal cells in samples of acute human wounds or in healthy skin. In contrast, basal keratinocytes at the migrating front of reepithelialization are the predominant source of collagenase during active wound repair (Fig. 6). Furthermore, collagenase expression by migrating keratinocytes is an invariable feature of a disrupted epidermis, as a consequence of normal wound healing by secondary intention, in ulceration resulting from a variety of disease processes, and in full-thickness burn wounds. Collectively, these observations implicate the keratinocyte as a major participant in the degradation of extracellular matrix during wound healing and suggest that fibroblasts, macrophages, and other cells within the dermis release collagenase only at certain stages of repair or in only some types of wounds.

5.2. Alterations in Cell–Matrix Interactions and Collagenase Production by Basal Keratinocytes

An interesting aspect of the epithelial expression of interstitial collagenase in wound skin is that the enzyme is not produced in nonulcerated samples. Also, *in situ* hybridization studies clearly show that only basal keratinocytes express collagenase mRNA and not the more differentiated cells of the stratum spinosum and stratum granulosum. The confinement of collagenase expression to the basal epidermal cells suggests that disruption of the basement membrane and subsequent exposure of keratinocytes to the underlying dermal stroma is apparently a critical determinant for the induction of epidermal collagenolytic activity. Basal keratinocytes normally rest on a basement membrane composed of various forms of laminin, entactin, proteoglycans, and type IV collagen (Stenn and Malhotra, 1992). In response to wounding, keratinocytes migrate from the edge of the wound under a provisional matrix of fibrin and

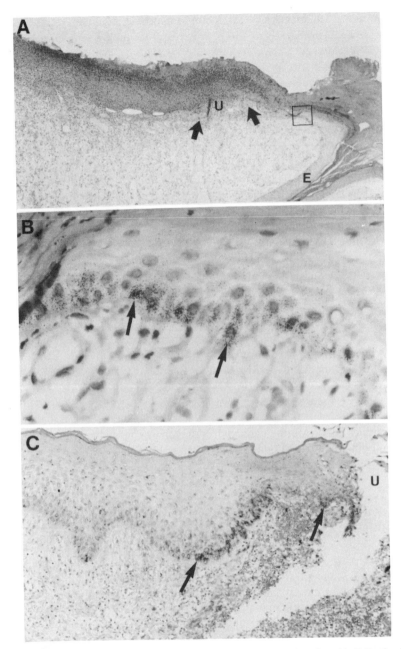

Figure 6. Collagenase is expressed by basal keratinocytes at the leading edge of reepithelialization in ulcers. (A and A′) Paired bright- and dark-field views of a sample of pyogenic granuloma hybridized with an ^{35}S-labeled antisense RNA probe specific for collagenase mRNA. An ulcerated area (U) is indicated, and collagenase-positive keratinocytes are seen on both sides of the ulcer (arrows). The intensity of the signal for collagenase mRNA diminishes with increasing distance from the ulcer. (B and B′) High magnification views of the area indicated by the box in panel A. Prominent collagenase expression is localized only to basal

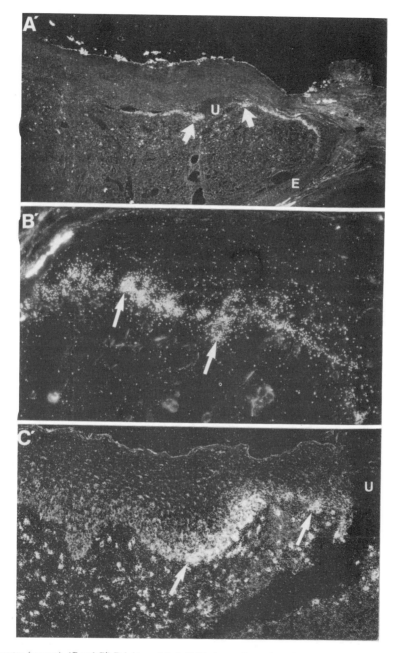

keratinocytes (arrows). (C and C′) Bright- and dark-field views of a section of a nonspecific ulcer hybridized for collagenase mRNA. Note the intense collagenase expression by basal keratinocytes (arrows) forming the leading edge of reepithelialization adjacent to the ulcer. In this specimen, collagenase is also expressed by underlying dermal fibroblasts. No specific signal was detected on any sample hybridized with a sense RNA probe (not shown).

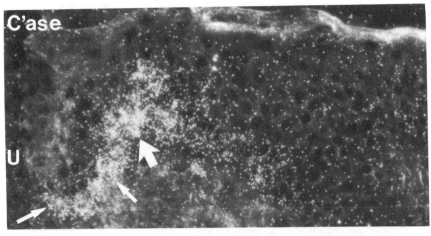

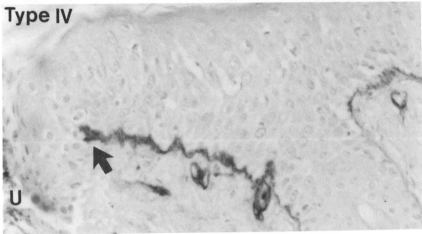

Figure 7. Keratinocytes expressing collagenase are not in contact with basement membrane. Serial sections from a specimen of a nonspecific chronic ulcer (U) were hybridized for collagenase (C'ase) or were immunostained for type IV collagen. The extent of basal lamina is marked by the large arrows. Most collagenase-positive keratinocytes (small arrows) are not in contact with basement membrane, but rather are migrating over the dermal matrix. Signal for collagenase mRNA diminishes rapidly in areas with intact basal lamina.

fibronectin (Clark *et al.*, 1982) and over or through the dermis, which includes structural macromolecules, such as type I collagen, microfibrils, and elastin, distinct from those in the basement membrane. Loss of contact with the basement membrane and establishment of new cell–matrix interaction with components of the dermal and provisional matrices may be a critical determinant that alters keratinocyte phenotype and induces collagenase production. Indeed, recent *in vivo* observations (Saarialho-Kere *et al.*, 1993b) show that collagenase-positive keratinocytes are not in contact with an intact basement membrane, as demonstrated by immunostaining for type IV col-

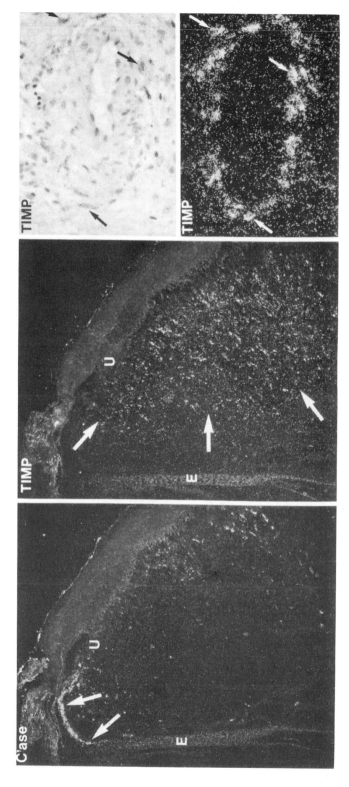

Figure 8. Collagenase and TIMP expression in ulcerated pyogenic granuloma. Shown are serial sections hybridized for collagenase (C'ase) or TIMP mRNAs. Collagenase-positive basal keratinocytes are detected only at the migrating front of epithelium (arrows). In contrast, TIMP-positive cells are found within the underlying granulation tissue (bordered by arrows), but not in epidermis.

lagen and laminin, and are migrating over the dermal wound matrix (Fig. 7). Reflecting this *in vivo* relationship between metalloproteinase expression and contact with the dermis, collagenase production is induced in primary keratinocytes grown on plates coated with native type I collagen, the most abundant component of dermal matrix. In contrast, components of the basement membrane and other proteins of the interstitial matrix do not affect collagenase expression (Saarialho-Kere *et al.*, 1993). In other studies, keratinocytes were shown to recognize and migrate on a type I collagen substratum, and that this interaction results in enhanced collagenase production (Petersen *et al.*, 1990). Collectively, these studies demonstrate a key role for type I collagen in initiating keratinocyte collagenase synthesis in the epithelial response to wounding.

The border between basement membrane and collagenase-positive basal keratinocytes is not always precise; however, the signal for collagenase mRNA decreases progressively within a few cells overlying the newly formed basement membrane (Fig. 8). This observation is consistent with the idea that cell–matrix interactions maintain the phenotype of basal keratinocytes in intact skin and indicates that collagenase expression ceases once the cell reestablishes contact with the basement membrane. Interestingly, laminin inhibits keratinocyte migration in culture (Woodley *et al.*, 1988), suggesting that this basement membrane protein may repress the phenotype of the activated keratinocyte once it reestablishes contact with basal lamina. In essentially all cell model studies, collagenase production is regulated at the level of transcription (Angel and Karin, 1992; Auble and Brinkerhoff, 1991; Mauviel *et al.*, 1992; Saarialho-Kere *et al.*, 1993b; S. Shapiro *et al.*, 1993), and collagenase mRNA has a half-life of about 6 hr (Saarialho-Kere *et al.*, 1993b). Thus, the weaker signal for collagenase mRNA in basal keratinocytes within the basement membrane border may represent transcripts that remain and are being degraded after gene expression has shut off.

In addition to encountering different ECM proteins, migrating keratinocytes also express a distinct pattern of matrix-binding integrins, and these may also be involved in regulation of collagenase production. Various groups have shown that integrins, such as $\alpha2\beta1$, $\alpha3\beta1$, and $\alpha6\beta4$, which are expressed on basal keratinocytes in intact skin, are also present in epidermal cells at the wound edge (Cavani *et al.*, 1993; Hertle *et al.*, 1992; Juhasz *et al.*, 1993; Larjava *et al.*, 1993). Migrating keratinocytes, however, selectively express additional integrins, such as $\alpha5\beta1$ and $\alpha v\beta3$, and these receptors are present on the same keratinocytes that express interstitial collagenase (Saarialho-Kere *et al.*, 1993b). The $\alpha5\beta1$ integrin is expressed in keratinocytes migrating out of skin explants (Guo *et al.*, 1991), and blocking this integrin receptor inhibits keratinocyte migration on fibronectin matrices (J. P. Kim *et al.*, 1992). Although the stimulus that induces integrin expression by keratinocytes is not known, TGFβ1, which would be released from degranulated platelets at the initiation of wound repair and from migratory and resident cells during other stages of healing, does up-regulate production of $\alpha5\beta1$ (Guo *et al.*, 1991).

Although the $\alpha5\beta1$ integrin recognizes fibronectin, which is present in both the provisional and dermal matrices but absent from the epidermal basement membrane, it is doubtful that this receptor mediates induction of collagenase expression by keratinocytes cultured on collagen. Most likely other integrins, such as the type I collagen-binding receptor, α2β1, which is constitutively expressed on keratinocytes (Hertle *et al.*, 1992; Pellegrini *et al.*, 1992; Symington *et al.*, 1993), participates in mediating induction of collagenase gene expression. Interestingly, basal keratinocytes constitutively express collagen-binding integrins, such as α1β1 and α2β1 (Hertle *et al.*, 1992; Juhasz *et al.*, 1993; Pellegrini *et al.*, 1992). The interaction of keratinocytes with the dermal matrix, and in particular type I collagen, may provide an early and critical signal to initiate the epithelial response to wounding. Binding of integrins with their matrix ligands activates intrinsic or nonreceptor tyrosine kinases (Burridge *et al.*, 1992; Shattil and Brugge, 1991; Zachary and Rozengurt, 1992), and recent findings demonstrated that collagen-mediated induction of collagenase by keratinocytes is dependent on tyrosine kinase activity (Sudbeck *et al.*, 1994). Since the epidermis is not normally in contact with type I collagen, it is tempting to speculate that the basal production of collagen-binding integrins, besides being involved in cell–cell and basement membrane interactions in the intact skin, keeps keratinocytes primed and ready to respond to injury. Furthermore, a mechanistic connection between collagenase expression and the recognition of type I collagen by the producing cell seems intuitive. After all, cells most likely do not release proteinases indiscriminately, especially an enzyme like interstitial collagenase that has such a defined substrate specificity, but rather rely on precise cell–matrix interactions to accurately remodel adjacent connective tissue.

5.3. Cytokines May Also Influence Collagenase Production by Keratinocytes

As stated above, collagenase expression is modulated by numerous proinflammatory mediators, such as IL-1 and TGF-α. Since many cytokines are present in the wound environment (Mauviel and Uitto, 1993) and because the epidermis is a source of many soluble mediators (Kupper, 1990; McKay and Leigh, 1991), expression of collagenase in migrating basal keratinocytes may be influenced by the presence of some or many of these factors. Indeed, TGF-α induces collagenolytic activity in epidermal rafts (Turksen *et al.*, 1991). In addition, contact with type I collagen is required but not sufficient for induction of collagenase by primary human keratinocytes (Sudbeck *et al.*, 1994). Serum must also be present, indicating that soluble factors are needed for collagenase production. It is possible that contact with dermal collagen is necessary to induce gene expression, whereas the pattern and quantity of cytokines regulate the net output of collagenase by keratinocytes. The overexpression of cytokines in chronic ulcers may lead to excess tissue remodeling and hindered repair (see Section 5.4).

5.4. Regulation of Collagenase Production in the Dermis

As stated, some expression of collagenase is detected in stromal cells in the vicinity of chronic ulcers (Saarialho-Kere *et al.*, 1992, 1993a). Typically, however, only a few scattered positive cells are present, and the signal per cell is much less than that detected in keratinocytes. Thus, collagenase expression in wound healing is primarily a response of the migrating epidermis. Dermal expression of collagenase, on the other hand, may play an important role at certain stages of wound repair, such as resolution of granulation and scar tissue, processes that may not be occurring in the chronic ulcers we have studied. In contrast to the idea that cell–matrix interactions are required for collagenase induction in keratinocytes, enzyme expression in stromal cells may be regulated by cytokines released by advancing granulation tissue (Mauviel, 1993; Porras-Reyez *et al.*, 1991). As mentioned above, up-regulation of collagenase gene expression by fibroblasts is mediated by various inflammatory agents including IL-1, TNF-α, and platelet-derived growth factor (PDGF) (Edwards *et al.*, 1987; Heckmann *et al.*, 1993; Unemori *et al.*, 1991b), and macrophage stimulation of collagenase biosynthesis is induced and stimulated by bacterial endotoxin (Saarialho-Kere *et al.*, 1993c) and by various cytokines. In human fibroblasts, EGF also induces expression of collagenase (Edwards *et al.*, 1987), and this cytokine may be released from platelets during wound healing (McKay and Leigh, 1991). Supportive of the requirement of inflammatory mediators for expression of collagenase by stromal cells, no expression of collagenase is seen in acute wounds or in samples with fibrotic ulcers but lacking any inflammation (Saarialho-Kere *et al.*, 1993a).

Cell–matrix interactions may also influence metalloproteinase production in dermal fibroblasts. Collagenase expression is markedly increased in dermal fibroblasts grown within collagen lattices (Mauch *et al.*, 1989). Werb *et al.* (1989) found that collagenase expression by synovial fibroblasts depends on the capacity to cross-link and cluster α5β1 receptors on the cell surface, and this effect is enhanced by exposure to tenascin (Tremble *et al.*, 1994). The fact that the invariable production of collagenase by migrating basal keratinocytes is usually not associated with expression by dermal cells (Saarialho-Kere *et al.*, 1992, 1993b) indicates that distinct mechanisms control this metalloproteinase in different compartments of the wound environment. Furthermore, the degradative activity of collagenase may be involved in distinct healing processes that are accomplished by the different cellular compartments. Keratinocytes may degrade dermal collagen to aid migration and promote reepithelialization, whereas stromal collagenase activity may affect tissue remodeling associated with granulation and scar formation.

5.5. Collagenase and TIMP-1 in Wound Healing

Collagenolytic activity is regulated in part by natural inhibitors, particularly TIMP-1. Although keratinocytes are capable of secreting TIMP-1 *in vitro* (Welgus and Stricklin, 1983), TIMP-1 mRNA does not colocalize with collagenase mRNA in migrating keratinocytes in chronic wounds (Saarialho-Kere *et al.*, 1992) and is detected only transiently at the edges of healing human burn wounds (Stricklin *et al.*, 1993). Typically, TIMP-1 is expressed by stromal or perivascular cells, usually away from sites of collagenase expression (Fig. 8) (Saarialho-Kere *et al.*, 1992). However, this feature was more consistent in the pyogenic granulomas we studied than in nonspecific ulcers, the former of which are characterized by extensive, proliferating blood vessels and where TIMP-1 mRNA was frequently observed in a perivascular distribution. This suggests that keratinocyte-derived collagenase is allowed to act without impedance from TIMP-1.

5.6. Stromelysins in Cutaneous Wound Healing

In addition to interstitial collagenase, we (H. G. W. and W. C. P.) have completed a thorough examination of the pattern of metalloproteinase expression in chronic wounds. Only a few matrix molecules are cleaved by collagenase, and in the setting of dermal wound healing, these are probably restricted to native types I and III collagen. Therefore, other MMPs capable of degrading fibronectin, laminin, type IV collagen, and glycosaminoglycans would be required for effective wound repair and tissue remodeling. Recent *in situ* hybridization findings suggest that stromelysins-1 and -2 are the major metalloproteinases responsible for performing these catalytic functions during wound repair (Saarialho-Kere *et al.,* 1994).

Interestingly, stromelysin-2 is expressed by the same population of migrating keratinocytes that expresses collagenase (Fig. 9). This enzyme is a distinct gene product that is closely related, both structurally and by substrate specificity, to stromelysin-1; but until now, its expression has been limited to a few cancer cells, and it has not been found in normal tissue (Murphy *et al.,* 1991; Sirum and Brinckerhoff, 1989). Unlike collagenase, however, stromelysin-2 expression is strictly confined to the epidermis and is not produced by any dermal or inflammatory cell in the wound environment. Stromelysin-1, as well, is expressed by the basal epidermis in chronic wounds, but the keratinocytes expressing this metalloproteinase are removed from the migrating front and are in contact with an intact basement membrane (Fig. 9). In further contrast to stromelysin-2, stromelysin-1 is abundantly expressed by dermal fibroblasts in the granulation tissue associated with wounds (Fig. 9). Because of its broad substrate specificity, stromelysin-1 may be an important enzyme in remodeling the dermal matrix during wound repair.

Since stromelysin-2 and interstitial collagenase are expressed by the same cells (Fig. 10), contact of actively migrating keratinocytes with the dermal matrix may influence production of both these enzymes; indeed, stromelysin-2 is expressed in keratinocytes cultured on a type I collagen substratum (Saarialho-Kere *et al.,* 1994). Furthermore, stromelysin-2-positive keratinocytes do not reside on an intact basement membrane, indicating that altered cell–matrix contacts may also be an important determinant in regulating expression of this metalloproteinase. Similar to collagenase (Fig. 10), the border between immunoreactive basement membrane and stromelysin-2-positive basal keratinocytes is not precise; however, the signal for stromelysin-2 mRNA decreases progressively within a few cells overlying the newly formed basement membrane (Saarialho-Kere *et al.,* 1994). This observation is consistent with the idea that cell–matrix interactions maintain the phenotype of basal keratinocytes in intact skin and indicates that stromelysin-2 expression ceases once the cell establishes contact with the basement membrane. Whatever does control stromelysin-2 production in keratinocytes, it is likely to be different from what regulates stromelysin-1 expression, especially in light of the distinct localization of these enzymes in the wound

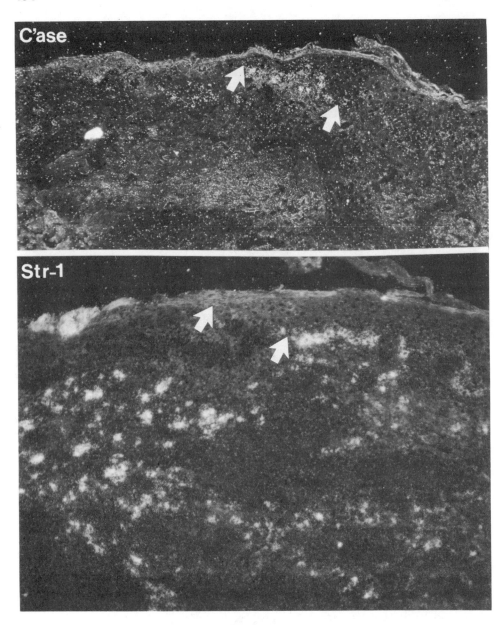

Figure 9. Collagenase and stromelysin-1 are expressed by distinct populations of basal keratinocytes. Serial sections of a nonspecific ulcer were processed for *in situ* hybridization for collagenase (C'ase) and stromelysin-1 (Str-1) mRNAs. The migrating front of the epidermis that is positive for collagenase mRNA is indicated by large arrows and is seen adjacent to an ulceration in the upper left corner. Stromelysin-1 mRNA was seen in the epidermis away from the migrating front of epithelium and in many dermal fibroblasts.

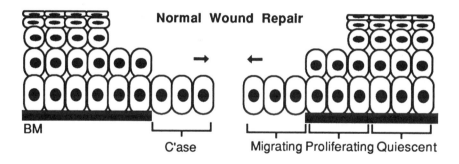

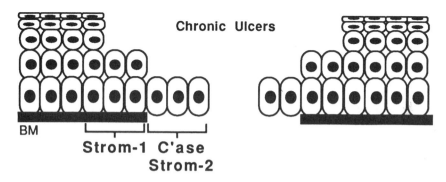

Figure 10. Spatial pattern of metalloproteinase expression in wounded epidermis. Collagenase (C'ase) is prominently and invariably expressed by migrating basal keratinocytes in all wounds, whether acute or chronic, characterized by disruption of the basement membrane (BM). Noticeably more collagenase is expressed by keratinocytes in chronic ulcers. Stromelysins 1 and 2 (Strom-1, Strom-2) are also expressed in the epidermis but by functionally distinct subpopulations of basal keratinocytes. In addition, the stromelysins are not expressed in all wounds, and thus, their proteolytic activity actually may be a detriment to proper wound healing in chronic ulcers.

environment. Although both enzymes share a high degree of amino acid sequence homology and similar proteolytic properties, the promoter regions of these two genes are quite disparate (Sirum and Brinckerhoff, 1989). Thus, whereas stromelysin-1 is synthesized by many cell types and is stimulated by a variety of cytokines, stromelysin-2 production is seemingly more limited and may be confined to epithelial cells (Windsor *et al.,* 1993). Furthermore, although the expression of both enzymes is induced by phorbol ester, TNF-α, and EGF, stromelysin-2 production is not influenced by IL-1, IL-6, or PDGF (Brinckerhoff *et al.,* 1992; Windsor *et al.,* 1993).

In contrast to collagenase and stromelysin-2, stromelysin-1-producing keratinocytes were in contact with an underlying basement membrane (Figs. 9 and 10). Since both resting keratinocytes in normal skin, which do not produce stromelysin-1, and stromelysin-1-positive keratinocytes in chronic ulcers reside on a basement membrane, the primary stimulus for stromelysin-1 expression in wound healing is probably not cellular interaction with a matrix molecule but rather exposure to a soluble factor. Although the identity of this inducing factor is not known, cytokines such as TNF-α (Grondahl-Hansen *et al.*, 1988), IL-1 (Golsen and Bauer, 1986), EGF (Gosline, 1978; Grant *et al.*, 1987), and PDGF (Corcoran *et al.*, 1992) stimulate stromelysin-1 expression in cultured fibroblasts and may also be active on keratinocytes. Alternatively, TGF-β1 may be involved. Interestingly, this cytokine down-regulates metalloproteinase expression in interstitial cells, such as fibroblasts and chondrocytes (Edwards *et al.*, 1987), but augments enzyme production by keratinocytes (Windsor *et al.*, 1993). However, since stromelysin-1 is expressed in both keratinocytes and dermal fibroblasts in the same samples, the production of TGF-β1 would have to be precisely localized and controlled to induce biosynthesis of stromelysin-1 by keratinocytes without inhibiting its production in the dermis.

5.7. Other Metalloproteinases

Other MMPs with broad catalytic activity, such as the 92-kDa and 72-kDa gelatinases and matrilysin, may be important in releasing keratinocytes from the basement membrane prior to lateral movement at the beginning of epithelial wound healing (Salo *et al.*, 1991), and both the 72-kDa and 92-kDa gelatinases are transiently seen in epidermal cells shortly after wounding (Salo *et al.*, 1994). In chronic wounds, however, these gelatinases are not actively synthesized by epidermal cells and are only occasionally expressed by either resident dermal or inflammatory cells (Saarialho-Kere *et al.*, 1993a). The metallogelatinases, especially 92-kDa gelatinase, may be secreted by certain inflammatory cells that migrate to wound sites, notably neutrophils (Hasty *et al.*, 1990), eosinophils (Ståhle-Bäckdahl *et al.*, 1994; Ståhle-Bäckdahl and Parks, 1993), and macrophages. Up-regulation of 92-kDa gelatinase is seen in the epithelial layer of healing rabbit corneal wounds (Fini *et al.*, 1992), and hence, the healing response in the cornea may be distinct from that in the skin.

5.8. Role of Metalloproteinases in Wound Repair

The invariant and prominent production of interstitial collagenase by basal keratinocytes in both acute and chronic wounds indicates that this metalloproteinase serves a critical and required role in reepithelialization rather than in dermal remodeling. As stated above, collagenase catalyzes the rate-limiting step in collagen degradation by making a single site-specific cleavage in the triple helical region of this abundant matrix molecule. At physiological temperature, these fragments are thermally unstable and denature into their constitutive polypeptide chains, forming gelatin peptides. Types I and III collagen are the major structural components of the dermal matrix and may serve as a substratum for keratinocyte migration. Since the $\alpha 2\beta 1$ integrin receptor for type I collagen binds native collagen with a higher affinity than it does for gelatin (Staatz *et al.*, 1989), the activity of interstitial collagenase by migrating keratinocytes may aid in dissociating the cell from the dermal matrix and thereby promote efficient locomotion over the dermal and provisional matrices. Thus, in the cutaneous wound-healing response, collagenase may serve a beneficial function, unlike its potentially destructive role in arthritis and vascular disease (Firestein *et al.,* 1991; Vine and Powell, 1991).

Similar to collagenase, stromelysin-2 may facilitate keratinocyte migration by degrading noncollagenous matrix molecules or by removing damaged basement membrane. It is also tempting to speculate that stromelysin-2 may be involved in the activation of cosecreted procollagenase (He *et al.,* 1989; Murphy *et al.,* 1987; Unemori *et al.,* 1991a). Since it is produced by proliferating cells (Fig. 10), stromelysin-1 is probably not involved in reepithelialization per se, but rather is needed for restructuring the newly formed basement membrane. In the dermis, collagenase and stromelysin-1 probably affect tissue repair at multiple stages, including remodeling during the formation and removal of granulation tissue and during the resolution of scar tissue. Furthermore, these two metalloproteinases may be needed for associated processes, such as angiogenesis and extravasation and migration of inflammatory cells.

Since epidermal expression of the two stromelysins was seen only in chronic wounds (Saarialho-Kere *et al.,* 1994), the production of these metalloproteinases may represent unregulated proteinase production that actually contributes to the inability of certain ulcers to heal. Appropriate and efficient reepithelialization may require the proper balance of proteinases and inhibitors (Fig. 10). In a chronic wound, however, overexpression of collagenase and the additional production of stromelysins may impair healing by destroying newly deposited matrix and cytokines and by disrupting cell–cell interactions. Further studies will be needed to accurately determine the role of the various metalloproteinases that are expressed in temporal and site-specific patterns during wound repair.

ACKNOWLEDGMENTS. The authors thank the Italian Association for Cancer Research (AIRC grant to P. M.), the National Institutes of Health (grants No. CA34282 to D. B. R.; HL-29594, HL-48762, and AM35805 to H. G. W. and W. C. P.) and the Dermatology Foundation (H. G. W. and W. C. P.) for their support. We are grateful to Drs. Ulpu Saarialho-Kere and Barry Sudbeck for their contributions.

References

Ågren, M. S., Taplin, C. J., Woessner, J. F., Eaglstein, W. H., and Mertz, P. M., 1992, Collagenase in wound healing: Effect of wound age and type, *J. Invest. Dermatol.* **99:**709–714.

Alessandri, G., Rajn, K., and Gullino, P. M., 1983, Mobilization of capillary endothelium *in vitro* induced by effectors of angiogenesis *in vivo, Cancer Res.* **43:**1790–1797.

Andrade-Gordon, P., and Strickland, S., 1986, Interaction of heparin with plasminogen activators and plasminogen: Effects on the activation of plasminogen, *Biochemistry* **25:**4033–4040.

Andreasen, P. A., Nielsen, L. S., Grøndahl-Hansen, J., Zenthen, J., Stephens, R., and Danø, K., 1984, Inactive proenzyme to tissue-type plasminogen activator from human melanoma cells, identified after affinity purification with a monoclonal antibody, *EMBO J.* **3:**51–56.

Andreasen, P. A., Sottrup-Jensen, L., Kjoller, L., Nykjaer, A., Moestrup, S. K., Petersen, C. M., and

Gliemann, J., 1994, Receptor-mediated endocytosis of plasminogen activators and activator/inhibitor complexes, *FEBS Lett.* **338:**239–245.

Angel, P., and Karin, M. I., 1992, Specific members of the jun protein family regulate collagenase expression in response to various extracellular stimuli, in: *Matrix Metalloproteinases and Inhibitors* (H. Birkedahl-Hansen, Z. Werb, H. G. Welgus, and H. E. van Wart, eds.), pp. 156–164, Gustav Fisher, Stuttgart.

Antalis, T. M., and Dickinson, J. L., 1992, Control of plasminogen-activator inhibitor type 2 gene expression in the differentiation of monocytic cells, *Eur. J. Biochem.* **205:**203–209.

Appella, E., Robinson, E. A., Ullrich, S. J., Stoppelli, M. P., Corti, A., Cassani, G., and Blasi, F., 1987, The receptor-binding sequence of urokinase. A biological function for the growth factor module of proteases, *J. Biol. Chem.* **262:**4437–4440.

Astedt, B., Lecander, I., Brodin, T., Ludblad, A., and Low, K., 1985, Purification of a specific placental plasminogen activator inhibitor by monoclonal antibody and its complex formation with plasminogen activator, *Thromb. Haemost.* **53:**122–125.

Auble, D. T., and Brinkerhoff, C. E., 1991, The AP-1 sequence is necessary but not sufficient for phorbol induction of collagenase in fibroblasts, *Biochemistry* **30:**4629–4635.

Ausprunk, D. H., and Folkman, J., 1977, Migration and proliferation of endothelial cells in preformed and newly formed blood vessels during tumor angiogenesis, *Microvasc. Res.* **14:**53–65.

Bacharach, E., Itin, A., and Keshet, E., 1992, *In vivo* patterns of expression of urokinase and its inhibitor PAI-1 suggest a concerted role in regulating physiological angiogenesis, *Proc. Natl. Acad. Sci. USA* **89:**10686–10690.

Baker, J. B., Low, D. A., Simmer, R. L., and Cunningham, D. D., 1980, Protease-nexin: A cellular component that links thrombin and plasminogen activator and mediates their binding to cells, *Cell* **21:**37–45.

Baragi, V., Fliszar, C. J., Conroy, M. C., Ye, Q.-Z., Shipley, J. M., and Welgus, H. G., 1994, Contribution of the C-terminal domain of metalloproteinases to binding by TIMP: C-terminal truncated stromelysin and matrilysin exhibit equally compromised affinities as compared to full-length stromelysin, *J. Biol. Chem.* **269:**12692–12697.

Bashkin, P., Doctrow, S., Klagsbrun, M., Svahn, C. M., Folkman, J., and Vlodavski, I., 1989, Basic fibroblast growth factor binds to subendothelial extracellular matrix and is released by heparitinase and heparin-like molecules, *Biochemistry* **28:**1737–1743.

Basilico, C., and Moscatelli, D., 1992, The FGF family of growth factors and oncogenes, *Adv. Cancer Res.* **59:**115–165.

Behrendt, N., Rønne, E., Ploug, M., Petri, T., Løber, D., Nielsen, L. S., Schleuning, W. D., Blasi, F., Appella, E., and Danø, K., 1990, The human receptor for urokinase plasminogen activator. NH_2-terminal sequence and glycosylation variants, *J. Biol. Chem.* **265:**6453–6460.

Behrendt, N., Ploug, M., Patthy, L., Houen, G., Blasi, F., and Danø, K., 1991, The ligand-binding domain of the cell surface receptor for urokinase-type plasminogen activator, *J. Biol. Chem.* **266:**7842–7847.

Behrendt, N., Ploug, M., Ronne, E., Høyer-Hansen, G., and Dano, K., 1993, Cellular receptor for urokinase-type plasminogen activator: Protein structure, *Methods Enzymol.* **223:**207–222.

Berkenpas, M. B., and Quigley, J. P., 1991, Transformation-dependent activation of urokinase-type plasminogen activator by a plasmin-independent mechanism: Involvement of cell surface membranes, *Proc. Natl. Acad. Sci. USA* **88:**7768–7772.

Birkedal-Hansen, H., Moore, W. G. I., Bodden, M. K., Windsor, L. J., Birkedal-Hansen, B., DeCarlo, A., and Engler, J. A., 1993, Matrix metalloproteinases: A review, *Crit. Rev. Oral Biol. Med.* **4:**197–250.

Bizik, J., Stephens, R. W., Grofova, M., and Vaheri, A., 1993, Binding of tissue-type plasminogen activator to human melanoma cells, *J. Cell Biochem.* **51:**326–335.

Bodden, M. K., Harber, G. J., Birkedal-Hansen, B., Windsor, L. J., Caterina, N. C. M., Engler, J. A., and Birkedal-Hansen, H., 1994, Functional domains of human TIMP-1 (tissue inhibitor of metalloproteinases), *J. Biol. Chem.* **269:**18943–18952.

Boone, T. C., Johnson, M. J., DeClerck, Y. A., and Langley, K. E., 1990, cDNA cloning and expression of a metalloproteinase inhibitor related to tissue inhibitor of metalloproteinases, *Proc. Natl. Acad. Sci. USA* **87:**2800–2804.

Bouma, B. N., and Griffin, J. H., 1978, Deficiency of factor XII-dependent plasminogen proactivator in prekallikrein-deficient plasma, *J. Lab. Clin. Med.* **91:**148–155.

Bowersox, J. C., and Sorgente, N., 1982, Chemotaxis of aortic endothelial cells in response to fibronectin, *Cancer Res.* **42:**2547–2551.

Brinckerhoff, C. E., Sirum-Connolly, K. L., Karmilowicz, M. J., and Auble, D., 1992, Expression of stromelysin and stromelysin-2 in rabbit and human fibroblasts, *Matrix* Suppl. **1:**165–175.

Brown, P. D., Kleiner, D. E., Unsworth, E. J., and Stetler-Stevenson, W. G., 1993, Cellular activation of the 72 kDa type IV procollagenase/TIMP-2 complex, *Kidney Int.* **43:**163–170.

Bu, G., Maksymovitch, E. A., and Schwartz, A. L., 1993, Receptor-mediated endocytosis of tissue-type plasminogen activator by low density lipoprotein receptor-related protein on human hepatoma HepG2 cells, *J. Biol. Chem.* **268:**13002–13009.

Buckley-Sturrock, A., Woodward, S. C., Senior, R. M., Griffin, G. L., Klagsbrun, M., and Davidson, J. M., 1989, Differential stimulation of collagenase and chemotactic activity in fibroblasts derived from rat wound repair tissue and human skin by growth factors, *J. Cell. Physiol.* **138:**70–78.

Bürk, R. R., 1973, A factor from a transformed cell line that affects cell migration, *Proc. Natl. Acad. Sci. USA* **70:**369–372.

Burridge, K., Turner, C. E., and Romer, L. H., 1992, Tyrosine phosphorylation of paxillin and pp125[FAK] accompanies cell adhesion to extracellular matrix: A role in cytoskeletal assembly, *J. Cell Biol.* **119:**893–903.

Bussolino, F., Di Renzo, M. F., Ziche, M., Bocchietto, E., Olivero, M., Naldini, L., Gaudino, G., Tamagnone, L., Coffer, A., and Comoglio, P. M., 1992, Hepatocyte growth factor is a potent angiogenic factor which stimulates endothelial cell motility and growth, *J. Cell Biol.* **119:**629–641.

Buttice, G., and Kurkinen, M., 1993, A polyomavirus enhancer A-binding protein-3 site and Ets-2 protein have a major role in the 12-*O*-tetradecanoylphorbol-13-acetate response of the human stromelysin gene, *J. Biol. Chem.* **268:**7169–7204.

Buttice, G., Quinones, S., and Kurkinen, M., 1991, The AP-1 site is required for basal expression but is not necessary for TPA-response of the human stromelysin gene, *Nucleic Acids Res.* **19:**3723–3731.

Carmeliet, P., Schoonjans, L., Kieckens, L., Ream, B., Degen, J., Bronson, R., De Vos, R., van den Oord, J. J., Collen, D., and Mulligan, R. C., 1994, Physiological consequences of loss of plasminogen activator gene function in mice, *Nature* **386:**419–425.

Carmichael, D. F., Sommer, A., Thompson, R. C., Anderson, D. C., Smith, C. G., Welgus, H. G., and Stricklin, G. P., 1986, Primary structure and cDNA cloning of human fibroblast collagenase inhibitor, *Proc. Natl. Acad. Sci. USA* **83:**2407–2411.

Carrel, R., and Travis, J., 1985, Alpha-1-antitrypsin and the serpins: Variation and countervariation, *Trends Biol. Sci.* **10:**20–25.

Carroll, P. M., Richards, W. G., Darrow, A. L., Wells, J. M., and Strickland, S., 1993, Preimplantation mouse embryos express a cell surface receptor for tissue-plasminogen activator, *Development* **119:**191–198.

Cavani, A., Zambruno, G., Marconi, A., Manca, V., Marchetti, M., and Giannetti, A., 1993, Distinctive integrin expression in the newly forming epidermis during wound healing in humans, *J. Invest. Dermatol.* **101:**600–604.

Chapman, H. A., and Stone, Jr., O. L., 1984, Cooperation between plasmin and elastase in elastin degradation by intact murine macrophages, *Biochem. J.* **222:**721–728.

Clark, R. A. F., Lanigan, J. M., DellaPelle, P., Manseau, E., Dvorak, H. F., and Colvin, R. B., 1982, Fibronectin and fibrin provide a provisional matrix for epidermal cell migration during wound re-epithelization, *J. Invest. Dermatol.* **79:**264–269.

Collier, I. E., Krasnov, P. A., Strongin, A. Y., Birkedal-Hansen, H., and Goldberg, G. I., 1992, Alanine scanning mutagenesis and functional analysis of the fibronectin-like collagen-binding domain from human 92 kDa type IV collagenase, *J. Biol. Chem.* **267:**6776–6781.

Colman, R. W., 1969, Activation of plasminogen by human plasma kallikrein, *Biochem. Biophys. Res. Commun.* **35:**273–279.

Corcoran, M. L., Stetler-Stevenson, W. G., Brown, P. D., and Wahl, L. M., 1992, Interleukin 4 inhibition of prostaglandin E2 synthesis block interstitial collagenase and 92-kDa type IV collagenase/gelatinase production by human monocytes, *J. Biol. Chem.* **267:**515–519.

Cubellis, M. V., Nolli, M. L., Cassani, G., and Blasi, F., 1986, Binding of single-chain prourokinase to the urokinase receptor of human U937 cells, *J. Biol. Chem.* **261:**15819–15822.

Cubellis, M. V., Wun, T. C., and Blasi, F., 1990, Receptor-mediated internalization and degradation of urokinase is caused by its specific inhibitor PAI-1, *EMBO J.* **9**:1079–1085.

Danø, K., Andreasen, P. A., Grøndahl-Hansen, J., Kristensen, B., Nielsen, L. S., and Skriver, L., 1985, Plasminogen activators, tissue degradation and cancer, *Adv. Cancer Res.* **44**:146–239.

DiMario, J., Buffinger, N., Yamada, S., and Strohman, R. C., 1989, Fibroblast growth factor in the extracellular matrix of dystrophic (mdx) mouse muscle, *Science* **244**:688–690.

Eaton, D. L., Scott, R. W., and Baker, J. B., 1984, Purification of human fibroblast urokinase proenzyme and analysis of its regulation by proteases and protease nexin, *J. Biol. Chem.* **259**:6241–6247.

Edlund, T., Ny, T., Ranby, M., Hedén, L. O., Palm, G., Holmgren, E., and Josephson, S., 1983, Isolation of cDNA sequences coding for a part of human tissue plasminogen activator, *Proc. Natl. Acad. Sci. USA* **80**:349–352.

Edwards, D. R., Murphy, G., Reynolds, J. J., Whitham, S. E., Docherty, A. J. P., Angel, P., and Heath, J. K., 1987, Transforming growth factor modulates the expression of collagenase and metalloproteinase inhibitor, *EMBO J.* **6**:1899–1904.

Ellis, V., Behrendt, N., and Danø, K., 1991, Plasminogen activation by receptor-bound urokinase. A kinetic study with both cell-associated and isolated receptor, *J. Biol. Chem.* **266**:12752–12758.

Ellis, V., Behrendt, N., and Dano, K., 1993, Cellular receptor for urokinase-type plasminogen activator: function in cell-surface proteolysis, *Methods Enzymol.* **223**:223–233.

Emonard, H. P., Remacle, A. G., Noel, A. C., Grimaud, J. A., Stetler-Stevenson, W. G., and Foidart, J. M., 1992, Tumor cell surface-associated binding site for the M(r) 72,000 type IV collagenase, *Cancer Res.* **52**:5845–5848.

Estreicher, A., Wohlwend, A., Belin, D., Schleuning, W. D., and Vassalli, J. D., 1989, Characterization of the cellular binding site for the urokinase-type plasminogen activator, *J. Biol. Chem.* **264**:1180–1189.

Estreicher, A., Mühlhauser, J., Carpentier, J. L., Orci, L., and Vassalli, J. D., 1990, The receptor for urokinase type plasminogen activator polarizes expression of the protease to the leading edge of migrating monocytes and promotes degradation of enzyme–inhibitor complexes, *J. Cell Biol.* **111**:783–792.

Felez, J., Chanquia, C. J., Fabregas, P., Plow, E. F., and Miles, L. A., 1993, Competition between plasminogen and tissue plasminogen activator for cellular binding sites, *Blood* **82**:2433–2441.

Fernandez, H. N., Henson, P. M., Otani, A., and Hugli, T. E., 1978, Chemotactic response to human C3a and C5a anaphylatoxins. I. Evaluation of C3a and C5a leukotaxis *in vitro* and under simulated *in vivo* conditions, *J. Immunol.* **120**:109–115.

Ferrara, N., Houck, K. A., Jakeman, L. B., Winer, J., and Leung, D. W., 1991, The vascular endothelial cell growth factor family of polypeptides, *J. Cell. Biochem.* **47**:211–218.

Fibbi, G., Ziche, M., Morbidelli, L., Magnelli, L., and Del Rosso, M., 1988, Interaction of urokinase with specific receptors stimulates mobilization of bovine adrenal capillary endothelial cells, *Exp. Cell Res.* **179**:385–395. [Published erratum in *Exp. Cell. Res.* **186**:196.]

Fini, M. E., Girard, M. T., and Matsubara, M., 1992, Collagenolytic/gelatinolytic enzymes in corneal wound healing, *Acta Ophthalmol.* **70**:26–33.

Firestein, G. S., Paine, M. M., and Littman, B. H., 1991, Gene expression (collagenase, tissue inhibitor of metalloproteinases, complement, and HLA-DR) in rheumatoid arthritis and osteoarthritis synovium. Quantitative analysis and effect of intraarticular corticosteroids, *Arthritis Rheum.* **34**:1094–1105.

Flaumenhaft, R., Moscatelli, D., and Rifkin, D. B., 1990, Heparin and heparan sulfate increase the radius of diffusion and action of basic fibroblast growth factor, *J. Cell Biol.* **111**:1651–1659.

Flaumenhaft, R., Abe, M., Mignatti, P., and Rifkin, D. B., 1992, Basic fibroblast growth factor-induced activation of latent transforming growth factor β in endothelial cells: Regulation of plasminogen activator activity, *J. Cell Biol.* **118**:901–909.

Freije, J. M. P., Diez-Itza, I., Balbin, M., Sanchez, L. M., Blasco, R., Tolivia, J., and Lopez-Otin, C., 1994, Molecular cloning and expression of collagenase-3, a novel human matrix metalloproteinase produced by breast carcinomas, *J. Biol. Chem.* **269**:16766–16773.

Gaire, M., Magbanua, Z., McDonnell, S., McNeil, L., Lovett, D. H., and Matrisian, L. M., 1994, Structure and expression of the human gene for the matrix metalloproteinase matrilysin, *J. Biol. Chem.* **269**:2032–2040.

Gherardi, E., Sharpe, M., and Lane, K., 1993, Properties and structure–function relationship of HGF-SF, *EXS* **65**:31–48.

Goldberg, G. I., Wilhelm, S. M., Kronberger, A., Bauer, E. A., Grant, G. A., and Eisen, A. Z., 1986, Human fibroblast collagenase. Complete primary structure and homology to an oncogene transformation-induced rat protein, *J. Biol. Chem.* **261:**6600–6605.

Golsen, J. B., and Bauer, E. A., 1986, Basal cell carcinoma and collagenase, *J. Dermatol. Surg. Oncol.* **12:**812–817.

Gordon, S., Unkeless, J. C., and Cohn, Z. A., 1974, Induction of macrophage plasminogen activator by endotoxin stimulation and phagocytosis, *J. Exp. Med.* **140:**995–1010.

Gosline, J. M., 1978, Hydrophobic interaction and a model for the elasticity of elastin, *Biopolymers* **17:**677–695.

Granelli-Piperno, A., Vassalli, J. D., and Reich, E., 1977, Secretion of plasminogen activator by human polymorphonuclear leukocytes. Modulation by glucocorticoids and other effectors, *J. Exp. Med.* **146:**1693–1706.

Grant, D. S., Kleinman, H. K., Goldberg, I. D., Bhargava, M. M., Nickoloff, B. J., Kinsella, J. L., Polverini, P., and Rosen, E. M., 1993, Scatter factor induces blood vessel formation *in vivo, Proc. Natl. Acad. Sci. USA* **90:**1937–1941.

Grant, G. A., Eisen, A. Z., Marmer, B. L., Roswit, W. T., and Goldberg, G. I., 1987, The activation of human skin fibroblast procollagenase. Sequence identification of the major conversion products, *J. Biol. Chem.* **262:**5886–5889.

Grobmyer, S. R., Kuo, A., Orishimo, M., Okada, S. S., Cines, D. B., and Barnathan, E. S., 1993, Determinants of binding and internalization of tissue-type plasminogen activator by human vascular smooth muscle and endothelial cells, *J. Biol. Chem.* **268:**13291–13300.

Grondahl-Hansen, J., Lund, L. R., Ralfkiaer, E., Ottevanger, V., and Danø, K., 1988, Urokinase- and tissue-type plasminogen activators in keratinocytes during wound reepithelization *in vivo, J. Invest. Dermatol.* **90:**790–795.

Guo, M., Kim, L. T., Akiyama, S. K., Gralnick, H. R., Yamada, K. M., and Grinnell, F., 1991, Altered processing of integrin receptors during keratinocyte activation, *Exp. Cell. Res.* **195:**315–322.

Hajjar, K. A., 1991, The endothelial cell tissue plasminogen activator receptor. Specific interaction with plasminogen, *J. Biol. Chem.* **266:**21962–21970.

Hajjar, K. A., 1993, Homocysteine-induced modulation of tissue plasminogen activator binding to its endothelial cell membrane receptor, *J. Clin. Invest.* **91:**2873–2879.

Hajjar, K. A., and Hamel, N. M., 1990, Identification and characterization of human endothelial cell membrane binding sites for tissue plasminogen activator and urokinase, *J. Biol. Chem.* **265**(5)**:**2908–2916.

Hajjar, K. A., Harpel, P. C., Jaffe, E. A., and Nachman, R. L., 1986, Binding of plasminogen to cultured human endothelial cells, *J. Biol. Chem.* **261:**11656–11662.

Hamilton, J. A., Whitty, G. A., Stanton, H., Wojta, J., Gallichio, M., McGrath, K., and Ianches, G., 1993a, Macrophage colony-stimulating factor and granulocyte-macrophage colony-stimulating factor stimulate the synthesis of plasminogen-activator inhibitors by human monocytes, *Blood* **82:**3616–3621.

Hamilton, J. A., Whitty, G. A., Wojta, J., Gallichio, M., McGrath, K., and Ianches, G., 1993b, Regulation of plasminogen activator inhibitor-1 levels in human monocytes, *Cell. Immunol.* **152:**7–17.

Hashimoto, K., Singer, K., Lide, W. B., Shafran, K., Webber, P., Morioka, S., and Lazarus, G. S., 1983, Plasminogen activator in cultured epidermal cells, *J. Invest. Dermatol.* **81:**424–429.

Hasty, K. A., Pourmotabbed, T. F., Goldberg, G. I., Thompson, J. P., Spinella, D. G., Stevens, R. M., and Mainardi, C. L., 1990, Human neutrophil collagenase. A distinct gene product with homology to other matrix metalloproteinases, *J. Biol. Chem.* **265:**11421–11424.

Hayakawa, T., Yamashita, K., Tanzawa, K., Uchujima, E., and Iwata, K., 1992, Growth-promoting activity of tissue inhibitor of metalloproteinase-1 (TIMP-1) for a wide range of cells, *FEBS Lett.* **298:**29–32.

He, C. S., Wilhelm, S. M., Pentland, A. P., Marmer, B. L., Grant, G. A., Eisen, A. Z., and Goldberg, G. I., 1989, Tissue cooperation in a proteolytic cascade activating human interstitial collagenase, *Proc. Natl. Acad. Sci. USA* **86:**2632–2636.

Hébert, C. A., and Baker, J. B., 1988, Linkage of extracellular plasminogen activator to the fibroblast cytoskeleton: Colocalization of cell surface urokinase with vinculin, *J. Cell. Biol.* **106:**1241–1248.

Heckmann, M., Adelmann, G. B. C., Hein, R., and Krieg, T., 1993, Biphasic effects of interleukin-1 alpha on dermal fibroblasts: Enhancement of chemotactic responsiveness at low concentrations and of mRNA expression for collagenase at high concentrations, *J. Invest. Dermatol.* **100:**780–784.

Heeb, M. J., Espana, F., Geiger, M., Collen, D., Stump, D. C., and Griffin, J. H., 1987, Immunological identity of heparin-dependent plasma and urinary protein C inhibitor and plasminogen activator inhibitor-3, *J. Biol. Chem.* **262**:15813–15816.

Heimark, R. L., Twardzik, D. R., and Schwartz, S. M., 1986, Inhibition of endothelial cell regeneration by type-beta transforming growth factor from platelets, *Science* **233**:1078–1080.

Hekman, C. M., and Loskutoff, D. J., 1985, Endothelial cells produce a latent inhibitor of plasminogen activators that can be activated by denaturants, *J. Biol. Chem.* **260**:11581–11587.

Hertle, M. D., Kubler, M.-D., Leigh, I. M., and Watt, F. M., 1992, Aberrant integrin expression during epidermal wound healing and in psoriatic epidermis, *J. Clin. Invest.* **89**:1892–1901.

Herz, J., Clouthier, D. E., and Hammer, R. E., 1992, LDL receptor-related protein internalizes and degrades uPA-PAI-1 complexes and is essential for embryo implantation, *Cell* **71**:411–421.

Howard, E. W., Bullen, E. C., and Banda, M. J., 1991, Regulation of the autoactivation of human 72 kDa progelatinase by tissue inhibitor of metalloproteinases-2, *J. Biol. Chem.* **266**:13064–13069.

Høyer-Hansen, G., Rønne, E., Solberg, H., Behrendt, N., Ploug, M., Lund, L. R., Ellis, V., and Danø, K., 1992, Urokinase plasminogen activator cleaves its cell surface receptor releasing the ligand binding domain, *J. Biol. Chem.* **267**:18224–18229.

Huhtala, P., Tuuttila, A., Chow, L. T., Lohi, J., Keski-Oja, J., and Tryggvason, K., 1991, Complete structure of the human gene for 92-kDa type IV collagenase, *J. Biol. Chem.* **266**:16485–16490.

Iadonato, S. P., Bu, G., Maksymovitch, E. A., and Schwartz, A. L., 1993, Interaction of a 39 kDa protein with the low-density-lipoprotein-receptor-related protein (LRP) on rat hepatoma cells, *Biochem. J.* **296**:867–875.

Ichinose, A., 1992, Multiple members of the plasminogen–apolipoprotein(a) gene family associated with thrombosis, *Biochemistry* **31**:3113–3118.

Ichinose, A., Kisiel, W., and Fjuikawa, K., 1984, Proteolytic activation of tissue plasminogen activator by plasma and tissue enzymes, *FEBS Lett.* **175**:412–418.

Isseroff, R. R., and Rifkin, D. B., 1983, Plasminogen is present in the basal layer of epidermis, *J. Invest. Dermatol.* **80**:297–299.

Isseroff, R. R., Fusenig, N. E., and Rifkin, D. B., 1983, Plasminogen activator in differentiating mouse keratinocytes, *J. Invest. Dermatol.* **80**:217–222.

Juhasz, I., Murphey, G. F., Yan, H.-C., Herlyn, M., and Albelda, S. M., 1993, Regulation of extracellular matrix proteins and integrin cell substratum adhesion receptors on epithelium during cutaneous wound healing *in vivo, Am. J. Pathol.* **143**:1458–1469.

Kanalas, J. J., and Makker, S. P., 1991, Identification of the rat Heymann nephritis autoantigen (gp330) as a receptor site for plasminogen, *J. Biol. Chem.* **266**:10825–10829.

Kawano, T., Morimoto, K., and Uemura, Y., 1970, Partial purification and properties of urokinase inhibitor from human placenta, *J. Biochem.* **67**:333–342.

Kim, J. P., Zhang, K., Kramer, R. H., Schall, T. J., and Woodley, D. T., 1992, Integrin receptors and RGD sequences in human keratinocyte migration: Unique anti-migratory function of a3bl epiligrin receptor, *J. Invest. Dermatol.* **5**:764–770.

Kim, K. J., Li, B., Houck, K., Winer, J., and Ferrara, N., 1992, The vascular endothelial growth factor proteins: Identification of biologically relevant regions by neutralizing monoclonal antibodies, *Growth Factors* **7**:53–64.

Kirchheimer, J. C., Nong, Y. H., and Remold, H. G., 1988, IFN-gamma, tumor necrosis factor alpha, and urokinase regulate the expression of urokinase receptors on human monocytes, *J. Immunol.* **141**:4229–4234.

Kobayashi, H., Schmitt, M., Goretzki, L., Chucholowski, N., Calvete, J., Kramer, M., Günzler, W. A., Jänicke, F., and Graeff, H., 1991, Cathepsin B efficiently activates the soluble and the tumor cell receptor-bound form of the proenzyme urokinase-type plasminogen activator (pro-uPA), *J. Biol. Chem.* **266**:5147–5152.

Kounnas, M. Z., Henkin, J., Argraves, W. S., and Strickland, D. K., 1993, Low density lipoprotein receptor-related protein/alpha 2-macroglobulin receptor mediates cellular uptake of prourokinase, *J. Biol. Chem.* **268**:21862–21867.

Kristensen, P., Larsson, L. I., Nielsen, L. S., Grøndahl-Hansen, J., Andreasen, P. A., and Danø, K., 1984, Human endothelial cells contain one type of plasminogen activator, *FEBS Lett.* **168**:33–37.

Kruithof, E. K. O., Vassalli, J. D., Schleuning, W. D., Mattaliano, R. J., and Bachman, F., 1986, Purification and characterization of a plasminogen activator inhibitor from the histiocytic lymphoma cell line U-937, *J. Biol. Chem.* **261**:11207–11213.

Kupper, T., 1990, The activated keratinocyte: A model for inducible cytokine production by non-bone marrow-derived cells in cutaneous inflammatory and immune responses, *J. Invest. Dermatol.* **94**(Suppl. 6):146S–149S.

Lacraz, S., Nicod, L., Galve-de Rochemonteix, B., Baumberger, C., Dayer, J.-M., and Welgus, H. G., 1992, Suppression of metalloproteinase biosynthesis in human alveolar macrophages by interleukin-4, *J. Clin. Invest.* **90**:382–388.

Lacraz, S., Isler, P., Welgus, H. G., and Dayer, J.-M., 1994a, Direct contact between T-lymphocytes and monocytes is a major pathway for the induction of metalloproteinase expression, *J. Biol. Chem.* **269**:22027–22033.

Lacraz, S., Nicod, L., Welgus, H. G., and Dayer, J.-M., 1995, IL-10 inhibits metalloproteinase and stimulates TIMP-1 production in human mononuclear phagocytes, *J. Clin. Invest.* in press.

Larjava, H., Salo, T., Haapasalmi, K., Kramer, R. H., and Heino, J., 1993, Expression of integrins and basement membrane components by wound keratinocytes, *J. Clin. Invest.* **92**:1425–1435.

Leco, K. J., Khokha, R., Pavloff, N., Hawkes, S. P., and Edwards, D. R., 1994, TIMP-3 is an extracellular matrix-associated protein with a distinctive pattern of expression in mouse cells and tissues, *J. Biol. Chem.* **269**:9352–9360.

Lee, S. W., Ellis, V., and Dichek, D. A., 1994, Characterization of plasminogen activation by glycosyl-phosphatidylinositol-anchored urokinase, *J. Biol. Chem.* **269**:2411–2418.

Li, W. H., and Graur, D., 1991, Evolution by gene duplication and exon shuffling. Exon shuffling, in: *Fundamentals of Molecular Evolution,* pp. 153–155, Sinauer Associates, Sunderland, Massachusetts.

Lipton, A., Klinger, I., Paul, D., and Holley, R. W., 1971, Migration of mouse 3T3 fibroblasts in response to a serum factor, *Proc. Natl. Acad. Sci. USA* **68**:2799–2801.

Loskutoff, D. J., and Edgington, T. S., 1977, Synthesis of a fibrinolytic activator and inhibitor by endothelial cells, *Proc. Natl. Acad. Sci. USA* **74**:3903–3907.

Low, D. A., Baker, J. B., Koonce, W. C., and Cunningham, D. D., 1981, Release of protease nexin regulates cellular binding, internalization, and degradation of serine proteinases, *Proc. Natl. Acad. Sci. USA* **78**:2340–2344.

Lu, H., Misrhahi, M. C., Krief, P., Soria, C., Soria, J., Mishal, Z., Bertrand, O., Perrot, J. Y., Li, H., Pujade, E., Bernadou, A., and Caen, J. P., 1988, Parallel induction of fibrinolysis and receptors for plasminogen and urokinase by interferon gamma on U937 cells, *Biochem. Biophys. Res. Commun.* **155**:418–422.

Lund, L. R., Rømer, J., Rønne, E., Ellis, V., Blasi, F., and Danø, K., 1991a, Urokinase-receptor biosynthesis, mRNA level and gene transcription are increased by transforming growth factor β1 in human A549 lung carcinoma cells, *EMBO J.* **10**:3399–3407.

Lund, L. R., Rønne, E., Roldan, A. L., Behrendt, N., Rømer, J., Blasi, F., and Danø, K., 1991b, Urokinase receptor mRNA level and gene transcription are strongly and rapidly increased by phorbol myristate acetate in human monocyte-like U937 cells, *J. Biol. Chem.* **266**:5177–5181.

Lyons, R. M., Keski-Oja, J., and Moses, H. L., 1988, Proteolytic activation of latent transforming growth factor-β from fibroblast conditioned medium, *J. Cell Biol.* **106**:1659–1665.

Lyons, R. M., Gentry, L. E., Purchio, A. F., and Moses, H. L., 1990, Mechanism of activation of latent recombinant transforming growth factor β-1 by plasmin, *J. Cell Biol.* **110**:1361–1367.

Maciag, T., 1984, Angiogenesis: The phenomenon, in: *Progress in Thrombosis and Haemostasis* (T. H. Spaet, ed.), pp. 167–182, New York, Grune and Stratton.

Mandle, R. J., and Kaplan, A. P., 1977, Hageman factor substrates. Human plasma prekallikrein: Mechanism of activation by Hageman factor and participation in Hageman factor-dependent fibrinolysis, *J. Biol. Chem.* **252**:6097–6104.

Marcotte, P. A., Kozan, I. M., Dorwin, S. A., and Ryan, J. M., 1992, The matrix metalloproteinase Pump-1 catalyzes formation of low molecular weight (pro)urokinase in cultures of normal human kidney cells, *J. Biol. Chem.* **267**:13803–13806.

Matrisian, L. M., 1990, Metalloproteinases and their inhibitors in matrix remodeling, *Trends Genet.* **6**:121–125.

Matrisian, L. M., 1992, The matrix-degrading metalloproteinases, *BioEssays* **14**:455–463.

Mauch, C., Adelmann, G. B., Hatamochi, A., and Krieg, T., 1989, Collagenase gene expression in fibroblasts is regulated by a three-dimensional contact with collagen, *FEBS Lett.* **250**:301–305.

Mauviel, A., 1993, Cytokine regulation of metalloproteinase gene expression, *J. Cell. Biochem.* **53**:288–295.

Mauviel, A., and Uitto, J., 1993, The extracellular matrix in wound healing: Role of the cytokine network, *Wounds* **5**:137–152.

Mauviel, A., Kähäri, V.-M., Evans, C. H., and Uitto, J., 1992, Transcriptional activation of fibroblast collagenase gene expression by a novel lymphokine, leukoregulin, *J. Biol. Chem.* **267**:5644–5648.

Mauviel, A., Halcin, C., Vasiloudes, P., Parks, W. C., Kurkinen, M., and Uitto, J., 1994, Uncoordinated regulation of collagenase, stromelysin, tissue inhibitor of metalloproteinases, and interleukin-8 genes by prostaglandin E2. Selective enhancement of collagenase gene expression in human dermal fibroblasts in culture, *J. Cell. Biochem.* **54**:465–472.

McKay, I. A., and Leigh, I. M., 1991, Epidermal cytokines and their roles in cutaneous wound healing, *Br. J. Dermatol.* **124**:513–518.

McNeill, H., and Jensen, P. J., 1990, A high-affinity receptor for urokinase plasminogen activator on human keratinocytes: Characterization and potential modulation during migration, *Cell Regul.* **1**:843–852.

Mignatti, P., and Rifkin, D. B., 1993, Biology and biochemistry of proteinases in tumor invasion, *Physiol. Rev.* **73**:161–195.

Mignatti, P., Robbins, E., and Rifkin, D. B., 1986, Tumor invasion through the human amniotic membrane: Requirement for a proteinase cascade, *Cell* **47**:487–498.

Mignatti, P., Tsuboi, R., Robbins, E., and Rifkin, D. B., 1989, *In vitro* angiogenesis on the human amniotic membrane: Requirement for basic fibroblast growth factor-induced proteinases, *J. Cell Biol.* **108**:671–682.

Mignatti, P., Mazzieri, R., and Rifkin, D. B., 1991, Expression of the urokinase receptor in vascular endothelial cells is stimulated by basic fibroblast growth factor, *J. Cell Biol.* **113**:1193–1201.

Miles, L. A., and Plow, E. F., 1985, Binding and activation of plasminogen on the platelet surface, *J. Biol. Chem.* **260**:4303–4311.

Miles, L. A., Dahlberg, C. M., Plescia, J., Felez, J., Kato, K., and Plow, E. F., 1991, Role of cell-surface lysines in plasminogen binding to cells: Identification of alpha-enolase as a candidate plasminogen receptor, *Biochemistry* **30**:1682–1691.

Miyazono, K., Hellman, U., Westedt, C., and Heldin, C. H., 1988, Latent high molecular weight complex of transforming growth factor β1, *J. Biol. Chem.* **263**:6407–6415.

Mizuno, K., and Nakamura, T., 1993, Molecular characteristics of HGF and the gene, and its biochemical aspects, *EXS* **65**:1–29.

Møller, L. B., Pöllanen, J., Rønne, E., Pedersen, N., and Blasi, F., 1993, N-linked glycosylation of the ligand-binding domain of the human urokinase receptor contributes to the affinity for its ligand, *J. Biol. Chem.* **268**:11152–11159.

Monsky, W. L., Kelly, T., Lin, C. Y., Yeh, Y., Stetler-Stevenson, W. G., Mueller, S. C., and Chen, W. T., 1993, Binding and localization of M(r) 72,000 matrix metalloproteinase at cell surface invadopodia, *Cancer Res.* **53**:3159–3164.

Montesano, R., Vassalli, J. D., Baird, A., Guillemin, R., and Orci, L., 1986, Basic fibroblast growth factor induces angiogenesis *in vitro*, *Proc. Natl. Acad. Sci. USA* **83**:7297–7301.

Montesano, R., Pepper, M. S., Möhle-Steinlein, U., Risau, W., Wagner, E. F., and Orci, L., 1990, Increased proteolytic activity is responsible for the aberrant morphogenetic behavior of endothelial cells expressing the middle T oncogene, *Cell* **62**:435–445.

Morton, P. A., Owensby, D. A., Wun, T. C., Billadello, J., and Schwartz, A. L., 1990, Identification of determinants involved in binding of tissue-type plasminogen activator–plasminogen activator inhibitor type 1 complexes to HepG2 cells, *J. Biol. Chem.* **265**:14093–14099.

Moscatelli, D., and Rifkin, D. B., 1988, Membrane and matrix localization of proteinases: A common theme in tumor cell invasion and angiogenesis, *Biochim. Biophys. Acta* **948**:67–85.

Moscatelli, D., Presta, M., and Rifkin, D. B., 1986, Purification of a factor from human placenta that stimulates capillary endothelial cell protease production, DNA synthesis and migration, *Proc. Natl. Acad. Sci. USA* **83**:2091–2095.

Moser, T. L., Enghild, J. J., Pizzo, S. V., and Stack, M. S., 1993, The extracellular matrix proteins laminin and fibronectin contain binding domains for human plasminogen and tissue plasminogen activator, *J. Biol. Chem.* **268:**18917–18923.

Muller, G., Behrens, J., Nussbaumer, U., Bohlen, P., and Birchmeyer, W., 1987, Inhibitory action of transforming growth factor β on endothelial cells, *Proc. Natl. Acad. Sci. USA* **84:**5600–5604.

Mullins, D. E., and Rohrlich, S. T., 1983, The role of proteinases in cellular invasiveness, *Biochim. Biophys. Acta* **695:**177–214.

Murphy, G., Cockett, M. I., Stephens, P. E., Smith, B. J., and Docherty, A. J., 1987, Stromelysin is an activator of procollagenase. A study with natural and recombinant enzymes, *Biochem. J.* **248:**265–268.

Murphy, G., Cockett, M. I., Ward, R. V., and Docherty, A. J. P., 1991, Matrix metalloproteinase degradation of elastin, type IV collagen and proteoglycan. A quantitative comparison of the activities of 95 kDa and 75 kDa gelatinases, stromelysins-1 and -2 and punctuated metalloproteinase (PUMP), *Biochem. J.* **277:**277–279.

Murphy, G., Allan, J. A., Willenbrock, F., Cockett, M. I., O'Connell, J. P., and Docherty, A. J. P., 1992a, The C-terminal domain in collagenase and stromelysin specificity, *J. Biol. Chem.* **267:**9612–9618.

Murphy, G., Ward, R., Gavrilovic, J., and Atkinson, S., 1992b, Physiological mechanisms for metalloproteinase activation, *Matrix* (Suppl.) **1:**224–230.

Murphy, G., Willenbrock, F., Ward, R. V., Cockett, M. I., Eaton, D., and Docherty, A. J. P., 1992c, The C-terminal domain of 72 kDa gelatinase A is not required for catalysis, but is essential for membrane activation and modulates interactions with tissue inhibitors of metalloproteinases, *Biochem. J.* **283:**637–641.

Murphy, G., Segain, J. P., O'Shea, M., Cockett, M., Ioannou, C., Lefebvre, O., Chambon, P., and Basset, P., 1993, The 28 kDa N-terminal domain of mouse stromelysin-3 has the general properties of a weak metalloproteinase, *J. Biol. Chem.* **268:**15435–15441.

Murphy, G., Nguyen, Q., Cockett, M. I., Atkinson, S. J., Allan, J. A., Knight, C. G., Willenbrock, F., and Docherty, A. J. P., 1994, Assessment of the role of the fibronectin-like domain of gelatinase A by analysis of a deletion mutant, *J. Biol. Chem.* **269:**6632–6636.

Nakamura, T., 1991, Structure and function of hepatocyte growth factor, *Prog. Growth Factor Res.* **3:**67–85.

Naldini, L., Tamagnone, L., Vigna, E., Sachs, M., Hartmann, G., Birchmeier, W., Daikuhara, Y., Tsubouchi, H., Blasi, F., and Comoglio, P. M., 1992, Extracellular proteolytic cleavage by urokinase is required for activation of hepatocyte growth factor/scatter factor, *EMBO J.* **11:**4825–4833.

Nielsen, L. S., Kellerman, G. M., Behrendt, N., Picone, R., Danø, K., and Blasi, F., 1988, A 55,000–60,000 M_r receptor protein for urokinase-type plasminogen activator: Identification in human tumor cell lines and partial purification, *J. Biol. Chem.* **263:**2358–2363.

Norris, D. A., Clark, R. A. F., Swigart, L. M., Huff, J. C., Westin, W. L., and Howell, S. E., 1982, Fibronectin fragments are chemotactic for human peripheral blood monocytes, *J. Immunol.* **129:**1612–1618.

O'Connell, J. P., Willenbrock, F., Docherty, A. J. P., Eaton, D., and Murphy, G., 1994, Analysis of the role of the COOH-terminal domain in the activation, proteolytic activity, and tissue inhibitor of metalloproteinase interactions of gelatinase B, *J. Biol. Chem.* **269:**14967–14973.

Odekon, L. E., Sato, Y., and Rifkin, D. B., 1992, Urokinase-type plasminogen activator mediates basic fibroblast growth factor-induced bovine endothelial cell migration independent of its proteolytic activity, *J. Cell. Physiol.* **150:**258–263.

O'Shea, M. O., Willenbrock, F., Williamson, R. A., Cockett, M. I., Freedman, R. B., Reynolds, J. J., Docherty, A. J. P., and Murphy, G., 1992, Site-directed mutations that alter the inhibitory activity of the tissue inhibitor of metalloproteinases-1: importance of the n-terminal region between cystein 3 and cystein 13, *Biochemistry* **31:**10146–10152.

Ossowski, L., Quigley, J. P., Kellerman, G. M., and Reich, E., 1973, Fibrinolysis associated with oncogenic transformation. Requirement of plasminogen for correlated changes in cellular morphology, colony formation in agar and cell migration, *J. Exp. Med.* **138:**1056–1064.

Overall, C. M., Wrana, J. L., and Sodek, J., 1991, Transcriptional and post-transcriptional regulation of 72-kDa gelatinase/type IV collagenase by transforming growth factor-beta 1 in human fibroblasts. Comparisons with collagenase and tissue inhibitor of matrix metalloproteinase gene expression, *J. Biol. Chem.* **266:**14064–14071.

Pellegrini, G., De Luca, M., Orecchia, G., Balzac, F., Cremona, O., Savoia, P., Cancedda, R., and Marchisio,

P. C., 1992, Expression, topography, and function of integrin receptors are severely altered in keratinocytes from involved and uninvolved psoriatic skin, *J. Clin. Invest.* **89:**1783–1795.

Pennica, D., Holmens, W. E., Kohr, W. J., Harkins, R. N., Vehar, G. A., Ward, C. A., Bennet, W. F., Yelverton, E., Seeburg, P. H., Heyneker, H. L., Goeddel, D. V., and Collen, D., 1983, Cloning and expression of human tissue-type plasminogen activator cDNA in *E. coli, Nature* **301:**214–221.

Pepper, M. S., Belin, D., Montesano, R., Orci, L., and Vassalli, J. D., 1990, Transforming growth factor-beta 1 modulates basic fibroblast growth factor-induced proteolytic and angiogenic properties of endothelial cells *in vitro, J. Cell Biol.* **111:**743–755.

Pepper, M. S., Ferrara, N., Orci, L., and Montesano, R., 1991, Vascular endothelial cell growth factor (VEGF) induces plasminogen activators and plasminogen activator inhibitor-1 in microvascular endothelial cells, *Biochem. Biophys. Res. Commun.* **181:**902–906.

Pepper, M. S., Ferrara, N., Orci, L., and Montesano, R., 1992a, Potent synergism between vascular endothelial growth factor and basic fibroblast growth factor in the induction of angiogenesis *in vitro, Biochem. Biophys. Res. Commun.* **189:**824–831.

Pepper, M. S., Matsumoto, K., Nakamura, T., Orci, L., and Montesano, R., 1992b, Hepatocyte growth factor increases urokinase-type plasminogen activator (u-PA) and u-PA receptor expression in Madin-Darby canine kidney epithelial cells, *J. Biol. Chem.* **267:**20493–20496.

Pepper, M. S., Sappino, A. P., Montesano, R., Orci, L., and Vassalli, J. D., 1992c, Plasminogen activator inhibitor-1 is induced in migrating endothelial cells, *J. Cell. Physiol.* **153:**129–139.

Pepper, M. S., Sappino, A. P., Stocklin, R., Montesano, R., Orci, L., and Vassalli, J. D., 1993, Up-regulation of urokinase receptor expression on migrating endothelial cells, *J. Cell Biol.* **122:**673–684.

Petersen, L. C., Lund, L. R., Nielsen, L. S., Danø, K., and Skriver, L., 1988, One chain urokinase-type plasminogen activator from human sarcoma cells is a proenzyme with little or no intrinsic activity, *J. Biol. Chem.* **263:**11189–11195.

Petersen, M. J., Woodley, D. T., Stricklin, G. P., and O'Keefe, E. J., 1990, Enhanced synthesis of collagenase by human keratinocytes cultured on type I or type IV collagen, *J. Invest. Dermatol.* **94:**341–346.

Peverali, F. A., Mandriota, S., Ciana, P., Marelli, R., Quax, P., Rifkin, D. B., Della Valle, G., and Mignatti, P., 1994, Tumor cells secrete an angiogenic factor that stimulates basic fibroblast growth factor and urokinase expression in vascular endothelial cells, *J. Cell. Physiol.* **161:**1–14.

Philips, M., Juul, A. G., and Thorsen, S., 1984, Human endothelial cells produce a plasminogen activator inhibitor and a tissue plasminogen activator–inhibitor complex, *Biochim. Biophys. Acta* **802:**99–110.

Picone, R., Kajtaniak, E. L., Nielsen, L. S., Behrendt, N., Mastronicola, M. R., Cubellis, M. V., Stoppelli, M. P., Pedersen, S., Danø, K., and Blasi, F., 1989, Regulation of urokinase receptors in monocyte-like U937 cells by the phorbol ester phorbol myristate acetate, *J. Cell Biol.* **108:**693–702.

Ploug, M., Rønne, E., Behrendt, N., Jensen, A. L., Blasi, F., and Danø, K., 1991, Cellular receptor for urokinase plasminogen activator. Carboxyl-terminal processing and membrane anchoring by glycosyl-phosphatidylinositol, *J. Biol. Chem.* **266:**1926–1933.

Plow, E. F., and Miles, L. A., 1990, Plasminogen receptors in the mediation of pericellular proteolysis, *Cell. Differ. Dev.* **32:**293–298.

Plow, E. F., Freaney, D. E., Plescia, J., and Miles, L. A., 1986, The plasminogen system and cell surfaces: Evidence for plasminogen and urokinase receptors on the same cell type, *J. Cell Biol.* **103:**2411–2420.

Pollack, R., and Rifkin, D. B., 1975, Actin-containing cables within anchorage-dependent rat embryo cells are dissociated by plasmin and trypsin, *Cell* **6:**495–506.

Pöllanen, J. J., 1993, The N-terminal domain of human urokinase receptor contains two distinct regions critical for ligand recognition, *Blood* **82:**2719–2729.

Pöllanen, J., Hedman, K., Nielsen, L. N., Danø, K., and Vaheri, A., 1988, Ultrastructural localization of plasma membrane-associated urokinase-type plasminogen activator at focal contacts, *J. Cell Biol.* **106:**87–95.

Ponting, C. P., Marshall, J. M., and Cederholm-Williams, S. A., 1992, Plasminogen: A structural review, *Blood Coagul. Fibrinolysis* **3:**605–614.

Porras-Reyez, B. H., Blair, H. C., Jeffrey, J. J., and Mustoe, T. A., 1991, Collagenase production at the border of granulation tissue in a healing wound: Macrophage and mesenchymal collagenase production *in vivo, Connect. Tissue Res.* **27:**63–71.

Postlethwaite, A. E., and Kang, A. H., 1976, Collagen and collagen peptide-induced chemotaxis of human blood monocytes, *J. Exp. Med.* **143**:1299–1307.

Quinn, C. O., Scott, D. K., Brinckerhoff, C. E., Matrisian, L. M., Jeffrey, J. J., and Partridge, N. C., 1990, Rat collagenase: Cloning, amino acid sequence comparison and parathyroid hormone regulation in osteoblastic cells, *J. Biol. Chem.* **265**:13521–13527.

Ramakrishnan, V., Sinicropi, D. V., Dere, R., Darbonne, W. C., Bechtol, K. B., and Baker, J. B., 1990, Interaction of wild-type and catalytically inactive mutant forms of tissue-type plasminogen activator with human umbilical vein endothelial cell monolayers, *J. Biol. Chem.* **265**:2755–2762.

Risch, J., Werb, Z., and Fukuyama, K., 1980, Effect of plasminogen and its activities on nuclear disintegration in newborn mice skin in culture, *J. Invest. Dermatol.* **174**:257 (Abstract).

Robbins, K. C., and Summaria, L., 1970, Human plasminogen and plasmin, *Methods Enzymol.* **19**:184–199.

Roberts, A. B., Sporn, M. B., Assoian, R. K., Smith, J. M., Roche, N. S., Wakefield, L. M., Heine, U. I., Liotta, L. A., Falanga, V., Kehrl, J. H., and Fauci, A. S., 1986, Transforming growth factor type β: Rapid induction of fibrosis and angiogenesis *in vivo* and stimulation of collagen formation *in vitro, Proc. Natl. Acad. Sci. USA* **83**:4167–4171.

Rodgers, W. H., Osteen, K. G., Matrisian, L. M., Navre, M., Giudice, L. C., and Gorstein, F., 1993, Expression and localization of matrilysin, a matrix metalloproteinase, in human endometrium during the reproductive cycle, *Am. J. Obstet. Gynecol.* **168**:253–260.

Rogelj, S., Klagsbrun, M., Atzmon, R., Kurokawa, M., Haimovitz, A., Fuks, Z., and Vlodavski, I., 1989, Basic fibroblast growth factor is an extracellular matrix component required for supporting the proliferation of vascular endothelial cells and the differentiation of PC12 cells, *J. Cell Biol.* **109**:823–831.

Rogister, B., Delree, P., Leprince, P., Martin, D., Sadzot, C., and Malgrange, B., 1993, Transforming growth factor beta as a neuronoglial signal during peripheral nervous system response to injury, *J. Neurosci. Res.* **34**:32–43.

Roldan, A. L., Cubellis, M. V., Masucci, M. T., Behrendt, N., Lund, L. R., Danø, K., Appella, E., and Blasi, F., 1990, Cloning and expression of the receptor for human urokinase plasminogen activator, a central molecule in cell surface, plasmin dependent proteolysis, *EMBO J.* **9**:467–474.

Rønne, E., Behrendt, N., Ellis, V., Ploug, M., Danø, K., and Høyer-Hansen, G., 1991, Cell-induced potentiation of the plasminogen activation system is abolished by a monoclonal antibody that recognizes the NH$_2$-terminal domain of the urokinase receptor, *FEBS Lett.* **288**:233–236.

Russell, M. E., Quertermous, T., Declerck, P. J., Collen, D., Haber, E., and Homcy, C. J., 1990, Binding of tissue-type plasminogen activator with human endothelial cell monolayers. Characterization of the high affinity interaction with plasminogen activator inhibitor-1, *J. Biol. Chem.* **265**:2569–2575.

Saarialho-Kere, U. K., Chang, E. S., Welgus, H. G., and Parks, W. C., 1992, Distinct localization of collagenase and TIMP expression in wound healing associated with ulcerative pyogenic granuloma, *J. Clin. Invest.* **90**:1952–1957.

Saarialho-Kere, U. K., Chang, E. S., Welgus, H. G., and Parks, W. C., 1993a, Expression of interstitial collagenase, 92 kDa gelatinase, and TIMP-1 in granuloma annulare and necrobiosis lipoidica diabeticorum, *J. Invest. Dermatol.* **100**:335–342.

Saarialho-Kere, U. K., Kovacs, S. O., Pentland, A. P., Olerud, J., Welgus, H. G., and Parks, W. C., 1993b, Cell–matrix interactions influence interstitial collagenase expression by human keratinocytes actively involved in wound healing, *J. Clin. Invest.* **92**:2858–2866.

Saarialho-Kere, U. K., Welgus, H. G., and Parks, W. C., 1993c, Distinct mechanisms regulate interstitial collagenase and 92 kDa gelatinase expression in human monocytic-like cells exposed to bacterial endotoxin, *J. Biol. Chem.* **268**:17354–17361.

Saarialho-Kere, U. K., Crouch, E. C., and Parks, W. C., 1995, The matrix metalloproteinase matrilysin is constitutively expressed in human exocrine epithelium, *J. Invest. Dermatol.* **105**:190–196.

Saarialho-Kere, U. K., Kovacs, S. O., Pentland, A. P., Parks, W. C., and Welgus, H. G., 1994, Distinct populations of keratinocytes express stromelysin-1 and -2 in chronic wounds, *J. Clin. Invest.* **94**:79–88.

Saarinen, J., Kalkkinen, N., Welgus, H. G., and Kovanen, P. T., 1994, Direct activation of human interstitial procollagenase (MMP-1) by human mast cell chymase in the presence of heparin, *J. Biol. Chem.* **269**:18134–18140.

Saksela, O., and Rifkin, D. B., 1988, Cell-associated plasminogen activation: Regulation and physiological functions, *Ann. Rev. Cell Biol.* **4:**93–126.

Saksela, O., and Rifkin, D. B., 1990, Release of basic fibroblast growth factor-heparan sulfate complexes from endothelial cells by plasminogen activator-mediated proteolytic activity, *J. Cell Biol.* **110:**767–775.

Saksela, O., Moscatelli, D., and Rifkin, D. B., 1987, The opposing effects of basic fibroblast growth factor and transforming growth factor beta on the regulation of plasminogen activator activity in capillary endothelial cells, *J. Cell Biol.* **105:**957–963.

Saksela, O., Moscatelli, D., Sommer, A., and Rifkin, D. B., 1988, Endothelial cell-derived heparan sulfate binds basic fibroblast growth factor and protects it from proteolytic degradation, *J. Cell Biol.* **107:**743–751.

Salo, T., Lyons, J. G., Rahemtulla, F., Birkedal-Hansen, H., and Larjava, H., 1991, Transforming growth factor-β1 up-regulates type IV collagenase expression in cultured human keratinocytes, *J. Biol. Chem.* **266:**11436–11441.

Salo, T., Mäkela, M., Kylmäniemi, M., Autio-Harmainen, H., and Larjava, H., 1994, Expression of matrix metalloproteinase-2 and -9 during early human wound healing, *Lab. Invest.* **70:**176–182.

Sato, H., Takino, T., Okada, Y., Cao, J., Shinagawa, A., Yamamoto, E., and Seiki, M., 1994, A matrix metalloproteinase expressed on the surface of invasive tumor cells, *Nature* **370:**61–65.

Sato, Y., and Rifkin, D. B., 1989, Inhibition of endothelial cell movement by pericytes and smooth muscle cells: Activation of a latent transforming growth factor β-1-like molecule by plasmin, *J. Cell Biol.* **109:**309–315.

Sato, Y., Tsuboi, R., Lyons, R. M., Moses, H. L., and Rifkin, D. B., 1990, Characterization of the activation of latent TGF-β by co-cultures of endothelial cells and pericytes or smooth muscle cells: A self-regulating system, *J. Cell Biol.* **111:**757–763.

Schneiderman, J., Sawdey, M., Craig, H., Thinnes, T., Bordin, G., and Loskutoff, D. J., 1993, Type 1 plasminogen activator inhibitor gene expression following partial hepatectomy, *Am. J. Pathol.* **143:**753–762.

Senior, R. M., Griffin, G. L., and Mecham, R. P., 1980, Chemotactic activity of elastin-derived peptides, *J. Clin. Invest.* **66:**859–862.

Shapiro, S., Doyle, G. A. D., Parks, W. C., Ley, T. J., and Welgus, H. G., 1993, Divergent mechanisms regulating the production of collagenase and TIMP in U937 cells: Evidence for involvement of delayed transcriptional activation and enhanced mRNA stability, *Biochemistry* **32:**4286–4292.

Shapiro, S. D., Campbell, E. J., Kobayashi, D. K., and Welgus, H. G., 1990, Immune modulation of metalloproteinase production in human macrophages. Selective pretranslational suppression of interstitial collagenase and stromelysin biosynthesis by interferon-γ, *J. Clin. Invest.* **86:**1204–1210.

Shapiro, S. D., Campbell, E. J., Kobayashi, D. K., and Welgus, H. G., 1991, Dexamethasone selectively modulates basal and lipopolysaccharide-induced metalloproteinase and tissue inhibitor of metalloproteinase production by human alveolar macrophages, *J. Immunol.* **146:**2724–2729.

Shapiro, S. D., Griffin, G. L., Gilber, D. J., Jenkins, N. A., Copeland, N. G., Welgus, H. G., Senior, R. M., and Ley, T. J., 1992, Molecular cloning, chromosomal localization and bacterial expression of a novel murine macrophage metalloelastase, *J. Biol. Chem.* **267:**4664–4671.

Shapiro, S. D., Kobayashi, D., Pentland, A. P., and Welgus, H. G., 1993a, Induction of macrophage metalloproteinases by extracellular matrix substrates: Evidence for enzyme- and substrate-specific responses involving prostaglandin-dependent mechanisms, *J. Biol. Chem.* **268:**8170–8175.

Shapiro, S. D., Kobayashi, D. K., and Ley, T. J., 1993b, Cloning and characterization of a unique elastolytic metalloproteinase produced by human alveolar macrophages, *J. Biol. Chem.* **268:**23824–23829.

Shattil, S. J., and Brugge, J. S., 1991, Protein tyrosine phosphorylation and the adhesive functions of platelets, *Curr. Opin. Cell Biol.* **3:**869–879.

Simon, D. I., Ezratty, A. M., Francis, S. A., Rennke, H., and Loscalzo, J., 1993, Fibrin(ogen) is internalized and degraded by activated human monocytoid cells via Mac-1 (CD11b/CD18): A nonplasmin fibrinolytic pathway, *Blood* **82:**2414–2422.

Sires, U. I., Griffin, G. L., Broekelmann, T. J., Mecham, R. P., Murphy, G., Chung, A. E., Welgus, H. G., and Senior, R. M., 1993, Degradation of entactin by matrix metalloproteinases. Susceptibility to matrilysin and identification of cleavage sites, *J. Biol. Chem.* **268:**2069–2074.

Sirum, K. L., and Brinckerhoff, C. E., 1989, Cloning of the genes for human stromelysin and stromelysin 2: Differential expression in rheumatoid synovial fibroblasts, *Biochemistry* **28:**8691–8698.

Skriver, L., Nielsen, L. S., Stephens, R., and Danø, K., 1982, Plasminogen activator released as inactive proenzyme from murine cells transformed by sarcoma virus, *Eur. J. Biochem.* **124**:409–414.

Staatz, W. D., Rajpara, S. M., Wayner, E. A., Carter, W. G., and Santoro, S. A., 1989, The membrane glycoprotein Ia-IIa (VLA-2) complex mediates the Mg^{+2}-dependent adhesion of platelets to collagen, *J. Cell Biol.* **108**:1917–1924.

Stack, M. S., and Pizzo, S. V., 1993, Modulation of tissue plasminogen activator-catalyzed plasminogen activation by synthetic peptides derived from the amino-terminal heparin binding domain of fibronectin, *J. Biol. Chem.* **268**:18924–18928.

Ståhle-Bäckdahl, M., and Parks, W. C., 1993, 92 kDa gelatinase is actively expressed by eosinophils and secreted by neutrophils in invasive squamous cell carcinoma, *Am. J. Pathol.* **142**:995–1000.

Ståhle-Bäckdahl, M., Inoue, M., Giudice, G. J., and Parks, W. C., 1994, 92 kDa gelatinase is produced by eosinophils at the site of blister formation in bullous pemphigoid and cleaves the extracellular domain of the 180 kDa bullous pemphigoid autoantigen (type XVII collagen), *J. Clin. Invest.* **93**:2202–2230.

Stecher, V. J., and Sorchin, E., 1972, The chemotactic activity of fibrin lysis products, *Int. Arch. Allergy Appl. Immunol.* **43**:879–886.

Stenn, K. S., and Malhotra, R., 1992, Epithelialization, in: *Wound Healing: Biochemical and Clinical Aspects* (I. K. Cohen, R. F. Diegelmann, and W. J. Linblad, eds.), pp. 115–127, Saunders, Philadelphia.

Stoppelli, M. P., Corti, A., Soffientini, A., Cassani, G., Blasi, F., and Assoian, R. K., 1985, Differentiation-enhanced binding of the amino-terminal fragment of human plasminogen activator to a specific receptor on U937 monocytes, *Proc. Natl. Acad. Sci. USA* **82**:4939–4943.

Stoppelli, M. P., Tacchetti, C., Cubellis, M. V., Corti, A., Hearing, V. J., Cassani, G., Appella, E., and Blasi, F., 1986, Autocrine saturation of pro-urokinase receptors on human A431 cells, *Cell* **45**:675–684.

Stricklin, G. P., Bauer, E. A., Jeffrey, J. J., and Eisen, A. Z., 1977, Human skin collagenase: Isolation of precursor and active forms from both fibroblast and organ cultures, *Biochemistry* **16**:1607–1615.

Stricklin, G. P., Li, L., Jancic, V., Wenczak, B. A., and Nanney, L. B., 1993, Localization of mRNAs representing collagenase and TIMP in sections of healing human burn wounds, *Am. J. Pathol.* **143**:1657–1666.

Sudbeck, B. D., Welgus, H. G., Parks, W. C., and Pentland, A. P., 1994, Collagen-mediated induction of collagenase by basal keratinocytes involves distinct intracellular pathways, *J. Biol. Chem.* **269**:30022–30029.

Summaria, L., Wohl, R. C., Boreisha, I. G., and Robbins, K. C., 1982, A virgin enzyme derived from human plasminogen. Specific cleavage of the arginyl-560-valyl peptide bond in the diisopropoxyphosphinyl virgin enzyme by plasminogen activators, *Biochemistry* **21**:2056–2059.

Symington, B. E., Takada, Y., and Carter, W. G., 1993, Interaction of integrins a3b1 and a2b1: Potential role in keratinocyte intercellular adhesion, *J. Cell Biol.* **120**:523–535.

Tashiro, K., Hagiya, M., Nishizawa, T., Seki, T., Shimonishi, M., Shimizu, S., and Nakamura, T., 1990, Deduced primary structure of rat hepatocyte growth factor and expression of the mRNA in rat tissues, *Proc. Natl. Acad. Sci. USA* **87**:3200–3204.

Tremble, P., Chiquet-Ehrismann, R., and Werb, Z., 1994, The extracellular matrix ligands fibronectin and tenascin collaborate in regulating collagenase gene expression in fibroblasts, *Mol. Biol. Cell* **5**:439–453.

Tsuboi, R., and Rifkin, D. B., 1990, Recombinant basic fibroblast growth factor stimulates wound healing in healing-impaired db/db mice, *J. Exp. Med.* **172**:245–251.

Tsuboi, R., Sato, Y., and Rifkin, D. B., 1990, Correlation of cell migration, cell invasion, receptor number, proteinase production, and basic fibroblast growth factor levels in endothelial cells, *J. Cell Biol.* **110**:511–517.

Tsuboi, R., Sato, C., Kurita, Y., Ron, D., Rubin, J. S., and Ogawa, H., 1993, Keratinocyte growth factor (FGF-7) stimulates migration and plasminogen activator activity of normal human keratinocytes, *J. Invest. Dermatol.* **101**:49–53.

Turksen, K., Choi, Y., and Fuchs, E., 1991, Transforming growth factor alpha induces collagen degradation and cell migration in differentiating human epidermal raft cultures, *Cell Regul.* **2**:613–625.

Unemori, E. N., Bair, M. J., Bauer, E. A., and Amento, E. P., 1991a, Stromelysin expression regulates collagenase activation in human fibroblasts, *J. Biol. Chem.* **266**:23477–23482.

Unemori, E. N., Hibbs, M. S., and Amento, E. P., 1991b, Constitutive expression of 92-kDa gelatinase (type

IV collagenase) by rheumatoid synovial fibroblasts and its induction in normal human fibroblasts by inflammatory cytokines, *J. Clin. Invest.* **88:**1656–1662.

Unkeless, J. C., Gordon, S., and Reich, E., 1974, Secretion of plasminogen activator by stimulated macrophages, *J. Exp. Med.* **139:**834–850.

Van Wart, H. E., and Birkedal-Hansen, H., 1990, The cysteine switch: A principle of regulation of metalloproteinase activity with potential applicability to the entire matrix metalloproteinase gene family, *Proc. Natl. Acad. Sci. USA* **87:**5578–5582.

Vassalli, J. D., Dayer, J. M., Wohlwend, A., and Belin, D., 1984, Concomitant secretion of prourokinase and of plasminogen activator-specific inhibitor by cultured human monocytes–macrophages, *J. Exp. Med.* **159:**1653–1668.

Verde, P., Stoppelli, M. P., Galeffi, P., Di Nocera, P., and Blasi, F., 1984, Identification and primary sequence of an unspliced human urokinase poly(A)+ RNA, *Proc. Natl. Acad. Sci. USA* **81:**4727–4731.

Vincenti, M. P., Coon, C. I., and Brinckerhoff, C. E., 1994, Regulation of collagenase gene expression by IL-1beta requires transcriptional and post-transcriptional mechanisms, *Nucleic Acid Res.* **22:**4818–4827.

Vine, N., and Powell, J. T., 1991, Metalloproteinases in degenerative aotic disease, *Clin. Sci.* **81:**233–239.

Vlodavsky, I., Fuks, Z., Bar-Ner, M., Ariav, Y., and Schirrmacher, V., 1983, Lymphoma cell-mediated degradation of sulfated proteoglycans in the subendothelial extracellular matrix: Relationship to tumor cell metastasis, *Cancer Res.* **43:**2704–2711.

Welgus, H. G., Jeffrey, J. J., and Eisen, A. Z., 1981, The collagen substrate specificity of human skin fibroblast collagenase, *J. Biol. Chem.* **256:**9511–9515.

Welgus, H. G., Campbell, E. J., Cury, J. D., Eisen, A. Z., Senior, R. M., Wilhelm, S. M., and Goldberg, G. I., 1990, Neutral metalloproteinases produced by human mononuclear phagocytes. Enzyme profile, regulation, and expression during cellular development, *J. Clin. Invest.* **86:**1496–1502.

Welgus, H. G., and Stricklin, G. P., 1983, Human skin fibroblast collagenase inhibitor: Comparative studies in human connective tissues, serum and amniotic fluid, *J. Biol. Chem.* **258:**12259–12264.

Werb, Z., Mainardi, C., Vater, C. A., and Harris, E. D., 1977, Endogenous activation of latent collagenase by rheumatoid synovial cells. Evidence for a role of plasminogen activator, *N. Engl. J. Med.* **296:**1017–1023.

Werb, Z., Banda, M. J., and Jones, P. A., 1980, Degradation of connective tissue matrices by macrophages. I. Proteolysis of elastin, glycoproteins and collagen by proteinases isolated from macrophages, *J. Exp. Med.* **152:**1340–1357.

Werb, Z., Tremble, P. M., Behrendtsen, O., Crowley, E., and Damsky, C. H., 1989, Signal transduction through the fibronectin receptor induces collagenase and stromelysin gene expression, *J. Cell. Biol.* **109:**877–889.

Windsor, L. J., Grenett, H., Birkedal-Hansen, B., Bodden, M. K., Engler, J. A., and Birkedal-Hansen, H., 1993, Cell type-specific regulation of SL-1 and SL-2 genes. Induction of the SL-2 gene but not the SL-1 gene by human keratinocytes in response to cytokines and phorbol esters, *J. Biol. Chem.* **268:**17341–17347.

Wittwer, A. J., and Sanzo, M. A., 1990, Effect of peptides on the inactivation of tissue plasminogen activator by plasminogen activator inhibitor-1 and on the binding of tissue plasminogen activator to endothelial cells, *Thromb. Haemost.* **64:**270–275.

Woessner, J. F., 1991, Matrix metalloproteinases and their inhibitors in connective tissue remodeling, *FASEB J.* **5:**2145–2154.

Woodley, D. T., Kalebec, T., Baines, A. J., Link, W., Prunieras, M., and Liotta, L., 1986, Adult human keratinocytes migrating over nonviable dermal collagen produce collagenolytic enzymes that degrade type I and type IV collagen, *J. Invest. Dermatol.* **4:**418–423.

Woodley, D. T., Bachmann, P. M., and O'Keefe, E. J., 1988, Laminin inhibits human keratinocyte migration, *J. Cell. Physiol.* **136:**140–146.

Yang, E. Y., and Moses, H. L., 1990, Transforming growth factor β1-induced changes in cell migration, proliferation, and angiogenesis in the chick chorioallantoic membrane, *J. Cell Biol.* **111:**731–741.

Zachary, I., and Rozengurt, E., 1992, Focal adhesion kinase (p125 FAC): A point of convergence in the action of neuropeptides, integrins, and oncogenes, *Cell* **71:**891–894.

Chapter 15

Proteoglycans and Their Role in Wound Repair

RICHARD L. GALLO and MERTON BERNFIELD

1. Introduction

Proteoglycans are a heterogeneous group of protein–carbohydrate complexes that are distinguished from all other macromolecules by bearing glycosaminoglycan (GAG) chains. These linear polysaccharide chains are highly polyanionic (due to sulfate and carboxylate residues), bear the highest charge density of any vertebrate macromolecule, and usually occupy a high proportion of the mass of the proteoglycans. The distinction is important because these properties differ from all other molecules in vertebrate tissues. Because of their chemical stability, the GAG chains have been known and well characterized for many years. Only relatively recently, with the identification of a large number of proteoglycan core proteins, have their roles in cellular behavior become apparent. The GAG chains play a primordial role in metazoans; they are produced by the simplest organisms and are synthesized very early during vertebrate development and by virtually every nucleated cell. While they have multiple potential functions, the explicit role of each GAG type depends both on its nature and on the core protein moiety to which the GAG chain is linked in a proteoglycan.

The core proteins of proteoglycans are diverse and heterogeneous, appear to share only their capacity to bear GAG chains, and represent the wide variety of the proteins found within secretory vesicles, at cell surfaces, and in the extracellular matrix. However, several proteoglycan gene families have evolved. These proteins, whose major function appears to be to carry GAG chains, also may have a variety of other domains. In general, the protein moiety of a proteoglycan is responsible for placing its GAG chain(s) in discrete locations at specific times. Thus, the regulation of expression of these proteins and their trafficking serves to regulate the availability of the GAG chains.

RICHARD L. GALLO • Department of Dermatology, Harvard Medical School, Boston Childrens' Hospital, Boston, Massachusetts 02115.   MERTON BERNFIELD • Joint Program in Neonatology, Harvard Medical School, Boston Children's Hospital, Boston, Massachusetts 02115.

The Molecular and Cellular Biology of Wound Repair (Second Edition), edited by Richard A. F. Clark. Plenum Press, New York, 1996

Proteoglycans are increasingly recognized as regulators of cell behavior. For example, proteoglycan GAG chains can organize macromolecular assemblies (e.g., the dermatan sulfate chains on decorin or on type IX collagen that organize collagens into fibrils), can enable various cellular effectors to act on cells (e.g., cell surface heparan sulfate that causes growth factors to interact with their signal-transducing receptors), or can enable cells to store secretory components (e.g., the binding of mast cell metalloproteases by intracellular heparin). These functions are in addition to the most prevalent use of GAGs in clinical practice, which is the regulation of blood coagulation.

Proteoglycans have been classified in several ways, with each classification mode emphasizing a particular aspect of proteoglycan biology or structure. The molecules have been classified according to GAG type, by their location in tissues and cells, by the gene family in which they are associated, or by their presumed function. The purpose of this chapter is to review the role of proteoglycans in wound repair. Thus, we will focus on the general nature of proteoglycans, the types of GAG chains they bear, and their functional significance for wound repair. Recent reviews should be consulted for more detailed information (e.g., Jackson *et al.*, 1991; Kjellen and Lindahl, 1991; Wight *et al.*, 1991; Hardingham and Fosang, 1992; Kreis and Vale, 1993; Kresse *et al.*, 1993; David, 1993).

2. The GAGs

Proteoglycans consist of a protein (known as the core protein) and one or more covalently linked GAG chain. The GAGs are polysaccharides consisting of repeating disaccharides in linear chains. In every case, the disaccharide is composed of a hexosamine alternating with an acidic sugar, either a uronic acid or a sulfated galactose. The chains are variable in size, ranging from as few as 10 to as many as 20,000 disaccharides. The chains are linear and have polarity, with the reducing end linked by a glycosidic bond to the core protein by a specific oligosaccharide, designated as a linker region, and with a nonreducing end that is distant from the core protein. The chains are not rigid in structure, and their flexibility in space depends on the nature of the sugars in the chains.

GAG chains are synthesized stepwise by an enzymatic addition of each alternating sugar from nucleotide sugar donors. These reactions yield a linear chain that is then enzymatically modified to yield the chains with their distinctive characteristics. In these modification reactions the product of one reaction is the substrate for the next. The initial polymerization reactions lead to repeating identical disaccharide units, but the subsequent modification reactions are often incomplete, yielding heterogeneity in the substituents. Thus, the structural complexity of the various GAGs is based on the specificity, rate, and availability of the modifying enzymes rather than on a preexisting template as with proteins, RNA, or DNA. This biosynthetic mechanism leads to considerable structural microheterogeneity among the GAG chains and can lead to confusion in nomenclature. In practice, however, the GAG chains are characterized by their behavior as polymers, including their electrophoretic or chromatographic mobility.

Table I. The Glycosaminoglycans of Proteoglycans

	Disaccharide composition		
	Hexosamine	Uronic acid or Gal	Distinguishing features
Hyaluronan	GlcN	GlcA	No core protein; unsulfated, fills extracellular spaces
Galactosaminoglycans			
Chondroitin sulfate	GalN	GlcA	Cartilage and cell surfaces; principally extracellular
Dermatan sulfate	GalN	GlcA/IdoA	
Heparan sulfates/heparin	GlcN	GlcA/IdoA	Binds proteins; principal intracellular GAG
Keratan sulfate	GlcN	Gal	Restricted distribution; skeletal and corneal forms

There are four types of vertebrate GAGs: hyaluronan, the galactosaminoglycans, the heparan sulfates, and the keratan sulfates. These GAGs differ primarily in the type of disaccharide but also in their size and extent of their modifications (see Table I).

2.1. Hyaluronan

Hyaluronan, formerly known as hyaluronic acid, is the simplest GAG, consisting of GlcNac alternating with GlcA. (The standard abbreviations are used for the sugars: Ac, acetyl; Gal, galactose; GalN, galactosamine; GlcN, glycosamine; GlcA, glucuronic acid; IdoA, iduronic acid; Xyl, xylose.) This polysaccharide is not linked covalently with protein, and thus, by definition, is not a proteoglycan; yet it is considered here because of its GAG nature and its importance in wound repair. Additional distinctions are that its disaccharide chains are not modified and contain no sulfate esters. These chains are very large, and a hyaluronan molecule can reach approximately 10 million Da in size. Hyaluronan is produced by most cell types, but it is especially abundant surrounding fibroblasts and other mesenchymal cells. Its synthesis also differs from the other GAGs; hyaluronan is synthesized by an enzyme complex at the plasma membrane through which the growing chain is extruded into the extracellular space (Prehm, 1983). This biosynthetic mechanism enables the hyaluronan to fill very large volumes as a single molecule. Hyaluronan is not readily degraded locally, but circulates within tissues via the lymph and is degraded by reticuloendothelial cells, principally of the liver (Roden *et al.,* 1989).

Hyaluronan is an exclusively extracellular macromolecule. Its chains are highly hydrated and occupy an extraordinary amount of extracellular space, imparting viscosity to tissues and fluids. Various ions and solutes can diffuse within these spaces, or hyaluronan can surround cells and cause them to aggregate. Indeed, cells bind to hyaluronan by several mechanisms, including cell surface receptors such as CD44, a transmembrane protein found on a variety of cell types. Hyaluronan is recognized by several proteins and proteoglycan core proteins, enabling it to organize these constituents within the extracellular matrix, e.g., the binding of aggrecan and link protein to hyaluronan organizes the proteoglycan aggregates within hyaline cartilage (Heinegard and Oldberg, 1989). Several other hyaluronan-binding proteins have been described, but their functions are not well defined.

2.2. Galactosaminoglycans

The galactosaminoglycans are chondroitin sulfate and dermatan sulfate. In these GAGs, the disaccharide consists of GalNac alternating with GlcA or IdoA. Chondroitin sulfate contains GlcA exclusively, whereas dermatan sulfate contains a proportion of its uronic acids as IdoA. These are sulfated GAG; the sulfate esters are on the C4 and C6 of the GalNac and often on the C-2 of the IdoA. These GAGs show extensive sequence heterogeneity because of variable extents of sulfation. Because of its IdoA content, dermatan sulfate can self-aggregate. These GAG chains are linked to a variety of core proteins via a tetrasaccharide linkage region, Glca-Gal-Gal-Xyl in which the Xyl is the reducing end. Each sugar in the linkage region is added by a distinctive enzyme residing in the Golgi apparatus (Vertel *et al.*, 1993). The GAG chain is then initiated by the addition of a GalNac residue to the GlcA at the nonreducing end and is extended by the sequential addition of GlcA and GalNAc residues from their respective nucleotide sugars. All subsequent modifications, including O-sulfations and the epimerization of GlcA to IdoA, occur on the preformed polysaccharide chain.

The galactosaminoglycans are produced by virtually all cell types, and a wide variety of core proteins contain these GAG chains (see Table II). The proteins are glycosylated via the linkage tetrasaccharide to a serine in distinctive Ser-Gly dipeptide sequences. The number of chains per protein can vary from one (e.g., in decorin) to several hundred (e.g., in aggrecan). These galactosaminoglycan chains can bind a variety of protein ligands, and the tissue and cellular distribution of these proteoglycans depend on the nature of their core protein.

2.3. Heparan Sulfates

The heparan sulfates, composed of heparin and heparan sulfate, are glucosaminoglycans. These are the most structurally complex GAGs, and there is no qualitative difference in structure between heparin and heparan sulfate. The disaccharide consists of GlcNac alternating with GlcA or IdoA. The acetyl substituent on the GlcN is frequently replaced with a sulfate, forming N-sulfated GlcN. In contrast to the β1–3, β1–4 linkages in hyaluronan, the galactosaminoglycans, and keratan sulfate, the sugars in heparan sulfates are linked β1–4, α1–4. These GAG chains are linked to proteins via the identical tetrasaccharide linkage region as described above for the galactosaminoglycans. Heparin consists primarily of N-sulfated regions enriched in IdoA and commonly O-sulfated at the C6 of the GlcN and the C2 of the IdoA. Heparan sulfate chains show greater complexity; N-sulfate, IdoA, and O-sulfated regions of

Table II. Major Types of Proteoglycans[a]

Location	Proteoglycan	GAG constituents[b]	Predominant cellular distribution	Potential function	Reference
Intracellular Vesicular	Serglycin	CS; multiple chains HS; multiple chains	Musocal mast cells Connective tissue, mast cells, and basophils	Storage and secretion of proteases and amines	Stevens and Austen (1989)
Cell surface Transmembrane	Syndecan family Syndecan-1 Syndecan-2 (fibroglycan) Syndecan-3 (N-syndecan) Syndecan-4 (amphiglycan, ryudocan)	 CS, HS; 2-5 chains HS HS CS, HS; 2-5 chains	 Epithelia Vascular endothelia Neural cells Ubiquitous	Coreceptors for growth factors, matrix components, other ligands; shed as paracrine effectors	Bernfield *et al.* (1992)
	NG2	CS; 2-3 chains	Neural and melanoma cells	Matrix adhesion	Nishiyama and Stallcup (1993)
GPI-linked	Glypican family Glypican Cerebroglycan OC1-5	 HS; 1-3 chains HS HS		As syndecan family, probably unique functions for the core proteins	David (1993)

(*continued*)

Table II. (*Continued*)

Location	Proteoglycan	GAG constituents[b]	Predominant cellular distribution	Potential function	Reference
Extracellular					
	Leucine-rich family				
	Decorin	DS; 1 chain	Fibroblasts	Organize type I collagen fibrils; core protein binds TGF-β, fibronectin	Scott (1993)
	Biglycan	DS; 2 chain	Fibroblasts	Bind TGF-β	Hildebrand *et al.* (1994)
	Fibromodulin	KS		Bind TGF-β	
	Lumican	KS		Bind TGF-β	Blochberger *et al.* (1992)
Basement membrane	Perlecan	HS; 1-3 chains	Epithelia, endothelia, fibroblasts	Organize basement membranes, bind bFGF	Iozzo (1994)
Hyaluronan-associated	Aggrecan family				
	Aggrecan	CS, KS; multiple chains	Chondrocytes	Bind water in cartilage; multiple domains	Neame (1993)
	Versican	CS; 12-15 chains	Fibroblasts	Bind water in interstitial matrix, multiple domains	Zimmerman and Ruoslahti (1989)
	Neurocan	CS; 1-3	Brain		Rauch *et al.* (1992)
	Brevican	CS	Brain		Yamada *et al.* (1994)
Others	Agrin	HS	Brain, neuromuscular junction	Aggregate acetylcholine receptors	Tsen *et al.* (1995)
	Testican	HS; CS 1 chain each	Testes		Silbert *et al.* (1995)

[a]Excludes part-time proteoglycans.
[b]CS is chondroitin sulfate; DS is dermatan sulfate; HS is heparan sulfate; KS is keratan sulfate; GPI is glycerylphosphatidylinositol.

approximately 12 to 30 disaccharides are separated by regions of variable size that contain GlcNAc and are not heavily sulfated (Turnbull and Gallagher, 1990). Thus, regions of very high charge density alternate with less acidic regions.

The heparan sulfates are on core proteins within vesicles, at the cell surface, and in the extracellular matrix. The major intracellular heparan sulfate is heparin, which is in serglycin, a proteoglycan within mast cells and basophils. Commercial heparin, clinically used as an anticoagulant, is a degradative product of mast cell heparin.

The heparan sulfates at cell surfaces and in the extracellular matrix are linked to a variety of proteins and consist of highly N-sulfated regions alternating with regions enriched in GlcNAc and GlcA. The size of the N-sulfated regions and of the chains themselves vary among cell types. The major role of these GAGs is the binding of proteins. Specific sequences are associated with the binding of specific ligands. For example, antithrombin-III selectively binds to a unique pentasaccharide sequence, accelerating the formation of a ternary complex with thrombin; this is the major basis of the anticoagulant activity of heparin (Lindahl *et al.*, 1980). Basic fibroblast growth factor (FGF) selectively binds to a penta- or heptasaccharide sequence, while a dodecasaccharide is needed to enable this growth factor to form a ternary complex with the FGF receptor at the cell surface, thus mediating a mitogenic signal (Maccarana *et al.*, 1993; Guimond *et al.*, 1993).

2.4. Keratan Sulfates

Keratan sulfate consists of a repeating disaccharide of GlcNac alternating with galactose. This GAG contains no uronic acids, but both the GlcNac and galactose may be O-sulfated at the 6 position. Keratan sulfate chains are linked to proteins via two distinct types of linkages. The first is similar to linkages in mucins in which the carbohydrate chain is O-linked to serine via a GalNac-containing hexasaccharide. This form of linkage is prevalent in skeletal keratan sulfate, as on aggrecan and versican. The other form of linkage is similar to those in typical glycoproteins in which the carbohydrate chain is N-linked to asparagines via a biantennary oligosaccharide. This form is prevalent in corneal keratan sulfate, as exists on fibromodulin and lumican. The physiological role of keratan sulfate at these sites is not well-defined.

3. Molecular Strategy of the Proteoglycans

The proteoglycans employ an economical strategy to exert their effects on cells. The core proteins provide attachment sites that dictate the nature of the GAG chain and direct the intracellular and extracellular trafficking that places the GAGs at the proper site and circumstance for their ligand binding. The various GAG chains bind a panoply of ligands. Once bound, these entities can then be stored or can interact with signal-transducing receptors at the cell surface, soluble proteins for endocytosis, enzymes, insoluble proteins for cell attachment or molecular organization, and a multitude of

Table III. Heparin–Heparan Sulfate Interactions with Protein Ligands Potentially Involved in Wound Repair[a]

	Protein ligand
Extracellular matrix components	Collagen types I, III, IV, V
	Fibronectin
	Laminin
	Pleiotropin
	Tenascin
	Thrombospondin
	Vitronectin
	wnt-1
Growth factors (GF)	Fibroblast GF family
	Hepatocyte GF/scatter factor
	Heparin-binding epidermal GF-like GF
	Platelet-derived GF
	Schwannoma derived GF
	Vascular endothelial GF
Growth factor binding proteins (BP)	Follistatin
	IGFBP-3
	TGF-β BP
Cytokines	IL-8
	Interferon-γ
	MIP-1β
Cell adhesion molecules	CD45
	L-selectin
	Mac-1
	N-CAM
	PECAM
Proteases	Elastase
	Thrombin
	Tissue plasminogen activator
Antiproteases	Antithrombin III
	Heparin cofactor II
	Leuserpin
	Plasminogen activator inhibitor-1
	Protease nexin I

[a]Modified from Bernfield *et al.* (1992) and Silbert *et al.* (1995), where the citations are listed.

other protein ligands (see Table III). Importantly, GAG binding of protein ligands is not highly specific for the protein, but is sufficiently high in affinity and specificity to be physiologically relevant.

The design of the GAG chains is tailored for binding ligands. The chains are flexible in space and can assume three-dimensional shapes that can be modified by the ligand. The binding results from the polyanionic character of the chains analogous to the induced-fit model of enzyme–substrate interactions. Binding of proteins is enhanced by the presence of IdoA residues, which impart conformational flexibility to dermatan sulfate and heparan sulfate chains (Casu *et al.*, 1988). Indeed, heparan sulfate and its heparin derivative bind a very large number of protein ligands because of the clusters of highly acidic residues in regions that are rich in IdoA. Heparin may show less selectivity because of its greater extent of sulfation (Spillman and Lindahl, 1994). The size, sequence, and extent of modifications of these chains on the same core protein vary between cell types. Thus, the nature of the GAG chain can be a differentiated characteristic that arises during development (Kato *et al.*, 1994). These considerations suggest that GAG chains can be informational. An argument for this idea is that when these chains are recognized by protein ligands, they produce a physiologically relevant informational signal.

The characteristics of the GAG-binding ligands are highly variable. Although a large number of proteins bind at physiological protein concentrations and under physiological ionic conditions, the amino acid sequences responsible for this binding vary widely (Spillman and Lindahl, 1994). These sequences are generally rich in lysine and arginine, but no unique or invariant sequences can be predicted with certainty to bind to any GAG. Indeed, the key amino acids may not be in a contiguous sequence. Nevertheless, the binding is relatively high affinity, ranging from 10^{-7} to 10^{-9} M. This scheme provides cells with a mechanism to snare a wide variety of physiological effectors without requiring that evolution generate multiple novel binding proteins that contain unique molecular features.

This economical strategy is most evident with heparan sulfate/heparin proteoglycans. Their extensive structural heterogeneity and presumed malleability underlie their ability to bind many proteins in a functionally relevant way (Table III). Their ligands include a large number of proteins thought to be involved in wound repair, including extracellular matrix components, growth factors and their binding proteins, cytokines, cell adhesion molecules, proteases, and antiproteases.

The heparan sulfate proteoglycans at the cell surface and in the extracellular matrix act as receptors or coreceptors for these ligands. The interaction of basic FGF with the cell has been best studied (see Turnbull *et al.*, 1992; Maccarana *et al.*, 1993; Guimond *et al.*, 1993; Spivak-Kroizman *et al.*, 1994). This growth factor binds at nanomolar affinities to both extracellular and cell surface heparan sulfate proteoglycans, which are substantially more abundant than the binding sites derived from FGF receptors. Once formed, the FGF–heparan sulfate complex can form a higher-affinity ternary complex with the FGF receptor, which, when occupied, initiates the intracellular signaling cascade. The interaction with the proteoglycan provides the cell with a way to regulate receptor signaling by means other than receptor expression and occupancy. This coreceptor mechanism could result in the inhibition and modulation of growth factor action that likely occurs during wound repair. Analogous coreceptor interactions occur with other extracellular effectors, e.g., fibronectin that may not induce a change in cellular behavior unless it engages cells by both its heparin-binding and integrin-binding domains (Woods and Couchman, 1994).

4. Proteoglycan Families

The proteoglycans are in specific gene families that encode proteins that consistently contain GAG when found in tissues. These proteins must be distinguished from those that can be isolated either with or without GAG chains; a variable proportion of such proteins exist as proteoglycan variants and are known as part-time proteoglycans. Part-time proteoglycans can be intracellular (e.g., Ia invariant chain), at the cell surface [e.g., thrombomodulin or transforming growth factor-β (TGF-β) type 3 receptor] or in the extracellular matrix (e.g., type IX collagen). It is sometimes difficult to ascertain which proteoglycans are "part-time." Usually, members of authentic proteoglycan core protein families have homologues that also bear GAG chains.

The core proteins of proteoglycans have evolved to contain GAG attachment sequences in addition to a variety of other functional domains. Each core protein predominantly bears either a single GAG type (e.g., perlecan contains exclusively heparan sulfate), or may contain multiple GAG types (e.g., aggrecan contains both chondroitin sulfate and keratan sulfate and syndecans-1 and -4 contain both heparan sulfate and chondroitin sulfate), or may differ in GAG type depending on the cell type (e.g., serglycin bears chondroitin sulfate in mucosal mast cells and heparan sulfate in connective tissue mast cells). The biosynthetic events responsible for dictating which of the GAG chains are placed on the tetrasaccharide linkage region are as yet not clear. Certain core protein sequences correlate with specific GAG types, but site-specific mutation studies have not shown that the sequences are uniquely selective for a GAG chain (Zhang and Esko, 1994). However, a core protein domain has been described that ensures that it will be consistently glycosylated with a GAG chain (Kokenyesi and Bernfield, 1994).

The major proteoglycan core protein families vary in their cellular locations. Indeed, because the nature of the core protein dictates its site within or outside the cell, a proteoglycan classification based on cellular location may be most relevant. These locations also give a clue to the nature of the GAG substituents and to their function. In general, intracellular and cell surface proteoglycans contain predominantly heparan sulfate. These proteins use their GAG chains to bind a variety of protein ligands, acting to stabilize the ligand to degradation, to present the ligand to another protein, or to maintain the ligand in an inactive or stored form. The functions of many of these proteoglycans are not yet clear. Extracellular proteoglycans contain predominantly dermatan sulfate and chondroitin sulfate; keratan sulfate is exclusively on extracellular proteoglycans. These proteoglycans are involved in organizing components of the extracellular matrix, including water and ions, the structural components of cartilage and basement membranes, and the architecture of a variety of collagenous fibrils.

5. Proteoglycans in the Early Phase of Wound Repair

The earliest phase of wound repair involves clot formation, cell adhesion, and inflammation. At the time of injury, vascular continuity is lost and blood extravasates into the extracellular space and begins to coagulate. Blood coagulation is prevented within the vascular space by the anticoagulant surface of endothelia, a property of the cell surface proteoglycans, including the syndecans (Kojima *et al.*, 1992b), and thrombomodulin, a part-time proteoglycan (Parkinson *et al.*, 1992). These act, in large part, by binding to antithrombin III, a protease inhibitor that reacts with serine proteases of the coagulation cascade. When antithrombin III binds to heparin or "heparinlike" regions on heparan sulfate, it changes conformation, enabling it to inactivate target proteases such as factor Xa and thrombin (Thunberg *et al.*, 1982; Lindahl *et al.*, 1979, 1984; Choay *et al.*, 1983; Danielsson *et al.*, 1986). Immediately following injury, coagulation and platelet adherence is promoted and antithrombin III activity is low. Platelets may augment this process through the release of heparitinase (Castellot *et al.*, 1982) and platelet factor 4 that inhibits the function of heparin *in vitro* (Ginsberg and Jaques, 1983). Several proteoglycans produced by endothelial cells are members of the syndecan family (Marcum and Rosenberg, 1984; Kojima *et al.*, 1992a,b; David *et al.*, 1992). However, only about 1–5% of the heparan sulfate chains bind to immobilized antithrombin III (Kojima *et al.*, 1992a), suggesting that endothelial cell surface proteoglycans have functions distinct from anticoagulant activity.

Proteoglycans may function as adhesion molecules during the inflammatory cell activation that arises immediately after coagulation in the early phase of wound repair. Cell surface proteoglycans such as the syndecans are also associated with cells of lymphoid and monocyte lineage (Sanderson *et al.*, 1989; Kim *et al.*, 1994). The roles for proteoglycans in these events are more speculative and based largely on the results of experiments performed *in vitro*. Macrophages regulate their expression of syndecan-1 during activation (Yeaman and Rapraeger 1993). Peritoneal macrophages express minimal amounts of cell surface syndecan-1 despite abundant syndecan-1 mRNA. Following stimulation with cAMP, these cells appear to release a posttranslational block and express syndecan-1 heparan sulfate proteoglycan at the cell surface. These activated macrophages may thus alter their responsiveness to extracellular matrix components or to heparin-binding growth factors. A similar phenomenon has been observed for cells of the B lineage by Sanderson *et al.* (1989). Syndecan-1 expression during the development of these cells correlates with differentiation and the capacity to circulate. B-cell precursors associated with the stromal matrix express syndecan-1 at the cell surface, lose expression when they mature to circulating B-cells, and reexpress cell surface syndecan-1 when they differentiate into plasma cells. Thus, the expression of syndecan-1, a receptor for several matrix components, is induced when these cells adhere to the stroma. Sanderson *et al.*, (1992b, 1994) have shown that the ability of these cells to migrate in a collagen matrix can be regulated by select forms of heparan sulfates.

Other glycosaminoglycans can regulate inflammatory cell function. For example, hyaluronan has been shown to affect the activity of neutrophils by decreasing adherence and chemotactic responsiveness (Forrester and Lackie, 1981; Forrester and Wilkinson, 1981). The mechanism of this inhibition is unclear, but may involve competitive inhibition of chemotactic factors or adhesion receptors with ligands present at the cell surface. One candidate for these receptors is the selectin class of adhesion molecules. This family of glycoproteins is expressed on neutrophil, lymphocyte, and endothelial cell surfaces and binds to specific ligands on the opposing cell membrane, mediating leukocyte movements (for review, see McEver, 1992). Several of these ligands are carbohydrate-based (Imai *et al.*, 1991; Lasky *et al.*, 1992), including heparan sulfate (Norgard-Sumnicht *et al.*, 1993).

6. Proteoglycans in the Development of Granulation Tissue

The role of proteoglycans in wound repair is based on observations of the synthesis and degradation of these molecules following the initiation of the early inflammatory response. During this phase of repair, cellular behaviors, including change in cell shape, proliferation, adhesion, and migration, must be coordinately regulated among the variety of cells within the epidermis and dermis. The initiation and control of these cellular behaviors lead to the events that characterize the repair such as angiogenesis, fibroblast proliferation, wound contraction, and reepithelialization. Proteoglycan expression is regulated during these events and, based on recent knowledge of their molecular interactions, are likely to participate in control of the repair process.

Studies on excisional wounds in adult skin performed over 40 years ago demonstrated the temporal expression of hyaluronan and sulfated GAGs in the skin following injury (Dunphy and Upuda, 1955). Hyaluronan is deposited early, reaches a peak, and then falls in the first few days when replaced by sulfated GAGs, presumably decorin, biglycan, and versican. Interestingly, the decrease in hyaluronan during adult wound repair is not seen when the repair of fetal skin was examined (Longaker *et al.*, 1991). In these studies, levels of hyaluronan remain markedly elevated in fetal skin. This persistently elevated level of hyaluronan distinguishes fetal from adult wound repair. Since fetal repair is also characterized by decreased inflammation and superior dermal matrix reorganization that lacks apparent scar formation, it has been proposed that hyaluronan is responsible for this result (Adzick and Longaker, 1992). Indeed, as noted earlier, hyaluronan can decrease neutrophil function, which may in turn influence matrix reorganization.

Proteoglycans are induced during wound repair. In the mouse, Elenius *et al.* (1991) have studied the expression of syndecan-1 and its transcripts. These studies demonstrated increased expression of syndecan-1 during the process of granulation tissue formation and in hyperproliferative keratinocytes distal from the wound edge. Recent analysis has shown that this induction is temporally regulated and specific to cell type and proteoglycan. In uninjured adult mouse or human skin, syndecan-1 and -4 are detected by immunostaining in the epidermis but not in the dermis. Following an incisional injury, syndecan-1 and -4 are induced in the dermis adjacent to the injury. Syndecan-1 is predominantly expressed on endothelia and syndecan-4 is expressed throughout the developing granulation tissue on endothelial cells and fibroblasts. This induction is transient; the expression of both syndecan-1 and -4 returns to baseline levels about 7 days following injury. In the epidermis, syndecan-1 is abundantly expressed on keratinocytes and its GAG chain size varies as cells stratify (Sanderson *et al.*, 1992a). Syndecan-4 is less abundantly expressed on keratinocytes and is not detectable on stratified cells. At the wound edge, however, keratinocytes that appear to be actively migrating across the fibrin clot show decreased syndecan-1 and -4 expression relative to keratinocytes distal from the injury, a pattern opposite to that seen in dermal cells. The loss of epithelial syndecan-1 expression has also been seen in corneal wounds (Grushkin-Lerner and Trinkaus-Randal, 1991) and is reminiscent of the loss of syndecan-1 expression seen in epithelia during organogenesis (reviewed in Bernfield *et al.*, 1992). Other proteoglycans such as decorin have also been shown to vary in expression during the wound repair process (Yeo *et al.*, 1991). Thus, as a group, several molecules must be considered when evaluating proteoglycan function during granulation tissue formation.

The significance of regulated expression of proteoglycans during wound repair is best understood in the context of the proposed function of these molecules. As discussed earlier, the GAG chains of proteoglycans bind to several components of the extracellular microenvironment. Binding of cell surface proteoglycans to components of the extracellular matrix such as collagens, fibronectin, and laminin would serve to mediate cell adhesion. Consistent with this hypothesis, syndecan-1 on mammary epithelial cells is associated with the cytoskeleton (Rapraeger *et al.*, 1986). Moreover, the cellular organelle involved in matrix adhesion of fibroblasts, the focal contact, contains both syndecan-1 (Yamagata *et al.*, 1993) and syndecan-4 (Woods and Couchman, 1994). These proteoglycans are induced during wound repair on fibroblasts and endothelial cells of the dermis, suggesting that they are involved in cell adhesion and migration of the principal cells within granulation tissue.

Other proteoglycans also interact with matrix molecules. The galactosaminoglycans, chondroitin sulfate and dermatan sulfate, bind matrix molecules such as fibronectin and laminin, but with less affinity than heparan sulfates (Rouslahti and Engvall, 1980; Brennan *et al.*, 1983). Galactosaminoglycans may function as inhibitors of adhesion as seen in studies of cell adhesion to fibronectin and in studies of cell migration (Brennan *et al.*, 1983; Knox and Wells, 1979; Rich *et al.*, 1981). It is conceivable that chondroitin sulfate or dermatan sulfate proteoglycans could be used to decrease adhesion, and thus facilitate migration during wound repair. Indeed, versican has been shown to have antiadhesive function (Yamagata *et al.*, 1993). Therefore, proteoglycans can exert fine control over cell adhesion to the matrix either through loss of expression of heparan sulfate proteoglycans or increased expression of antiadhesive galactosaminoglycan-containing proteoglycans.

Experimental evidence from numerous laboratories has suggested that several growth factors require heparin or heparan sulfate proteoglycans to exert their effect on the induction of cell growth (Rapraeger *et al.*, 1991; Yayon *et al.*, 1991). The mechanism by which this occurs is thought to depend on establishment of a ternary complex at the cell surface between the heparin-binding growth factor, its selective high-affinity signaling receptor, and a heparan sulfate proteoglycan. Growth factors known to depend on or be influenced by heparin or heparan sulfate are listed in Table III and include members of the FGF family such as FGF-2 (Klagsburn and Baird, 1991) and FGF-7 (Reich-Slotkey *et al.*, 1994), and vascular endothelial growth factor (Gitay-Goren *et al.*, 1992; Soker *et al.*, 1993; Tessler *et al.*, 1994). These growth factors have been shown to be regulated during wound repair and to influence its outcome (McGee *et al.*, 1988; Werner *et al.*, 1992, 1994; Brown *et al.*, 1992). The induction of expression of syndecan-1 and syndecan-4 during wound repair is coincident with expression of these growth factors. Heparan sulfate proteoglycans such as the syndecans can therefore act as a common molecule through which multiple growth factors could be regulated. In some *in vitro* cell systems, and in the rabbit ear chamber model of angiogenesis, perlecan has been shown to function as a positive signal for proliferation in response to FGF-2, while syndecan-1 may have a negative influence (Aviezer *et al.*,

1994). Thus, multiple proteoglycans are available during wound repair to influence growth factor responsiveness. This degree of fine control and coordinated influences is entirely consistent with a functional role for proteoglycans during wound repair.

The multiple binding interactions and functional influences of proteoglycans have directed considerable interest toward understanding systems that regulate their expression. Cells such as macrophages have abundant mRNA for syndecan-1 core protein synthesis but little intact proteoglycan until stimulated by cAMP, thus illustrating the potential for posttranscriptional regulation of these genes (Yeaman and Rapraeger, 1993). Indeed, the growth factor TGF-β can regulate expression of chondroitin and dermatan sulfate proteoglycans (Bassols and Massagué, 1988) and enhance addition of chondroitin sulfate chains to the syndecan-1 core protein (Rapraeger, 1989). FGF-2 (Elenius *et al.*, 1992) and increased culture density (Lories *et al.*, 1992) can increase the expression of syndecan-1 on fibroblasts. In addition, a 39-amino-acid proline- and arginine-rich peptide found in wound fluid and derived from wound neutrophils induces syndecan-1 and -4 on confluent mesenchymal cells (Gallo *et al.*, 1994). Therefore, syndecans and other proteoglycans are likely to be under the influence of multiple control mechanisms during wound repair. Clearly, understanding these systems will have important implications for wound repair through their capacity to influence inflammation, epithelial proliferation, angiogenesis, and fibrosis.

7. Conclusions

Recent advances in our understanding of the proteoglycans has highlighted the importance of these ubiquitous molecules. These are highly efficient molecules that place GAG chains at sites where these information-rich polysaccharides can interact with a large number of cellular effectors. Recapitulating the strict regulation of proteoglycans seen in embryogenesis (reviewed in Bernfield *et al.*, 1992), the repairing wound demonstrates complex control over proteoglycan expression. Cells within the wound environment utilize proteoglycans in a variety of specific roles that include growth factor receptor, matrix anchor, adhesion molecule, and ligand for proteases and protease inhibitors. To understand the role of proteoglycans in wound repair, these molecules must be evaluated as distinct factors with potentially opposing functions. With this specific knowledge, proteoglycans promise to be a useful tool in the control of cell behaviors such as those critical to the successful repair of a wound.

References

Adzick, N. S., and Longaker, M. T., 1992, *Fetal Wound Healing,* Elsevier, New York.

Aviezer, D., Hecht, D., Safran, M., Elsinger, M., David, G., and Yayon, A., 1994, Perlecan: Basal lamina proteoglycan, promotes basic fibroblast growth factor-receptor binding, mitogenesis, and angiogenesis, *Cell* **79:**1005–1013.

Bassols, A., and Massagué, J., 1988, Transforming growth factor β regulates the expression and structure of extracellular matrix chondroitin/dermatan sulfate proteoglycans, *J. Biol. Chem.* **263:**3039–3045.

Bernfield, M., Kokenyesi, R., Kato, M., Hinkes, M. T., Spring, J., Gallo, R. L., and Lose, E. J., 1992, Biology of the syndecans: A family of transmembrane heparan sulfate proteoglycans, *Annu. Rev. Cell Biol.* **8:**365–393.

Blochberger, T. C., Vergnes, J. P., Hempel, J., and Hassell, J. R., 1992, cDNA to chick lumican (corneal keratan sulfate proteoglycan) reveals homology to the small interstitital proteoglycan gene family and expression in muscle and intestine, *J. Biol. Chem.* **267:**347–352.

Brennan, M. J., Oldberg, A., Hayman, E. G., and Ruoslahti, E., 1983, Effect of a proteoglycan produced by rat tumor cells on their adhesion to fibronectin–collagen substrata, *Cancer Res.* **43:**4302–4307.

Brown, L. F., Kiang-Tech, Y., Berse, B., Tet-Kin, Y., Senger, D. R., Dvorak, H. F., and Van De Water, L., 1992, Expression of vascular permiability factor (vascular endothelial growth factor) by epidermal keratinocytes during wound healing, *J. Exp. Med.* **176:**1375–1379.

Castellot, J. J., Favreau, L. V., Karnovsky, M. J., and Rosenberg, R. D., 1982, Inhibition of vascular smooth muscle cell growth by endothelial cell-derived heparin: Possible role of a platelet endoglycosidase, *J. Biol. Chem.* **257:**11256–11260.

Casu, B., Petitou, M., Provasoli, M., and Sinay, P., 1988, Conformational flexibility: A new concept for explaining binding and biological properties of iduronic acid-containing glycosaminoglycans, *Trends Biochem. Sci.* **13:**221–225.

Choay, J., Pititous, M., Lormeau, J. E., Sinay, P., Lasu, B., and Gatti, G., 1983, Structure–activity relationship in heparin: A synthetic pentasaccharide with high affinity for antithrombin III and eliciting high anti-factor Xa activity, *Biochem. Biophys. Res. Commun.* **116:**492–499.

Danielsson, Å., Raub, E., Lindahl, U., and Björk, I., 1986, Role of ternary complexes, in which heparin binds both antithrombin and proteinase, in the acceleration of the reactions between antithrombin and thrombin or factor Xa, *J. Biol. Chem.* **261:**15467–15473.

David, G., 1993, Integral membrane heparan sulfate proteoglycans, *FASEB J.* **7:**1023–1030.

David, G., van der Schueren, B., Marynen, P., Cassiman, J-J., and van den Berghe, H., 1992, Molecular cloning of amphiglycan, a novel integral membrane heparan sulfate proteoglycan expressed by epithelial and fibroblastic cells, *J. Cell Biol.* **118:**961–969.

Dunphy, J. E., and Upuda, K. N., 1955, Chemical and histochemical sequences in the normal healing of wounds, *N. Engl. J. Med.* **253:**847–851.

Elenius, K., Vainio, S., Laato, M., Salmivirta, M., Theslef, I., and Jalkanen, M., 1991, Induced expression of syndecan in healing wounds, *J. Cell Biol.* **114:**585–595.

Elenius, K., Määttä, A., Salmivirta, M., and Jalkanen, M., 1992, Growth factors induce 3T3 cells to express bFGF-binding syndecan, *J. Biol. Chem.* **267:**6435–6441.

Forrester, J. V., and Lackie, J. M., 1981, Effect of hyaluronic acid on neutrophil adhesion, *J. Cell. Sci.* **50:**329–344.

Forrester, J. V., and Wilkinson, P. C., 1981, Inhibition of leukocyte locomotion by hyaluronic acid, *J. Cell Sci.* **48:**315–331.

Gallo, R. L., Ono, M., Povsic, T., Page, C., Eriksson, E., Klagsbrun, M., and Bernfield, M., 1994, Syndecans, cell surface heparan sulfate proteoglycans, are induced by a proline-rich antimicrobial peptide from wounds, *Proc. Natl. Acad. Sci. USA* **91:**11035–11039.

Ginsberg, M. H., and Jaques, B. C., 1983, Platelet membrane proteins, in: *Measurements of Platelet Function* (L. A. Harker and T. S. Zimmerman, eds.), pp. 158–176, Churchill-Livingston, Edinburgh.

Gitay-Goren, H., Soker, S., Vlodavsky, I., and Neufeld, G., 1992, The binding of vascular endothelial growth factor to its receptors is dependent on cell surface-associated heparin-like molecules, *J. Biol. Chem.* **267:**6093–6098.

Gruskin-Lerner, L. S., and Trinkaus-Randall, V., 1991, Localization of integrin and syndecan *in vivo* in a corneal epithelial abrasion and keratectomy, *Curr. Eye Res.* **10:**75–85.

Guimond, S., Maccarana, M., Olwin, B. B., Lindahl, U., and Rapraeger, A. C., 1993, Activating and inhibitory heparin sequences for FGF-2 (basic FGF): Distinct requirements for FGF-1, FGF-2 and FGF-4, *J. Biol. Chem.* **268:**23906–23914.

Hardingham, T. E., and Fosang, A. J., 1992, Proteoglycans: Many forms and many functions, *FASEB J.* **6:**861–870.

Heinegard, D., and Oldberg, A., 1989, Structure and biology of cartilage and bone matrix noncollagenous macromolecules, *FASEB J.* **3:**2042–2051.

Hildebrand, A., Romaris, M., Rasmussen, L. M., Heinegard, K., Twardzik, D. R., Border, W. A., and Ruoslahti, E., 1994, Interaction of the small interstitial proteoglycans biglycan, decorin and fibromodulin with transforming growth factor β, *Biochem. J.* **302:**527–534.

Imai, Y., Singer, M. S., Fennie, C., Lasky, L. A., and Rosen, S. D., 1991, Identification of a carbohydrate-based endothelial ligand for a lymphocyte homing receptor, *J. Cell Biol.* **113:**1213–1221.

Iozzo, R. V., 1994, Perlecan: A gem of a proteoglycan, *Matrix Biol.* **14:**203–208.

Jackson, R. L., Busch, S. J., and Cardin, A. D., 1991, Glycosaminoglycans: Molecular properties, protein interactions, and role in physiological processes, *Physiol. Rev.* **71:**481–539.

Kato, M., Wang, H., Bernfield, M., Gallagher, J. T., and Turnbull, J. E., 1994, Cell surface syndecan-1 on distinct cell types differs in fine structure and ligand binding of its heparan sulfate chains, *J. Biol. Chem.* **269:**18881–18890.

Kim, C. W., Goldberger, O. A., Gallo, R. L., and Bernfield, M., 1994, Members of the syndecan family of heparan sulfate proteoglycans are expressed in distinct cell-, tissue-, and development-specific patterns, *Mol. Biol. Cell* **5:**797–805.

Kjellen, L., and Lindahl, U., 1991, Proteoglycans: Structures and interactions, *Annu. Rev. Biochem.* **60:**443–475.

Klagsbrun, M., and Baird, A., 1991, A dual receptor system is required for basic fibroblast growth factor activity, *Cell* **67:**229–231.

Knox, P., and Wells, P., 1979, Cell adhesion and proteoglycans, *J. Cell Sci.* **40:**77–88.

Kojima, T., Leone, C., Marchildon, G. A., Marcum, J. A., and Rosenberg, R. D., 1992a, Isolation and characterization of heparan sulfate proteoglycans produced by cloned rat microvascular endothelial cells, *J. Biol. Chem.* **267:**4859–4869.

Kojima, T., Shworak, N. W., and Rosenberg, R. D., 1992b, Molecular cloning and expression of two distinct cDNA-encoding heparan sulfate proteoglycan core proteins from a rat endothelial cell line, *J. Biol. Chem.* **267:**4870–4877.

Kokenyesi, R., and Bernfield, M., 1994, Core protein structure and sequence determine the site and presence of heparan sulfate and chondroitin sulfate on syndecan-1, *J. Biol. Chem.* **269:**12305–12309.

Kreis, T., and Vale, R., 1993, *Guidebook to the Extracellular Matrix and Adhesion Proteins,* Oxford University Press, New York.

Kresse, H., Hausser, H., and Schonherr, E., 1993, Small proteoglycans, *Experientia* **49:**403–416.

Lasky, L. A., Singer, M. S., Dowbenko, D., Imai, Y., Henzel, W. J., Grimley, C., Fennie, C., Gillet, N., Watson, S. R., and Rosen, S. D., 1992, An endothelial ligand for 1-selectin is a novel mucin-like molecule, *Cell* **69:**927–938.

Lindahl, U., Bäckström, G., Höök, M., Thunberg, L., Fransson, L. °., and Linker, A., 1979, Structure of the antithrombin-binding site in heparin, *Proc. Natl. Acad. Sci. USA* **76:**3198–3202.

Lindahl, U., Backstrom, G., Thunberg, L., and Leder, I. G., 1980, Evidence for a 3-O-sulfated D-glucosamine residue in the antithrombin-binding sequence of heparin, *Proc. Natl. Acad. Sci. USA* **77:**6551–6555.

Lindahl, U., Thunberg, L., Bäckström, G., Riesenfeld, J., Nordling, K., and Björk, I., 1984, Extension and structural variability of the antithrombin-binding sequence in heparin, *J. Biol. Chem.* **259:**12368–12376.

Lories, V., Cassiman, J-J., van den Berghe, H., and David, G., 1992, Differential expression of cell surface heparan sulfate proteoglycans in human mammary epithelial cells and lung fibroblasts, *J. Biol. Chem.* **267:**1116–1122.

Maccarana, M., Casu, B., and Lindahl, U., 1993, Minimal sequence in heparin/heparan sulfate required for binding of basic fibroblast growth factor, *J. Biol. Chem.* **268:**23898–23905.

Marcum, J. A., and Rosenberg, R. D., 1984, Anticoagulantly active heparin-like molecules from vasular tissue, *Biochemistry* **23:**1730–1737.

McEver, R. P., 1992, Leukocyte-endothelial cell interactions, *Curr. Opin. Cell Biol.* **4:**840–849.

McGee, G. S., Davidson, J. M., Buckley, A., Sommer, A., Woodward, S. C., Aquino, A. M., Barbour, R., and Demitrium, A. A., 1988, Recominant basic fibroblast growth factor accelerates wound healing, *J. Surg. Res.* **45:**145–153.

Neame, P. J., 1993, Extracellular matrix of cartilage: Proteoglycans, in: *Joint Cartilage Degradation: Basic and Clinical Aspects* (J. E. Woessner and D. S. Howell, eds.), pp. 109–138, Marcel Dekker, New York.

Nishiyama, A., and Stallcup, W. B., 1993, Expression of NG2 proteoglycan causes retention of type VI collagen on the cell surface, *Mol. Biol. Cell* **4:**1097–1108.

Norgard-Sumnicht, K. E., Varki, N. M., and Varki, A., 1993, Calcium-dependent heparin-like ligands for L-selectin in nonlymphoid endothelial cells, *Science* **261:**480–483.

Parkinson, J. F., Koyama, T., Bang, N. Y., and Preissner, K. T., 1992, Thrombomoduline: An anticoagulant cell surface proteoglycan with physiologically relevant glycosaminoglycan moiety, *Adv. Exp. Med. Biol.* **313:**177–188.

Prehm, P., 1983, Synthesis of hyaluronate in differentiated teratocarcinoma cells: Mechanism of chain growth, *Biochem J.* **211:**191–198.

Rapraeger, A., 1989, Transforming growth factor (type β) promotes the addition of chondroitin sulfate chains to the cell surface proteoglycan (syndecan) of mouse mammary epithelia, *J. Cell Biol.* **109:**2509–2518.

Rapraeger, A., Jalkanen, M., and Bernfield, M., 1986, Cell surface proteoglycan associates with the cytoskeleton at the basolateral cell surface of mouse mammary epithelial cells, *J. Cell Biol.* **103:**2683–2696.

Rapraeger, A. C., Krufka, A., and Olwin, B. B., 1991, Requirement of heparan sulfate for bFGF-mediated fibroblast growth and myoblast differentiation, *Science* **252:**1705–1708.

Rauch, U., Karthikeyan, L., Maurel, P., Margolis, R. U., and Margolis, R. K., 1992, Cloning and primary structure of neurocan, a developmentally regulated, aggregating chondroitin sulfate proteoglycan of brain, *J. Biol. Chem.* **267:**19536–19547.

Reich-Slotky, R., Bonneh-Barkay, D., Shaoul, E., Bluma, B., Svahn, C. M., and Ron, D., 1994, Differential effect of cell-associated heparan sulfates on the binding of keratinocyte growth factor (KGF) and acidic fibroblast growth factor to the KGH receptor, *J. Biol. Chem.* **269:**32279–32285.

Rich, A. M., Pearlstein, E., Weissman, G., and Hoffstein, S. T., 1981, Cartilage proteoglycans inhibit fibronectin-mediated adhesion, *Nature* **293:**224–226.

Roden, L., Campbell, P., Fraser, J. R. E., Laurent, T. E., Perloft, H., and Thompson, J. N., 1989, Enzymatic pathways of hyaluronan catabolism, in: *The Biology of Hyaluronan* (D. Evered and J. Whelan, eds.), pp. 60–86, Ciba Foundation Symposium, 143, Chichester, Wiley.

Ruoslahti, E., and Engvall, E., 1980, Complexing of fibronectin, glycosaminoglycans and collagen, *Biochim. Biophys. Acta* **631:**350–358.

Sanderson, R. D., Lalor, P., and Bernfield, M., 1989, B lymphocytes express and lose syndecan at specific stages of differentiation, *Cell Reg.* **1:**27–35.

Sanderson, R. D., Hinkes, M. D., and Bernfield, M., 1992a, Syndecan-1, a cell-surface proteoglycan, changes in size and abundance when keratinocytes stratify, *J. Invest. Derm.* **99:**1–7.

Sanderson, R. D., Sneed, T., Young, L., Sullivan, G., and Lander, A., 1992b, Adhesion of B lymphoid (MPC-11) cells to type I collagen is mediated by the integral membrane proteoglycan, syndecan, *J. Immunol.* **148:**3902–3911.

Sanderson, R. D., Turnbull, J. E., Gallagher, J. T., and Lander, A. D., 1994, Fine structure of heparan sulfate regulates syndecan-1 function and cell behavior, *J. Biol. Chem.* **269:**13100–13106.

Scott, J. E., 1993, Proteoglycan-fibrillar collagen interactions in tissues: Dermatan sulfate proteoglycan as a tissue organizer, in: *Dermatan Sulphate Proteoglycans: Chemistry, Biology, Chemical Pathology* (J. E. Scott, ed.), pp. 165–181, Portland Press, London.

Silbert, J. L., Bernfield, M., and Kokenyesi, R., 1995, Proteoglycans: A very special class of glycoproteins, in: *Glycoproteins* (J. Montreuil, H. Schachter, and J. F. G. Vliegenthart, eds.), Elsevier, Amsterdam, in press.

Soker, S., Svahn, C. M., and Neufeld, G., 1993, Vascular endothelial growth factor is inactivated by binding to α₂-macroglobulin and the binding is inhibited by heparin, *J. Biol. Chem.* **268:**7685–7691.

Spillman, D., and Lindahl, U., 1994, Glycosaminoglycan–protein interactions: A question of specificity, *Curr. Opin. Struct. Biol.* **4:**677–682.

Spivak-Kroizman, T., Lemmon, M. A., Dikic, I., Ladbury, J. E., Pinchasi, D., Huang, F., Jaye, M., Crumley, G., Schlessinger, J., and Lax, I., 1994, Heparin-induced oligomerization of FGF molecules is responsible for FGf receptor dimerization, activation and cell proliferation, *Cell* **79:**1015–1024.

Stevens, R. L., and Austen, K. F., 1989, Recent advances in the cellular and molecular biology of mast cells, *Immunol. Today* **10:**381–386.

Tessler, S., Rockwell, P., Hicklin, D., Cohen, T., Levi, B-Z., Witte, L., Limischka, I. R., and Neufeld, G.,

1994, Heparin modulates the interaction of VEGF$_{165}$ with soluble and cell associated *flk-1* receptors, *J. Biol. Chem.* **269:**12456–12461.

Thunberg, L., Backstron, G., and Lindahl, U., 1982, Further characterization of the antithrombin binding sequence in heparin, *Carbohydr. Res.* **100:**393–410.

Tsen, G., Halfter, W., Kröger, S., and Cole, G. J., 1995, Agrin is a heparan sulfate proteoglycan, *J. Biol. Chem.* **270:**3392–3399.

Turnbull, J. E., and Gallagher, J. E., 1990, Molecular organization of heparan sulfate from human skin fibroblasts, *Biochem J.* **265:**715–724.

Turnbull, J. E., Fernig, D. G., Ke, Y., Wilkinson, M. C., and Gallagher, J. T., 1992, Identification of the basic fibroblast growth factor binding sequence in fibroblast heparan sulfate, *J. Biol. Chem.* **267:**10337–10341.

Vertel, B. M., Walters, L. M., Flay, N., Kearns, A. E., and Schwartz, N. B., 1993, Xylosylation is an endoplasmic reticulum to Golgi event, *J. Biol. Chem.* **268:**11105–11112.

Werner, S., Peters, K. G., Longaker, M. T., Fuller-Pace, F., Banda, M. J., and Williams, L. T., 1992, Large induction of keratinocyte growth factor expression in the dermis during wound healing, *Proc. Natl. Acad. Sci. USA* **89:**6896–6900.

Werner, S., Smola, H., Liao, X., Longaker, M. T., Krieg, T., Hofschneider, P. H., and Williams, L. T., 1994, The function of KGF in morphogenesis of epithelium and reepithelialization of wounds, *Science* **266:**819–822.

Wight, T. N., Heinegard, D. K., and Hascall, V. C., 1991, Proteoglycans: Structure and function, in: *Cell Biology of Extracellular Matrix* (E. D. Hay, ed.), pp. 45–78, Plenum Press, New York.

Woods, A., and Couchman, J. R., 1994, Syndecan-4 heparan sulfate proteoglycan is a selectively enriched and widespread focal adhesion components, *Mol. Biol. Cell* **5:**183–192.

Yamada, H., Watanabe, K., Shimonaka, M., and Yamaguchi, Y., 1994, Molecular cloning of brevican, a novel brain proteoglycan of the aggrecan/versican family, *J. Biol. Chem.* **269:**10119–10126.

Yamagata, M., Saga, S., Kato, M., Bernfield, M., and Kimata, K., 1993, Selective distributions of proteoglycans and their ligands in pericellular matrix of cultured fibroblasts. Implications for their roles in cell-substratum adhesion, *J. Cell Sci.* **106:**55–65.

Yayon, A., Klagsbrun, M., Esko, J. D., Leder, P., and Ornitz, D. M., 1991, Cell surface, heparin-like molecules are required for binding of basic fibroblast growth factor to its high affinity receptor, *Cell* **64:**841–848.

Yeaman, C., and Rapraeger, A. C., 1993, Post-transcriptional regulation of syndecan-1 expression by cAMP in peritoneal macrophages, *J. Cell Biol.* **122:**945–950.

Yeo, T-K., Brown, L., and Dvorak, H. F., 1991, Alterations in proteoglycan synthesis common to healing wounds and tumors, *Am. J. Pathol.* **138:**1437–1450.

Zhang, L., and Esko, J. D., 1994, Amino acid determinants that drive heparan sulfate assembly in a proteoglycan, *J. Biol. Chem.* **269:**19295–19299.

Zimmerman, D. R., and Ruoslahti, E., 1989, Multiple domains of the large fibroblast proteoglycan versican, *EMBO J.* **8:**2975–2981.

Chapter 16

Collagens and the Reestablishment of Dermal Integrity

BEATE ECKES, MONIQUE AUMAILLEY, and THOMAS KRIEG

1. Introduction

Skin contains a large number of different morphological structures that are composed of various extracellular matrix (ECM) components (Table I). Following tissue injury and destruction, ECM restoration has to be achieved by a controlled *de novo* synthesis as well as degradation of damaged ECM molecules. Although the ECM contains a large number of glycoproteins, those belonging to the family of collagens probably play the most important role, since they not only provide the structural scaffold of the tissue but also regulate many cellular functions.

For many years, collagens were defined as extracellular molecules with a very typical helicoidal structure (Gross, 1956; Ramachandran and Reddi, 1976); however, today the definition of collagens as well as the classification turned out to be more complex. This is due to the fact that most of the connective tissue molecules are chimeras sharing one or more collagenous and noncollagenous structural domains, suggesting that they may have evolved by combinations of genes selected from a relatively small repertoire (Engel, 1991; Baron *et al.*, 1991). These domains are often not exclusively found in the ECM, and the collagenous domain is also found in molecules that do not belong to connective tissue, such as the macrophage scavenger receptor or the C1q component of complement (Table II). In addition, some molecules are classified as collagens although they contain large noncollagenous domains representing up to 90% of their molecular mass, or they are not strictly extracellular but rather anchored within the cell membrane.

BEATE ECKES and THOMAS KRIEG • Department of Dermatology, University of Cologne, D-50924 Cologne, Germany. MONIQUE AUMAILLEY • Institute of Protein Biology and Chemistry, CNRS, 69367 Lyon cedex 7, France.

The Molecular and Cellular Biology of Wound Repair. (Second Edition), edited by Richard A. F. Clark. Plenum Press, New York, 1996

Table I. Morphological Structures and Molecular Composition

Structure	Constituent
Thick/thin fibrils	Collagens I, III, V, XII, XIV
Microfilaments (100 nm)	Collagen VI
Elastin-associated microfibrils	Fibrillin
Reticular fibers	Collagens V, III ?
	Fibronectin ?
Anchoring fibrils	Collagen VII
Anchoring filaments	Laminins 5 and 6

1.1. Structure of Collagens

Collagens are homo- or heterotrimeric molecules involving several α chains that are distinct gene products. Altogether, 32 different α chains have been identified and at least 17 are expressed in skin (Fig. 1). Common to all α chains is the repetitive motif Gly-X-Y, which allows folding into a triple helix. According to the primary structure of α chains and their assembly into collagen molecules, 19 different collagen types have been identified and characterized in vertebrates. They differ in length and number of collagenous domains and in their content of noncollagenous regions. Based on their ultrastructural organization, collagens can be subdivided into two main subfamilies: the fibril-forming and the nonfibrillar collagens (van der Rest and Garrone, 1992).

The interstitial collagens I, II, III, V, and XI are synthesized in precursor forms from which large polypeptides are cleaved off after secretion of the molecules. The resulting molecule has a 300-nm helical rod with short noncollagenous sequences at the N- and C-terminal ends (Prockop *et al.,* 1979; Peltonen *et al.,* 1985; Mayne and Burgeson, 1987; Linsenmayer *et al.,* 1993; Morris and Bächinger, 1987). These molecules assemble head-to-tail longitudinally and aggregate laterally in the characteristic quarter-staggered manner to form fibrils (Fig. 2). There was a long debate as to whether fibrils are formed by a single or several different collagen types; it is now well established that fibrils are heteropolymers of various collagen types (Henkel and Glanville, 1982; Birk *et al.,* 1988; Mendler *et al.,* 1989). Probably collagens V and XI represent the central core when collagen I, II, or III make most of the mass of the

Table II. Molecules with Collagenous Triple
Helical Domains

C1q (complement system)
Acetylcholine esterase
Mannose binding protein
Lung surfactant
Macrophage scavenger receptor

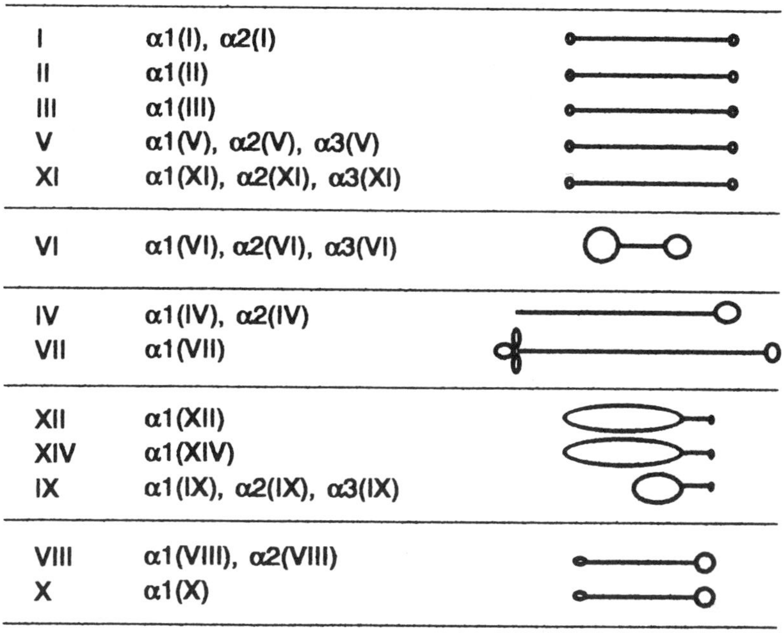

I	α1(I), α2(I)	
II	α1(II)	
III	α1(III)	
V	α1(V), α2(V), α3(V)	
XI	α1(XI), α2(XI), α3(XI)	
VI	α1(VI), α2(VI), α3(VI)	
IV	α1(IV), α2(IV)	
VII	α1(VII)	
XII	α1(XII)	
XIV	α1(XIV)	
IX	α1(IX), α2(IX), α3(IX)	
VIII	α1(VIII), α2(VIII)	
X	α1(X)	

Figure 1. Structural models of different collagens. Triple helical domains are represented by straight lines, noncollagenous regions by circles and oblong shapes.

fibrils. The FACIT collagens (fibril-associated collagens with interrupted triple helices) as well as some proteoglycans are attached to the outer surface of the fibrils (Fig. 3) (Keene *et al.*, 1991; Font *et al.*, 1993; Vogel *et al.*, 1984).

Collagen VI is a ubiquitous component of extracellular matrices (Timpl and Engel, 1987; Hessle and Engvall, 1984; von der Mark *et al.*, 1984). It is a heterotrimer composed of three polypeptide chains: α1(VI) and α2(VI), with a molecular mass of 140 kDa, and an α3 chain of 260 kDa (Trueb and Winterhalter, 1986; Colombatti *et al.*, 1987). The three chains are assembled into a dumbbell-shaped molecule with a short triple helical domain of 105 nm and with globular domains at both ends. These make up for about two thirds of the entire molecule. The globular domains consist of repetitive motifs with homology to the A domain of von Willebrand factor (Chu *et al.*, 1990). The helical portions of two collagen VI monomers associate in an antiparallel fashion and two dimers associate into tetramers, which in turn form end-to-end aggregates that constitute the microfibrillar structure with a 110-nm periodicity (Timpl and Engel, 1987).

The triple helical domain of the FACIT collagens is short (30 nm) and, in addition, interrupted. Most of these molecules (> 90%) are constituted by noncollagenous regions at the amino-terminal end (van der Rest and Garrone, 1992). These noncollagenous domains are formed by several single or successive motifs with homology to the type III repeats of fibronectin interspaced by von Willebrand factor A motifs

Supramolecular assemblies

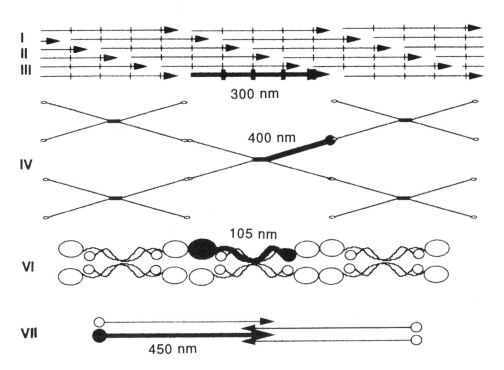

Figure 2. Higher-order structural arrangement of some collagens. Within each structure, one monomer is depicted by bold lines.

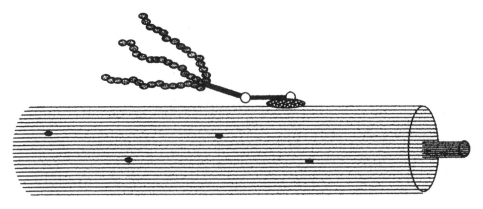

Figure 3. Model for heterotypic collagen fibril. Fibrillar collagens (e.g., type I) arranged around a central core of collagen type V constitute the main body of the fibril. Attached to the surface are FACIT collagens, e.g., type XII.

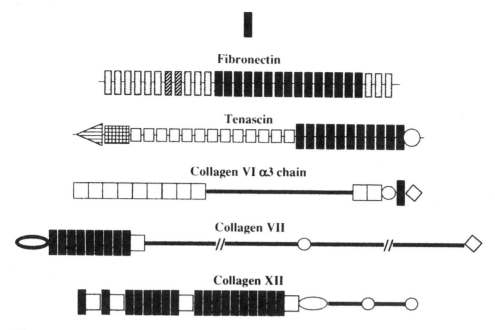

Figure 4. Modular composition of selected ECM macromolecules depicting fibronectin type III homology regions (dark boxes).

(Yamagata *et al.,* 1991; Trueb and Trueb, 1992). The number of these motifs can vary considerably (Fig. 4). No homotypical assemblies have yet been described for collagens IX, XII, and XIV, and they are probably involved in heterotypical suprastructures with the interstitial collagens, which they decorate (Keene *et al.,* 1991; Eyre *et al.,* 1987). The function of the FACIT collagens is therefore thought to be the contribution to the spatial organization of the interstitial collagen fibrils via further interaction with proteoglycans (Font *et al.,* 1993; Shaw and Olsen, 1991; Brown *et al.,* 1993).

The typical collagenous components of the dermoepidermal junction are collagens IV, VII, and XVII, with collagen IV being the most abundant. Collagens IV and VII contain long and flexible helicoidal domains of 400 and 470 nm, respectively, while collagen XVII contains 13 short collagen domains located exclusively in the carboxy-terminal extracellular portion of the molecule (Timpl, 1993; Burgeson, 1993; Guidice *et al.,* 1992). Collagens IV and VII form specific and distinct supramolecular assemblies, while that of collagen XVII has not yet been fully characterized. Collagen IV is composed of two $\alpha 1(IV)$ chains and one $\alpha 2(IV)$ chain. The molecules associate into dimers by interaction between the carboxy-terminal noncollagenous domains (NC1) and into tetramers by interactions between four overlapping amino-terminal extremities. This assembly leads to the formation of an irregular polygonal mesh stabilized by additional lateral aggregation of part of the collagen IV major triple helices. This meshwork constitutes the scaffold of the basement membrane and provides an anchoring substrate for other basement membrane molecules and adjacent cells. There is also evidence that the collagen IV network is directly connected to the suprabasal anchoring filaments via interaction with the amino-terminal globular domains of collagen VII, with the rest of the molecule being involved in collagen VII dimer formation and lateral aggregation of the triple helices to form the anchoring fibrils (Burgeson, 1993).

1.2. Biosynthesis of Collagens

Biosynthesis of collagen is complex and involves several posttranslational modifications that are catalyzed by specific enzymes (Peltonen *et al.*, 1985) (Table III). This has been characterized in detail mainly for the interstitial collagen types and has been reviewed extensively in many previous publications (Prockop *et al.*, 1979; Kivirikko *et al.*, 1990; Kivirikko and Myllylä, 1985). The modifications include hydroxylation of prolyl and lysyl residues and glucosylation and galactosylation of lysyl and hydroxylysyl residues. After chain selection, posttranslational modification, and triple helix formation, the molecules are secreted into the extracellular space where procollagen peptides are removed by the action of N- and C-procollagen peptidases. Later, the molecules aggregate into fibrils that are then stabilized by intermolecular cross-link formation, which is catalyzed by lysyl oxidase. Regulation of collagen deposition is controlled at various levels. The promotor regions of the interstitial collagens have been characterized in detail, and several stimulatory as well as inhibitory sequences have been identified (Karsenty *et al.*, 1991; Goldberg *et al.*, 1992; Maity *et al.*, 1992; Routeshouser and de Crombrugghe, 1992). A detailed characterization is also available for the bidirectional promotor controlling gene expression of collagen IV (Pöschl *et al.*, 1988). Several cytokines have been shown to modulate collagen gene expression but also to influence mRNA stability (Penttinen *et al.*, 1988; Kähäri *et al.*, 1990; Postlethwaite *et al.*, 1988). Regulation of collagen production and deposition can also involve posttranslational modifications as well as collagen degradation (Fig. 5).

2. Control of Collagen Synthesis during Wound Healing

Collagen gene expression as well as its biosynthesis have been investigated in detail during the different phases of wound healing (Clark, 1985; Clore *et al.*, 1979). In

Table III. Posttranslational Modifications

Targets	Enzymes
Hydroxylation of prolyl residues	Prolyl hydroxylase
Hydroxylation of lysyl residues	Lysyl hydroxylase
Glycosylation	Glucosyl transferase
	Galactosyl transferase
Triple helix formation	*cis-trans* Isomerase
Cleavage of procollagen peptides	N- and C-terminal peptidases
Oxidation of lysyl and OH-lysyl residues	Lysyl oxidase

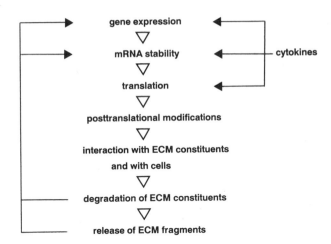

Figure 5. Regulation of connective tissue deposition. Diagrammatic representation of the different stages involved in the control of ECM deposition. Closed arrows indicate modulatory activity and feedback control mechanisms by cytokines and matrix cleavage and degradation products, respectively.

the initial steps, collagen III and fibronectin are deposited, and later on, collagen III is gradually replaced by collagen I. When total newly synthesized collagen was measured after injection of radioactively labeled proline, accumulation of collagen was found for about 3 weeks after wounding (Madden and Peacock, 1971). Although synthesis remained activated after this time, reduced deposition is probably caused by an equilibrium between synthesis and degradation. Using collagen-I-specific antibodies in experimentally induced granulation tissue, collagen I deposition was observed at day 7 (Mäkelä and Vuorio, 1986). These studies were corroborated by extraction of mRNA from the tissue, where also an early induction of transcript levels for $\alpha 1$(III) was observed; $\alpha 1$(I) was induced after a few days and down-regulated in later stages (Oono et al., 1993).

By in situ hybridization studies, collagen mRNA transcript levels could be associated with morphological structures (Scharffetter et al., 1989b). The $\alpha 1$(I) mRNA was initially found in the deep layers of granulation tissue only. By days 6–13, induced transcript levels were also found in the upper layers. By days 13–23, synthesis in the lower levels of the granulation tissue was down-regulated and high amounts of transcripts were only found within the subepidermal layers. After day 26, only faint labeling was observed (Fig. 6). Similar data have been found for $\alpha 1$(III) using in situ hybridization techniques (Oono et al., 1993).

Whereas synthesis and degradation of the major interstitial collagens have been investigated in detail, much less is known concerning the metabolism of the minor collagens. For some, like collagen VI, limited information is available. The genes for the $\alpha 1$(VI) and $\alpha 2$(VI) chains were both located on chromosome 21, and the $\alpha 3$(VI) gene maps on chromosome 2 (Weil et al., 1988). Several in vitro experiments demonstrated that various conditions, including the exposure to cytokines, e.g., interferon-γ (Heckmann et al., 1989) or transforming growth factor-β, can differentially regulate

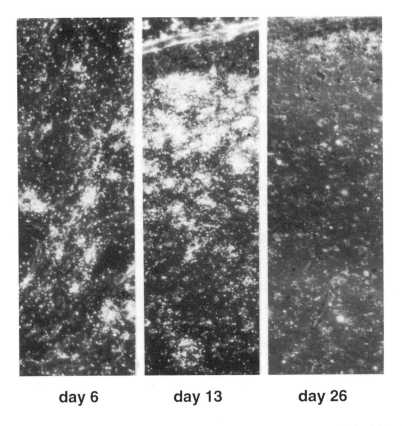

day 6   **day 13**   **day 26**

Figure 6. (A) Time course of collagen α1(I) expression in healing wounds in a rat model. Spatial distribution of collagen α1(I) transcripts was studied during a 4-week period postwounding by *in situ* hybridization to granulation tissue biopsies (Scharffetter *et al.,* 1989b). At day 6, most fibroblastic cells throughout the tissue express collagen-I-specific transcripts, whereas at day 13, synthesis is confined to the upper dermis; after 3–4 weeks, transcripts accumulate in a dense restricted area located directly subepidermally. (B) Quantitation of grain density by image analysis for signals obtained at days 6 and 21, indicating even distribution of collagen transcripts in upper and lower dermis in early stages of wound healing and spatial restriction to upper layers in later phases.

gene expression of these α chains. Interestingly, several sets of experiments also indicated that the stability of the newly synthesized molecules is only given when a heterotrimeric assembly is formed. Biosynthesis of the α3(VI) chain therefore is thought to be the rate-limiting step in the formation of microfibrils. During wound healing, collagen VI is not expressed by myofibroblasts but in endothelial and fi-broblastlike cells (Oono *et al.,* 1993). In early phases of wound healing, α3(VI) as well as the α1(VI) and α2(VI) chains are expressed in a similar manner as those of collagen I and III. Also, the location of gene expression is similar between all these collagen types. In later phases, however, the α3(VI) chain is the first chain that is down-regulated, whereas synthesis of the α1(VI) and α2(VI) chains is still induced. In analogy to various *in vitro* experiments (Heckmann *et al.,* 1989), it may be postulated

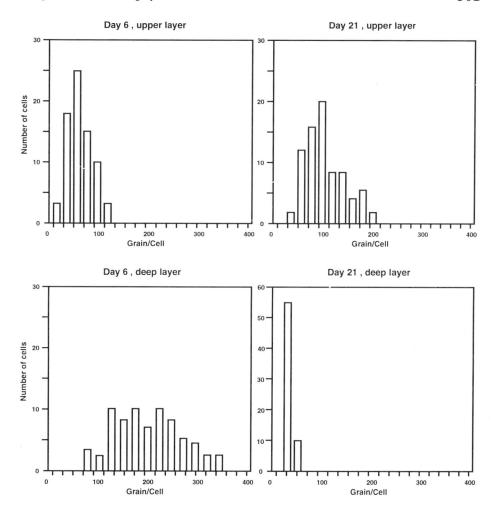

that this down-regulation of the α3(VI) mRNA may be sufficient for a down-regulation of total collagen VI protein synthesis. Since collagen VI has several important biological functions, including its cell-binding activity (Pfaff *et al.,* 1993), regulation of collagen VI might play a key role in controlling matrix organization in the restoration of dermal integrity.

2.1. Cytokines and Collagen Synthesis

A large amount of information is available concerning the influence of cytokines, which are released during tissue injury, on *de novo* formation of collagen (Mauch *et al.,* 1994). Most of these data come from *in vitro* experiments; some, however, are obtained by a time-dependent follow-up of cytokine expression on the mRNA and/or protein levels and collagen production *in vivo*. Combination of both sets of data provides a basis for gaining insight into the complex network of cytokine activities.

The transforming growth factor-β (TGF-β) family includes various types of cytokines with a molecular mass of about 25 kDa (Roberts and Sporn, 1990). They are synthesized in a precursor form by different cells, and, following activation, bind to receptor molecules on target cells (e.g., fibroblasts). TGF-β1 and -2 have been reported to have profound influences on ECM metabolism by activation of the biosynthesis of several proteins (Ignotz *et al.*, 1987; Varga *et al.*, 1987; Shah *et al.*, 1992; Kulozik *et al.*, 1990; Roberts *et al.*, 1986), as well as by the inhibition of proteolytic activities (Overall *et al.*, 1989; Matrisian, 1990). The molecular mechanism of the activation of collagen α1(I) transcription has been studied in detail and was found to involve defined sequences in the promotor region of collagen α1(I) gene (Rossi *et al.*, 1988). *In situ* hybridization analysis has demonstrated a coexpression of TGF-β1 and collagen α1(I) mRNA in early stages of wound healing (Quaglino *et al.*, 1991; Peltonen *et al.*, 1991).

Also, interleukin-1 (IL-1) modulates procollagen I and III expression in human fibroblasts in a dose-dependent manner. Low concentrations of IL-1 cause stimulation, whereas high concentrations lead to an inhibition of the expression of both collagens in cultured fibroblasts (Heckmann *et al.*, 1993). Similarly, in two- as well as in three-dimensional culture systems, IL-4 was found to enhance collagen α1(I) mRNA levels, as well as biosynthesis of type I collagen (Gillery *et al.*, 1992). The activity of these stimulatory cytokines is counteracted by interferon-γ (Varga *et al.*, 1990) and tumor necrosis factor-α. These cytokines can either alone or in combination lead to a down-regulation of collagen α1(I) transcription (Scharffetter *et al.*, 1989a). It is not yet clear whether regulation takes place strictly at the transcriptional level or whether changes in mRNA stability are involved as well.

2.2. Regulation of Collagen Synthesis by ECM

Regulation of collagen synthesis by cytokines is only one mechanism operating during wound healing. Whereas during many years the ECM was thought to have primarily structural functions and to provide the mechanical scaffold for cells and tissues, it is now clear that specific structural domains of ECM constitutents are endowed with various biological activities. In particular, ECM serves as a reservoir for growth factors and as a source of informations for cells. Cell adhesion-promoting activity has been characterized for collagens I (Davis, 1992), IV (Aumailley and Timpl, 1986; Vandenberg *et al.*, 1991), V (Ruggiero *et al.*, 1994), and VI (Pfaff *et al.*, 1993; Aumailley *et al.*, 1989). Interestingly, cell adhesion to collagens is dependent on the helical conformation of the molecules. However, denatured collagens I, V, and VI, but not collagen IV, can also induce cell adhesion, although by different molecular mechanisms. The fact that collagens can interact differently with cells when in native or denatured forms is particularly interesting in the context of wound healing, since according to their conformational state, collagens fulfill different biological functions. The influence of collagenous molecules on cellular activities might also be mediated by collagen peptides (Katayama *et al.*, 1993; Fouser *et al.*, 1991; Wiestner *et al.*, 1979). In particular, N-propeptides of collagen were found to have feedback regulation effects on collagen biosynthesis in fibroblasts. Degradation of collagen by bacterial collagenase results in small peptides that were found to have high chemotactic activity for fibroblasts and other cells (Postlethwaite *et al.*, 1978). In addition, the large cleavage products generated by the activity of mammalian collagenase possess chemotactic activity (Albini and Adelmann-Grill, 1985). Besides fibroblasts, attachment of other cells to collagen can modulate inflammation and wound healing considerably; e.g., attachment of neutrophils and granulocytes facilitates the respiratory burst releasing proteases and cytokines that are required to coordinate the various events in early phases of tissue repair (Werb and Gordon, 1975).

A great deal of information has been derived from cells grown in a three-dimensional collagenous matrix. This allows the influence of collagen on the cell behavior under *in vivo*-like conditions (Bell *et al.*, 1979) to be studied. Besides collagen, other substances have been used to form three-dimensional lattices. These include artificial polymers (Fleischmajer *et al.*, 1991), fibrin (Gillery *et al.*, 1989), agarose, alginate (Schlumberger *et al.*, 1989), or extracts from a basement membrane-producing tumor (Vukicevic *et al.*, 1992). Although fibroblasts are capable of using all these polymeric structures for attachment and spatial support, most studies have been carried out using collagen I matrices. It is reasonably abundant in skin and tendons, providing the sources from which collagen I can be extracted under nondenaturing conditions and subsequently purified according to well-established procedures (Bell *et al.*, 1979). The influence of this collagen I matrix may easily be altered by including other ECM constituents, e.g., other collagen types (Ehrlich, 1988) including fibril-associated collagens, which have profound effects on the supramolecular fiber organization, as well as fibronectin or proteoglycans (Cidadao, 1989; Docherty *et al.*, 1989; Guidry and Grinnell, 1987).

The communication between cells and extracellular macromolecules is mediated by specific receptors located on the fibroblast surface. These receptors include different classes of proteins, e.g., the syndecans. The major receptors for ECM proteins, however, are integrins (Hynes, 1992). Integrins are a large family of heterodimeric molecules. They are composed by two noncovalently associated subunits, α and β (Fig. 7). Nineteen heterodimers resulting from the combination of 8 β and 14 α subunits have been identified. Each integrin has a large extracellular domain, a short transmembrane segment, and a short intracellular tail (except for the $\beta4$ subunit, which has a cytosolic domain of >1000 amino acids) (Hogervorst *et al.*, 1990). The extracellular domain of the α subunits contains three to five motifs with homology to the EF-hands of calmodulin and parvalbumin, representing potential cation-binding sites that are required for dimer formation and ligand-binding functions of the receptor (Hynes, 1992).

The intracellular domains are thought to interact with cytoskeletal proteins such as talin and α-actinin, which in turn can bind vinculin and the actin cytoskeleton (Juliano and Haskill, 1993). Integrins can therefore transmit information directly from the ECM to the cell nucleus. The exact mechanism used for this signal transduction is still unclear. There is evidence, however, that phosphorylation of cytoskeleton-associated proteins by tyrosine (Schaller *et al.*, 1992) and/or serine/threonine kinases could be involved.

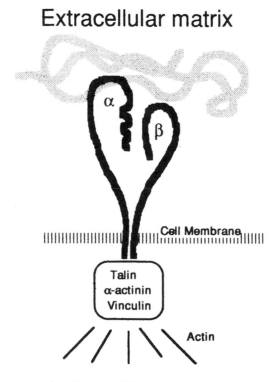

Figure 7. Schematic representation of the potential interaction between ECM and cytoskeletal components via integrins.

Using three-dimensional systems, various cells have been investigated and were found to be modulated in their shape, their differentiation status, or in distinct biosynthetic functions. In the following, selected examples of such matrix-induced regulation will be discussed.

Synthesis of collagen VII by keratinocytes was studied (Contard *et al.,* 1993) at an ultrastructural level using keratinocytes seeded on top of a fibroblast-derived matrix. Deposition of collagen VII and formation of anchoring fibrils detectable by electron microscopy were shown to be a long-term process requiring 5–6 weeks of culture and to be dependent on the presence of a mesenchymal matrix, on epithelial–mesenchyme interactions and serum components. More detailed information regarding the interactions between epithelial and mesenchymal cells has come from studies (König and Bruckner-Tuderman, 1991, 1994) employing cocultures of keratinocytes and fibroblasts in a two-chamber system, indicating that collagen VII synthesis by keratinocytes can be strongly enhanced by the presence of fibroblasts. The presence of keratinocytes cannot, however, induce elevated collagen VII synthesis in fibroblasts that are capable of synthesizing minimal amounts of collagen VII in this system. In search of an activating, diffusible factor, TGF-β2 was found to be a major but probably not the only stimulator of collagen VII synthesis for keratinocytes in coculture. These studies were further extended to an organotypic skin model with keratinocytes on top of a devitalized dermis and confirmed the importance of TGF-β2 and of ECM contact in stimulating the synthesis and organized deposition of collagen VII at the dermoepidermal interface.

Composition and organization of ECM previously had been shown to exert profound effects on the regulation of capillary endothelial cells. Specifically, collagen IV and laminin were shown to modulate endothelial cell migration, proliferation, and multicellular organization during angiogenesis (Montesano *et al.,* 1983). However, striking results were obtained by embedding endothelial cells within a matrix reconstituted from mainly type I collagen. Cells were evenly interspersed within the lattice, which they contracted much like dermal fibroblasts do. In addition, extensive complex tubelike structures started to form. These structures labeled intensely with antibodies to collagen V, indicative of enhanced synthesis and deposition of this collagen type. Collagens I and III were negative. This tube morphogenesis and collagen V production was enhanced by addition of TGF-β and could not be demonstrated in cultures of endothelial cells lacking the three-dimensional contact with matrix (Madri *et al.,* 1988).

In light of recent reports, these results may be interpreted as laminin and collagen playing an indirect role in these phenomena. ECM molecules can be thought to serve as substrates for anchoring growth factors such as TGF-β. In addition, anchoring of growth factors to ECM protects the growth factors from degradation on one hand, and facilitates interactions of growth factors with their corresponding receptors on the cell membrane on the other (crinopexie).

Further evidence showing matrix-mediated effects on collagen V was derived from chicken embryo corneal fibroblasts responsible for synthesizing the corneal stroma consisting mainly of collagen I and of lesser but significant amounts of collagen V (Linsenmayer *et al.,* 1984). All cells in a population were shown to produce both collagen types, indicating the absence of subpopulations responsible for the exclusive synthesis of only one collagen type. In all culture conditions tested, corneal fibroblasts synthesized collagens I and V in a stable ratio of 4:1. However, only culture within a three-dimensional collagen gel resulted in the assembly of heterotypic fibrils that are characteristic of uniform, small-diameter corneal fibrils (McLaughlin *et al.,* 1989).

In contrast to the information discussed above, more knowledge has accumulated regarding modulation of biosynthesis of the abundantly present fibril-forming collagens I and III. When fibroblasts or smooth muscle cells are seeded into these three-dimensional lattices, the lattices are contracted until a dense connective tissue network is formed. This contraction is thought to represent wound contraction and is dependent on a properly functioning protein biosynthesis as well as an intact intracellular cytoskeleton. Seeding dermal fibroblasts of different donor species within a three-dimensional collagenous matrix leads to coordinated down-regulation by 90–95% of collagen α1(I), α2(I), and α1(III) chains within 1–2 days (Mauch *et al.,* 1988). The rate of reduction depends on the number of cells embedded into the matrix. Reduction was shown to take place both at the protein and the mRNA levels, indicating pretranslational control mechanisms.

In comparison to monolayer cultures, cleavage of procollagen peptides was more efficient in the three-dimensional culture. Analysis of collagen distribution among the different compartments (cells, medium, matrix) revealed that in the lattice culture more than 90% was associated with the matrix, and 55% of this was converted to mature collagen. In contrast, cells in monolayer cultures secrete 75% into the medium, of which 85% remains in the procollagen form. There was no evidence for enhanced collagen degradation in lattice cultures, as judged by comparable amounts of dialyzable hydroxyproline in both monolayer and lattice cultures.

Reduction in α1(I) procollagen mRNA levels by more than 90% of monolayer values was observed using *in situ* hybridization as well as Northern and dot blot analysis. In order to understand the underlying molecular mechanisms leading to repressed steady-state mRNA levels, different control mechanisms can be envisioned. These include (1) differential expression or binding activity of nuclear proteins interacting with the collagen promotor, (2) differential methylation of regulatory sequences of the collagen gene, (3) differential transcription rates, (4) differences in transcript stability, and (5) differences in signal transduction mediated by collagen-binding integrins. Whereas only preliminary information is available regarding the first two mechanisms, more insight could be obtained into the latter three.

Rates of transcription of the human α1(I) collagen gene were assessed in primary human fibroblasts by nuclear run-on analysis (Eckes *et al.,* 1993). In agreement with decreased steady-state transcript levels, *de novo* synthesis of this mRNA was found to be decreased. However, the observed reduction by 50%, in comparison to monolayer cultures, could not account for the more severe reduction in steady-state levels. Therefore, rates of collagen mRNA decay in the two culture systems were compared. In contrast to fibronectin transcripts, which decayed at a comparable rate in monolayer and lattice cultures, α1(I) procollagen mRNA is approximately 50% less stable in fibroblasts cultured in three-dimensional contact with a collagenous matrix than in monolayer culture. These results indicate that the matrix environment triggers transcriptional as well as posttranscriptional control mechanisms leading to downregulation of collagen transcript levels.

Down-regulation of collagens I and III as described above occurs in freely contracting collagen lattices. In contrast, maintaining the matrix under tension, i.e., by fixing the rims of the gel, thereby prohibiting radial gel contraction, yielded different results (Grinnell, 1994), implicating the importance of mechanical forces and the role of mechanotransducers in gene regulation (Ingber and Folkman, 1989). In the attached matrices, fibroblasts assume an elongated, bipolar shape and they are oriented perpendicular to the plane of the force. In contrast to fibroblasts in freely contracting gels that stop proliferating, in the attached system proliferation proceeds. The force generated by the cells can be measured experimentally. It is maximal after 6 hr of culture and is on the order of 0.25 g/10^6 cells (Lambert *et al.,* 1992). In addition, biosynthetic capacity is modulated differently in the attached matrices. Here, synthesis of α1(I) and α2(I) collagen transcripts remains nearly constant over a period of 5 days, whereas all three chains of collagen VI are increased relative to the levels detected in freely contracting gels.

Since the interaction between fibroblasts and the surrounding collagenous matrix involves integrins of the β1 family, expression of these receptors was studied. Immunoprecipitation of total cellular $\alpha 1$, $\alpha 2$, $\alpha 3$, $\alpha 5$, $\alpha 6$, αv, and $\beta 1$ chains showed that in freely contracting lattices, $\alpha 2$ was increased, whereas the others remained unchanged (Klein *et al.*, 1991). Results obtained using tumor cells (rhabdomyosarcoma) confirmed the role of $\alpha 2\beta 1$ in mediating matrix contraction: rhabdomyosarcoma cells are inherently unable to contract a collagen matrix, but acquire this capability upon transfection of the $\alpha 2$ integrin gene (Schiro *et al.*, 1991; Chan *et al.*, 1992).

However, $\alpha 2\beta 1$ integrin is only one of at least two distinct receptors for collagen I expressed on the surface of fibroblasts. Studies are now in progress attempting to relate binding of different integrins to the identical ligand with distinctly different biochemical responses. Preliminary results suggest that binding of collagen I to $\alpha 2\beta 1$ is not responsible for transducing the signal "reduce synthesis of collagen I" into the nucleus, but that another receptor carries this function (Langholz *et al.*, 1995). Intracellular signaling has been shown to involve phosphorylation of intracellular proteins, e.g., pp125[FAK], itself a tyrosine kinase located in focal adhesions (Roeckel and Krieg, 1994), as well as the activity of phospholipase C (Lambert *et al.*, 1993) and protein kinase C (Guidry, 1993).

2.3. The ECM as Reservoir for Growth Factors

ECM molecules including collagens do not only regulate cellular activities by direct binding to membrane bound integrins. These multifunctional molecules have been shown to specifically bind growth factors including basic FGF or TGF-β (Nathan and Sporn, 1991; Ruoslahti and Yamaguchi, 1991; Witt and Lander, 1994). This binding can activate the cytokines (bFGF) or inactivate them. The ECM therefore acts as an important reservoir for these mediators that can be rapidly released locally at sites of tissue injury. This mechanism has been identified for heparan sulfate proteoglycan and also noncollagenous glycoproteins of ECM. Direct information for collagen is not yet available; however, by analogy this could be an important function for regulating the reestablishment of dermal integrity after injury.

References

Albini, A., and Adelmann-Grill, B. C., 1985, Collagenolytic cleavage products of collagen type I as chemoattractants for human dermal fibroblasts, *Eur. J. Cell Biol.* **36:**104–107.

Aumailley, M., and Timpl, R., 1986, Attachment of cells to basement membrane collagen type IV, *J. Cell Biol.* **103:**1569–1575.

Aumailley, M., Mann, K., von der Mark, H., and Timpl, R., 1989, Cell attachment properties of collagen VI and Arg-Gly-Asp dependent binding to its $\alpha 2$ (VI) and $\alpha 3$ (VI) chains, *Exp. Cell Res.* **181:**463–474.

Baron, M., Norman, D. G., and Campbell, I. D., 1991, Protein modules, *Trends Biochem. Sci.* **16:**13–17.

Bell, E., Ivarsson, B., and Merrill, C., 1979, Production of a tissue-like structure by contraction of collagen lattices by human fibroblasts of different proliferative potential *in vitro, Proc. Natl. Acad. Sci. USA* **76:**1274–1278.

Birk, D. E., Fitch, J. M., Barbiarz, J. P., and Linsenmayer, T. F., 1988, Collagen type I and V are present in the same fibril in the avian corneal stroma, *J. Cell Biol.* **106:**999–1008.

Brown, J. C., Mann, K., Wiedemann, H., and Timpl, R., 1993, Structure and binding properties of collagen type XIV isolated from human placenta, *J. Cell Biol.* **120:**557–567.

Burgeson, R. E., 1993, Type VII collagen, anchoring fibrils and epidermolysis bullosa, *J. Invest. Dermatol.* **101:**252–255.

Chan, B. M., Kassner, P. D., Schiro, J. A., Byers, H. R., Kupper, T. S., and Hemler, M. E., 1992, Distinct cellular functions mediated by different VLA integrin alpha subunit cytoplasmic domains, *Cell* **68:**1051–1060.

Chu, M. L., Zhan, R. Z., Pan, T. C., Stokes, D., Kuo, H. J., Glanville, R. W., Mayer, U., Mann, K., Deutzmann, R., and Timpl, R., 1990, Mosaic structure of globular domains in the human type VI collagen α3 chain: Similarity to von Willebrand factor, fibronectin, actin, salivary proteins, and aprotinin type protease inhibitors, *EMBO J.* **9:**385–393.

Cidadao, A. J., 1989, Interactions between fibronectin, glycosaminoglycans and native collagen fibrils—An EM study in artificial three-dimensional extracellular matrices, *Eur. J. Cell Biol.* **48:**303–312.

Clark, R. A. F., 1985, Cutaneous tissue repair: Basic biological considerations, *J. Am. Acad. Dermatol.* **13:**701–725.

Clore, J. N., Cohen, I. K., and Diegelmann, R. F., 1979, Quantitation of collagen types I and III during wound healing in rat skin, *Proc. Soc. Exp. Biol. Med.* **161:**337–340.

Colombatti, A., Bonaldo, P., Ainger, K., Bressan, G. M., and Volpin, D., 1987, Biosynthesis of chick type VI collagen: Intercellular assembly and molecular structure, *J. Biol. Chem.* **262:**14454–14460.

Contard, P., Bartel, R. L., Jacobs, L., Perlish, J., MacDonald, E. D., Handler, L., Cone, D., and Fleischmajer, R., 1993, Culturing keratinocytes and fibroblasts in a three-dimensional mesh results in epidermal differentiation and formation of a basal lamina-anchoring zone, *J. Invest. Dermatol.* **100:**35–39.

Davis, G. E., 1992, Affinity of integrin for damaged extracellular matrix αvβ3 binds to denatured collagen type I through RGD sites, *Biochem. Biophys. Res. Commun.* **182:**1025–1031.

Docherty, R., Forrester, J. V., Lackie, J. M., and Gregory, D. W., 1989, Glycosaminoglycans facilitate the movement of fibroblasts through three-dimensional collagen matrices, *J. Cell Sci.* **92:**263–270.

Eckes, B., Mauch, C., Hüppe, G., and Krieg, T., 1993, Down-regulation of collagen synthesis in fibroblasts within three-dimensional collagen lattices involves transcriptional and posttranscriptional mechanisms, *FEBS Lett.* **318:**129–133.

Ehrlich, H. P., 1988, The modulation of contraction of fibroblast populated collagen lattices by types I, II, and III collagen, *Tissue Cell* **20:**47–50.

Engel, J., 1991, Common structural motifs in proteins of the extracellular matrix, *Curr. Opin. Cell Biol.* **3:**779–785.

Eyre, D. R., Apon, S., Wu, J. J., Ericsson, L. H., and Walsh, K. A., 1987, Collagen type IX: Evidence for covalent linkages to type II collagen in cartilage, *FEBS Lett.* **220:**337–341.

Fleischmajer, R., Contard, P., Schwartz, E., MacDonald, E. D., Jacobs, L., and Sakai, L. Y., 1991, Elastin-associated microfibrils (10 nm) in a three-dimensional fibroblast culture, *J. Invest. Dermatol.* **97:**638–643.

Font, B., Aubert-Foucher, E., Goldschmidt, D., Eichenberger, D., and van der Rest, M., 1993, Binding of collagen XIV with the dermatan sulfate side chain of decorin, *J. Biol. Chem.* **268:**25015–25018.

Fouser, L., Sage, E. H., Clark, J., and Bornstein, P., 1991, Feedback regulation of collagen gene expression: A Trojan horse approach, *Proc. Natl. Acad. Sci. USA* **88:**10158–10162.

Gillery, P., Bellon, G., Coustry, F., and Borel, J. P., 1989, Cultures of fibroblasts in fibrin lattices—Models for a study of metabolic activities of the cells in physiological conditions, *J. Cell. Physiol.* **140:**483–490.

Gillery, P., Fertin, C., Nicolas, J. F., Chastang, F., Kalis, B., Banchereau, J., and Maquart, F. X., 1992, Interleukin-4 stimulates collagen gene expression in human fibroblast monolayer cultures. Potential role in fibrosis, *FEBS Lett.* **302:**231–234.

Goldberg, H., Helaakoski, T., Garrett, L. A., Karsenty, G., Pellegrino, A., Lozano, G., Maity, S., and de Crombrugghe, B., 1992, Tissue-specific expression of the mouse α2 (I) collagen promoter, *J. Biol. Chem.* **267:**19622–19630.

Grinnell, F., 1994, Fibroblasts, myofibroblasts, and wound contraction, *J. Cell Biol.* **124:**401–404.

Gross, J., 1956, The behavior of collagen as a model in morphogenesis, *J. Biophys. Biochem. Cytol.* (Suppl.) **2:**261–294.

Guidice, G. J., Emery, D. J., and Diaz, L. A., 1992, Cloning and primary structural analysis of the bullous pemphigoid autoantigen BP 180, *J. Invest. Dermatol.* **99**:243–250.

Guidry, C., 1993, Fibroblast contraction of collagen gels requires activation of protein kinase C, *J. Cell. Physiol.* **155**:358–367.

Guidry, C., and Grinnell, F., 1987, Heparin modulates the organization of hydrated collagen gels and inhibits gel contraction by fibroblasts, *J. Cell Biol.* **104**:1097–1103.

Heckmann, M., Aumailley, M., Hatamochi, A., Chu, M. L., Timpl, R., and Krieg, T., 1989, Down-regulation of α3 (VI) chain expression by γ-interferon decreases synthesis and deposition of collagen type VI, *Eur. J. Biochem.* **182**:719–726.

Heckmann, M., Adelmann-Grill, B. C., Hein, R., and Krieg, T., 1993, Biphasic effects of interleukin-1α on dermal fibroblasts: Enhancement of chemotactic responsiveness at low concentrations and of mRNA expression for collagenase at high concentrations, *J. Invest. Dermatol.* **100**:780–784.

Henkel, W., and Glanville, R. W., 1982, Covalent cross-linking between molecules of type I and type III collagen, *Eur. J. Biochem.* **122**:205–213.

Hessle, H., and Engvall, E., 1984, Type VI collagen: Studies on its localization, structure and biosynthetic form with monoclonal antibodies, *J. Biol. Chem.* **259**:3955–3961.

Hogervorst, F., Kuikman, I., von dem Borne, A. E., and Sonnenberg, A., 1990, Cloning and sequence analysis of β4 cDNA: An integrin subunit that contains a unique 118 kd cytoplasmic domain, *EMBO J.* **9**:765–770.

Hynes, R. O., 1992, Integrins: Versatility, modulation, and signaling in cell adhesion, *Cell* **69**:11–25.

Ignotz, R. A., Endo, T., and Massague, J., 1987, Regulation of fibronectin and type I collagen mRNA levels by transforming growth factor-β, *J. Biol. Chem.* **262**:6443–6446.

Ingber, D. E., and Folkman, J., 1989, Mechanochemical switching between growth and differentiation during fibroblast growth factor-stimulated angiogenesis *in vitro:* Role of extracellular matrix, *J. Cell Biol.* **109**:317–330.

Juliano, R. L., and Haskill, S., 1993, Signal transduction from the extracellular matrix, *J. Cell Biol.* **120**:577–585.

Kähäri, V. M., Chen, Y. Q., Su, N. W., Ramirez, F., and Uitto, J., 1990, TNF-α and interferon-γ suppress the activation of human type I collagen gene expression by TGF-β, *J. Clin. Invest.* **86**:1489–1495.

Karsenty, G., Ravazzolo, R., and de Crombrugghe, B., 1991, Purification and functional characterization of a DNA-binding protein that interacts with a negative element in the mouse α1 (I) collagen promoter, *J. Biol. Chem.* **266**:24842–24848.

Katayama, K., Armendariz-Borunda, J., Raghow, R., Kang, A. H., and Seyer, J. M., 1993, A pentapeptide from type I procollagen promotes extracellular matrix production, *J. Biol. Chem.* **268**:9941–9944.

Keene, D. R., Lunstrum, G. P., Morris, N. P., Stoddard, D. W., and Burgeson, R. E., 1991, Two type XII-like collagens localized to the surface of banded collagen fibrils, *J. Cell Biol.* **113**:971–978.

Kivirikko, K. I., and Myllylä, R., 1985, Posttranslational processing of procollagens, *Ann. NY Acad. Sci.* **460**:187–201.

Kivirikko, K. I., Helaakoski, T., Tasanen, K., Vuori, K., Myllylä, R., Parkkonen, T., and Pihlajaniemi, T., 1990, Molecular biology of prolyl 4-hydroxylase, in: *Structure, Molecular Biology and Pathology of Collagen* (R. Fleischmajer, B. R. Olsen, and K. Kühn, eds.), pp. 132–142, Academic Press, New York.

Klein, C. E., Dressel, D., Steinmayer, T., Mauch, C., Eckes, B., Krieg, T., Bankert, R. W., and Weber, L., 1991, Integrin α2β1 is up-regulated in fibroblasts and highly aggressive melanoma cells in three-dimensional collagen lattices and mediates the reorganization of collagen I fibrils, *J. Cell. Biol.* **115**:1427–1436.

König, A., and Bruckner-Tuderman, L., 1991, Epithelial–mesenchymal interactions enhance expression of collagen VII *in vitro, J. Invest. Dermatol.* **96**:803–808.

König, A., and Bruckner-Tuderman, L., 1994, Transforming growth factor-β promotes deposition of collagen VII in a modified organotypic skin model, *Lab. Invest.* **70**:203–209.

Kulozik, M., Hogg, A., Lankat-Buttgereit, B., and Krieg, T., 1990, Co-localization of transforming growth factor β2 with α1(I) procollagen mRNA in tissue sections of patients with systemic sclerosis, *J. Clin. Invest.* **86**:917–922.

Lambert, C. A., Soudant, E. P., Nusgens, B. V., and Lapiere, Ch. M., 1992, Pretranslational regulation of the extracellular matrix macromolecules and collagenase expression in fibroblasts by mechanical forces, *Lab. Invest.* **66:**444–451.

Lambert, C. A., Martens, H., Nusgens, B. V., and Lapiere, Ch. M., 1993, Regulation of collagenase and COL1A1 genes, but not β-actin gene in fibroblast populated collagen gels is mediated via tyrosine kinases and phospholipase C, *J. Invest. Dermatol.* **100:**436 (Abstract).

Langholz, O., Roeckel, D., Mauch, C., Kozlowska, E., Bank, I., Krieg, T., and Eckes, B., 1995, Collagen and collagenase gene expression in three-dimensional collagen lattices are differentially regulated by α1β1 and α2β1 integrins, *J. Cell Biol.*, in press.

Linsenmayer, T. F., Fitch, J. M., and Mayne, R., 1984, Extracellular matrices in the developing avian eye. Type V collagen in corneal and noncorneal tissues, *Invest. Ophthalmol. Vis. Sci.* **25:**41–47.

Linsenmayer, T. F., Gibney, E., Igoe, F., Gordon, M. F., Fitch, J. M., Fessler, L. I., and Birk, D. E., 1993, Type V collagen: Molecular structure and fibrillar organization of the chicken α1 (V) NH$_2$-terminal domain, a putative regulator of corneal fibrillogenesis, *J. Cell Biol.* **121:**1181–1189.

Madden, J. W., and Peacock, E. E., 1971, Studies on the biology of collagen during wound healing. III. Dynamic metabolism of scar collagen and remodeling of dermal wounds, *Ann. Surg.* **174:**511–522.

Madri, J. A., Pratt, B. M., and Tucker, A. M., 1988, Phenotypic modulation of endothelial cells by transforming growth factor-β depends upon the composition and organization of the extracellular matrix, *J. Cell Biol.* **106:**1375–1384.

Maity, S. N., Sinha, S., Routeshouser, E. C., and de Crombrugghe, B., 1992, Three different polypeptides are necessary for DNA binding of the mammalian heteromeric CCAAT binding factor, *J. Biol. Chem.* **267:**16574–16580.

Mäkelä, J. K., and Vuorio, E., 1986, Type I collagen messenger RNA levels in experimental granulation tissue and silicosis in rats, *Med. Biol.* **64:**15–22.

Matrisian, L. M., 1990, Metalloproteinases and their inhibitors in matrix remodeling, *Trends Genet.* **6:**121–125.

Mauch, C., Hatamochi, A., Scharffetter, K., and Krieg, T., 1988, Regulation of collagen synthesis in fibroblasts within a three-dimensional collagen gel, *Exp. Cell Res.* **178:**493–503.

Mauch, C., Oono, T., Eckes, B., and Krieg, T., 1994, Cytokines and wound healing, in: *Epidermal Growth Factors and Cytokines* (T. A. Luger and T. Schwarz, eds.), pp. 325–344, Marcel Dekker, New York.

Mayne, R., and Burgeson, R. E. (eds.), 1987, *Structure and Function of Collagen Types,* Academic Press, Orlando, Florida.

McLaughlin, J. S., Linsenmayer, T. F., and Birk, D. E., 1989, Type V collagen synthesis and deposition by chicken embryo corneal fibroblasts *in vitro, J. Cell Sci.* **94:**371–379.

Mendler, M., Eich-Bender, S. G., Vaughan, L., Winterhalter, K. H., and Bruckner, P., 1989, Cartilage contains mixed fibrils of collagen types II, IX and XI, *J. Cell Biol.* **108:**191–197.

Montesano, R. L., Orci, L., and Vassalli, P., 1983, *In vitro* rapid organization of endothelial cells into capillary-like networks is promoted by collagen matrices, *J. Cell. Biol.* **97:**1648–1652.

Morris, N. P., and Bächinger, H. P., 1987, Type XI collagen is a heterotrimer with the composition (1α, 2α, 3α) retaining non triple helical domains, *J. Biol. Chem.* **262:**11345–11350.

Nathan, C., and Sporn, M., 1991, Cytokines in context, *J. Cell Biol.* **113:**981–986.

Oono, T., Specks, U., Eckes, B., Majewski, S., Hunzelmann, N., Timpl, R., and Krieg, T., 1993, Expression of type VI collagen mRNA during wound healing by *in situ* hybridization, *J. Invest. Dermatol.* **100:**329–334.

Overall, C. M., Wrana, J. L., and Sodek, J., 1989, Independent regulation of collagenase, 72 kDa progelatinase and metalloproteinase inhibitor (TIMP) expression in human fibroblasts by transforming growth factor-β, *J. Biol. Chem.* **264:**1860–1869.

Peltonen, L., Halila, R., and Ryhänen, L., 1985, Enzymes converting procollagens to collagens, *J. Cell. Biochem.* **28:**15–21.

Peltonen, J., Hsiao, L. L., Jaakkola, S., Sollberg, S., Aumailley, M., Timpl, R., Chu, M. L., and Uitto, J., 1991, Activation of collagen gene expression in keloids. Colocalization of type I and VI collagen and transforming growth factor β mRNAs, *J. Invest. Dermatol.* **97:**240–248.

Penttinen, R. P., Kobayaski, S., and Bornstein, P., 1988, TGF-β increases mRNA for matrix proteins both in the presence and in the absence of changes in mRNA stability, *Proc. Natl. Acad. Sci. USA* **85:**1105–1108.

Pfaff, M., Aumailley, M., Specks, U., Knolle, J., Zerwes, H. G., and Timpl, R., 1993, Integrin and Arg-Gly-Asp dependence of cell adhesion to the native and unfolded triple helix of collagen type VI, *Exp. Cell Res.* **206:**161–166.

Pöschl, E., Pollinger, R., and Kühn, K., 1988, The genes for the α1 (IV) and α2 (IV) chains of human basement membrane collagen type IV are arranged head-to-head and separated by a bidirectional promoter of unique structure, *EMBO J.* **7:**2687–2695.

Postlethwaite, A. E., Seyer, J. M., and Kang, A. H., 1978, Chemotactic attraction of human fibroblasts to type I, I and III collagen and collagen-derived peptides, *Proc. Natl. Acad. Sci. USA* **75:**871–874.

Postlethwaite, A. E., Raghow, R., Stricklin, G. P., Poppleton, A., Seyer, J. M., and Kang, A. H., 1988, Modulation of fibroblast functions by interleukin-1 procollagen messenger RNAs and stimulation of other functions but not chemotaxis by human recombinant interleukin-1α and β, *J. Cell Biol.* **106:**311–318.

Prockop, D. J., Kivirikko, K. I., Tuderman, L., and Guzman, N. A., 1979, The biosynthesis of collagen and its disorders, *N. Engl. J. Med.* **301:**13–23.

Quaglino, D., Nanney, L. B., Ditesheim, J. A., and Davidson, J. M., 1991, Transforming growth factor-β stimulates wound healing and modulates extracellular matrix gene expression in pig skin: Incisional wound model, *J. Invest. Dermatol.* **97:**34–42.

Ramachandran, G. N., and Reddi, A. H. (eds.), 1976, *Biochemistry of Collagen,* Plenum Press, New York.

Roberts, A. B., and Sporn, M. B., 1990, The transforming growth factor-βs, in: *Handbook of Experimental Pharmacology.* Vol. 95, I: *Peptide Growth Factors and Their Receptors* (M. B. Sporn and A. B. Roberts, eds.), pp. 419–472, Springer, Berlin.

Roberts, A. B., Sporn, M. B., Assoian, R. K., Smith, J. M., Roche, N. S., Wakefield, L. M., Heine, U. I., Liotta, L. A., Falanga, V., Kehrl, J. H., and Fauci, A. S., 1986, Transforming growth factor type β: Rapid induction of fibrosis and angiogenesis *in vivo* and stimulation of collagen formation *in vitro, Proc. Natl. Acad. Sci. USA* **83:**4167–4171.

Roeckel, D., and Krieg, T., 1994, Three-dimensional contact with type I collagen mediates tyrosine phosphorylation in primary human fibroblasts, *Exp. Cell Res.* **211:**42–48.

Rossi, P., Karsenty, G., Roberts, A. B., Roche, N. S., Sporn, M. B., and de Crombrugghe, B., 1988, A nuclear factor I binding site mediates the transcriptional activation of a type I collagen promoter by transforming growth factor-β, *Cell* **52:**405–414.

Routeshouser, E. C., and de Crombrugghe, B., 1992, Purification of BBF, a DNA-binding protein recognizing a positive *cis*-acting element in the mouse α1(III) collagen promoter, *J. Biol. Chem.* **267:**14398–14404.

Ruggiero, F., Champliaud, M. F., Garrone, R., and Aumailley, M., 1994, Interactions between cells and collagen V molecules or single chains involve distinct mechanisms, *Exp. Cell Res.* **210:**215–222.

Ruoslahti, E., and Yamaguchi, Y., 1991, Proteoglycans as modulators of growth factor activities, *Cell* **64:**867–869.

Schaller, M. D., Borgman, C. A., Cobb, B. S., Vines, R. R., Reynolds, A. B., and Parsons, J. T., 1992, pp125[FAK], a structurally distinctive protein tyrosine kinase associated with focal adhesions, *Proc. Natl. Acad. Sci. USA* **89:**5192–5196.

Scharffetter, K., Heckmann, M., Hatamochi, A., Mauch, C., Stein, B., Riethmüller, G., Ziegler-Heitbrock, H. W. L., and Krieg, T., 1989a, Synergistic effect of tumor necrosis factor alpha and interferon-γ on collagen synthesis of human skin fibroblasts *in vitro, Exp. Cell Res.* **181:**409–419.

Scharffetter, K., Stolz, W., Lankat-Buttgereit, B., Hatamochi, A., Söhnchen, R., and Krieg, T., 1989b, Localization of collagen α1 (I) gene expression during wound healing by *in situ* hybridization, *J. Invest. Dermatol.* **93:**405–412.

Schiro, J. A., Chan, B. M. C., Roswit, W. T., Kassner, P. D., Pentland, A. P., Hemler, M. E., Eisen, A. Z., and Kupper, T. S., 1991, Integrin α2β1 (VLA-2) mediates reorganization and contraction of collagen matrices by human cells, *Cell* **67:**403–410.

Schlumberger, W., Thie, M., Rauterberg, J., Kresse, H., and Robenek, H., 1989, Deposition and ultrastructural organization of collagen and proteoglycans in the extracellular matrix of gel-cultured fibroblasts, *Eur. J. Cell Biol.* **50:**100–110.

Shah, M., Foreman, D. M., and Ferguson, M. W., 1992, Control of scarring in adult wounds by neutralizing antibody to transforming growth factor β, *Lancet* **339:**213–214.

Shaw, L. M., and Olsen, B. J., 1991, FACIT collagens: Diverse molecular bridges in extracellular matrices, *Trends Biochem. Sci.* **16**:191–194.

Timpl, R., 1993, Structure and biological activity of basement membrane proteins, *Eur. J. Biochem.* **180**:487–502.

Timpl, R., and Engel, J., 1987, Type VI collagen, in: *Structure and Function of Collagen Types* (R. Mayne and R. E. Burgeson, eds.), pp. 105–143, Academic Press, New York.

Trueb, B., and Winterhalter, K. H., 1986, Type VI collagen is composed of a 200 kd and two 140 kd subunits, *EMBO J.* **5**:2815–2819.

Trueb, J., and Trueb, B., 1992, Type XIV collagen is a variant of undulin, *Eur. J. Biochem.* **207**:549–557.

Vandenberg, P., Kern, A., Ries, A., Luckenbill-Edds, L., Mann, K., and Kühn, K., 1991, Characterization of a type IV collagen major cell binding site with affinity to the $\alpha 1\beta 1$ and the $\alpha 2\beta 1$ integrins, *J. Cell Biol.* **113**:1475–1483.

van der Rest, M., and Garrone, R., 1992, The collagen family of proteins, *FASEB J.* **5**:2814–2823.

Varga, J., Rosenbloom, J., and Jimenez, S. A., 1987, Transforming growth factor-β causes a persistent increase in steady state amounts of type I and type III collagen and fibronectin mRNAs in normal human dermal fibroblasts, *Biochem. J.* **247**:597–604.

Varga, J., Olsen, A., Herhal, J., Constantine, G., Rosenbloom, J., and Jimenez, S. A., 1990, Interferon-γ reverses the stimulation of collagen but not fibronectin gene expression by TGF-β in normal human fibroblasts, *Eur. J. Clin. Invest.* **20**:487–493.

Vogel, K. G., Paulsson, M., and Heinegard, D., 1984, Specific inhibition of type I and type II collagen fibrillogenesis by the small proteoglycan of tendon, *Biochem. J.* **223**:587–597.

von der Mark, H., Aumailley, M., Wick, G., Fleischmajer, R., and Timpl, R., 1984, Immunochemistry, genuine size and tissue localization of collagen VI, *Eur. J. Biochem.* **142**:493–502.

Vukicevic, S., Kleinman, H. K., Luyten, F. P., Roberts, A. B., Roche, N. S., and Reddi, A. H., 1992, Identification of multiple active growth factors in basement membrane Matrigel suggests caution in interpretation of cellular activity related to extracellular matrix components, *Exp. Cell Res.* **202**:1–8.

Weil, D., Mattei, M. G., Passarge, E., Long, N. V., Pribula-Conway, D., Mann, K., Deutzmann, R., Timpl, R., and Chu, M. L., 1988, Cloning and chromosomal localization of human genes encoding the three chains of type VI collagen, *Am. J. Hum. Genet.* **42**:435–445.

Werb, Z., and Gordon, S., 1975, Elastase secretion by stimulated macrophages. Characterization and regulation, *J. Exp. Med.* **142**:361–377.

Wiestner, M., Krieg, T., Hörlein, D., Glanville, R. W., Fietzek, P., and Müller, P. K., 1979, Inhibiting effect of procollagen peptides on collagen biosynthesis in fibroblast cultures, *J. Biol. Chem.* **254**:7016–7023.

Witt, D. P., and Lander, A. D., 1994, Differential binding of chemokines to glycosaminoglycan subpopulations, *Curr. Biol.* **4**:394–400.

Yamagata, M., Yamada, K. M., Yamada, S. S., Shinomura, T., Tanaka, H., Nishida, Y., Obara, M., and Kimata, K., 1991, The complete primary structure of type XII collagen shows a chimeric molecule with reiterated fibronectin type III motifs, von Willebrand factor A motifs, a domain homologous to a noncollagenous region of type IX collagen, and short noncollagenous domains with an Arg-Gly-Asp site, *J. Cell Biol.* **115**:209–221.

Chapter 17

The Dermal–Epidermal Basement Membrane Zone in Cutaneous Wound Healing

JOUNI UITTO, ALAIN MAUVIEL, and JOHN McGRATH

1. Introduction

The formation and repair of the functional extracellular matrix as part of the wound healing process requires coordinate expression of a repertoire of related and unrelated genes, including those encoding matrix proteins and proteolytic and regulatory enzymes. In addition, the formation of a functionally organized basement membrane between the epidermis and the dermis is essential for the integrity of the skin to allow its function as a protective organ. This chapter will highlight the recent progress made in understanding the biochemistry and molecular biology of the cutaneous basement membrane zone.

2. Structural Features of the Cutaneous Basement Membrane Zone

The basement membrane zone (BMZ) of the skin consists of a large number of distinct structural macromolecules that form an intricate attachment zone at the dermal–epidermal interface (Fig. 1). It is now apparent that the presence of sufficient quantities of these macromolecules and their discrete intermolecular interactions are necessary for the functional integrity of the BMZ. Electron microscopic examination of the cutaneous BMZ has revealed distinct, morphologically recognizable features that reflect the presence of some 20 different gene products assembled into the structural elements (Fig. 2). On the epidermal side of the cutaneous BMZ, one can recognize

JOUNI UITTO, ALAIN MAUVIEL, and JOHN McGRATH • Departments of Dermatology and Cutaneous Biology, and Biochemistry and Molecular Biology, Jefferson Medical College, and Section of Molecular Dermatology, Jefferson Institute of Molecular Medicine, Thomas Jefferson University, Philadelphia, Pennsylvania 19107.

The Molecular and Cellular Biology of Wound Repair (Second Edition), edited by Richard A. F. Clark. Plenum Press, New York, 1996

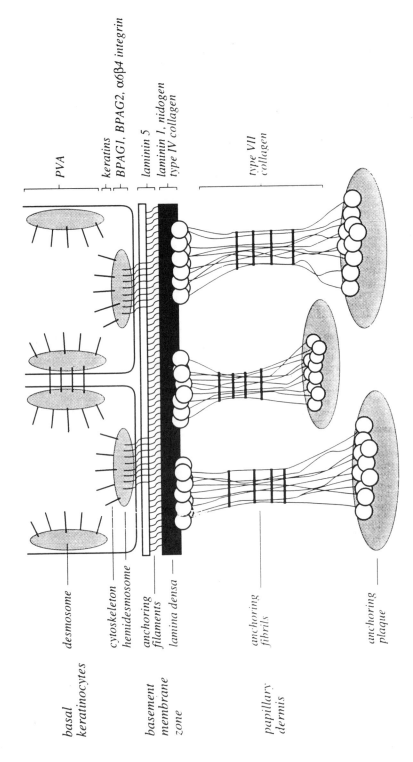

Figure 1. Schematic representation of the complexity of the cutaneous basement membrane zone. This figure depicts the presence of basal keratinocytes overlying the papillary dermis, with the basement membrane separating these two compartments, as shown on the left. The presence of ultrastructurally recognizable morphologic structures are indicated on the left side of the figure, and the individual protein components comprising these structures are indicated on the right. (Modified from Uitto and Christiano, 1992).

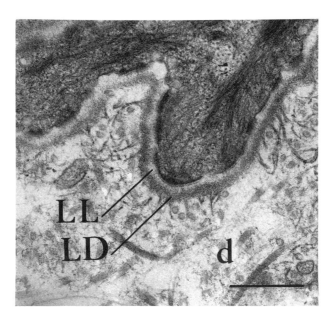

Figure 2. Transmission electron microscopy of the cutaneous basement membrane zone. Note the presence of a basement membrane consisting of two layers, lamina lucida (LL) and lamina densa (LD), separating the epidermis (e) from the underlying dermis (d). Bar: 0.5 μm.

hemidesmosomes, attachment structures that reside at the basal keratinocyte–lamina lucida interface. The hemidesmosomes consist of an intracellular plaque and a trans-membrane component that form a link securing the attachment of basal keratinocytes to the adjacent dermal–epidermal BMZ (Figs. 1 and 2). Immunoelectron microscopy has topographically localized at least three protein components to the hemidesmosomes, that is, bullous pemphigoid antigens 1 and 2, and the α6β4 integrin. The bullous pemphigoid antigen 1 (BPAG-1), a 230-kDa noncollagenous protein, is the major autoantigen in bullous pemphigoid, a blistering autoimmune disease (Stanley *et al.,* 1981; Stanley, 1989). Another hemidesmosomal protein is bullous pemphigoid antigen 2 (BPAG-2), a 180-kDa collagenous protein, also known as type XVII collagen (Li *et al.,* 1993), which serves as the pathogenetic autoantigen in patients with bullous pemphigoid (Diaz *et al.,* 1990; Liu *et al.,* 1993, 1995). BPAG-2 is also the autoantigen in herpes gestationis, a blistering disease associated with pregnancy (Morrison *et al.,* 1988; Giudice *et al.,* 1993). Several lines of evidence summarized below indicate that the two bullous pemphigoid antigens are clearly distinct gene products. The third protein associated with the hemidesmosomal complexes is the α6β4 integrin, a cell-surface-binding protein characteristically expressed in epithelial cells (Buck and Hor-witz, 1987; Albelda and Buck, 1990; Larjava *et al.,* 1990; Sonnenberg *et al.,* 1990, 1991; Jones *et al.,* 1991). The α6β4 integrin has been shown to be polarized to the underside of epidermal keratinocytes, on the surface apposed to the basement mem-brane, where it has been postulated to mediate the attachment of hemidesmosomes to the underlying BMZ (Peltonen *et al.*, 1989; DeLuca *et al.*, 1990; Kurpakus *et al.*, 1991; Sonnenberg *et al.*, 1991). Thus, the hemidesmosomal proteins form a network that extends from the intracellular milieu of the basal keratinocytes to the extracellular space and physically secures the attachment of the epidermis to the underlying basement membrane (Fig. 1).

The core of the cutaneous BMZ consists of a basement membrane, which by conventional transmission electron microscopy can be divided into two distinct layers. The upper, electron-lucent layer is known as the lamina lucida, and the lower portion, an electron-dense layer, is known as the lamina densa (Fig. 2). It has been suggested, however, that this ultrastructural appearance of two distinct layers is an artifact due to fixation for electron microscopy (Eady, 1988). Nevertheless, within the lamina lucida, there are clearly recognizable fine filamentous structures, anchoring filaments, which are most numerous subjacent to the hemidesmosomes (Fig. 3). It appears, therefore,

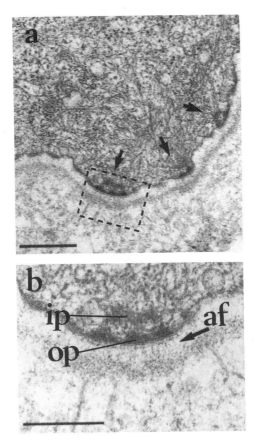

Figure 3. Electron microscopic visualization of the hemidesmosome-anchoring filament complexes at the cutaneous basement membrane zone. (a) Note the presence of clearly recognizable hemidesmosomes (arrows). (b) Enlargement of the area outlined in (a) reveals that the hemidesmosomes consist of an inner plaque (ip) and an outer plaque (op). Underneath the hemidesmosomes are anchoring filaments (af, arrow) which traverse the lamina lucida to the electron dense lamina densa. Bars: a. 0.5 μm, b. 0.25 μm.

that the hemidesmosomes and anchoring filaments form a continuous adhesion complex.

The presence of the hemidesmosomal component and the anchoring filament proteins are characteristic of stratifying squamous epithelia, such as those present in the skin, upper gastrointestinal tract, and the cornea of the eye. In contrast, some of the components of the cutaneous BMZ, such as type IV collagen, laminin 1, and nidogen, are ubiquitously expressed in basement membranes throughout the body, and these proteins are also present as major structural components of the cutaneous basement membrane (Olsen *et al.,* 1989a,b).

In the lower portion of the cutaneous BMZ, distinct attachment structures, anchoring fibrils, can be detected (Figs. 2 and 4). These fibrils extend from the lamina densa of the basement membrane to the papillary dermis. Anchoring fibrils may either loop to reattach to the lamina densa or the lower portion may attach to electron-dense, basement membrane-like structures known as anchoring plaques (Keene *et al.,* 1987; Burgeson *et al.,* 1990; Uitto and Christiano, 1992; Burgeson, 1993). Anchoring fibrils can be recognized ultrastructurally by their characteristic banding pattern, and the centrosymmetrical arrangement suggests that the ends of type VII collagen attach to other basement membrane components (Fig. 4). Immunoelectron microscopic, cell biological, and protein structural analyses have suggested that type VII collagen is the major, if not the exclusive, component of anchoring fibrils (see Section 3.3.3.).

In summary, the dermal–epidermal basement membrane forms a structural attachment zone which on its upper portion serves as an attachment site for basal keratinocytes through the formation of hemidesmosome-anchoring filament complexes, while the lower portion of the cutaneous BMZ is securely attached to the underlying dermis by anchoring fibrils. Thus, preservation of the structural features and controlled synthesis of these BMZ components, followed by their aggregation into a delicate

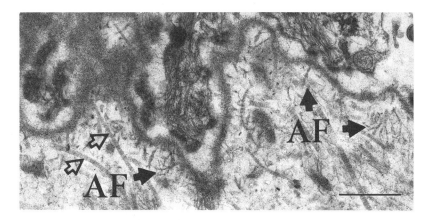

Figure 4. Demonstration of the presence of anchoring fibrils in the sublamina densa space. Note the presence of multiple anchoring fibrils extending from the lamina densa to the upper papillary dermis (AF, solid arrows). These fibrils are clearly distinct from the broad, banded dermal collagen fibers consisting of type I and III collagens (open arrows). Bar: 0.5 μm.

network, are critical steps toward securing the integrity of the dermal–epidermal junction during fetal skin development and during reparative wound healing processes.

This chapter will highlight recent advances made in understanding the structure and function of the cutaneous BMZ components, the regulation of the corresponding gene expression by cytokines, and their relevance to normal and compromised wound healing.

3. Molecular Biology of the Cutaneous BMZ Components

As indicated above, the cutaneous BMZ consists of a large number of proteins that represent as many as 20 distinct gene products. Since these proteins form a complex, highly cross-linked network within the intact skin, the dissection of the structural features of the human proteins was initially hampered by the insolubility of the individual components. Recently, however, significant progress toward understanding the structural features of the BMZ proteins has been made through molecular cloning of the corresponding genes and their complementary DNA sequences (Uitto and Christiano, 1992). The information derived from the nucleotide sequencing has allowed the primary and secondary structures of these protein components to be deduced and predictions to be made about their molecular organization, based on the characteristics of the putative peptide sequences.

As examples of the tremendous progress made in understanding the molecular biology of the individual components of the cutaneous BMZ, this section will highlight the cloning of the genes encoding the hemidesmosomal protein, BPAG-1, the anchoring filament protein, laminin 5, and the anchoring fibril protein, type VII collagen, respectively.

3.1. Bullous Pemphigoid Antigen 1: A Hemidesmosomal Protein

BPAG-1 is a major component of the hemidesmosomes and is also known as the 230-kDa bullous pemphigoid antigen (Stanley *et al.*, 1981). Immunofluorescence studies have indicated that the 230-kDa bullous pemphigoid antigen is the major component of the intracellular hemidesmosomal plaque (Mutasim *et al.*, 1985; Mueller *et al.*, 1989). Due to difficulties in isolating the BPAG-1 protein, initial studies did not yield complete information about its structural features. However, recent cloning of the full-length human BPAG-1 cDNA sequences has subsequently allowed a detailed structural analysis.

3.1.1. cDNA Cloning

Isolation of a partial, approximately 2-kilobase (kb), cDNA corresponding to the BPAG-1 sequences was first reported by Stanley *et al.* (1988). These authors isolated a cDNA from a cDNA expression library by an immunoscreening technique utilizing antiserum from a patient with bullous pemphigoid containing relatively high titers of immunoglobulin G (lgG) antibodies recognizing BPAG-1 epitopes. For the purpose of performing potential linkage studies in families with heritable blistering skin diseases (Christiano and Uitto, 1992), oligomer primers were subsequently synthesized using the published sequence information to amplify an approximately 450-basepair (bp) segment of the BPAG-1 mRNA by reverse transcriptase–polymerase chain reaction (RT-PCR) (Sawamura *et al.*, 1990, 1991b). This PCR-generated cDNA was subsequently used to screen a human keratinocyte cDNA library. The initial screening resulted in the isolation of five positive clones, the largest one being ~2.2 kb in size. Subsequently, the use of a variety of techniques, including screening of the cDNA library with a synthetic oligonucleotide corresponding to the 5' end of the cDNA and PCR amplification mRNA sequences, resulted in the isolation of overlapping cDNAs that were demonstrated to correspond to the entire full-length human BPAG-1 (Sawamura *et al.*, 1991a).

3.1.2. Protein Structure

The primary sequence of the 230-kDa BPAG, deduced from the open reading frame of the full-length cDNA, was shown to consist of a polypeptide of 2649 amino acids (Fig. 5). A hydrophilicity blot of the entire sequence revealed a large central

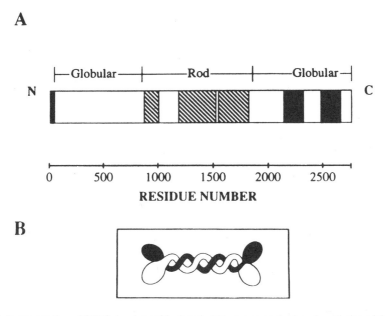

Figure 5. Representation of BPAG-1 polypeptide domain (A), and computer-based prediction of the protein structure (B), based on deduced primary amino acid sequence. (A) The individual polypeptide demonstrates the presence of three subdomains characterized by heptad repeats (▨) and two subdomains containing 38-residue repeats (■). (B) Modeling of the protein homodimer structure depicts a central coiled-coil rod flanked by globular end domains. (Modified from Sawamura *et al.*, 1991a.)

portion, consisting of ~1000 amino acid residues, to be predominantly hydrophobic, while the peptide sequences toward both ends were in general more hydrophilic. The putative primary structure of the BPAG-1 protein was also subjected to computer search for internal repeats as well as for homologies with other protein sequences in the GenBank (Sawamura *et al.,* 1991a). Such searches revealed several interesting features within the primary structure of the predicted polypeptide. First, a 38-residue repeat structure was detected in two subdomains, each consisting of four copies of the repeat, at the carboxyl end of the molecule (Fig. 5A). These repeat domains demonstrated homology with a similar, 38-residue repeat found in desmoplakin I, a desmosomal protein implicated in binding of intermediate filaments to the plasma membrane. This structural similarity between BPAG-1 and desmoplakin I suggested a functional role for BPAG-1 as a hemidesmosomal attachment site for intermediate filaments. In fact, this suggestion is consistent with the intracellular location of BPAG-1 within the basal keratinocytes as a major component of the hemidesmosomal plaque.

A second interesting feature of the deduced protein structure was the presence of an internal heptad repeat that was detected in three separate subdomains within the central portion of the BPAG-1 polypeptide (Fig. 5). Within the repeating substructure $(abcdefg)_n$, the amino acid residues at the a and d positions were predominantly apolar (Sawamura *et al.,* 1991a). Computer analysis predicted that the region containing the heptad repeat had a high probability (~81%) of having an α-helical conformation. This pattern of regular repeats, which is found, for example, in myosin, suggested that BPAG-1 may form a double-stranded coiled-coil structure in the central portion of the protein, and this central core would then be flanked by globular domains at both ends, as predicted from the primary sequence (Fig. 5B).

3.1.3. Genomic Organization

To elucidate the exon–intron organization of the human BPAG-1 gene, a λFIX genomic DNA library was screened with cDNAs corresponding to different parts of the full-length mRNA (Fig. 6). Isolation of seven overlapping λ clones allowed elucidation of the entire gene, which was shown to span ~20 kb of genomic sequence (Tamai *et al.,* 1993). Identification of the intron–exon borders revealed 22 distinct exons, which varied from 78 to 2810 bp in size. Interestingly, the 3′ end of the gene corresponding to the 38-amino acid residue repeat region was encoded by a single, relatively large exon (number 22). The seven exons in the region encoding the three α-helical coiled-coil subdomains varied from 84 to 2730 bp in size. However, two of the subdomains were encoded by a single exon (number 21) without interruption, while the most upstream part of the 3′ subdomain was encoded by three separate exons (numbers 15–17). The region corresponding to the amino-terminal globular domain consisted of 15 distinct exons varying from 78 to 591 bp in size (Fig. 6). These observations indicate that different protein domains within the 230-kDa BPAG are encoded by gene segments depicting distinct intron–exon organizations (Tamai *et al.,* 1993). These arrangements may relate to the stability of the gene with potential for rearrangements, the occurrence of which could conceivably result in heritable diseases.

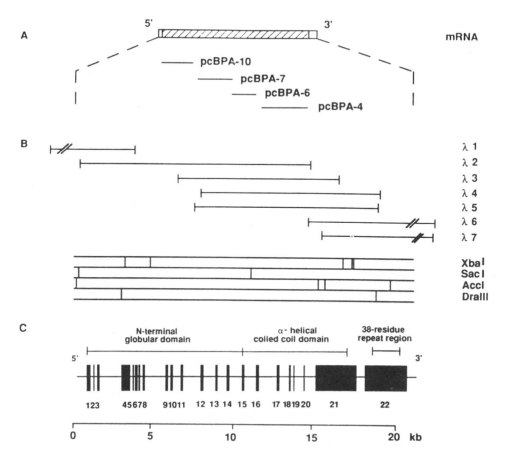

Figure 6. Cloning of the human BPAG-1 gene. (A) Four separate complementary DNAs (pcBPA) corresponding to different segments of the full-length human BPAG-1 mRNA were used to screen a genomic DNA library. (B) Seven distinct genomic clones (λ1–7), varying from ~9 to ~14 kb in size, were isolated. Restriction enzyme digestions with XbaI, SacI, AccI, and DraIII endonucleases allowed alignment of the clones. The segment corresponding to the coding region of the BPAG-1 mRNA spans ~20 kb of genomic DNA. (C) Comparison of the cDNA with the corresponding genomic sequences identified 22 distinct exons within the gene. These exons are depicted by vertical bars, the introns are shown by horizontal lines. The gene segments corresponding to specific domain structures (see Figure 5) are indicated above the exons. (Modified from Tamai *et al.*, 1993.)

3.1.4. Tissue Specificity of Gene Expression

To examine the tissue-specific expression of the BPAG-1 gene, extensive analysis at the mRNA level has been performed (Sawamura *et al.*, 1991b; Tamai *et al.*, 1994a,b). First, Northern analyses were performed with RNA isolated from a variety of cultured cells, including normal human epidermal keratinocytes, HeLa cells, as well as a human oral epidermoid carcinoma cell line (KB) and human transformed amniotic epithelial

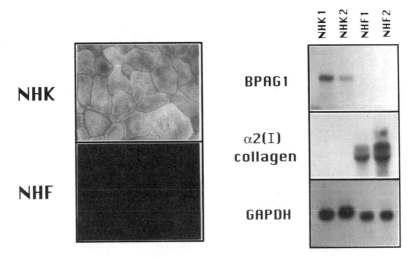

Figure 7. Demonstration of selective expression of the BPAG-1 gene, as determined at the protein level by indirect immunofluorescence (A) or at the mRNA level by Northern hybridization (B). (A) Normal human keratinocytes (NHK) or normal human fibroblasts (NHF) were incubated in parallel, and the presence of BPAG-1 epitopes was detected by immunostaining with a monoclonal antibody. Note the detectable immunosignal only in NHK cultures, while NHF are entirely negative. (B) Two different cell strains of NHK or NHF were incubated in culture, total RNA was isolated, and Northern analyses were performed with BPAG-1 and type I collagen $\alpha 2(I)$ cDNAs. Note the clearly detectable BPAG-1 mRNA transcripts in NHK1 and 2 cultures, while NHF1 and 2 are entirely negative. In contrast, the $\alpha 2(I)$ collagen gene is expressed only in NHF cultures. Hybridization with GAPDH, a ubiquitously expressed housekeeping gene, was used as a control. (Reproduced from Tamai *et al.*, 1995b.)

cells (WISH), which were shown by immunofluorescence staining to express BPAG-1 (see Fig. 7A). An mRNA transcript was detected only in normal human epidermal keratinocytes cultured in serum-free medium containing low concentrations (0.15 mM) of Ca^{2+}, and not in cultured human dermal fibroblasts examined in parallel (Fig. 7B). Instead, type I collagen $\alpha 2(I)$ mRNA transcripts were readily detected in cultured fibroblasts but were absent in keratinocytes. Thus, the expression of the BPAG-1 gene appears to be largely restricted to keratinocytes, particularly those with the mitotic, undifferentiated phenotype characteristic of basal cells.

To examine the regulatory elements potentially responsible for tissue-specific expression of the gene at the transcriptional level, fragments of DNA corresponding to the 5'-flanking region of the gene were isolated (Tamai *et al.*, 1993). These fragments were then ligated to the bacterial chloramphenicol acetyl transferase (CAT) reporter gene, and the constructs were used for transient cell transfections of cultured keratinocytes, as well as of cell types that did not demonstrate expression of the gene at the mRNA level, as determined by Northern analyses, such as dermal fibroblasts (Tamai *et al.*, 1994a, 1995b). The 5'-flanking region of the gene, upstream from the ATG initiation codon for translation, was found to contain several putative transcriptional response elements. Specifically, these included three motifs potentially conferring keratinocyte-specific expression to the gene. The presence of such elements was ini-

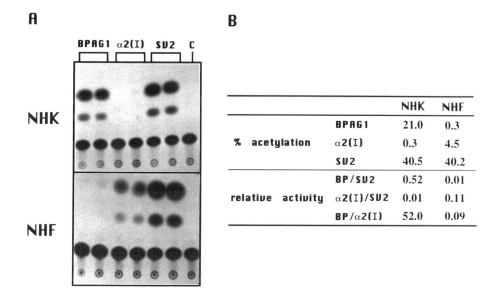

Figure 8. Transient transfections of normal human keratinocytes (NHK) or normal human fibroblasts (NHF) in culture with BPAG-1 and α2(I) collagen promoter/CAT reporter gene constructs. (A) Assay of CAT activity by thin layer chromatography demonstrating separation of acetylated forms of [^{14}C] chloramphenicol from the unacetylated forms. Note the clearly detectable CAT activity in NHK cultures transfected with BPAG promoter/CAT construct, while the α2(I) collagen promoter construct was not expressed. In contrast, NHF cultures clearly expressed the collagen promoter, while BPAG1 promoter activity was essentially undetectable. (B) Calculation of percent acetylation of [^{14}C] chloramphenicol used as substrate for the CAT assay, allowed expression of the relative promoter activity in NHK and NHF cultures. The results indicated that BPAG-1 promoter activity in NHK cultures is ~50-fold higher than the expression of the α2(I) collagen promoter, attesting to the specificity of the expression. SV2 viral promoter/CAT construct was used as a positive control; lane C contains a negative control containing protein extract from cells without transfection. For technical details, see Tamai *et al.*, 1995b. (Reproduced from Tamai *et al.*, 1995b.)

tially suggested by the high level of expression of a promoter–CAT construct in normal human epidermal keratinocytes, as compared to a low level of expression in fibroblasts (Fig. 8). Careful analysis of the potential motifs within the promoter region indicated that so-called CK-8-mer, which has previously been identified in a number of keratin genes as well as in the gene for human involucrin (Blessing *et al.*, 1987, 1989; Cripe *et al.*, 1987), was not functional in the human BPAG-1 gene, as determined by gel mobility shift assays with nuclear proteins isolated from cultured keratinocytes or fibroblasts (Tamai *et al.*, 1993, 1994a). In contrast, a putative AP-2 binding sequence, which was designated as keratinocyte responsive element 2 (KRE-2) at position -1786 to -1778, was shown to confer partial keratinocyte-specific expression to the gene (Tamai *et al.*, 1994a). This conclusion was based on several lines of evidence. First, 5′ deletion clones that eliminated this region resulted in a dramatic reduction in the level of expression of the promoter–CAT construct in keratinocytes, but did not alter the low level of expression noted in fibroblasts (Fig. 9). However, cloning of the KRE-2 sequence in front of the truncated promoter–CAT construct restored the high level of

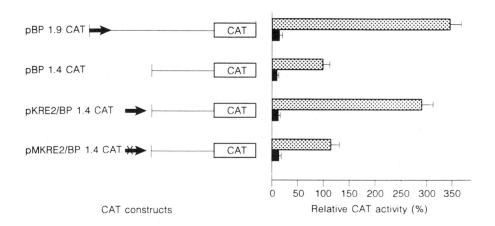

Figure 9. Demonstration that keratinocyte response element 2 provides tissue specificity to the BPAG-1 gene expression. The construct pBP1.9 CAT containing the KRE2 sequence (→) at position -[1878 to 1888], and a truncated construct pBP1.4 CAT devoid of KRE2 were transfected into normal human keratinocytes (■) or normal human fibroblasts (■) in culture. Note the 3.5-fold activity with the pBP1.9 CAT construct in keratinocytes as compared to pBP1.4 CAT, while no significant difference in fibroblast cultures was noted. Cloning of the KRE2 element in front of a pBP1.4 CAT restored the high level of expression of keratinocytes, but not in fibroblasts. A mutated construct, pMKRE2/BP1.4 CAT, in which two C→A substitutions were made in the KRE2 sequence, showed activity similar to that of pBP1.4 CAT. (Reproduced from Tamai *et al.,* 1993.)

expression in keratinocytes. Furthermore, introduction of specific mutations into the KRE-2 sequence abolished its ability to restore the keratinocyte-specific activity (Fig. 9). Second, gel mobility shift assays utilizing nuclear protein extracts from normal keratinocytes identified a unique DNA-binding protein, keratinocyte transcriptional protein 1 (KTP-1), which was then shown to bind to KRE-2 (Tamai *et al.,* 1994a). This *trans*-acting factor was further characterized by competition assays with consensus AP-2 oligomers, by UV cross-linking studies, and by Southwestern analyses. The results indicated that KTP-1 is not present in human skin fibroblasts or HeLa cells, which do, however, contain the genuine AP-2. Subsequent UV cross-linking studies and Southwestern analyses suggested that KTP-1 binds to DNA as a single polypeptide of ~110 kDa. Collectively, these data indicated that KTP-1 is a novel DNA-binding protein, clearly distinct from AP-2, and this protein may be responsible for the basal keratinocyte-specific expression of the BPAG-1 gene (Tamai *et al.,* 1994a). It was noted, however, that elimination of the KRE-2 sequence from the promoter region did not completely abolish the tissue specificity of BPAG-1 gene expression in keratinocytes (Tamai *et al.,* 1994a). This observation suggested the possibility that additional regulatory elements may contribute to the basal keratinocyte specific expression of this gene.

Subsequently, another *cis*-element, keratinocyte-responsive element 3 (KRE-3), at position -216 to -197 of the human BPAG-1 gene, was demonstrated to be a major

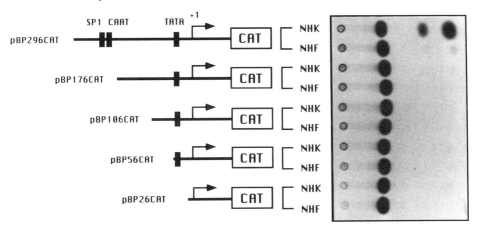

Figure 10. Demonstration of the presence of a keratinocyte responsive element, KRE3 in the promoter region -[296 to -176] of the human BPAG-1 promoter. BPAG-1 promoter region/CAT constructs were transfected to normal human keratinocytes (NHK) or normal human fibroblasts (NHF) in culture. Note the clearly detectable promoter activity in NHK cultures with the construct pBP296 CAT. 5′ deletions of this construct abolished the high level of expression in NHK. The presence of SP1, CAAT, and TATA consensus sequences are shown upstream. The transcription initiation site of the gene is indicated by +1. (Reproduced from Tamai *et al.*, 1995b.)

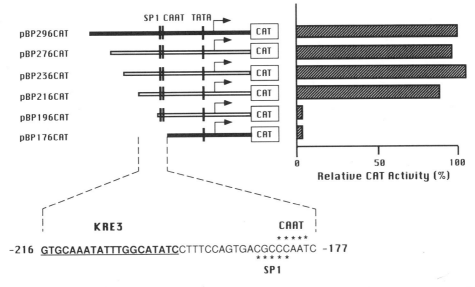

Figure 11. Identification of the keratinocyte responsive element 3 (KRE3) by 5′-deletion constructs of BPAG-1/CAT reporter gene constructs. Six constructs depicted in the figure were transfected into normal human keratinocytes in culture, and CAT activity was determined as described in Figures 8–10. Note that deletion of the segment between nucleotides -216 and -177 abolishes the CAT activity. Nucleotide sequencing of this region reveals the presence of the KRE3 segment (underlined) upstream from SP1 and CAAT consensus sequences (asterisks). (Reproduced from Tamai *et al.*, 1995b.)

sequence element conferring tissue-specific expression to the BPAG-1 gene (Tamai *et al.*, 1995b). Specifically, a promoter–CAT construct containing this element had ~50-fold higher expression than a similar 5'-deletion construct devoid only of this sequence, when tested in transient transfections of cultured human keratinocytes (Fig. 10). Again, however, there was no effect on the low baseline level of expression in cultured skin fibroblasts. Careful analysis of KRE3 revealed that it consisted of a palindromic sequence, 5'-CAAATATTTG-3', and mutations in this sequence significantly reduced the promoter activity (Tamai *et al.*, 1995b) (Fig. 11). Gel mobility shift assays with an oligomer-containing KRE-3 sequence demonstrated binding activity with nuclear proteins isolated from keratinocytes. One of the DNA–protein complexes was clearly specific, since competition with >12.5-fold excess of the unlabeled oligomer resulted in disappearance of this band (Tamai *et al.*, 1995b). No specific binding activity was noted with nuclear proteins extracted from fibroblasts. Thus, KRE-3 appears to serve as the binding site for keratinocyte-specific *trans*-activating factor(s), and KRE-3, together with KRE-2, may thus confer the tissue-specific expression to the BPAG-1 gene.

3.1.5. Gene Mapping

To determine the genomic location of the human BPAG-1 gene, chromosomal *in situ* hybridizations with radioactively labeled cDNAs were performed (Sawamura *et al.*, 1990, 1992). Results indicated that the human BPAG-1 gene resides at the chromosomal locus 6p11–6p12 (Fig. 12). This conclusion was supported by hybridizations with a panel of human × rodent hybrid cell DNA, which indicated concordance with the human chromosome 6.

3.2. Bullous Pemphigoid Antigen 2: A Hemidesmosomal Transmembrane Protein, Type XVII Collagen

BPAG-2, the 180-kDa bullous pemphigoid antigen, is a collagenous hemidesmosomal protein, which has recently been designated as type XVII collagen (Giudice *et al.*, 1991; Li *et al.*, 1991, 1993). In fact, type XVII collagen consists of 13 distinct collagenous domains in its carboxy-terminal segment and of a large globular domain in the amino-terminal portion (Fig. 13). Sequence analyses have also revealed the presence of a membrane-associated domain, which initially suggested, on the basis of computer predictions, that type XVII collagen is a transmembrane protein (Li *et al.*, 1991). Subsequent studies involving epitope mapping have demonstrated that this protein is indeed a transmembrane protein in an unusual type II topography (Giudice *et al.*, 1992; Hopkinson *et al.*, 1992; Li *et al.*, 1992). Specifically, the carboxy-terminal portion of the molecule containing the collagenous domains is located in the extracellular space embedded in the dermal–epidermal basement membrane, potentially serving as an attachment site for other BMZ molecules. In contrast, the amino-terminal noncollagenous domain resides intracellularly within the hemidesmosomes and may associate

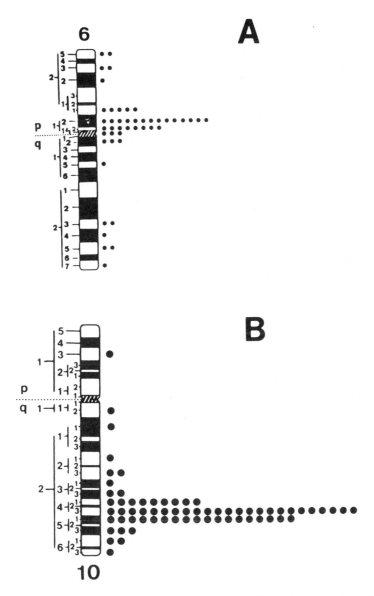

Figure 12. Idiograms of human chromosomes 6 and 10 indicating different chromosomal loci for BPAG-1 (A) and BPAG-2 (B), as revealed by chromosomal *in situ* hybridizations with the corresponding cDNA probes. For experimental details, see original data in Sawamura *et al.,* 1991b and Li *et al.,* 1991. (Reproduced from Sawamura *et al.,* 1992.)

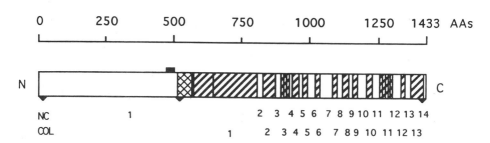

Figure 13. Representation of the α1(XVII) polypeptide, the 180-kD bullous pemphigoid antigen, as deduced from cloned complementary DNAs. The polypeptide consists of a total of 1,433 amino acids. The 5′ segment (NC1) is characterized by a 573-amino acid noncollagenous sequence. The carboxyl-terminal domain consists of 13 collagenous segments (COL1–13), which are interrupted by noncollagenous segments (NC2–13). The most carboxyl-terminal segment within the NC-1 domain is characterized by the presence of 8 heptad repeats (▧). Computer analysis of the primary sequence identifies a membrane-associated segment (■) as well as three potentially antigenic sites (▼). (Reproduced from Li *et al.*, 1993.)

directly with the 230-kDa bullous pemphigoid antigen, BPAG-1, within the hemidesmosomal plaque.

3.2.1. Demonstration That BPAG-1 and BPAG-2 Are Distinct Gene Products

As indicated above, the sera from patients with bullous pemphigoid (BP), a blistering skin disease, have been shown to contain autoantibodies that recognize at least two autoantigens. By Western immunoblotting analyses, one of these antigens has been shown to be a 230-kDa protein that is recognized by autoantibodies in the sera from the majority of patients with BP. In addition, antibodies recognizing a 180-kDa protein have been shown to be present in the sera of some patients with BP and also in patients with herpes gestationis (Morrison *et al.*, 1988; Giudice *et al.*, 1993). Immunofluorescence data have demonstrated that these antigens reside within the cutaneous BMZ, and furthermore, immunoelectron microscopy has localized them to the hemidesmosomes.

Previously, there has been some uncertainty as to the precise structural relationship between these two autoantigens (Labib, 1991). This issue was complicated by the fact that different authors have assigned somewhat different molecular weights for these proteins. However, recent data on the cloning and characterization of the BPAG-1 and BPAG-2 genes have unequivocally demonstrated that these proteins are two distinct gene products without structural homology. The evidence to support this conclusion has come from at least three different approaches (Sawamura *et al.*, 1992). First, cloning of the full-length cDNA sequences corresponding to these two proteins has allowed the entire primary amino acid sequence for both of them to be deduced, and comparison of these sequences has not revealed any homology (Sawamura *et al.*, 1992; Giudice *et al.*, 1992; Li *et al.*, 1993). In fact, BPAG-2 is a collagenous protein with multiple segments characterized by repeating Gly-X-Y sequences, while BPAG-1 is a noncollagenous protein.

Second, Northern hybridization of RNA from cells expressing both proteins, such as cultured epidermal keratinocytes, has revealed the presence of mRNA transcripts of two distinct sizes. Specifically, the BPAG-1 protein is encoded by an ~9-kb mRNA,

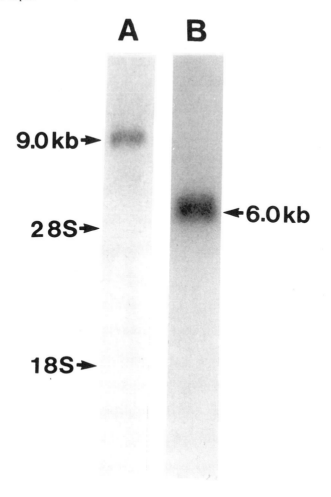

Figure 14. Demonstration that BPAG-1 and BPAG-2 are encoded by mRNA transcripts of different sizes. Normal human keratinocyte RNA was used for Northern analyses with hybridizations with a human BPAG-1 (lane A) or BPAG-2 (lane B) cDNAs. Note the different sizes of the BPAG-1 and BPAG-2 mRNAs are 9.0 and 6.0 kb, respectively. (Reproduced from Sawamura *et al.,* 1992.)

whereas BPAG-2 sequences are contained within an ~6-kb transcript (Fig. 14). Although the size difference in mRNA could theoretically be explained by alternative splicing of the primary pre-mRNA transcripts, the presence of multiple transcription initiation sites within a single gene or by differential utilization of 3' polyadenylation signals, the absence of any sequence homology at the cDNA level mitigates these possibilities.

Third, chromosomal *in situ* hybridizations with BPAG-1 and BPAG-2 cDNAs revealed that the genes for these two bullous pemphigoid antigens reside in different chromosomes within the human genome (Fig. 12). Specifically, as noted above, BPAG-1 resides in the short arm of human chromosome 6, corresponding to the region 6p11–12 (Sawamura *et al.,* 1990), whereas BPAG-2 resides in the long arm of chromosome 10 at the locus 10q24.3 (Li *et al.,* 1991).

This demonstration that the 230-kD and 180-kD bullous pemphigoid antigens are clearly distinct gene products (BPAG-1 and BPAG-2) illustrates the potential of molecular biology in elucidating the structural components of the cutaneous basement membrane zone (Uitto and Christiano, 1992).

3.3. Laminin 5: An Anchoring Filament Protein

As indicated above, electron microscopic examination of the cutaneous BMZ reveals the presence of structures known as anchoring filaments, which traverse from the hemidesmosomes across the lamina lucida to the lower portion of the dermal–epidermal basement membrane. Laminin 5 is a recently characterized BMZ component that appears to be an integral component of the anchoring filaments.

3.3.1. Structural Features

Extensive biochemical studies at the protein level have established that laminin 5 is a member of the laminin superfamily (Beck *et al.*, 1990; Tryggvason, 1993; Timpl and Brown, 1994; Burgeson *et al.*, 1994). There are currently as many as seven different laminin isoforms that have characteristic tissue distribution and specific subunit composition (Table I). Each of the laminin isotypes is a heterotrimer consisting of subunit polypeptides, nine of which have been characterized and shown to be distinct gene products (Table II).

Table I. Summary of Laminin Isoforms, Their Chain Composition and Tissue Distribution[a]

Laminin isoforms	Chain composition	Previous name	Tissue distiribution
Laminin 1	α1β1γ1	EHS laminin	Kidney, testis, meninges, neuroretina, some regions of brain, blood vessels of muscle and skin
Laminin 2	α2β1γ2	Merosin	Striated muscle, peripheral nerve, choroid plexus, meninges, developing and regenerating liver, thymus, testis, placental trophoblast
Laminin 3	α1β2γ1	S-laminin	Synapses at neuromuscular junctions, perineurium in perhipheral nerves, certain blood vessels and glomeruli in the kidney
Laminin 4	α2β2γ1	Merosin/S-laminin	Myotendinous junction, placental trophoblast
Laminin 5	α3β3γ2	Nicein/Kalinin/ Epiligrin	Skin, amnion, larynx, esophagus, stomach, small intestine, bladder, pancreas
Laminin 6[b]	α3β1γ1	K-laminin	Same as for laminin 5
Laminin 7[b]	α3β2γ1	KS-laminin	Unknown

[a]Modified from Burgeson *et al.* (1994).
[b]The identity of α chains in laminins 6 and 7 with the α3 chain of laminin 5 is not firmly established.

Table II. Genes Encoding Laminin Polypeptide Chains and Their Chromosomal Location[a]

Chain	Previous name	Gene	Chromosomal locus[b]	References
α1	A, Ae	LAMA1	18p11.3	Nagayoshi *et al.* (1989)
α2	M, Am	LAMA2	6q22-23	Vuolteenaho *et al.* (1994)
α3	200 kDa[c]	LAMA3	18q11.2	Ryan *et al.* (1994)
α4		LAMA4	6q21	Richards *et al.* (1994)
β1	B1, B1e	LAMB1	7q22	Pikkarainen *et al.* (1987)
β2	S, B1s	LAMB2	3p21	Wewer *et al.* (1994)
β3	140 kDa[c]	LAMB3	1q32	Vailly *et al.* (1994b)
γ1	B2, B2e	LAMC1	1q25-31	Fukushima *et al.* (1988)
γ2	B2t[c]	LAMC2	1q25-31	Vailly *et al.* (1994b)

[a]Modified from Burgeson *et al.* (1994).
[b]Chromosomal location determined for human genes.
[c]Identified in laminin 5.

The most extensively studied laminin isoform, laminin 1, was initially isolated from the murine Engelbreth-Holm-Swarm tumor, hence the original name EHS-laminin (Timpl and Brown, 1994). Laminin 1 is composed of three genetically distinct, but structurally related polypeptides, the α1, β1, and γ1 chains, with molecular weights of 400, 215, and 200 kDa, respectively. These chains form an asymmetric cross-shaped structure with three short arms and a long arm (Fig. 15). The amino-termini of the α, β, and γ subunits extend separately to form the three short arms, which consist of several subdomains. In particular, subdomains VI and IV form globular structures, while subdomains V and III consist of epidermal growth factor (EGF)-like repeats, each containing cysteine residues. The stem of the long arm is formed by subdomains II and I, and the three chains fold together into a coiled-coil structure (Fig. 15). The β chain contains a cysteine-rich, so-called α subdomain between the major domains II and I. The carboxy-terminal third of each α chain forms a large globular domain at the end of the long arm.

Laminin 5 was first recognized by a monoclonal antibody, GB3, and was designated as BM600; this protein was subsequently renamed nicein (Verrando *et al.*, 1988). At the same time, a BMZ protein, kalinin, was independently isolated (Rousselle *et al.*, 1991; Marinkovich *et al.*, 1992b). Subsequent immunofluorescent and biochemical analysis have clearly established that nicein and kalinin polypeptides are identical products of the same genes (Marinkovich *et al.*, 1993). Furthermore, antibody approaches have identified a third anchoring filament protein, epiligrin (Carter *et al.*, 1991). This protein appears to be similar, if not identical, with kalinin/nicein or at least contains this protein complexed with other components of the cutaneous BMZ (Gil *et al.*, 1994). Consequently, the protein nicein/kalinin/epiligrin was renamed laminin 5 (Burgeson *et al.*, 1994).

Laminin 5, as do all members of the laminin family, consists of three individual subunits—α3, β3, and γ2 chains—which assemble into a trimeric cross-shaped structure (Fig. 15). Biosynthesis of the α3, β3, and γ2 polypeptides of laminin 5 occurs in

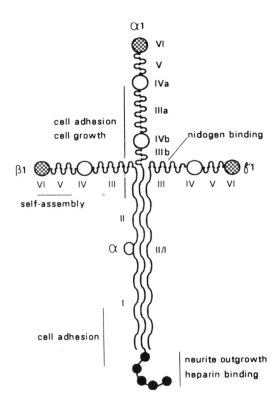

the basal keratinocytes, where they form a heterotrimer, followed by secretion and a two-step proteolytic processing. The cellular form of laminin 5 is ~460 kDa in size and is secreted by keratinocytes into the cell culture medium (Marinkovich *et al.*, 1992a). This particular form of laminin 5 appears to facilitate attachment and adhesion of proliferating and migrating cells. The cellular form is subsequently converted in the medium to a ~440-kDa intermediate protein as a result of processing of the α3 chain. Furthermore, the ~440-kDa form is converted to the ~400-kDa protein by subsequent processing of the γ2 chain. It has been suggested that the latter form of laminin 5 facilitates epithelial–mesenchymal cohesion *in vivo*. Thus, each of the different forms of laminin 5 have divergent and important roles in maintaining the adhesion zone between the epidermis and dermis.

Biologically, laminin 5 is suggested to be involved in the attachment of keratinocytes to the basement membrane. Certain sites of the molecule are capable of interacting with receptors on the keratinocyte surface, while other regions appear to reside within the lamina densa, thus providing cell–substrata adhesion. Specifically, laminin 5 has been shown to bind α3β1 integrin, which is located apart from the hemidesmosomes in the plasma membrane of basal keratinocytes (Carter *et al.*, 1990). Laminin 5 has also been shown to interact with the hemidesmosomal protein, α6β4 integrin, which is expressed by epithelial cells on the side apposed to the basement membrane (Peltonen *et al.*, 1989; DeLuca *et al.*, 1990; Stepp *et al.*, 1990; Kurpakus *et al.*, 1991;

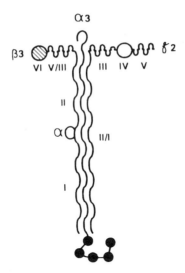

Figure 15. Schematic representation of the structures of laminin 1 (left) and laminin 5 (right). Laminin 1 is composed of α1, β1 and γ1 polypeptides, while the corresponding polypeptides in laminin 5 are α3, β3 and γ2. The structural domains of the individual polypeptides in both laminins are indicated by Roman numerals (I–VI), and functional domains of laminin 1 are described. (The figure was kindly provided by Dr. Leena Pulkkinen, Jefferson Medical College.)

Sonnenberg *et al.*, 1991). Thus, laminin 5 serves as a specific ligand for the extracellular portion of α6β4 integrin.

3.3.2. cDNA Cloning

Precise elucidation of the laminin 5 protein structure has been advanced by cloning of the corresponding DNA sequences. Specifically, the entire γ2 and β3 chain cDNA sequences have been published (Kallunki *et al.*, 1992; Gerecke *et al.*, 1994; Vailly *et al.*, 1994a). Also, partial sequence to the α3 polypeptide has been reported (Ryan *et al.*, 1994). This information has allowed the primary sequences of the corresponding polypeptides to be deduced. cDNA cloning of these polypeptides has also facilitated structural analysis of different laminin subunits, and specifically cloning of laminin 5 polypeptides (α3, β3, and γ2 chains) has allowed detailed comparison with the corresponding laminin 1 polypeptide sequences (α1, β1, and γ1 chains). For example, recent cloning of the β3 chain cDNA and genomic DNA sequences has allowed direct comparison of the structures of the corresponding genes (LAMB1 and LAMB3), as well as direct comparison of the β1 and β3 chain protein domains (Pulkkinen *et al.*, 1995).

3.3.3. Gene Structure

The classic laminin 1 has the chain composition α1, β1, and γ1 (Table I). The β1 chain consists of 1765 amino acids, encoded by a 5.6-kb mRNA, and the corresponding gene, LAMB1, has been mapped to human chromosome 7q22 (Table II). Since the LAMB3 gene was subsequently mapped to a different chromosome, 1q25 (Vailly *et al.*,

1994b), it was of interest to compare the structural organization of the LAMB1 and LAMB3 genes. Direct comparison indicated that LAMB3, ~29 kb in size, is considerably more compact than LAMB1, which extends over 80 kb of genomic sequences (Pikkarainen *et al.*, 1987). The LAMB1 gene consists of 34 exons, while LAMB3 comprises 23 exons (Fig. 16). This difference reflects the fact that LAMB1 encodes 1765 amino acids, as compared to 1172 amino acids encoded by LAMB3, explaining the difference in the overall size of these two genes. In addition, the sizes of some of the introns in LAMB1 are considerably larger, >50 kb in size, as compared with those in LAMB3. However, the sizes of individual exons in certain regions of these two genes are comparable. For example, exons 3–10, which encode domains VI and V of both polypeptides, show considerable similarity in their sizes (Fig. 17). The largest exon in LAMB3 (exon 14) is 379 bp in size, and the corresponding exon in LAMB1 (exon 25) is 370 bp, both encoding the border of domains III and II. Strikingly, the sizes of exons 18–21 in LAMB3-encoding domain I sequences are precisely the same size as exons 29–32 in LAMB1 corresponding to the same peptide domain (Pulkkinen *et al.*, 1995). Comparison of LAMB1 and LAMB3 exons indicates that the coding regions corresponding to the 3′ end of domain V, the entire domain IV, and the 5′ end of domain III are missing from the LAMB3 gene (Fig. 17). Thus, although there is a deletion of a segment from the short arm of the β3 chain in comparison to the β1 chain, there is considerable conservation in the structural organization of the segments of these genes within the existing domains in the β1 and β3 polypeptides (Pulkkinen *et al.*, 1995).

The observed conservation of exon–intron organization between the LAMB1 and LAMB3 genes reflects the similarity of the overall structure, as suggested by the predicted amino acid sequence. Domains VI and V/III of β3 have approximately 42–54% homology with the β1 chain, while domains I and III in the β3 chain have about 20% homology with the corresponding sequences in the β1 chains, respectively (Pulkkinen *et al.*, 1995). The sequence conservation, although somewhat low, in domains I and II is not particularly surprising, since these domains containing the heptad repeats are responsible for the coiled-coil structure that initiates and stabilizes the interactions between the three laminin subunits (Engel, 1992). One would predict that the length of these domains could vary only slightly, if at all, without a negative effect on the molecular stability.

The sizes of exons encoding the short arm domains are less conserved, possibly reflecting the increased divergence of the amino acid sequences in this segment of β1 and β3 chains (Pulkkinen *et al.*, 1995). In particular, the truncated structure of the β3 chain is considerably different from the structure of the β1 chain. In addition, it has been suggested that laminin 5 interacts with other basement membrane molecules via mechanisms that are different from those of laminin 1. Specifically, laminin 1 has been suggested to self-assemble and to interact with a variety of other laminins to form networks, and these interactions are mediated by domain VI of the short arm of the β1 chain (Schittny and Yurchenco, 1990; Yurchenco and Schittny, 1990). In contrast, laminin 5 is thought to form covalent interactions with laminin 6 and laminin 7, and these interactions may involve the short arm of the β3 chain (Marinkovich *et al.*, 1992b).

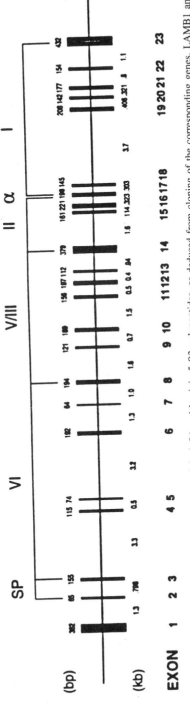

Figure 16. Comparison of the organization of laminin 1 β1 and laminin 5 β3 polypeptides, as deduced from cloning of the corresponding genes, LAMB1 and LAMB3. Note that LAMB1 consists of 34 exons, while LAMB3 has 23 exons. The central portion corresponding to β1 polypeptide has been deleted from β3 polypeptide, as shown by dotted lines. The sizes of the remaining exons, as indicated in bp within each exon, are well conserved. (Reproduced from Pulkkinen *et al.,* 1995.)

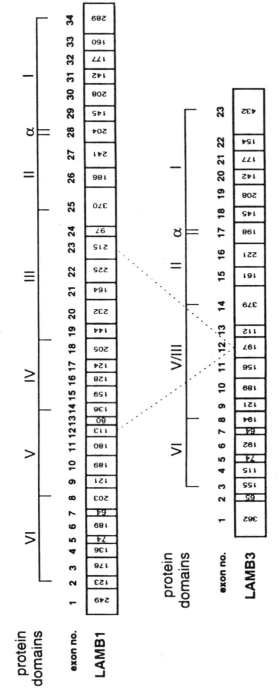

Figure 17. Intron–exon organization of the human LAMB3 gene. The gene consists of 23 exons within ~29 kb of genomic DNA. The sizes of the exons (bp) are indicated above each exon, and the approximate sizes of the introns (kb) are shown below. The domain organization of the β3 chain of laminin 5 encoded by distinct exons is indicated above the gene structure. SP, signal peptide. Roman numerals and α refer to protein domains as shown in Figure 15. (Modified from Pulkkinen *et al.*, 1995.)

3.4. Type VII Collagen: The Anchoring Fibril Protein

The collagens constitute a family of closely related yet genetically distinct proteins, and 19 different collagen types have been identified in vertebrate tissues (Kivirikko, 1993). Thus far, 11 of these genetically distinct collagens have been detected in the skin. Among the collagens, type VII collagen has a relatively limited tissue distribution, as it is found predominantly in the BMZ of stratifying squamous epithelia, i.e., in the skin, mucous membranes, and the cornea of the eye (Burgeson, 1993; Wetzels *et al.*, 1991). Type VII collagen has been demonstrated by immunolocalization studies to be a component of anchoring fibrils, morphologically distinct attachment structures extending from the lower portion of the basement membrane to the papillary dermis (Fig. 18).

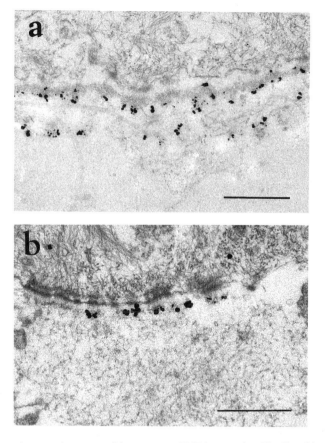

Figure 18. Immunoelectron microscopy of the cutaneous BMZ in normal and healing skin with a monoclonal antibody recognizing the amino-terminal noncollagenous (NC-1) domain of type VII collagen. (a) In normal skin, specific labeling is noted at both ends of the anchoring fibrils, one being adjacent to the lamina densa, the lower part being within the upper dermis. (b) During the early stages of wound healing, specific labeling is noted only in the lamina densa region, while the lower level of labeling by comparison with normal skin is absent, reflecting a gradual reassembly of anchoring fibrils, during tissue repair.

3.4.1. Structural Features

Type VII collagen was initially isolated from extracts of human amniotic membranes by limited pepsin proteolysis and was designated as "long-chain collagen," since it has an unusually long triple-helical region of ~450 nm (Bentz *et al.*, 1983; Burgeson, 1993). Extensive biochemical analyses at the protein level suggested that the tissue form of trimeric type VII collagen consists of only one type of α-chain, α1(VII), and that the triple-helical domain contains interchain disulfide bonds (Morris *et al.*, 1986). These studies also predicted the presence of a pepsin-sensitive, nonhelical site close to the center of the triple helix (Seltzer *et al.*, 1989). Subsequent studies examined biosynthetic products of cells that express type VII collagen, such as epidermal keratinocytes, transformed human epithelial cell lines (KB and WISH), and, to a lesser degree, dermal fibroblasts (Ryynänen *et al.*, 1991, 1992; König and Bruckner-Tuderman, 1992). These studies indicated that type VII collagen is synthesized as a procollagen, each pro-α1(VII) chain being ~350 kDa in size.

Type VII collagen is the major constituent of anchoring fibrils within the cutaneous BMZ (Sakai *et al.*, 1986; Keene *et al.*, 1987). The assembly of anchoring fibrils involves several discrete steps following the secretion of the type VII procollagen molecules into the extracellular space (Fig. 19). The assembly of the anchoring fibrils is initiated by formation of antiparallel dimers linked through their overlapping carboxy-terminal ends (Burgeson, 1993). This association is apparently stabilized by the formation of interchain disulfide bonds, and subsequently, part of the carboxy-terminal domain is proteolytically cleaved (Fig. 19). A large number of these antiparallel dimer molecules then laterally associate to form the anchoring fibril structures that can be recognized at the sublamina densa space by electron microscopy through their characteristic banding pattern (see Figs. 2, 4, and 19). In fact, the presence of the characteristic banding pattern of segment-long spacing aggregates of type VII collagen initially led to identification of type VII collagen as the major component of the anchoring fibrils (Bruns, 1969).

3.4.2. Molecular Cloning of Type VII Collagen Sequences

Elucidation of the primary sequence of pro-α1(VII) chain has been greatly enhanced by cloning of the corresponding complementary and genomic DNA sequences (Parente *et al.*, 1991; Christiano *et al.*, 1994b,c). The information on amino acid sequences as deduced from the nucleotide sequence has confirmed that each of the α chains of type VII collagen is composed of a central collagenous segment characterized by the presence of repeating Gly-X-Y sequences (Fig. 20). Examination of the amino acid sequence deduced from the full-length cDNA corresponding to the α1(VII) chain has revealed, however, that the collagenous domain is interrupted at 19 locations by imperfections or interruptions in the Gly-X-Y amino acid repeat sequence (Christiano *et al.*, 1994b). The largest one of these interruptions is a 39-amino acid noncollagenous segment close to the center of the triple-helical domain (Fig. 20). The presence of this noncollagenous segment was predicted at the protein level by pepsin digestion, which indicated cleavage of the triple-helix into two large fragments under nondenaturing conditions (Seltzer *et al.*, 1989). The interruptions within the triple-helical domain of

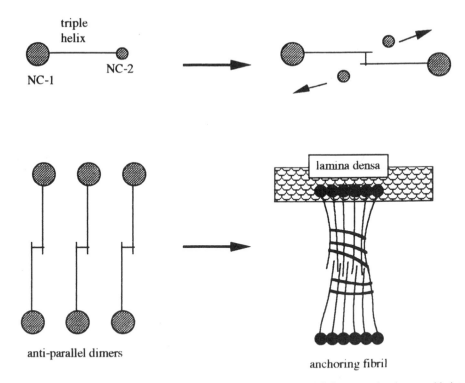

Figure 19. Schematic representation of type VII collagen molecules and their supramolecular assembly into anchoring fibrils. Type VII collagen molecules consist of a central collagenous domain flanked by amino-terminal noncollagenous (NC-1) and carboxy-terminal noncollagenous (NC-2) domains. In the extracellular space, two procollagen molecules align to form antiparallel dimers, which are stabilized by the formation of intermolecular disulfide bonds. Part of the NC-2 domain is proteolytically removed to yield stable type VII collagen antiparallel dimers. A large number of these dimer molecules laterally associate to form anchoring fibrils, which extend from the lamina densa to the papillary dermis.

type VII have been suggested to provide flexibility to the anchoring fibrils. The central collagenous segment is flanked by noncollagenous domains at both ends of the poly-peptide. The large amino-terminal NC-1 domain, ~145 kDa, consists of submodules with homology to known adhesive proteins, including the most amino-terminal seg-ment with homology to cartilage matrix protein, followed by nine consecutive fi-bronectin type III (FN-III)-like domains, and a segment with homology to the A domain of von Willebrand factor (Christiano *et al.,* 1992; Gammon *et al.,* 1992) (Fig. 20). The carboxy-terminal noncollagenous NC-2 end is relatively small, ~17 kDa, when calculated from the primary sequence, and contains a segment with homology to the Kunitz protease inhibitor molecule (Greenspan, 1993). It is unclear, however, whether this segment serves as an active inhibitor of proteases.

3.4.3. Domain Organization

Homology searches with GenBank sequences showed that the nine consecutive fibronectin type III-like repeat sequences have overall homology averaging ~23%.

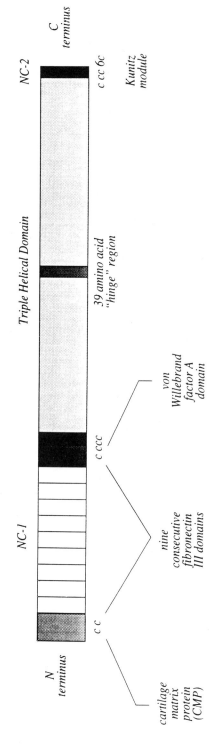

Figure 20. Domain organization of type VII collagen polypeptides, as deduced from cloned cDNA sequences. Note the central collagenous domain, which is interrupted by noncollagenous imperfections, including a 39-amino acid segment in the middle of the triple helix. The collagenous domain is flanked by the amino-terminal, noncollagenous NC-1 domain, which consists of submodules with homology to known adhesive proteins: cartilage matrix protein; fibronectin type III-like domains; von Willebrand factor A domain-like sequence; and by the carboxy-terminal NC-2 domain, which contains a Kunitz protease inhibitorlike module. Note the presence of multiple cysteine (c) residues.

Individually, these domains consist of ~90 amino acids in size and comprise two halves, each encoded by a separate exon, characteristic for the fibronectin type III domain (Christiano *et al.*, 1992). The A exon of each pair encodes 39–55 amino acids, with a conserved tryptophan residue in the approximate center of the sequence. The B exon of each pair encodes 43–50 amino acids, and a conserved octet sequence in the center of the B module is observed. This exon arrangement is also characteristic of fibronectin type III repeats, and has been found in a large number of unrelated proteins, including fibronectin, tenascin, and leukocyte antigen-related protein, among others. In addition, recent analyses of different collagens have shown the presence of fibronectin III repeats in the α3 chain of type VI collagen and the α1 chain of type XII collagen (see Christiano *et al.*, 1992).

Upstream of the FN-III domains, a 123-amino-acid segment with ~21% homology to the cartilage matrix protein sequences was found (Fig. 20). Previous comparisons have suggested that cartilage matrix protein has functional homology with von Willebrand factor (Shelton-Inloes *et al.*, 1986). Both of these sequences are characterized by the presence of two conserved cysteine residues near the edges of each domain. In addition to the cysteine residues identified at the end regions of the von Willebrand factor A-like domains, two additional cysteines were identified in the region just preceding the Gly-X-Y collagenous sequence. This 65-amino-acid segment also contains 8 prolyl residues, and this cysteine/proline-rich region is encoded by two exons.

3.4.4. Molecular Interactions

The chimeric NC-1 domain of type VII collagen consists of subdomains homologous to known adhesive proteins, suggesting that the amino-terminal end may be involved in protein–protein interactions. Specifically, it has been suggested that the NC-1 domain of type VII collagen may facilitate the binding of anchoring fibrils to other macromolecular components of the basement membrane in the lamina densa and within the anchoring plaques (Uitto *et al.*, 1992; Burgeson, 1993). Such association would stabilize the adhesion of the basement membrane to the underlying dermis, thus providing integrity to the cutaneous BMZ. This prediction has been supported by the demonstration of a large number of mutations in the type VII collagen gene in families with dystrophic forms of the inherited blistering skin disorder, epidermolysis bullosa, manifesting with fragility of the cutaneous BMZ and compromised epidermal wound healing (see Section 5).

The anchoring fibrils span from the lower part of the cutaneous basement membrane to the upper papillary dermis where they associate with structures known as anchoring plaques (Burgeson *et al.*, 1990). Anchoring plaques are basement membranelike islands within the papillary dermis, containing type IV collagen and laminin 1, ubiquitously expressed components of all basement membranes. Some of the anchoring fibrils adopt a U-shaped configuration, in which both ends containing the NC-1 domains are inserted into the basal lamina. Such anchoring fibril loops entrap the broad, banded interstitial collagen fibrils that consist primarily of type I and type III collagens. These types of supramolecular interactions allow the anchoring fibrils to stabilize the binding of the lower portion of the basement membrane to the upper part of the papillary dermis. Thus, type VII collagen plays a critical role in stabilizing the attachment at the lower portion of the cutaneous basement membrane to the underlying dermis.

3.4.5. Cloning of the COL7A1 Gene

Recently, the gene for type VII collagen (COL7A1) has been cloned, and the entire exon–intron organization characterized (Christiano *et al.*, 1994b,c). The gene is a remarkably compact, yet complex, arrangement of introns and exons. Specifically, the gene spans only ~32 kb of genomic DNA, yet it consists of 118 separate exons, the largest number of exons in any previously reported gene (Christiano *et al.*, 1994c). The gene has been mapped to the short arm of human chromosome 3, corresponding to the band 3p21 (Parente *et al.*, 1991).

4. Cytokine Modulation of BMZ Gene Expression

Several lines of evidence indicate that the different phases of the wound-healing process all potentially involve cytokines and growth factors (Mauviel and Uitto, 1993). Following injury, tissue repair takes place in an organized sequential manner (see Chapter 1, this volume). The initial fibrin clot formation is rapidly followed by invasion of the injured tissue by mononuclear cells and granulocytes, followed by angiogenesis, granulation tissue formation, fibroblast proliferation, matrix deposition, and tissue remodeling. The final step, maturation of the scar tissue, which involves the contraction of the dermal collagen network and squamous differentiation of the keratinocytes to allow reepithelialization, leads to wound closure. The inflammatory cells that gather at the wound site at the beginning of the repair process secrete a variety of soluble mediators, cytokines, and growth factors with pleiotropic activities, which include the induction of fibroblast proliferation and chemotaxis, and modulation of extracellular matrix gene expression (Mauviel and Uitto, 1993).

4.1. Proinflammatory Cytokines

Although extensive *in vitro* studies have shown that a variety of inflammatory signals produced by circulation-derived cells in tissue are capable of modulating the metabolism of extracellular matrix, only a few of these studies deal with the regulation of the expression of basement membrane components. Among the numerous soluble factors released by inflammatory cells during the wound-healing process, a number of them have been the subject of intensive characterization of their biological functions; these include proinflammatory cytokines such as interleukin-1 (IL-1) and tumor necrosis factor-α (TNF-α), several growth factors, and in particular transforming growth factor-β (TGF-β), as well as interferon-γ (IFN-γ), a T-cell-derived cytokine with potent antiviral and antiproliferative properties (for review, see Dinarello, 1993; Vilcek and Lee, 1991; Mauriel and Uitto, 1993). Although expressed by fibroblasts and keratinocytes, IL-1 and TNF-α are predominantly produced by activated macrophages, and these two cytokines play crucial roles during inflammatory processes; however, little is known about the effect of these two cytokines on the regulation of basement membrane gene expression. We have recently provided evidence that both cytokines are potent enhancers of type VII collagen gene expression in fibroblast cultures (Mauviel *et al.*, 1994). Also, we have shown, using human keratinocytes in culture, that TNF-α can selectively inhibit the expression of the subunits of laminin 5, whereas it does not appreciably affect the expression of the β3 or α2 subunits (Korang *et al.*, 1995).

4.2. Up-Regulatory Cytokines

TGF-β, the most extensively studied growth factor, belongs to a family of polypeptides secreted by a variety of cell types (Massagué, 1990; Moses *et al.*, 1990; Wahl, 1992). Aside from its immunoregulatory functions, it has been shown to stimulate extracellular matrix deposition by mechanisms that involve both enhanced matrix synthesis and reduced matrix degradation. With regard to the basement membrane components, TGF-β has been shown to enhance the expression of both BPAG-1 and BPAG-2 in epidermal keratinocytes (Sollberg *et al.*, 1992b). It is also a potent inducer of fibronectin and collagen type IV in HT1080 fibrosarcoma cells in culture, whereas it has no effect on the laminin β1 and γ1 chain or nidogen expression in the same cell type (Kähäri *et al.*, 1991). TGF-β has also been shown to enhance the expression of the type VII collagen gene in both fibroblasts and keratinocytes in culture (Ryynänen *et al.*, 1991; König and Bruckner-Tuderman, 1992; Mauviel *et al.*, 1994). Interestingly, TGF-β synergizes with TNF-α and other proinflammatory cytokines to induce the expression of type VII collagen gene expression in dermal fibroblasts in culture (Figs. 21 and 22) (Mauviel *et al.*, 1994).

We have recently demonstrated that TGF-β, as measured either at the mRNA or protein levels, is also capable of up-regulating the expression of the genes encoding the three subunits of laminin 5, α3, β3, and γ2 (Korang *et al.*, 1995). This up-regulation of laminin 5 gene expression was observed in human basal keratinocytes and HaCaT cells in culture. In addition, TGF-β has been shown to stimulate the expression of keratinocyte integrins either *in vitro* or *in vivo* during reepithelialization of cutaneous wounds (Heino *et al.*, 1989; Sollberg *et al.*, 1992a; Gailit *et al.*, 1994). Specifically, TGF-β has been shown to increase the expression of β4, β5, and αV integrins, but to have little or no effect on β1 integrin expression *in vivo* (Gailit *et al.*, 1994). Also, the expression of cell surface integrins α5 and αV was increased after TGF-β treatment. *In vivo*, during reepithelialization of cutaneous wounds, the expression of TGF-β was coordinated with increased expression of these integrin subunits in migrating keratinocytes in the adjacent epidermis. Therefore, it appears that TGF-β may induce epidermal keratinocytes to express integrins during reepithelialization, therefore facilitating the migratory component of reepithelialization.

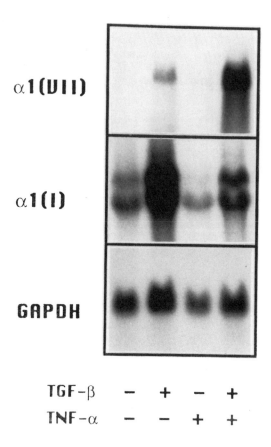

α1(UII)

α1(I)

GAPDH

| TGF-β | – | + | – | + |
| TNF-α | – | – | + | + |

Figure 21. Demonstration that TGF-β and TNF-α synergistically enhance type VII collagen gene expression, while TNF-α counteracts the TGF-β activation of type I collagen gene expression. Normal human fibroblasts in culture were incubated with (+) or without (−) TGF-β (5 ng/ml) or TNF-α (10 ng/ml), as indicated. Total RNA was isolated, and Northern hybridizations were performed with α 1(VII) and α 1(I) collagen cDNAs. GAPDH cDNA was used for control hybridizations. Note that TGF-β markedly enhances the expression of both type VII and type I collagen mRNAs. Combination of TGF-β and TNF-α further enhances type VII collagen gene expression, while the effect of TGF-β on the α1(I) mRNA levels is abrogated by TNF-α. (Reproduced from Mauviel *et al.*, 1994.)

4.3. Interferon-γ

Contrasting with the stimulatory effects of TGF-β on BPAG-1 and BPAG-2 gene expression (Sollberg *et al.*, 1992b), IFN-γ was recently shown to inactivate the transcription of the 230-kDa BPAG gene (BPAG-1). This inhibitory effect of IFN-γ was specific for BPAG-1 and did not affect the expression of BPAG-2 (Tamai *et al.*, 1995a). Inhibition of BPAG-1 gene expression was demonstrated at the mRNA level by North ern analysis, at the protein level by indirect immunofluorescence using a monoclonal antibody recognizing human BPAG-1, and by transient cell transfections of human keratinocytes with BPAG-1 promoter–CAT reporter gene constructs. Reduced tran-

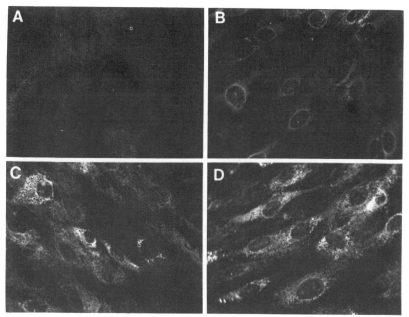

Figure 22. Indirect immunofluorescence staining of keratinocytes with a monoclonal antitype VII collagen antibody of normal keratinocyte cultures (A), or after incubation with TNF-α (B), TGF-β (C), or combination of these two growth factors (D). Note that TNF-α and TGF-β (B and C) alone enhance type VII collagen expression, and combination of these growth factors (D) further increases the protein expression. (Reproduced from Mauviel *et al.*, 1994.)

scription of the BPAG-1 gene by INF-γ was also demonstrated by *in vitro* nuclear run-on assays (Tamai *et al.*, 1995a).

4.4. The Cytokine Network

The use of highly purified or recombinant growth factors and cytokines in *in vitro* experiments provides a convenient, carefully controlled system to analyze the effect of a single factor on the expression of extracellular matrix and BMZ genes. It should be noted, however, that these conditions do not necessarily reflect *in vivo* situations where a variety of cytokines and growth factors may be present simultaneously at the sites of inflammation and the healing wound. Up to now, however, relatively little has been known about the interactions of cytokines that modulate the expression of the genes encoding basement membrane components. As indicated in Section 4.2, TGF-β and TNF-α, which often antagonize each other in regulating the expression of extracellular matrix genes, such as those encoding type I collagen, elastin, and fibronectin (Mauviel and Uitto, 1993), can have additive or synergistic effects to stimulate the expression of basement membrane protein genes such as that encoding type VII collagen (Mauviel *et al.*, 1994).

4.5. Role of Matrix Metalloproteases

Aside from direct modulation of the extracellular matrix and BMZ gene expression, cytokines and growth factors can also regulate the deposition of these macromolecules through the modulation of the expression of matrix metalloproteinases (MMPs) (Mauviel, 1993). Three classes of MMPs have been determined, as defined by their specific substrates: collagenases, stromelysins, and gelatinases (Table III) (Mauviel, 1993). Each member of the MMP family share strong structural homologies at both amino acid and DNA sequence levels. They are secreted as zymogens and require activation for proteolytic activity. Collagenases degrade exclusively fibrillar components of the extracellular matrix, such as type I and III collagens, whereas stromelysins and gelatinases have a broad spectrum of activity and can degrade numerous macromolecules such as fibronectin, elastin, gelatins, type IV collagen, laminins, and proteoglycans (Table III). Therefore, these last two classes of MMPs can degrade basement membranes. The gene structure of stromelysin 1 (MMP-3) has been exten-

Table III. The Family of Matrix Metalloproteinases

MMP No.	Names	Mol. wt. (kDa)	ECM substrates	Distribution
	Collagenases			
1	Interstitial collagenase (human/rabbit)	52.5	Fibrillar collagens (I, II, III, V, VII, X)	Connective tissue cells Monocyte/macrophages Endothelial cells
8	Neutrophil collagenase	75	Fibrillar collagens (I, II, III) Gelatin	Neutrophils (PMN)
	Gelatinases			
2	72-kDa gelatinase (gelatinase A) Type IV collagenase	72	Gelatin Collagens IV, V, VII, X, XI FN, elastin	Most cell types Tumor cells
9	92-kDa gelatinase (gelatinase B)	92	Same as gelatinase A	Connective tissue cells Monocyte/macrophages Tumor cells
	Stromelysins			
3	Stromelysin-1 (rat transin)	57/60	Proteoglycans, FN Laminin Collagens III, IV, V, IX Gelatins Activates MMP-1	Same MMP-1 Tumor cells
10	Stromelysin-2 (rat transin-2)	53	Same as MMP-3 but lower activity	Macrophages Tumor cells
7	Matrilysin (pump-1)	28	Gelatins FN, proteoglycans Collagen IV?, elastin	Immature monocytes Connective tissue cells Tumor cells
	Others			
11	Stromelysin-3	54.6[a]	Unknown	Stromal cells of tumors
12	Metalloelastase	21	Elastin, FN	Macrophages

[a]The molecular weight has been estimated from the cDNA.

sively studied together with the regulation of its expression. Proinflammatory cytokines, such as IL-1 and TNF-α, have been shown to stimulate the expression of stromelysin by two distinct mechanisms: (1) increased promoter activity leading to transcriptional activation of the gene, and (2) increased mRNA stability. On the other hand, TGF-β has been shown to reduce stromelysin expression and activity in cultured cells. This inhibition results from both reduced stromelysin gene expression at the transcriptional level and increased tissue inhibitor of metalloproteinases (TIMP) gene expression (Edwards *et al.,* 1987).

TGF-β has been shown to up-regulate both 92-kDa and 72-kDa gelatinase activities in cultured fibroblasts and keratinocytes, thus contrasting with its inhibitory action on MMP-3 gene expression (Overall *et al.,* 1991; Salo *et al.,* 1991). The effect of TGF-β on gelatinase gene expression appears to result from both elevation of the corresponding mRNA levels through transient activation of transcription, as well as from increased stability of the mRNA.

Contrasting with the up-regulatory effect of TGF-β on gelatinase gene expression, IL-4, a potent suppressor of monocytic functions as well as a B-cell growth factor, suppresses the synthesis of 92-kDa gelatinase by human alveolar macrophages without affecting the expression of TIMP (Lacraz *et al.,* 1992). The effect of IL-4 on MMP gene expression takes place at the pretranslational level by reduction of the corresponding mRNA steady-state levels, which may be a prostaglandin E_2-mediated, cAMP-dependent mechanism (Corcoran *et al.,* 1992). IFN-γ also suppresses constitutive expression of macrophage 92-kDa gelatinase (Shapiro *et al.,* 1990).

5. Relevance of BMZ Regeneration to Wound Healing

It is clear that the presence of functionally intact BMZ components, which assemble into discrete dermal–epidermal attachment structures, is necessary for the integrity of the skin as a protective organ. During reparative wound-healing processes, the synthesis of individual BMZ components, under the regulation of cytokines and growth factors, has to be balanced with ongoing degradation and turnover. Furthermore, the expression of individual genes has to be coordinated to allow assembly of the subunit polypeptides and component proteins into supramolecular structures. Thus, there are several critical steps necessary for the successful progression of epidermal wound healing. The importance of individual components of the BMZ in providing integrity to the dermal–epidermal junction is illustrated by the group of heritable diseases, collectively known as epidermolysis bullosa, which manifest with fragility and compromised wound healing of the skin.

5.1. Epidermolysis Bullosa: A Prototypic Heritable Disease of Compromised Wound Healing

Epidermolysis bullosa (EB) is a group of genodermatoses characterized by fragility and easy blistering of the skin and the mucous membranes (Fine *et al.,* 1991;

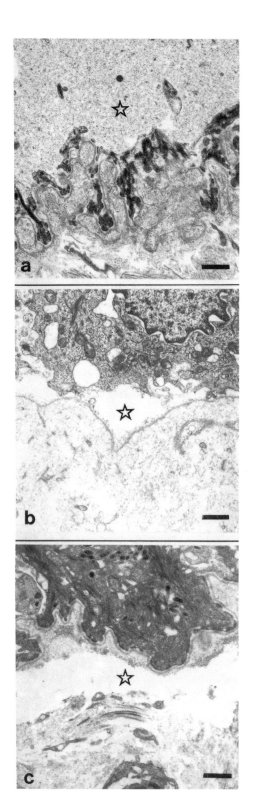

Figure 23. Demonstration of the level of tissue separation in different forms of epidermolysis bullosa, the blister cavity being indicated by a star (☆). (a) In the simplex forms of EB, blister formation occurs within the basal keratinocytes. (b) In the junctional forms, tissue cleavage occurs within the lamina lucida of the dermal–epidermal basement membrane, with intact lamina densa being recognized at the floor of the blister. (c) In the dystrophic forms of EB, blister formation occurs within the papillary dermis just below the lamina densa. Bars, 1 μm.

Eady *et al.,* 1994; Uitto and Christiano, 1994; Uitto *et al.,* 1994). Although the initial blistering is in response to mechanical trauma, healing of erosions is compromised by continued fragility of the skin, and in some cases, by excessive granulomatous or fibrotic tissue response. Recent advances in the molecular genetics of EB, with identification of specific mutations in different forms of this disease, have provided evidence for the functional importance of different BMZ components in providing integrity to the dermal–epidermal junction (Uitto and Christiano, 1992).

In the simplest form of classification, EB can be divided into three major categories on the basis of clinical observations and the level of blister formation, as determined by transmission electron microscopy (Fig. 23). In the simplex forms of EB, blister formation occurs at the level of basal keratinocytes, and erosions generally heal without significant scarring. In the junctional forms of EB, tissue separation occurs within the dermal–epidermal basement membrane, at the level of the lamina lucida. In the dystrophic forms of EB, tissue separation occurs below the basement membrane within the papillary dermis at the level of anchoring fibrils, and healing of the lesion results in extensive scarring.

5.2. EB Simplex

Recent studies have clearly established that the simplex forms of EB, usually inherited in an autosomal dominant fashion, are due to mutations in the genes encoding the basal keratins, KRT5 and KRT14 (Bonifas *et al.,* 1991; Coulombe *et al.,* 1991; Epstein, 1992). Most of these mutations are single-base substitutions that result in the formation of a perturbed keratin intermediate filament network by dominant negative interference. As a consequence, the basal keratinocytes are fragile, and minor trauma to the skin leads to basal cell cytolysis and blister formation just above the basement membrane. The consequences of these single-base substitutions have also been verified through generation of transgenic mice with keratin mutations, which similarly demonstrate blister formation at the level of basal keratinocytes (Vassar *et al.,* 1991). Furthermore, *in vitro* reconstitution of the mutated keratin filaments containing the amino acid substitutions have unequivocally demonstrated the role of these mutations as the cause of tissue fragility in the simplex forms of EB (Fuchs and Coulombe, 1992).

5.3. Junctional EB

The junctional forms of EB are characterized by tissue separation within the dermal–epidermal basement membrane (Fig. 22), and the lamina lucida protein, laminin 5, was initially implicated as the candidate protein in most cases, based on immunofluorescence analysis (Verrando *et al.,* 1991; Meneguzzi *et al.,* 1992; Baudoin *et al.,* 1994). However, evidence for BPAG-2 as the candidate gene has also been presented (Jonkman *et al.*, 1995). Recent cloning of the laminin 5 polypeptides, α3, β3, and γ2 chains, has allowed initiation of mutation detection in the corresponding genes, LAMA3, LAMB3, and LAMC2, respectively (Uitto *et al.*, 1994). Amplification of genomic DNA or mRNA sequences by PCR, followed by the application of electrophoretic mutation scanning techniques (Ganguly *et al.*, 1993), has revealed a large number of mutations in the three genes encoding each of the constitutive polypeptides of laminin 5 (Pulkkinen *et al.*, 1994a,b; Vailly *et al.*, 1995a,b; McGrath *et al.*, 1995a,b; Kivirikko *et al.*, 1995). Most of these mutations are premature termination codons, predicting synthesis of truncated, nonfunctional polypeptides. Since synthesis of a full-length polypeptide of each laminin 5 subunit is a prerequisite for the formation of the functional heterotrimeric protein molecule, such mutations could explain the absence of laminin 5 epitopes from the skin in affected individuals. Interestingly, specific mutations have also been detected in the gene encoding the β4 subunit of the α6β4 integrin in a patient with a junctional EB variant, which in addition to characteristic cutaneous lesions has pyloric atresia (Vidal *et al.*, 1995). Furthermore, we have recently disclosed mutations in the BPAG-2 gene in patients with generalized atrophine benign epidermolysis bullosa, a variant of nonlethal junctional EP (McGrath *et al,.* 1995c). These observations on junctional EB attest to the importance of hemidesmosome-anchoring filament proteins, including laminin 5, BPAG-2, and the α6β4 integrin, in providing integrity to keratinocyte–basement membrane adhesion.

5.4. Dystrophic EB

The dystrophic forms of EB, which are characterized by abnormalities in the anchoring fibrils, including altered morphology, paucity or even complete absence, are due to mutations in type VII collagen (Bruckner-Tuderman *et al.*, 1989; Christiano *et al.*, 1993, 1994d; McGrath *et al.*, 1993; Uitto and Christiano, 1994; Uitto *et al.*, 1994, 1995). The most severe forms of recessive dystrophic EB, the mutilating Hallopeau-Siemens type (HS-RDEB), is characterized by excessive scarring in the affected areas of skin. In this form of EB, the predominant genetic lesion is a premature termination codon of translation in the type VII collagen gene, predicting a truncated, nonfunctional polypeptide (Christiano *et al.*, 1994a, 1995; Hilal *et al.*, 1993; Hovnanian *et al.*, 1994). In addition, since the premature termination codons of translation frequently lead to reduced levels of mRNA transcripts containing the mutation (Urlaub *et al.*, 1989; McIntosh *et al.*, 1993; Belgrader and Maquat, 1994), these patients' cells are unable to synthesize any normal type VII collagen molecules and to assemble functional anchoring fibrils. This finding would explain the extreme fragility and poor epidermal wound healing in these patients. The extensive scarring in HS-RDEB patients may reflect overproduction of collagens type I and III by dermal fibroblasts, but the precise cellular mechanisms resulting in scarring and complicating the wound-healing processes are currently unknown. Nevertheless, the illustration of mutations in the type VII collagen gene highlights the importance of this protein in the formation of anchoring fibrils during fetal skin development and in reparative wound healing processes.

6. BMZ Repair Processes in Normal Wound Healing

Following tissue injury and disruption of the basement membrane, repair processes take place in a sequential, organized, and specific manner. The orchestration of the deposition of the various components within the BMZ is under the control of a variety of cytokines and growth factors (Mauviel and Uitto, 1993; Mauviel *et al.*, 1993; Chapter 1, this volume). *In vitro* experiments have shown that TGF-β, for example, can stimulate the expression of cell surface integrins, therefore allowing migration of keratinocytes on the wound bed. Also, TGF-β has been shown to up-regulate the expression of core components of the basement membrane, type IV collagen, and laminin 1 genes (Kähäri *et al.*, 1991). It also induces the expression of cell–matrix attachment molecules, such as α6β4 integrin and BPAGs. These molecules, linked to the hemidesmosomal structures, allow the attachment of basal keratinocytes to the underlying basement membrane. It should be noted that the basement membrane is a polarized structure, attached to the dermis by means of anchoring fibrils composed of type VII collagen. The expression of these macromolecules is strongly up-regulated *in vitro* by TGF-β and, in the case of type VII collagen, by proinflammatory cytokines such as TNF-α and IL-1. It therefore could be anticipated that a concerted action of various cytokines and growth factors during the wound-healing process will accelerate the deposition of attachment structures in the BMZ, leading to stabilization of the newly repaired tissue.

The assembly of supramolecular attachment structures during cutaneous wound healing is preceded by activation of the synthesis of the corresponding protein components either by migrating keratinocytes or adjacent fibroblasts. Such delay in the assembly of the attachment structures is illustrated by analyses of type VII collagen gene expression during the wound-healing processes. Specifically, as shown in Fig. 24A, type VII collagen epitopes demonstrate a linear immunofluorescent staining pattern at the dermal–epidermal junction in normal human skin. In wounded skin, the corresponding epitopes are present during the early stages of wound healing predominantly within the follicular epidermis, while immunostaining is reduced and patchy in the interfollicular epidermis (Fig. 24B). Examination of the skin at the same stage of wound healing by immunoelectron microscopy, utilizing a monoclonal antibody that recognizes the NC-1 domains of type VII collagen, demonstrates the presence of epitopes only in a local distribution adjacent to the basement membrane at the level of the regenerating lamina densa (Fig. 18B). This is in a distinct contrast with the immunoelectron microscopy appearances of normal wounded skin, which demonstrate two distinct layers of gold particles, corresponding to NC-1 domains adjacent to the lamina densa at one end of the anchoring fibrils and within the papillary dermis that correspond to the anchoring plaques at the other end. Similarly, staining of normal human skin with anti-type IV collagen antibodies reveals a linear staining pattern at the

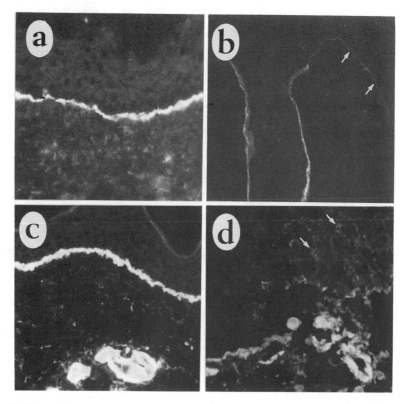

Figure 24. Immunofluorescence staining of normal human skin (a and c) or wounded skin during early stages of repair (b and d) with antibodies recognizing type VII collagen (a and b) or type IV collagen (c and d) epitopes. Note the presence of a linear staining pattern for both of these collagens at the dermal–epidermal junction of the normal skin, while the protein expression at the early stages of wound healing is patchy (b and d, arrows). However, in the case of type VII collagen (b), bright staining persists within the follicular basement membrane. In the case of type IV collagen (c and d), intradermal blood vessels are recognized by bright immunofluorescent staining. Also, note the presence of weak, yet clearly detectable, immunosignal within the basal keratinocytes.

dermal–epidermal junction. In addition to the cutaneous BMZ, other basement membranes, including those associated with dermal blood vessels, are also recognized (Fig. 24C). During the early stages of epidermal wound healing, type IV collagen epitopes are present at the dermal–epidermal junction in a discontinuous and markedly attenuated pattern. These findings are corroborated by electron microscopy (Fig. 18B), which also ultrastructurally demonstrates a patchy, poorly organized lamina densa. It is of interest to note that weak, yet clearly detectable, immunostaining is found in association with epidermal keratinocytes during the early wound healing (Fig. 24D). These observations suggest that epidermal keratinocytes are responsible for the synthesis of type IV collagen. This observation is consistent with the previous demonstration that human epidermal keratinocytes in culture are capable of expressing the type IV collagen genes (Olsen and Uitto, 1989). The delay between the activation of gene expression of individual BMZ components and the assembly of recognizable attachment structures may explain the continued fragility of the epidermis observed clinically for extended periods after apparently complete reepithelialization of the wounds.

7. Future Prospects

Further understanding of the regulatory role of growth factors and cytokines *in vitro,* as well as development of *in vivo* models of wound healing controllable in a spatially and temporally regulated manner, is essential to determine the exact significance of various mechanistic steps at the different stages of repair of the dermal–epidermal junction. These studies will eventually lead to the development of cytokine/ growth factor-based pharmacological treatments to correct or accelerate the wound repair processes or counteract age-associated alterations of skin integrity such as increased fragility and impaired wound-healing tendency as observed in the elderly individuals (Chen *et al.,* 1994).

ACKNOWLEDGMENTS. The authors thank Tamara Alexander for superb assistance in preparation of this chapter. The original studies by the authors summarized in this chapter were supported by the United States Public Health Service, National Institutes of Health grants P01-AR38923 and R01-AR41439. Dr. Mauviel was recipient of a Research Career Development Award from the Dermatology Foundation. Dr. McGrath was supported by Research Fellowships from the Dermatology Foundation and the Dystrophic Epidermolysis Bullosa Research Association of the United Kingdom.

References

Albelda, S. M., and Buck, C. A., 1990, Integrins and other cell adhesion molecules, *FASEB J.* **4**:2668–2880.

Baudoin, C., Miquel, C., Blanchet-Bardon, C., Gambini, G., Meneguzzi, G., and Ortonne, J.-P., 1994, Herlitz junctional epidermolysis bullosa keratinocytes display heterogeneous defects of nicein/kalinin gene expression, *J. Clin. Invest.* **93**:862–869.

Beck, K., Hunter, I., and Engel, J., 1990, Structure and function of laminin: Anatomy of a multidomain glycoprotein, *FASEB J.* **4**:148–160.

Belgrader, P., and Maquat, L. E., 1994, Nonsense but not missense mutations can decrease the abundance of nuclear mRNA for the mouse major urinary protein, while both types of mutations can facilitate exon-skipping, *Mol. Cell. Biol.* **14**:6326–6336.

Bentz, H., Morris, N. P., Murray, L., Sakai, L. Y., Hollister, D. W., and Burgeson, R. E., 1983, Isolation and partial characterization of a new human collagen with an extended triple helical structural domain, *Proc. Natl. Acad. Sci. USA* **80**:3168–3172.

Blessing, M., Zentgraf, H., and Jorcano, J. L., 1987, Differentially expressed bovine cytokeratin genes. Analysis of gene linkage and evolutionary conservation of 5'-upstream sequences, *Eur. Mol. Biol. Organ. J.* **6**:567–575.

Blessing, M., Jorcano, J. L., and Franke, W. W., 1989, Enhancer elements directing cell-type-specific expression of cytokeratin genes and changes of the epithelial cytoskeleton by transfections of hybrid cytokeratin genes, *Eur. Mol. Biol. Organ. J.* **8**:117–126.

Bonifas, J. M., Rothman, A. L., and Epstein, Jr., E. H., 1991, Epidermolysis bullosa simplex: Evidence in two families for keratin gene abnormalities, *Science* **254**:1202–1205.

Bruckner-Tuderman, L., Mitsuhashi, Y., Schnyder, U. W., and Bruckner, P., 1989, Anchoring fibrils and type VII collagen are absent from skin in severe recessive dystrophic epidermolysis bullosa, *J. Invest. Dermatol.* **93**:3–9.

Bruns, R. R., 1969, A symmetrical extracellular fibril, *J. Cell Biol.* **42**:418–430.

Buck, C. A., and Horwitz, 1987, Cell surface receptors for extracellular matrix molecules, *Annu. Rev. Cell. Biol.* **3**:179–205.

Burgeson, R. E., 1993, Type VII collagen, anchoring fibrils, and epidermolysis bullosa, *J. Invest. Dermatol.* **101**:252–255.

Burgeson, R. E., Lunstrum, G. P., Rokosova, B., Rimberg, C. S., Rosenbaum, L. M., and Keene, D. R., 1990, The structure and function of type VII collagen, *Ann. NY Acad. Sci.* **580**:32–43.

Burgeson, R. E., Chiquet, M., Deutzmann, R., Ekblom, P., Engel, J., Kleinman, H., Martin, G. R., Meneguzzi, G., Paulsson, M., Sanes, J., Timpl, R., Tryggvason, K., Yamada, Y., and Yurchenco, P. D., 1994, A new nomenclature for laminins, *Matrix Biol.* **14**:209–211.

Carter, W. G., Kaur, P., Gil, S. G., Gahr, P. J., and Wayner, E. A., 1990, Distinct functions for integrins $\alpha3\beta1$ in focal adhesions and $\alpha6\beta4$/bullous pemphigoid antigen in a new stable anchoring contact (SAC) of keratinocytes: Relation to hemidesmosomes, *J. Cell Biol.* **111**:3141–3154.

Carter, W. G., Ryan, M. C., and Gahr, P. J., 1991, Epiligrin, a new cell adhesion ligand for integrin $\alpha3\beta1$ in epithelial basement membranes, *Cell* **65**:599–610.

Chen, Y. Q., Mauviel, A., Ryynänen, J., Sollberg, S., and Uitto, J., 1994, Type VII collagen gene expression by human skin fibroblasts and keratinocytes in culture. Influence of donor age and cytokine responses, *J. Invest. Dermatol.* **102**:205–209.

Christiano, A. M., and Uitto, J., 1992, Polymorphism of the human genome: Markers for genetic linkage analyses in heritable diseases of the skin, *J. Invest. Dermatol.* **99**:519–523.

Christiano, A. M., Rosenbaum, L. M., Chung-Honet, L., Parente, M. G., Woodley, D. T., Pan, T.-C., Zhang, R. Z., Chu, M.-L., Burgeson, R. E., and Uitto, J., 1992, The large non-collagenous domain (NC-1) of type VII collagen is amino terminal and chimeric. Homology to cartilage matrix protein, the type III domain of fibronectin and the A domain of von Willebrand factor, *Hum. Molec. Genet.* **7**:475–481.

Christiano, A. M., Greenspan, D. S., Hoffman, G. G., Zhang, X., Tamai, Y., Lin, A. N., Dietz, H. C., Hovnanian, A., and Uitto, J., 1993, A missense mutation in type VII collagen in two affected siblings with recessive dystrophic epidermolysis bullosa, *Nature Genet.* **4**:62–66.

Christiano, A. M., Anhalt, G., Gibbons, S., Bauer, E. A., and Uitto, J., 1994a, Premature termination codons in the type VII collagen gene (COL7A1) underlie severe, mutilating recessive dystrophic epidermolysis bullosa, *Genomics* **21**:160–168.

Christiano, A. M., Greenspan, D. S., Lee, S., and Uitto, J., 1994b, Cloning of human type VII collagen. Complete primary sequence of the $\alpha1$(VII) chain and identification of intragenic polymorphisms, *J. Biol. Chem.* **269**:20256–20262.

Christiano, A. M., Hoffman, G. G., Chung-Honet, L. C., Lee, S., Cheng, W., Uitto, J., and Greenspan, D. S., 1994c, Structural organization of the human type VII collagen gene (COL7A1), comprised of more exons than any previously characterized gene, *Genomics* **21**:169–179.

Christiano, A. M., Ryynänen, M., and Uitto, J., 1994d, Dominant dystrophic epidermolysis bullosa: Identification of a glycine-to-serine substitution in the triple-helical domain of type VII collagen, *Proc. Natl. Acad. Sci. USA* **91**:3549–3553.

Christiano, A. M., Suga, Y., Greenspan, D. S., Ogawa, H., and Uitto, J., 1995, Premature termination codons on both alleles of the type VII collagen gene (COL7A1) in three brothers with recessive dystrophic epidermolysis bullosa, *J. Clin. Invest.* **95**:1328–1334.

Corcoran, M. L., Stetler-Stevenson, W. G., Brown, P. D., and Wahl, L. M., 1992, Interleukin-4 inhibition of prostaglandin E_2 synthesis blocks interstitial collagenase and 92-kDa type IV collagenase/gelatinase production by human monocytes, *J. Biol. Chem.* **267**:515–519.

Coulombe, P. A., Hutton, M. E., Letai, A., Hebert, A., Paller, A. S., and Fuchs, E., 1991, Point mutations in human keratin 14 genes of epidermolysis bullosa simplex patients: Genetic and functional analyses, *Cell* **66**:1301–1311.

Cripe, T. P., Haugen, T. H., Turk, J. P., Tabatabai, F., Schmid, P. G., II, Durst, M., Gissman, L., Roman, A., and Turek, L. P., 1987. Transcriptional regulation of the human papilloma virus-16 E6-E7 promoter by a keratinocyte-dependent enhancer, and by viral E2 *trans*-activator and repressor gene products: Implications for cervical carcinogenesis, *Eur. Mol. Biol. Organ. J.* **6**:3745–3753.

DeLuca, M., Tamura, R. M., Kajiji, S., Bondanza, S., Rossino, P., Cancedda, R., Marchisio, P. C., and Quaranta, V., 1990, Polarized integrin mediates human keratinocyte adhesion to basal lamina, *Proc. Natl. Acad. Sci. USA* **87**:6888–6892.

Diaz, L., Ratrie, III, H., Saunders, W., Futamara, S., Squiquera, H. L., Anhalt, G. J., and Giudice, G. J., 1990, Isolation of a human epidermal cDNA corresponding to the 180-kd autoantigen recognized by bullous pemphigoid and herpes gestationis sera: Immunolocalization of this protein to the hemidesmosome, *J. Clin. Invest.* **86**:1088–1094.

Dinarello, C. A., 1993, The role of interleukin-1 in disease, *N. Engl. J. Med.* **328**:106–113.

Eady, R. A. J., 1988, The basement membrane. Interface between the epithelium and the dermis: Structural features, *Arch. Dermatol.* **124**:709–712.

Eady, R. A. J., McGrath, J. A., and McMillan, J., 1994, Ultrastructural clues to genetic disorders of skin: The dermal–epidermal junction, *J. Invest. Dermatol.* **103**:13S–18S.

Edwards, D. R., Murphy, G., Reynolds, J. J., Whitham, S. E., Docherty, A. J. P., Angel, P., and Heath, J. K., 1987, Transforming growth factor-beta modulates the expression of collagenase and metalloprotease inhibitor, *Eur. Mol. Biol. Organ. J.* **6**:1899–1904.

Engel, J., 1992, Laminins and other strange proteins, *Biochemistry* **31**:10643–10651.

Epstein, Jr., E. H., 1992, Molecular genetics of epidermolysis bullosa, *Science* **256**:799–803.

Fine, J.-D., Bauer, E. A., Briggaman, R. A., Carter, D. M., Eady, R. A. J., Esterly, N. B., Holbrook, K. A., Hurwitz, S., Johnson, L., Lin, A., Pearson, R., and Sybert, V. P., 1991, Revised clinical and laboratory criteria for subtypes of epidermolysis bullosa, *J. Am. Acad. Dermatol.* **24**:119–135.

Fuchs, E., and Coulombe, P. A., 1992, Of mice and men: Genetic skin diseases of keratin, *Cell* **69**:899–902.

Fukushima, Y., Pikkarainen, T., Kallunki, T., Eddy, R. L., Byers, M. G., Hayley, L. L., Henry, W. M., Tryggvason, K., and Shows, T. B., 1988, Isolation of human laminin B2 (LAMB2) cDNA clone and alignment of the gene to the chromosome 1q25–31 region, *Cytogen. Cell Genet.* **48**:137–141.

Gailit, J., Welsh, M. P., and Clark, R. A. F., 1994, TGF-β stimulates expression of keratinocyte integrins during re-epithelialization of cutaneous wounds, *J. Invest. Dermatol.* **102**:221–227.

Gammon, W. R., Abernethy, M. L., Padilla, K. M., Prisayanh, P. S., Cook, M. E., Wright, J., Briggaman, R. A., and Hunt, S. W., 1992, Noncollagenous (NC1) domain of collagen VII resembles multidomain adhesion proteins involved in tissue-specific organization of extracellular matrix, *J. Invest. Dermatol.* **99**:691–696.

Ganguly, A., Rock, M. J., and Prockop, D. J., 1993, Conformation-sensitive gel electrophoresis for rapid detection of single-base differences in double-stranded PCR products and DNA fragments: Evidence for solvent-induced bends in DNA heteroduplexes, *Proc. Natl. Acad. Sci. USA* **90**:10325–10329.

Gerecke, D. R., Wagman, D. W., Champliaud, M. F., and Burgeson, R. E., 1994, The complete primary structure for a novel laminin chain, the laminin Blk chain, *J. Biol. Chem.* **269**:11073–11080.

Gil, S. A., Brown, T. A., Ryan, M. C., and Carter, W. G., 1994, Junctional epidermolysis bullosa: Defects in expression of epiligrin/nicein/kalinin and integrin-β4 that inhibit hemidesmosome formation, *J. Invest. Dermatol.* **103**:315–385.

Giudice, G. J., Squiquera, H. L., Elias, P. M., and Diaz, L. A., 1991, Identification of two collagen domains within the bullous pemphigoid autoantigen, BP180, *J. Clin. Invest.* **87**:734–738.

Giudice, G. J., Emery, D. J., and Diaz, L. A., 1992, Cloning and primary structural analysis of the bullous pemphigoid autoantigen, BP180, *J. Invest. Dermatol.* **99**:243–250.

Giudice, G. J., Emery, D. J., Zelickson, B. D., Anhalt, G. J., Liu, Z., and Diaz, L. A., 1993, Bullous pemphigoid and herpes gestationis autoantibodies recognize a common non-collagenous site of the BP180 ectodomain, *J. Immunol.* **151**:5742–5750.

Greenspan, D. S., 1993, The carboxy-terminal half of type VII collagen, including the non-collagenous NC-2 domain and intron/exon organization of the corresponding region of the COL7A1 gene, *Hum. Mol. Genet.* **2**:273–278.

Heino, J., Ignotz, R. A., Hemler, M. E., and Massague, J., 1989, Regulation of cell adhesion receptors by transforming growth factor-β. Concomitant regulation of integrins that share a common β1 subunit, *J. Biol. Chem.* **264:**380–388.

Hilal, H., Rochat, A., Duquesnoy, P., Blanchet-Bardon, C., Wechsler, J., Martin, N., Christiano, A. M., Barrandon, Y., Uitto, J., Goossens, M., and Hovnanian, A., 1993, A homozygous insertion-deletion in the type VII collagen gene (COL7A1) in Hallopeau-Siemens dystrophic epidermolysis bullosa, *Nature Genet.* **5:**287–293.

Hopkinson, S. B., Riddelle, K. S., and Jones, J. C. R., 1992, Cytoplasmic domain of the 180-kDa bullous pemphigoid antigen, a hemidesmosomal component: Molecular and cell biologic characterization, *J. Invest. Dermatol.* **99:**264–270.

Hovnanian, A., Hilal, L., Blanchet-Bardon, C., de Prost, Y., Christiano, A. M., Uitto, J., and Goossens, M., 1994, Recurrent nonsense mutations within type VII collagen in patients with severe mutilating recessive dystrophic epidermolysis bullosa, *Am. J. Hum. Genet.* **55:**289–296.

Jones, J. C. R., Kurpakus, M. A., Cooper, H. M., and Quaranta, V., 1991, A function for the integrin α6β4 in the hemidesmosome, *Cell Regul.* **2:**437–438.

Jonkman, M. F., de Jong, M. C. J. M., Heeres, K., Pas, H. H., Bruins, S., van der Meer, J. B., Niessen, C. M., and Sonnenberg, A., 1995, 180-kD bullous pemphigoid antigen (BPAG2) is not expressed in generalized atrophic benign epidermolysis bullosa, *J. Clin. Invest.* **95:**1345–1352.

Kähäri, V.-M., Peltonen, J., Chen, Y. Q., and Uitto, J., 1991, Differential modulation of basement membrane gene expression in human fibrosarcoma HT-1080 cells by transforming growth factor-β1: Enhanced type IV collagen gene expression correlates with altered culture phenotype of the cells, *Lab. Invest.* **64:**807–818.

Kallunki, P., Sainio, K., Eddy, R., Byers, M., Kallunki, T., Sariola, H., Beck, K., Hirvonen, H., Shows, T. B., and Tryggvason, K., 1992, A truncated laminin chain homologous to the B2 chain: Structure, spatial expression and chromosomal assignment, *J. Cell Biol.* **119:**679–693.

Keene, D. R., Sakai, L. Y., Lunstrum, G. P., Morris, N. P., and Burgeson, R. E., 1987, Type VII collagen forms an extended network of anchoring fibrils, *J. Cell Biol.* **104:**611–621.

Kivirikko, K. I., 1993, Collagens and their abnormalities in a wide spectrum of diseases, *Ann. Med.* **25:** 113–126.

Kivirikko, S., McGrath, J. A., Aberdam, D., Ciatti, S., Baudoin, C., Dunnill, M. G. S., McMillan, J. R., Eady, R. A. J., Ortonne, J.-P., Meneguzzi, G., Uitto, J., and Christiano, A. M., 1995, A homozygous nonsense mutation in the α3 chain gene of laminin 5 (LAMA3) in lethal (Herlitz) junctional epidermolysis bullosa, *Hum. Mol. Genet.* **4:**959–962.

König, A., and Bruckner-Tuderman, L., 1992, Transforming growth factor-β stimulates collagen VII expression by cutaneous cells *in vitro, J. Cell Biol.* **117:**679–685.

Korang, K., Christiano, A. M., Uitto, J., and Mauviel, A., 1995, Differential cytokine regulation of the genes encoding the three subunits of laminin 5 (LAMA3, LAMB3, LAMC2) in epidermal keratinocytes, *FEBS Lett.,* **368:**556–558.

Kurpakus, M. A., Quaranta, V., and Jones, J. C. R., 1991, Surface relocation of alpha6beta4 integrins and assembly of hemidesmosomes in an *in vitro* model of wound healing, *J. Cell. Biol.* **115:**1737–1750.

Labib, R. S., 1991, Heterogeneity of the bullous pemphigoid antigen: What is true and what is artifact? *J. Invest. Dermatol.* **96:**527–529.

Lacraz, S., Nicod, L., Galve-de Rochemonteix, B., Baumberger, C., Dayer, V.-M., and Welgus, H. G., 1992, Suppression of metalloproteinase biosynthesis in human alveolar macrophages by interleukin-4, *J. Clin. Invest.* **90:**382–388.

Larjava, H., Peltonen, J., Akiyama, S. K., Yamada, S. S., Gralnick, H. R., Uitto, J., and Yamada, K. M., 1990, Novel function for β-1 integrins in keratinocyte cell–cell interactions, *J. Cell Biol.* **110:**803–815.

Li, K., Sawamura, D., Giudice, G. J., Diaz, L. A., Mattei, M.-G., Chu, M.-L., and Uitto, J., 1991, Genomic organization of collagenous domains and chromosomal assignment of human 180-kD bullous pemphigoid antigen (BPAG2), a novel collagen of stratified squamous epithelium, *J. Biol. Chem.* **266:**24064–24069.

Li, K., Giudice, G. J., Tamai, K., Do, H. C., Sawamura, D., Diaz, L. A., and Uitto, J., 1992, Cloning of partial cDNAs for mouse 180-kDa bullous pemphigoid antigen (BPAG2), a highly conserved collagenous protein of the cutaneous basement membrane zone, *J. Invest. Dermatol.* **99:**258–263.

Li, K., Tamai, K., Tan, E. M. L., and Uitto, J., 1993, Cloning of type XVII collagen. Complementary and genomic DNA sequences of mouse 180-kDa bullous pemphigoid antigen (BPAG2) predict an interrupted collagenous domain, a transmembrane segment, and unusual features in the 5'-end of the gene and the 3'-untranslated region of the mRNA, *J. Biol. Chem.* **268:**8825–8834.

Liu, Z., Diaz, L. A., Troy, J. L., Taylor, A. F., Emery, D. J., Fairley, J. A., and Giudice, G. J., 1993, A passive transfer model of the organ-specific autoimmune disease, bullous pemphigoid, using antibodies generated against the hemidesmosomal antigen, BP180, *J. Clin. Invest.* **92:**2480–2488.

Liu, Z., Giudice, G. J., Swartz, S. J., Fairley, J. A., Till, G. O., Troy, J. L., and Diaz, L. A., 1995, The role of complement in experimental bullous pemphigoid, *J. Clin. Invest.* **95:**1539–1544.

Marinkovich, M. P., Lunstrum, G. P., and Burgeson, R. E., 1992a, The anchoring filament protein kalinin is synthesized and secreted as a high molecular weight precursor, *J. Biol. Chem.* **267:**17900–17906.

Marinkovich, M. P., Lunstrum, G. P., Keene, D. R., and Burgeson, R. E., 1992b, The dermal–epidermal junction of human skin contains a novel laminin variant, *J. Cell Biol.* **119:**695–703.

Marinkovich, M. P., Verrando, P., Keene, D. R., Meneguzzi, G., Lunstrum, G. P., Ortonne, J. P., and Burgeson, R. E., 1993, The basement membrane proteins kalinin and nicein are structurally and immunologically homologous, *Lab. Invest.* **69:**295–299.

Massagué, J., 1990, The transforming growth factor-β family, *Annu. Rev. Cell Biol.* **6:**597–641.

Mauviel, A., 1993, Cytokine regulation of metalloproteinase gene expression, *J. Cell Biochem.* **53:**288–295.

Mauviel, A., and Uitto, J., 1993, The extracellular matrix in wound healing: Role of the cytokine network, *Wounds* **5:**137–152.

Mauviel, A., Chen, Y. Q., Evans, C. H., Dong, W., and Uitto, J., 1993, Transcriptional interactions of transforming growth factor beta (TGF-β) with pro-inflammatory cytokines, *Curr. Biol.* **3:**822–831.

Mauviel, A., Lapière, J.-C., Halcin, C., Evans, C. H., and Uitto, J., 1994, Differential cytokine regulation of type I and type VII collagen gene expression in cultured human dermal fibroblasts, *J. Biol. Chem.* **269:**25–28.

McGrath, J. A., Ishida-Yamamoto, A., O'Grady, A., Leigh, I. M., and Eady, R. A. J., 1993, Structural variations in anchoring fibrils in dystrophic epidermolysis bullosa: Correlation with type VII collagen expression, *J. Invest. Dermatol.* **100:**366–372.

McGrath, J. A., McMillan, J. R., Dunnill, M. G. S., Pulkkinen, L., Christiano, A. M., Rodeck, C. H. J., Eady, R. A. J., and Uitto, J., 1995a, Genetic basis of lethal junctional epidermolysis bullosa in an affected fetus: Implications for prenatal diagnosis in one family, *Prenat. Diagn.* **15:**647–654.

McGrath, J. A., Pulkkinen, L., Christiano, A. M., Leigh, I. M., Eady, R. A. J., and Uitto, J., 1995b, Altered laminin 5 expression due to mutations in the gene encoding the β3 chain (LAMB3) in generalized atrophic benign epidermolysis bullosa, *J. Invest. Dermatol.* **104:**467–474.

McGrath, J. A., Gatalica, B., Christiano, A. M., Li, K., Orvaribe, K., McMillan, J. R., Eady, R. A. J., and Uitto, J., 1995c, Mutations in the 180-kD bullous pemphigoid antigen (BPAG2), a hemidesmosomal transmembrane collage (COL17A1), in generalized atrophic benign epidermolysis bullosa, *Nature Genet.* **11:**83–86.

McIntosh, I., Hamosh, A., and Diez, H., 1993, Nonsense mutations and diminished mRNA levels, *Nature Genet.* **4:**219.

Meneguzzi, G., Marinkovich, M. P., Aberdam, D., Pisani, A., Burgeson, R., and Ortonne, J.-P., 1992, Kalinin is abnormally expressed in epithelial basement membranes of Herlitz's junctional epidermolysis bullosa patients, *Exp. Dermatol.* **1:**221–229.

Morris, N. P., Keene, D. R., Glanville, R. W., Bentz, H., and Burgeson, R. E., 1986, The tissue form of type VII collagen is an antiparallel dimer, *J. Biol. Chem.* **261:**5638–5644.

Morrison, L. H., Labib, R. S., Zone, J. J., Diaz, L. A., and Anhalt, G. J., 1988, Herpes gestationis autoantibodies recognize a 180 kD human epidermal antigen, *J. Clin. Invest.* **81:**2023–2026.

Moses, H. L., Yang, E. Y., and Pietenpol, J. A., 1990, TGF-β stimulation and inhibition of cell proliferation. New mechanistic insights, *Cell* **63:**245–247.

Mueller, S. V., Klaus-Kovtun, V., and Stanley, J. R., 1989, A 230-kD basic protein is the major bullous pemphigoid antigen, *J. Invest. Dermatol.* **92:**33–38.

Mutasim, D. F., Takashi, Y., Labib, R. S., Anhalt, G. J., Patel, H. P., and Diaz, L. A., 1985, A pool of bullous pemphigoid antigen(s) is intracellular and associated with the basal cell cytoskeleton–hemidesmosome complex, *J. Invest. Dermatol.* **84:**47–53.

Nagayoshi, T., Mattei, M.-G., Passage, E., Knowlton, R., Chu, M.-L., and Uitto, J., 1989, Human lamin A chain (LAMA) gene: Chromosomal mapping to locus 18p11.3, *Genomics* **5:**932–935.

Olsen, D. R., and Uitto, J., 1989, Differential expression of type IV procollagen and laminin genes by fetal *vs.* adult skin fibroblasts in culture: Determination of subunit mRNA steady-state levels, *J. Invest. Dermatol.* **93**:127–131.

Olsen, D. R., Nagayoshi, T., Fazio, M., Mattei, M.-G., Passage, E., Weil, D., Chu, M.-L., Deutzmann, R., Timpl, R., and Uitto, J., 1989a, Human nidogen: cDNA cloning, cellular expression, and mapping of the gene to chromosome 1q43, *Am. J. Human Genet.* **44**:876–885.

Olsen, D. R., Nagayoshi, T., Fazio, M., Peltonen, J., Jaakkola, S., Sanborn, D., Sasaki, T., Kuivaniemi, H., Chu, M.-L., Deutzmann, R., Timpl, R., and Uitto, J., 1989b, Human laminin: Cloning and sequence analysis of cDNAs encoding A, B1 and B2 chains, and the expression of corresponding genes in human skin and cultured cells, *Lab. Invest.* **60**:772–782.

Overall, C. M., Wrana, J. L., and Sodek, J., 1991, Transcriptional and posttranscriptional regulation of 72-kDa gelatinase/type IV collagenase by transforming growth factor-β1 in human fibroblasts, *J. Biol. Chem.* **266**:14604–14071.

Parente, M. G., Chung, L. C., Ryynänen, J., Woodley, D. T., Wynn, K. C., Bauer, E. A., Mattei, M.-G., Chu, M.-L., and Uitto, J., 1991, Human type VII collagen: cDNA cloning and chromosomal mapping of the gene, *Proc. Natl. Acad. Sci. USA* **88**:6931–6935.

Peltonen, J., Larjava, H., Jaakkola, S., Gralnick, H., Akiyama, S. K., Yamada, K. M., and Uitto, J., 1989, Localization of integrin receptors for fibronectin, collagen and laminin in human skin: Variable expression in basal and squamous cell carcinomas, *J. Clin. Invest.* **84**:1916–1922.

Pikkarainen, T., Eddy, R., Fukushima, Y., Byers, M., Shows, T., Pihlajaniemi, T., Saraste, M., and Tryggvason, K., 1987, Human laminin β1 chain, a multidomain protein with gene (LAMB1) locus into q22 region of chromosome 7, *J. Biol. Chem.* **262**:10454–10462.

Pulkkinen, L., Christiano, A. M., Airenne, T., Haakana, H., Tryggvason, K., and Uitto, J., 1994a, Mutations in the γ2 chain gene (LAMC2) of kalinin/laminin 5 in the junctional forms of epidermolysis bullosa, *Nature Genet.* **6**:293–298.

Pulkkinen, L., Christiano, A. M., Gerecke, D., Burgeson, R. E., Pittelkow, M. R., and Uitto, J., 1994b, A homozygous nonsense mutation in the β3 chain gene of laminin 5 (LAMB3) in Herlitz junctional epidermolysis bullosa, *Genomics* **24**:357–360.

Pulkkinen, L., Gerecke, D. R., Christiano, A. M., Wagman, D. W., Burgeson, R. E., and Uitto, J., 1995, Cloning of the β3 chain gene (LAMB3) of human laminin 5, a candidate gene in junctional epidermolysis bullosa, *Genomics* **25**:192–198.

Richards, A. J., Al-Imara, L., Carter, N. P., Lloyd, J. C., Leversha, M. A., and Pope, F. M., 1994, Localization of the gene (LAMA4) to chromosome 6q21 and isolation of a partial cDNA encoding a variant laminin A chain, *Genomics* **22**:237–239.

Rousselle, P., Lunstrum, G. P., Keene, D. R., and Burgeson, R. E., 1991, Kalinin: An epithelium-specific basement membrane adhesion molecule that is a component of anchoring filaments, *J. Cell Biol.* **114**:567–576.

Ryan, M. C., Tizard, R., VanDevanter, D. R., and Carter, W. G., 1994, Cloning of the LamA3 gene encoding the α3 chain of the adhesive ligand epiligrin, *J. Biol. Chem.* **269**:22779–22787.

Ryynänen, J., Sollberg, S., Olsen, D. R., and Uitto, J., 1991, Transforming growth factor-β up-regulates type VII collagen gene expression in normal and transformed epidermal keratinocytes in culture, *Biochem. Biophys. Res. Commun.* **180**:673–680.

Ryynänen, J., Sollberg, S., Parente, M. G., Chung, L. C., Christiano, A. M., and Uitto, J., 1992, Type VII collagen gene expression by cultured human cells and in fetal skin. Abundant mRNA and protein levels in epidermal keratinocytes, *J. Clin. Invest.* **89**:163–168.

Sakai, L. Y., Keene, D. R., Morris, N. P., and Burgeson, R. E., 1986, Type VII collagen is a major structural component of anchoring fibrils, *J. Cell Biol.* **103**:1577–1586.

Salo, T., Lyons, J. G., Rahemtulla, F., Birkedal-Hansen, H., and Larjava, H., 1991, Transforming growth factor-β1 up-regulates type IV collagenase expression in cultured human keratinocytes, *J. Biol. Chem.* **266**:11436–11441.

Sawamura, D., Nomura, K., Sugita, Y., Mattei, M.-G., Chu, M.-L., Knowlton, R., and Uitto, J., 1990, Bullous pemphigoid antigen (BPAG1): cDNA cloning and mapping of the gene to the short arm of human chromosome 6, *Genomics* **8**:722–726.

Sawamura, D., Li, K.-H., Chu, M.-L., and Uitto, J., 1991a, Human bullous pemphigoid antigen (BPAG1): Amino acid sequence deduced from cloned cDNAs predict biologically important peptide segments and protein domains, *J. Biol. Chem.* **266**:17784–17790.

Sawamura, D., Li, K.-H., Nomura, K., Sugita, Y., Christiano, A., and Uitto, J., 1991b, Bullous pemphigoid antigen: cDNA cloning, cellular expression, and evidence for polymorphism of the human gene, *J. Invest. Dermatol.* **96:**908–915.

Sawamura, D., Li, K., and Uitto, J., 1992, 230-kD and 180-kD bullous pemphigoid antigens are distinct gene products, *J. Invest. Dermatol.* **98:**942–943.

Schittny, J. C., and Yurchenco, P. D., 1990, Terminal short arm domains of basement membrane laminin are critical for its self-assembly, *J. Cell Biol.* **110:**825–832.

Seltzer, J. L., Eisen, A. Z., Bauer, E. A., Morris, N. P., Glanville, R. W., and Burgeson, R. E., 1989, Cleavage of type VII collagen by interstitial collagenase and type IV collagenase (gelatinase) derived from human skin, *J. Biol. Chem.* **264:**3822–3826.

Shapiro, S. D., Campbell, E. J., Kobayashi, D. K., and Welgus, H. G., 1990, Immune modulation of metalloproteinase production in human macrophages. Selective pre-translational suppression of interstitial collagenase and stromelysin biosynthesis by gamma interferon, *J. Clin. Invest.* **86:**1204–1210.

Shelton-Inloes, B. B., Titani, K., and Sadler, J. E., 1986, cDNA sequences for human von Willebrand factor reveal five types of repeated domains and five possible protein sequence polymorphisms, *Biochemistry* **25:**3164–3171.

Sollberg, S., Peltonen, J., and Uitto, J., 1992a, Differential expression of laminin isoforms and β4 integrin epitopes in the basement membrane zone of normal human skin and basal cell carcinomas, *J. Invest. Dermatol.* **98:**864–870.

Sollberg, S., Ryynänen, J., Olsen, D. R., and Uitto, J., 1992b, Transforming growth factor-β up-regulates the expression of the genes for β4 integrin and bullous pemphigoid antigens (BPAG1 and BPAG2) in normal and transformed human keratinocytes, *J. Invest. Dermatol.* **99:**409–414.

Sonnenberg, A., Linders, C. J. T., Modderman, P. W., Damsky, C. H., Aumailley, M., and Timpl, R., 1990, Integrin recognition of different cell-binding fragments of laminin (P1, E3, E8) and evidence that α6β1 but not α6β4 functions as a major receptor for fragment E8, *J. Cell Biol.* **110:**2145–2155.

Sonnenberg, A., Calafat, J., Janssen, H., Daams, H., van der Raaij-Helmer, L. M., Falcioni, R., Kennel, S. J., Aplin, J. D., Baker, J., Loizidou, M., and Garrod, D., 1991, Integrin α6β4 complex is located in hemidesmosomes, suggesting a major role in epidermal cell–basement membrane adhesion, *J. Cell Biol.* **113:**907–917.

Stanley, J. R., 1989, Pemphigus and pemphigoid as paradigms of organ-specific, autoantibody-mediated diseases, *J. Clin. Invest.* **83:**1443–1448.

Stanley, J. R., Hawley-Nelson, P., Yuspa, S. H., Shevach, E. M., and Katz, S. I., 1981, Characterization of bullous pemphigoid antigen: A unique basement membrane protein of stratified squamous epithelia, *Cell* **24:**897–903.

Stanley, J. R., Tanaka, T., Mueller, S., Klaus-Kovtun, V., and Roop, D., 1988, Isolation of complementary DNA for bullous pemphigoid antigen by use of patients' autoantibodies, *J. Clin. Invest.* **82:**1864–1870.

Stepp, M. A., Spurr-Michaud, S., Tisdale, A., Elwell, J., and Gipson, I. K., 1990, α6β4 Integrin heterodimer is a component of hemidesmosomes, *Proc. Natl. Acad. Sci. USA* **87:**8970–8974.

Tamai, K., Sawamura, D., Do, H. C., Li, K., Tamai, Y., and Uitto, J., 1993, The human 230-kDa bullous pemphigoid antigen gene (BPAG1): Exon–intron organization and identification of regulatory *cis*-elements in the promoter region, *J. Clin. Invest.* **92:**814–822.

Tamai, K., Li, K., and Uitto, J., 1994a, Identification of a DNA binding protein (keratinocyte transcriptional protein-1) recognizing a keratinocyte-specific regulatory element in the 230-kDa bullous pemphigoid antigen gene, *J. Biol. Chem.* **269:**493–502.

Tamai, K., Sawamura, D., Do, H. Y. C., Li, K., and Uitto, J., 1994b, Molecular biology of the 230-kDa bullous pemphigoid antigen. Cloning of the gene (BPAG1) and its tissue specific regulation, *Dermatology* **189**(suppl.1):27–33.

Tamai, K., Li, K., Silos, S., Rudnicka, L., Hashimoto, T., Nishikawa, T., and Uitto, J., 1995a, Interferon-γ mediated inactivation of transcription of the 230-kDa bullous pemphigoid antigen gene (BPAG1) provides novel insight into keratinocyte differentiation, *J. Biol. Chem.* **270:**392–396.

Tamai, K., Silos, S. A., Li, K., Korkeela, E., Ishikawa, H., and Uitto, J., 1995b, Tissue-specific expression of the 230-kDa bullous pemphigoid antigen gene (BPAG1). Identification of a novel keratinocyte regulatory *cis*-element, KRE3, *J. Biol. Chem.* **270:**7609–7614.

Timpl, R., and Brown, J. C., 1994, The laminins, *Matrix Biol.* **14:**275–281.

Tryggvason, K., 1993, The laminin family, *Curr. Opin. Cell Biol.* **5:**877–882.

Uitto, J., and Christiano, A. M., 1992, Molecular genetics of the cutaneous basement membrane zone. Perspectives on epidermolysis bullosa and other blistering skin diseases, *J. Clin. Invest.* **90**:687–692.

Uitto, J., and Christiano, A. M., 1994, Molecular basis of the dystrophic forms of epidermolysis bullosa: Mutations in the type VII collagen gene, *Arch. Dermatol. Res.* **287**:16–22.

Uitto, J., Chung-Honet, L. C., and Christiano, A. M., 1992, Molecular biology and pathology of type VII collagen, *Exp. Derm.* **1**:2–11.

Uitto, J., Pulkkinen, L., and Christiano, A. M., 1994, Molecular basis of the dystrophic and junctional forms of epidermolysis bullosa: Mutations in the type VII collagen and kalinin (laminin 5) genes, *J. Invest. Dermatol.* **103**:39S–46S.

Uitto, J., Hovnanian, A., and Christiano, A. M., 1995, Premature termination codon mutations in the type VII collagen gene (COL7A1) underlie severe recessive dystrophic epidermolysis bullosa, *Proc. Assoc. Am. Phys.* **107**:245–252.

Urlaub, G., Mitchell, P. J., Ciudad, C. J., and Chasin, L. A., 1989, Nonsense mutations in the dihydrofolate reductase gene affect RNA processing, *Mol. Cell Biol.* **9**:2868–2880.

Vailly, J., Verrando, P., Champliaud, M. F., Gerecke, D., Wagman, D. W., Baudoin, C., Aberdam, D., Burgeson, R., Bauer, E., and Ortonne, J.-P., 1994a, The 100-kDa chain of nicein/kalinin is a laminin B2 chain variant, *Eur. J. Biochem.* **219**:209–218.

Vailly, J., Szepetowski, P., Pedeutour, F., Mattei, M. G., Burgeson, R. E., Ortonne, J.-P., and Meneguzzi, G., 1994b, The genes for nicein/kalinin 125-kDa and 100-kDa subunits, candidate for junctional epidermolysis bullosa, map to chromosome 1q32 and 1q25–31, *Genomics* **21**:286–288.

Vailly, J., Pulkkinen, L., Christiano, A. M., Tryggvason, K., Uitto, J., Ortonne, J.-P., and Meneguzzi, G., 1995a, Identification of a homozygous exon skipping mutation in the LAMC2 gene in a patient with Herlitz's junctional epidermolysis bullosa, *J. Invest. Dermatol.* **104**:434–437.

Vailly, J., Pulkkinen, L., Miquel, C., Christiano, A. M., Gerecke, D. R., Burgeson, R. E., Uitto, J., Ortonne, J.-P., and Meneguzzi, G., 1995b, Identification of a one basepair deletion in exon 14 of the LAMB3 gene in a patient with Herlitz junctional epidermolysis bullosa, and prenatal diagnosis in a family at risk for recurrence, *J. Invest. Dermatol.* **104**:462–466.

Vassar, R., Coulombe, P. A., Degenstein, L., Albers, K., and Fuchs, E., 1991, Mutant keratin expression in transgenic mice causes marked abnormalities resembling a human genetic skin disease, *Cell* **64**:365–380.

Verrando, P., Pisani, A., and Ortonne, J.-P., 1988, The new basement membrane antigen recognized by the monoclonal antibody GB3 is a large size glycoprotein: Modulation of its expression by retinoic acid, *Biochim. Biophys. Acta* **942**:45–56.

Verrando, P., Blanchet-Bardon, C., Pisani, A., Thomas, L., Cambazard, F., Eady, R. A. J., Schofield, O., and Ortonne, J.-P., 1991, Monoclonal antibody GB3 defines a widespread defect of several basement membranes and keratinocyte dysfunction in patients with lethal junctional epidermolysis bullosa, *Lab. Invest.* **64**:85–92.

Vidal, F., Aberdam, D., Christiano, A. M., Uitto, J., Ortonne, J.-P., and Meneguzzi, G., 1995, Mutations in the gene for the integrin β4 subunit are associated with junctional epidermolysis bullosa with pyloric atresia, *Nature Genet.* **10**:229–234.

Vilcek, J., and Lee, T. H., 1991, Tumor necrosis factor: New insights into the molecular mechanisms of its multiple actions, *J. Biol. Chem.* **266**:7313–7316.

Vuolteenaho, R., Nissinen, M., Sainio, K., Byers, M., Eddy, R., Hirvonen, H., Shows, T. B., Sariola, H., Engvall, E., and Tryggvason, K., 1994, Human laminin M chain (merosin): Complete primary structure, chromosomal assignment, and expression of the M and A chain in human fetal tissue, *J. Cell Biol.* **124**:381–394.

Wahl, S. M., 1992, Transforming growth factor beta (TGF-β) in inflammation: A cause and a cure, *Clin. Immunol.* **12**:61–74.

Wetzels, R. H. W., Robben, H. C. M., Leigh, I. M., Schaafsma, H. E., Vooijs, G. P., and Ramaekers, F. C. S., 1991, Distribution patterns of type VII collagen in normal and malignant human tissues, *Am. J. Pathol.* **139**:451–459.

Wewer, U. M., Gerecke, D. R., Durkin, M. E., Kurtz, K. S., Mattei, M.-G., Champliaud, M.-F., Burgeson, R. E., and Albrechtsen, R., 1994, Human β2 chain of laminin (formerly S chain): cDNA cloning, chromosomal localization, and expression in carcinomas, *Genomics* **24**:243–252.

Yurchenco, P. D., and Schittny, J. C., 1990, Molecular architecture of basement membrane, *FASEB J.* **4**:1577–1590.

Chapter 18

Fetal Wound Healing and the Development of Antiscarring Therapies for Adult Wound Healing

R. L. McCALLION and M. W. J. FERGUSON

1. Introduction

Scarring is an important clinical problem, often resulting in adverse effects on function and growth as well as an undesirable cosmetic appearance. Adult wound healing is characterized by acute inflammation, contraction, and collagen deposition, responses likely to have been optimized for rapid wound closure and minimizing infection. Similar processes may also result in fibrotic diseases that are common in many areas of medicine and surgery. Abdominal surgery often leads to intraperitoneal fibrous adhesions, while fibrotic retinopathy in diabetes, pulmonary fibrosis, and hepatic cirrhosis are significant medical problems. A major medical objective is therefore the reduction, and ideally the prevention, of scarring.

Despite extensive research into wound healing and scarring, there is still no consensus on the definition of a scar, nor an accurate, quantifiable means of measuring scarring, either clinically or histologically. A cutaneous scar may be defined as the macroscopic disturbance of normal skin structure and function arising as a consequence of wound repair. These macroscopic changes are due to alterations in the epidermal, dermal, and subcutaneous tissues (Ferguson *et al.*, 1995). As yet, there have not been any good correlations made between the degree of scarring and the associated histological, cellular, and molecular changes within the wounded tissue. These are important issues that need to be addressed if the degree of scarring is to be accurately reported and meaningful comparisons are to be made between the results obtained by different investigators.

Extensive experimental evidence has demonstrated that the embryo and early

R. L. McCALLION and M. W. J. FERGUSON • School of Biological Sciences, University of Manchester, Manchester M13 9PT, England.

The Molecular and Cellular Biology of Wound Repair (Second Edition), edited by Richard A. F. Clark. Plenum Press, New York, 1996

fetus respond to wounding in a way that is fundamentally different from the adult. In general, fetal wound healing occurs with the absence of scar formation. There are many differences between fetal and adult wound healing. In the fetus, rapid wound closure is achieved without scab formation in a warm, sterile fluid environment. The characteristic inflammation seen in adult wounds is absent in the fetus, the ratio of different cytokines is altered, and their levels generally reduced. The extracellular matrix of the fetal wound is rich in hyaluronic acid, matrix deposition is rapid and organized in structure, more similar to unwounded skin, sometimes leading to the complete restitution of normal dermal architecture. The causative factors in scar-free fetal wound healing need to be defined, since not all differences between adult and fetal healing may be significant in terms of scarring. Not all fetal wounds heal without a scar and there are a number of important parameters that together determine if a fetal wound will heal in a scar-free fashion.

1.1. Species

Many different species of animal have been used to investigate fetal wound healing, including the rat (Goss, 1977; Robinson and Goss, 1981; Roswell, 1984), opossum (Ferguson and Howarth, 1991), rabbit (Adzick *et al.*, 1985b; Krummel *et al.*, 1987), sheep (Burrington, 1971; Longaker *et al.*, 1990b; Stern *et al.*, 1993), monkey (Lorenz *et al.*, 1993), and mouse (Martin and Lewis, 1992; McCluskey *et al.*, 1993; Whitby and Ferguson, 1991a), and differences between species are pronounced. Nowhere is this more evident than in the case of fetal rabbits, which do not have the ability to contract their wounds, in contrast to most other species (Somasundaram and Prathap, 1970).

1.2. Gestational Age

Gestational age is important in determining if a fetal wound will scar. Generally, early embryos do not scar, while late-gestation fetuses heal with adultlike scar formation. In the rat (term = 20 days), the transition from the scar-free to the scarring phenotype occurs between days 18 and 19 of gestation (Ihara *et al.*, 1990), while in the sheep (term = 145 days) this transition occurs between days 100 to 120, around the middle of the third trimester (Longaker *et al.*, 1990b). The change from scar-free to scarring phenotype is more gradual in primates and spans several weeks' gestation. Lorenz and colleagues (1993) demonstrated that rhesus monkeys (term = 165 days) healed with complete restoration of normal dermal architecture at 75 days gestation but after 107 days (early 3rd trimester), with adultlike scar formation. These authors described a "transition wound" between 85 and 100 days gestation in the rhesus monkey, in which wounds healed with an absence of sebaceous glands and hair follicles, but with reticular dermal collagen architecture, similar to unwounded dermis.

This gradual transition from the scar-free to scarring phenotype has also been described *exutero* in the opossum marsupial model. Armstrong and Ferguson (1995) have described microscopic changes in the wounded dermis of opossum pouch young that became more pronounced as the age of the pouch young at wounding increased. These changes reflect a transition from no visible scar to the frank scarring of the adult animal. The marsupial studies also highlight two important findings: first, evidence of scar growth, and second, the problem of definition of a scar. Minor histological changes in the restituted wounded dermis were present but did not result in a macroscopically visible mark. Should this be considered a scar? How much dermal disruption is required to produce a macroscopically visible or a function-impairing scar? This is a very pertinent question when considering the efficacy of potential antiscarring therapies.

1.3. Tissue Damage

Most investigations into the importance of gestational age on fetal scarring have utilized a defined, incisional dermal wound. However, Horne and colleagues (1992) employed excisional wounds in their studies of fetal lambs. They reported scarring of lamb fetal wounds at gestational ages of 75, 90, and 120 days. This apparent anomaly demonstrates the critical importance of the severity of the wound on the scarring phenotype. If the degree of inflammation and the amount of dermal disruption are important determinants of scar formation, this would suggest a correlation between the extent of tissue damage and the gestational age at which the wound is inflicted to achieve scar-free healing. The greater the degree of damage, the earlier one needs to operate to achieve scar-free healing.

1.4. Body Site

The site of the body at which a wound is made is another significant factor in determining if scar-free healing will occur in the fetus. Early gestation skin incisions in the sheep fetus will heal without scarring, while similar wounds in fetal diaphragm, stomach, and peritoneum all heal with scar formation (Adzick *et al.*, 1985a; Longaker *et al.*, 1991c). There are similar site-specific variations in the degree of scarring in the adult human dermis.

There are therefore a multiplicity of factors in fetal wound healing that account for the apparently conflicting results obtained by different investigators. It is certain, however, that under defined conditions, early fetal dermal wounds heal without scarring. A thorough understanding of the mechanisms that underlie fetal scar-free healing is central to the goal of manipulation of the adult wound to achieve scar-free healing. Those differences between fetal and adult wound healing that are directly responsible for the scar-free phenotype need to be elucidated and separated from those that may be inconsequential. This chapter summarizes research into these differences and discusses the implications for the manipulation of adult wound healing to achieve the ultimate goal of scar-free healing.

2. Fetal Environment

One of the most fundamental differences between adult and fetal wound healing is the environment in which the wound heals. Fetal skin is perfused with fetal serum and bathed in warm, sterile amniotic fluid, rich in growth factors, hyaluronic acid, fibronectin, and other factors. The question therefore arises, is the unique fetal environment responsible for scar-free healing?

Typically, fetal arterial PO_2 is 20–25 mmHg, significantly hypoxic when compared to the adult (Nelson, 1976). This may cause a degree of dependence on anaerobic metabolism in the fetus and a reduced use of oxygen in fetal wound healing (Burrington, 1971). Studies in adult wound healing have stressed the importance of a sufficiently high oxygen concentration. Hypoxia in adult wound healing results in impaired leukocyte function, delayed healing, and increased incidence of wound infection. Oxygen is essential for the hydroxylation of proline and lysine during collagen synthesis, and supplemental oxygen has been shown to stimulate collagen deposition by adult fibroblasts (Hunt et al., 1961). However, tissue hypoxia seems to be important in the induction of angiogenesis, the activation of certain growth factors, and the stimulation of fibroblast and macrophage metabolism and division (Tahery and Lee, 1989). The relative dependence of fetal and adult wound healing on the degree of tissue oxygenation have not yet been determined, and the apparent paradox of accelerated wound healing in a scar-free fashion by the fetus, in a relatively hypoxic environment, remains an area for further study.

Fetal serum is known to have a different composition than that of the adult. The level of insulinlike growth factor II (IGF-II) is high in fetal serum and tissue and the levels decline postnatally (Estes et al., 1991; Lee et al., 1991). In rat serum, IGF-II concentrations were reported to be 20- to 100-fold higher in fetal rats than in adults (Moses et al., 1980). The profiles of other growth factors in fetal serum are also likely to be different from those of the adult due to the growth and differentiation status of the fetus (Lorenz and Adzick, 1993). Indeed, the ability of fetal calf serum to support the growth and maintenance of cells in culture has long been known, yet the precise components of the serum that are important remain to be elucidated. It is possible that the relative levels of individual growth factors or other serum components may have some influence on the degree of scarring.

The extracellular matrix of normal fetal dermis and fetal wounds is rich in hyaluronic acid (HA), a glycosaminoglycan found at high concentrations in regions of tissue repair, proliferation, and regeneration. However, the factors that cause the persistently high levels of HA in the fetus have not yet been fully described. Fetal serum is rich in an HA-stimulating factor, thought to be ubiquitous, reaching a peak at 40% of gestation time; this factor may control HA deposition (Longaker et al., 1989a). Fetal serum contains a factor that causes fibroblasts to migrate more quickly in vitro, and this acceleration of fibroblast migration was not due to an increased mitogenicity of the serum. This factor and the HA-stimulating factor described in fetal serum may be identical or related substances (Longaker et al., 1989b). The presence of this factor in fetal serum may facilitate or cause the rapid healing of fetal wounds, since increased levels of HA may provide a matrix that favors rapid cell migration.

The fetus is surrounded by amniotic fluid, known to be rich in nutrients, growth factors, and other substances important in fetal development. The characteristics of this fluid environment may play some role in the scar-free healing of the fetus. Amniotic fluid contains an HA-stimulating factor, similar to that in fetal serum, as well as a high concentration of HA itself (Longaker *et al.,* 1990a).

Fetal wounds heal without the formation of a fibrin clot, and it is thought that the absence of a scab is probably the result of healing in a fluid environment (Somasundaram and Prathap, 1972). It is not yet clear what role, if any, the presence of a fibrin clot may have in the adult scarring phenotype, but it is possible that the clot may provide a reservoir of growth factors important in scarring. Decreased fetal fibrin clot formation may be due to a combination of an immature fetal coagulation cascade and the fibrinolytic activity of amniotic fluid.

Gao and colleagues (1994) have investigated the effects of amniotic fluid on proteases *in vitro* and showed that human amniotic fluid enhanced collagenase activity. However, the activities of hyaluronidase, elastase, and cathespin B were all inhibited. The authors suggest that amniotic fluid may play an important role in fetal scar-free healing by regulating these matrix-degrading enzymes. These findings may in part explain the increases in HA found in fetal wound healing, due to the suppression of hyaluronidase activity, and increased collagenase activity may have a role to play in the superior collagen organization seen in fetal wound healing.

Amniotic fluid has also been found to have a role in wound contraction, although the results obtained by different workers often appear to be contradictory. Studies carried out by Longaker and colleagues (1991a) have demonstrated that excisional fetal lamb wounds contract *in utero,* and exposure to amniotic fluid does not significantly retard wound contraction. However, Hallock and colleagues (1988) examined fetal rats and stated that excisional wounds do not contract in the presence of amniotic fluid. Contrary to this report, Ihara and Motobayashi (1992) observed wound contraction in fetal rats of 16 days' gestation. However, this study was performed *in vitro* and this may account for the differing results observed. Excisional fetal rabbit wounds do not contract when exposed to amniotic fluid (Somasundaram and Prathap, 1970); but when excluded from the fluid by application of a silastic patch, wound contraction occurred rapidly (Somasundaram and Prathap, 1972). Rabbit amniotic fluid is, however, very different in composition from that of most other mammals, due to a different placentation mechanism, and it contains large amounts of high-molecular-weight proteins such as immunoglobulins. Further *in vitro* studies have demonstrated that human amniotic fluid, of specific gestational age (21 weeks), can inhibit contraction of a fibroblast-populated collagen lattice (FPCL) (Wider *et al.,* 1993). However, sheep amniotic fluid (80–85 days) enhances the contraction of lattices populated with fetal lamb fibroblasts (Burd *et al.,* 1991a).

The effects of amniotic fluid on wound contraction are demonstrably species-dependent, and the particular composition of amniotic fluid from different species may explain these apparent anomalies. Rittenberg and colleagues (1991) have demonstrated a 40,000 molecular weight protein in 125-day sheep amniotic fluid that stimulates FPCL contraction, and studies are underway to characterize a small-molecular-weight component of rabbit amniotic fluid that inhibits FPCL contraction (Longaker *et al.,* 1991c).

Another difference between the adult and fetal healing environment is that the fetal wound heals under sterile conditions. Frantz and co-workers (1993) have shown that despite the immaturity of the immune system, fetal rabbits are capable of mounting an acute inflammatory response to bacteria introduced into the wound via a polyvinyl alcohol (PVA) sponge. They reported that implants containing live bacteria resulted in fibroplasia and neovascularization, a typical adultlike response. These authors therefore concluded that the sterile environment appeared to play a role in scar-free fetal wound healing.

Despite these findings, the question arises, is the unique fetal environment the cause of scar-free fetal healing? There is evidence that intrinsic effects, even in the adult, are more important than environment, since cells that display some embryonic characteristics, such as hair follicles and oral mucosa, exhibit scar-free healing when wounded in the adult (Jahoda and Oliver, 1984). The most compelling evidence that the unique fetal environment is an epiphenomenon and that the scar-free healing phenotype arises from intrinsic differences comes from experiments on the pouch young of marsupials.

The Brazilian gray short-tailed opossum, *Monodelphis domesticus,* is a small, pouchless marsupial that is easy to breed in captivity. Marsupials have the advantage that the embryos are born at a very early developmental stage, equivalent to a 6-week gestation human fetus, in the case of *Monodelphis* (Morykwas *et al.,* 1991), and these "embryos" are externally accessible for repeated wound healing or other studies. In marsupial pouch young the immune system is poorly developed, as is the case for eutherian mammalian embryos, but the skin is well developed and keratinized. The cranial end of the embryo is more differentiated than the caudal, which allows attachment to the nipple and suckling (Ferguson and Howarth, 1991). The use of *Monodelphis* as a wound-healing model allows the investigation of specific intrinsic factors important in fetal wound healing, while exposed to an adult external environment.

Investigations into wound healing in *Monodelphis* carried out by Ferguson and Howarth (1991) revealed fundamental differences in healing between day-2 pouch young compared to day-28 pouch young and adults. The day-2 pouch young demonstrated typical scar-free fetal wound-healing features including rapid reepithelialization, minimal inflammatory and angiogenic responses and lack of fibrosis. Day-2 pouch young *Monodelphis,* however, did show scab formation and wound contraction. Day-28 pouch young healed their wounds like adult *Monodelphis,* resulting in scar formation. Armstrong and Ferguson (1995) confirmed these findings and described a transitionary phase, between pouch days 4 and 9, when macroscopic scarring could not be seen, but microscopic dermal collagen fiber disruption was evident. The transition from the scar-free to scarring phenotype was coincident with increasing numbers of inflammatory cells at the wound site and development of the adipose layer and differentiation of the dermis. The proposal that sterile conditions are responsible for the minimal inflammation in fetal wounds seems unlikely, since *Monodelphis* pouch young do not reside in a sterile environment, yet little inflammation is present in the pouch day-4 wounds. The lack of inflammation is more likely due to the immaturity of the immune cells and their ability to respond to different stimuli.

To further determine the relative importance of intrinsic and extrinsic factors in fetal scar-free healing, a number of studies have utilized grafts of fetal skin in the adult environment and *vice versa*. Longaker and colleagues (1994) transplanted adult (maternal) sheep skin onto the backs of fetal lambs before 77 days gestation, at which time the transplants would not be rejected. Forty days after transplantation, incisional wounds were made in the adult skin graft and the surrounding fetal tissue. Late gestation (120 days) fetal skin was also transplanted and wounded in the same way and control (fetal–fetal) transplants were also wounded. Wounds in the adult skin graft healed with scar formation in the fetal environment, while fetal–fetal transplants did not scar. This illustrates that the differentiated skin of the adult cannot be modulated to heal in a scar-free fashion by exposure to amniotic fluid or perfusion by fetal serum. The skin of late-gestation sheep grafted onto a younger embryo also healed with scar formation. The cause of scarring in the adult skin wound was not therefore a barrier effect of the keratinized adult epidermis shielding the graft from the influence of amniotic fluid.

The converse of this experiment was also carried out by Lorenz and co-workers (1992) in which fetal skin was transplanted into an adult environment. Grafts of human fetal skin, of varying gestational ages, were placed on adult athymic mice, in either a cutaneous or subcutaneous position. Scarring was observed in all fetal skin grafts in a cutaneous position, regardless of fetal skin gestational age. However, the grafts placed in a subcutaneous pocket all healed in a scar-free manner. This scarless healing of fetal skin in an adult, subcutaneous environment again illustrates that continuous perfusion by fetal serum and immersion in amniotic fluid is not prerequisite for scar-free wound repair. The nonscarring phenotype of the fetus is intrinsic to fetal skin, although the location of the graft on the adult recipient is crucial. It had previously been known that human fetal skin transplanted cutaneously onto nude mice and exposed to air demonstrated a marked acceleration of differentiation, while skin transplanted to a subcutaneous location differentiated at a rate comparable to that *in utero* (Lane *et al.,* 1989). The more rapid rate of tissue differentiation when fetal skin is exposed to a cutaneous environment is offered as a possible explanation for why fetal skin of the same gestational age will heal in a scar-free fashion when transplanted subcutaneously, but will scar in a cutaneous location.

In summary, current evidence suggests that the fetal environment is not an essential component of scar-free healing and that the degree of differentiation of the fetus is the most important determinant in scarring. It has been shown that advancing tissue maturity, associated with increasingly complex dermal structure, cellular differentiation, and maturity of the immune system all correlate with the diminished ability to heal without scarring with increasing gestational age.

3. Extracellular Matrix

The regulation of cell function by the extracellular matrix (ECM) is a fundamental mechanism that controls cell behavior and phenotype. Interactions between the individual components of the ECM and specific cell surface molecules can initiate a cascade of signal transduction events leading to many varied cellular responses. The ECM is also important as a reservoir for growth factors and cytokines, and components of the ECM can also interact with cytokines in a synergistic or antagonistic fashion.

Collagen is the principal component of the ECM and may be fibrillar, such as types I, II, and III, nonfibrillar, such as type IV, or fibril-associated collagens with interrupted triple helices (FACIT) (Vuorio and Crombrugghe, 1990; Kuhn, 1987; Shaw and Olsen, 1991). The different types of collagen may be present in varying amounts, depending on tissue type. Fibronectin is a 540-kDa glycoprotein dimer of two similar polypeptide chains, with some variation induced by alternative splicing. Fibronectin is a component of the fibrous ECM and basement membrane and also exists as a soluble dimer in plasma. It is an extensively studied adhesive protein component of the ECM, involved in cell attachment and chemotaxis (Clark, 1988). Laminin, a 500-kDa glyco-protein composed of three chains is a major component of basement membranes. It exhibits many biological activities, including chemotaxis, cell proliferation, and attach-ment (Raghow, 1994). Tenascin is a very large glycoprotein (almost 2000 kDa) with restricted distribution in normal tissue. It is made up of three repeating molecules, which include 13 epidermal growth factorlike repeats. The principal effect of tenascin in the ECM may be antiadhesive, interfering with fibronectin action (Chiquet-Ehrismann *et al.*, 1988). Proteoglycans are made up of a core polypeptide to which linear gly-cosaminoglycans (GAG) are covalently bonded and are the most abundant nonfibrillar component of the ECM (Jackson *et al.*, 1991; Wight *et al.*, 1992). GAG chains may consist of chondroitin sulfate, heparan sulfate, dermatan sulfate, or keratan sulfate, while hyaluronic acid is not attached to a polypeptide. Decorin and fibromodulin interact with fibrillar collagens and can inhibit fibrillogenesis *in vitro* (Hedbom and Heinegard, 1989). Decorin may also competitively inhibit transforming growth factor-beta (TGF-β) receptor interactions (Border and Ruoshlati, 1992), while TGF-β type III receptor (beta glycan) is another proteoglycan that may be required for TGF-β signaling (Attisano *et al.*, 1994).

Many studies have shown striking differences between the ECM of fetal and adult wounds, particularly in the restoration of normal collagen fiber orientation and tissue architecture in scarless fetal wounds. Differences in ECM profile between adult and fetal wounds may lead to changes in the migration and orientation of infiltrating cells and influences on fibrillogenesis, which may determine whether a wound heals in a scarred or scar-free manner. Hence, the constituents of the ECM and the kinetics of their deposition in wounding are important parameters to be studied in the search for the causes of the fetal scar-free phenotype.

3.1. Collagen

Collagen is the major structural protein of scar tissue in postnatal animals. Many studies have focused on collagen deposition using different species, wound models, and varying age of animal. In the past, there has been some controversy as to whether collagen is deposited at all in fetal wounds. Rowsell (1984) reported the absence of collagen deposition in fetal rat wounds. However, trichrome staining was the only detection method used, which is a relatively insensitive technique. Data obtained by Krummel and colleagues (1987) supported this finding, detecting no collagen in PVA–silastic implants in fetal rabbit wounds (Merkel *et al.*, 1988). Adzick and colleagues

(1985b) used a rabbit wound model and detected greater hydroxyproline accumulation in a Gortex sponge in fetal wounds than neonatal or adult wounds, indicating much more abundant collagen deposition in fetal wounds. Siebert *et al.* (1990) subsequently confirmed these findings using the same model, but found the differential between adult and fetal wound hydroxyproline deposition to be smaller than reported by Adzick and co-workers. However, wound implants may themselves influence ECM deposition and provoke a "foreign body" inflammatory response; how well they mimic wound healing events is debatable. It is now accepted that collagen is deposited in fetal wounds, at a greater rate than in the adult, and in a normal reticular pattern rather than the scarred pattern of the adult wound.

Burd and colleagues (1990) confirmed collagen deposition in fetal sheep wounds and demonstrated quantitative differences in fetal and adult collagens deposited in the wounds. Collagen, as demonstrated by trichrome staining, was evident in PVA implants in both adult and fetal (75–100 days gestation) wounds from 5 to 20 days postwounding. Biochemical analysis of hydroxyproline content showed an increase in fetal and adult wounds over 15 days postwounding. Thereafter, adult levels fell, whereas 75- and 100-day postwounding fetal levels continued to rise. Also detected was an increase in the proportion of type I collagen in the implants between 10 and 20 days postwounding in both fetal and adult wounds, but by 20 days postwounding the quantity of collagen decreased in both wound types. Interestingly, by 20 days postwounding the adult wounds had greatest tensile strength, yet contained the least collagen. These differences in tensile strength between adult and fetal tissue were also seen in the rabbit (Julia *et al.*, 1993). The adult wound had greater absolute tensile strength, but the proportional strengths between adult and fetal normal skin and adult and fetal wounds were the same, around 20% of normal showing equivalent wound tensile strength relative to normal tissue in both adult and fetal wounds. These authors concluded that there was no correlation between absolute collagen content and wound tensile strength, but that the pattern of collagen deposition in the fetal wound resulted in an improved cosmetic appearance.

Frantz and colleagues (1992) quantified the rate of collagen synthesis in fetal and adult rabbit wounds. In adults there was a preferential stimulation of collagen synthesis compared with noncollagenous protein synthesis, from 5 to 10 days after wounding. However, in the fetus, both collagen and noncollagenous protein synthesis were elevated for the first 5 days postwounding. These workers attributed the delay in onset of collagen synthesis in the adult to a number of factors. Adult fibroblasts are relatively sparse and in a quiescent state in normal dermis and must be activated. They require time for proliferation and migration into the wound before collagen synthesis can begin, steps that typically take 2 to 3 days. A significant cell density in the wound space is also required for maximal collagen synthesis. However, fetal fibroblasts exist in a relatively active state and may require no specific stimulation to up-regulate collagen synthesis. Fetal fibroblasts also reside in an ECM rich in HA, which facilitates the migration and proliferation of these cells at the wound site (Depalma *et al.*, 1989). Fetal fibroblasts are capable of proliferating and synthesizing collagen simultaneously *in vitro*, whereas adult fibroblasts maximally produce collagen when not dividing (Graham *et al.*, 1984).

The regulation of collagen synthesis in fetal and adult rabbit wounds has also been investigated using *in situ* hybridization to localize collagen mRNA (Nath *et al.*, 1994b). Fetal cells at the wound site had an overall increased message for type I collagen compared to surrounding tissue cells, with levels peaking 3–5 days postwounding. In adult wounds, fibroblast number increased by 7 days postwounding and again, total message for type I collagen was increased compared to normal skin. These results correlate well with those obtained by other workers and emphasize the accelerated rate of collagen deposition that occurs in fetal wounds. Morphometric analysis of the *in situ* hybridization signal revealed that fetal wounds accumulate type I collagen by increasing the number of cells within the wound, not by up-regulation of gene transcription, whereas adult wounds showed both fibroblast migration and up-regulation of pro-collagen type I mRNA. There may be many reasons why the fetus does not rely on gene induction to deposit wound collagen, including the possibility that a stimulus for supramaximal collagen synthesis (e.g., TGF-β) is not present or is present in an inactive form.

Collagen fibrillogenesis is a process that is as yet incompletely understood. It is known that collagen types I, III, and V are present within the same fibrils (Birk *et al.*, 1988; Keene *et al.*, 1987) and that interactions between the different collagen types can alter fibril size *in vitro* (Birk *et al.*, 1990). Proteoglycans can also alter fibril size (Scott, 1988), as can fibronectin (Speranz *et al.*, 1987); thus, fibrillogenesis may be influenced by the fetal ECM environment. Fetal tissues contain a higher proportion of type III collagen compared to type I (Merkel *et al.*, 1988; Epstein, 1974), and it is known that in the rat the adult ratio of collagen types do not become established until 10–15 days postpartum (Hallock *et al.*, 1993). This observation has also been made in mice (Boon *et al.*, 1992) and opossum (Morykwas *et al.*, 1991). It is thought that the ratio of type III to type I collagen may influence collagen fiber size. However, it is known that fetal wounds heal with scarring long before the transition from fetal to adult type III to I collagen ratios, so it is unlikely that the differences in these collagen ratios are of major importance in the scar-free fetal phenotype.

Whitby and Ferguson (1991a) carried out a detailed study of ECM deposition in fetal, neonatal, and adult lip wounds in mice. These authors did not use a wound implant model, which may in itself influence deposition of ECM, but determined the spatial and temporal distribution of various ECM components in the lip wound by immunohistochemistry. This method does not allow precise quantification of the amounts of each ECM constituent, but presents the least artificial wound model. These authors found that collagen types I, III, IV, V, and VI were present in wounds from animals of all ages, but both the timing and pattern of collagen deposition varied. In the fetus and neonate, collagen was detected by 48 hr postwounding, but did not appear in the adult wounds until 5 days postwounding. All five types of collagen examined appeared almost simultaneously. The major difference between collagens in the fetal and adult wounds was in the organization of the collagen fibrils. In the fetus, collagen was deposited in a reticular fashion, indistinguishable from the surrounding dermis, but adult collagen fibrils were deposited in dense parallel bundles, typical of scar tissue. This illustrates the critical difference in control and patterning of collagen fibrillogenesis between fetal and adult wounds. Similar results were also observed in sheep wounds (Longaker *et al.*, 1990b).

3.2. Fibronectin

Fibronectin (Fn) is expressed at high levels during wound healing (Kurkinen *et al.*, 1980; Grinnell *et al.*, 1981), before the appearance of collagen, and is a major component of the primary ECM during tissue repair. The Fn at the wound site is derived from two sources: plasma Fn (pFn) in the exudate from damaged blood vessels, present in the α-granules of platelets, and cellular Fn (cFn), which is synthesized locally at the wound site (Clark *et al.*, 1983). Fn is involved in the migration of all the major cell types into the wound site. Fibroblasts and keratinocytes are known to be stimulated to migrate by Fn (Donaldson and Mahan, 1983; Takashima *et al.*, 1986). Fn is chemo-attractant to endothelial cells (Bowersox and Sorgente, 1982) and macrophages (Norris *et al.*, 1982). Fn opsonizes tissue debris after injury (Martin *et al.*, 1988) and may also act as a provisional matrix for ECM assembly (McDonald, 1988). Fn is a constituent of the ECM early in embryonic development (Duband *et al.*, 1987), where its principal role is also in mediating cell migration and adhesion (Duband *et al.*, 1988). Antibodies to Fn or to cell surface receptors of the integrin family can block cell migration when injected into the intact embryo (Boucaut *et al.*, 1984; Poole and Thiery, 1986; Bonner-Frazer, 1985, 1986).

Differences in the deposition of Fn in fetal and adult wound healing have been investigated in full-thickness linear incisions, made on adult rabbits and fetuses of 24 days' gestation (Longaker *et al.*, 1989c). Wounds were harvested between 4 and 24 hr postwounding and the total Fn was detected by immunohistochemistry. Fn was initially associated with the fibrin clot, and deposition occurred with clot formation, the first event in adult wound healing. In the fetus, Fn was first detected 4 hr after wounding, becoming more prominent until the staining was very strong and covered the wound edge by 24 hr. This is in contrast to the adult wound, in which initial deposition of Fn was not detected until 12 hr postwounding. By 24 hr postwounding, staining for Fn was also marked in the adult, throughout the clot and adjacent to the wound surface. No collagen was present in any of the wounds at this time point. These authors stated that the earlier deposition of Fn in the fetal wound may provide an earlier signal for cell migration and hence underlie the rapid reepithelialization of the fetal wound compared to the adult.

Fn deposition was also studied in fetal, adult, and neonatal mouse lip wounds by immunohistochemistry. Whitby and Ferguson (1991a) showed that Fn was present at the surface of the wounds in all age groups by 1 hr postwounding. This staining persisted up to 72 hr in 16- and 18-day gestation fetal wounds, but was present up to 7 days in neonatal and adult wounds. Hence, in the mouse, the timing and pattern of Fn deposition was similar in the fetal and adult wounds, but the signal disappeared much earlier in the fetus. The decrease in staining for Fn appeared concomitant with the increase in collagen deposition.

Antibody staining methods do not always allow the precise discrimination of the type of Fn present at the wound site. Multiple isoforms of Fn may be present and these may play different roles in wound healing. Fn has three spliced segments: EIIIA and EIIIB are part of a series of type 3 repeats, termed "extra domains" because they are either completely included or excluded in the mature molecule (Schwarzbauer *et al.*, 1983, 1987; Tamkun *et al.*, 1984); the third segment, V (variable), contains several internal splice sites in humans, but only one in the rat, giving rise to three possible forms, i.e., included, partially excluded, or completely excluded (Magnuson *et al.*, 1991).

ffrench-Constant and colleagues (1989) investigated alternative Fn splicing in normal rat skin and after wounding, using *in situ* hybridization with probes that recognize all forms of Fn, or those that were specific to different spliced variants. In the normal rat skin, the majority of Fn mRNA lacks the EIIIA and EIIIB domains, but includes the V segment. Fn mRNA was detected mainly in cells within the reticular dermis, but was absent from cells of the papillary dermis and epidermis. After wounding, total Fn mRNA was detected by day 1 postwounding in cells that formed a band immediately beneath the clot. Many more positive cells were seen in the surrounding dermis and panniculus carnosus compared to normal skin. This pattern of labeling became more intense until 7 days postwounding, when the granulation tissue was still positive, but the labeling was less extensive in the neighboring dermis and muscle. By 14 days postwounding, the labeling within the granulation tissue was weaker and had returned to normal in the surrounding tissue. When the splicing of the Fn mRNA was examined, it was found that the V segment was included, as for normal skin; however, EIIIA and EIIIB positive variants were present, largely restricted to the granulation tissue itself. EIIIB positive isoforms were less abundant than EIIIA positive isoforms. The observed *in situ* staining patterns may reflect a splicing difference in which EIIIA and EIIIB are included in Fn wound mRNA to a much greater degree than in normal tissue.

These authors had previously shown that both EIIIA and EIIIB are present in the early chicken embryo (ffrench-Constant and Hynes, 1988) and become spliced out in tissue-specific patterns once embryogenesis and organogenesis near completion (ffrench-Constant and Hynes, 1989). Since the patterns of Fn splicing are highly conserved among species, the authors suggest that it is likely that the rat embryo contains predominantly EIIIA and EIIIB positive Fn and that the results of the wound healing investigation demonstrate a return to the embryonic pattern of splicing. The return to an embryonic pattern of splicing after injury in the rat has also been demonstrated after induction of experimental hepatic fibrosis by ligation of the biliary duct (Jarnagin *et al.*, 1994). Here, the EIIIA segment of Fn was biologically active and may mediate the conversion of lipocytes to myofibroblasts in hepatic fibrosis.

The mechanisms that control the patterns of Fn splicing are poorly understood. However, TGF-β, an important cytokine in wound healing, increases the synthesis of total Fn and its receptor (Ignotz *et al.*, 1987; Ignotz and Massague, 1987) and also increases the *in vitro* inclusion of EIIIA into Fn (Balza *et al.*, 1988). It therefore seems likely that TGF-β and other growth factors may play a role in alternative Fn splicing.

Wounding of tissue results in the reappearance of embryonic spliced Fn, which may be more functionally appropriate in terms of wound repair. ffrench-Constant and co-workers (1989) suggest that the localization of EIIIA and EIIIB positive spliced variants at the base and sides of wounds is consistent with a role for these forms in epithelial cell migration, but also speculate roles for spliced variants in cell proliferation, differentiation, and chemotaxis. The normal existence of embryonic splicing in the fetal wound may contribute to the accelerated healing observed, since there is no requirement for the type of Fn to be switched. Moreover, after wounding, the predominant form of Fn in the adult is pFn, released from platelets, which may be a less efficient Fn for wound healing. It takes approximately 4 days for the embryonic type Fn to appear in significant amounts in adult wounds (Shaw and Olsen, 1991). Hence, the different isoforms of Fn present at the time of wounding may be important in influencing the rate of healing.

3.3. Tenascin

Tenascin (Tn) expression in normal adult tissues is greatly restricted, but high transient expression of Tn has been found coincident with actively migrating or proliferating cells during wound healing and also during embryonic morphogenesis and oncogenesis (Mackie *et al.*, 1988; Erickson and Bourdon, 1989; Daniloff *et al.*, 1989; Chiquet-Ehrismann, 1990). Tn has been ascribed both adhesive and antiadhesive properties. The addition of soluble Tn to cultured tumor cells resulted in partial detachment, loss of intercellular contacts, and inhibition of cell migration and spreading on basal lamina (Chiquet-Ehrismann, 1990). Tn-coated substrates retard the attachment and/or spreading of a number of cell types on Fn, laminin, and basal lamina (Erickson and Bourdon, 1989; Lightner and Erickson, 1990; Chiquet-Ehrismann, 1990). Some investigators have proposed a functional antagonism between Tn and Fn that is now well accepted (Lotz *et al.*, 1989; Chiquet-Ehrismann *et al.*, 1988). Others have provided evidence for adhesive domains in Tn that interact with an RGD (Arg-Gly-Asp) sequence-sensitive integrin or chondroitin sulfate proteoglycan (Mackie *et al.*, 1988). One exception to the limited distribution of Tn in adult tissue is in the thymus, which expresses significant amounts, and it has been reported that Tn can block T-cell proliferation and activation in this organ (Ruegg *et al.*, 1989). The immunosuppressive effect of Tn has been confirmed by other authors (Hemesath *et al.*, 1994), who state that Tn may modulate the ability of Fn to facilitate T-cell activation or migration, and as such may be important in resolution of the inflammatory response.

Tn is a family of related ECM proteins, consisting of Tn-C, Tn-R, and Tn-X. The original Tn is Tn-C, or cytotaxin, and the name continues to be used without a letter in research concerning only Tn-C (Erickson, 1993). Mice have been genetically engineered to knockout the Tn-C gene, and surprisingly have developed normally, showing no apparent defect (Saga *et al.*, 1992). This lack of phenotype does not necessarily indicate that Tn has no important function, but that the function may be subtle. The Tn knockout, however, does cast some doubt on the functions ascribed to the molecule. It was reported that these Tn knockout mice healed their wounds normally, but much more detailed study is required to fully elucidate the effects of the absence of Tn-C.

Very few studies have been carried out on Tn in fetal wound healing. Whitby and Ferguson (1991a) examined the distribution of Tn in mice lip wounds in their detailed study of the differences in ECM deposition in fetal neonatal and adult wounds. In the fetus, Tn was present at the wound surface within 1 hr postwounding. By 24 hr, the staining for Tn was more prominent and was present in the mesenchyme adjacent to the wound site. However, Tn staining was found to be most intense near the basement membrane of the epidermis. The staining for Tn persisted in the fetus at 48 hr postwounding, but became very sparse by 72 hr, and by 5 days after wounding, the amount of Tn had returned to normal, unwounded levels. This temporal distribution contrasted significantly with that observed in the adult, although the spatial pattern was similar. In the adult wound, Tn was not detectable until 24 hr postwounding, but was most intense after 48 hr. At 5 days postwounding, Tn expression was present diffusely and still detectable 12 days after wounding. The expression in the neonate was approximately midway between the fetal and adult patterns, with Tn expression first evident at 12 hr postwounding and reverting to normal by 5 days postwounding. The time at which Tn was first detected paralleled the rate of reepithelialization and wound closure, which was most rapid in the fetus and slowest in the adult. Similar results were obtained in healing fetal and adult sheep wounds (Whitby *et al.*, 1991). Tn may play an important role in wound closure by reducing cell adhesion through cellular interactions with Fn, hence initiating cell migration. The rapid appearance of Tn in fetal wounds may underlie their rapid reepithelialization when compared to adults. Rapid deposition of Tn in the fetal wound may also attenuate proinflammatory signals and so contribute to the poor inflammatory infiltrate seen in fetal wounds.

3.4. Proteoglycans and Glycosaminoglycans

The major glycosaminoglycans (GAGs) in skin are hyaluronic acid (HA), dermatan sulfate (DS), and chondroitin sulfate (CS), with a minor contribution from heparan sulfate (HS) (Mast *et al.*, 1992a). HA does not have a protein core, but the other GAGs are covalently linked to proteins in the form of proteoglycans. HA is the most extensively studied GAG in fetal and adult wound healing. As yet, little is known about the possible role of other GAGs and proteoglycans in fetal wound healing.

HA is a high-molecular-weight, nonsulfated GAG, consisting of alternating residues of N-acetylglucosamine and glucuronic acid. It is present in the ECM of most animal tissues and is found at highest concentrations in soft connective tissues (Laurent, 1987). HA is a linear polysaccharide, although its macromolecular structure is that of an expanded coil, which makes it susceptible to enzymatic degradation, resulting in the release of low-molecular-weight oligosaccharides.

It has long been known that the ECM becomes enriched with HA coincident with episodes of rapid cellular migration and proliferation. This has been well documented in developing, regenerating, and remodeling tissues and in tissues undergoing tumor cell invasion (Toole, 1991; Knudson *et al.*, 1989). HA is involved in the detachment process within the cell cycle that allows cells to move, and the burst in HA synthesis before mitosis allows cells to become detached from neighboring cells (Turley and

Torrence, 1984; Tomida *et al.*, 1974; Mian, 1986; Brecht *et al.*, 1986). HA is also known to inhibit cell differentiation (Kujawa and Tepperman, 1983; Kujawa *et al.*, 1986). It is thought to exert effects on the behavior of cells in several different ways. It is believed to stimulate cell proliferation and migration by creating a low-resistance hydrated matrix that removes cells from contact inhibition and restricted mobility (Brecht *et al.*, 1986). HA-rich matrices are also thought to bind growth factors, thus influencing cell growth and differentiation by changing the local concentration of these cytokines (Ruoslahti and Yamaguchi, 1991). It is also believed to regulate cell behavior by the provision of ligands for cell attachment and motility (Turley *et al.*, 1991), mainly via CD44, identified as the principal cell surface receptor (Aruffo *et al.*, 1990).

HA is also thought to play an important role in angiogenesis. An ECM rich in HA is known to inhibit blood vessel formation in chick embryo limb buds and within granulation tissue (Balazs and Darzynkiewicz, 1973; Feinberg and Beebe, 1983). However, HA oligosaccharides, considered to be degradation products of HA, stimulate angiogenesis in the chick chorioallantoic membrane assay (West *et al.*, 1985). This stimulation of angiogenesis is associated with a marked up-regulation of collagen synthesis, suggesting that endothelial cells may require type I collagen as a substrate for cell migration (Kumar *et al.*, 1992).

HA affects collagen synthesis in a number of systems. Chandrakasan and colleagues (1986) showed that HA stimulated the deposition of type III collagen by cultured human fibroblasts. Addition of HA to fetal rabbit fibroblasts *in vitro* significantly stimulated collagen deposition (Mast *et al.*, 1993). Scott and Hughes (1986) studied the early development of chick and bovine tendons and found that fetal collagen fibrils were smaller in diameter when HA was more abundant and that fibril size increased concomitantly with decreasing HA concentrations. Fetal skin contains much higher levels of HA than adult skin. The ECM of fetal wounds is rich in HA compared to the transient up-regulation of HA at the adult wound site, a finding observed in the fetal rabbit (Depalma *et al.*, 1989; Mast *et al.*, 1991; Stern *et al.*, 1992) and fetal sheep (Chiu *et al.*, 1990; Longaker *et al.*, 1991b).

Estes *et al.* (1993) used Hunt-Schilling chambers to measure the HA content of fetal lamb wound fluid at 75, 100, and 120 days gestation and compared the levels to those found in adult sheep. HA levels were highest in wound fluid from lambs of 75 days gestation, peaking at 7 days postwounding. Wound fluid from 100-day gestation animals showed a similar profile, but with slightly lower HA concentrations at each time point. However, in the 120-day gestation animals, the HA levels detected were significantly lower than in the younger fetuses and by 14 days postwounding had decreased to near undetectable levels, contrasting with the high levels still present at this time in the 75- and 100-day-old fetuses. They suggested that high and prolonged levels of HA in fetal wounds promoted scar-free healing, and indeed the predicted decrease in wound HA content with increasing fetal age correlated with the onset of scarring in the 120-day-old fetus. They also investigated the hyaluronic acid stimulating activity (HASA) present in the wound fluid and found a direct correlation between HASA and HA levels in the wound, with HASA virtually absent in the adult wound fluid. HASA, found in fetal serum, amniotic fluid, and urine (Longaker *et al.*, 1990a) may be a mechanism by which HA levels remain high in the fetus. The levels of GAGs in fetal, newborn, and adult sheep have also been investigated by Knight and colleagues (1994), using PVA implants. These authors found that in uninjured skin, the contribution of HA to the total GAG content fell progressively with increasing fetal maturity. However, they reported no decrease in the wound HA content with gestational age. The reasons for this anomaly are not apparent, but may be related to differences in the HA detection method or in the type of wound model used.

In adult wound repair, HA is the first macromolecule to appear following formation of the platelet plug, but this elevation is transient (Dunphy and Upuda, 1955) and the decrease in HA content of adult wounds corresponds to the appearance of hyaluronidase at the wound site (Bertolami and Dunoff, 1978). Fibrinogen from a number of mammalian species, including man, specifically binds HA (Le Boeuf *et al.,* 1986; Frost and Weigel, 1990). It has been postulated that HA is brought to the wound site by fibrinogen in the circulation and platelets and that fibrin in the clot binds the HA to create a fibrin–HA matrix. This feature of the early ECM of the wound may make the clot more porous, increasing water retention and allowing cells to infiltrate the clot (Weigel *et al.,* 1986).

HA concentration may be an important factor in determining the inflammatory response at the wound site. Dillon and colleagues (1994) have examined the adherence of fetal lymphocytes to a number of ECM components, particularly HA. Fetal lymphocytes were capable of binding a number of glycoproteins, including fibronectin, vitronectin, and collagen I and III, early in their ontogeny. However, these cells showed no binding to HA, the most abundant component of fetal wound ECM. These authors propose that the minimal inflammatory response seen in fetal wound healing may be due in part to the inability of lymphocytes to adhere to HA.

Degradation of HA and other GAGs by the addition of crude hyaluronidase to implants in fetal skin led to a direct increase in the numbers of inflammatory cells, fibroblast proliferation, angiogenesis, and collagen deposition (Mast *et al.,* 1992b). However, as noted earlier, implants may represent more of a foreign body reaction rather than modeling the wound healing response. Therefore, it is still not known whether degradation of HA in fetal wounds causes them to heal with a scar. There are several possible mechanisms whereby high levels of HA could contribute to fetal scar-free healing. The inhibition of cell differentiation by high levels of HA may keep cells in a "fetallike" state, allowing wound healing to occur by a process more similar to regeneration than repair. The less resistant HA-rich matrix could allow quicker cell infiltration into the wound site and high HA levels may also enhance cell proliferation. The high HA concentration in the fetal wound may also influence the amount and type of collagen deposited and the nature of the fibrils formed, thus resulting in a more reticular pattern of collagen deposition. The lack of HA degradation products at the fetal wound could contribute to the low levels of angiogenesis observed, while inhibition of lymphocyte infiltration by HA is one of the possible mechanisms by which inflammation at the fetal wound site could be reduced.

Proteoglycans (PGs) and GAGs have a ubiquitous distribution and their pattern and composition are different between species, tissue types of the same species, and even within the same tissues. PG function may be modulated by its particular GAG component, by the core protein component, or by both. Many functions are related to the binding of PGs or GAGs to other molecules, especially proteins (Sames, 1994). Many PGs are constituents of the ECM, including the large PGs versican, aggrecan, and perlecan and the small PGs biglycan, decorin, and fibromodulin. Some PGs, such as syndecan, have membrane-embedded core proteins. Small, large, and very large PGs may bind to collagen fibrils, maintaining space between them, resulting in tissue hydration and creating channels for the movement of water-soluble materials (Scott, 1992). GAG chains are large, extended structures with highly charged sulfate and carboxylate groups. The high negative charge of PGs attracts counter ions and the osmotic imbalance caused by a high local concentration of ions draws water from surrounding areas, hence, PGs keep the matrix hydrated (Hardingham and Bayliss, 1990).

Decorin associates with type I and II collagen, both *in vitro* and *in vivo* in fetal and adult skin (Fleischamajer *et al.,* 1991; Scott, 1988). As decorin binds to the surface of collagen fibrils, the assembly of individual collagen molecules is delayed, resulting in reduction of the final diameter of the collagen fibers (Vogel *et al.,* 1984; Vogel and Trotter, 1987). Schonherr *et al.* (1995) have mapped the collagen fibrillogenesis ability, TGF-β binding, collagen fiber binding, and endocytotic functions of the decorin molecule to different domains within its protein core. Fibromodulin also binds to collagens I and II and inhibits fibrillogenesis *in vitro* (Hedbom and Heinegard, 1989). It binds collagen *in vivo* at different sites than decorin and may play a similar role to decorin in the ECM, by organizing fibril formation (Hedlund *et al.,* 1994). Biglycan is also present in the skin, but has not been described as associating with collagen fibrils (Fleischamajer *et al.,* 1991) or to show any specific effect on collagen fibril formation *in vitro* (Hedbom and Heinegard, 1989). Biglycan also has an abundant pericellular distribution in the layers of the epidermis, but little biglycan has been detected in the dermis (Tan *et al.,* 1993). In fetal tissue, biglycan has also been identified at the cell surface, in a similar distribution to that seen in adult skin (Bianco *et al.,* 1990).

In addition to their effects on ECM structure, PGs are also important mediators of growth factor activity. They may function as receptors for growth factors, as protectors and storage ligands, or as inactivators. The role of heparan sulfate PGs in binding fibroblast growth factors (FGFs) and other heparan-binding growth factors is well documented (Ruoslahti and Yamaguchi, 1991). Binding of growth factors to PGs in the ECM or at the cell surface may protect them from degradation and provide local reservoirs of growth factors (Damon *et al.,* 1989). The local release of heparin (free HS) during acute inflammation may displace HS-bound growth factors such as FGFs (Thompson *et al.,* 1990). Proteolytic cleavage of the PG may also mobilize PG fragments with the growth factors still attached (Saksela and Rifkin, 1990). Therefore, acute inflammation and proteolytic activity on wounding may result in a rapid availability of growth factors *in situ,* to help promote early events in wound repair. FGF-2 must be bound to HS side chains of a PG or to heparin in order to bind to its cell surface receptor (Yayon *et al.,* 1991). A HS PG has also been identified as a FGF receptor, with higher affinity for FGF-1 than FGF-2 (Sakaguchi *et al.,* 1991). It is clear that the modulation of HS and HS PGs are important in the action of FGFs and other growth factors.

TGF-β also binds to PGs, but is distinct from the HS-binding growth factors in that it binds to the core protein rather than to the GAG side chain. One of the PGs that TGF-β binds to is betaglycan, the type III TGF-β receptor (Andres *et al.,* 1989). Betaglycan is able to bind all three mammalian forms of TGF-β with relatively high affinity (Massague *et al.,* 1990; Andres *et al.,* 1991). Membrane-anchored betaglycan plays a critical role in TGF-β action by binding TGF-β and presenting it to the signaling type II receptor (Lopez-Casillas *et al.,* 1993). The extracellular domain of betaglycan can be shed by cells in a soluble form that can act as an antagonist to TGF-β binding and actions (Lopez-Casillas *et al.,* 1994). Hence, the relative ratio of membrane bound to soluble betaglycan *in vivo* may be an important determinant of TGF-β activity. Another PG that can bind TGF-β is decorin, and this binding leads to the neutralization of TGF-β activity (Yamaguchi *et al.,* 1990). Hildebrand and colleagues (1994) found that decorin, biglycan, and fibromodulin could all bind to TGF-β with slightly different affinities. These authors demonstrated that fibromodulin was a more effective binder of TGF-β than decorin or biglycan, and that this was the only one of the three PGs that could bind, even slightly, the latent TGF-β complex. ECM-bound decorin and fibromodulin may compete with receptors for TGF-β, sequestering it into the matrix, and hence controlling its bioavailability *in vivo.*

Growth factors can also affect the rate of PG synthesis. TGF-β1 markedly increases the expression of mRNA for biglycan and versican by normal human skin fibroblasts and inhibits the expression of decorin mRNA (Kahari *et al.,* 1991). This response was also seen in human gingival fibroblasts, although the gingival cells were more responsive to TGF-β1 in terms of biglycan expression. Analysis of the production of ^{35}S-labeled PGs confirmed the stimulation of biglycan and versican production and the inhibition of decorin synthesis. The enhancement of biglycan and versican was found to be coordinate with up-regulation of type I procollagen gene expression, and the authors suggested a possible role for these PGs in the activation of collagen formation stimulated by TGF-β1. In contrast to TGF-β1, FGF-2 was found to upregulate the expression of decorin mRNA in normal skin and in cultured dermal fibroblasts, the level of versican was not changed, and the amount of mRNA for biglycan was reduced, concomitant with a reduction in type I procollagen gene expression (Tan *et al.,* 1993). The relative levels of PG induced by growth factors may influence the rate at which collagen is deposited in wound repair. Divergent results have been reported on the effects of TGF-β1 on decorin expression. Border and colleagues (1990b) documented an up-regulation of both decorin and biglycan synthesis by TGF-β1 in kidney mesangial cells. It appears that the control of decorin synthesis is cell type specific and the molecular events controlling this expression are not yet fully understood.

There have been few studies documenting PG alterations in fetal and adult wound healing and fibrosis. Yeo and colleagues (1991) demonstrated increased decorin production in adult guinea pig wounds at 7 days postwounding by immunohistochemical and biochemical techniques. Increases in both decorin and biglycan expression have also been reported in experimentally induced liver fibrosis (Meyer *et al.,* 1992). The expression of the cell surface PG syndecan has been studied during the healing of cutaneous wounds in adult mice (Elenius *et al.,* 1991). There was an enhanced expression of syndecan in proliferating and migrating cells of the epidermis and hair follicles and a limited expression on the surface of vascular endothelial cells within the granulation tissue. In their study of the ECM of fetal, neonatal, and adult lip wounds in mice, Whitby and Ferguson (1991a) documented some changes in the levels of the GAGs HS and CS. These authors reported no differences between fetal, neonatal, and adult wounds in terms of HS localization. HS staining in normal skin was restricted to the basement membranes of the epidermis and oral mucosa, and at the wound site staining was also restricted to reforming epithelial basement membranes. There were, however, differences in the localization of CS staining between the fetus and the adult. CS was present diffusely in the fetal dermis but absent from neonatal and adult dermis, except for small areas around hair follicles and oral mucosal glands. At no time point studied was CS seen in the wounds of neonatal and adult mice, but in the fetal wound CS was detected at the wound site by 20 hr postwounding and had returned to its normal distribution by 48 hr postwounding. The absence of CS staining in adult wounds contrasts with the up-regulation of CS in adult guinea pig wounds reported by Yeo and co-workers (1991). The differences in the reported results are likely due to differences in antibody affinities and the masking of CS binding sites by other GAG side chains.

4. Fibroblasts

Fibroblasts are critical to the wound-healing process, since these cells are responsible for the deposition and remodeling of most of the new ECM. The process of fibroblast recruitment to the wound site and ECM proliferation has been termed *fibroplasia* (Clark, 1993). The study of fibroplasia is complex since the phenotype of the fibroblast changes during wound healing, as does the composition of the ECM that surrounds it. Growth factors are essential for the stimulation of fibroplasia and the wound environment stimulates fibroblasts to assume a number of distinct phenotypes.

Welch and colleagues (1990) have documented fibroblast behavior in porcine, full-thickness adult skin wounds. No fibroblasts were found at the wound site in the first 3 days after wounding, despite the presence of chemoattractant cytokines and the provisional matrix of the fibrin clot. However, during this time period, the fibroblasts in the subcutaneous septa below the wound proliferated in response to factors released during the initial injury. At day 4 postwounding, the fibroblasts migrated into the wound site, and by 7 days postwounding had completely filled the dermal defect. It was postulated that proliferation of fibroblasts was required for phenotypic change to migratory cells and that this change may involve development of a motor apparatus, loss of receptors for attachment to the ECM, and the expression of new receptors necessary for motility.

After migration into the wound space the fibroblasts assume a synthetic phenotype and begin to synthesize collagen and Fn. Welch and co-workers judged collagen production to be maximal at 7 days postwounding, and at this time the actin fibrils within the fibroblasts were seen to condense into bundles. Fetal fibroblasts can both proliferate and synthesize collagen simultaneously, in contrast to adult fibroblasts (Graham *et al.,* 1984). This difference in fibroblast phenotype may directly contribute to the faster rate of fetal healing. Welch *et al.* (1990) also observed the formation of intercellular junctions and cell matrix connections with the characteristics of Fn-rich links called fibronexis at 7 days postwounding. By 9 days postwounding, the synthetic phenotype had largely been replaced by fibroblasts containing tightly bundled actin fibrils, known as myofibroblasts, which aligned themselves across the wound. These myofibroblasts may bring about wound contraction, which occurs from 7 to 14 days after wounding.

Distinct pathways and stages of cellular differentiation tend to be associated with specific cell surface antigen expression. Rettig and colleagues (1988) first described cell surface antigens that are differentially expressed by mesenchymal cells during normal development, proliferation, or malignant transformation. One such antigen, identified by a monoclonal antibody termed F19, was found to be expressed in normal fetal tissue, tumor tissue-cultured fibroblasts, and in adult wounds, but was not normally expressed in adult connective tissues. These workers suggested that F19 is a cell surface marker for actively proliferating mesenchymal cells and that its expression may be induced by growth factors or during malignant transformation. The presence of F19 antigen was subsequently confirmed to be present on the surface of adult fibroblasts from wounds, 7 to 21 days postwounding, but absent from normal adult skin fibroblasts (Garin-Chesa *et al.*, 1990). This antigen has now been cloned and identified as fibroblast activation protein α (FAP-α) and the authors speculate that the distinct phenotype, FAP-α^+, reflects differences in fibroblast activation state and function (Scanlan *et al.*, 1994). This work suggests that normal fetal fibroblasts are phenotypically different from normal adult fibroblasts and that adult fibroblasts must switch their phenotype during wound healing to express more fetallike characteristics. The need to activate this switch may be one reason for the initial lag period during normal adult wound healing, which is absent in fetal wound healing. Distinct fetal fibroblast phenotypes may be important determinants of fetal scar-free healing.

Schor and colleagues (1985) demonstrated that fetal fibroblasts will migrate into a three-dimensional collagen gel, whereas very few normal adult fibroblasts will do so. These authors proposed that a novel protein, migration stimulation factor (MSF), produced by fetal fibroblasts was the factor responsible for the migratory capacity of the fetal cells. They also found this factor to be secreted by tumor fibroblasts and by fibroblasts of the adult oral mucosa (Schor *et al.*, 1988). MSF stimulated the migration of normal adult fibroblasts in culture, but the adult cells themselves did not secrete the protein. It is well documented that oral mucosa heals more rapidly than skin, with much reduced scarring, and it has been suggested that these cells represent a fetallike subpopulation of adult fibroblasts (Sloan, 1991). MSF may increase cell migration by stimulation of the production of HA by fibroblasts (Schor *et al.*, 1989). Addition of TGF-β1 to MSF-induced migrating fibroblasts inhibited MSF-induced cell migration and HA synthesis (Ellis *et al.*, 1992). TGF-β3, at low concentrations, showed some synergistic effects with MSF on HA synthesis and cell migration, but at high concentrations was antagonistic (Ellis, 1993). MSF has also been detected in adult wound fluid and may be synthesized by a small subpopulation of adult fibroblasts in response to wounding (Picardo *et al.*, 1992). Secretion of MSF during wound healing may be a signal for migration of fibroblasts into the wound, effected by increased HA synthesis

(Picardo *et al.*, 1992). In the fetal wound, high levels of MSF may accelerate fibroblast migration and maintain the high concentration of HA within the wound ECM, but in the adult wound, high levels of TGF-β1 (not found in fetal wounds) may quickly antagonize the action of any MSF present. The restriction of fibroblast motility and low HA concentrations in the adult wound may directly contribute to the scarring adult phenotype. This theory is further supported by the observation that adult oral mucosa, which secretes high levels of MSF and has many fetallike characteristics, does not scar. MSF may be similar if not identical to a domain of Fn and may be related to HASA detected in amniotic fluid and fetal serum, which has been suggested to be an important factor in maintaining the high concentrations of HA found in fetal tissue (Longaker *et al.*, 1990a).

There is still some dispute as to the mechanism of wound contraction and conflicting reports on wound contraction in the fetus. Gabbiani *et al.* (1972) first described the change in phenotype of fibroblasts in granulation tissue, with the expression of α-smooth muscle actin (ASMA), the actin isoform present in vascular smooth muscle cells. These cells were termed myofibroblasts and have also been observed to be permanently present in fibrocontractive diseases, such as hepatic cirrhosis and renal and pulmonary fibrosis (Schurch *et al.*, 1992). The current prevailing hypothesis is that the myofibroblast is responsible for wound contraction. Stress fibers are presumed to be the force-generating element and these forces are thought to be transduced to the ECM through the fibronexus. However, the importance of the myofibroblast in wound contraction is disputed. An alternative hypothesis has been proposed by Erlich and Rajartnum (1990), who postulated that individual fibroblasts reorientate the collagen fibrils associated with them and this "stressed" matrix transmits the force necessary for wound closure. Erlich (1988) stated that the control of wound contraction was linked to the composition of the ECM and that a type III collagen matrix contracted faster than a matrix composed of type I collagen. Gross and colleagues (1995) have proposed another mechanism of wound closure. These authors removed the central granulation tissue and epidermis from adult, full-thickness porcine wounds daily and found that these wounds still closed at a normal rate. The authors concluded that wound closure is brought about by the fibroblasts at the wound margins, which undergo a form of directed migration.

Not all fetal wounds undergo contraction and this appears to be species dependent. Rabbit fetal wounds apparently do not contract until after birth (Somasundaram and Prathap, 1970), whereas fetal sheep wounds are known to contract *in utero* (Hallock *et al.*, 1988). Ihara and Motobayashi (1992) found that fetal rat skin closes an open wound at 16 days gestation. They stated that the covering of the wound derives from "spreading" of the peripheral full-thickness skin surrounding the wound and no granulation tissue is required for wound closure, an observation similar to that made by Gross and colleagues (1995) in adult wounds. Martin and Lewis (1992), in contrast to other workers, suggested that the epidermis was important in fetal wound closure. They studied wound closure in fetal chickens and mice and proposed wound closure by a "purse string" mechanism, based on the observation of a continuous ring of actin cables within epithelial cells at the rim of the wounds. However, this method of wound closure is disputed by Gross and co-workers (1995), who found that wound closure was not dependent on epidermal contraction and occurred at an unchanged rate, even if the actin cable was cut.

Excisional wounds made on sheep fetuses at 100 days gestation do contract and contain myofibroblasts (Longaker *et al.,* 1991a). Estes and co-workers (1994) described the development and ultrastructural features of myofibroblasts in fetal sheep wounds from 75 to 120 days gestation. ASMA staining was absent from normal skin and wounds of 75-day gestation animals, except in vascular smooth muscle cells and pericytes. However, by 100 days gestation, substantial ASMA staining was observed, increasing from 3 to 7 days postwounding. ASMA staining was also strongly positive in the wounds of sheep of 120 days gestation. When the wound fibroblasts were cultured *in vitro,* all expressed ASMA, including those from 75-day gestation animals, indicating that the fibroblasts from the early gestation fetuses have the capability of expressing ASMA but lack the necessary signals to do so *in vivo.* Transition from scar-free healing to scarring in the sheep coincides with the appearance of ASMA expression by wound fibroblasts and the lack of myofibroblasts in nonscarring fetuses may be due to the absence of an appropriate inducing stimulus, such as TGF-β.

The differences in phenotype between adult and fetal fibroblasts may have a significant influence on the speed of wound healing and subsequent scarring. Fetal fibroblasts are active, relatively homogeneous and may be able to respond more quickly to wounding, whereas adult cells must undergo proliferation and phenotypic change. The causes and consequences of wound contraction are not yet clear, but it is possible that the mechanism of wound contraction has a direct bearing on the degree of scarring. Fetal fibroblasts may lack the necessary signals, possibly from growth factors, to change to a myofibroblast phenotype, and hence wound closure may proceed differently from that of the adult animal.

5. Inflammation and Growth Factors

One of the most fundamental differences between adult and fetal wound healing is the relative lack of an inflammatory response in the fetus and correspondingly low levels of inflammatory growth factors at the fetal wound site. In adult wound healing, one of the initial events is platelet degranulation, resulting in immediate release of a number of growth factors, including platelet-derived growth factor (PDGF), epidermal growth factor (EGF), and TGF-β. Neutrophils are immediately attracted to the wound site, where they phagocytose bacteria and damaged tissue. Macrophage infiltration follows and neutrophils are injested by the invading macrophages. TGF-β directly stimulates the influx of both neutrophils and macrophages (Wahl *et al.,* 1987). Macrophages at the wound site are activated by the growth factors released by degranulating platelets to produce a wide variety of growth factors themselves, including interleukin-1 (IL-1), FGF-2, PDGF, TGF-α, and TGF-β, forming a positive feedback loop for the recruitment of more macrophages to the wound site. Leibovich and Ross (1975) demonstrated severe inhibition of tissue debridement, delayed fibroblast proliferation, and general retardation of wound healing when adult wounds were depleted of monocytes and macrophages. TGF-β and PDGF are potent chemoattractants for fibroblasts and stimulate them to proliferate and produce new ECM constituents. These stimulated fibroblasts also produce TGF-β and PDGF in an autocrine feedback loop, similarly to macrophages (Pierce *et al.,* 1991).

Angiogenesis is a key component of granulation tissue formation and is stimulated particularly by vascular endothelial growth factor (VEGF), FGF-1, FGF-2, TGF-α, and TGF-β. The FGFs act directly on endothelial cells to stimulate their migration, proliferation, and tubule formation (Gospodarowicz *et al.,* 1987), while TGF-β appears to act indirectly, by the recruitment of more macrophages to the wound site, which in turn secrete angiogenic substances (Kiritsy and Lynch, 1993). PDGF-BB, secreted by macrophages and endothelial cells, may also play an important role in neovascularization, by inducing endothelial cell chemotaxis (Fiegel *et al.,* 1991). Concurrent with granulation tissue formation and inflammation is the process of reepithelialization. EGF and TGF-α directly increase the rate of reepithelialization, while keratinocyte growth factor (FGF-7) affects the rate of reepithelialization by the stimulation of keratinocyte proliferation (Brown *et al.,* 1986; Schultz *et al.,* 1987). Epithelial cells themselves are an important source of growth factors, which may function in an autocrine fashion on the epithelial cells themselves or in a paracrine manner to influence granulation tissue formation (Stadnyk, 1994). Indeed, it has been suggested that epithelial cells may be a more important source of PDGF and TGF-β than macrophages in the first few days after wounding (Antoniades *et al.,* 1991; Kane *et al.,* 1991). The factors that bring about resolution of the inflammatory response and cessation of granulation tissue formation have not been studied in detail. The production of "anti-inflammatory" cytokines such as IL-10, which has been shown to inhibit TGF-β synthesis (Van Vlasselaer *et al.,* 1994), may be significant in switching off the inflammatory response, and collagens can attenuate the response of dermal fibroblasts to TGF-β (Clark *et al.,* 1995). Gabianni and colleagues (Desmouliere *et al.,* 1995) have demonstrated that fibroblasts are removed from the wound site by apoptosis, and this may be another important mechanism in the resolution of the wound-healing response.

Fetal wound healing is fundamentally different from wound healing in the adult, since little inflammatory response is required for repair. A direct consequence of the lack of inflammation in the fetus may be the absence of scar tissue formation. A growing body of evidence directly implicates high levels of specific growth factors and TGF-β1 and -β2 in particular, with scarring during wound healing and with other forms of fibrosis. Three isoforms of TGF-β are present in mammals, which have similar activities *in vitro* but display marked differences in their spatial and temporal distribution during wound healing. In adult wounds, TGF-β2 and -β3 were prevalent by 24 hr after wounding and were strongly associated with the epidermis. In contrast, TGF-β1 was not detected at significant levels in the dermis until 5 days postwounding, when reepithelialization was complete (Levine *et al.,* 1993). These results support individual *in vivo* functions of TGF-β isoforms in wound repair. TGF-β1 is the isoform most commonly associated with fibrosis, although isoform-specific effects have not yet been widely investigated. *In vitro,* TGF-β1 has been shown to function both as an agonist and antagonist of cell proliferation and inflammation; however, it consistently acts on cells to induce deposition of ECM constituents (Roberts and Sporn, 1990).

Excessive accumulation of ECM is the chief pathological feature of fibrotic disease, and abnormal organization of ECM in wound healing results in scarring.

The exogenous addition of TGF-β1 accelerates the rate of healing in the rat, by increasing the rate of accumulation of total wound collagen (Sporn *et al.*, 1983), resulting in greater tensile strength of the wounds at early time points and a greater influx of monocytes and fibroblasts into the wound site (Mustoe *et al.*, 1987). Exogenously applied TGF-βs stimulate the expression of endogenous TGF-β1, partly by increased recruitment and/or activation of macrophages and partly by positive induction of the TGF-β1 promoter (Schmid *et al.*, 1993). The fibrotic potential of TGF-β has been shown when levels remain elevated. Intravenous injections of TGF-β in the rat produce marked fibrosis of the kidney and liver and also at the injection site (Terrell *et al.*, 1993). A marked systemic fibrosis was also observed in nude mice when TGF-β1 was applied intraperitoneally for 10 days (Zugmaier *et al.*, 1991). High levels of TGF-β have also been implicated in a wide variety of human fibrotic diseases, including glomerulonephritis (Yoshioka *et al.*, 1993), diabetic nephropathy (Yamamoto *et al.*, 1993), and lung fibrosis (Broeklmann *et al.*, 1991).

The distribution of growth factors in fetal wounds was investigated by Whitby and Ferguson (1991b), who studied the localization of PDGF, TGF-β, and FGF-2 in fetal, neonatal, and adult wounds, using immunocytochemical techniques. PDGF was detected at the wound site of fetal, neonatal, and adult wounds within 1 hr of wounding. However, the rate of clearance of PDGF in the different wounds studied varied with age. In the fetal wound, PDGF was no longer detected 48 hr postwounding; in the neonate, staining had disappeared by 72 hr postwounding; but in the adult, PDGF was still detected at 72 hr after wounding but was not present after 120 hr. TGF-β1 and -β2 were not detected within the fetal wound or at the wound margins at any of the time points studied, but were seen at 1, 6, and 12 hr postwounding at the wound surface and within the clot of both neonatal and adult wounds. Fetal wounds lack a fibrin clot and exhibit poor platelet degranulation. The initial release of growth factors on wounding and the storage of such factors in the clot prior to release is markedly different in the fetus compared to the adult. FGF-2 was not detected in the fetal wound, but was observed in both adult and neonatal wounds at 1 and 6 hr after wounding. The authors suggested that the continuing presence of PDGF at the adult wound site may be due to synthesis of this growth factor by cells such as macrophages recruited to the wound site but absent from fetal wounds. The lack of TGF-β in the fetal wound may also be due to the lack of an inflammatory response, a lack of degranulating platelets in the fetal wound, or a relative absence of TGF-β from fetal platelets. They also suggested that the absence of FGF-2 may be secondary to absence of an inflammatory response in the fetal wound. Lack of staining for TGF-β and bFGF does not mean an absolute absence of these growth factors from the fetal wound, but rather reduced levels below the detection limits of the antibodies employed. These results clearly illustrate major differences in the relative levels of these growth factors in adult and fetal wounds.

The lack of TGF-β1 and -β2 in fetal wounds was supported by studies carried out on the rabbit by Nath and colleagues (1994a). The distribution of TGF-β1 in the wounds of fetal mice was also investigated by Martin and colleagues (1993) using *in situ* hybridization techniques. Transcripts for TGF-β1 synthesis were found to be rapidly induced in epithelial cells at the wound margins within 1 to 3 hr of wounding and by 3 to 6 hr in the mesenchyme of the wound bed. No TGF-β3 induction was observed in the wounds, and the levels of TGF-β2 could not be determined due to limitations of the model used. TGF-β1 protein was detected by immunocytochemistry using an isoform-specific antibody, within 1 hr of wounding. However, the clearance of the staining was rapid, so that by 18 hr postwounding the levels of TGF-β1 had returned to near background. This was in contrast to the adult wound where sustained levels of TGF-β1 protein were detected (Kane *et al.*, 1991). Martin and colleagues (1993) suggested that the rapid appearance of TGF-β1 protein at the fetal wound site may not be due solely to gene expression, but to posttranslational release of intracellular stores. They postulate that the fetal environment may provide a richer source of TGF-β regulatory molecules than found in the adult and these may limit the availability of TGF-β within the fetal wound. Martin and co-workers (Hopkinson-Woolley *et al.*, 1994) confirmed the lack of macrophage infiltration into the fetal wound, the cell type that is the most potent and sustained source of TGF-β. Differences in the reported levels of TGF-β in fetal wounds may be explained by the use of antibodies, often with unreported specificities, which may detect latent or active TGF-β or both forms of the molecule. In general, it is accepted that TGF-β is transiently expressed in fetal wounds but is present at higher levels and for prolonged periods in adult wounds. However, in both fetus and adult, the levels of active TGF-β and their relative distributions are unknown.

Krummel and colleagues (1988) investigated the effects of exogenous application of TGF-β1 on the response of 24-day gestation fetal rabbits to wounding. PVA sponges containing TGF-β were implanted subcutaneously into the fetus and the histological responses documented. At 7 days postwounding, a grossly fibrotic reaction was observed in the fetus in response to TGF-β, with marked fibroblast proliferation and collagen accumulation, reactions not observed in response to the implant alone. These observations demonstrate that TGF-β can generate a fibrotic response and that the fetus is capable of responding to TGF-β. Durham and co-workers (1989) investigated the binding of TGF-β to embryonic (14 days gestation), fetal (24 days gestation), and adult rabbit fibroblasts in order to test if the observed *in vivo* effect of exogenous TGF-β in the fetal rabbit is due to a direct influence on the fibroblast. These authors demonstrated that embryonic and fetal fibroblasts possessed cell surface receptors for TGF-β, as did the adult cells. Furthermore, they described a progressive reduction in the numbers of these cell surface receptors during development. Lack of TGF-β-mediated fibrosis in fetal wound healing is therefore not due to the inability of the fetal fibroblast to respond to TGF-β. Endogenous TGF-β at the fetal wound site may be in the form of an inactive precursor, may be quickly inactivated, or may not be present in adequate quantities or for sufficient length of time to mediate a scarring response. The TGF-β receptor isoform profile (Derynck, 1994) of the fetal fibroblasts may also be different to that of the adult, so directing different responses to TGF-β isoforms.

Macrophage recruitment during wound healing in the fetal mouse has been investigated using a macrophage-specific monoclonal antibody (Hopkinson-Woolley *et al.*, 1994). Embryos were wounded at gestation times of 11.5 to 14.5 days. Macrophages were not recruited to the wound site in the early embryo during wound closure, and the onset of macrophage recruitment occurred at gestation day 14.5. This recruitment was coincident with the onset of scarring in the fetal mouse. The ability of gestation day 11.5 macrophages to respond to chemotactic stimuli was investigated by inflicting burn wounds, to produce localized cell necrosis (Hopkinson-Woolley *et al.*, 1994). In these animals, a significant number of macrophages were recruited to the burn wound site. This illustrates that embryonic macrophages are capable of mounting an inflammatory response in the presence of appropriate stimuli. Hence, the severity of wounding is an important determinant in the degree of inflammatory response seen in the fetus and therefore the degree of scarring. A similar correlation between the ability to mount an inflammatory response and the onset of scarring was observed in the pouch young of the marsupial *Monodelphis domesticus* (Armstrong and Ferguson, 1995).

It is evident that in fetal wounds, the absence of scarring correlates with a poor inflammatory response and that scar-free healing may be a consequence of low levels of active TGF-β1 and -β2 at the wound site. The presence of other growth factors at the fetal wound site have not yet been extensively investigated, but these levels may also have significant effects on scarring, either directly influencing collagen deposition or indirectly as integral members of the complicated cascade of growth factors directing the initiation and maintenance of the inflammatory response and fibroplasia.

6. Manipulation of Adult Wound Healing to Reduce Scarring

Research into the fundamental differences between adult wound healing and scar-free fetal healing have suggested several areas in which the adult wound could be manipulated to resemble more closely particular characteristics of the fetus, in the hope of reducing scarring within the adult wound. The major areas in which attention has been focused are modulation of the inflammatory response and growth factors in the adult wound and manipulation of ECM constituents toward a more fetallike profile.

6.1. Manipulation of the ECM

Fibronectin is deposited more quickly in fetal wound healing than in adult tissue repair (Longaker *et al.*, 1989c). Cheng and colleagues (1988) found that application of exogenous Fn to adult rat wounds accelerated the rate of closure of the wounds. They suggested that exogenously applied Fn may be a source of extra chemotactic fragments and may accelerate macrophage and fibroblast accumulation at the site of the injury. These results suggest that faster Fn deposition may be an important mechanism by which the fetus effects rapid reepithelialization. However, no advantageous effects on scarring of the adult wound were reported and the wounds were only examined for a short time after wounding (Cheng *et al.*, 1988).

The most striking difference between the ECM of fetal and adult wounds is the high concentration of HA at the fetal wound site; hence, a number of workers have attempted to reduce scarring after adult wounding by the exogenous application of HA.

Abatangelo and colleagues (1983) studied the rate of healing of wounds from normal and alloxan-diabetic rats, when treated with topical applications of HA. These wounds were not full thickness, but scratch incisions, which did not result in scarring. These authors reported a marked acceleration of the normally reduced rate of healing of the diabetic rats but little effect on the healing rate of normal control animals.

Hellstrom and Laurent (1987) applied 1% HA to the edges of rat tympanic membrane perforations and found that the healing rate was accelerated from closure in an average of 12 days to healing in 7 days. In the HA-treated wounds, the epithelium bridged the gap in advance of the connective tissue layer, and the authors suggested that exogenous HA may facilitate this earlier cell movement. Further studies indicated that HA application also markedly affected the structural quality of the healed membrane, with beneficial effects on scarring seen within 1 month (Laurent et al., 1988). The closure time of perforations and reduction in membrane scarring correlated with the concentration of HA applied.

The HA applied to wounds in these and other studies was extracted from rooster comb and is known not to be pure HA but a complex mixture. Burd and colleagues (1989) examined the composition of HA extracted from human skin. These workers found that HA was associated with collagen and sugars such as glucose and mannose. They reported that HA may also be complexed with elastin or other microfibrilar proteins. These authors demonstrated that tissue-extracted HA inhibited fibroblast replication in culture, whereas recombinant HA purified by fermentation did not, and the effect of tissue-extracted HA could not be abolished by the addition of hyaluronidase (Burd et al., 1991b). They proposed that HA–protein complexes play a significant role in the in vivo organization of scar tissue. They suggested that HA may act as a protective carrier, delivering growth factors and other proteins to the wound site. Fermentation-produced HA added exogenously to adult rat cutaneous incisional wounds had little effect on scarring (R. L. McCallion et al., unpublished observations).

6.2. Manipulation of Growth Factor Profile

Fetal wound healing does not elicit a major inflammatory response and as such is fundamentally different from the process of wound healing in the adult. Leibovich and Ross (1975) attempted to make the adult wound response more fetallike by inhibiting the infiltration of monocytes and macrophages into the wound site. However, wound healing was significantly retarded by these treatments, and it was concluded that the presence of some monocytes and macrophages is required for the process of adult wound repair. Other general anti-inflammatory therapies have been equally unsuccessful in reducing scarring. In our laboratory, suramin, an antiparasitic drug known to inhibit the binding of a number of growth factors (e.g., TGF-β, PDGF, bFGF, EGF) to their receptors, was applied intradermally to rat wounds (Chamberlain et al., 1995). The drug was found to have significant side effects and to inhibit wound repair. No benefit was conferred in terms of scar reduction. The general lack of success of anti-inflammatory agents in the reduction of scarring suggests that the causes of scarring are more complex than a simple consequence of the general inflammatory response. The alteration of specific growth factor levels may be more important in terms of scar prevention.

The significance of TGF-β in scarring and fibrosis has been documented by many workers. Studies have been carried out in our laboratory to determine the contribution of TGF-β to the scarring process, to establish the significance of the different isoforms of TGF-β within the wound, and to investigate if a reduction in the levels of TGF-β at the wound site reduces scarring in adult wounds. In the initial experiments, a polyclonal neutralizing antibody to TGF-β was used to lower the concentration of active TGF-β at the wound site. The antibody neutralized all isoforms of TGF-β but with different affinities so that β1 and β2 were bound to the greatest extent, but little TGF-β3 activity was neutralized. Neutralizing antibody was injected intradermally into adult rat wounds, just prior to wounding and on days 1 and 2 postwounding. A significant effect on scarring was noted by 42 days postwounding when antibody-treated wounds displayed much reduced scarring, with collagen fiber orientation much more similar to unwounded dermis than the parallel, dense, aligned collagen fibers seen in untreated scarred wounds (Shah *et al.,* 1992). Neutralizing antibody-treated wounds had a significantly reduced monocyte and macrophage profile compared to control wounds and substantially fewer new blood vessels. At 7 days postwounding, the antibody-treated wounds contained less collagen and fibronectin than untreated wounds, but measurements of tensile strength revealed no significant differences in the breaking strength of antibody-treated and control wounds. The lack of adverse effect of reduced collagen content on wound strength in the treated wounds was thought to be due to the more normal reticular pattern of collagen fiber orientation in these wounds. The neutralizing antibody must be administered at the time of wounding, or shortly thereafter, for significant anti-scarring effects to be observed. Neutralizing the TGF-β levels immediately after wounding may prevent the induction of TGF-β amplification at the wound site. This could result from modulation of monocyte and macrophage recruitment, inhibition of the TGF-β promotor, or reduction of the amount of TGF-β stored in the fibrin clot and elsewhere at the wound site. Neutralizing TGF-β may also reduce the levels of plasminogen activator inhibitor, and hence the relative levels of plasminogen activator and plasmin may be increased, leading to increased fibrinolysis. This would make the fibrin scaffold less compact and may facilitate fibroblast migration, and thus a more reticular orientation of collagen deposition (Shah *et al.,* 1994). Application of neutralizing antibodies to TGF-β have also proved effective in the reduction of fibrosis in conditions such as glomerulonephritis (Border *et al.,* 1990a) and pulmonary fibrosis (Giri *et al.,* 1993).

Isoform-specific neutralizing antibodies were also applied to wounds to determine the contribution of each TGF-β isoform to the scarring response (Shah *et al.,* 1995). The addition of neutralizing antibody to TGF-β2 alone had little effect on inflammatory cell infiltration, angiogenesis, or the resultant scar. Antibody to TGF-β1 alone resulted in some reduction of monocytes and macrophages at the wound site, but only marginal effects on scarring. Both TGF-β1 and -β2 must be neutralized to significantly reduce scarring in adult rodent cutaneous wounds. This indicates a synergistic effect of neutralization of both TGF-β1 and -β2 on scarring. Neutralizing antibodies to growth factors may have varying effects, depending on their concentration relative to the ligand, their affinity for the ligand, and the number of epitopes that they recognize. Generally, a low-affinity antibody or one at low concentration may have an agonistic effect similar to a binding protein, whereas a high-affinity antibody or one at a high concentration usually has neutralizing activity. It is likely that the size of the immune complex formed is critical in determining whether the antibody–growth factor complex is cleared or circulates as a carrier complex. A similar phenomenon has been described in mice with antibodies to interleukin-6 (Heremans *et al.*, 1992; Martens *et al.*, 1993). This makes antibody therapy in wound healing an attractive therapeutic option since different antibodies to TGF-β could be used to either accelerate wound healing or to reduce scarring.

Surprisingly, exogenous addition of the growth factor TGF-β3 at low concentrations was found to have marked antiscarring activity, similar to the effects of neutralizing TGF-β1 and -β2. However, angiogenesis was markedly increased by application of TGF-β3 in contrast to the neutralizing antibody treatment. Addition of TGF-β3 to adult rodent wounds resulted in a decreased monocyte and macrophage profile, reduced Fn deposition at 7 and 14 days postwounding, and a marked improvement in the orientation and organization of the collagen deposited (Shah *et al.*, 1995). This work clearly demonstrates isoform-specific differences in the roles of TGF-β in wound healing and cutaneous scarring. *In vivo,* TGF-β3 may act as a negative regulator of inflammation and of TGF-β1 and TGF-β2 at the wound site. Decreasing the levels of TGF-β1 and TGF-β2 by the addition of neutralizing antibody may result in a concomitant increase of TGF-β3 levels. Scar prevention evidently depends on a subtle alteration in the relative levels of TGF-β isoforms and it is the ratio of TGF-β3 compared to TGF-β1 and -β2 that is important in cutaneous scarring. This conclusion was supported by the finding that exogenous application of a panspecific antibody to TGF-β, which neutralized all isoforms equally, had no advantageous effect on scarring (Shah *et al.*, 1995). Preliminary experiments of neutralizing PDGF and adding exogenous FGF-2 at the time of wounding have shown some advantageous effects on subsequent scarring (Shah *et al.*, 1992), reinforcing the hypothesis that it is the relative ratios of growth factors that are important in scarring.

To investigate the importance of the relative levels of TGF-β isoform transcription or translation in scarring, antisense oligonucleotides to TGF-β1, -β2, or -β3 and appropriate sense or scrambled controls were applied to adult rat wounds (Chamberlain and Ferguson, 1995). *In vitro,* the antisense oligonucleotides inhibited 70% of TGF-β protein production in an isoform-specific fashion in a target cell line. Intradermal injection of the oligonucleotides before wounding was critical, since loading of the cells was achieved by exploiting the transient permeability of the cells at the wound edge upon incision. Antisense oligonucleotides to TGF-β1 and -β2 reduced cutaneous scarring, but to a much reduced extent when compared to either neutralizing antibody or exogenous TGF-β3 application. This may be expected, since antisense oligonucleotides can have no effect on the activation of TGF-β isoforms or on their release from cellular or ECM stores.

TGF-βs are secreted associated with a latency-associated peptide (LAP), as part of a biologically inactive complex (latent TGF-β) unable to interact with cell surface TGF-β receptors. Two of the three oligosaccharide side chains of the LAP have mannose-6-phosphate (M6P) as their terminal residues. M6P binds specifically and with high affinity to the cation-independent M6P/insulinlike growth factor II (IGF II) receptor and active TGF-β does not contain M6P residues (Purchio *et al.*, 1988). Binding of the inactive TGF-β–LAP complex to the M6P–IGF II receptor appears to be required for activation of the latent TGF-β complex (Dennis and Rifkin, 1991). This activation of TGF-β also requires the presence of plasmin, plasminogen activator (Dennis and Rifkin, 1991), and transglutaminase (Kojima *et al.*, 1993). Studies carried out in our laboratory, utilizing rat and pig wounds, have investigated the use of M6P as an antiscarring agent. M6P was applied to the wounds by intradermal injection, in similar regimens to the application of neutralizing antibodies. M1P was administered as a control carbohydrate, which binds to the mannose receptor, as does M6P, but does not bind to the M6P receptor. At early time points, accelerated collagen deposition was observed in wounds treated with both M6P and M1P. Fewer monocytes and macrophages infiltrated the M6P-treated wounds at 7 days postwounding compared to M1P-treated and control wounds. By 40 days postwounding, the dermal collagen architecture of the M6P-treated wounds was similar to unwounded skin and a reduction in scarring was evident, an effect not observed when M1P was applied (McCallion *et al.*, manuscript in preparation). The mechanism of action of M6P in reducing scarring may involve (1) inhibition of activation of latent TGFβ, (2) reduction in the cellular sequestration of platelet-released latent TGF-β by cells at the wound site, (3) inhibition of formation of the large latent TGF-β complex by prevention of latent TGF-β-binding protein polymerization to the LAP at the M6P receptor, and (4) induction of changes in the number or state of activation of inflammatory cells recruited to the wound site. The diverse activities of M6P may result from binding to both the mannose (early modulation of collagen synthesis) and M6P receptors (antiscarring), making it an extremely attractive therapeutic agent for wound repair.

Collectively, these adult manipulation studies suggest that there is a large therapeutic window for improving the quality of adult wound repair by the prevention of scarring. Adult wound healing may have been optimized by evolutionary selection for speed of healing under dirty conditions. Consequently, the inflammatory response and cytokine profile is excessive, resulting in scarring. With contemporary hygiene and wound care, it becomes possible to manipulate this "overdrive" response and so markedly influence the quality of wound repair, without adversely affecting wound strength or the speed of repair. Such manipulations may include alterations of the concentration of a number of active growth factors, probably soon after injury. Identification of such factors will be facilitated by further careful comparisons between molecular mechanisms of scar-free fetal wound healing and scar-forming adult wound healing. Another great challenge for adult wound healing is the discovery of therapies that will improve the quality of an existing scar.

References

Abatangelo, G., Martelli, M., and Vecchia, P., 1983, Healing of hyaluronic acid enriched wounds: Histological observations, *J. Surg. Res.* **35**:410–416.

Adzick, N. S., Outwater, K. M., Harrison, M. R., Davies, P., Glick, P. L., deLorimer, A. A., and Reid, L. M., 1985a, Correction of congenital diaphragmatic hernia *in utero*. IV An early gestational age fetal lamb model for pulmonary vascular morphometric analysis, *J. Pediatr. Surg.* **20**:673–680.

Adzick, N. S., Harrison, M. R., Glick, P. L., Beckstead, J. H., Villa, R. L., Schevenstuhl, H., and Goodson, W. H., 1985b, Comparison of fetal, newborn and adult wound healing by histologic, enzyme-histochemical and hydroxyproline determination, *J. Pediatr. Surg.* **20**:315–319.

Andres, J. L., Stanley, K., Cheifetz, S., and Massague, J., 1989, Membrane-anchored and soluble forms of betaglycan, a polymorphic proteoglycan that binds transforming growth factor, *J. Cell Biol.* **109**:3137–3145.

Andres, J. L., Ronnstrand, L., Cheiftz, S., and Massague, J., 1991, Purification of the TGFβ binding proteoglycan betaglycan, *J. Biol. Chem.* **266**:23282–23287.

Antoniades, H. N., Galanolpoulos, T. G., Neville-Golden, J., Kiritsy, C. P., and Lynch, S. E., 1991, Injury induces *in vivo* expression of platelet-derived growth factor (PDGF) and PDGF mRNAs in skin epithelial cells and PDGF mRNAs in connective tissue fibroblasts, *Proc. Natl. Acad. Sci. USA* **88**:565–569.

Armstrong, J. R., and Ferguson, M. W. J., 1995, Ontogeny of the skin and the transition from scar-free to scarring phenotype during wound healing in the pouch young of a marsupial *Monodelphis domestica*, *Dev. Biol.* **169**(1):242–260.

Aruffo, A., Stamenkovic, I., Melnick, M., Underhill, C. B., and Seed, B., 1990, CD44 is the principle cell surface receptor for hyaluronate, *Cell* **61**:1303–1313.

Attisano, L., Wrana, J. L., Lopes-Casillas, F., and Massague, J., 1994, TGFβ receptors and actions, *Biochim. Biophys. Acta.* **1222**:71–80.

Balazs, E. A., and Darzynkiewicz, Z., 1973, The effect of hyaluronic acid on fibroblasts, mononuclear phagocytes and lymphocytes, in: *Biology of the Fibroblast* (E. Kulonen and J. Pikkarainen, eds.), pp. 237–252, Academic Press, New York.

Balza, E., Borsi, L., Allemanni, G., and Zardi, L., 1988, Transforming growth factor-β regulates the levels of different fibronectin isoforms in normal human cultured fibroblasts, *FEBS Lett.* **228**:42–44.

Bertolami, C. N., and Dunoff, R. B., 1978, Hyaluronidase activity during open wound healing in rabbits: A preliminary report, *J. Surg. Res.* **25**:256–259.

Bianco, P., Fisher, L. W., Young, M. F., Termine, J. D., and Robey, P. G., 1990, Expression and localisation of the two small proteoglycans biglycan and decorin in developing human skeletal and non-skeletal tissues, *J. Histochem. Cytochem.* **38**:1549–1563.

Birk, D. E., Fitch, J. M., Barbiaz, J. P., and Linsenmayer, T. F., 1988, Collagen type I and type V are present in the same fibril in the avian corneal stroma, *J. Cell Biol.* **106**:999–1008.

Birk, D. E., Fitch, J. M., Babiarz, J. R., Doane, K. J., and Linsenmayer, T. M., 1990, Collagen fibrillogenesis *in vitro*: Interaction of types I and V collagen regulates fibril diameter, *J. Cell Sci.* **95**:649–657.

Bonner-Frazer, M., 1985, Alterations in neural crest migration by a monoclonal antibody that affects cell adhesion, *J. Cell Biol.* **101**:610–617.

Boon, L., Manicourt, D., Marbaix, E., Vandenabeele, M., and Vanwijck, R., 1992, A comparative analysis of surgical cleft lip corrected *in utero* and in neonates, *Plast. Reconstr. Surg.* **89**:11–17.

Border, W. A., and Ruoshlati, E., 1992, Transforming growth factor-β in disease: The dark side of tissue repair, *J. Clin. Invest.* **90**:1–5.

Border, W. A., Okuda, S., Languino, L. R., Sporn, M. B., Ruoslahti, E., 1990a, Suppression of experimental glomerulophritis by anti-serum against transforming growth factor-β, *Nature* **346**:371–374.

Border, W. A., Okuda, S., Languino, L. R., and Ruoslahti, E., 1990b, Transforming growth factor beta regulates production of proteoglycans by mesangial cells, *Kidney Int.* **37**:689–695.

Boucaut, J. C., Darriebere, J., Boulekbache, H., and Thiery, J. P., 1984, Prevention of gastrulation but not neuralation by antibodies to fibronectin in amphibian embryos, *Nature* **307**:364–367.

Bowersox, J. C., and Sorgente, N., 1982, Chemotaxis of aortic endothelial cells in response to fibronectin, *Cancer Res.* **42**:2547–2551.

Brecht, M., Mayer, U., and Schlosser, E., 1986, Increased hyaluronic acid synthesis is required for fibroblast detachment and mitosis, *Biochem. J.* **239**:445–450.

Broeklmann, T. J., Limper, A. M., Colby, T. V., and McDonald, J. A., 1991, Transforming growth factor β1 is present at sites of extracellular matrix gene expression in human pulmonary fibrosis, *Proc. Natl. Acad. Sci. USA* **88**:6642–6646.

Brown, G. L., Curtsinger, L., Brightwell, J. R., Ackerman, D. M., Tobin, G. R., Polk, H. C., George-Nascimento, C., Valenzuela, P., and Schultz, G. S., 1986, Enhancement of epidermal regeneration by biosynthetic epidermal growth factor, *J. Exp. Med.* **163**:1319–1324.

Burd, D. A. R., Siebert, J. W., Ehrlich, H. P., and Garg, H. G., 1989, Human skin and post-burn hyaluronan: Demonstration of the association with collagen and other proteins, *Matrix* **9**:322–327.

Burd, D. A. R., Longaker, M. T., Adzick, N. S., Harrison, M. R., and Erlich, H. P., 1990, Fetal wound healing in a large animal model: The deposition of collagen is confirmed, *Br. J. Plast. Surg.* **43**:571–577.

Burd, D. A. R., Longaker, M. T., Rittenberg, T., Adzick, N. S., Harrison, M. R., and Erlich, H. P., 1991a, *In vitro* foetal wound contraction: The effect of amniotic fluid, *Br. J. Plast. Surg.* **44**:302–305.

Burd, D. A. R., Greco, R. M., Regauer, S., Longaker, M. T., Siebert, J. W., and Garg, H. G., 1991b, Hyaluronan and wound healing: A new perspective, *Br. J. Plast. Surg.* **44**:579–584.

Burrington, J. D., 1971, Wound healing in the fetal lamb, *J. Pediatr. Surg.* **6**:523–528.

Chamberlain, J., and Ferguson, M. W. J., 1995, Use of antisense oligonucleotides to TGFβ in adult wound repair, *J. Invest. Dermatol.,* in press.

Chamberlain, J., Shah, M., and Ferguson, M. W. J., 1995, The effect of suramin on healing adult rodent dermal wounds, *J. Anat.* **186**:87–96.

Chandrakasan, G., Rutka, J., and Stern, R., 1986, Hyaluronic acid stimulates collagen synthesis and levels of type III collagen in cultures of human fibroblasts (Abstract), *J. Cell Biol.* **103**:252.

Cheng, C. Y., Martin, D. E., Leggett, C. G., Reece, M. C., and Reese, A. C., 1988, Fibronectin enhances healing of excised wounds in rats, *Arch. Dermatol.* **124**:221–225.

Chiquet-Ehrismann, R., 1990, What distinguishes tenascin from fibronectin? *FASEB J.* **4**:2598–2604.

Chiquet-Ehrismann, R., Kalla, P., Pearson, C. A., Beck, K., and Chiquet, M., 1988, Tenascin interferes with fibronectin action, *Cell* **53**:383–390.

Chiu, E. S., Longaker, M. T., and Adzick, N. S., Stern, M., Harrison, M. P., and Stern, R., 1990, Hyaluronic acid patterns in fetal and adult wound fluid, *Surg. Forum* **41**:636–639.

Clark, R. A. F., 1988, Potential roles of fibronectin in cutaneous wound repair, *Arch. Dermatol.* **124**:201–206.

Clark, R. A. F., 1993, Regulation of fibroplasia in cutaneous wound repair, *Am. J. Med. Sci.* **306**:42–48.

Clark, R. A. F., Winn, H. J., Dvorak, H. F., and Colvin, R. B., 1983, Fibronectin beneath re-epithelialising epidermis *in vivo,* sources and significance, *J. Invest. Dermatol.* **80**:26–30S.

Clark, R. A. F., Nielsen, L. D., Welch, M. P., and McPherson, J. M., 1995, Collagen matrices attenuate the collagen-synthetic response of cultured fibroblasts to TGFβ, *J. Cell Sci.* **108**:1251–1261.

Damon, D. H., Lobb, R. R., D'Amore, P. A., and Wagner, J. A., 1989, Heparin potentiates the action of acidic fibroblast growth factor by prolonging its biological half life, *J. Cell Physiol.* **138**:221–226.

Daniloff, J. K., Crossin, K. L., Pincon-Raymond, M., Murawsky, M., Rieger, F., and Edelman, G. M., 1989, Expression of cytotactin in the normal regenerating neuromuscular system, *J. Cell Biol.* **108**:625–635.

Dennis, P. A., and Rifkin, D. B., 1991, Cellular activation of latent TGFβ requires binding to the cation-independent M6P/IGF II receptor, *Proc. Natl. Acad. Sci. USA* **88**:580–584.

Depalma, R. L., Krummel, T. M., Durham, L. A., Michna, B. A., Thomas, B. L., Nelson, J. M., and Diegelmann, R. F., 1989, Characterisation and quantitation of wound matrix in the fetal rabbit, *Matrix* **9**:224–231.

Derynck, R., 1994, TGFβ receptor-mediated signalling, *Trends Biochem. Sci.* **19**:548–553.

Desmouliere, A., Redard, M., Darby, I., and Gabbiani, G., 1995, Apoptosis mediates the decrease in cellularity during the transition between granulation tissue and scar, *Am. J. Pathol.* **146**:56–66.

Dillon, P. W., Keefer, K., Blackburn, J. H., Houghton, P. E., and Krummel, T. M., 1994, The extracellular matrix of the fetal wound: Hyaluronic acid controls lymphocyte adhesion, *J. Surg. Res.* **57**:170–173.

Donaldson, D. J., and Mahan, J. T., 1983, Fibrinogen and fibronectin as substrates for epidermal cell migration during wound closure, *J. Cell Sci.* **62**:117–127.

Duband, J. L., Darriebere, T., Boucaut, J. C., Boulekbache, H., and Thiery, J. P., 1987, Regulation of development by the extracellular matrix, in: *Cell Membranes: Methods and Reviews* (E. L. Elson, W. A. Frazier, and L. Glaser, eds.), pp. 1–53, Plenum Press, New York.

Duband, J. L., Dufor, S., and Thiery, J. P., 1988, Extracellular matrix–cytoskeleton interactions in locomoting embryonic cells, *Protoplasma* **145:**112–119.

Dunphy, J. E., and Upuda, K. N., 1955, Chemical and histochemical sequences in the normal healing of wounds, *N. Engl. J. Med.* **235:**847–851.

Durham, L. A., Krummel, T. M., Cawthorn, J. W., Thomas, B. L., and Diegelmann, R. F., 1989, Analysis of transforming growth factor beta receptor binding in embryonic, fetal and adult rabbit fibroblasts, *J. Pediatr. Surg.* **24:**784–788.

Elenius, K., Vainio, S., Laato, M., Samivirta, M., Thesleff, I., and Jalkanen, M., 1991, Induced expression of syndecan in healing wounds, *J. Cell Biol.* **114:**585–595.

Ellis, I., Grey, A. M., Schor, A. M., and Schor, S. L., 1992, Antagonistic effects of TGF-β1 and MSF on fibroblast migration and hyaluronic acid synthesis: Possible implications for dermal wound healing, *J. Cell Sci.* **102:**447–456.

Ellis, I. R., 1993, *Migration Stimulating Factor: Biochemical Characterisation, Mode of Action,* PhD thesis, University of Manchester.

Epstein, E. H., 1974, (alpha 1 (3)) Human skin collagen. Release by pepsin digestion and preponderance in fetal life. *J. Biol. Chem.* **249:**3225–3231.

Erickson, H. P., 1993, Tenascin-C, tenascin-R and tenasin-R and tenascin-X: A family of talented proteins in search of functions, *Curr. Opin. Cell Biol.* **5:**869–876.

Erickson, H. P., and Bourdon, M. A., 1989, Tenascin: An extracellular matrix protein prominent in specialised embryonic tissues and tumours, *Annu. Rev. Cell Biol.* **5:**71–92.

Erlich, H. P., 1988, Wound closure: Evidence of co-operation between fibroblasts and collagen matrix, *Eye* **2:**149–157.

Erlich, H. P., and Rajartnum, J. M. B., 1990, Cell locomotion forces versus cell contraction forces for collagen lattice contraction: *In vitro* model of wound contraction, *Tissue Cell* **22:**407–417.

Estes, J. H., Spencer, E. M., Longaker, M. T., and Adzick, N. S., 1991, Insulin-like growth factor II in ovine wound fluid. Evidence for developmental regulation, *Surg. Forum* **42:**659–661.

Estes, J. M., Adzick, N. S., Harrison, M. R., Longaker, M. T., and Stern, R., 1993, Hyaluronate metabolism undergoes an ontogenic transition during fetal development: Implications for scar-free wound healing, *J. Pediatr. Surg.* **28:**1227–1231.

Estes, J. M., Van de Berg, J. S., Adzick, N. S., MacGillivray, T. E., Desmouliere, A., and Gabbiani, G., 1994, Phenotypic and functional features of myofibroblasts in sheep wounds, *Differentiation* **56:**173–181.

Feinberg, R. N., and Beebe, D. L., 1983, Hyaluronate in vasculogenesis, *Science* **220:**1177–1179.

Ferguson, M. W. J., and Howarth, G. F., 1991, Marsupial models of scarless fetal wound healing, in: *Fetal Wound Healing,* 1st ed. (N. S. Adzick and M. T. Longaker, eds.), pp. 92–125, Elsevier, Holland.

Ferguson, M. W. J., Shah, M., Armstrong, J., Whitby, D. J., and Longaker, M. T., 1995, Scar formation. The spectral nature of fetal and adult wound repair, *Plas. Reconstr. Surg.,* in press.

ffrench-Constant, C., and Hynes, R. O., 1988, Patterns of fibronectin gene expression and splicing during cell migration in chicken embryos, *Development* **104:**369–382.

ffrench-Constant, C., and Hynes, R. O., 1989, Alternative splicing of fibronectin is temporally and spatially regulated in the chicken embryo, *Development* **106:**375–388.

ffrench-Constant, C., Van De Water, L., Dvorak, H. F., and Hynes, R. O., 1989, Reappearance of an embryonic pattern of fibronectin splicing during wound healing in the adult rat, *J. Cell Biol.* **109:**903–914.

Fiegel, V. D., Penner, B. G., Wohl, R. C., and Knighton, D. R., 1991, PDGF-BB induces wound capillary endothelial cell chemotaxis, Wound Healing Society Programme Abstracts, no. 1.

Fleischamajer, R., Fisher, L. W., MacDonald, E. D., Jacobs, L., Perlish, J. S., and Termine, J. D., 1991, Decorin interacts with fibrillar collagen of embryonic and adult human skin, *J. Struct. Biol.* **106:**82–90.

Frantz, F. W., Diegelmann, R. F., Mast, B. A., and Cohen, K., 1992, Biology of fetal wound healing: Collagen biosynthesis during dermal repair, *J. Pediatr. Surg.* **27:**945–949.

Frantz, F. W., Bettinger, D. A., Haynes, J. H., Johnson, D. E., Harvey, K. H., Dalton, H. P., Yager, D. R., Diegelmann, R. F., and Cohen, I. K., 1993, Biology of fetal repair: The presence of bacteria in fetal wounds induces an adult-like healing response, *J. Pediatr. Surg.* **78:**428–434.

Frost, S. J., and Weigel, P. H., 1990, Binding of hyaluronic acid to mammalian fibrinogens, *Biochim. Biophys. Acta* **1034:**39–45.

Gabbiani, G., Hirschel, B. J., Ryan, G. B., Statkov, P. R., and Majno, G., 1972, Granulation tissue as a contractile organ: A study of structure and function, *J. Exp. Med.* **135:**719–734.

Gao, X. X., Devoe, L. D., and Given, K. S., 1994, Effects of amniotic fluid on proteases—A possible role of amniotic fluid in fetal wound healing, *Ann. Plast. Surg.* **33:**128–134.

Garin-Chesa, P., Old, L. J., and Rettig, W. J., 1990, Cell surface glycoprotein of reactive stromal fibroblasts as a potential antibody target in human epithelial cancers, *Proc. Natl. Acad. Sci. USA* **87:**7235–7239.

Giri, S. N., Hyde, D. M., and Hollinger, M. A., 1993, Effect of antibody to transforming growth factor-β on bleomycin induced accumulation of lung collagen in mice, *Thorax* **48:**959–966.

Gospodarowicz, D., Ferrara, N., Schweigere, L., and Neufeld, G., 1987, Structural characterisation and biological functions of fibroblast growth factor, *Endocrinol. Rev.* **8:**95–114.

Goss, A. N., 1977, Intrauterine healing of fetal rat oral mucosal, skin and cartilage wounds, *J. Oral Pathol.* **6:**35–38.

Graham, M. F., Diegelmann, R. F., and Cohen, I. K., 1984, An *in vitro* model of fibroplasia: Simultaneous quantification of fibroblast proliferation, migration and collagen synthesis, *Proc. Soc. Exp. Med.* **176:**302–308.

Grinnell, F., Billingham, R. E., and Burgess, L., 1981, Distribution of fibronectin during wound healing *in vivo*, *J. Invest. Dermatol.* **76:**181–189.

Gross, J., Farinelli, W., Sadow, P., Anderson, R., and Bruns, R., 1995, On the mechanism of skin wound contraction. A granulation tissue knockout with a normal phenotype, *Proc. Natl. Acad. Sci. USA* **92:**5982–5986.

Hallock, G., Rice, D. C., Merkel, J. R., and DiPaolo, B. R., 1988, Analysis of collagen content in the fetal wound, *Ann. Plast. Surg.* **21:**310–315.

Hallock, G. G., Merkel, J. R., Rice, D. C., and DiPaolo, B. R., 1993, The ontogenic transition of collagen deposition in rat skin, *Ann. Plast. Surg.* **30:**239–243.

Hardingham, T. E., and Bayliss, M. T., 1990, Proteoglycans of articular cartilage changes in ageing and joint disease, *Semin. Arthritis Rheum. Suppl.* **1:**12–33.

Hedbom, E., and Heinegard, D., 1989, Interaction of a 59 kDa connective tissue matrix protein with collagen I and II, *J. Biol. Chem.* **264:**6898–6905.

Hedlund, H., Mengarelli-Widholm, S., Heinegard, D., Reinholt, F. P., and Svensson, O., 1994, Fibromodulin distribution and association with collagen, *Matrix Biol.* **14:**227–232.

Hellstrom, S., and Laurent, C., 1987, Hyaluronan and healing of tympanic membrane perforations: An experimental study, *Acta. Otolaryngol.* **S442:**54–61.

Hemesath, T. J., Marton, L. S., and Stefansson, K., 1994, Inhibition of T cell activation by the extracellular matrix protein tenascin, *J. Immunol.* **152:**5199–5207.

Heremans, H., Dillen, C., Put, W., Vandamme, J., and Billiau, A., 1992, Protective effect of anti-interleukin (IL) -6 antibody against endotoxin, associated with paradoxically increased IL-6 levels, *Eur. J. Immunol.* **22:**2395–2401.

Hildebrand, A., Romaris, M., Rasmussen, L. M., Heinegard, D., Twardzik, D. R., Border, W. A., and Ruoslahti, E., 1994, Interaction of small interstitial proteoglycans biglycan, decorin and fibromodulin with transforming growth factor-beta, *Biochem. J.* **302:**527–534.

Hopkinson-Woolley, J., Hughes, D., Gordon, S., and Martin, P., 1994, Macrophage recruitment during limb development and wound healing in the embryonic and fetal mouse, *J. Cell Sci.* **107:**1159–1167.

Horne, R. S. C., Hurley, J. V., Crowe, D. M., Ritz, M. H., McCO'Brien, B., and Arnold, L. I., 1992, Wound healing in fetal sheep: a histological and electron microscope study, *Br. J. Plast. Surg.* **45:**333–345.

Hunt, T. K., Zederfeldt, B., and Goldstick, T. K., 1961, Oxygen and wound healing, *Am. J. Surg.* **118:**521–525.

Ignotz, R. A., and Massague, J., 1987, Cell adhesion protein receptors as targets for transforming growth factor β action, *Cell* **51:**189–197.

Ignotz, R. A., Endo, T., and Massague, J., 1987, Regulation of fibronectin and type I collagen mRNA levels by transforming growth factor β, *J. Biol. Chem.* **262:**6443–6446.

Ihara, S., and Motobayashi, Y., 1992, Wound closure in fetal rat skin, *Development* **114:**573–582.

Ihara, S., Motobayashi, Y., Nagao, E., and Kistler, A., 1990, Ontogenic transition of wound healing pattern in rat skin occurring at the rat fetal stage, *Development* **110:**671–680.

Jackson, R. L., Busch, S. J., and Cardin, A. L., 1991, Glycosaminoglycans: Molecular properties, protein interactions and role in physiological processes, *Physiol. Rev.* **71:**481–539.

Jahoda, C. A. B., and Oliver, R. F., 1984, Histological studies of the effects of wounding vibrissa follicles in the hooded rat, *J. Embryol. Exp. Morphol.* **83:**95–108.

Jarnagin, W. R., Rockey, D. C., Koteliansky, V. E., Wang, S.-S., and Bissell, D. M., 1994, Expression of variant fibronectins in wound healing: Cellular source and biological activity of the EIIIA segment in rat hepatic fibrogenesis, *J. Cell Biol.* **127:**2037–2048.

Julia, M. V., Albert, A., Morales, L., Miro, D., Sancho, M. A., and Garcia, X., 1993, Wound healing in the fetal period: The resistance of the scar to rupture, *J. Pediatr. Surg.* **28:**1458–1462.

Kahari, V. M., Larjava, H., and Uitto, J., 1991, Differential regulation of extracellular matrix proteoglycan gene expression, *J. Biol. Chem.* **266:**10608–10615.

Kane, C. J. M., Mansbridge, J. N., Hebda, P. A., and Hanawalt, P. C., 1991, Direct evidence for spatial and temporal regulation of transforming growth factor β1 expression during cutaneous wound healing, *J. Cell Physiol.* **148:**157–173.

Keene, D. R., Sakai, L. Y., Bachinger, H. P., and Burgeson, R. E., 1987, Type III collagen can be present on banded collagen fibrils regardless of fibril diameter, *J. Cell Biol.* **105:**2393–2402.

Kiritsy, C. P., and Lynch, S. E., 1993, Role of growth factors in cutaneous wound healing: A review, *Crit. Rev. Oral Biol. Med.* **4:**729–760.

Knight, K. R., Horne, R. S. C., Lepore, D. A., Kumta, S., Ritz, M., Hurley, J. V., and McCO'Brien, B., 1994, Glycosaminoglycan composition of uninjured skin and of scar tissue in fetal, newborn and adult sheep, *Res. Exp. Med.* **194:**119–127.

Knudson, W., Biswas, C., Li, X. Q., Nemec, R. E., and Toole, B. P., 1989, The role and regulation of tumour-associated hyaluronan, in: *The Biology of Hyaluronan,* (CIBA Foundation Symposium), (D. Evered and J. Whelan, eds.), pp. 150–169, John Wiley and Sons, Chichester, England.

Kojima, S., Nara, K., and Rifkin, D. B., 1993, Requirement for transglutaminase in the activation of latent transforming growth factor β in bovine endothelial cells, *J. Cell Biol.* **121:**439–448.

Krummel, T. M., Nelson, J. M., Diegelmann, R. F., Linblad, W. J., Salzberg, A. M., Greenfield, L. J., and Cohen, I. K., 1987, Fetal response to injury in the rabbit, *J. Pediatr. Surg.* **22:**640–644.

Krummel, T. M., Michna, B. A., Thomas, B. L., Sporn, M. B., Nelson, J. M., Salzberg, I. K., and Diegelmann, R. F., 1988, Transforming growth factor beta (TGF-β) induces fibrosis in a fetal wound model, *J. Pediatr. Surg.* **23:**647–652.

Kuhn, K., 1987, The classical collagens: Types I, II and III, in: *Structure and Function of Collagen Types* (R. Mayne and R. E. Burgeson, eds.), pp. 1–42, Academic Press, New York.

Kujawa, M. J., and Tepperman, K., 1983, Culturing chick muscle cells on glycosaminoglycan substrates: Attachment and differentiation, *Dev. Biol.* **99:**277–286.

Kujawa, M. J., Pechak, D. G., Fiszman, M. Y., Caplan, A. I., 1986, Hyaluronic acid bonded to cell culture surfaces inhibits the program of myogenesis, *Dev. Biol.* **113:**10–16.

Kumar, S., Kumar, P., Ponting, J. M., Sattar, A., Rooney, P., Pye, D., and Hunter, R. D., 1992, Hyaluronic acid promotes and inhibits angiogenesis, in: *Angiogenesis in Health and Disease* (M. E. Maragoudakis, P. Lelkes, and P. M. Gullino, eds.), pp. 253–263, Plenum Press, New York.

Kurkinen, M., Vaheri, A., Roberts, P. J., and Stenman, S., 1980, Sequential appearance of fibronectin and collagen in experimental granulation tissue, *Lab. Invest.* **43:**47–51.

Lane, A. T., Scott, G. A., and Day, K. H., 1989, Development of fetal skin transplanted onto nude mice, *J. Invest. Dermatol.* **93:**787–791.

Laurent, T. C., 1987, Biochemistry of hyaluronan, *Acta Otolaryngol.* (Stockh.) **442(Suppl):**7–24.

Laurent, T. C., Hellstrom, S., and Fellenius, E., 1988, Hyaluronan improves the healing of experimental membrane perforations. *Arch. Otolaryngol. Head Neck Surg.* **114:**1435–1441.

Le Boeuf, R. D., Raja, R. H., Fuller, G. M., and Weigel, P. H., 1986, Human fibrinogen specifically binds hyaluronic acid, *J. Biol. Chem.* **261:**12586–12592.

Lee, W. H., Bowsher, R. R., Apathy, J. M., Smith, M. M., and Henry, D. P., 1991, Measurement of insulin-like growth factor II in physiological fluids and tissues. II Extraction and quantification in rat tissues, *Endocrinology* **128**:815–822.

Leibovich, S. J., and Ross, R., 1975, The role of the macrophage in wound repair. A study with hydrocortisone and anti-macrophage serum, *Am. J. Pathol.* **78**:71–100.

Levine, J. H., Moses, H. L., Gold, L. I., and Nanney, L. B., 1993, Spatial and temporal patterns of immunoreactive transforming growth factor β1, β2 and β3 during excisional wound repair, *Am. J. Pathol.* **143**:368–380.

Lightner, V. A., and Erickson, H. P., 1990, Binding of hexabrachion (tenascin) to the extracellular matrix and substratum and its effect on cell adhesion, *J. Cell Sci.* **95**:263–277.

Longaker, M. T., Harrison, M. R., Crombleholme, T. M., Langer, J. C., Decker, M., Verrier, E. D., Spendlove, R., and Stern, R., 1989a, Studies in fetal wound healing: I A factor in fetal serum that stimulates deposition of hyaluronic acid, *J. Pediatr. Surg.* **24**:789–792.

Longaker, M. T., Harrison, M. R., Langer, J. C., Crombleholme, T. M., Verrier, E. D., Spendlove, R., and Stern, R., 1989b, Studies in fetal wound healing: II A fetal environment accelerates fibroblast migration *in vitro*, *J. Pediatr. Surg.* **24**:793–798.

Longaker, M. T., Whitby, D. J., Ferguson, M. W. J., Harrison, M. R., Crombleholme, T. M., Langer, J. C., Cochrum, K. C., Verrier, E. D., and Stern, R., 1989c, Studies in fetal wound healing: III Early deposition of fibronectin distinguishes fetal from adult wound healing, *J. Pediatr. Surg.* **24**:799–805.

Longaker, M. T., Adzick, N. S., Hall, J. L., Stair, S. E., Crombleholme, T. M., Duncan, B. W., Bradley, S. M., Harrison, M. R., and Stern, R., 1990a, Studies in fetal wound healing: VII Fetal wound healing may be modulated by hyaluronic acid stimulating activity in amniotic fluid, *J. Pediatr. Surg.* **25**:430–433.

Longaker, M. T., Whitby, D. J., Adzick, N. S., Crombleholme, T. M., Langer, J. C., Duncan, B. W., Bradley, S. M., Stern, R., Ferguson, M. W., and Harrison, M. R., 1990b, Studies in fetal wound healing. VI Second and early third trimester fetal wounds demonstrate rapid collagen deposition without scar formation, *J. Pediatr. Surg.* **25**:63–68.

Longaker, M. T., Burd, A. R., Gowen, A. H., Yen, T. S. B., Jennings, R. W., Duncan, B. W., Harrison, M. R., and Adzick, N. S., 1991a, Midgestational excisional fetal lamb wounds contract *in utero*, *J. Pediatr. Surg.* **26**:942–948.

Longaker, M. T., Chiu, E. S., Adzick, N. S., Stern, M., Harrison, M. R., and Stern, R., 1991b, Studies in fetal wound healing. A prolonged presence of hyaluronic acid characterizes fetal wound healing, *Ann. Surg.* **213**:292–296.

Longaker, M. T., Whitby, D. J., Jennings, R. W., Duncan, B. W., Ferguson, M. W., Harrison, M. R., and Adzick, N. S., 1991c, Fetal diaphragmatic wounds heal with scar formation, *J. Surg. Res.* **50**:375–385.

Longaker, M. T., Whitby, D. J., Ferguson, M. W. J., Lorenz, H. P., Harrison, M. R., and Adzick, N. S., 1994, Adult skin wounds in the fetal environment heal with scar formation, *Ann. Surg.* **219**:65–72.

Lopez-Casillas, F., Wrana, J. L., and Massague, J., 1993, Betaglycan presents ligand to the TGF-β signalling receptor, *Cell* **73**:1435–1444.

Lopez-Casillas, F., Payne, H. M., Andres, J. L., and Massague, J., 1994, Betaglycan can act as a dual modulator of TGF-β access to signalling receptors: Mapping of ligand binding and GAG attachment sites, *J. Cell Biol.* **124**:557–568.

Lorenz, P. H., and Adzick, N. S., 1993, Scarless skin wound repair in the fetus, *West. J. Med.* **159**:350–355.

Lorenz, H. P., Longaker, M. T., Perocha, L. A., Jennings, R. W., Harrison, M. R., and Adzick, N. S., 1992, Scarless wound repair: A human fetal skin model, *Development* **114**:253–259.

Lorenz, H. P., Whitby, D. J., Longaker, M. T., and Adzick, N. S., 1993, Fetal wound healing: The ontogeny of scar formation in the non-human primate, *Ann. Surg.* **217**:391–396.

Lotz, M. M., Burdsal, C. A., Erickson, H. P., and McClay, D. R., 1989, Cell adhesion to fibronectin and tenascin: Quantitative measurements of initial binding and subsequent strengthening response, *J. Cell Biol.* **109**:1795–1805.

Mackie, E. J., Halfter, W., and Liverani, D., 1988, Induction of tenascin in healing wounds, *J. Cell. Biol.* **107**:2757–2767.

Magnuson, V. L., Young, M., Schattenberg, D. G., Mancini, M. A., Chen, B., Steffensen, B., and Klebe, R. J., 1991, The alternative splicing of fibronectin pre-mRNA is altered during ageing and in response to growth factors, *J. Biol. Chem.* **266**:14654–14662.

Martens, E., Dillen, C., Put, W., Heremans, H., Vandamme, J., and Billiau, A., 1993, Increased circulating interleukin-6 (IL-6) activity in endotoxin-challenged mice pretreated with anti-IL-6 antibody is due to IL-6 accumulated in antigen-antibody complexes, *Eur. J. Immunol.* **23**:2026–2029.

Martin, D. E., Reece, M. C., Maher, J. E., and Reese, A. C., 1988, Tissue debris at the injury site is coated by plasma fibronectin and subsequently removed by tissue macrophages, *Arch. Dermatol.* **124**:226–229.

Martin, P., and Lewis, J., 1992, Actin cables and epidermal movement in embryonic wound healing, *Nature* **360**:179–183.

Martin, P., Dickson, M. C., Millan, F. A., and Akhurst, R. J., 1993, Rapid induction and clearance of TGF-β1 is an early response to wounding in the mouse embryo, *Dev. Genet.* **14**:225–238.

Massague, J., Cheiftetz, S., Boyd, F. T., and Andres, J. L., 1990, TGF-beta receptors and TGF-beta binding proteoglycans: Recent progress in identifying their functional properties, *Ann. NY Acad. Sci.* **593**: 59–72.

Mast, B. A., Flood, L. C., Haynes, J. H., DePalma, R. L., Cohen, I. K., Dieglmann, R. F., and Krummel, T. M., 1991, Hyaluronic acid is a major component of the matrix of fetal rabbit skin and wounds: Implications for healing by regeneration, *Matrix* **11**:63–68.

Mast, B. A., Dieglemann, R. F., Krummel, T. M., and Cohen, I. K., 1992a, Scarless wound healing in the mammalian fetus, *Surg. Gynecol. Obstet.* **174**:441–451.

Mast, B. A., Haynes, J. H., Krummel, T. M., Diegelmann, R. F., and Cohen, I. K., 1992b, *In vivo* degradation of fetal wound hyaluronic acid results in increased fibroplasia, collagen deposition and neovascularisation, *Plast. Reconstr. Surg.* **89**:503–509.

Mast, B. A., Diegelmann, R. F., Krummel, T. M., and Cohen, I. K., 1993, Hyaluronic acid modulates proliferation, collagen and protein synthesis of cultured fetal fibroblasts, *Matrix* **13**:441–446.

McCluskey, J., Hopkinson-Wooley, J., Luke, B., and Martin, P., 1993, A study of wound healing in the E11.5 mouse embryo by light and electron microscopy, *Tissue Cell* **25**:173–181.

McDonald, J. A., 1988, Extracellular matrix assembly, *Annu. Rev. Cell Biol.* **4**:183–207.

Merkel, J. R., DiPaolo, B. R., Hallock, G. G., and Rice, D. C., 1988, Types I and types III collagen content of healing wounds in fetal and adult rats, *Proc. Soc. Exp. Biol. Med.* **187**:493–497.

Meyer, D. H., Krull, N., Dreher, K. L., and Gressner, A. M., 1992, Biglycan and decorin gene expression in normal and fibrotic rat liver: Cellular localization and regulatory factors, *Hepatology* **16**:204–216.

Mian, N., 1986, Analysis of cell-growth-phase-related variations in hyaluronate synthase activity of isolated plasma membrane fractions of cultured human skin fibroblasts, *Biochem. J.* **237**:333–342.

Morykwas, M. J., Ditesheim, J. A., Ledbetter, M. S., Crook, E., White, W. L., Jennings, D. A., and Argenta, L. C., 1991, *Monodelphis domesticius:* A model for early developmental wound healing, *Ann. Plast. Surg.* **4**:327–331.

Moses, A. C., Nissley, P. S., Short, P. A., Rechler, M. M., White, R. M., Knight, A. B., and Higa, O. Z., 1980, Increase levels of multiplication stimulating activity, an insulin-like growth factor in fetal rat serum, *Proc. Natl. Acad. Sci. USA* **77**:3649–3653.

Mustoe, T. A., Pierce, G. F., Thomason, A., Gramates, P., Sporn, M. B., and Deuel, T. F., 1987, Accelerated healing of incisional wounds in rats induced by transforming growth factor β, *Science* **237**:1333–1336.

Nath, R. K., LaRegina, M., Markham, H., Ksander, G. A., and Weeks, P. M., 1994a, The expression of transforming growth factor beta in fetal and adult rabbit skin wounds, *J. Pediatr. Surg.* **29**:416–421.

Nath, R. K., Parks, W. C., Mackinnon, S. E., Hunter, D. A., Markham, H., and Weeks, P. M., 1994b, The regulation of collagen in fetal skin wounds: mRNA localization and analysis, *J. Pediatr. Surg.* **29**:855–862.

Nelson, N. M., 1976, Respiration and circulatory changes before birth, in: *The Physiology of the Newborn Infant,* 4th ed. (C. A. Smith and N. M. Nelson, eds.), pp. 75–121, C. C. Thomas, Springfield, Massachusetts.

Norris, D. A., Clark, R. A. F., Swigart, L. M., Huff, J. C., Weston, W. L., and Howell, S. E., 1982, Fibronectin fragment(s) are chemotactic for human peripheral blood monocytes, *J. Immunol.* **129**:1612–1618.

Picardo, M., McGurk, M., Schor, S. L., Grey, A. M., and Ellis, I., 1992, Identification of migration stimulating factor in wound fluid, *Exp. Mol. Pathol.* **57**:8–21.

Pierce, G. F., Vandeberg, J., Rudolph, R., Tarpley, J., and Mustoe, T., 1991, Platelet derived growth factor-BB and transforming growth factor beta-1 selectively modulate glycosaminoglycans, collagen and myofibroblasts in excisional wounds, *Am. J. Pathol.* **138**:629–646.

Poole, T. J., and Thiery, J. P., 1986, Antibodies and a synthetic peptide that block cell–fibronectin adhesion arrest neural crest migration *in vivo, Prog. Clin. Biol. Res.* **217B:**235–238.

Purchio, A. F., Cooper, J. A., Brunner, A. M., Lioubin, M. N., Gentry, L. E., Kovacina, K. S., Roth, R. A., and Marquardt, H., 1988, Identification of mannose-6-phosphate in two asparagine-linked sugar chains of recombinant transforming growth factor-β1 precursor, *J. Biol. Chem.* **263:**14211–14215.

Raghow, R., 1994, The role of extracellular matrix in post inflammatory wound healing and fibrosis, *FASEB J.* **8:**823–831.

Rettig, W. J., Garin-Chesa, P., Beresford, H. R., Oettgen, H. F., Melamed, M. R., and Old, L. J., 1988, Cell surface glycoproteins of human sarcomas: Differential expression in normal and malignant tissues and cultured cells, *Proc. Natl. Acad. Sci. USA* **85:**3110–3114.

Rittenberg, T., Longaker, M. T., Adzick, N. S., and Erlich, H. P., 1991, Sheep amniotic fluid has a protein factor which stimulates human fibroblast populated collagen lattice contraction, *J. Cell Physiol.* **149:**444–450.

Roberts, A. B., and Sporn, M. B., 1990, The transforming growth factors TGF-β, in: *Peptide Growth Factors and Their Receptors,* Vol. 95 *Handbook of Experimental Pathology* (M. B. Sporn and A. B. Roberts, eds.), pp. 419–472, Springer-Verlag, New York.

Robinson, B. W., and Goss, A. N., 1981, Intrauterine healing of fetal rat cheek wounds, *Cleft Palate J.* **18:**251–255.

Roswell, A. R., 1984, The intrauterine healing of fetal muscle wounds: experimental study in the rat, *Br. J. Plast. Surg.* **37:**635–642.

Ruegg, C. R., Chiquet-Ehrismann, R., and Alkan, S. S., 1989, Tenascin, an extracellular matrix protein exerts immunomodulatory activities, *Proc. Natl. Acad. Sci. USA* **86:**7437–7441.

Ruoslahti, E., and Yamaguchi, Y., 1991, Proteoglycans as modulators of growth factor activities, *Cell* **64:**867–869.

Saga, Y., Yagi, T., Ikawa, Y., Sakakura, T., and Aizawa, S., 1992, Mice develop normally without tenascin, *Genes Dev.* **6:**1821–1831.

Sakaguchi, K., Yanagishita, M., Takeuchi, Y., and Aurbach, G. D., 1991, Identification of heparan sulfate proteoglycans as a high affinity receptor for acidic fibroblast growth factor (aFGF) in a parathyroid cell line, *J. Biol. Chem.* **266:**7270–7278.

Saksela, O., Rifkin, D. B., 1990, Release of basic fibroblast growth factor–heparan sulfate complexes from endothelial cells by plasminogen activator-mediated proteolytic activity, *J. Cell Biol.* **110:**767–775.

Sames, K., 1994, Introduction: Biochemistry of proteoglycans and glycosaminoglycans, in: *The Role of Proteoglycans and Glycosaminoglycans in Ageing. Interdisciplinary Topics in Gerontology* (H. P. von Hahn, ed.), pp. 1–17, S. Karger, Basel, Switzerland.

Scanlan, M. J., Mohan Raj, B. K., Calvo, B., Garin-Chesa, P., Sanz-Moncasi, M. P., Healey, J. H., Old, L. J., and Rettig, W. J., 1994, Molecular cloning of fibroblast activation protein a, a member of the serine protease family selectively expressed in stromal fibroblasts of epithelial cancers, *Proc. Natl. Acad. Sci. USA* **91:**5657–5661.

Schmid, P., Kunz, S., Cerletti, N., McMaster, G., and Cox, D., 1993, Injury induced expression of TGF-β1 mRNA is enhanced by exogenously applied TGFβs, *Biochem. Biophys. Res. Commun.* **194:**399–406.

Schonherr, E., Hausser, H., Beavan, L., and Kresse, H., 1995, Decorin-type I collagen-interaction: Presence of separate core protein binding domains, *J. Biol. Chem.* **270:**8877–8883.

Schor, S. L., Schor, A. M., Rushton, G., and Smith, L., 1985, Adult, fetal and transformed fibroblasts display different migratory phenotypes on collagen gels: Evidence for an isoform transition during fetal development, *J. Cell Sci.* **73:**221–234.

Schor, S. L., Schor, A. M., Grey, A. M., and Rushton, G., 1988, Fetal and cancer patient fibroblasts produce an autocrine migration stimulating factor not made by normal adult cells, *J. Cell Sci.* **90:**391–399.

Schor, S. L., Schor, A. M., Grey, A. M., Chen, J., Rushton, G., Grant, M., and Ellis, I., 1989, Mechanisms of action of the migration stimulating factor (MSF) produced by fetal and cancer patient fibroblasts: Effect on hyaluronic acid synthesis, *In Vitro Cell. Dev. Biol.* **25:**737–746.

Schultz, G. S., White, M., Mitchell, R., Brown, G., Lynch, J., Twardzick, D. R., and Todaro, G. J., 1987, Epithelial wound healing enhanced by transforming growth factor a and vaccinia growth factor, *Science* **235:**350–352.

Schurch, W., Seemayer, T. A., and Gabbiani, G., 1992, Myofibroblasts, in: *Histology for Pathologists* (S. S. Sternberg, ed.), pp. 109–144, Raven Press, New York.

Schwarzbauer, J. E., Tamkun, J. W., Lemiscka, I. R., and Hynes, R. O., 1983, Three different fibronectin mRNA arise by alternative splicing within the coding region, *Cell* **35**:421–431.

Schwarzbauer, J. E., Patel, R. S., Fonda, D., and Hynes, R. O., 1987, Multiple sites of alternative splicing of the rat fibronectin gene transcript, *EMBO J.* **6**:2573–2580.

Scott, J. E., 1988, Proteoglycan–fibrillar collagen interactions, *Biochem. J.* **252**:313–323.

Scott, J. E., 1992, Supramolecular organization of extracellular matrix glycosaminoglycans, *in vitro* and in the tissues, *FASEB J.* **6**:2639–2645.

Scott, J. E., and Hughes, E. W., 1986, Proteoglycan–collagen relationships in developing chick and bovine tendons. Influence of physiological environment, *Connect. Tiss. Res.* **14**:267–278.

Shah, M., Foreman, D. M., and Ferguson, M. W. J., 1992, Control of scarring in adult wounds by neutralising antibody to TGF-β, *Lancet* **339**:213–214.

Shah, M., Foreman, D. M., and Ferguson, M. W. J., 1994, Neutralising antibody to TGFβ1,2 reduces cutaneous scarring in adult rodents, *J. Cell Sci.* **107**:1137–1157.

Shah, M., Foreman, D. M., and Ferguson, M. W. J., 1995, Neutralisation of TGF-β1 and TGF-β2 or exogenous addition of TGF-β3 to cutaneous rat wounds reduces scarring, *J. Cell Sci.* **108**:15–17.

Shaw, L. M., and Olsen, B. R., 1991, FACIT collagens: Diverse molecular bridges in extracellular matrices, *Trends Biochem. Sci.* **16**:191–194.

Siebert, J. W., Burd, D. A. R., McCarthy, J., and Erlich, H. P., 1990, Fetal wound healing: A biochemical study of scarless healing, *Plast. Reconst. Surg.* **85**:495–504.

Sloan, P., 1991, Current concepts in the role of fibroblasts and extracellular matrix in wound healing and their relevance to oral implantology, *J. Dent.* **19**:107–109.

Somasundaram, K., and Prathap, K., 1970, Intra-uterine healing of skin wounds in rabbit wounds in rabbit fetuses, *J. Pathol.* **100**:81–86.

Somasundaram, K., and Prathap, K., 1972, The effect of exclusion of amniotic fluid on intra-uterine healing of skin wounds in rabbit fetuses, *J. Pathol.* **107**:127–130.

Speranz, M. L., Valentini, G., and Calligaro, A., 1987, Influence of fibronectin on fibrillogenesis of type I and type III collagen, *Coll. Rel. Res.* **7**:115–123.

Sporn, M. B., Roberts, A. B., Schull, J. M., Smith, J. M., Ward, J. M., and Sodek, J., 1983, Polypeptide transforming growth factors isolated from bovine sources and used for wound healing *in vivo*, *Science* **219**:1329–1331.

Stadnyk, A. W., 1994, Cytokine production by epithelial cells, *FASEB J.* **8**:1041–1047.

Stern, M., Schmid, B., Dodson, T. B., Stern, R., and Kaban, K. B., 1992, Fetal cleft lip repair in rabbits: Histology and role of hyaluronic acid, *J. Oral Maxillofac. Surg.* **50**:263–268.

Stern, M., Dodson, T. B., Longaker, M. T., Lorenz, H. P., Harrison, M. R., and Kaban, L. B., 1993, Fetal cleft lip repair in lambs: Histologic characteristics of the healing wound, *Int. J. Oral Maxillofac. Surg.* **22**:371–374.

Tahery, T. J., and Lee, D. A., 1989, Review: Pharmacologic control of wound healing in glaucoma filtration surgery, *J. Ocular Pharmacol.* **5**:155–179.

Takashima, A., Billingham, R. E., and Grinnell, F., 1986, Activation of rabbit keratinocyte receptor function *in vivo* during wound healing, *J. Invest. Dermatol.* **86**:585–590.

Tamkun, J. W., Schwarzbauer, J. E., and Hynes, R. O., 1984, A single rat fibronectin gene generates three different mRNAs by alternative splicing of a complex exon, *Proc. Natl. Acad. Sci. USA* **81**:5140–5144.

Tan, E. M. L., Hoffren, J., Rouda, S., Greenbaum, S., Fox, J. W., Moore, J. H., and Dodge, G. R., 1993, Decorin, versican and biglycan gene expression by keloid and normal dermal fibroblasts: Differential regulation by basic fibroblast growth factor, *Exp. Cell Res.* **209**:200–207.

Terrell, T. G., Working, P. K., Chow, C. P., Green, C. P., and Green, J. D., 1993, Pathology of recombinant human transforming growth factor-β1 in rats and rabbits, *Int. Rev. Exp. Pathol.* **34**:43–67.

Thompson, R. W., Whalen, G. F., Saunders, K. B., Hores, T., and D'Amore, P. A., 1990, Heparin-mediated release of fibroblast growth factor-like activity into the circulation of rabbits, *Growth Factors* **3**:221–229.

Tomida, M., Koyama, H., and Ono, T., 1974, Hyaluronic acid synthetase in cultured mammalian cells producing hyaluronic acid. Oscillatory change during the growth phase and suppression by 5-bro-modeoxyuridine, *Biochem. Biophys. Acta* **338:**352–363.

Toole, B. P., 1991, Proteoglycans and hyaluronan in morphogenesis and differentiation, in: *Cell Biology of the Extracellular Matrix* (E. D. Hay, ed.), pp. 305–339, Plenum Press, New York.

Turley, B. P., and Torrance, J., 1984, Localization of hyaluronate-binding protein on motile and non-motile fibroblasts, *Exp. Cell Res.* **161:**17–28.

Turley, E. A., Austen, L., Vandeligt, K., and Clary, C., 1991, Hyaluronan and a cell associated hyaluronan binding protein regulates the locomotion of Ras-transformed cells, *J. Cell Biol.* **112:**1041–1047.

Van Vlasselaer, P., Borremans, B., van Gorp, U., Dasch, J. R., and DeWaal-Malefyt, R., 1994, Interleukin-10 inhibits transforming growth factor-β (TGF-β) synthesis required for osteogenic commitment of mouse bone marrow cells, *J. Cell Biol.* **124:**569–577.

Vogel, K. G., and Trotter, J. A., 1987, The effect of proteoglycans on the morphology of collagen fibrils formed *in vitro, Collagen Rel. Res.* **7:**105–114.

Vogel, K. G., Paulsson, M., and Heinegard, D., 1984, Specific inhibition of type I and type II collagen fibrillogenesis by the small proteoglycan of tendon, *Biochem. J.* **223:**587–597.

Vuorio, E., and Crombrugghe, B., 1990, The family of collagen genes, *Annu. Rev. Biochem.* **59:**837–872.

Wahl, S. M., Hunt, D. A., Wakefield, N., McCartney-Francis, L. M., Wahl, A. B., Roberts, A. B., and Sporn, M. B., 1987, Transforming growth factor-β (TGF-β) induces monocyte chemotaxis and growth factor production, *Proc. Natl. Acad. Sci. USA* **84:**5788–5792.

Weigel, P. H., Fuller, G. M., and Le Boeuf, R. D., 1986, A model for the role of hyaluronic acid and fibrin in the early events during the inflammatory response and wound healing, *J. Theor. Biol.* **119:**219–234.

Welch, M. P., Odland, G. F., and Clark, R. A. F., 1990, Temporal relationships of F-actin bundle formation, collagen and fibronectin matrix assembly and fibronectin receptor expression during wound contraction, *J. Cell Biol.* **110:**133–145.

West, D. C., Hampson, I. N., Arnold, F., and Kumar, S., 1985, Angiogenesis induced by degradation products of hyaluronic acid, *Science* **228:**1324–1326.

Whitby, D. J., and Ferguson, M. W. J., 1991a, The extracellular matrix of lip wounds in fetal, neonatal and adult mice, *Development* **112:**651–668.

Whitby, D. J., and Ferguson, M. W. J., 1991b, Immunolocalization of growth factors in fetal wound healing, *Dev. Biol.* **147:**2207–2215.

Whitby, D. J., Longaker, M. T., Harrison, M. R., Adzick, N. S., and Ferguson, M. W. J., 1991, Rapid epithelialisation of fetal wounds is associated with the early deposition of tenascin, *J. Cell Sci.* **99:** 583–586.

Wider, T. M., Yager, J. S., Rittenberg, T., Hugo, N. E., and Erlich, P., 1993, The inhibition of fibroblast-populated collagen lattice contraction by human amniotic fluid: A chronologic examination, *Plast. Reconstruc. Surg.* **91:**1287–1293.

Wight, T. N., Kinsella, M. G., and Qwarnstrom, E. E., 1992, The role of proteoglycans in cell adhesion, migration and proliferation, *Curr. Opin. Cell Biol.* **4:**793–801.

Yamaguchi, Y., Mann, D. M., and Ruoslahti, E., 1990, Negative regulation of transforming growth factor-beta by the proteoglycan decorin, *Nature* **346:**281–284.

Yamamoto, T., Nakamura, T., Noble, N. A., Ruoslahti, E., and Border, W. A., 1993, Expression of transforming growth factor β is elevated in human and experimental diabetic nephropathy, *Proc. Natl. Acad. Sci. USA* **90:**1814–1818.

Yayon, A., Klagsbrun, M., Esko, J. D., Leder, P., and Ornitz, D. M., 1991, Cell surface heparin like molecules are required for binding of basic fibroblast growth factor to its high affinity receptor, *Cell* **64:**841–848.

Yeo, T. K., Brown, L., and Dvork, H. L., 1991, Alterations in proteoglycan synthesis common to healing wounds and tumors, *Am. J. Pathol.* **138:**437–1450.

Yoshioka, K., Takemura, T., Murakami, K., Okada, M., Hino, S., Miyamoto, H., and Maki, S., 1993, Transforming growth factor β and mRNA in glomeruli in normal and diseased human kidneys, *Lab. Invest.* **68:**154–163.

Zugmaier, G., Paik, S., Wilding, G., Knabbe, C., Bano, M., Lupu, R., Deschauer, B., Simpson, S., Dickson, R. B., and Lippman, M., 1991, Transforming growth factor-β1 induces cachexia and systemic fibrosis without an anti-tumor effect in nude mice, *Cancer Res.* **51:**3590–3594.

Index